DESIGN MANUAL FOR TRANSISTOR CIRCUITS

DESIGN MANUAL
FOR TRANSISTOR CIRCUITS

John M. Carroll
Managing Editor, ELECTRONICS

McGRAW-HILL BOOK COMPANY, INC.
New York   Toronto   London   1961

DESIGN MANUAL FOR TRANSISTOR CIRCUITS

Copyright © 1961 by the McGraw-Hill Book Company, Inc. Printed in the United States of America. All rights reserved. This book, or parts thereof, may not be reproduced in any form without permission of the publisher.

Library of Congress Catalog Card Number: 61-10714

10144

PREFACE

This book is designed as a complete manual for the designer of transistor circuits. The main part of the book includes 128 articles carefully selected from nearly 300 articles on transistors and semiconductor devices published in *Electronics* magazine during 1958, 1959, and 1960. The articles were selected because they seemed useful to the greatest number of transistor circuit designers and because they seemed to make the greatest contribution to the art of transistor circuit design.

The articles are grouped into 21 chapters for ease of reference. The chapters deal with basic circuits such as amplifiers, oscillators, and power supplies, as well as with specific applications such as radio, F-M and television receivers, test instruments, industrial and radiation measuring instruments, medical equipment, and computer circuits. There are many basic articles describing circuit design techniques as well as articles giving circuits for specific applications. In nearly all cases, all component values are given and the transistors used are commercially available.

In this book, so many articles on circuits for missiles and satellites appeared that it seemed advisable to divide them into two chapters: one on telemetering circuits and one on guidance circuits. Similarly, the number of digital computing circuits was so great that it was decided to have a basic chapter on counting circuits and three chapters on computer applications covering switching and control circuits, memory circuits, and circuits for input and output devices. The field of industrial electronics is covered in chapters on solid-state switching, servomechanisms, and measuring instruments. The new tunnel diode is covered in an article describing the theory of the device, typical circuits, and applications.

For newcomers to the field of transistor circuit design as well as those who want a chance to brush up or to check up on some elementary point, this book begins with a chapter on the transistor itself. This chapter was written especially for this book. It covers semiconductor materials, forward and reverse biased *p-n* junctions, transistor action, transistor load lines, hybrid matrix parameters, equivalent-T circuit, high-frequency transistors, power transistors, and special devices such as unijunction transistors, controlled rectifiers, thyristors, unipolar transistors, and integrated semiconductor circuits.

The early chapters of the book deal with problems common to all transistor circuit applications. Several popular design charts and nomographs are reproduced here. Knotty problems are dealt with, such as thermal design problems and how to ease transistor operating loads.

This book should prove useful as a self-study text for designers getting into transistor circuit design work, because of the wealth of basic circuit design information. The book should prove invaluable also as a reference for the experienced designer looking for a specific circuit for some special application.

The author thanks Donald Hall of Texas Instruments Incorporated, Dallas, Texas for his help in revising Chapter 1.

John M. Carroll

CONTENTS

Preface — v

Chapter 1. SEMICONDUCTOR DEVICES — 1
Appendix: Matrix Algebra — 20
Bibliography — 22

Chapter 2. CIRCUIT DESIGN TECHNIQUES — 23
Q-Multiplier for Audio Frequencies *G. B. Miller* — 25
Tube-Transistor Hybrids Afford Design Economy *G. A. Dunn and N. C. Hekimian* — 28
Easing Transistor Loads *W. F. Saunders, Jr.* — 31
Starved Transistors Raise D-C Input Impedance *B. M. Bramson* — 32
Bias Method Raises Breakdown Point *A. Somlyody* — 34
Tunnel Diode—Circuits and Applications *I. A. Lesk, N. Holonyak, Jr. and U. S. Davidsohn* — 36
Diode Reduces Cutoff Current Drift *H. H. Hoge* — 41
Diodes Offset Silicon Transistor Heat Drift *D. H. Bryan* — 42

Chapter 3. CIRCUIT DESIGN CHARTS — 43
High-Frequency Cutoff Nomograph *H. E. Schauwecker* — 45
Rapid Conversion of Hybrid Parameters *S. Sherr* — 46
Thermal Design Chart *O. D. Hawley and M. Kato* — 48
Finding Dissipation of Switching Transistors *D. W. Boensel* — 50
Switch-Time Nomograph *T. A. Prugh* — 52

Chapter 4. AMPLIFIERS — 53
One-Transistor "Push-Pull" *J. A. Worcester* — 55
Designing Audio Power Amplifiers *M. B. Herscher* — 56
Designing High-Quality Audio Amplifiers *R. Minton* — 60
Feedback Design for Amplifier Stages *T. R. Hoffman* — 62
Amplifier with 100-Megacycle Bandwidth *J. C. de Broekert and R. M. Scarlett* — 65
Single-Ended Amplifiers for Class-B Operation *H. C. Lin and B. H. White* — 68
Amplifier Using Multiple Feedback *H. Lefkowitz* — 70
D-C Operational Amplifier with Transistor Chopper *W. Hochwald and F. H. Gerhard* — 72
D-C Amplifier for Rugged Use in Field *W. G. Van Dorn* — 75
Low-Pass Filter for Subaudio Frequencies *R. C. Onstad* — 76
Audio Volume Compressor *E. C. Miller* — 79
Amplifier Design Method *V. R. Latorre* — 80

Chapter 5. OSCILLATORS 81

Transistorized F-M Oscillator *P. W. Wood*	83
Graphical Design of Oscillators *W. R. McSpadden and E. Eberhard*	84
Solid-State Generator of Microwave Power *M. M. Fortini and J. Vilms*	88
Design of High-Power Oscillators *W. E. Roach*	90
Transfluxor Oscillator Gives Drift-Free Output *R. J. Sherin*	94
Synthesizing Timbre for Musical Tones *W. S. Pike and C. N. Hoyler*	96

Chapter 6. POWER SUPPLIES 99

Rectifier Gives D-C of Either Polarity *R. R. Bockemuehl*	101
Design of Power Converters *T. R. Pye*	102
Line Voltage Control Uses Zener Diodes *R. A. Greiner*	105
Designing Highly Stable Power Supplies *E. Baldinger and W. Czaja*	106
Equations for Power-Supply Design *T. Hamm, Jr.*	110
Inverse Feedback Stabilizes Dry-Cell Current Sources *G. E. Fasching*	113
Constant-Current-Coupled Power Supply *E. Gordy and P. Hasenpusch*	114
Inverters for Fluorescent Lamps *L. J. Gardner*	116
Solid-State Generator Regulator for Automobiles *L. D. Clements*	118
Linear Circuits Regulate Solid-State Inverter *R. Wileman*	121
Designing D-C to A-C Converters *S. Schenkerman*	124
Efficient Photoflash Power Converter *R. J. Sherin*	127
Fast Response Overload Protection *F. W. Kear*	128

Chapter 7. COUNTING CIRCUITS 129

Steering Circuits Control Reversible Counters *R. D. Carlson*	131
Choosing Transistors for Monostable Multivibrators *J. R. Kotlarski*	134
Increasing Counting System Reliability *H. A. Kampf*	136
Binary Circuits Count Backwards or Forwards *H. J. Weber*	138
Triggered Bistable Circuits *J. B. Hangstefer and L. H. Dixon, Jr.*	140
Pulse Sorting with Transistors and Ferrites *J. H. Porter*	142
Transistor Drives Cold-Cathode Counter *H. Sadowski and M. E. Cassidy*	144
Transistor Switch Design *A. Gill*	146
Insuring Reliability in Time-Delay Multivibrators *P. E. Harris*	147
Feedback Stabilizers Flip-Flop *P. Chelik*	148
Magnetic Core Operates Counter *E. H. Sommerfield*	150

Chapter 8. RADIO RECEIVERS 151

Eyeglass Radio Receiver *H. F. Cooke*	153
Design of Reflexed Radio Receivers *J. Waring*	154
European Designs for High-Frequency Radio Receivers *R. Shah*	157
Special Circuits for Radio Receivers *W. E. Sheehan and W. H. Ryer*	160
Design of Automobile Broadcast Receivers *R. A. Santilli and C. F. Wheatley*	162
F-M Tuner Uses Four Transistors *H. Cooke*	166
Wideband F-M with Capacitance Diodes *C. Arsem*	168

Chapter 9. TELEVISION RECEIVERS 171

Tuners for Portable Television Receivers *V. Mukai and P. V. Simpson*	173
Designing Television Tuners with Mesa Transistors *H. F. Cooke*	176
Horizontal Deflection Circuits for Television Receivers *M. Fischman*	182
Television Sound Detector Uses Drift Transistor *M. Meth*	186
Low-Distortion Television Monitor Amplifier *H. J. Paz*	189

Chapter 10. COMMUNICATIONS APPLICATIONS 193

Carrier Transmission for Closed-Circuit Television *L. G. Schimpf*	195
Mobile Radio System Provides 920 Channels *F. Brauer and D. Kammer*	198
Amplifier for 16-MM Sound Movie Camera *E. M. Tink*	202
Portable Multiplexer for Telephone Communications *P. W. Kiesling, Jr.*	204

Versatile F-M Transducer *C. S. Burrus*	207
How to Construct a Miniature F-M Transmitter *D. E. Thomas and J. M. Klein*	208

Chapter 11. TEST INSTRUMENTS **211**

Generator for Pulse Circuit Design *L. Neumann*	213
Battery-Operated Transistor Oscilloscope *O. Svehaug and J. R. Kobbe*	216
Automatic Measurement of Transistor Beta (h_{FE}) *E. P. Hojak*	220
Generating Accurate Sawtooth and Pulse Waveforms *C. A. Von Uriff and R. W. Ahrons*	222
Clock Source for Electronic Counters *T. F. Marker*	225
Zero-Crossing Synchronizes Wavetrain Outputs *J. A. Wereb, Jr.*	226
Measuring Alpha-Cutoff Frequency *G. I. Turner*	228
Multiple-Waveform Generator *J. E. Curry*	229
Simplified Design of a Sweep Generator *H. P. Brockman*	230

Chapter 12. SOLID-STATE SWITCHING AND CONTROL **231**

Industrial Temperature Controller *H. Sutcliffe*	233
Solid-State Thyratron Switches Kilowatts *R. P. Frenzel and F. W. Gutzwiller*	234
Time Delay for Industrial Control *L. Szmauz and H. Bakes*	238
Amplifiers for Nuclear Reactor Control *E. J. Wade and D. S. Davidson*	240

Chapter 13. SERVOMECHANISMS **243**

Servo Preamplifiers Using Direct-Coupled Transistors *A. N. Desautels*	245
Constant-Current Technique Cuts Servo Response Time *L. H. Dulberger*	246
Reducing Relay Servo Size *S. Shenfeld*	249
Controlled Rectifiers Drive A-C and D-C Motors *W. R. Seegmiller*	252

Chapter 14. INDUSTRIAL MEASURING INSTRUMENTS **255**

Amplifiers for Strain-Gages and Thermocouples *R. S. Burwen*	257
Electronic Judging of Fast-Moving Sports Contests *W. R. Durrett*	260
Balloon-Borne Circuits Sort High-Altitude Cosmic Rays *D. Enemark*	262
Monitoring Radioisotope Tracers for Industry *F. E. Armstrong and E. A. Pavelka*	266
Radioactive Tracers Find Jet Fuel Flow Rates *J. D. Keys and G. F. Alexander*	268
Solid-State Photocell Sees through Haze *P. Weisman and S. L. Ruby*	270

Chapter 15. MEDICAL INSTRUMENTS **273**

Circuit Substitutes for Larynx	275
Instrument Monitors Blood Pressure *O. Z. Roy and J. R. Charbonneau*	276
Impedance Measurements of Living Tissue *S. Bagno and F. M. Liebman*	278
Detecting Foetal Heart Sounds *T. I. Humphreys*	280
Electronic Tonometer for Glaucoma Diagnosis *R. S. Mackay and E. Marg*	282
Artificial Neuron Uses Transistors	284

Chapter 16. MISSILE AND SATELLITE TELEMETRY **285**

Data Conversion Circuits for Earth-Satellite Telemetry *D. N. Carson and S. K. Dhawan*	287
Infrared Communications Receiver for Space Vehicles *W. E. Osborne*	290
Multiplexing Techniques for Satellite Applications *O. B. King*	292
Circuits for Space Probes *R. R. Bennett, G. J. Gleghorn, L. A. Hoffman, M. G. McLeod and Y. Shibuya*	297

Chapter 17. MISSILE AND AIRCRAFT GUIDANCE **301**

Solid-State Guidance for Able-Series Rockets *R. E. King and H. Low*	303
Switching Circuits for Missile Countdowns *D. W. Boensel*	310
Designing Safety into Automatic Pilot Systems *C. W. McWilliams*	307
Simplified Controls for Target Drones *G. B. Herzog*	313

Chapter 18. RADAR, SONAR AND BEACONS 317

Dual Conversion for Marker-Beacon Receivers *R. G. Erdmann*   319
Low-Power Sweep Circuits for Radar Indicators *C. E. Veazie*   322
Using Magnetic Circuits to Pulse Radar Sets *A. Krinitz*   324
Portable Depth Finder for Small Boats *H. C. Single*   326
Receiver for Marker-Beacon Use *F. P. Smith*   328
Radio Beacon Helps Locate Aircraft Crashes *D. M. Makow*   331
Distress Transmitter Uses Hybrid Circuit *H. B. Weisbecker*   334

Chapter 19. DIGITAL COMPUTER SWITCHING CIRCUITS 335

Frequency Control of Magnetic Multivibrators *W. A. Geyger*   337
Computer Switching with High-Power Transistors *J. S. Ronne*   340
Alarm Circuit Warns of Faults in Digital Circuits *S. Fierston*   344
Transistors Provide Computer Clock Signals *S. Schoen*   346

Chapter 20. DIGITAL COMPUTER MEMORY CIRCUITS 349

A Digital Recorder Holds Data after Shock *C. P. Hedges*   351
Clock Track Recorder for Memory Drum *A. J. Strassman and R. E. Keeter*   358
Compact Memories Have Flexible Capacities *D. Haagens*   354

Chapter 21. DATA PROCESSING INPUT-OUTPUT CIRCUITS . . . 361

Character Generator for Digital Computers *E. D. Jones*   363
Encoder Measures Random-Event Time Intervals *R. J. Kelso and J. C. Groce*   366
Solid-State Digital Code-to-Code Converter *R. Wasserman and W. Nutting*   370
Pulse-Height-to-Digital Signal Converter *W. W. Grannemann, C. D. Longerot,*
  R. D. Jones, D. Endsley, T. Summers, T. Lommasson, A. Pope and D. Smith   374

Index   377

Chapter 1
SEMICONDUCTOR DEVICES

In the past decade a whole new world of circuit design has been opened by the transistor and related semiconductor devices. In addition to replacing electron tubes in many applications, the ever-increasing members of the solid state family have permitted the circuit designer to create new types of systems, previously unattainable. Features such as small size, low power consumption, and lack of warm-up time are being designed into many circuits.

Transistors normally operate from a few microwatts to several watts of power. Newer techniques make possible power outputs of tens of watts at several megacycles. Other semiconductor devices can switch kilowatts of power at lower frequencies.

Basic Materials

The operation of transistors and other semiconductor devices depends on the control of impurities in a nearly perfect crystal lattice structure. Semiconductors are elements whose electrical conductivity lies between metals (good conductors) and nonmetals (insulators). Pure single-crystal semiconductors behave as excellent insulators. Controlled amounts of impurities allow the conductivity to be varied to suit the needs of the transistor designer. The elements most used for transistors are silicon and germanium, both from Group IV of the periodic table, but transistors have been fabricated using combinations of Groups III and V such as gallium arsenide and indium antimonide.

A semiconductor free from impurities is called intrinsic. In this form its resistance is high, making it an excellent insulator. Intrinsic material has impurity levels of less than one foreign atom for 100,000,000,000 atoms of the basic material. The fabrication of semiconductor devices starts with intrinsic material into which is introduced a controlled amount of impurities to obtain the desired characteristic.

The semiconductor material is used in single crystal form. The atoms are arranged in a diamond-shaped lattice, the bonds between each two atoms formed by a pair of electrons as shown in Fig. 1.

The semiconductors have four electrons in the outer, or valence, band, each normally forming bonds with adjacent atoms in the crystal structure. When all the bonds are filled, no electrons are available for conduction, hence the single-crystal semiconductor material makes a good insulator. Any impurity upsets some of these bonds, leaving free electrons or holes for conduction.

Two types of impurities may be added to the intrinsic material. If atoms of a material containing only three electrons in the valence band are present, an atom adjacent to it has an electron not in a bond and thus relatively free to move about. This leaves a vacancy or "hole" shown in Fig. 2, which can be filled by an electron from another atom causing an apparent movement of the hole. The effects of hole conduction appear as those of a positive particle with characteristics similar to an electron. Materials of this type appear in Group III of the periodic table, examples being boron, indium, aluminum, and gallium. Such impurities are called acceptors or p-type impurities because they give rise to conduction by positive particles.

The second type of impurity is one containing five electrons in the valence band. In a semiconductor crystal four of these enter into bonds, the other being easily dislodged to allow conduction as illustrated in Fig. 3. These materials appear in Group V of the periodic table, examples being phosphorous, antimony, and arsenic. These impurities are called donors or n-type because of their conduction by negative particles (electrons).

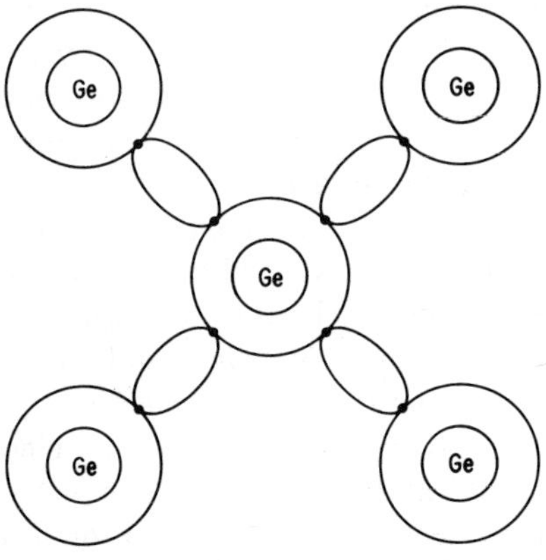

FIG. 1—Bonds between germanium atoms in crystal lattice are formed by electron pairs

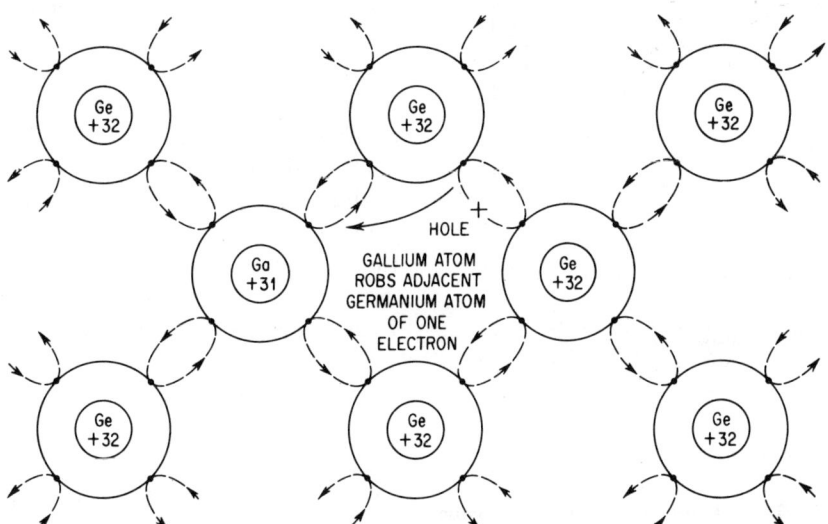

FIG. 2—Germanium crystal lattice with gallium atom, and acceptor or p-type impurity

It is possible to have both types of conduction in the same area of a semiconductor crystal. If the predominant type of conduction is by electrons, then holes are called minority carriers. The reverse also is true: if holes predominate, electrons are the minority carriers. If equal numbers of carriers are present, the material appears much the same as intrinsic, except that, at any temperature, the conductivity is higher because of thermally generated carriers. Examples of minority carrier conduction will be mentioned later in describing the functioning of several types of semiconductor devices.

The simplest form of carrier motion in a semiconductor is diffusion. This type of motion is random in direction, similar to what happens when an ink drop is placed in a glass of water. Because diffusion is a comparatively slow phenomenon, transistors which depend solely on it for their performance are confined to low-frequency operation.

If an electric field is present, the charged carriers are influenced by it to move in a direction and at an increased rate determined by the field. This effect, called drift, increases the high-frequency performance, providing one of the advantages of mesa and drift transistors over more conventional structures.

Diode Action

The simplest active semiconductor device is the diode. Composed of two regions, one n-type and one p-type, the diode operates on principles outlined in the preceding section.

The junction between the p-region and the n-region is one of two types, either a sharp change from one type to another or a gradual change. The alloy diode, made by alloying a metal containing one type of impurity onto a wafer of the other type, has a sharp division or junction.

Other diode types, such as grown or diffused, have gradual or graded junctions. This difference in junctions, as will be seen later, influences diode behavior, particularly with reverse voltage applied.

Barrier Potential

When a piece of n-type semiconductor material is brought into intimate contact with a piece of p-type semiconductor material, a p-n junction, or diode, is formed.

Superficially it might seem that the excess electrons in the n-type material would immediately diffuse across the p-n junction and neutralize the excess holes in the p-type material.

However, this is not the case. The excess positive charge on the n-type impurity atoms near the junction sets up a strong positive electrical field that repels the holes in the p-type material on the other side. The excess negative charge on the p-type impurity atoms near the junction sets up a strong negative field that repels the electrons in the n-type material across the junction. The net result of these two opposing fields is a potential hill or barrier as shown in Fig. 4(A). Very few electrons and holes acquire sufficient thermal energy to get over this potential hill with no potential applied externally. The few holes that diffuse into the n-region equal the few electrons that diffuse into the p-region and there is no net current flow across the p-n junction.

Reverse Biased Diode

If the negative terminal of a battery is connected to the p-region of the diode and the positive terminal is connected to the n-region of the diode, the diode is said to be in the reverse biased condition, Fig. 4(C). The battery potential adds to that of the hill, and current flow is cut off except for the small reverse saturation current I_s.

As the potential across a reverse biased diode is increased, junction breakdown eventually occurs. See Fig. 5. There are two physical mechanisms that can cause junction breakdown, avalanche effect, and Zener effect.

In the avalanche effect, the high reverse field caused by back biasing the diode accelerates thermally produced free electrons and holes, imparting to them sufficient energy so that they liberate other electrons and holes in pairs when they collide with atoms of the crystal lattice. The liberated electrons and holes collide with other

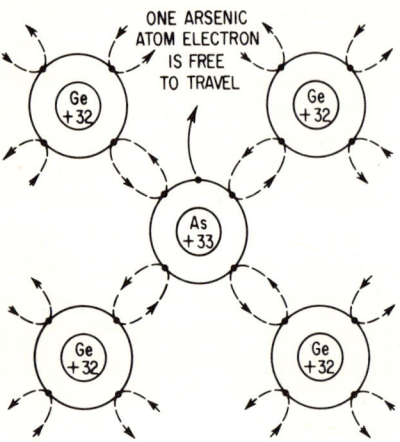

FIG. 3—Germanium crystal lattice with arsenic atom, a donor or n-type impurity

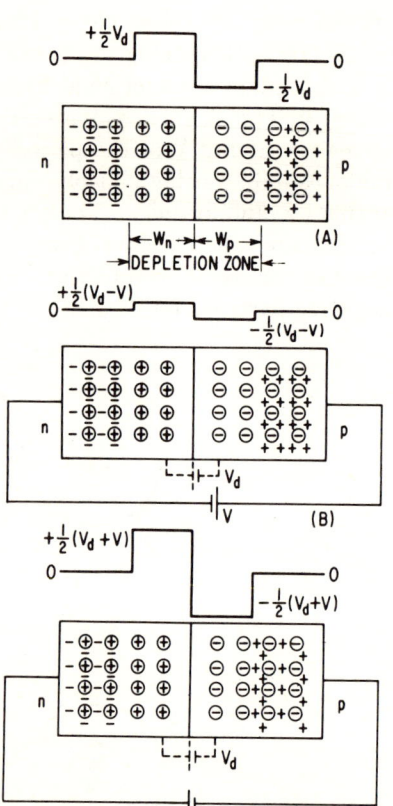

FIG. 4—A p-n junction potential barrier (A). Effects of forward biasing (B). Effects of reverse biasing (C)

atoms forming still more hole-electron pairs. This charge-carrier multiplication is known as avalanche effect. The current in the reverse direction builds up rapidly. Similarly, when the high reverse field is removed, recombination of the hole-electron pairs or diode recovery occurs rapidly. Avalanche diodes used as switching devices in digital computers have been made to operate in tenths of millimicroseconds (nanoseconds).

The electric field (created by the back-biasing potential) can also become the source of energy that makes additional conduction electrons available. The field actually pulls electrons out of their covalent bonds. This mechanism, which also causes current to increase rapidly, is known as the Zener effect.

The reverse breakdown characteristic of diodes is used in voltage regulators. These diodes, designed to have closely controlled breakdown voltages, are used as voltage references and by themselves to provide a constant voltage source. Two breakdown diodes back-to-back provide voltage regulation in either direction.

Forward Biased Diode

If the positive terminal of a battery is connected to the *p*-region of the diode and the negative terminal to the *n*-region, the battery's potential will subtract from the negative potential existing at the *p*-region side of the junction and from the positive potential existing at the *n*-region side of the junction. See Fig. 4(B).

The junction is now forward biased. Current will flow and increase in magnitude as the potential applied to the diode increases (Fig. 5). The magnitude of this forward or majority-carrier current is given by

$$I = I_s \exp(eV/kT)$$

where I_s is the reverse saturation current in amperes, e is the electronic charge (1.602×10^{-19} coulomb), k is Boltzmann's constant (1.38×10^{-23} joule per degree Kelvin), and T is the junction temperature in degrees Kelvin.

As the voltage is increased in the forward direction, forward current increases. The charge carriers become more numerous. The electrons travel faster and collide with other semiconductor atoms, knocking still more electrons out of their covalent bonds. There are higher I^2R losses in the material due to increased current flow. This heat adds energy to the semiconductor atoms and helps free more valence electrons for conduction. Thus the resistance of the diode drops sharply and current increases rapidly. This condition can permanently damage a semiconductor device.

Diode Capacitance

A reverse biased diode exhibits a property known as junction capacitance. How this junction capacitance develops can be seen from the definition of capacitance as a dependence of charge concentration on applied voltage or

$$C = Q/V$$

where C is capacitance in farads, Q is charge in coulombs, and V is potential in volts. However, in a reverse biased diode, the junction capacitance is not constant but depends upon the instantaneous magnitude of the applied voltage. Thus the junction capacitance is actually an incremental capacitance.

$$\Delta C_j = q/\Delta V$$

where Δ indicates a "small change in."

In a reverse biased diode, depletion regions exist on either side of the *p-n* junction. These are regions from which charge carriers have been repelled by the junction potential itself and by the potentials applied to the diode terminals.

The width of the depletion layer depends upon the net voltage applied across the junction barrier. As the voltage increases, more holes are pulled out of the *p*-depletion region toward the negative terminal of the battery. Likewise, more electrons are pulled out of the *n*-depletion regions toward the positive terminal of the battery. Whether the wider portion of the depletion region exists in the *n* or *p* side depends upon which has the higher resistivity. The higher the resistivity, the wider the portion of the depletion region.

Increasing potential across the junction barrier increases depletion layer width, analogously to a parallel-plate capacitor

$$C_j = \epsilon A / W_d$$

where A is the junction cross section area, ϵ is the permittivity of the junction material (permittivity equals dielectric constant times the permittivity of free space, 8.855×10^{-12} farad per meter), and W_d is the depletion layer width in meters. The larger the applied voltage, the wider the depletion layer and the lower the junction capacitance per unit area.

The property of junction capacitance is made use of in the voltage variable capacitor, a specially designed semiconductor diode. The concept of junction capacitance is also important in understanding the high-frequency behavior of transistors.

The magnitude of junction capacitance may be determined from the composition of the semiconductor material, the voltage across the *p-n* barrier, and the diode construction. For a germanium alloy diode:

$$C_j = \left(\frac{\epsilon}{2\mu\rho}\right)^{1/2} (V_j)^{-1/2}$$

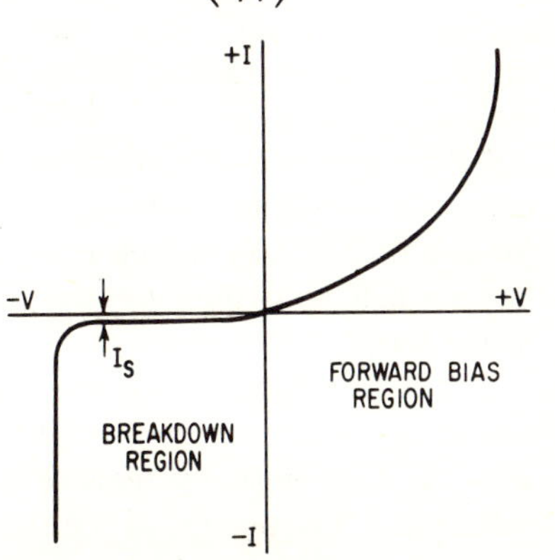

FIG. 5—Voltage-current relationships in a semiconductor junction diode

where ϵ is the permittivity of the diode material. For germanium, the permittivity is 1.4×10^{-12} farad per centimeter. Quantity μ is the electron mobility in the diode material (3,900 cm/sec/v-cm for germanium). Resistivity ρ varies from 1 to 45 ohm-cm for germanium (up to 240,000 ohm-cm for silicon). Resistivity values depend in either case upon the concentration of impurities. The barrier or net junction potential V_j is given by:

$$V_j = V_d - V$$

where V_d is the built-in potential difference across the p-n junction and V is the externally applied potential taken as positive in the direction of forward current flow. Note that increasing the potential across a reverse biased diode decreases its junction capacitance.

The manner in which the junction capacitance varies with reverse voltage is determined by the junction type. For sharp junctions, such as alloyed ones, the capacitance is proportional to $1/\sqrt{V_j}$, and for graded junctions, such as diffused, to $1/\sqrt[3]{V_j}$.

Transistor Action

The junction transistor results from a proper combination of two diodes in a single-crystal structure. The common region is made thin and called the base. One junction called the emitter is forward biased, the other called the collector, reverse biased.

The emitter provides carriers which are injected into the base region. Most of these carriers cross the base region, either by diffusion alone or by diffusion aided by a drift field, and enter the collector depletion region. They are swept into the collector region by the field and flow out the collector terminal.

For a combination of reasons, not all the emitter current injected into the base actually reaches the collector. The portion of the emitter current that reaches the collector is given by alpha, or h_{fb}, the short-circuit forward-current amplification factor in the common-base connection. In a junction transistor h_{fb} varies from -0.9 to -0.997. The higher h_{fb}, generally, the better the transistor. Under normal conditions h_{fb} is always less than unity. An average value is -0.98.

Current Amplification

Several effects go to determine the value of h_{fb}, the common-base short-circuit current gain. They are summarized in the equation

$$h_{fb} = \beta \gamma \delta$$

where β is the transport factor, γ is the emitter efficiency, and δ is the collector multiplication factor.

Emitter Efficiency

The first component of h_{fb} to be considered is γ, the emitter efficiency. This is a measure of the net forward current flowing across the emitter-base junction. The forward bias on the junction permits electrons (*npn* case) from the emitter to flow into the base. This flow constitutes the emitter current that heads for the collector. However, the forward bias also permits holes to flow from the base to the emitter. This hole current makes no contribution to collector current. The ratio of emitter-junction electron current to total emitter junction current (electrons and holes) is γ, the emitter efficiency. Gamma is a number less than, but very close to, unity.

Transport Factor

In a typical junction transistor, β is a number less than, but close to, unity. The quantity $(1 - \beta)$ is a measure of the portion of the emitter current (electrons for *npn* units, holes for *pnp* units) that recombines in the base region and consequently never reaches the collector.

Considering an *npn* example: some of the electrons injected from the *n*-type emitter region recombine with holes, the majority current carriers in the *p*-type base region. The charge carriers that participate in these recombinations make up the base current of the transistor. Base current flows out the base and never reaches the collector.

In a good transistor, beta is made as high as possible. This is done by keeping the base region as narrow as possible to decrease the possibility of recombination. Also, the transistor material is kept as free as possible of unwanted impurities. Impurities, such as boron in germanium, form recombination centers in which an electron may be trapped and held until it recombines with a hole.

Collector Multiplication

The quantity δ (delta) is the collector multiplication factor. It is either equal to, or slightly more than, unity. As electrons are accelerated toward the collector, they collide with and ionize semiconductor atoms in the depletion region, setting free additional electrons to add to the total collector potential.

Variation of Current Gain

The current gain may be seen in Fig. 6 to be dependent upon the collector potential. At very low values of collector potential V_c, the collector-base junction is insufficiently back-biased and some collector electrons flow into the base region (*npn* case) reducing net collector current and consequently the value of h_{fb}. As V_c is increased h_{fb} reaches its normal value. A further increase in V_c causes h_{fb} to approach unity almost asymptotically. In this region as h_{fb} approaches unity β increases as V_c sweeps electrons more and more rapidly across the base region permitting less recombination. As V_c is increased still further the influence of δ, the collector multiplication factor, becomes predominant and h_{fb} increases rapidly to values greater than unity.

Emitter Base Breakdown

The base-emitter junction of the transistor forms a diode. It is normally forward biased, but in switching and some audio amplifier applications the diode may be reverse biased. It is important that the diode be capable of withstanding this reverse voltage. Alloy transistors usually have emitter-base diode breakdowns in the order of one-half to one-fourth that of the collector-base diode. Grown junction transistors often have emitter-base diode breakdowns of 1 to 5 volts. Diffused transistors usually have emitter-base breakdowns in the 3- to 10-volt region. The manufacturer specifies breakdown and often a leakage current at some fraction of the breakdown voltage.

The collector-base diode has the same sort of characteristics as the emitter-base diode, but the breakdowns are higher, somewhere in the 10- to 200-volt range. The breakdown voltage and one or more values of leakage current associated with this diode are normally supplied by the manufacturer.

Collector-Base and Collector-Emitter Breakdown

An important transistor parameter is the collector junction reverse breakdown potential BV. Since, for reasons to be developed later, transistors are usually operated with signal applied to the base and the emitter grounded, junction breakdown potential is measured in this, the common-emitter connection. Junction breakdown potential measured this way is designated BV_{CER}. Since junction breakdown is not sharply defined, a value of collector reverse current is specified at which breakdown is considered to have taken place. For diffused transistors, the voltage-current relationship exhibits a negative resistance region as shown in Fig. 7. The effect of external emitter-base resistance is also shown. Since the open-base breakdown, BV_{CEO}, is the most stringent rating, it is usually given by the manufacturer. Care should be exercised, because exceeding this breakdown can sometimes result in destruction of the transistor.

Transit Time Effects

Since charge carriers usually move across the base region of the transistor by the process of diffusion they describe random, dog-leg paths rather than straight lines. It takes a significant amount of time for the carriers to cross from emitter to collector. Thus, conditions inside the transistor are such that the charge carriers are unable to move fast enough to follow the rapid variations of a high-frequency signal, in the same manner as they follow the variations of a low-frequency signal of the same amplitude. Indeed, the shorter the time that minority carriers spend in the base region, the better the transistor. Not only do transit time effects restrict the operating frequency of the transistor but when minority carriers spend a relatively long time in the base region, more opportunities exist for recombination, which also tends to reduce transistor gain even at lower frequencies.

To get the charge carriers across the base faster, either the collector can exert a stronger pull or the base region can be made smaller. The latter measure is used in surface barrier transistors.

Alternatively, a nonuniform field of increasing attraction throughout the base region can be used to pull charge carriers faster and faster toward the collector. The nonuniform accelerating field is used in drift and gaseous diffusion transistors.

Also, the charge carriers can be forced to follow a straight line and not wander throughout the base region. This restriction of charge carrier diffusion is accomplished electrically in the tetrode transistor and physically in the mesa transistor.

Some or all these methods of coping with high frequency and charge carrier recombination effects can be combined. For example, a gaseous diffusion transistor inherently provides a narrow base region and permits use of high collector potential as well as supplying a nonuniform accelerating field. The gaseous diffusion manufacturing technique is then used to make transistors of the mesa shape to get the maximum high-frequency performance.

The inability of charge carriers to follow the variations in an applied higher-frequency signal is known as transit time effect and also occurs in vacuum-tube amplifiers. Transit time begins to take effect when the time it takes for charge carriers to diffuse across the base region becomes a significant part of the input-signal frequency cycle. The result of transit time effect is that h_{fb} at high frequencies drops off relative to h_{fb} at low frequencies. The low-frequency value of the short-circuit common-base forward-current gain is usually written h_{fbo} in discussions of high-frequency effects.

Also, h_{fb} lags h_{fbo} by a phase angle that increases with signal frequency. This lagging phase angle represents the phase angle of the input-signal cycle corresponding to the time it takes for charge carriers to diffuse across the base region.

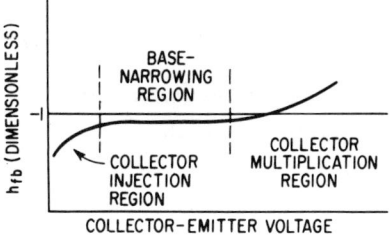

FIG. 6—Variation of transistor bias with collector-emitter potential

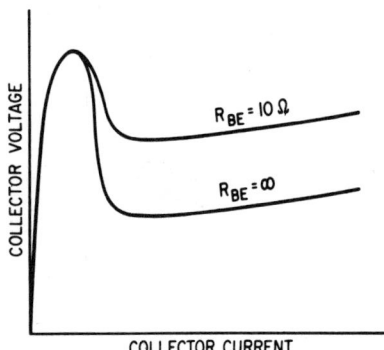

FIG. 7—Voltage-current relationship for diffused transistors shows a negative-resistance region

Both magnitude and angle variations of h_{fb} are illustrated in Fig. 8.

Effects on Current Gain

The frequency dependence of h_{fb} at high frequencies is given by

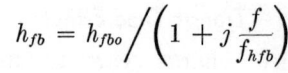

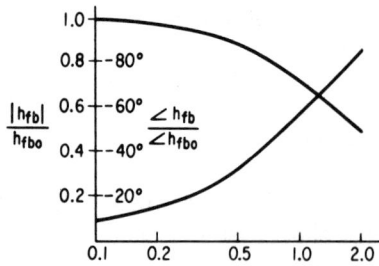

FIG. 8—Change in phase and magnitude of h_{FE} with frequency (normalized with respect to alpha cutoff frequency) for common-base connection.

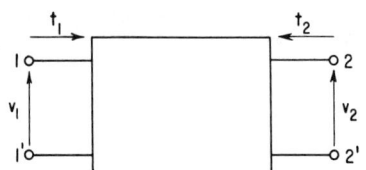

FIG. 9—Four-terminal representation of a transistor

$$h_{fb} = h_{fbo} \bigg/ \left(1 + j\frac{f}{f_{hfb}}\right)$$

where h_{fbo} is the value of alpha at audio frequencies, f is the operating frequency, and f_{hfb} is the current-gain cutoff frequency. The current gain cutoff frequency is defined as the frequency at which h_{fb} has dropped to 0.707 times its low-frequency value or, h_{fbo}. This drop off in current gain corresponds to a loss in power gain of 3 db.

Quantity h_{fb} is a phasor quantity. It has magnitude and direction.

Consider a transistor with a low-frequency current gain of 0.98 when operated at its alpha cutoff frequency

$$h_{fb} = 0.98/(1 + j1)$$
$$= 0.707 \times 0.98 \underline{/-45°}$$

This result demonstrates that the forward current gain is lower at high frequencies and that the current coming out of the transistor lags the current going in since

$$I_c = h_{fb} I_e$$

For the example given, the phase lag is 45 degrees. More accurate formulas, however, disclose that the phase angle is more like 50 degrees. This phase shift in current indicates transit time effect. It is apparent that the designer is faced, at high frequencies, not only with a loss in gain but also with troublesome feedback effects.

The common-emitter connection is even more sensitive to undesirable high-frequency effects on current gain than is the common-base connection. The common-emitter cutoff frequency is given by

$$f_{he} = f_{hb}(1 - h_{fbo})$$

and the phase and magnitude of h_{fe} are given by

$$h_{fe} = h_{feo} \bigg/ \left(1 + j\frac{f}{f_{hfe}}\right)$$

The Transistor as a Circuit Element

The preceding sections have considered the transistor as a physical entity, but the user is interested in the transistor primarily as a circuit element. Many of the techniques of network design can be applied to transistor circuits. The following section will treat the transistor as an element of a circuit, giving information useful to the circuit designer.

One convenient way to represent the transistor is shown in Fig. 9, where a four-terminal network is used. Various means exist of relating the four external voltages and currents to the internal network. Each way, however, requires four network parameters to be specified. These four parameters may take the form of resistances, admittances, or dimensionless ratios.

Hybrid Matrix Parameters

Today most circuit design is done with hybrid matrix or h-parameters. There are four h-parameters for each of the three basic transistor amplifier connections. Measuring techniques are demonstrated in Fig. 10.

The first h-parameter is h_{11}, the input impedance in ohms of a four-terminal network with the two

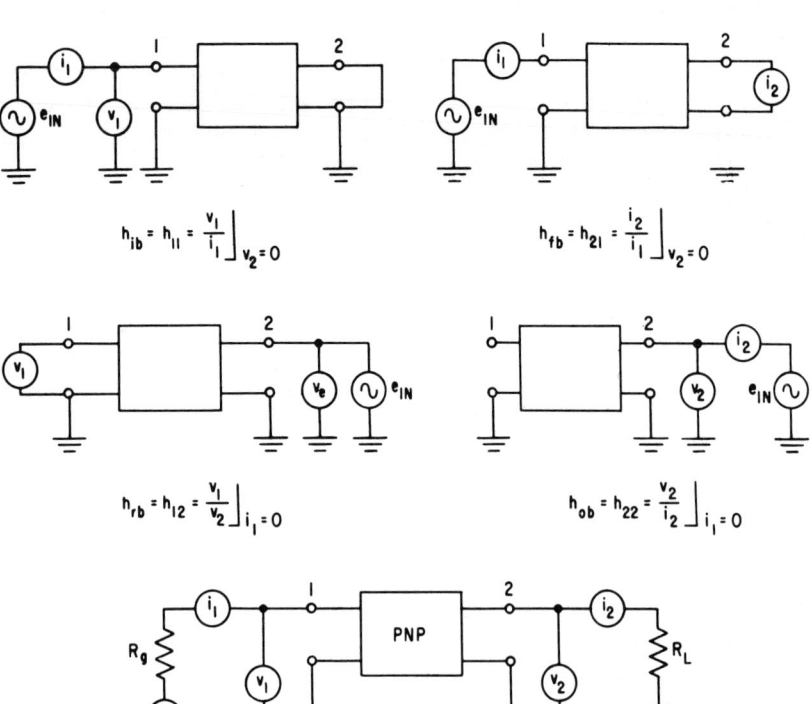

FIG. 10—Measuring techniques for h-matrix parameters

8

output terminals short circuited (Fig. 10, top, left). In transistor manuals the input impedance in the common-base connection is usually given as

$$h_{ib} = h_{11} = \left.\frac{v_1}{i_1}\right|_{v_2=0}$$

In the Institute of Radio Engineers nomenclature for h-parameters there are two subscripts for each hybrid matrix parameter symbol. The first subscript may be i, o, f, or r, meaning respectively: input, output, forward, or reverse. The second subscript may be b, e, or c, pertaining respectively to the common-base, common-emitter, or common-collector connection.

The second hybrid parameter is the forward current transfer ratio, h_{21}. It is a dimensionless ratio. In the common-base connection, it is h_{fb} or α. In the common-emitter connection h_{fe} is equal to α_{ce} or β referred to in various writings. Transistor manuals usually give the value of $h_{fe} = \alpha_{ce}$. This can readily be converted into h_{fb} by

$$h_{fb} = -h_{fe}/(1 + h_{fe})$$

The forward-current transfer ratio is measured with the output circuit shorted (Fig. 10, top, right) and is given by

$$h_{fb} = h_{21} = \left.\frac{i_2}{i_1}\right|_{v_2=0}$$

The third h-parameter is h_{12}, the reverse voltage feedback ratio. It is also dimensionless. Transistor manuals usually quote the common-base value of this ratio. The ratio is measured with the input circuit open (Fig. 16, bottom, left)

$$h_{rb} = h_{12} = \left.\frac{v_1}{v_2}\right|_{i_1=0}$$

The last h-parameter is the output conductance, h_{22}, measured in mhos. It is usually quoted for the common-base connection and is measured with the input circuit open (Fig. 10, bottom, right)

$$h_{ob} = h_{22} = \left.\frac{i_2}{v_2}\right|_{i_1=0}$$

The term "hybrid" derives from the fact that these parameters are given in ohms, mhos, and two dimensionless ratios, not all in one compatible set of units such as the equivalent-T parameters, which are all in ohms.

Some typical values for common-emitter h-parameters are: $h_{ie} = 1,500$ ohms; $h_{fe} = 50$; $h_{re} = 11 \times 10^{-4}$ and $h_{oe} = 50 \times 10^{-6}$ mho. Some typical values for common-base h-parameters are: $h_{ib} = 40$ ohms; $h_{fb} = 0.98$; $h_{rb} = 4 \times 10^{-4}$; and $h_{ob} = 1 \times 10^{-6}$ mho.

Equivalent-T Circuit

For transistor circuit design work, it is sometimes convenient to be able to represent a transistor by an equivalent electrical circuit. The equivalent-T presents a good picture of what a transistor looks like electrically. The equivalent-T circuit of a common-base transistor amplifier is shown in Fig. 11(A).

Quantity r_e is the internal emitter resistance, on the order of 20 to 150 ohms for junction transistors; r_b is the internal base resistance on the order of 500 to 1,000 ohms; and r_c is the internal collector resistance, from 1 to 4 megohms. Quantity $r_m i_e$ corresponds to voltage generator μe_g in vacuum-tube analysis; r_m, the mutual resistance, is equal to $-h_{fb} r_c$. The lower-case letters indicate equivalent circuit components within the transistor itself. Quantities R_e and R_l represent external emitter and load resistors.

The foregoing parameters are measured by considering the transistor as a four-pole network, Fig. 11(B). Input voltage and current, designated v_1 and i_1, respectively are measured at audio frequencies. As indicated in Fig. 12, these measurements are made between emitter and base. Output current and voltage, designated v_2 and i_2,

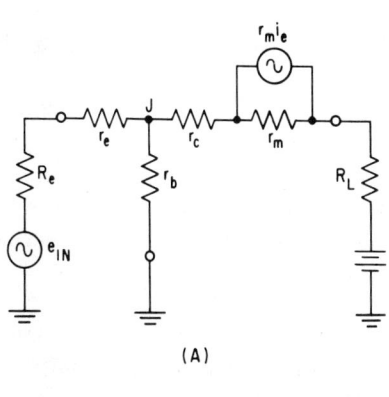

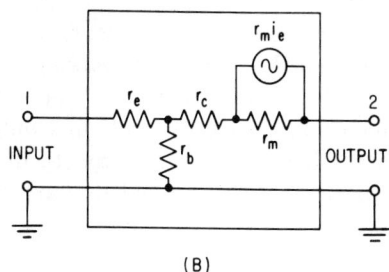

FIG. 11—Transistor equivalent-T circuit (A), and four-pole network representation (B)

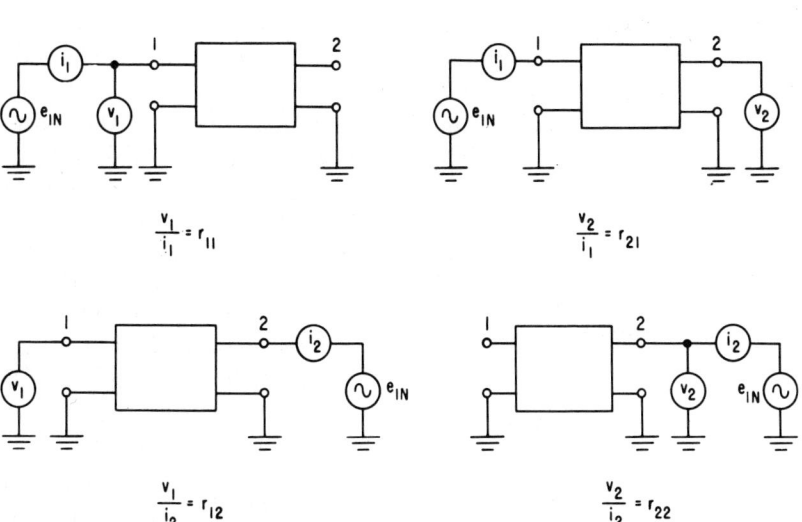

FIG. 12—Measuring techniques for obtaining equivalent-T circuit parameters. Measurements are taken at audio frequencies

are measured between collector and base.

From the measurements indicated in Fig. 12, the following relationships are obtained:

$$r_{11} = v_1/i_1$$
$$r_{12} = v_1/i_2$$
$$r_{21} = v_2/i_1$$
$$r_{22} = v_2/i_2$$

The equivalent-T circuit components r_e, r_b, r_c, and r_m are then obtained from solution of the following equations:

$$r_{11} = r_e + r_b$$
$$r_{12} = r_b$$
$$r_{21} = r_m + r_b$$
$$r_{22} = r_c + r_b$$

However, in designing transistor circuits the important parameters are the input resistance, R_i; the output resistance R_o; and the voltage gain A_v. These circuit parameters may be derived from the equivalent-T transistor representation as will be shown later.

High-frequency Equivalent-T Circuit

A high-frequency equivalent-T circuit of a common-base transistor amplifier is shown in Fig. 13. The equivalent-T circuit reveals that both the emitter and collector internal resistances are shunted by capacitances. The input capacitance is associated with the emitter-base junction. It is the diffusion capacitance C_D and is characteristic of forward-biased semiconductor diodes. The output capacitance is the collector capacitance C_c and is characteristic of reverse biased semiconductor diodes.

Collector Capacitance

The collector capacitance C_c is variously known as junction

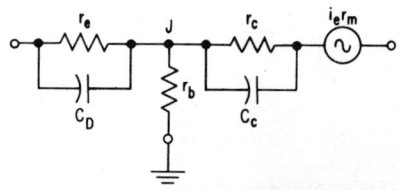

FIG. 13—High-frequency equivalent-T circuit

capacitance, barrier capacitance, space-charge capacitance, transition capacitance, and depletion-layer capacitance. Evaluation of collector junction capacitance is done as described in sections on the p-n junction. In junction transistors C_c ranges from 0.5 to 50 micromicrofarads (picofarads).

Diffusion Capacitance

An appreciable capacitance, called diffusion capacitance, shunts the emitter-base junction.

The diffusion capacitance cannot readily be given a physical meaning similar to the parallel-plate capacitor. Diffusion capacitance does, nevertheless, relate to an incremental change in charge concentration with applied voltage.

In a *pnp* transistor, the diffusion capacitance may be considered the storage capacitance for holes or minority carriers piled up in the base region before diffusing on to the collector. Consider the minority carrier distribution in the base region. The distribution decreases almost linearly from a high value near the emitter junction to nearly zero at the edge of the collector depletion region. The amount of charge stored in this region depends upon the mean distance a hole travels before recombination. This diffusion capacitance or storage capacitance for holes is given by

$$C_D = L_p^2 g/D_p$$

where L_p is the mean distance in meters that a hole travels before recombination, D_p is the diffusion constant for holes (0.5 m²/sec for holes in germanium, *pnp* case; 1.0 m²/sec for electrons in germanium, *npn* case), and g is the dynamic conductance of the emitter-base diode i_e/v_{be} in mhos.

Dimensionally: L_p is in meters; D_p is in meters²/second and g is in mhos (ampere/volt or coulomb/volt-second). Therefore

$$C_D = (\text{meter}^2) \frac{\text{coulomb}}{\text{volt-sec}} \frac{\text{second}}{(\text{meter}^2)}$$

which is the familiar relationship for capacitance: $C = q/v$. For audio transistors C_D can run as high as 5,000 micromicrofarads.

Capacitance Effects Compared

Consider a *pnp* transistor designed for low frequency amplifier service. Equivalent circuit values are given by: $C_D = 5,000$ pf; $C_c = 50$ pf; $r_b = 100$ ohms; $r_e = 20$ ohms; and $r_c = 1$ megohm. At an input frequency of 84 kc, the emitter resistance is shunted by a capacitive reactance of

$$X_c = 1/2\pi f C = 380 \text{ ohms}$$

Thus the input arm of the equivalent-T circuit has an impedance of

$$Z_e = r_e j X_c/(r_e + j X_c)$$
$$= 19.8 \underline{/5°} \text{ ohms}$$

The capacitive reactance of the output arm of the equivalent-T circuit is

$$X_c = 1/2\pi f C = 38,000 \text{ ohms}$$

and the impedance of the output arm is

$$Z_c = 38,000 \underline{/89°} \text{ ohms}$$

Thus a very large value of diffusion capacitance shunting a small resistance has little effect on the emitter arm, while a small value of collector capacitance shunting the large collector resistance has an important effect indeed on the collector arm of the circuit.

Theory indicates several courses to take in designing transistors for high frequency use. Recalling that

$$C_c = A\epsilon/W_d$$

it may be seen that making a transistor physically small can improve high-frequency performance by reducing the collector area and the base width. However, making the transistor small decreases its power-handling ability by providing only a small volume of material to dissipate the heat created by current flow.

Collector Potential

Increasing collector potential is another way to increase high-frequency response. Higher inverse voltage applied to the collector-base diode of a junction transistor increases the width of the collector depletion region. Collector junction capacitance is reduced as depletion layer width is increased.

Transistor Gain

Probably the most important attribute of any active electronic device is gain, or the ability to transform a small signal into a larger one. The transistor is capable of gain at very low power levels, with good efficiency. At the microwatt and milliwatt levels, absence of a filament makes the over-all circuit efficiency higher than that obtainable with a vacuum tube.

One characteristic of the transistor which sometimes causes difficulty in circuit design is the finite power gain. A vacuum tube operated Class A with the grid negatively biased is, for all practical purposes, an infinite-power-gain device. A voltage on the grid can control large amounts of output power in the plate circuit.

A transistor, however, always has a finite input resistance, making some input power necessary for its proper operation. This finite power gain prevents the practice common in vacuum tube circuits of considering voltage gain and power gain as synonymous. To design accurately a transistor circuit, power gain must be taken into account.

The two constituents of power gain in the transistor are current gain and voltage gain. Voltage gain is the ability of an active element to accept a small input voltage and produce a larger voltage at the output terminals. Current gain, likewise, is the ability to accept a small input current and produce a larger output current. In the transistor a small voltage across the low input resistance produces a larger voltage across a high load resistance. This may or may not represent power gain, depending on the resistance and voltage gain. A small current flowing into the low input resistance of a common emitter stage produces a larger current through a larger load resistance, giving power gain.

It may seem odd to speak of a "gain" of less than unity. However, h_{fb} measures only forward current gain. The resistance of the forward-biased emitter diode circuit is a few hundred ohms or less and the resistance of the reverse biased collector diode circuit is a megohm or more. This means there can be large voltage and power gains.

Consider a grounded-base transistor amplifier with $h_{fb} = -0.96$, collector current $I_c = 20$ ma, a change in emitter current $i_e = 1$ μa, and emitter resistance $r_e = 500$ ohms. Short-circuit voltage gain across the transistor alone, disregarding the rest of the circuit is approximately

$$A_v = \frac{e_{out}}{e_{in}}$$

Since $h_{fb} = \frac{i_c}{i_e}$, emitter current may be found to be 20.8 ma. Thus

$$A_v = \frac{i_c r_c}{i_e r_e}$$
$$= \frac{0.96 \times 1 \times 10^6 \times 10^{-6}}{1 \times 500 \times 10^{-6}}$$
$$= 1,920$$

considering power again

$$A_p = \frac{P_{out}}{P_{in}} = \frac{e_{out} i_{out}}{e_{in} i_{in}}$$
$$= h_{fb} A_v = 1,843$$

Expressed in decibels

$$A_p \text{ (db)} = 10 \log 10 \frac{P_{out}}{P_{in}} = 32.6 \text{ db}$$

Even though the current gain of a junction transistor is less than unity, large voltage and power gains may be achieved with these devices.

Transistor Characteristic Curves

Of the several graphical representations of transistor parameters, the one most commonly encountered is the average collector characteristic. Here the collector-to-emitter voltage is displayed along the abscissa. It is conventional to represent the absolute magnitude of collector voltage in all cases even though the collector voltage increases negatively for *pnp* transistors. Likewise, the collector current, usually expressed in milliamperes, is shown along the ordinate increasing from bottom to top ignoring the fact that collector current increases negatively for *pnp* units. Base current is shown as the running parameter.

Note in Figs. 14 and 15 that the transistor collector characteristic resembles the plate characteristic curves of a pentode vacuum tube.

Similarly to vacuum-tube curves,

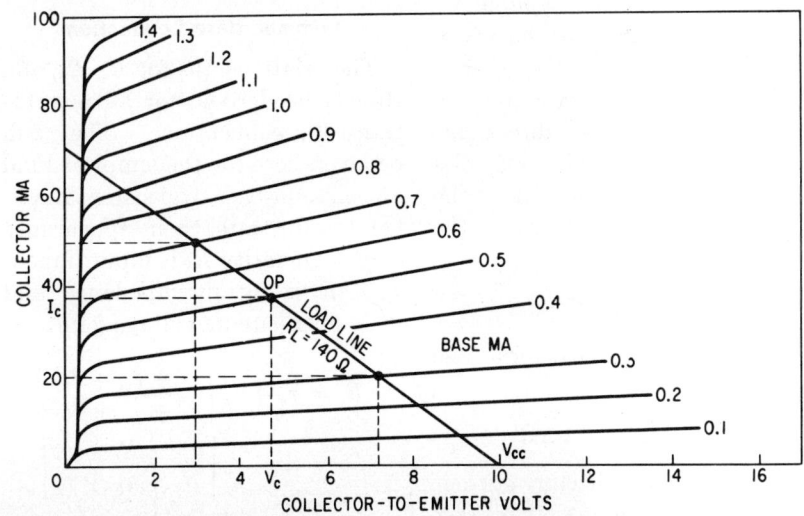

FIG. 14—Average collector characteristics for an RCA type 2N647 *npn* audio power amplifier

11

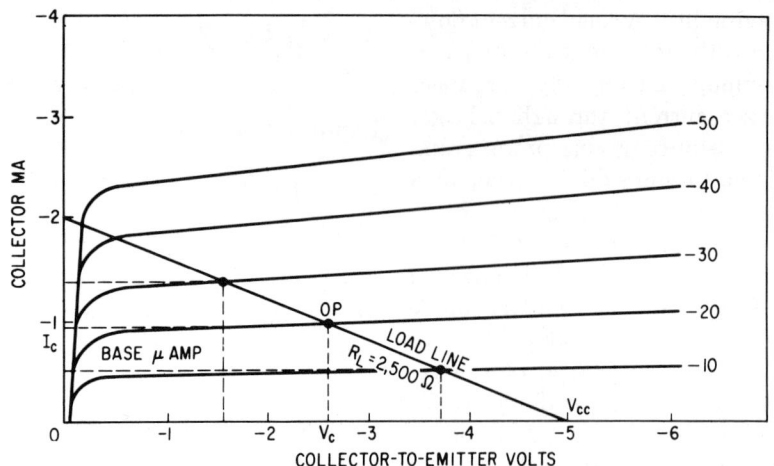

FIG. 15—Average collector characteristics for an RCA type 2N331 *pnp* audio amplifier

transistor characteristic curves are useful in the design of amplifiers. A load line may be constructed by establishing point V_{CC}, the collector supply potential on the abscissa where $I_C = 0$. Thus with almost no collector current flowing, practically the entire collector supply potential appears between collector and emitter of the transistor. Actually with no base current flowing, a small amount of collector current I_{CEO} does flow. This is known as the collector-emitter cutoff current and is usually only a few microamperes in magnitude. It is dependent on the collector-base leakage current, I_{CBO}, and the current gain of the transistor. The load line is established by drawing it through point $I_C = 0$ and the desired operating point (I_C, V_C).

For a type 2N647 *npn* audio power amplifier whose characteristics are shown in Fig. 14, assume a collector supply potential of 10 volts. The desired direct-current operating point is $V_C = 5$ volts and $I_C = 35$ ma. The slope of the load line is given by the ratio of the two zero intercepts:

$$R_L = 10 \text{ volts}/70 \text{ ma}$$

or 140 ohms. Note that for a base current swing of

$$0.7 - 0.3 = 0.4 \text{ ma}$$

there results a collector current swing $50 - 20 = 30$ ma. This represents a current gain of 75.

For a type 2N331 *pnp* audio amplifier whose characteristics are shown in Fig. 15, assume a collector supply potential of -5 volt. The operating point, OP, is $V_C = 2.5$ volts, $I_C = 0.8$ ma. Slope of the load line is

$$R_L = -5 \text{ volt}/-2 \text{ ma}$$
$$= 2,500 \text{ ohms}$$

Note that a base current swing of 10 microamperes results in a collector current swing of 1 milliampere, a current gain of 100.

Amplifier Configurations

The transistor has been considered alone, but it must be used in a circuit. The following sections show how the circuit performance is affected by the transistor connections.

Common-Base Connection

The relationships for R_i, R_o, and A_v can be derived for all possible transistor connections. They will be given here for the simplest kind of common-base transistor amplifier to illustrate their appearance only. Quantity R_i is input impedance, R_o is output impedance, and A_v is open-circuit voltage gain

$$R_i = r_e + r_b \left[\frac{r_c - r_m + R_L}{r_b + r_c + R_L} \right]$$

$$R_o = r_c - r_b \left[\frac{r_m - R_e - r_e}{R_e + r_b + r_e} \right]$$

$$A_v = \frac{r_m + r_b}{R_e + r_b + r_e}$$

In a typical common-base junction transistor amplifier, R_E the external emitter resistance is 500 ohms; R_L the external collector load resistor is about 100,000 ohms.

Figure 16(A), shows that a common-base transistor amplifier is electrically similar to a grounded-grid vacuum-tube amplifier. The common-base transistor amplifier provides current gain equal to h_{fb}.

A set of typical values for a common-base amplifier are current gain $h_{fb} = -0.98$; voltage gain $A_v = 150$; output impedance $R_o = 1$ megohm; and input impedance $R_i = 30$ ohms.

There is no voltage phase reversal in the common-base junction transistor amplifier. Although the gain in the common-base connection is not so high as in the common-emitter connection, the common-base connection is frequently used because it is not as sensitive to variations in transistor parameters as a common-emitter amplifier. In the common-base connection, less critical selection of transistors is called for than is required when the common-emitter connection is used.

A common-base amplifier requires a low driving resistance. Thus, expensive step-down transformers often must be used to interconnect cascaded common-base amplifiers since the output resistance of a transistor amplifier is high. However, when a transistor amplifier must operate from a low-impedance input, the common-base connection is ideal.

Common-Collector Amplifier

The common-collector transistor amplifier is also known as an emitter follower. Figure 16(A), shows that it corresponds to the grounded-plate vacuum-tube amplifier or cathode follower. A common-collector amplifier has a low output resistance and high input resistance. Voltage gain is less than unity and current gain is given by

$$h_{fc} = \left. \frac{\Delta I_e}{\Delta I_b} \right|_{V_c = 0} = \frac{1}{1 + h_{fb}}$$

There is no voltage phase reversal. Common-collector amplifiers are frequently used as isolation amplifiers or as impedance matching devices.

Common-Emitter Amplifier

A transistor operated common-emitter corresponds to a grounded-cathode vacuum-tube amplifier, Fig. 16(B). Current gain is given by

$$h_{fe} = \frac{\Delta I_c}{\Delta I_b}\bigg|_{V_c=0} = \frac{-h_{fb}}{1 + h_{fb}}$$

Typical values of h_{fe} run from 10 to 300. Typical values of input resistance R_i are 1,000 to 2,000 ohms, but can be from 100 to 100,000 ohms. A typical value of output resistance R_o is 50,000 ohms but the output resistance can vary from 5,000 to 500,000 ohms.

There is a voltage phase reversal in the common-emitter junction transistor connection. A typical value of voltage gain is 300.

Because of significant current gain and high voltage gain, the common-emitter amplifier can provide high power gain. Since the input resistance is relatively high, conventional resistance-capacitance coupling can be used rather than the special matching transformers required for common-base amplifiers.

In some books on transistor circuits, the short-circuit forward-current amplification factor in the grounded-emitter connection is called beta. However, since beta is also used to designate the transport factor, the symbol h_{fe} is preferred to denote common-emitter current gain.

Transistor handbooks often quote h_{fe} rather than h_{fb} since the common-emitter connection is most often used in practical amplifier design. In this case h_{fb} can be found from

$$h_{fb} = \frac{-h_{fe}}{1 + h_{fe}}$$

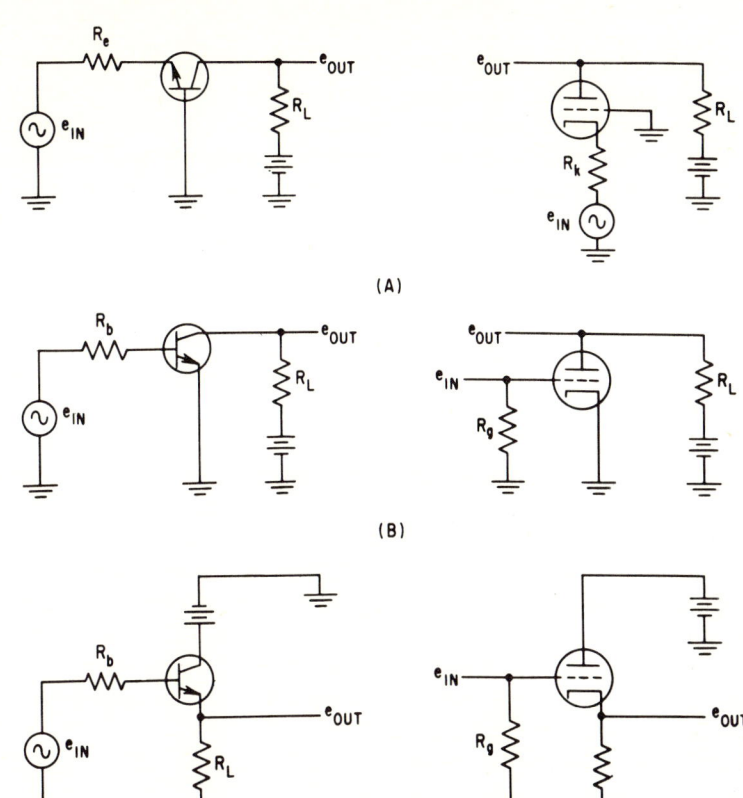

FIG. 16—Transistor amplifiers and corresponding vacuum-tube amplifiers. Grounded-base and grounded-grid amplifiers (A). Grounded-emitter and grounded-cathode amplifiers (B). Grounded-collector (emitter follower) and grounded-plate (cathode follower) amplifiers (C)

Bias Circuits

Heretofore transistor circuits have been depicted with fixed battery bias applied to the collector, base, and emitter terminals. In practice the required bias voltages are usually obtained by voltage divider action from a common supply.

The common-emitter base-input circuit is the most used transistor amplifier connection. The bias required at the base may be obtained by a resistor from base to the collector supply as shown in Fig. 17(A). Neglecting the small voltage drop across the forward-biased base-emitter junction:

$$I_b = V/R_1$$

and collector current is given by

$$I_c = h_{fe}I_b = h_{fe}V/R_1$$

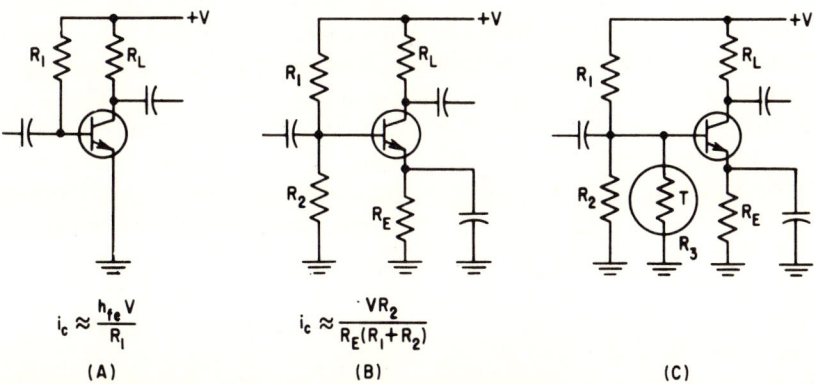

FIG. 17—Biasing circuits: uncompensated (A); emitter resistor compensation (B); and thermistor compensation (C)

The common-emitter connection, however, is susceptible to variations in transistor characteristics. These characteristics not only vary between different transistors but are dependent upon the h_{fe} stability and repeatability from unit to unit of the particular transistor type selected.

Thermal effects can be a problem also. As the ambient temperature at which the transistor is operated increases, the number of available charge carriers increases, causing an increase in the collector current. This increase in collector current means increased collector heat dissipation and increased opportunity for collision of charge carriers and semiconductor atoms. As a result of these collisions still more charge carriers are made available. This effect is a cumulative process and can lead to a condition called collector runaway, which can permanently damage the semiconductor device. Thermal effects are more pronounced with germanium units than with silicon units because of the lower energy gap of germanium.

Stability Factor

A measure of the susceptibility of a transistor amplifier circuit to the kinds of thermal instability described above is S, the stability factor. This factor might more properly be called the instability factor. Stability factor S is defined as

$$S = \frac{\Delta I_c}{\Delta I_{CBO}}$$

which is the ratio of a change in collector current to a change in the thermally dependent collector cut-off current. For a common-emitter stage the factor S is approximately equal to

$$S \approx h_{fe} \frac{1 + R_E/R_B}{1 + h_{fe} R_E/R_B}$$

where R_E is the external emitter resistor and R_B is the external base resistor. For R_E much larger than R_B, S is nearly equal to unity. This is the most favorable condition for thermal stability. For R_E much smaller than R_B, S nearly is equal to h_{fe}, the least favorable condition for thermal stability. However, a small value of R_B has the undesirable effect of shunting the input to the transistor making less signal power available. A large value of R_E robs power from the collector battery and lowers the collector-emitter voltage thus reducing the available amplifire output. Therefore, a compromise must be achieved between stable amplifier operation and higher amplifier performance.

A biasing arrangement for use with an emitter resistor and a voltage divider in the base circuit is shown in Fig. 17(B). Here collector current is given approximately by

$$I_c \approx V R_2 / R_E (R_1 + R_2)$$

where R_1 and R_2 make up the base-circuit voltage divider. Resistor R_2 is chosen to be 5 to 10 times larger than R_E. Typical values of R_E are between 500 and 1,000 ohms. Resistor R_1 is chosen so that voltage divider R_1 and R_2 sets the desired value of base current bias.

One additional way of reducing thermal instability is to shunt R_2 with thermistor R_3 as shown in Fig. 17(C). The thermistor has a high resistance at design temperature. As the temperature increases, collector current increases as previously described. However, increasing temperature causes the thermistor resistance to decrease. The result is that the voltage divider now reduces the base-emitter voltage and consequently the base current bias. This bias reduction changes the operating point of the transistor, reducing the collector current and thereby compensating for the thermally induced rise in collector current.

Manufacturing Techniques

Two general manufacturing methods are used to make nearly all of the transistors now in use. These methods are assembling or growing. A gaseous diffused transistor can broadly be considered an assembled type.

An alloy-junction *pnp* transistor is assembled by placing a pellet of indium on either side of a thin slab of *n*-type single-crystal germanium. The assembly is then heated until the indium begins to alloy with the germanium. Upon cooling, recrystallization forms the *p*-type emitter and collector regions in the original germanium lattice structure. Although both *pnp* and *npn* transistors can be made by alloying, alloy junction transistors are usually of the *pnp* type. The collector dot is larger than the emitter dot.

There are many methods for making grown junction transistors—double doping, rate growing, and melt-back (or melt-quench) being three. In the double-coping method, a single germanium crystal is grown from a melt of lightly doped *n*-type germanium. An indium pellet is dropped into the melt to form the *p*-type base region. Then the melt is heavily doped to form the second *n*-region. The resulting crystal is an *npn* sandwich which is sliced and diced into bars. The tiny germanium bars are etched to reveal the three regions and three separate gold leads are welded to the emitter, base, and collector.

In the rate-growing method, a crystal is grown from a melt containing both *n* and *p*-type impurities. However, the rate at which the crystal is pulled out of the melt is cycled. Since the *n* and *p*-type impurities do not diffuse through the molten germanium at the same rate, several layers of alternately *p* and *n*-type semiconductor are formed in the crystal. Thus several junctions are available and many more transistors can be made from a single crystal in the rate-growing process than in the double-doping process.

In the melt-quench or melt-back method a long slender rod of double-doped semiconductor material is prepared. The temperature is cycled to permit the rod alternately to melt and to freeze. Because of the different diffusion rates of the n and p-type impurities, a p-n junction is formed during each melt-freeze cycle.

The most versatile method of building transistors involves the diffusion of impurities from the gaseous state into single-crystal slabs. Because the time necessary to do this is measured in hours, a high degree of control can be achieved.

The method has several variations leading to single, double, and even triple diffused types. The simplest process consists of passing the "carrier" gas containing the desired impurities over a high-resistivity slab of semiconductor to form the collector-base junction. A second such diffusion produces the emitter-base diode. These diffusions may be combined into one by choosing impurities which allow the base diffusion to move into the slab faster than the emitter diffusion.

The advantages of the diffusion method are uniformity, large numbers of units produced simultaneously, and good control for high-frequency performance. Large-area junctions can be produced in this manner also, making power transistors feasible. Most new transistor types being developed today use some variation of the basic diffusion process.

Once the basic three layer "sandwich" is formed, contacts are alloyed to the emitter and base regions, the excess material etched away, and the collector connected to the "header" or bottom of the transistor can. A completed transistor of this type is shown in Fig. 18(A). Because of the shape this type is called a "mesa" transistor after the mesas of the western deserts.

Another variation of the diffused process is the planar transistor shown in Fig. 18(B). In this type the junctions are formed and a complete transistor exists without etching. This allows somewhat better protection of the transistor surface, resulting in lower values of leakage currents. This type of construction is presently restricted to silicon transistors.

Surface Barrier Transistor

Since base-region minority carrier transit time restricts transistor high-frequency operation, performance can be improved by reducing the width of the base region. A narrow base region is achieved in the surface barrier transistor shown in Fig. 19. The base region is a thin wafer of n-type germanium or silicon. The base and collector impurities are introduced in the form of a salt solution. The salt solution is sprayed onto the base region in a high-velocity stream of small cross section. The electrical potentials applied to the salt stream and wafer initially are such that the salt stream erodes away the base wafer. When the wafer has been made thin enough by erosion, the potential of the salt stream is suddenly reversed and the impurity metal is plated out of the salt solution into the surface of the base wafer. When the leads are welded to the plating, the impurities alloy with the semiconductor material forming the emitter and collector regions. Frequency response well over 100 megacycles has been achieved with surface barrier transistor amplifiers.

Various combinations of the above methods may be used for transistor production. One example of this is the alloy-diffused transistor. Here the emitter-base diode is made by diffusion and the collector base diode by alloying. The drift transistor shows the collector-base diode made by diffusion and the emitter-base diode made by alloying. The grown-diffused transistor utilizes diffusion of impurities in the melt during crystal growing to obtain a graded impurity distribution in the base region. The electro chemical-etch tech-

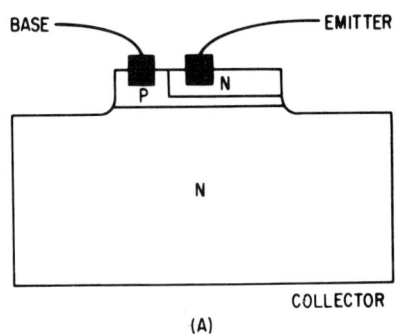

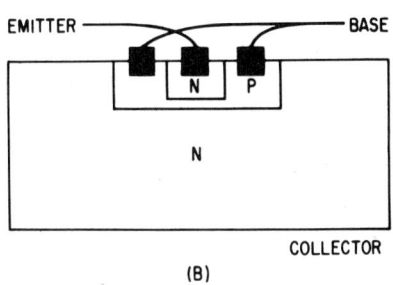

FIG. 18—Mesa transistor (A) and planar transistor (B)

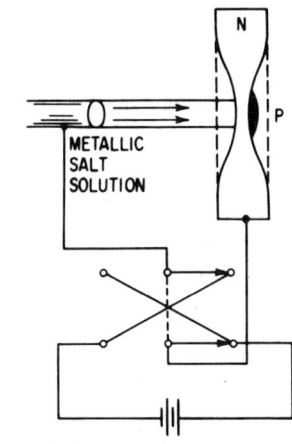

FIG. 19—Manufacturing technique for making surface-barrier transistors

nique has been combined with diffusion to produce a micro alloy-diffused base transistor. Many other combinations or variations are possible, but the above illustrates some which have been used in commercial production.

Figure 20(A) shows that the paths of minority carriers diffusing through the base region of a junction transistor are not straight lines from emitter to collector. The carriers start out headed in nearly all directions and their mean path to the collector is considerably

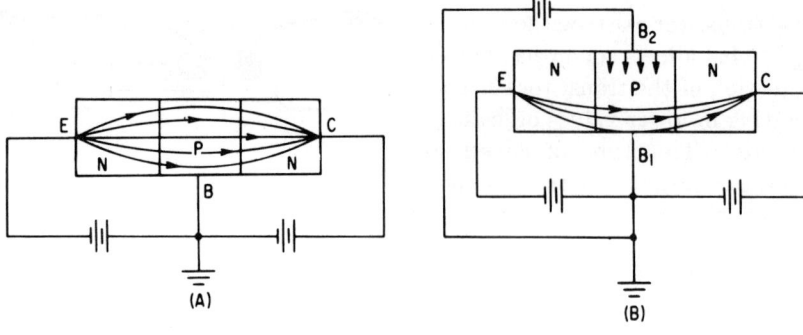

FIG. 20—Diffusion paths in *npn* junction transistor (A). Tetrode transistor action (B).

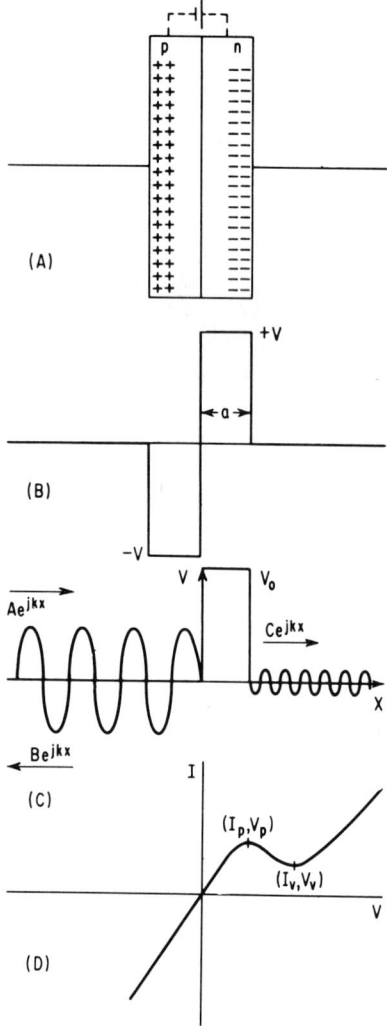

FIG. 21—Theory of the tunnel diode: distribution of free charges in a *p-n* junction (A); potential hill formed by space charge left on crystal lattice (B); electron colliding with one-dimensional square potential barrier (C); and voltage-current characteristic of a tunnel diode (D).

more than the straight line distance from emitter to collector.

However, if the minority charge carriers could be forced to follow a direct path through the base from emitter to collector, both the diffusion and collector capacitances would be reduced. The junction capacitance is reduced by forcing the minority carriers to follow a straight line from emitter to collector, reducing the effective base area. When the electron paths are straightened out, minority carriers remain in the base region for a shorter period of time. This reduces the opportunities for recombination and also reduces the diffusion capacitance. These effects may be seen in Fig. 20(B).

The second base connection is made to the top of the base region. A negative potential is impressed on the second base connection. The negative field repels the electrons crossing the base region. (Almost all tetrodes are *npn* junction transistors.) The electrons are forced to follow a nearly straight line across the bottom of the base region from emitter to collector. This restriction on electron path reduces transit time and permits higher frequency operation. Because most of the operation takes place near the base-1 lead, the effective base resistance is also reduced. Four-terminal transistors of this type are called tetrode transistors. With tetrodes, operation to 70 Mc and above can be achieved.

Special Semiconductor Devices

The emphasis given solid state research by the transistor has resulted in numerous other devices. Some of these have not been proven by time, while others are gaining widespread use. Many have been incorporated in circuits with transistors, others have led to development of circuit designs utilizing their peculiar properties.

The devices described in the following section make use of various phenomena of the solid state, some of them known for many years, others discovered in the continual research accelerated by the invention of the transistor.

Tunnel Diode

The Esaki or tunnel diode is a two-terminal semiconductor device that exhibits a negative-resistance characteristic. This property makes the tunnel diode useful as the active element of amplifiers, oscillators, and multivibrators as well as in such diode applications as detectors, switches, and computer logic elements. The tunnel diode features low noise operation and rapid switching. The tunnel diode takes its name from the tunnel-effect in quantum mechanics described by McColl in the early 1930's. Operation of the tunnel diode was described by the Japanese physicist Leo Esaki in 1958.

One form of the device consists of a heavily doped *p-n* junction diode constructed so as to have a sharp transition from the *p* to *n* regions. One diode terminal is a metal plate bonded to the semiconductor. The other terminal is bonded wire.

When the tunnel diode is biased in the reverse direction, there is heavy reverse current flow because of the high level of impurities. With no applied bias potential, there is initially a small flow of electrons and holes across the junction. These charge carriers neutralize each other and leave space charge regions in the crystal lattice structure from which they came: a positive space charge in the *n* region and negative space charge in the *p* region. See Fig. 21(A). Because the semiconductor material is heavily doped, the space charge potential is high. Because the p-n transition is abrupt, the potential hill is steep as shown in Fig. 21(B).

Classical theory predicts that a large forward bias must be applied to overcome the junction space-charge potential and cause appreciable junction current flow in the forward direction.

However, in the tunnel diode, appreciable currents (up to 5 amp in some units) can flow with very small values of forward bias. This current flow is explained by the so-called tunnel effect. The behavior of an electron at the junction corresponds to the quantum mechanical problem of a particle approaching a one-dimensional square potential barrier of height U_o and thickness a [Fig. 21(C)]. The motion of the particle is given by

$$S(x) = \exp(j2\pi px/h)$$

where j is the complex operator -1, p is momentum, and h is Planck's constant (6.625×10^{-34} joule sec). The notation exp means that whatever follows is an exponent of the Naperian logarithm base e or 2.718. For x less than or equal to 0

$$S(x) = A \exp jkx + B \exp(-jkx)$$

and for x equal to or greater than a

$$S(x) = C \exp jkx$$

This shows that there is a transmitted wave and a reflected wave. (In quantum mechanics it is not unusual to consider an electron as either a wave or a particle.) The situation is analogous to the behavior of microwave energy half a wavelength back from an open circuit in waveguide or to the behavior of light waves impinging on a semiopaque barrier where some energy manages to get through although most of the energy is reflected.

Therefore, even at low forward bias potentials, an appreciable forward current can flow in the tunnel diode. However, this current comes largely from local charge carriers. As forward bias is increased, local charge-carrier depletion soon sets in and the Esaki or tunnel current decreases. As forward bias is increased further, however, the junction space charge barrier is overcome and normal forward conduction results.

The U-shaped voltage-current characteristic shown in Fig. 21(D) illustrates the negative resistance property that makes it possible for the tunnel diode to amplify. This negative resistance characteristic of the tunnel diode resembles the dynatron plate characteristic of the vacuum-tube tetrode.

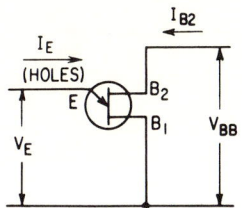 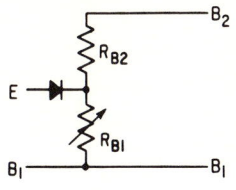

FIG. 22—Double-base diode or unijunction transistor, left, and its equivalent circuit, right

Double-base Diode

The double-base diode or unijunction transistor is a three-terminal semiconductor device that provides a highly stable negative resistance effect useful in oscillators and in bistable pulse circuits such as flip-flops, multivibrators, trigger, and sweep circuits.

The double-base diode is a silicon bar with two ohmic contacts B_1 and B_2 and a single rectifying contact E. See Fig. 22. The rectifying contact is close to base B_2. With no emitter current flowing, the semiconductor bar acts as a voltage divider $R_{B1} - R_{B2}$. The resistance of the voltage divider can range from 5,000 and 10,000 ohms. A typical value for resistance R_{B1} with $I_E = 0$ is 4,600 ohms.

Under these conditions a voltage V_{BB} appears at the emitter. When V_E is less than V_{BB}, the diode is reverse biased and no emitter current flows. When V_E is more than V_{BB}, the diode is forward biased and emitter current flows. The emitter current consists of holes flowing from E to B_1. The hole current flow results in an equal number of electrons in the E to B_1 region. The end result is that I_E increases and V_E decreases giving a stable negative resistance characteristic. In a typical double-base diode, R_{B1} is 40 ohms for

$$I_E = 50 \text{ ma}$$

Silicon Controlled Rectifier

The silicon controlled rectifier is a regenerative switch whose operation is similar to that of a thyratron or gas-filled electron tube. The silicon controlled rectifier can have a 1-microsecond response time. Some silicon controlled rectifiers can handle 50 amperes continuously. At least one type of silicon controlled rectifier can switch a kilowatt of power or more.

A silicon controlled rectifier or $pnpn$ (anode = p) transistor is shown schematically in Fig. 23(A) and physically in Fig. 23(B).

With reverse voltage, that is with the cathode positive, the con-

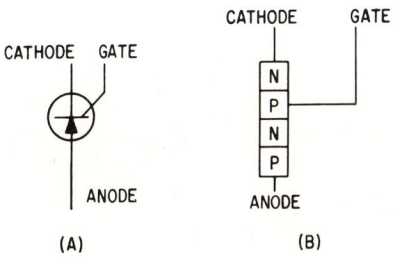

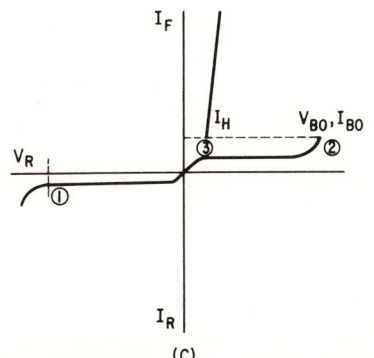

FIG. 23—Silicon controlled rectifier schematic (A), physical representation (B), and operating characteristics (C)

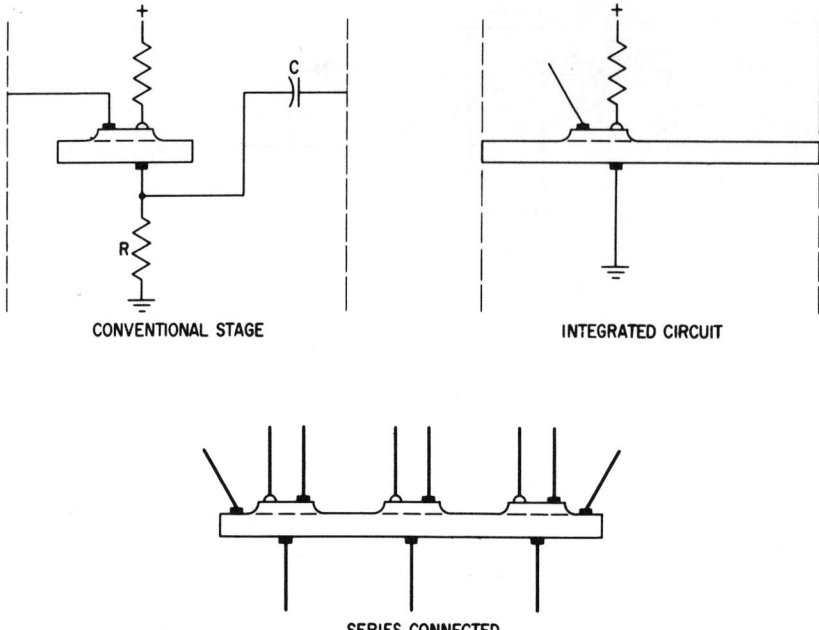

FIG. 24—Use of integrated semiconductor devices to make a digital computer shift register: conventional building block using thyristor, single integrated circuit, and three such integrated circuit stages in series

trolled rectifier blocks the flow of current [point 1 in Fig. 23(C)]. With forward bias, the anode positive with respect to the cathode, the controlled rectifier still blocks the flow of cathode-to-anode current up to the breakdown voltage point V_{BO} (point 2). At a value of anode voltage equal to V_{BO}, the resistance of the controlled rectifier drops almost instantaneously to a very low value.

Cathode current is limited only by the external voltage and the circuit impedance (point 3). At positive anode to cathode voltages less than V_{BO}, the controlled rectifier can be switched to its highly conductive state by applying a low level of gate-to-cathode current.

The controlled rectifier can be switched off by reducing the cathode current to a value less than I_H or holding current. This reduction in cathode current can be achieved by dropping the anode voltage to zero or less, as occurs once every cycle when an alternating voltage is applied to the anode.

Another way to switch off the controlled rectifier is to divert the cathode current around the controlled rectifier for a few microseconds.

Integrated Thyristor Shift Register

Figure 24 illustrates development of an integrated shift register used in digital computers. The active component is a thyristor, a three-terminal semiconductor device which displays a negative resistance characteristic and thus can be used to store one binary digit (one = ON; zero = OFF). This digit or ON signal may be shifted to an adjacent register stage turning that stage ON. The digit shift is accomplished by turning all thyristors in the register OFF simultaneously. Meanwhile, from the stage that was ON originally, a pulse is sent through an R-C circuit physically built into the integrated shift register toward the adjacent shift register stage eventually turning that stage ON. The R-C circuit must have a time constant long enough to store the pulse while all the register stages are being cleared or turned OFF.

In the integrated circuit shown in Fig. 24, the R-C circuit becomes a germanium bar. The shift pulse sends minority carriers from the ON

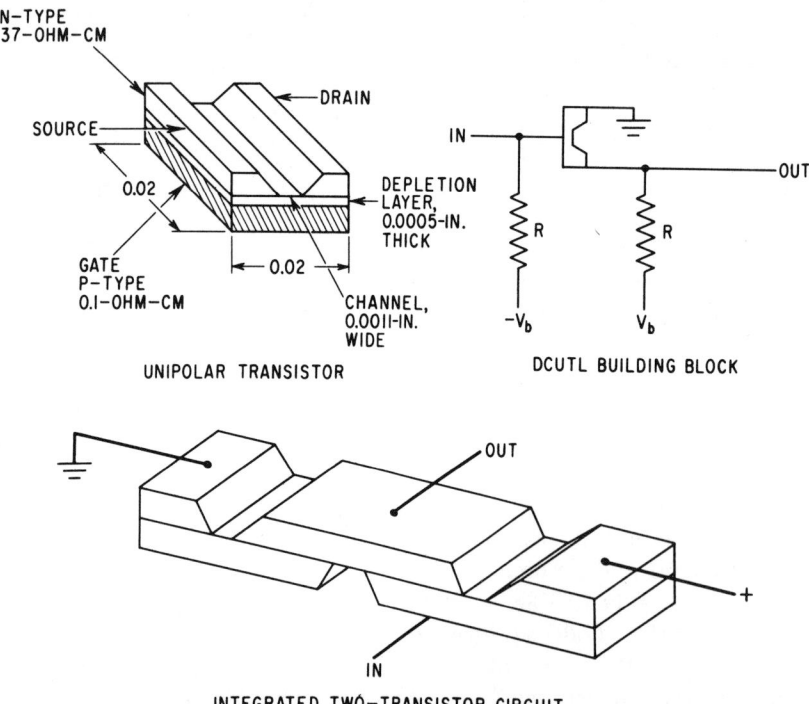

FIG. 25—Use of integrated semiconductor devices to construct direct-coupled unipolar-transistor logic circuits (DCUTL): structure of unipolar transistor, schematic of DCUTL circuit, and two-stage integrated semiconductor circuit

stage down the bar. The carriers are collected at the end of the bar after a time equal to their transit time and are used to trigger the next stage ON. Thus the semiconductor bar functions as the R-C delay line.

Unipolar Transistor Switching

The structures shown in Fig. 25 go to make up a direct-coupled unipolar transistor logic circuit or DCUTL. This circuit is used in digital computers and switching circuits.

A unipolar transistor device has three terminals. These terminals are called the gate, source, and drain. Current flows from the source to the drain and the amount of flow is controlled by the gate.

The unipolar transistor has a *p*-type gate contact. The *p-n* junction is reverse biased resulting in the depletion layer shown. The *n* region has two contacts: the source and the drain. When a gate pulse is applied, the depletion region grows until it encroaches on the current channel region between source and drain making the channel narrower. Now the resistance from source to drain becomes of the order of 10 to 100 megohms.

When the reverse bias is reduced, the depletion layer retreats from the channel region and the resistance from source to drain is reduced to approximately 5,000 ohms. At the same time the gate is insulated from the source-drain by the reverse biased *p-n* junction. In silicon, the value of this insulating resistance is approximately 100 megohms.

The unipolar transistor is used as a voltage-controlled relay. Typical high-frequency cutoff is about 10 Mc.

The basic DCUTL is shown schematically. It consists of a unipolar transistor with input and output load resistors. When several DCUTL building blocks are used, the output resistance of one can serve as the input resistor of the stage following.

Since the value of the output resistors is close to the channel resistance of the unipolar transistor, an additional unipolar transistor may be substituted for the output resistor. The channel width of this additional unipolar transistor is made slightly smaller than it would be normally to obtain the correct value of output resistance for the preceding stage. The unipolar transistor resistance is highly nonlinear, which ensures positive logic switching action in computers and switching circuits.

Semiconductor Solid Circuits

One of the most intriguing developments in the field of semiconductors has been the monolithic circuit. These are called integrated semiconductor devices and Solid Circuit* Semiconductor Networks. It has been shown that semiconductor material can be used as a resistor, capacitor, amplifier, and switch. As yet, it has not been possible to devise semiconductor inductance.

Within a single block of semiconductor material can exist all the necessary components for a great many electronic circuits. Even inductance can be achieved by vacuum deposition of a thin metallic film on one of the surfaces of the block. A whole unit of equipment could conceivably be devised without the use even of interconnecting metal leads.

The main advantage of inte-

* Trademark of Texas Instruments, Inc.

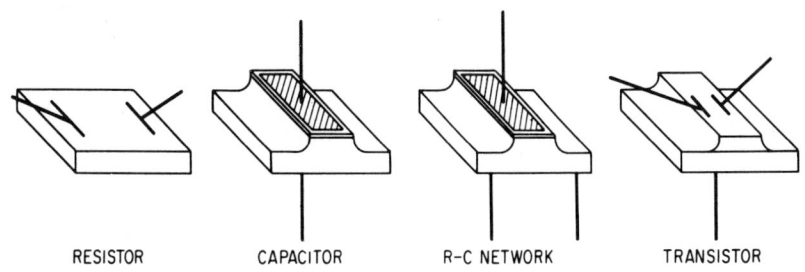

FIG. 26—Development of basic elements for Texas Instruments Incorporated Semiconductor Solid Circuits: resistor, capacitor, resistance-capacitance network, and transistor

grated circuits is that they permit true microminiaturization of electronic equipment. Equivalent packing densities of 100,000,000 conventional parts per cubic foot have been realized in single circuits: about 8 million parts per cubic foot now seems practical in electronic equipment.

Figure 26 illustrates the development of several circuit elements. Resistors are formed by applying ohmic or nonrectifying contacts to a semiconductor wafer. Resistance value is determined from

$$R = \rho l/A$$

where ρ is the resistivity of the material usually silicon, l is the length, and A the cross-sectional area of path. Silicon resistors have low noise characteristics, good linearity, positive temperature coefficients with little variation of resistance at high temperatures, and good stability with passage of time.

Thermistor

One of the simplest semiconductor devices is the thermistor, a resistor whose resistance decreases with increasing temperature. Because increasing current flow causes a rise in temperature in the thermistor due to I^2R losses, increasing current flow has the effect of decreasing the resistance of a thermistor.

This resistance decreases with increasing temperature because the increasing temperature adds energy to the semiconductor atomic systems and makes more conduction electrons available. The property

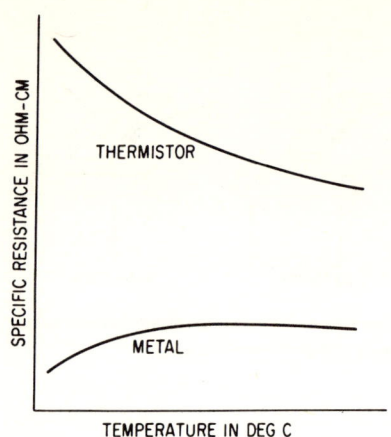

FIG. 27—Variation of resistance with temperature for thermistors and for metals

of decreasing resistance with increasing temperature is called a negative temperature coefficient of resistance and is illustrated in Fig. 27. Most metals have a positive temperature coefficient of resistance: their resistances increase with increasing temperature. Positive-temperature-coefficient thermistors are available also.

The resistance variation in a thermistor with temperature over the thermistor's operating range is given by

$$R = R_0 \exp \beta(1/T - 1/T_0)$$

where R is the final resistance of the thermistor in ohms at temperature T in degrees Kelvin. (The Kelvin, or absolute temperature scale, is a centigrade scale which begins at zero or -273.2 C equal to -459.6 F.) Beta is a constant depending upon the composition of the thermistor. Initial conditions are resistance R_0 and temperature T_0.

Thermistors may be made from mixed oxides such as titanium dioxide (TiO_2), magnesium oxide (MgO), nickel oxide (NiO), manganese oxide (Mn_2O_3), cobalt oxide (CO_2O_3) and iron oxide (Fe_3O_4).

Thermistors are used to measure heat or electrical power. They are especially useful as temperature compensating resistors in transistor amplifiers.

APPENDIX
Matrix Algebra

The "matrix" part of the full name of the h-parameters comes from the fact that useful design equations are derived by setting up the h-parameters in an algebraic form called a matrix:

$$\begin{bmatrix} v_1 \\ i_2 \end{bmatrix} = \begin{bmatrix} h_{11} & h_{12} \\ h_{21} & h_{22} \end{bmatrix} \begin{bmatrix} i_i \\ v_2 \end{bmatrix}$$

In matrix algebra two matrices are multiplied by multiplying term by term each row of one into each column of another in the order shown by the arrows above. Doing this, the following formulas are derived

$$v_1 = h_{11}i_1 + h_{12}v_2 \qquad (1)$$
$$i_2 = h_{21}i_1 + h_{22}v_2 \qquad (2)$$

Two additional formulas may be derived by a process known as taking the inverse of the matrix. A matrix may be inverted by using the formula

$$\alpha_{nm}^{-1} = \frac{1}{\Delta}\frac{\partial \Delta}{\partial \alpha_{mn}}$$

Quantity α refers to every term or "cell" of the matrix. The subscripts nm and mn are the numerical subscripts used in the matrix. Quantity α_{nm}^{-1} is the general term of the inverted matrix. Operator ∂ indicates a partial derivative. Quantity Δ is the determinant of the matrix. The determinant of a 2×2 matrix is the difference of the diagonal products of the matrix:

$$\begin{bmatrix} h_{11} & h_{12} \\ h_{21} & h_{22} \end{bmatrix}$$

$$\Delta = h_{11}h_{22} - h_{12}h_{21}$$

Consider the upper left cell of the matrix:

$$h_{12}^{-1} = \frac{1}{\Delta} \frac{\partial (h_{11}h_{22} - h_{12}h_{21})}{\partial h_{21}}$$

In carrying out the partial differentiation,

$$h_{12}^{-1} = -\frac{h_{12}}{\Delta}$$

After we have inverted the *h*-parameter matrix we premultiply (multiplier on the left) each side of the matrix equation by the inverse of the *h*-parameter matrix. Recalling that the product of any matrix and its inverse is the identity matrix (equivalent to unity), the matrix equation now becomes

$$\begin{bmatrix} i_1 \\ v_2 \end{bmatrix} = \begin{bmatrix} \dfrac{h_{22}}{\Delta} & -\dfrac{h_{12}}{\Delta} \\ -\dfrac{h_{21}}{\Delta} & \dfrac{h_{11}}{\Delta} \end{bmatrix} \begin{bmatrix} v_1 \\ i_2 \end{bmatrix}$$

By multiplying out the new matrix equation, two additional formulas can be derived.

$$i_1 = \frac{v_1 h_{22}}{\Delta} - \frac{i_2 h_{12}}{\Delta} \tag{3}$$

$$v_2 = \frac{i_2 h_{11}}{\Delta} - \frac{v_1 h_{21}}{\Delta} \tag{4}$$

By looking at the external portions of the transistor circuit only and replacing the transistor by its equivalent current and/or voltage generators as provided by Thévenin's theorem, we can write two more formulas:

$$v_2 = -i_2 R_L \tag{5}$$
$$v_1 = i_1 R_B \tag{6}$$

where R_L is the load resistor and R_E or R_B, as the case may be, is the input resistor.

	HYBRID MATRIX (h) PARAMETERS		
	COMMON BASE	COMMON EMITTER	COMMON COLLECTOR
	(circuit with R_e, V_{IN}, V_{OUT}, R_L)	(circuit with R_b, V_{IN}, V_{OUT}, R_L)	(circuit with R_b, V_{IN}, V_{OUT}, R_L)
VOLTAGE GAIN $A_v = \dfrac{V_{OUT}}{V_{IN}}$	$\dfrac{-h_{fb} R_L}{h_{ib} + \Delta R_L}$	$\dfrac{(\Delta + h_{fb}) R_L}{h_{ib} + \Delta R_L}$	$\dfrac{(1 - h_{rb}) R_L}{h_{ib} + R_L}$
CURRENT GAIN $A_i = \dfrac{i_{OUT}}{i_{IN}}$	$\dfrac{-h_{fb}}{h_{ob} R_L + 1}$	$\dfrac{\Delta + h_{fb}}{h_{ob} R_L + M}$	$\dfrac{1 - h_{rb}}{h_{ob} R_L + M}$
INPUT RESISTANCE $R_i = \dfrac{v_{IN}}{i_{IN}}$	$\dfrac{h_{ib} + \Delta R_L}{1 + h_{ob} R_L}$	$\dfrac{\Delta R_L + h_{ib}}{h_{ob} R_L + M}$	$\dfrac{h_{ib} + R_L}{h_{ob} R_L + M}$
OUTPUT RESISTANCE $R_o = \dfrac{v_{OUT}}{i_{OUT}}$	$\dfrac{h_{ib} + R_e}{\Delta + h_{ob} R_e}$	$\dfrac{h_{ib} + M R_b}{\Delta + h_{ob} R_b}$	$\dfrac{h_{ib} + M R_b}{1 + h_{ob} R_b}$

$\Delta = h_i h_o - h_f h_r \qquad M = \Delta + h_f - h_r + 1$

There are four important quantities of interest in transistor amplifier design: the current gain A_i, the voltage gain A_v, the input circuit resistance R_i, and the output circuit resistance R_o. These may be obtained from the relationships

$$A_i = i_2/i_1$$
$$A_v = v_2/v_1$$
$$R_i = v_1/i_1$$
$$R_o = v_2/i_2$$

No simplifying assumptions such as shorting or open-circuiting the input or output to the amplifier may be used here.

In the grounded-base circuit, the equation for current gain can be derived using Eqs. 2 and 5.

$$A_i = -h_{fb}/(h_{ob}R_L + 1)$$

The equation for voltage gain is derived by using Eqs. 4 and 5.

$$A_v = -h_{fb}R_L/(h_{ib} + \Delta R_L)$$

where

$$\Delta = h_{ib}h_{ob} - h_{fb}h_{rb}$$

The equation for the input resistance may be derived using Eqs. 1, 3, and 5.

$$R_i = (h_{ib} + \Delta R_L)/(1 + h_{ob}R_L)$$

The equation for the output resistance is derived using Eqs. 2, 4, and 6.

$$R_o = (h_{ib} + R_E)/(\Delta + h_{ob}R_E)$$

J. Carroll, "Modern Transistor Circuits," McGraw-Hill Book Company, Inc., New York, 1959.

J. Carroll, "Transistor Circuits and Applications," McGraw-Hill Book Company, Inc., New York, 1957.

A. Coblenz and H. L. Owens, "Transistors: Theory and Applications," McGraw-Hill Book Company, Inc., New York, 1956.

D. Dewitt and A. L. Rossoff, "Transistor Electronics," McGraw-Hill Book Company, Inc., New York, 1957.

G. P. Harnwell, "Principles of Electricity and Electromagnetism," McGraw-Hill Book Company, Inc., New York, 1949.

A. E. Hayes, Transistor Formulas Use h-Matrix Parameters, *Electronics*, p. 81, Feb. 28, 1958.

Jack S. Kilby, Semiconductor Solid Circuits, *Electronics*, p. 110, Aug. 7, 1959.

R. C. Lyman and C. I. Jones, Electroluminescent Panels for Automatic Displays, *Electronics*, p. 44, July 10, 1959.

Millimicrosecond Switch, *Electronics*, p. 26, Sept. 19, 1958.

J. Millman, "Vacuum Tube and Semiconductor Electronics," McGraw-Hill Book Company, Inc., New York, 1958.

"RCA Transistors and Semiconductor Diodes," Radio Corporation of America, Somerville, N.J., 1957.

V. Rojansky, "Introductory Quantum Mechanics," Prentice-Hall, Inc., Englewood Cliffs, N.J., 1938.

T. R. Scott, "Transistors and Other Crystal Valves," Macdonald and Evans, Ltd., London, 1955.

M. J. Sinnott, "The Solid State for Engineers," John Wiley & Sons, Inc., New York, 1957.

F. E. Terman, "Electronics and Radio Engineering," McGraw-Hill Book Company, Inc., New York, 1955.

F. E. Terman, "Radio Engineers' Handbook," McGraw-Hill Book Company, Inc., New York, 1943.

"Transistor Fundamentals and Applications," Radio Corporation of America, Camden, N.J., 1958.

"Transistor Manual," General Electric Company, Syracuse, N.Y., 1958.

Tunnel Diode: Big Impact? *Electronics*, p. 61, Aug. 7, 1959.

"2N647 Junction Transistor," Radio Corporation of America, Somerville, N.J., 1958.

"2N331 Junction Transistor," Radio Corporation of America, Somerville, N.J., 1958.

J. T. Wallmark and S. M. Marcus, Semiconductor Devices for Microminiaturization, *Electronics*, p. 35, June 26, 1959.

"Classification of Junction Transistors," Texas Instruments, Incorporated, Dallas, Tex., August, 1959.

Chapter 2
CIRCUIT DESIGN TECHNIQUES

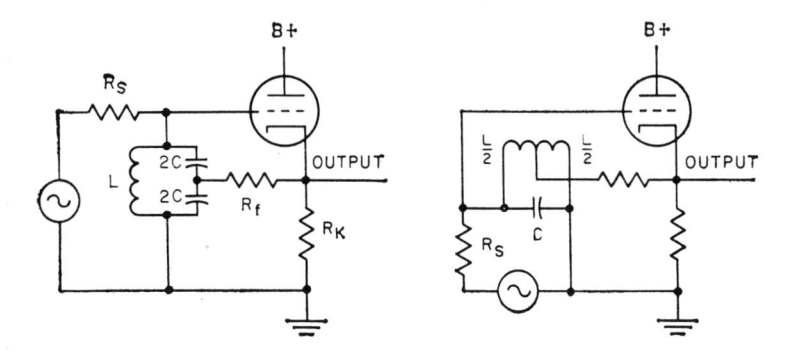

FIG. 1—Vacuum-tube Q-multiplier circuits use center-tapped coil or capacitor divider

Q-Multiplier for Audio Frequencies

High selectivity and stability may be provided in audio-frequency equipment that must be portable, or in which power is at a premium, by use of transistorized Q-multiplier circuit. Series-resonant circuit is applied to variable-selectivity a-f amplifier and multichannel selective-calling unit

By G. B. MILLER   The British Thomson-Houston Co., Ltd., Rugby, England

POSITIVE FEEDBACK to increase the selectivity of a tuned circuit has been put to use on a sound engineering basis only in recent years. The literature gives all the design data for tube-operated Q-multipliers.[1] This article indicates what modifications must be made when transistors are used.

Theory

Figure 1 shows the basic Q-multiplier circuit for either a center-tapped capacitor or coil. The selectivity of this stage is determined by

$$Q_{eff}/Q_o = R_f/(R_f - \tfrac{1}{4}R_d) \quad (1)$$

where Q_{eff} is the effective Q of the coil circuit, Q_o is the Q of the coil at the resonant frequency ω_o and $R_d = \omega_o L Q_o$. The effective Q, and thus the selectivity of the circuit, increases as R_f approaches $\tfrac{1}{4} R_d$. When R_f equals $\tfrac{1}{4} R_d$ the effective Q becomes infinite and the amplifier is unstable; oscillations occur for all values of R_f less than $\tfrac{1}{4} R_d$.

Equation 1 is valid only on the assumptions that the input impedance of the tube is infinite, the output impedance is negligibly small and $g_m R_k$ is much greater than unity.

Figure 2A shows the form taken by the circuit when a transistor replaces the tube. Resistors R_1 and R_2 provide bias to the base of the transistor and C_c prevents the bias from being shorted out by the coil. Since the input impedance to the transistor will not be infinite, Eq. 1 cannot be used as it is.

The value of R_d as used in Eq. 1 is the dynamic impedance of the parallel-tuned circuit at resonance and is purely resistive. The input resistance at XX in Fig. 2A, (with R_f and the tuned circuit disconnected is given by the following:

$$R_g = 1/(1/R_1 + 1/R_2 + 1/R_t) \quad (2)$$

where $R_t = \beta R_k$, the input impedance of the grounded-collector transistor; β is the grounded-emitter current-amplification factor.

The circuit of Fig. 2A can then be replaced by that of Fig. 2B in which the transistor is considered ideal, with infinite input impedance, and L and C are pure reactances. Resistance R_g can be treated as a damping resistor which lowers the Q and reduces R_d to a lower value R'_d where

$$R'_d = R_g R_d/(R_g + R_d) \quad (3)$$

Using R'_d in place of R_d in Eq. 1

$$\frac{Q_{eff}}{Q_o} = \frac{R_f}{R_f - \tfrac{1}{4}R'_d} \quad (4)$$

Derivation of Eq. 4 is not dependent upon the transistor having infinite input impedance. Since the grounded collector stage also has

25

negligibly small output impedance and $g_m R_k \gg 1$ is easily obtained, Eq. 4 can be used for the circuit of Fig. 2A. This equation may be used for circuits using either tubes or transistors.

Stability

The formulas derived with regard to the stability of tube-type Q multipliers all involve Q_o. The effect of the finite input impedance of the circuit using a transistor causes a reduction in the coil Q; this reduction must be taken into account in assessing the stability of the circuit.

The greater the Q multiplication required to achieve a specified selectivity, the lower the stability will be. For maximum stability the value of Q_o should be as high as possible; for a given coil this requires that R_g be large. Unfortunately, maintenance of the correct operating conditions with changes of temperature requires that R_2 be as small as possible. The design must therefore be a compromise between these two conflicting requirements.

Temperature Effects

A change of temperature shifts the operating point of the transistor and causes a change in the current amplification factor β.

Normal methods of temperature stabilization cannot be used without seriously affecting the amount of stable Q multiplication which can

Table 1—Conditions for Circuit Instability

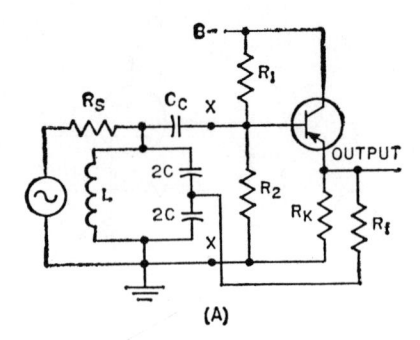

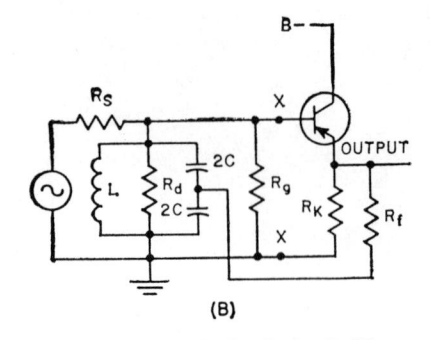

Freq. (cps)	L (h)	C (μf)	R_L (ohms)	R_s (ohms)	$R_f = \frac{1}{4}R'_d$ (ohms×1,000) Calc	Meas
190	6.8	0.1	840	0	6.8	6.8
				82	6.6	6.6
				180	6.45	6.3
				235	6.3	6.23
				500	5.7	5.63
216.5	4.5	0.12	790	0	5.5	5.48
				82	5.23	5.15
				180	5	4.85
				235	4.58	4.57
				500	4.25	4.2
235	3.8	0.12	1,255	0	3.88	3.85
				82	3.73	3.68
				180	3.62	3.6
				235	3.48	3.47
				500	3.12	3.1

FIG. 2—Transistor equivalent of vacuum-tube version (A) and idealized circuit (B)

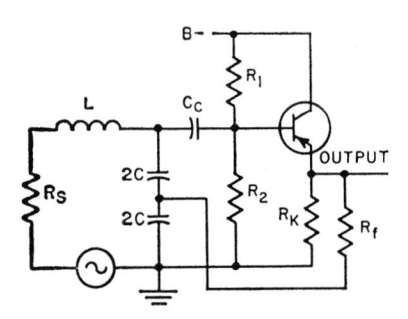

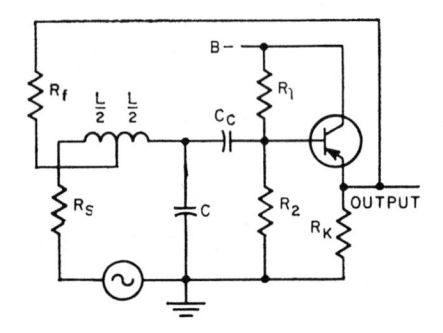

FIG. 3—Series-tuned circuits overcome shortcomings of parallel-tuned versions

be obtained. It has been found desirable to design the stage for the largest signal possible consistent with a minimum value of R_g. This minimum is easily derived from Eq. 3 and the stability requirement that $Q_{eff}/Q_o < \frac{1}{2} g_m R_k$. Since Q_{eff} is usually specified and $g_m R_k$ is known, Q_o can be calculated.

Let the Q of the coil used in the tuned circuit be $Q_o = R_d/\omega_o L$. When R_g is shunted across the coil the Q is reduced to $Q'_o = R'_d/\omega_o L$. Substituting for R_d and R'_d in Eq. 3 gives

$$R_{g\,min} = \omega_o L Q_o Q'_o / (Q_o - Q'_o) \quad (5)$$

If the transistor stage is designed for the largest signal possible consistent with the value of R_g given by Eq. 5 and then operated at a much lower signal level than it is designed for, a reasonable shift in operating point can take place without the transistor introducing distortion.

Referring to Eq. 2, R_g will be reasonably independent of R_t if $R_t \gg R_1$ and R_2; this is the case for most transistors if R_k is kept reasonably high. The value of β decreases with increasing temperature and thus R_g will also decrease with temperature.

A reduction of R_g causes a reduction in R'_d and this reduces the ratio Q_{eff}/Q_o. An increase in temperature will not therefore lead to instability but will reduce the Q-multiplication obtained. If this reduction is unacceptable, it is necessary to allow R_f to decrease with temperature.

Parallel Operation

When two or more selective amplifiers tuned to independent frequencies are to be operated from the same signal source, as in frequency-selective calling equipment, the parallel-tuned configuration shown in Fig. 2 is not suitable. Each tuned circuit tends to inject signals into the adjoining circuits, reducing adjacent channel rejection.

This difficulty has been overcome by the configuration shown in Fig. 3. The input is series-resonant, as seen from the signal source, and parallel-resonant, as seen by the input to the transistor. In this arrangement, each tuned circuit at its resonant frequency effectively shunts the input to all the other tuned circuits greatly reducing the breakthrough of signals.

Resistance R_s in Fig. 3 is the signal source resistance; it is ef-

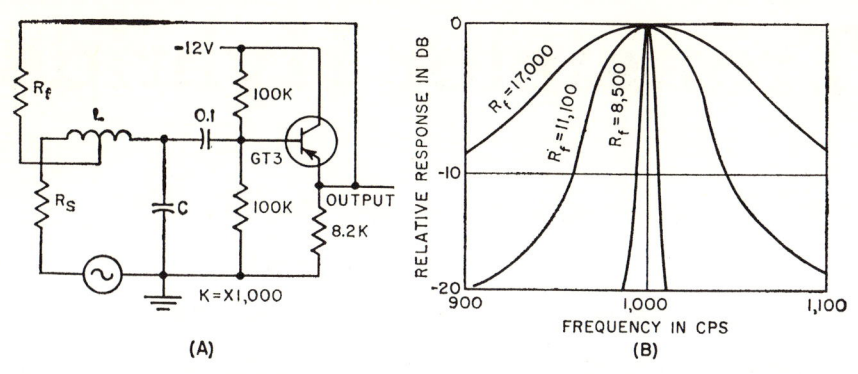

FIG. 4—Varying R_f of selective a-f amplifier (A) changes circuit Q-multiplication (B)

fectively in series with the coil and must be treated as part of the coil resistance when calculating the value of Q_o. Thus, $Q_o = \omega_o L/(R_s + R_L)$ where R_L is the resistance of the coil at the resonant frequency.

Practical Circuits

Figure 4A shows an audio-frequency selective circuit suitable for use either as a c-w note filter or as one channel in a multichannel frequency-selective amplifier.

When $R_f = \frac{1}{4} R'_d$, Eq. 4 shows that the circuit is unstable and will oscillate. The value of resistor R_f to give this condition were calculated for four different coils and for various values of generator resistance R_s. These calculated values were then compared with the actual measured values of R_f in the circuit for each condition and coil. The results, in Table I, indicate a close correlation between calculated and measured values.

The coils in Table I were built up into a three-channel a-f selective amplifier and connected as shown in Fig. 5A; Fig. 5B shows the response of each filter. The unit was driven from a generator with an impedance of 500 ohms. In comparison with a similar parallel-tuned unit, the adjacent channel rejection is much better.

With a minor modification, the circuit shown in Fig. 4A, can be used as a variable selective c-w filter; R_f should be replaced by a resistor and potentiometer in series. The resistor and potentiometer are each made equal to the minimum resistance needed for the maximum required selectivity. This minimum value is calculated from Eq. 4, but if maximum possible selectivity is wanted, this value should be made equal to $\frac{1}{4} R'_d + 1$ percent. The 1-percent margin is usually sufficient to ensure that the circuit does not oscillate, but the value may have to be adjusted. The potentiometer acts as a selectivity control, with maximum selectivity corresponding to minimum potentiometer resistance; Fig. 4B shows a typical set of selectivity curves for various positions of the potentiometer.

The maximum Q-multiplication that can be achieved with the transistor Q-multipliers has not been fully determined. However, a Q of 1,000 at 200 cps was obtained without any difficulty using a coil with a Q of 10.

Factor R_t appearing in the expression for R_g covers a multitude of troubles, especially when the transistor circuit is to be used at frequencies approaching its cutoff frequency; for audio frequencies it is generally sufficient to treat R_t as being purely resistive, but at higher frequencies the complete expression must be used and account must be taken of the input capacitance.

Performance

An experimental parallel-tuned circuit set up for maximum selectivity at 200 kc gave an overall bandwidth of 300 cps; although it was completely stable against normal temperature and voltage fluctuations, it was possible to shock excite it into oscillation after which it would continue to oscillate. Tests indicate that it is possible to do this at all frequencies with both tube and transistor Q-multipliers; it is believed that this is due simply to driving the tube or transistor into a region of nonlinearity.

When the Q-multiplication is restricted to less than $\frac{1}{2} g_m R_k$ the shock excitation instability does not occur. If this restriction on the amount of multiplication is observed and, in addition, the input level is kept small enough to ensure that the transistor is not overloaded, trouble from this form of instability should not be experienced.

A temperature run was made on the unit shown in Fig. 4A with the temperature cycled from 20 to 60 C and back. Between 20 and 47 C, no measurable change in effective Q was detected, but from 47 to 60 there was a reduction of Q multiplication to $\frac{2}{3}$ of its initial value.

The resonant frequency of the filter was not affected when the transistor temperature was raised from 20 to 60 C.

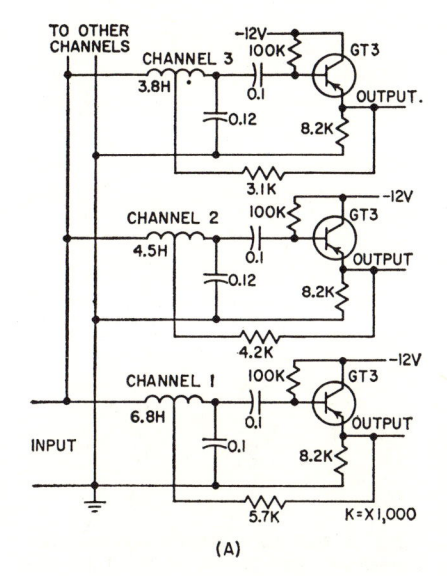

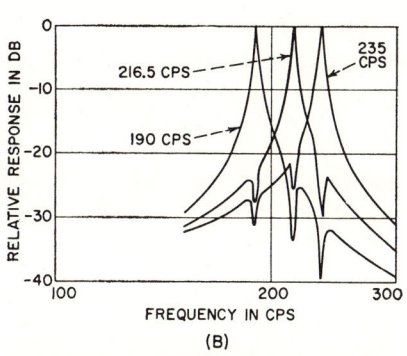

FIG. 5—Three channels of multichannel selective a-f amplifier (A) use coil and capacitor combinations in Table I for staggered resonant frequencies (B)

Reference
(1) H. E. Harris, Simplified Q Multiplier, ELECTRONICS, p 130, May 1951.

Tube-Transistor Hybrids

Design technique is described for combining transistors and vacuum tubes in a single circuit, resulting in reduction of power consumption, bulk and cost while improving reliability. Emphasis is on hybrid design of regenerative circuits such as a bistable cathode follower and four-stage ring counter

By G. A. DUNN and N. C. HEKIMIAN, Department of Defense, Washington, D. C.

TRANSISTORS AND VACUUM TUBES have frequently been combined in one device to produce so-called hybrid equipment, but little attention has been given to the combination of these two elements within individual circuits.[1,2]

Advantages

Although in recent years transistors have become increasingly popular as circuit elements, there are still many applications in which characteristics attributable only to vacuum tubes are required. For example the high input impedance available in vacuum tubes is often difficult and expensive to obtain in transistor circuits. In addition, the complete absence of reverse current, the minor dependence of their characteristics upon ambient temperature and the higher permissible voltage swings often dictate the use of vacuum tubes. On the other hand, greater circuit economy, reduced power consumption, simplified wiring, and smaller bulk and weight tend to make the transistor an attractive element. Further, certain transistor applications using complementary *pnp-npn* combinations are impossible with the conventional vacuum tube since the latter has no *pnp* equivalent.

Fortunately, transistor voltages are generally of the same order of magnitude as the grid-cathode voltages of most vacuum tubes. As a result, it is usually possible by judicious selection of bias voltages to operate the transistor without an auxiliary power supply. Where certain base voltages are required, these can be obtained through relatively low current dividers across a single supply. A great many variations can be obtained by changing the reference ground. Furthermore, it is frequently possible to use the complementary type of transistor by providing an appropriate change in biasing.

Regenerative Circuits

Some of the most attractive features of hybrid circuits are obtained by forming regenerative connections to enable the device to work as a switch, oscillator, pulse generator or other element exhibiting a negative resistance characteristic.

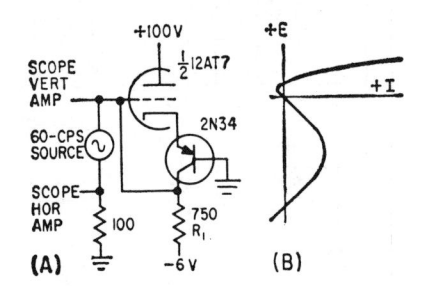

FIG. 1—Test circuit (A) produces negative resistance characteristic (B) on oscilloscope

The basic requirements for an inherently bistable device are that the magnitude of current gain and voltage gain each be greater than unity and that the phase shift be zero so that regenerative action is obtained when the output-input path is closed.

Since the normal static bias conditions for a conventional vacuum tube require that the grid be negative with respect to the cathode, it is possible to effect a great economy in circuit elements and a general reduction in complexity by employing the grid-cathode bias as at least part of the supply voltage for the transistor. It is also possible to arrange the circuits so that in the on condition, when both the transistor and the tube are conducting, the grid voltage remains negative to minimize grid current flow. This technique reduces triggering difficulties that arise from the lowered grid-circuit impedance in the on condition normally associated with the saturating type of Eccles-Jordan flip-flop.

The 60-cps test circuit shown in Fig. 1A may be used to display the static negative resistance characteristic. The negative supply voltage is made large enough to assure cut-off of the tube when the transistor is nonconducting; the plate supply voltage is so chosen that excessive tube current is not drawn even with zero grid voltage, and collector resistor R_1 is selected to assure saturation of the transistor when the tube current at zero grid voltage is impressed on the emitter.

Negative Resistance

The resulting trace on the scope, as shown in Fig. 1B, is a display of generator voltage as a function of generator current. The upper and lower portions of the curve have a positive slope, while the middle portion has a negative slope, indicating a negative resistance region. The voltage axis where the generator current is zero intersects the curve at three places. The upper intersec-

Afford Design Economy

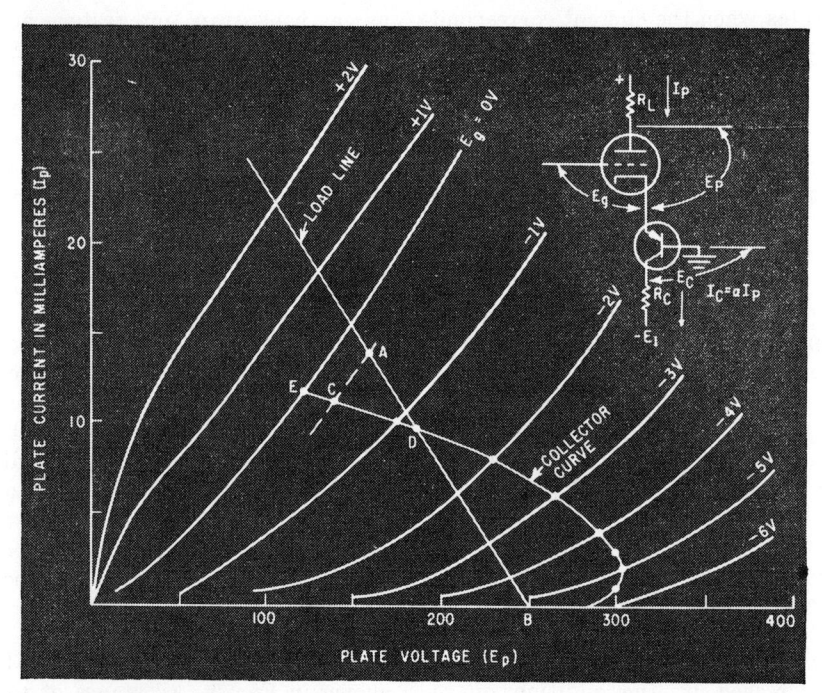

FIG. 2—Collector voltage is plotted as a function of plate current on triode plate characteristics to determine circuit parameters

plate load resistor. Intersections of the collector voltage curve and the plate load line represent mutually satisfactory conditions of plate current, grid voltage and collector voltage with no external input signal applied. The intersection at point B indicates that a stable off condition exists; the negative supply voltage is sufficient to maintain the tube at or near cutoff. The second intersection of the collector curve and load line at point D is of primary interest, and is referred to as the critical point.

It will be recalled that for the circuit to be stable in the on condition, the transistor should be in saturation. The emitter-to-collector voltage of a saturated transistor is generally a few tenths of a volt and can be represented by point C on the collector curve. Point C also represents the approximate collector current as read on the plate current scale. Since the stable operating point for a triode must lie on its load line and since the grid voltage is equal to the transistor emitter-collector voltage, the on condition for the tube must be at point A. The difference in current between points A and C is then the value of the saturation base current. It is evident that for the circuit to be stable in the on condition, the critical point must lie in the negative grid voltage region of the tube curve.

Preventing Damage

The saturation base current for low-power transistors should be restricted to not more than a few milliamperes to prevent damage. Although a plate load resistor is not required for bistable operation, it serves to minimize variations in saturation base current from changes in tube characteristics and power supply voltages.

The minimum trigger potential necessary to turn the circuit on or off is that which causes the plate current to rise above or fall below, respectively, the current at the critical point. Thus, the difference in grid voltage between points B

tion indicates that a stable on-condition exists; the center intersection occurs in the negative resistance region and is therefore unstable; and the lower intersection indicates that a stable off-condition exists, the voltage at this point being essentially the negative supply voltage.

Inspection of the circuit reveals that the tube provides current gain while the transistor provides voltage gain, both at zero phase shift. The overall voltage gain of the hybrid circuit is of primary concern and it can be shown to be $A_v = \mu a R_c/r_p = g_m a R_c$, where μ, r_p and g_m are the tube parameters, and a is the common base current gain of the transistor. If a plate load resistor R_L is used, the quantity $g_m' = \mu/(r_p + R_L)$ should be substituted for g_m. The input impedance of the transistor is considerably smaller than $r_p/(1 + \mu)$ and its effect is therefore neglected.

Design Procedure

A convenient procedure to follow when determining the necessary circuit parameters to make the circuit bistable is to plot the collector voltage as a function of plate current on the triode plate characteristics;

the values of grid voltage E_g are collector voltage coordinates since for this circuit the grid and collector voltages are approximately equal. The equation of the collector voltage to ground is $E_c = E_1 + a I_p R_c$, where E_1 is always negative and it is assumed that the base-to-emitter voltage of the transistor is zero. The collector voltage curve obtained in this fashion is a function of plate current only and no direct relationship exists between this curve and the plate voltage abscissa of the tube characteristics.

Figure 2 shows such a plot of the collector voltage. The straight line on this figure is the load line for a

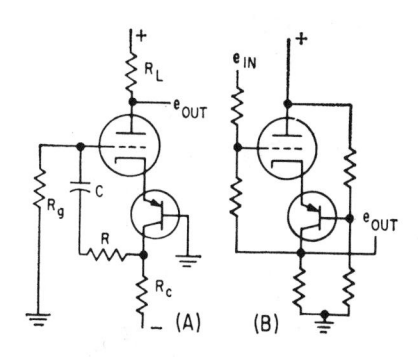

FIG. 3—Hybrid single-shot or free-running multivibrator (A) and bistable cathode follower (B)

29

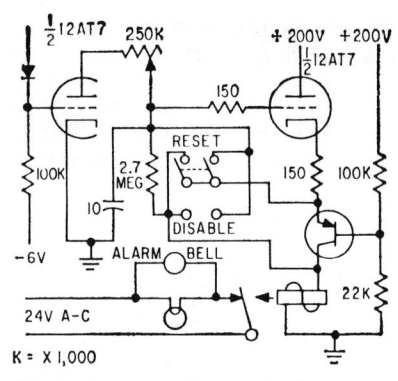

FIG. 4—Memory and alarm circuit accumulates predetermined number of pulses, then switches off until reset

and D is the minimum turn-on potential, and the voltage difference between D and A is the minimum turn-off potential.

The same procedure can be followed when designing a single-shot or free-running circuit. The critical point for the single-shot circuit occurs in the negative grid region just as with the bistable circuit, and for the free-running circuit it must occur in the positive grid region.

The bistable hybrid circuit does not lend itself well to binary counting applications since it requires alternate pulses of opposite polarity to turn it on and off. However it may be employed in storage or matrix circuits, or as a replacement for gas tubes. The inclusion of a plate load resistor provides a means of obtaining a high voltage swing which can be used to operate neon lamps and other circuits requiring high-voltage drive.

A normally-on single-shot or free-running square-wave generator is shown in Fig. 3A. The particular mode of operation is determined by whether or not the transistor saturates when the current corresponding to approximately zero grid voltage is flowing. The negative supply voltage for either mode of operation should be somewhat greater than the cut-off bias of the tube. A relatively fast rise and decay time can be achieved by making the loop gain large.

The single-shot circuit is conveniently triggered by a negative pulse coupled through a diode to either the grid or the collector.

Grid Current

The period of the free-running circuit is normally considerably shorter than the off period because the grid is driven positive with respect to the cathode during the on condition, resulting in grid current flow and subsequent capacitor discharge. This period can be greatly increased by the addition of resistor R in series with capacitor C.

One variation, obtained by changing the reference from ground to a higher potential, is shown in Fig. 3B. The new reference is obtained from the divider across the supply voltage and ground. The resulting circuit may be considered a bistable cathode follower. When the tube and transistor are in the on state the tube is a normally functioning cathode follower with a load resistance comprised of the base divider in parallel with the collector resistor. The transistor is nothing more than a closed switch. If the stage is momentarily turned off it remains in this state if the base divider network is properly designed. This circuit has been used as a boot-strap integrator and switch combination. A practical version is shown in Fig. 4.

This circuit is part of a memory and alarm system devised to accumulate pulses over a long period of time as represented by the time constant $R_g C_g / (1 - A)$, where A is the gain of the cathode follower. It has the property that when the integrated number of errors in a given time exceeds a predetermined level the circuit switches off and will not come back on unless reset. By placing an alarm relay in the collector of the transistor in addition to or in place of the collector load resistor and arranging alarm contacts that are normally closed, a fail-safe alarm circuit is obtained.

Ring Counter

A four-stage ring counter is shown in Fig. 5. All stages of the ring are identical and similar to the bistable circuit. Assume that the first stage is conducting; the transistor is then saturated and the tube bias is largely determined by the voltage drop across cathode resistor R_2. The voltage drop across R_1 causes diode D_1 to conduct, which raises the bias of the second stage and charges the priming capacitor of that stage.

A negative trigger pulse causes the conducting stage to be cut off and, when the pulse is removed the charge of the priming capacitor causes the second stage to turn on. Thus, each succeeding stage is turned on each time a negative trigger is applied. The width of the trigger pulse must be less than the discharge time of the priming circuit or the priming charge will be lost before the trigger pulse is removed.

The circuit as shown was found to count reliably at rates up to about 500 kc, with a trigger amplitude of about 4.5 v. The highest counting rate achieved was in excess of 750 kc.

REFERENCES

(1) R. N. Mital, Series Connected Transistor Amplifier, U. S. Patent No. 2,801,298.
(2) C. A. Bergfors, Transistor Trigger Circuit With Tube Controlling Emitter, U. S. Patent No. 2,825,806.

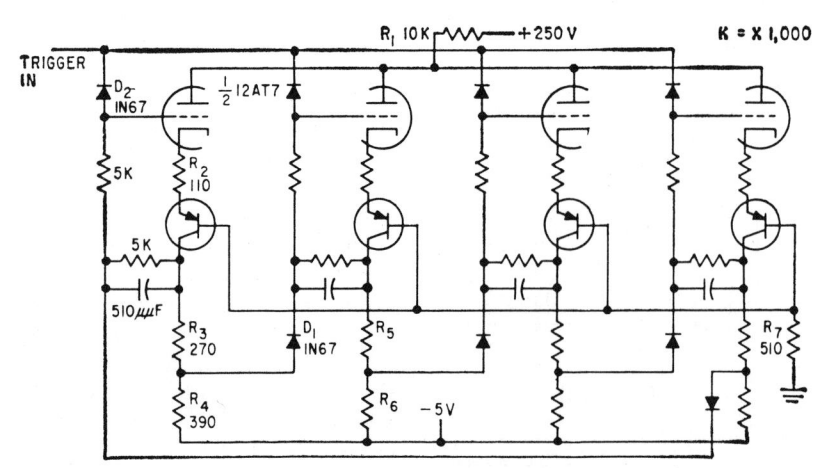

FIG. 5—Schematic of four-stage ring counter. All stages are identical

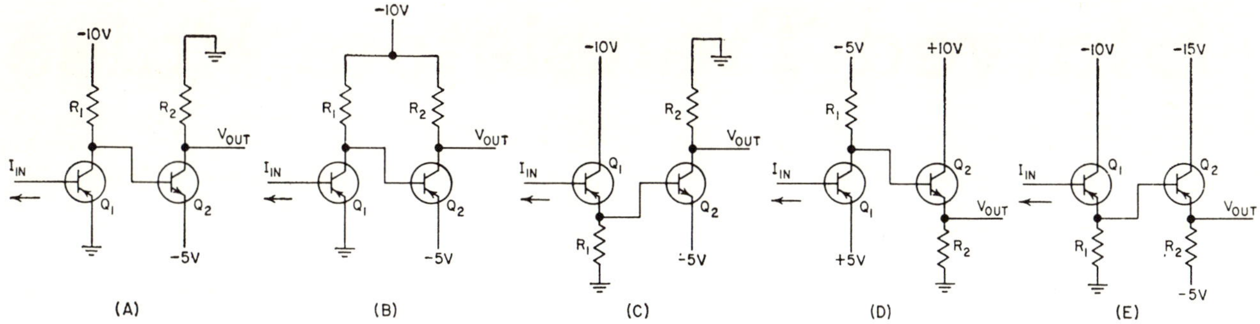

FIG. 1—Typical d-c transistor amplifiers. In (A), large values of I_{in} can burn out Q_2. Other circuits effectively limit current

Easing Transistor Loads

Direct-coupled transistor amplifiers are often driven to overload during normal operation. Some simple design rules presented here limit the current and power dissipation to safe limits when the input stage is overloaded

By WILLIAM F. SAUNDERS, III General Engineer, Link Aviation, Inc., Hillcrest, N. Y.

DESIGN OF DIRECT-COUPLED transistor amplifiers must include a means for limiting power dissipation in the transistors during amplifier overloads when every stage may be cut off or saturated with large currents. Considering current and power dissipation of each transistor, with the preceding stage in both extreme conditions, produces a direct-coupled amplifier that will always operate within the ratings of the transistors.

Figure 1A shows a possible two-stage, direct-coupled transistor amplifier. Assume that Q_1 can be driven either to cutoff or to saturation by I_{in} and examine the base current of Q_2 under these conditions. When I_{in} is so small that Q_1 is cut off, then Q_2 is also cut off and the ratings of Q_2 are not exceeded. For normal operation with V_{out} between 0 and −5 volts, the collector voltage of Q_1 is fixed at approximately −5 volts and the collector current divides between R_1 and the base of Q_2. Any increase in I_{in} appears in the collector of Q_1 multiplied by the current gain of Q_1. This increase in current flows directly into the base of Q_2 since the current through R_1 remains constant. Thus, the base current and power dissipation of Q_2 can reach large values which may cause the destruction of Q_2. This configuration would be satisfactory only if I_{in} were limited.

Other Methods

Another two-stage transistor amplifier is considered in Fig. 1B. If I_{in} is large, the increased collector current of Q_1 flows through R_1 making the collector voltage of Q_1 more positive, cutting off Q_2. For an I_{in} so small that Q_1 is cut off, the base current of Q_2 is limited by the current through R_1. This current is approximately equal to the normal collector current of Q_1, and usually other design requirements such as amplification and normal operating currents result in an R_1 large enough so the maximum ratings of Q_2 are not exceeded when Q_1 is cut off, even though Q_2 is saturated. Thus, the maximum current in Q_2 is independent of the maximum I_{in} and Q_2 is protected from the effects of extremely large input signals. A common-collector circuit shown in Fig. 1C, has no excessive current or power dissipation in Q_2 when Q_1 is overloaded, provided R_1 is large.

The criteria for either configuration can be written as two rules relating the configuration and transistor type (pnp, npn). First, a common-emitter stage must be followed by a similar transistor.

Second, a common-collector stage must be followed by a transistor of the complementary type. These two rules are completely general and always provide protection provided R_1 is large.

The low current gain of common-base stages will usually eliminate any saturation problems. For the same reason, it is not normally advantageous to use common-base stages in direct-coupled amplifiers.

Another design procedure restricts the operation of a critical stage to only one of the overloaded conditions. This may be accomplished in an operational amplifier by limiting the input signal to only one polarity. In Fig. 1D and 1E the maximum available base voltage of Q_2 is insufficient to cause large currents in Q_2 even when Q_1 is overloaded. This is not true if the +5 and +10 volt supplies are interchanged in Fig. 1D or if the −10 and −15 volt supplies are interchanged in Fig. 1E.

Starved Transistors Raise

Bootstrapped-collector circuit uses starved transistor to provide 500-megohm d-c input resistance with 100-v input signal; a-c resistance is even higher. Circuit operates without periodic adjustment in temperatures up to 60 C. Design criteria and transistor behavior under starved conditions are detailed

By B. M. BRAMSON, Project Engineer, Baird-Atomic, Inc., Cambridge, Mass.

PHOTOCONDUCTIVE DEVICES and a wide variety of other measuring circuits require transistor buffer stages having high d-c input resistance as well as high a-c input impedance.

The a-c aspect seems to have received the main attention to date and input impedances up to 1,000 megohms have been obtained.[1,2] On the d-c side, 0.4-megohm input resistance has been reported[3] for a system of short-term stability and relatively low input signal.

Input-Current Limitations

As the input resistance depends on the d-c current drawn by the first stage, this current must be kept to as low a value as possible.

The input resistance for the circuit of Fig. 1 is

$$R = V_{sig}/I_s$$

where I_s is the current drawn by the first stage and has a minimum value that is dictated by several considerations. D-c beta falls off continuously as I_c is reduced and there is no advantage in operating the transistor at a beta of nearly unity. Also, there seems to be a minimum I_c for proper operation whose value at this time is based on intuition, caution and suspicion. A third factor is temperature range, which affects both I_{co} and beta.

About the most that circuitry can expect to do is to permit the initial base current to be the minimum practical, not allowing it to increase appreciably with increasing input signal. This is equivalent to saying that I_s for an optimum circuit should remain practically constant regardless of signal input. Therefore, from Eq. 1, R is almost proportional to V_{sig}; this is basic feature of such circuits.

As I_s flows in signal source resistance, R_s in Fig. 1, it produces a spurious voltage drop V_r that is reflected in output voltage V_o of the system.

The percentage inaccuracy this introduces is

$$[V_r/(V_{sig} + V_r)]\,100$$

As V_r remains of the same order

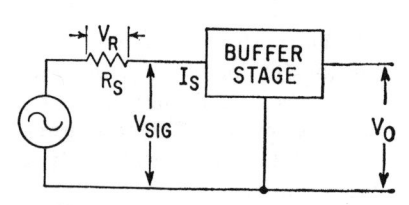

FIG. 1—Basic input configuration

independently of V_{sig} the voltage-transfer accuracy is roughly proportional to signal input voltage when $R > R_s$. This is another basic feature of such arrangements.

First-Stage Transistors

Some investigations on low-current operation of silicon transistors were made a few years ago.[4] The semiconductor art has since advanced to the stage where transistors are available with a 25 C d-c beta greater than 20 at $I_c = 10\ \mu a$, 60 C I_{co} less than 100 millimicroamp at $V_c = 3$ v, 25 C I_{co} approximately 5 millimicroamps and a-c small signal beta higher than the d-c beta. Two commercially available transistors with these characteristics are the X284 and the ST1028. The circuitry to be described was developed about these units.

Probably the simplest circuit for obtaining high d-c input resistance with transistors is the grounded collector in Fig. 2, operating under starved conditions. For maximum d-c input resistance, regardless of signal level, the d-c voltage drop across R_1 should be high compared to the maximum signal swing. The input current will then only swing by a small percentage and hence remain closer to the practical minimum.

With this circuit, a-c input impedances of 100 megohms may be obtained at audio frequencies. This is possible because of the considerable increase in collector impedance Z_c that occurs at low I_c. The input impedance almost equals that of the collector capacitance at frequencies below beta cutoff and is much increased at low I_c. Units in the 2N-336 class give around 5-kc cutoff and the 2N338 type around 30 kc at 10 μa of I_c.

Collector breakdown voltage limits the signal swing obtainable with this method.

It is sometimes desirable to cascade another grounded-collector stage to that in Fig. 2; R_1 is then

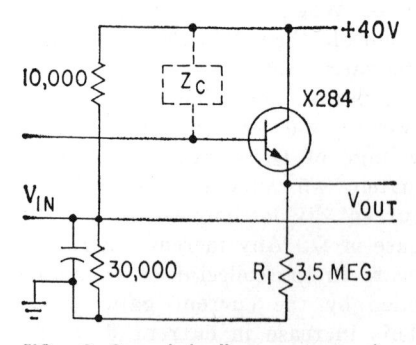

FIG. 2—Grounded-collector circuit for small signals operates under starved conditions for high input impedance

D-C Input Impedance

disconnected. The 10 μa emitter current of the first stage will then be just right for normal operation of the second transistor.

Large-Signal Circuit

In the circuit of Fig. 3A, the collector of the first stage is bootstrapped to the outer emitter. By this means the first-stage collector-emitter swing is so reduced that a V_c of 5 v accommodates a 100-v signal. This configuration also raises the effective first-stage base-to-collector impedance, giving considerably higher a-c input impedance to the system.

A curve of d-c input resistance obtained with a circuit of this type is shown in Fig. 3B. The input current varied from about 0.2 ma at 100-v signal to 0.1 ma at 10 v. This represents a resistance variation from 500 megohms at 100 v to 100 megohms at 10-v input signal.

The a-c input impedance will be higher than 10 megohms at 10 or 20 kc and in the 1,000-megohm region at low frequencies.

Maximum Signal Swing

If R_p is the parallel value of R_4 and R_5 in Fig. 3A, the relationship $R_p/(R_p + R_3)$ equal to or greater than $V_{a-c, p}/(V_{d-c})$ must be satisfied where $V_{a-c, p}$ = peak a-c signal voltage and V_{d-c} = d-c signal voltage. Otherwise peak clipping will result due to C_1 in accordance with r-c coupling theory.

For best circuit performance the d-c voltage across Zener diodes SV8 and SV11 should remain constant regardless of signal level. Therefore, the d-c current through them should remain as constant as possible. One simple way to insure this without the use of high d-c supply voltages is to use a floating supply XY. This approach has the further advantage that changes in V_{XY} and components cannot change the d-c operating point of the 953 emitter, a condition which can be objectionable at low signal levels.

The d-c current through the SV8 and SV11 Zener diodes should be at least 100 μa and preferably about 1 ma so that $(V_{XY}{-20})/(R_4 + R_5) \geqq$ 100 μa, where V_{XY} is the floating power supply voltage and the voltage drop across the SV8 and SV11 diodes is 20 v. Capacitors C_2 and C_3 maintain the voltage across the SV8 and SV11 diodes under a-c conditions where these diodes would be driven below their minimum operating current.

The minimum instantaneous signal that should be applied to the circuit during operation is about 2 v, the approximate voltage required to overcome the forward voltage drop in the three series-connected emitter junctions.

Potentiometer R_6 is set to meet this requirement for the maximum swing that is anticipated. The zero offset so created is easily corrected at the output if required.

Allowance

The collector current of the 2N-338 in Fig. 3A is selected to be high compared to the highest I_{co} that is anticipated for the 953. This current is established by the value of R_1. If this condition did not appertain, then the circuit would become inoperative at high values of I_{co} of the 953. Similarly, R_2 is selected so that the X284 takes a greater emitter current than any I_{co} expected from the 2N338.

A screen around the first stage components, connected to the output terminal, will assist considerably in keeping the input impedance high at higher frequencies.

For low ambient temperatures where beta might fall off somewhat, the transistors can be housed in a small temperature-controlled oven that will insure they will not fall beneath 25 C.

When using silicon transistors at low currents to obtain high d-c and a-c input resistance the following points may be helpful: d-c beta will increase with temperature and may increase or decrease by a small amount with life; I_{co} may increase by four times or so during the life of the transistor and also increase rapidly with temperature.

There is no fixed ratio between 1-ma beta and, say, 10-μa beta for all units of a particular type; therefore, the beta of each transistor must be tested separately at low current.[5]

The author thanks H. L. Aronson for his many valuable suggestions.

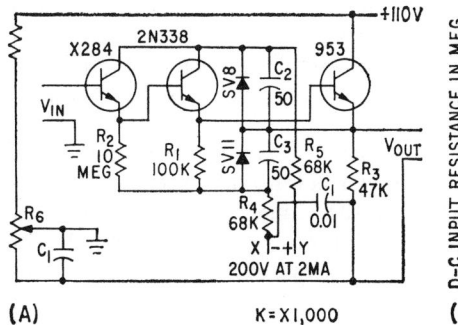

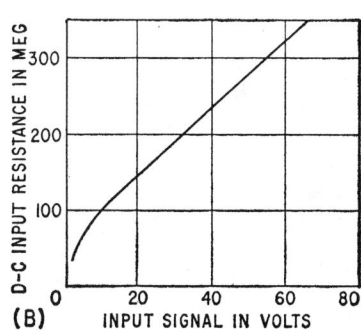

FIG. 3—Bootstrapping collector of first stage reduces collector-emitter swing and increases base-collector impedance (A); d-c input resistance varies fairly linearly with input signal level (B)

REFERENCES

(1) I. F. Barditch and J. D. Sullivan, Transistor Impedance Changer, *Electr Indus*, p. 77, Jan. 1958.
(2) P. J. Anzalone, A High Impedance Transistor Circuit, *Electr Design*, p. 38, June 1, 1957.
(3) D. Schuster, D-C Transistor Amplifier for High Impedance Input, ELECTRONICS, p 64, Feb. 28, 1958.
(4) E. Keonjian, Micro-Power Operation of Silicon Transistors, *Tele-Tech*, p 76, May 1956.
(5) W. M. Webster, The Variation of Junction Transistor Current Amplification Factor with Emitter Current, *Proc IRE*, p 194, June 1954.

Bias Method Raises

Reverse-biasing technique, which permits transistors to switch voltages higher than their collector-to-emitter rating, can be applied to many switching problems. It is now being used in Trixies, modular units for switching Nixie tube cathodes

By ARPAD SOMLYODY,
Circuit Design Analyst, Electronic Tube Div., Burroughs Corp., Plainfield, N. J.

USING TRANSISTORS above their rated collector-to-emitter voltage suggests many possibilities in on-off control circuits. In the application to be described, the technique permits use of lower cost transistors to drive Nixie numerical indicator tubes. The same approach could be used to increase output of more costly transistors having higher voltage ratings.

There are many applications in which medium-voltage, low-speed switching by transistors would be desirable. Examples include flip-flops yielding higher voltage outputs and gating circuits, as well as activation of gas discharge tubes. However, because of cost of transistors with higher collector-to-emitter voltage ratings, other switching methods are often used.

Reverse Biasing

Low-power, audio-type transistors are generally not expensive, but they tend to have low collector-to-emitter voltage ratings. For example, characteristics of the Sylvania 1750 25-volt transistor are listed in Tables I and II. Although maximum collector-to-base rating is 40 volts, it is useable only up to 25 volts in a common-emitter circuit because maximum collector-to-emitter rating is only 25 volts.

More careful analysis reveals that with the base-to-emitter junction reverse biased, collector-to-emitter breakdown voltage increases above the 40-volt collector-to-base rating. Since transistors in switching circuits are normally operated in either the saturated or off condition, the SYL 1750 can be used in 40-volt switching circuits.

A sample quantity of SYL 1750 transistors were tested with reverse bias applied to the base-to-emitter junction. Collector breakdown potentials were found to fall in the range from 45 to 60 volts with respect to the emitter. An output voltage swing of the same order

Table I—Absolute Maximum Ratings of SYL 1750 at 25 C

Collector-to-base volts	40 v
Collector-to-emitter volts	25 v
Collector current	100 ma
Power dissipation	150 mw
Junction temperature	85 C

can be realized if the base is reverse biased whenever the collector rises to its breakdown potential.

The collector of the SYL 1750 transistor can be operated at these high potentials provided that rated dissipation of the transistor is never exceeded. Dissipation can be limited by two methods: using a suitable constant-current collector supply or providing a sufficiently high load impedance in conjunction with a high-voltage supply. Also, a nonlinear load can be used, as in driving the gas indicator tubes.

Application to Nixie Tubes

The reverse-biasing technique has been successfully used in driving the ten cathodes of the Nixie numerical indicator tube using the basic circuit in Fig. 1. The Nixie

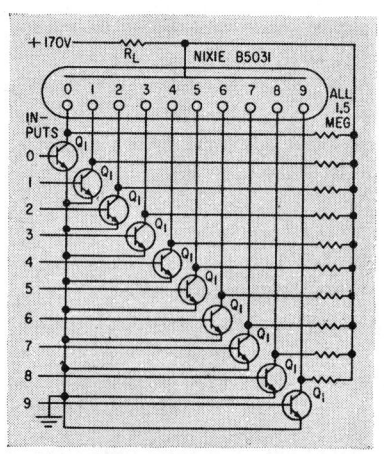

FIG. 1—Transistor bias V_{EB} of −1.5 v (I_B of −150 μa) enables Sylvania 1750 transistors to switch Nixie tube cathodes. Drive current I_B is 200 μa

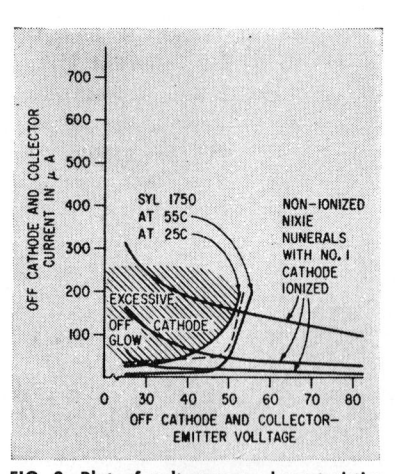

FIG. 2—Plot of volt-ampere characteristics of reverse-biased SYL 1750 transistor and Nixie tube cathodes shows how transistors can be used to switch cathodes

Breakdown Point

Table II—Electrical Characteristics of SYL 1750 at 25 C

Characteristic	Min	Max	Unit
I_{CBO} at $V_{CB}=40$ v, $I_E=0$	—	10	µa
I_{CER} at $V_{CE}=25$ v, $R_{BE}=10$ K	—	100	µa
I_{EBO} at $V_{EB}=10$ v, $I_c=0$	—	10	µa
h_{fe}	15	125	—
h_{ie} at $V_{CE}=6$ v, $I_c=1$ ma, $F=1$ kc	0.8 K	4 K	ohms
fh_{fe} at $V_{CE}=6$ v, $I_c=1$ ma	10	10	kc

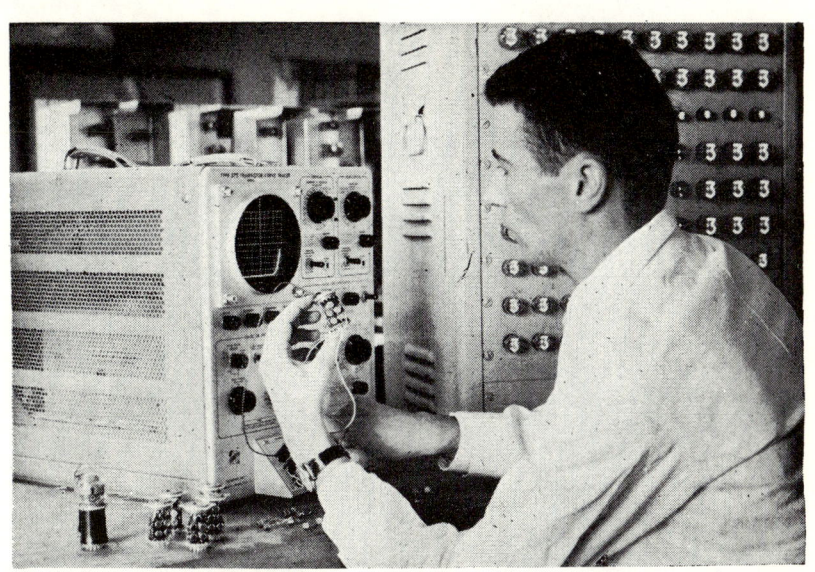

THE FRONT COVER. Modular Trixie using ten reverse-biased transistors to switch Nixie tube cathodes is tested by author

tube has ten cathodes in the form of numerals 0 through 9, together with a common anode, all sealed in a neon gas atmosphere. Application of a negative voltage to any one cathode with respect to the anode causes that cathode to glow in the shape of its particular numeral. The potential difference required between anode and selected cathode is ordinarily 150 volts. However, by using the prebiasing technique, input required on the cathode is reduced to between 40 and 60 volts (Prebiasing consists of holding the nine off cathodes at a potential about 40 to 60 volts above that of the on cathode.)

To adapt this readout device to low-voltage transistor counting and logic circuits, it became necessary to use buffer amplifiers for each of its ten cathodes. Since the cost of one buffer amplifier is multiplied by ten for each decade of readout, the use of high-voltage npn switching transistors is impractical for most commercial applications. However, by combining the Nixie tube cathode characteristics with the observed collector characteristics of reverse-biased SYL 1750 transistors a means was found of using these transistors for Nixie tube activation without sacrifice of reliability.

Characteristics

Figure 2 indicates a portion of the volt-ampere characteristics of some of the off or non-ionized cathodes of the Nixie tube when the number one cathode is grounded and the anode has about 150 volts applied to it. Operating the off cathodes in the higher current region of their characteristics (shaded part of Fig. 2) should be avoided. As potential on these cathodes is lowered, increasing off numeral glow becomes objectionable, causing difficulty in distinguishing the on numeral.

Figure 2 also shows the collector volt-ampere characteristics of SYL 1750 transistors in the reverse-biased condition at 25 C. When it is superimposed on the Nixie tube curves, the intersection of these curves establishes the average spread of operating points of the Nixie tube when used in this circuit. Transistor dissipation at these intersections does not even approach maximum rating of 150 mw.

As shown in Fig. 2, as temperature is increased, the transistor characteristic curve shifts vertically upward. If the transistor characteristics taken at 55 C are superimposed on the Nixie tube characteristics, some intersections fall inside the region where objectionable background occurs. The increase in I_{CBO} and leakage currents are directly responsible for the rise.

To obviate this condition, a 1.5-megohm resistor can be connected from each collector to the 150-volt point, as shown in Fig. 1. The current through these resistors shifts the transistor curve vertically down, allowing all intersections to lie outside the undesired region.

Trixie Modules

With the addition of the resistors, the reverse-biased circuit becomes applicable to driving Nixie tubes over the temperature range from −30 to 55 C. Such a circuit has been evaluated and is being produced. The units, called Trixies, are modular in design and are available in two forms: One operates standard and supersize and the other miniature Nixie tubes.

Success of the reverse-biasing technique here opens the way for further developments in circuit design requiring higher than rated collector-to-emitter voltages.

35

Tunnel Diode—

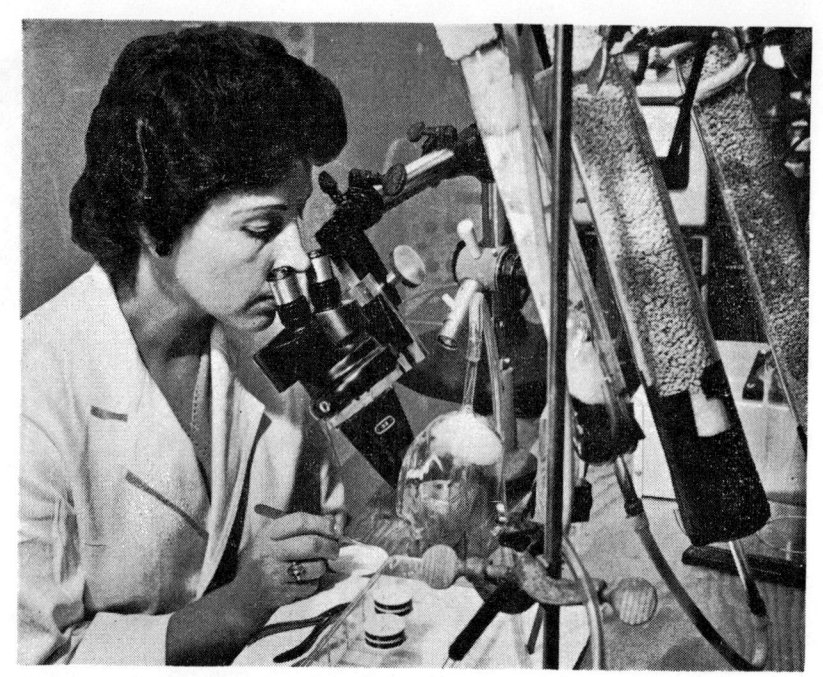

THE FRONT COVER—Technician prepares to attach leads to tunnel diode with aid of micromanipulator. Diode is assembled in bell jar containing inert gas

TUNNEL DIODE PARAMETERS

Parameter values for typical experimental tunnel diodes, with 1 ma peak currents, are $C = 5$ μμf, $-g = 1/150$ mho, $R_s = 1$ ohm and $L_s = 2$ n μh. In general, to find the parameters for other values of current (I_p) in ma., use $C \approx 5 I_p$ μμf, $g = I_p/150$ mho, $R_s = 1/I_p$ ohm and $L_s \approx 2$ n μh (controlled by case design).

TUNNEL DIODES are characterized by a negative conductance region that results when the current falls from an excessively high value at very low forward voltages to a value somewhat above that of a normal p-n junction at a higher forward voltage[1,2,3,4]. Hence, the tunnel diode tends to be a high current, low-voltage device with a large negative conductance (ELECTRONICS, p 54, Nov. 6, 1959). These features make tunnel diodes useful in many types of circuits, several of which are described in this article.

There are two requirements that must be met if a p-n junction is to be made into a tunnel diode. The first is that the junction (space charge region) must be narrow—on the order of 150 A. This means that the transition from p-type to n-type must be abrupt. The second requirement is that both the p-type and n-type regions must be degenerate; that is, the Fermi level must be within the conduction band on the n-type side and within (or very close to) the valence band on the p-type side, or the reverse. For a good tunnel diode, this requires very high impurity concentrations—greater than 5×10^{19} per cubic centimeter for silicon and greater than 2×10^{19} per cubic centimeter for germanium.

Silicon tunnel diodes at liquid helium temperatures (4.2 K) exhibit an anomalous behavior in the tunnel region. Four nodes exist in this region where the slope of the current-voltage characteristic decreases and then increases. This effect has been observed in a variety of tunnel diodes and due to phonon interaction in the tunneling process[5].

One of the tunnel diode's important features is its resistance to nuclear radiation. Preliminary experimental results have shown tunnel diodes to be at least ten times more resistant to nuclear radiation than transistors.

Transistors and other semiconductor devices are degraded by radiation because lifetime is lowered, resistivity changes (usually rises), and noise and transients are produced by ionization in the bulk and on the surface (see p 55). The tunnel diode does not depend upon carrier injection or extraction to produce its negative resistance region,

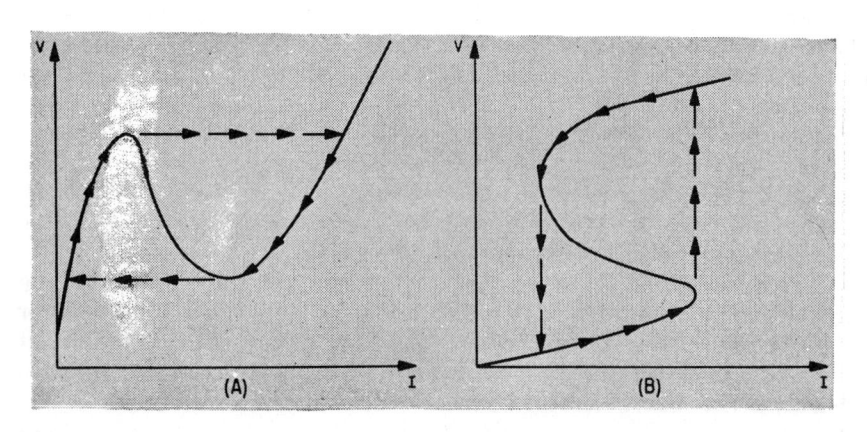

FIG. 1—(Left) Negative resistance characteristics for n-type (A) and s-type (B) devices

FIG. 2—Tunnel diode equivalent circuit has junction capacitance, C. The G is given by low-frequency characteristic. Semiconductor and leads give L and R_S

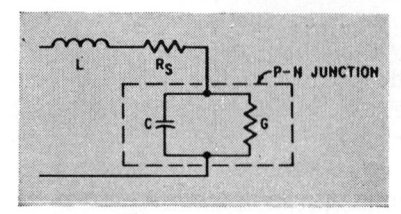

Circuits and Applications

Characteristics of tunnel diodes make them useful in oscillators, sweep circuits, detectors, multivibrators and amplifiers. Here are some typical circuits

By I. A. LESK, N. HOLONYAK, JR. and U. S. DAVIDSOHN
Semiconductor Products Dept., General Electric Co., Syracuse, N. Y.

so lifetime is a minor consideration.

Resistivity will not change until very large dosages have been applied because it is so low initially on both sides of the junction. Also, since the tunnel diode can work to such high temperatures, it can be heated during irradiation to anneal damage as it occurs. Germanium tunnel diodes have a practical upper temperature limit of 200 C (ELECTRONICS p 43, Oct. 30, 1959). Tunnel diodes should be much less affected by ionizing transients because they are quite insensitive to surface changes and to minority carriers produced by ionizing radiation for example, light.

Tunnel Diode Combinations

Simple combinations of tunnel diodes can be used to give some novel circuit characteristics. A symmetrical switch can be made by connecting two tunnel diodes in opposition. If many tunnel diodes with slightly different peak currents are connected in series, an n-stage stepping device results. If two groups of such tunnel diodes are placed in opposition, a symmetrical n-stage stepping device results. This type of device would find application in such areas as memories and analog-to-digital converters.

If a tunnel diode is placed in series with a backward diode (a diode with no tunnel diode characteristic during forward biasing), the negative conductance region is shifted to a high voltage. This could also be done with a normal diode, in which case the inverse characteristics would be rectifying

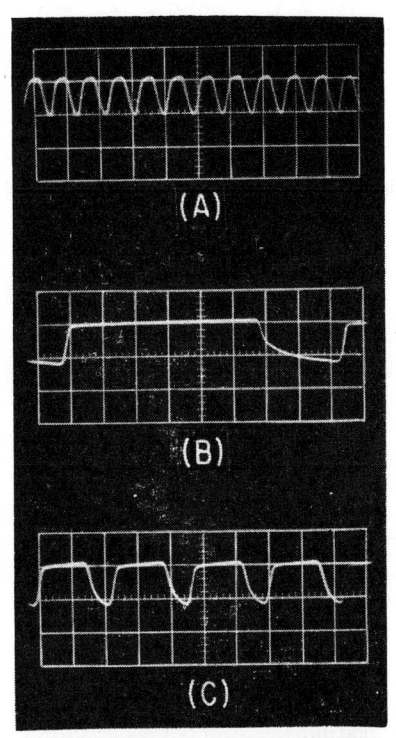

FIG. 3—Waveforms for tunnel diode relaxation oscillator with series inductance of 4.7 μh (A), 68 μh (B) and 15 μh (C). Scale is 0.1 μsec/cm

instead of conducting. If two such series combinations are placed in series opposition, a symmetrical higher voltage switch is obtained. Each additional diode connected in an opposing section adds a state to each polarity.

Relaxation Oscillator

Distinction is made between voltage-stable and current-stable negative resistance devices by calling the former an s characteristic and the latter an n characteristic. These characteristics are shown in Fig. 1, with arrows indicating the idealized

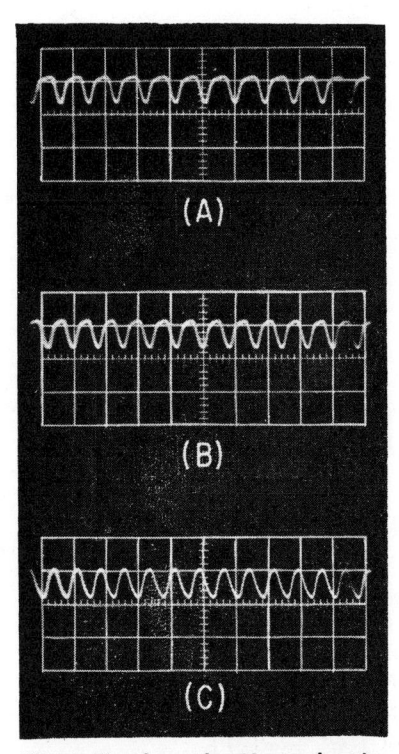

FIG. 4—Waveforms for 12-mc relaxation oscillator with 5 μh series inductor show effects of R_s of 8 ohms (A), 25 ohms (B) and 50 ohms (C). Scale is 0.1 μsec/cm

direction of rotation in the current-voltage plane when the device is operated as a relaxation oscillator.

The unijunction and avalanche transistors, four-layer diodes and controlled rectifiers are current-stable, while the tunnel diode is an example of the less common voltage-stable device. The type of stability is defined with respect to that ordinate which intercepts the device characteristic at one point only.

From considerations of tunneling and p-n junction theory, and practical fabrication limitations, the equivalent circuit of a tunnel diode

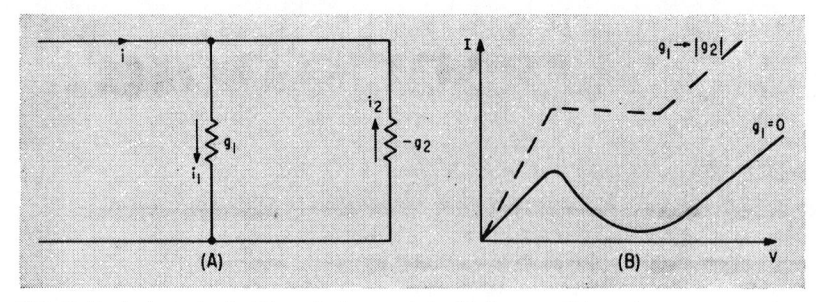

FIG. 5—Equivalent circuit (A) and characteristic (B) for parallel conductance operation. Input characteristic follows dotted line as g_1 approaches g_2

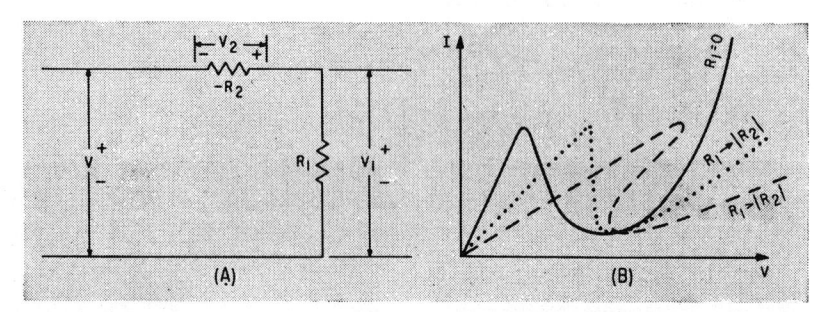

FIG. 6—Equivalent circuit (A) and characteristic (B) for series resistance operation. Dotted curve results when R_1 approaches R_2

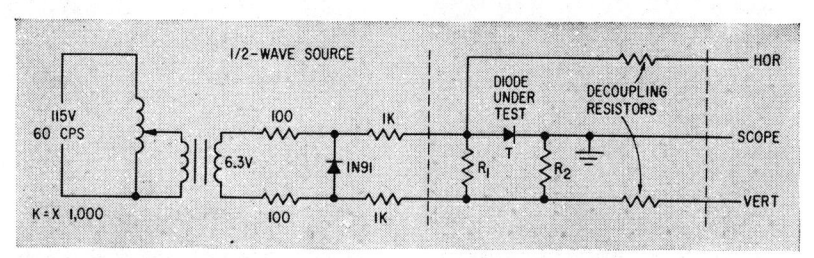

FIG. 7—Value of R_1+R_2 is critical in tunnel diode current-voltage curve tracer. Too small a value leads to circuit oscillation

may be drawn as shown in Fig. 2. For small signal operation, $Z_{in} = [R_s + G/(G^2 + \omega^2 C^2)] + j[\omega L - \omega C/(G^2 + \omega^2 C^2)]$

For operation in the negative conductance region, G is defined as $-g$. Negative input resistance will occur as long as R_s is less than $g/(g^2 + \omega^2 c^2)$. Letting ω_c be the frequency above which a negative input resistance no longer exists, then $\omega_c = (g/c)[(1/g R_s) - 1]$.

Waveforms for a relaxation oscillator using a series inductor and a voltage source are shown in Fig. 3. Since R_s includes the source resistance, the waveform is controlled by varying R_s so as to make ω_c approach ω, in which region the oscillations become sinusoidal. Oscillation is stopped by making ω_c less than ω. Effects of R_s are shown in Fig. 4.

Consider the parallel arrangement of a conductance and a negative conductance shown in Fig. 5A. Here $Z_{in} = 1/(g_1 - g_2)$ and $v_{in} = i/(g_1 - g_2)$, while the current amplification $i_1/i = 1/(1 - g_2/g_1)$.

To stabilize the device in the negative conductance region requires that $g_1 > |g_2|$ and as g_1 approaches $|g_2|$ the input current-voltage characteristic follows the dotted line in Fig. 5B. A series connection is shown in Fig. 6A. Here $Z_{in} = R_1 - R_2$ and $i_{in} = v/(R_1 - R_2)$ while the voltage amplification $v_1/v = 1/(1 - R_2/R_1) = 1/(1 - g_1/g_2)$.

As R_1 approaches $|R_2|$ the dotted curve shown in Fig. 6B, results. For $R_1 > |R_2|$, the dashed curve shows the current-voltage characteristic.

It is apparent that different results are obtained when operating into a series or parallel circuit. Thus, for a harmonic-rich oscillator a series voltage mode is best, while for amplification current coupling should be used. The circuits for a mixer oscillator have been constructed (ELECTRONICS, p 43, Oct. 30, 1959).

Sweep Circuits and Curve Tracers

The usual form of a curve tracer is illustrated in Fig. 7. One must be particularly careful to avoid distributed capacitance across the diode or oscillations will be introduced. The value of $R_1 + R_2$ is also critical. Where $(R_1 + R_2) > 1/|g|$ the device can only switch and the s characteristic will not be seen (Fig. 8). (The negative conductance region is not seen in Fig. 8 because the relatively high input impedance of the curve tracer results in a voltage jump at the peak and valley points.)

Too small a value of $(R_1 + R_2)$ (or R_s) can also lead to oscillations because some parasitic inductance, as well as capacitance, cannot be avoided. As both the circuit Q and ω_c are increased with decreasing R_s there exists a minimum R_s below which the oscillations cannot be avoided.

The sweep circuit can be used to much greater advantage than is immediately apparent. As has been mentioned, the presence of oscillation can easily be seen. In addition, the circuit can be used to obtain visual indications of other r-f signals.

Figure 9 shows the pip or beating between the self-oscillating tunnel diode and a high-frequency injected signal. Since the diode voltage (and negative conductance and capacitance) is changing at a 60-cps rate, the self oscillating frequency is being frequency modulated.

Figure 10 shows the result of mixing where the signal injected is the output of the input stage of a superregenerative receiver (a series of r-f bursts). This allows using a tunnel diode as a relatively narrow-band (10-mc at 100-mc) spectrum analyzer.

The averaged-out effects of r-f input are shown in Fig. 11 where the effect on the apparent characteristics of modulated r-f is illustrated. When two r-f signals are mixed in a tunnel diode, the negative conductance amplifies the resultant output as well. This leads to application in "reflexed" receiver circuits.

The effect of voltage on frequency of oscillation leads to the usual f-m oscillator applications. By impressing an audio signal current across R_s, 100-mc f-m oscillators have been built. The signals are amplitude modulated as well, but another tunnel diode can be used as an amplitude-limiter to eliminate this effect.

Detectors

Both tunnel diodes and backward diodes have such a low crossover voltage that an immediate application is evident as a low-level r-f detector and voltmeter rectifier. When the circuit shown in Fig. 12 was used, some effects were observed which may lead to further possible applications.

Figure 13 shows the d-c output as a function of r-f input. In some instances, one mv of r-f gave 50 mv of d-c output. Most significant is the r-f level at which the effect occurs. The backward diode presents the same phenomenon, but to a lesser extent. This may be because the nature of the characteristic in the crossover region of both will produce a gradual buildup of charge on the capacitor with time, thereby leading to d-c voltages which can exceed the peak r-f swing.

The very large d-c output after an r-f threshold has been reached suggests applications in amplified or augmented agc control circuits. The fact that the d-c output is nonlinearly related to the r-f amplitude may limit straight rectifier-detector applications. However, it should be possible to parallel the diode with a suitable resistor so that the conductance in the active region of both is slightly positive, thereby approaching ideal rectifier characteristics.

In addition, the characteristics displayed in Fig. 13 have two straight-line segments on a log/log plot. This suggests some additional uses in analog circuits.

Oscillators

To date tunnel diodes have been operated as sinusoidal oscillators in a transmission-line structure up to fundamental frequencies of kmc. It is noted that as nonlinear analysis shows, the device can generate so many harmonics and subharmonics, as well as all the cross products of these frequencies, determining what is the fundamental frequency can be a problem.

Figure 10 should not, therefore, be taken to mean that the tunnel diode was oscillating at 365 mc but only that the diode was detecting it. The fact that the local oscillator can pull the tunnel diode oscillator when using superheterodyne detection techniques and that the negative conductance may cause an absorption wavemeter to oscillate (or the tunnel diode oscillator may track the detector when using tuning stubs), demands some strenuous countermeasures to avoid confusion. Using suitable attenuators to decouple the oscillating and measuring circuits will prove helpful.

Strong harmonics of a 300-mc oscillator have been observed to greater than 2 kmc (frequency being limited by the detection system at hand). Some of this energy may of course be coupled into a resonant cavity and utilized.

Relaxation oscillators have been built and observed to 10 mc. Since the oscilloscope used has only a 30-mc bandwidth, odd harmonics of any frequency greater than 10 mc would be very rapidly attenuated. However, there is no reason to suppose that this is by any means a

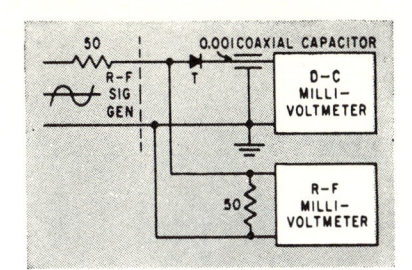

FIG. 12—R-f rectifier test circuit makes use of tunnel diode

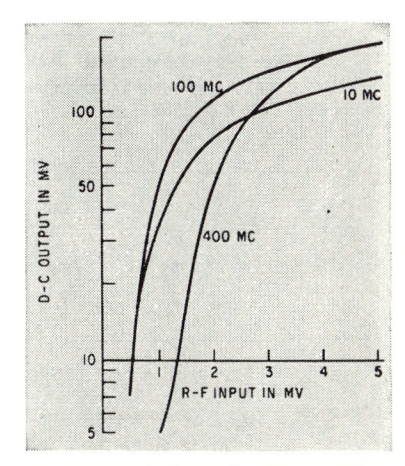

FIG. 13—Graph shows d-c output as function of r-f input for tunnel diode low-level rectifier

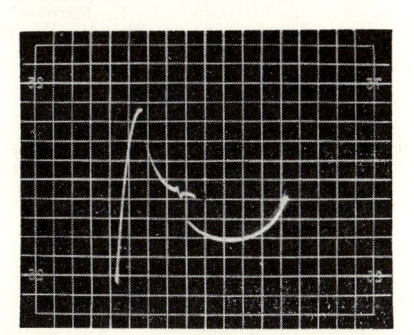

FIG. 8—Typical tunnel diode characteristic is shown with 0.2ma/div vertical scale and 0.1 v/div horizontal scale

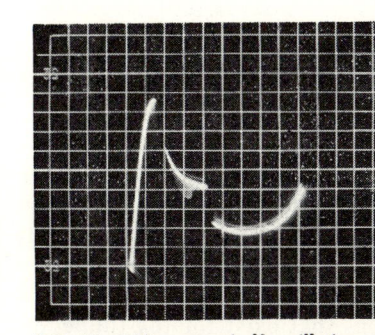

FIG. 10—X-Y sweep (self-oscillating with 10 μμf across diode) shows beat with 365-mc output of receiver

FIG. 9—X-Y sweep (self-oscillating with 10 μμf across diode) shows pip with 730-mc r-f signal

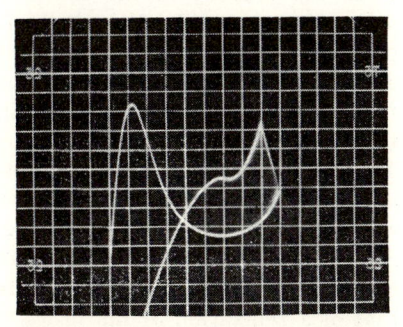

FIG. 11—Change in apparent characteristics with 70-mv, r-f (200 mc), 100-percent, 1-kc modulation

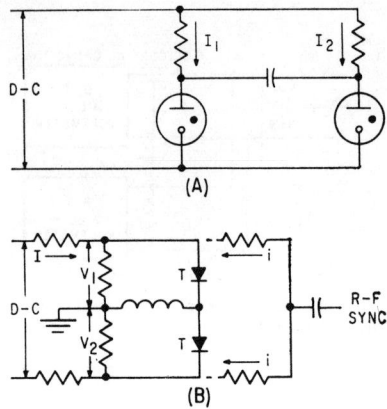

FIG. 14—Gas tube free-running multivibrator circuit (A) is basis for tunnel diode dual circuit (B)

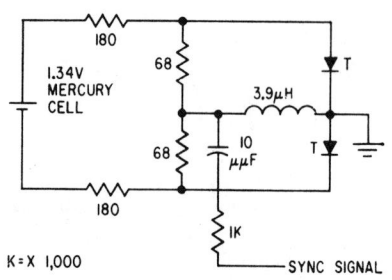

FIG. 15—Ten-mc multivibrator uses two 1 ma, −0.01-mho tunnel diodes

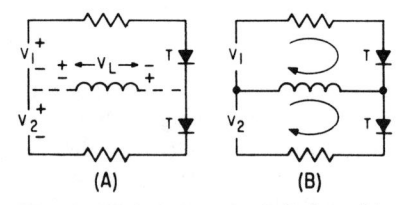

FIG. 16—Free-running tunnel diode multivibrator equivalent circuit during switching (A) and dwell (B) periods

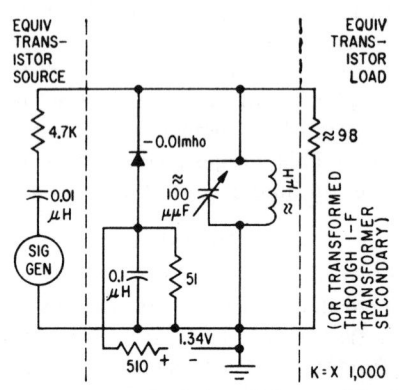

FIG. 17—Audio-modulated, 1-mc tuned oscillator circuit withstands wide temperature range

FIG. 18—Tunnel diode 445-kc amplifier has approximately 20-db gain

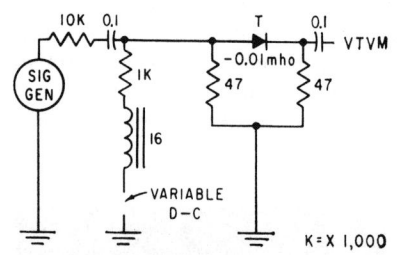

FIG. 19—Tunnel diode audio amplifier

limit. In fact from the mathematics and the harmonic content of our 300-mc oscillator, this too is likely to be of the relaxation type of operation.

Coupled Pairs

The relaxation oscillator leads one naturally to apply two such units as some form of multivibrator. From an analysis of the familiar n type free-running neon tube multivibrator shown in Fig. 14, the obvious tunnel diode dual circuit was obtained.

We have found little difficulty in constructing the 10-mc free running multivibrator shown in Fig. 15. Pulling effects have been observed on this multivibrator with 200-mc sync signals, but we cannot be sure of the effects since the scope used will not see the sync signal. However this leads one to expect that 30 and even 50-mc operation is possible.

Figure 16 illustrates the two conditions that exist for the tunnel diode multivibrator circuit. The first condition (A) is during switching when the current in the inductor does not change. Condition (B) is the dwell period when this current does change. From (A) it can be seen that the current change in both tunnel diodes must be the same. This change together with the voltage appearing across the inductor steers the direction of switching. Since current in the diodes can change, the gyration is not quite that of a relaxation oscillator.

From Fig. 6 it can be seen that sensitivity can be improved at the expense of frequency response by having R_s approach R_g. The operating time will be affected by the d-c bias level.

By suitably increasing R_s and adjusting the bias conditions the free-running multivibrator can be converted into a stable flip-flop for counting and logic circuits.

Transistors are obvious drivers for tunnel diodes and the reverse is also true. Many such applications can be visualized in logic and switching circuits. Small-signal combined circuits can be anticipated where the best features of both are used to maximum advantage, or where the interaction of both will produce unique applications impossible for either alone.

To demonstrate the versatility of tunnel diodes in oscillators, as well as their temperature range and freedom from surface effects, the audio modulated 1-mc tuned oscillator shown in Fig. 17 was made. The silicon tunnel diode, with no surface protection, may be dipped in liquid nitrogen (77 K), placed in a furnace at 300 C, and immersed in acid, the effect being only a minor change in oscillator and modulation frequencies.

Use of tunnel diodes in amplifiers is shown in Figs. 18 and 19.

The authors wish to acknowledge the contributions of R. N. Hall, J. J. Tiemann and C. S. Kim, and the assistance of G. K. Wessel, H. A. Jensen, S. F. Bevacqua and Mrs. M. Roehrig.

REFERENCES

(1) L. Esaki, Phys Rev, **109**, 603 (1958).
(2) T. Yajima and L. Esaki, J Phys Soc Jap, 13, p 1281, 1958.
(3) H. S. Sommers, Jr., *Proc IRE*, 47, p 1201, 1959.
(4) I. A. Lesk, N. Holonyak, Jr., U. S. Davidsohn, and M. W. Aarons, Germanium and Silicon Tunnel Diodes—Design, Operation, and Application, *1959 Wescon Conv Rec*.
(5) N. Holonyak, Jr., I. A. Lesk, R. N. Hall, J. J. Tiemann, and H. Ehrenreich, Direct Observation of Phonons During Tunneling in Narrow Junction Diodes, *Phys Rev Letters*, Aug. 15, 1959.

BIBLIOGRAPHY

The following papers were presented at the 1959 IRE-AIEE Solid State Devices Research Conference, Cornell University, June 17, 1959:

R. L. Batdorf, An Esaki Type Diode In InSb
H. S. Sommers, Jr., and H. Nelson, Tunnel Diodes as High Frequency Devices.
R. N. Hall and J. H. Racette, Tunnel Diodes in III-V Semiconductors.
J. J. Tiemann and R. L. Watters, Noise Considerations of Tunnel Diode Amplifiers.
N. Holonyak, Jr. and I. A. Lesk, Anomalous Behavior of Silicon Tunnel Diodes at Low Temperatures.

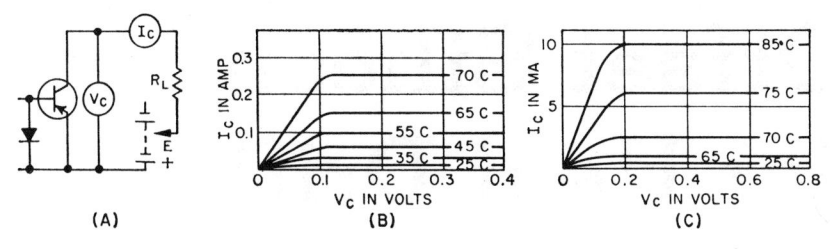

FIG. 1—Improvement in thermal drift brought about by diode compensation (A) is evidenced by drift characteristic of 2N290 transistor alone (B) and with GEX541 diode connected across emitter-base junction (C)

FIG. 2—Conventional (A) and compensated class-B push-pull amplifier (B)

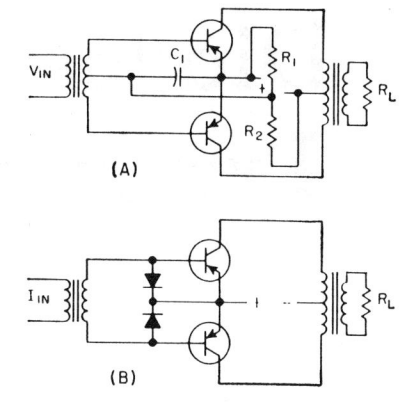

Diode Reduces Cutoff Current Drift

Amplified thermal I_{co} variations in grounded-emitter amplifiers can be compensated by connecting diode, having similar collector-base junction saturation-current characteristics, across transistor's base-emitter junction

By HENRI H. HOGE Chief Engineer, Advanced Research Associates, Inc., Kensington, Md.

USE OF RESISTORS to compensate a transistor d-c amplifier's drift current simply restricts the transistor's operating range, leaving the drift current still present and uncompensated. The compensation method to be described does not include any linear elements to restrict the transistor's dynamic range. It has effected negligible drift when incorporated in 10 to 20-w amplifiers.

In the common-emitter configuration, collector-current drift is a result of the variations of the I_{co}, flowing through the emitter-base junction, being amplified by a factor of beta. Figure 1 illustrates the way in which a diode can be connected across the transistor emitter-base junction to compensate for the amplified leakage current. The diode must have a junction area whose saturation-current characteristic is similar to that of the transistor's collector-base junction. Also the diode must be located as near to the transistor as possible so they will both be at the same operating temperature (transistors have been constructed with the diode in the same housing).

Push-Pull Circuits

When transistors with this compensation are used in push-pull class-B amplifiers, the unbalance or drift is approximately equal to the difference between the I_{cbo} of the two halves of the circuit. Compensated transistors of this type should be fed from a current source, with the attendent advantage that the base-emitter potentials need only be high enough to allow the necessary current to be fed through the base-emitter diode. In this way the additional capacitance of the diodes does not greatly affect the alpha-cutoff frequency.

In the conventional class-B transformer-coupled amplifier shown in Fig. 2A divider R_1R_2 reduces the overall efficiency of the stage and also necessitates the addition of C_1 to reduce the effects of R_1. The transistors in this circuit are driven from a voltage source which means that the off transistor is back-biased. This back-biasing not only sweeps all the carriers from the base region but also causes I_{beo} to flow. This of course has to be made up on the next change of polarity before transistor action can occur, manifesting itself as crossover distortion.

The circuit in Fig. 2B uses the compensated transistors fed by a current source. The off transistor is back-biased only by the forward drop of the compensating diode. Crossover distortion is reduced, the resistors and capacitor can be eliminated increasing efficiency.

41

Diodes Offset Silicon Transistor Heat Drift

By DAVID H. BRYAN

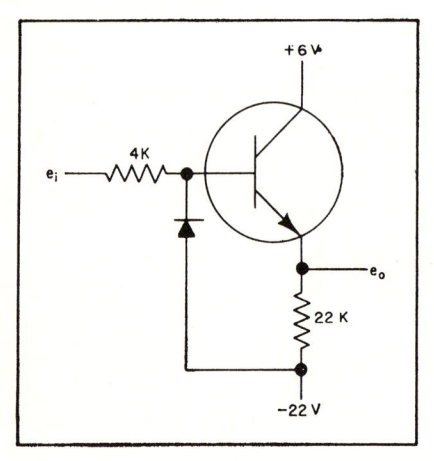

FIG. 1—Germanium diode with 4,000-ohm input resistor compensates drift in silicon transistor amplifier

IN BOTH germanium and silicon transistors, increased temperature decreases resistance of both the collector-to-base and emitter-to-base junctions. For germanium the change in the collector junction is sufficiently large to account for practically all heat drift effects. However, in silicon the change in collector-to-base junction resistance is practically negligible. The change in resistance of the emitter-to-base junction accounts for most of heat-drift effects. As a matter of fact the change in emitter-to-base resistance with changes in temperature is greater for silicon than for germanium because of the higher resistivity of silicon.

There are several methods of compensating heat drift in silicon transistor d-c amplifiers.

For an *npn* silicon emitter follower with 1-ma emitter current, output rises about 100-mv as room temperature rises to 55 C. This drift can be offset by adding a 4,000 ohms in series with the base and a back-biased germanium diode in parallel with the base. Over this temperature range diode current changes about 25 microamperes. This current through 4,000 ohms lowers voltage on the base 100 mv to provide compensation.

Using this network, the output variation with heat was found to be steady to within 10 mv. Since input impedance of the emitter follower is high, the 4,000-ohm resistor has little effect. The shunt diode offers about one-megohm resistance, which is also not significant.

A similar arrangement of compensation can be used for an amplifier circuit using emitter-resistor degeneration. If too much compensation results it is better to bleed part of the diode current to ground than to reduce the drop across the diode. This is because diode behavior is inclined to change operating point at low voltages but levels off at 20 volts or more. Hence more consistent behavior among diodes is obtained if the drop is maintained at twenty volts.

Applying this method to grounded-emitter transistor stages is not so effective unless several silicon diodes are used in series. In this type circuit the diode is in series with the collector and load resistor and is forward biased. The output can be thought of as the output of a two-input adding circuit comprising two diodes that drift the same amount with heat changes. However, the effect of each diode is opposite in sign and hence they tend to cancel.

Where a back-biased diode is used, it should be regarded as supplying a compensating current. This diode must be germanium since the leakage of silicon is negligible for these applications. On the other hand, when a forward-biased diode is used, silicon is a little better because there is more variation with heat. This diode should be regarded as a voltage source that provides a compensating voltage.

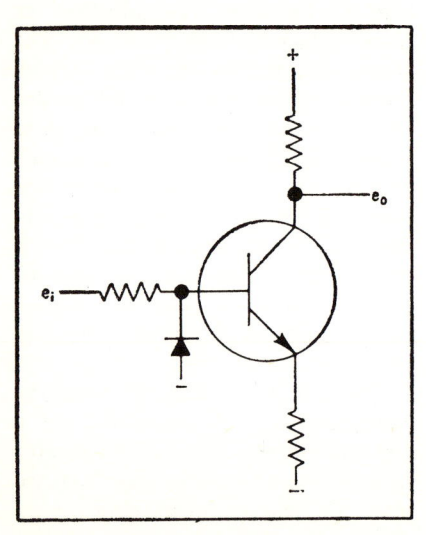

FIG. 2—If too much compensation results in this circuit, part of diode current can be bled to ground

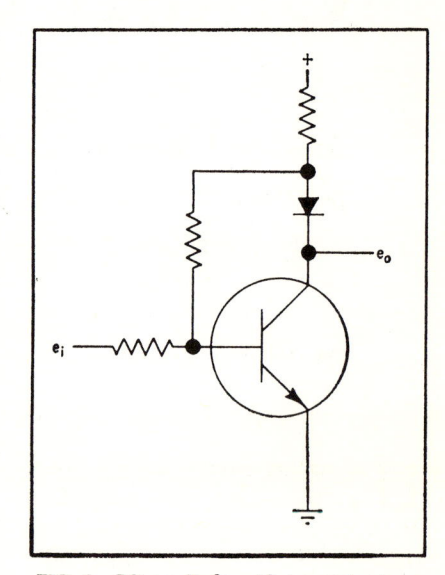

FIG. 3—Silicon diode with transistor junction form adder circuit to offset temperature drift

Chapter 3
CIRCUIT DESIGN CHARTS

High-Frequency Cutoff Nomograph

When either alpha cutoff frequency f_α or maximum oscillation frequency f_{max} are specified, chart permits easy conversion from one to the other. With value of f_{max} known, maximum power gain at any frequency can be found

By H. E. SCHAUWECKER Gilfillan Bros. Inc., Los Angeles, California

TRANSISTOR ALPHA CUTOFF frequency which results from the finite diffusion time of charge carriers through the base region is one of two characteristics that limit high-frequency performance. The frequency characteristic of alpha is represented approximately by

$$\alpha = \alpha_o/(1 + j\omega/\omega_a)$$

Collector capacitance C_c, the capacitance across the collector to base junction, also limits the maximum frequency of operation.

For most transistors, alpha is approximately equal to unity and maximum oscillating frequency is approximated by

$$f_{max} = (f\alpha/25.1 R_B C_c)^{1/2}$$

or in terms of f_α

$$f\alpha = 25 R_b C_c (f_{max})^2$$

Since some manufacturers specify $f\alpha$ and others f_{max}, conversion from one to the other is desirable.

Practical Examples

When f_α is 20 mc, R_B is 40 ohms and C_c is 10 μμf, f_{max} is desired. A straight line is drawn between the two points $f_\alpha = 20$ mc and $R_B C_c = 400$ μμsec, or 0.4 millimicroseconds. The value for f_{max} is read from the left-hand scale as 44 mc.

A transistor with an f_{max} of 10 mc and an $R_B C_c$ product of 2,000 μμsec or 2 mμsec is available. It is desired to determine the beta cutoff frequency for this transistor for which f_α is required. Referring to the nomogram, a straight line is drawn through $f_{max} = 10$ mc and $R_B C_c = 2$ mμsec. The line intersects the center scale at $f_\alpha = 5$ mc.

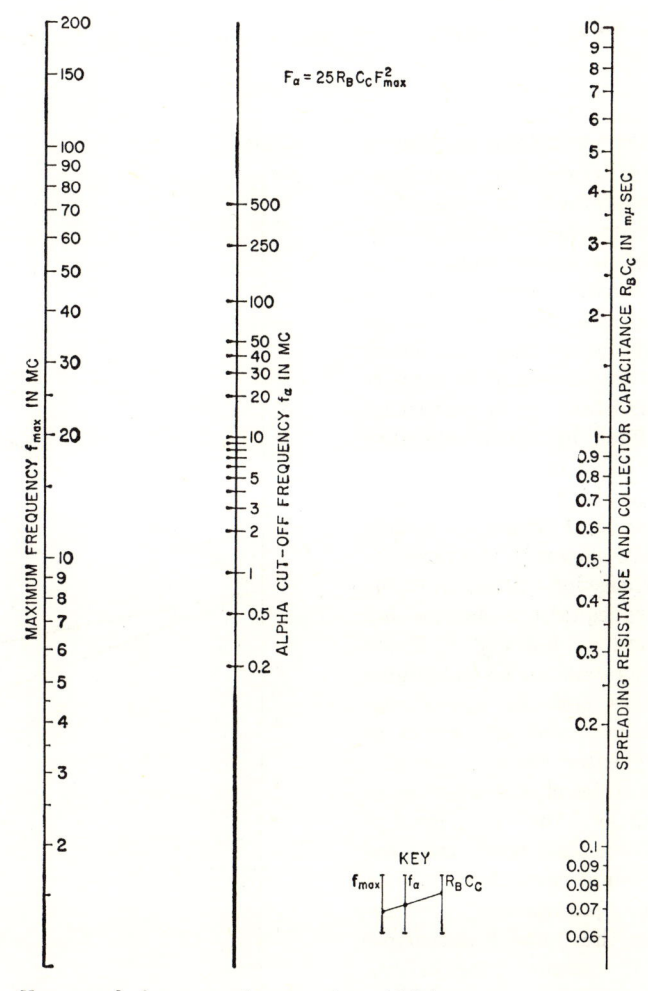

Nomograph for converting transistor high-frequency parameters

Rapid Conversion of Hybrid Parameters

Chart and nomograph simplify conversion of grounded-base transistor parameters, which are usually used on manufacturers' specification sheets, to grounded-emitter form. In most cases, common-collector parameters are then either identical to common-emitter parameters or can be obtained from them by the simple process of subtraction

By SOL SHERR Section Head, General Precision Laboratory, Inc., Pleasantville, New York

CIRCUIT DESIGNERS frequently find transistor specifications contained in manufacturers' data sheets to be in only one set of parameter values, usually the common-base representation. To use the transistor in one of the other two representations, it is necessary to convert the parameters. This conversion is simplified by the accompanying chart and nomograph.

Relationships between common-base and common-emitter values are shown in the approximate expressions listed in Table I.[1] These equations assume $h_{12b} \ll 1$ and $h_{11b} h_{22b} \ll 1$. These conditions exist in all acceptable transistors and the approximations do not generally result in errors greater than 1 percent.

The e subscript denotes common-emitter, the b subscript denotes common-base and the number subscripts denote four-terminal parameters.[2] Interchanging the e and b subscripts results in common-base parameters in terms of common-emitter parameters.

Another set of symbols employs only letter subscripts and is related to the number subscripts

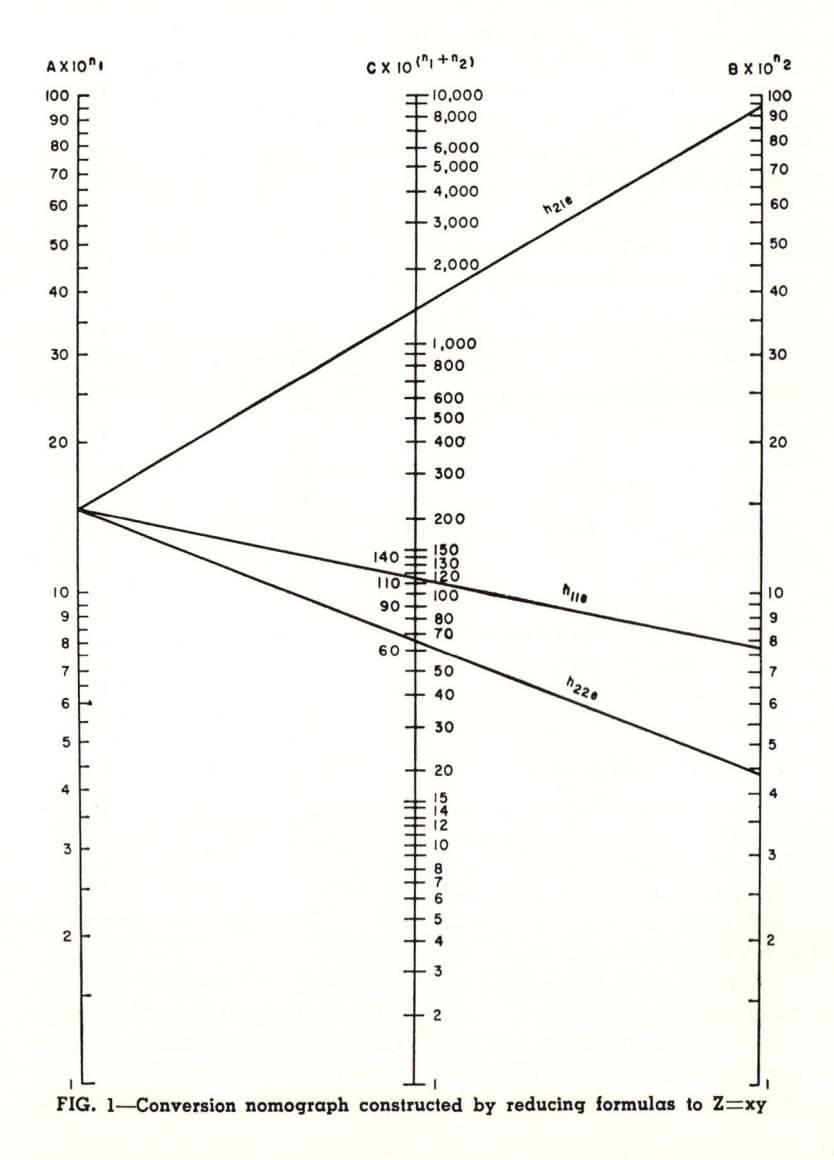

FIG. 1—Conversion nomograph constructed by reducing formulas to $Z = xy$

as shown in Table II.[2] It is easier to use the letter notation for many purposes, while the number notation will appeal to those accustomed to working with matrix notation.

Using the relationship $X = 1/(1 + h_{21b})$, all the expressions in Table I except that for h_{21e} reduce to the form $Z = xy$ and the nomograph of Fig. 1 is simple to construct for three cases. The value of X is found from Fig. 2 where it is plotted for value of h_{21b} ranging from 0.9 to 0.999.

Example

The conversion procedure can be illustrated by finding the grounded-emitter parameters for a transistor whose grounded-base parameters are $h_{11b} = 78$, $h_{21b} = 0.932$ and $h_{22b} = 4.4$.

To find h_{11e}, first find the value of X from Fig. 2 using $h_{21b} = 0.932$. Then on Fig. 1, draw a line from $X = 14.7$ on scale A through scale C to the appropriate value on scale B for h_{11b} (78). From scale C, h_{11e} is found to be 115, which agrees closely with the calculated value of 114.6.

If the values on scales A and B are multiplied by 10, the answer on scale C must be multiplied by each 10 or by a total factor of 100.

To find h_{21e}, use the previously found value of $X = 14.7$. Draw a line from this value on scale A to the value of h_{21b} (0.932) on scale B; h_{21e} is found to be 13.6 on scale C. The calculated value of h_{21e} is 13.7.

The procedure for finding h_{22e} is similar; however, the value of h_{22b} (4.4) is used on scale B. The resultant value from scale C is 6.5 compared to the calculated value of 6.47.

To find h_{12e}, first find h_{22e} as indicated above. Draw a line from h_{22e} on scale A to h_{11b} on scale B; the resultant on scale C is the value of the bracketed term in the relationship $h_{12e} = [h_{11b} h_{22b}/(1 + h_{21b})] - h_{12b}$. The value of h_{12e} is obtained by subtracting h_{12b} from the resultant on scale C.

Common Collector Values

The common collector h parameters can be found from the common-emitter parameters by using the relationships shown in Table III.[3] Two parameters are identical and the other two require only simple subtractions. The formulas for common collector in terms of common base are presented for completeness in Table IV.[1] These equations are based on the same approximations as those in Table I.

Parameters h_{11c} and h_{22c} may be found from the nomograph in the same manner as the common-emitter values, while h_{21c} may be found as x directly from Fig. 1.

Table I—Common-Emitter Parameters in Terms of Common-Base Parameters

$$h_{11e} = \frac{h_{11b}}{(1 + h_{21b})}$$

$$h_{12e} = \frac{h_{11b} h_{22b}}{(1 + h_{21b})} - h_{12b}$$

$$h_{21e} = \frac{-h_{21b}}{(1 + h_{21b})}$$

$$h_{22e} = \frac{h_{22b}}{(1 + h_{21b})}$$

Table II—Letters and Number Subscript Relationships

$h_{11e} = h_{ie}$	$h_{11b} = h_{ib}$
$h_{12e} = h_{re}$	$h_{12b} = h_{rb}$
$h_{21e} = h_{fe}$	$h_{21b} = h_{fb}$
$h_{22e} = h_{oe}$	$h_{22b} = h_{ob}$

Table III—Common-Collector Parameters in Terms of Common-Emitter Parameters

$h_{11c} = h_{11e}$
$h_{12c} = 1 - h_{12e} \approx 1$
$h_{21c} = -(1 + h_{21e})$
$h_{22c} = h_{22e}$

Table IV—Common-Collector Parameters in Terms of Common-Base Parameters

$$h_{11c} = \frac{h_{11b}}{(1 + h_{21b})}$$

$$h_{12c} = 1 - \frac{h_{11b} h_{22b}}{1 + h_{21b}} + h_{12b} \approx 1$$

$$h_{21c} = -\frac{1}{(1 + h_{21b})}$$

$$h_{22c} = \frac{h_{22b}}{(1 + h_{21b})}$$

It is also possible to find common-base parameters in terms of common-emitter parameters. The procedures are exactly the same, with the exception that the common-emitter parameters are used for Fig. 1 and 2.

Figure 2 may also be used to multiply any two numbers together, or to take the square root of any number.

REFERENCES

(1) L. P. Hunter, "Handbook of Semiconductor Electronics", McGraw-Hill Book Co., Inc., New York, 1957.
(2) IRE Standards on Letter Symbols for Semiconductor Devices, *Proc IRE*, July 1956.
(3) E. K. Novak, Transistor Parameter Conversion Charts Tables, *Electronic Design*, July 15, 1957.

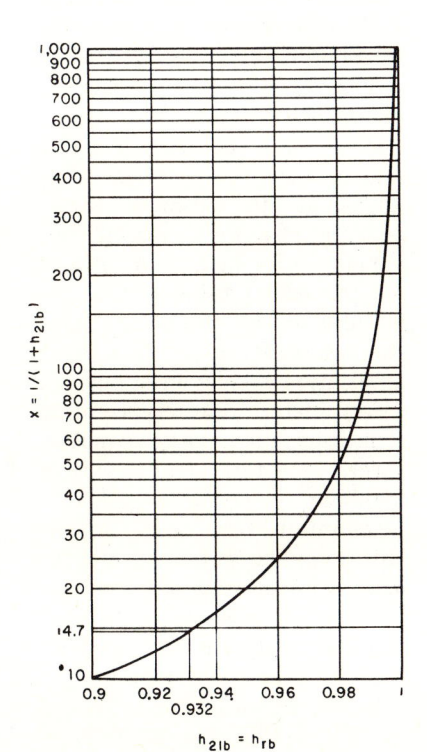

FIG. 2—Chart provides factor X for nomograph

Thermal Design Chart

Nomograph enables circuit designers to determine safety factor in terms of power dissipation and thermal resistance when using power transistors

By O. D. HAWLEY and M. KATO,
Nortronics, A Division of Northrop Corp., Hawthorne, Calif.

A DIFFICULT PROBLEM confronting the designer of transistorized equipment is making calculations necessary to insure conservative thermal design. This problem becomes acute when dealing with power transistors.

Since many designs are a matter of cut and try, thermal calculations are tedious, and sometimes difficult. Actual power ratings are dependent on temperature and a realistic concept of safety factor is needed. This article explains the ratings in terms presented on the manufacturers' data sheets. An expression for safety factor is developed. A nomograph is also presented which implements the necessary thermal calculations and can be used to determine the safety factor inherent in any design.

Junction Temperature

When the manufacturer's rating for maximum junction temperature is exceeded, the life of a transistor may be shortened even if the junction is not immediately destroyed. The junction temperature is related to ambient temperature T_a, power dissipated P_c and the thermal resistance R_t from the ambient to junction by the expression, $T_j = T_a + P_c R_t$, where temperature is in degrees C, power in watts and thermal resistance in deg C/watt. This may be described by reference to Fig. 1.

With no power dissipated in the transistor, the junction temperature is the same as ambient temperature (point A). When power is applied and dissipation in the transistor increases, the temperature also increases along line AO whose slope is equal to the thermal resistance from the junction to the ambient medium.

The majority of power tran-

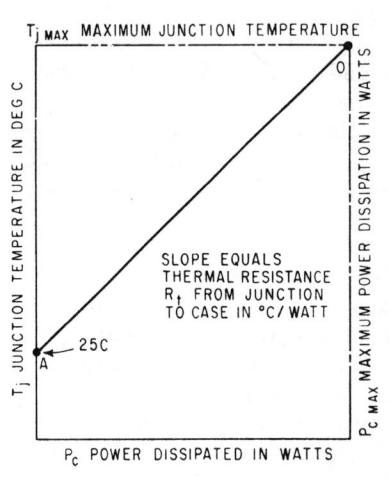

FIG. 1—Temperature-dissipation diagram showing maximum ratings usually supplied by transistor manufacturer's data

sistors carry two maximum ratings. Both maximum junction temperature and maximum power dissipation are usually specified. These are shown on the temperature-dissipation diagram in Fig. 1.

In most power transistors, the line AO connecting the point corresponding to zero dissipation at 25 C with the point corresponding to maximum power at maximum junction temperature has a slope equal to the minimum thermal resistance from the junction to the case. In most practical applications, with convection cooling, thermal resistance from junction to the ambient medium is 5 to 30 times the thermal resistance from junction to case.

Safety Factor

The concept of safety factor is based on the general engineering definition, $SF = S_u/S_w$, where SF is the safety factor, S_u is the ultimate stress, and S_w the working stress.

The safety factor is the ratio of ultimate stress to working stress. The allowable junction temperature rise over 25 C is analogous to ultimate stress since a rise in excess of this value results in a short life if not immediate failure. This figure divided by the actual junction temperature rise in a given application results in a realistic safety factor. A reference of 25 C is used since many manufacturers rate their transistors at maximum power when the case is maintained at 25 C. Apparently, higher power cannot be achieved even with a cooler case because of mechanical stresses induced by the thermal gradient between case and junction.

Safety factor can be expressed by $SF = (T_{j\max} - 25)/(T_j - 25)$, where $T_{j\max}$ is the maximum allowable junction temperature

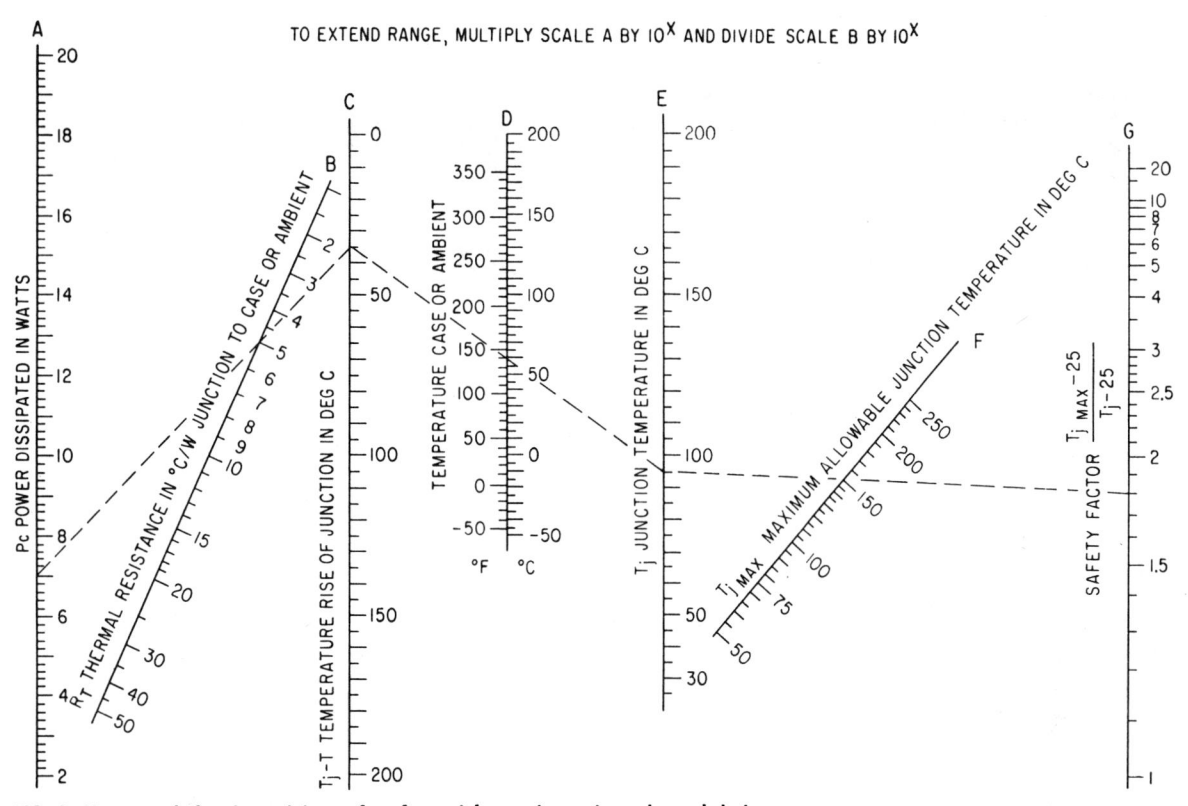

FIG. 2—Nomograph for determining safety factor inherent in a given thermal design

in deg C, and T_j is the actual junction temperature in deg C.

Another method of expressing the concept of safety factor is to use the maximum allowable power as the ultimate stress and the working power plus the power equivalent of case temperature as the working stress. This results in the expression $SF = P_{cmax}/\{P_c + [(T_c - 25)/R_t]\}$ where P_{cmax} is the maximum allowable power dissipation in watts at a case temperature of 25 C, P_c is the actual power dissipation in watts, and T_c is the case temperature in deg C.

Using Nomograph

To solve the equations relating thermal resistance, power dissipated, ambient temperature, junction temperature and safety factor, a nomograph has been developed and is shown in Fig. 2. The following example illustrates the use of the nomograph.

(1) Determine the power being dissipated in the transistor by measurement in the actual circuit. In the example shown, this is assumed to be 7 watts.

(2) Determine the thermal resistance from junction to ambient medium. The thermal resistance from case to ambient can be determined by measuring the change in case temperature for a known change in power dissipated. The thermal resistance from junction to ambient medium is the sum of the thermal resistance just determined and the thermal resistance from junction to case given in the manufacturer's data sheet. In the example, the total thermal resistance is 5 deg C/watt.

(3) Draw a line from the known power dissipated on scale A through the thermal resistance value on scale B. The extension of this line intersects scale C at the rise of the junction temperature above ambient. This is 35 C.

(4) Draw a line through the point just located on scale C and the ambient temperature on scale D. The extension of this line intersects scale E at the actual junction temperature (95 C).

(5) From this point, draw a line through the maximum allowable junction temperature on scale F. The value used in the example is 150 C. Extension of this line intersects scale G at the inherent safety factor. This turns out to be 1.8.

Range of the nomograph can be extended by multiplying scale A by some power of 10 and dividing scale B by an equal quantity, or by multiplying B by some power of 10 and dividing A by an equal quantity.

The nomograph can also be used to solve the required thermal resistance for a given safety factor, ambient temperature, power dissipated, and maximum allowable junction temperature.

By considering scale B as thermal resistance from junction to case and scale D as case temperature, calculations can be made based on case temperature using similar procedures.

Finding Dissipation of Switching Transistors

By **DONALD W. BOENSEL**, Member of Technical Staff, Space Electronics Corp., Glendale, Calif.

AN IMPORTANT problem in using transistors as switches is to determine maximum performance as a function of parameters such as load power, switching frequency and ambient temperature. In any case, collector junction temperature usually dictates maximum performance. Factors contributing directly to a rise in junction temperature are components of dissipated power resulting from such imperfect switch characteristics as OFF leakage current, ON saturation resistance, and non-zero turn-on and turn-off times.

To calculate transistor switching dissipation, assume that a generalized switching circuit may take the form shown in Fig. 1A and that the idealized output voltage and current waveforms are adequately represented by those shown in Fig. 1B. The subsequent analysis assumes the following: input dissipation is negligible; R_{sat} is determined by I_+ and load current V_c/R_L; turn-on and turn-off times (δ) are related to I_+' and I_-'; input peaking transients essentially reach completion regardless of drive frequency.

Also assume that $R_L \gg R_{sat}$,
$I_{co(max)} \ll V_c/R_L$,
$\beta = \alpha/(1-\alpha) = \beta_o/(1 + S/\omega_\beta)$
$I_{co}(T_j) = I_{co(25C)} 2^{K_L(T_j - 25)}$

The thermal resistance from collector junction to ambient consists of ρ_d (the resistance from junction to heat sink), and ρ_s (the resistance from sink to ambient). For most applications, the sum of these, ρ_t, can be treated as constant. Thus

$$T_j = T_{amb} + \rho_t P_c \quad (1)$$

Average collector dissipation per cycle is given by the sum of four components. This expression

$$P_c = \frac{1}{\Delta}(E_{on} + E_{off}) + \frac{V_c I_{co}(T_j)}{2} + \left(\frac{V_c}{R_L}\right)^2 \frac{R_{sat}}{2} \quad (2)$$

assumes that Δ, the period of the excitation, is much greater than the sum of δ_{on} and δ_{off} and that the drive is symmetrical (*off* interval = *on* interval). Energies dissipated during turn-on and turn-off are E_{on} and E_{off} respectively.

In any switching circuit designed to achieve reasonably effi-

Table I—Symbols

R_{sat}	saturation resistance
I_+	steady-state *on* base current
I_+'	peak turn-on current
I_-'	peak turn-off current
I_{co}	reverse saturation current
β_o	l-f common-emitter gain
S	$j\omega$, at the frequency of β
ω_β	beta cutoff f in rad/sec
T_j	collector temp in deg C
T_{amb}	ambient temp in deg C
K_L	0.07 for Ge, 0.11 for Si
ρ_t	deg C/w
P_c	collector power dissipation, w
P_L	peak load power, w

cient operation, input peaking is used to improve switching times; this condition means that $\beta_o I_+'$ and $\beta_o I_-'$ are $\gg V_c/R_L$.

Thus transition times are

$$\delta_{on} \cong [(V_c/R_L)/(\beta_o I_+')](1/\omega_\beta) \quad (3A)$$
$$\delta_{off} \cong [(V_c/R_L)/(\beta_o I_-')](1/\omega_\beta) \quad (3B)$$

During transitions, therefore, collector current and voltage can be represented by simple linear time-varying expressions. For turn-on

$$i_c = [(V_c/R_L)/(\delta_{on})]t \quad (4A)$$
$$v_{ce} = V_c(1 - t/\delta_{on}) \quad (4B)$$

For turn-off:

$$i_c = (V_c/R_L)(1 - t/\delta_{off}) \quad (5A)$$
$$v_{ce} = (V_c/\delta_{off})t \quad (5B)$$

When integrated over the transition intervals, Eq. 3, 4 and 5 lead to

$$E_{on} = [V_c^3/(6 R_L^2 \omega_\alpha)]/I_+'$$

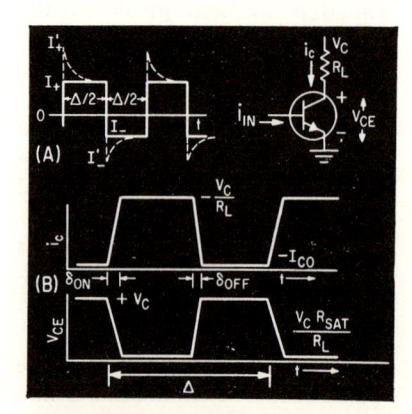

FIG. 1—Switching circuit (A) and assumed output waveforms (B)

$E_{off} = [V_c^3/(6R_L^2\omega_\alpha)]/I_-'$

where $\omega_\alpha = \beta_o \omega_\beta$ is the normal alpha cut-off frequency in radians/sec.

Although there are a number of possible combinations of knowns and unknowns in Eq. 1, it is practical to treat only one representative set. One important situation is determining the trade-off between operating frequency and load power. A particularly convenient form of Eq. 1 for such a determination is

$$T_j = T_{amb} + [P_L^2(K_1 f + K_2) + K_3 2^{K_L(T_j - 25)}] \quad (6)$$

where

$$K_1 = [(1/I_+') + (1/I_-')][\rho_t/6V_c\omega_\alpha] \quad (7)$$
$$K_2 = \rho_t R_{sat}/2V_c^2 \quad (8)$$
$$K_3 = \rho_t V_c I_{co}(25\text{C})/2 \quad (9)$$

and variables are in v, amp, ohms, deg C radians/sec and cps.

Ordinarily, either a maximum junction temperature is specified for a particular device or a lower temperature limit is chosen for safety or reliability so that T_j is known. This, of course, also determines $I_{co}(T_j)$. Although it is apparent from Eq. 6, 7, 8 and 9 that V_c can be optimized to give a maximum fP_L, such a consideration is beyond the scope of this treatment. It will therefore be assumed that some arbitrary V_c is known, allowing calculations of K_1, K_2, and K_3.

The nomogram is an aid in calculating the bracketed terms in Eq. 6. Scales (1), (2), and (3) are used to find $K_3 2^{K_L(T_j-25)}$ and scales (3), (4), and (5) to find $P_L^2 K_2$. The product $P_L^2 K_1 f$ is found with scales (5), (6), (7), (8), and (9). Although the product scales (3) and (9) have been limited to temperature terms from 1 to 100 C, scaling can be used to accommodate different P_L^2 or f ranges. Specifically for an n-decade increase in P_L^2, multiply (5) by 10^n and (6) by 10^{-n}. For an m-decade increase in f, multiply (8) by 10^m and (6) and (7) by 10^{-m}.

As an example of using the nomogram, consider the following problem.

A silicon transistor has $\omega_\alpha = 10^6$, $I_{co}(25\text{ C}) = 1$ ma, $\beta_o = 100$, $R_{sat} = 0.1$ ohm and $\rho_t = 2$ deg/w. Maximum T_j is 95 C and maximum ambient is 55 C. Determine the maximum average switched power obtainable from a 28-v supply at 100 kc with I_+' and I_-' limited to 1 amp apiece. Solving Eq. 7, 8 and 9
$K_1 = 2.4 \times 10^{-8}$, $K_2 = 1.3 \times 10^{-4}$ and $K_3 = 3 \times 10^{-2}$.

As shown by the dotted lines on Fig. 2, for

$$T_{jmax} - 25 = 70\text{ C}, \quad K_3 2^{K_L(T_j-25)} = 7\text{ C}.$$

Since K_2 is extremely small in this case, assume that it makes a negligible contribution to junction temperature. The maximum contribution of $P_L^2 K_1 f$ is obtained from

$$P_L^2 K_1 f \leq T_j - T_{amb} - K_3 2^{K_L(T_j-25)}$$
$$\leq 70 - 55 - 7 \leq 8.$$

The settings on scales (6), (7), (8), and (9) reveal that the maximum attainable power $= (3 \times 10^3)^{\frac{1}{2}} = 55$ w, or an average load power of 27.5 w.

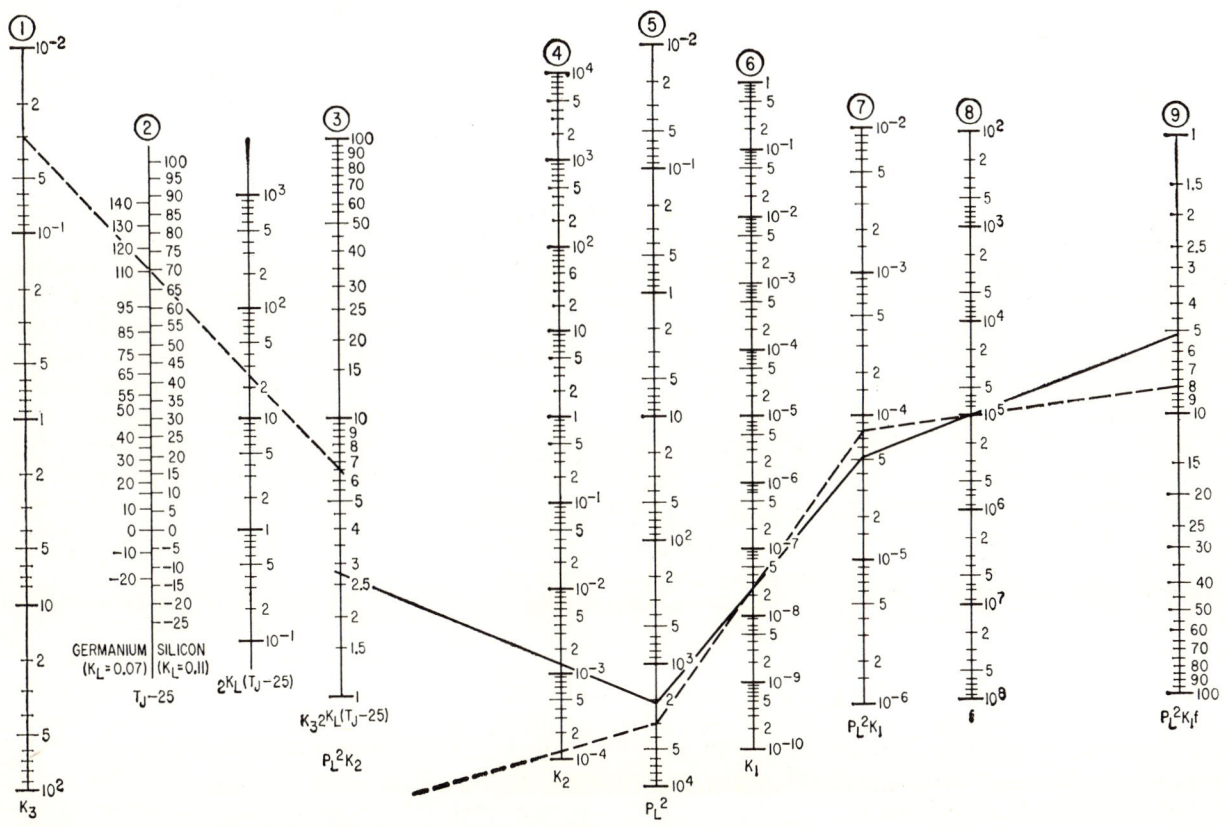

FIG. 2—Solid lines are for same problem shown by dotted lines, with $R_{sat} = 1$ ohm

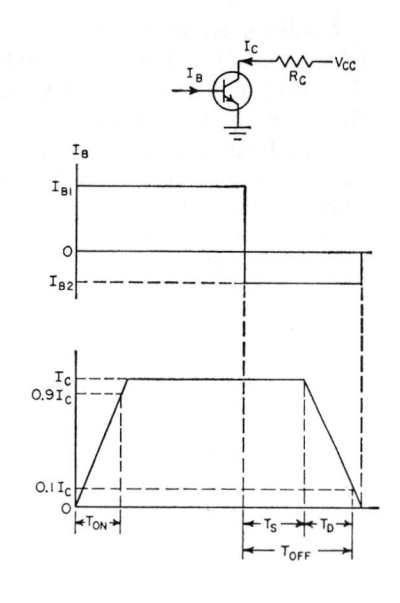

FIG. 1—Common-emitter switching circuit and waveform parameters

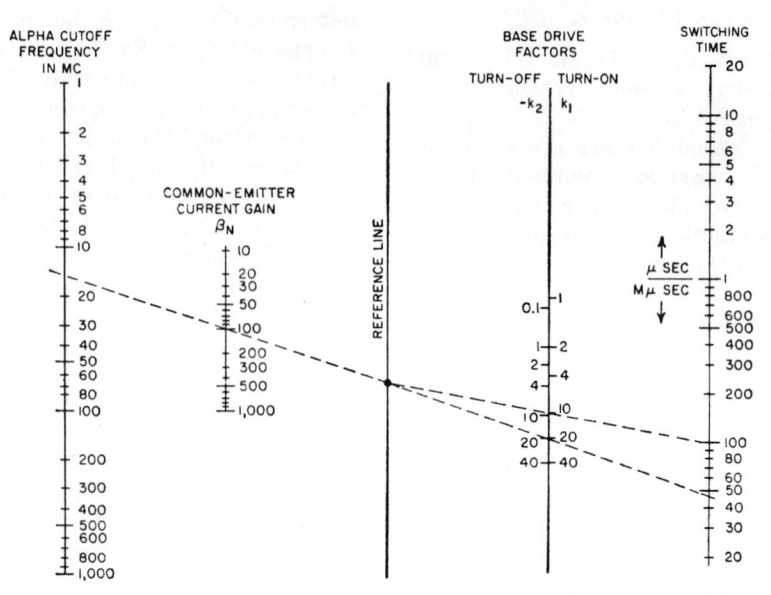

FIG. 2—Both rise and decay times are shown on nomograph scale at far right. Dashed lines show examples described

Switch-Time Nomograph

When common-emitter transistors are used as electronic switches, rise and decay times at turn-on and turn-off can be readily determined through use of the nomograph. Formula for calculation of storage time is also presented

By T. A. PRUGH Diamond Ordnance Fuze Laboratories, Washington, D. C.

TRANSISTORS in the common-emitter configuration are used in many circuits as switching elements. Switching times have been calculated in terms of basic transistor parameters and circuit conditions.[1] The results are of the general form $T = A (\ln B)$, where A is a function of the transistor and B is a function of both circuit conditions and transistor parameters.

The circuit and switching waveforms are shown in Fig. 1. Turn-off time T_{off} comprises the storage time T_s and the decay time T_d. The turn-on time simplifies to:

$$T_{on} = \{1/[(1-\alpha_n)\omega_n]\} \ln[k_1/(k_1-0.9)],$$

where α_n is the common-base short circuit current gain in the normal direction, ω_n the angular cutoff frequency of α_n and k_1 the ratio of base current I_{b1} used to turn on a transistor in a particular application to that base current I_c/β_n just necessary to saturate the transistor. Quantity I_c is the limiting value of collector current V_{cc}/R_c, and β_n is the common-emitter short circuit current gain in the direction $\alpha_n/(1-\alpha_n)$.

Decay time is:

$$T_d = \{1/[(1-\alpha_n)\omega_n]\} \ln [(k_2-1)/(k_2-0.1)],$$ where $k_2 = I_{b2}/(I_c/\beta_n)$.

Parameter k_2 is negative since the turn-off base current must be opposite to the collector current.

The nomograph, of Fig. 2 provides the turn-on and decay times. As an example, assume a transistor and circuit with $f_{ab} = 16$ mc, $\omega_n = 2\pi f_{ab} = 100 \times 10^6$, $\alpha_n = 0.99$, $(1-\alpha_n) = 0.01$, $\beta_n = 99$ and $I_c = 1,000$ μa. Also assume a turn-on base current I_{b1} of 100 μa and a turn-off base current I_{b2} of −200 microamperes.

Calculating $k_1 = 9.9$ and $k_2 = -19.8$, on the nomograph draw a straight line through $f_{ab} = 16$ mc and $\beta_n = 99$ to locate a point on the reference line. From this point draw another straight line through $k_1 = 9.9$ or $k_2 = -19.8$ to obtain the corresponding switching time. In this case the turn-on time is 100 mμsec and the decay time is 47 mμsec.

Storage time T_s is more involved and cannot be computed by a single nomograph. It is

$$T_s = \{(\omega_n + \omega_i)/[\omega_n \omega_i (1-\alpha_n \alpha_i)]\} \times \ln[(k_1-k_2)/(1-k_2)],$$

where the parameters with the i subscripts are measured with the emitter and collector connections interchanged.

REFERENCE
(1) J. J. Ebers and J. L. Moll, Large-Signal Behaviour of Junction Transistors, *Proc IRE*, p 1761, Dec. 1954.

Chapter 4
AMPLIFIERS

One-Transistor "Push-Pull"

Output circuit uses single transistor to approximate push-pull class-B audio output. Variable bias is derived from audio output. Unusual circuit configuration eliminates need for input transformer.

By JOSEPH A. WORCESTER, Radio Receiver Department, General Electric Co., Utica, New York

OPERATIONAL advantages of an audio push-pull class-B output stage are achieved by using a single transistor in a sliding class-A output circuit. Battery drain and power output of the sliding class-A circuit approximate the values obtained from the push-pull output. In addition to the cost reduction resulting from elimination of one transistor, a further saving is obtained by elimination of the input transformer normally used. Figure 1 shows the basic class-B output system.

Circuit Description

The circuit used to replace the class-B output is shown in Fig. 2. Starting with a conventional class-A stage, the fixed bias is lowered until the collector current is reduced to quiescent proportions. A variable opening bias derived from the output audio signal is then added to achieve essentially class-B operation.

Quiescent current conditions are established by the resistors R_1 and R_2. The primary and secondary of the output transformer T_1 are stacked to provide maximum a-c voltage for bias production. This a-c voltage is applied through C_1 to germanium diode D_1. The rectified a-c is filtered by C_2 and the resulting negative d-c is used to open up the transistor in accordance with the syllabic content of the output signal.

Proper operation of this circuit requires some attention to design details. The first few cycles of a syllable which produces the opening bias are clipped to some extent as the transistor is still operating under quiescent conditions. Distortion from this cause is not troublesome if the time constant of the diode output circuit, R_2C_2, is made just large enough to prevent feedback of low frequency audio signals. So that the bias circuit operates without observable delay from a syllabic standpoint, this time constant should not be much longer than required.

The diode is biased so that it starts to conduct just before the maximum undistorted output permitted by quiescent current alone is approached. If the diode operates later than this, distortion results. If it opens substantially sooner, the transistor is open more than required to handle the audio signal, causing excessive battery drain.

Diode bias is obtained by dividing the total resistance used to provide transistor quiescent current between R_1 and R_2. If R_1 is made relatively smaller and R_2 increased, the anode of the diode will be moved closer to ground potential and the cut-off bias will be increased since the diode emitter returns to B+ through R_3.

Biasing Resistance

Distribution of the quiescent biasing resistance between R_1 and R_2 is an initial circuit design consideration and, once determined, further adjustment or selection is not necessary to account for variations in transistors and diodes. Such is not the case for resistor R_3. By varying R_3 it is possible to regulate the percentage of the total control voltage developed by the diode that is utilized to open the transistor. The larger this resistor is made, the smaller the potential applied to the transistor. This adjustment is important since the gain of production transistors varies over such wide limits. For a low limit transistor, the a-c voltage in the collector circuit will be low and nearly all the developed control potential may be required to provide sufficient opening bias. This would necessitate a low value of R_3.

On the other hand, very high gain units will produce control bias greatly in excess of requirements and R_3 will have to be increased substantially. In practice values of R_3 are selected to provide the proper battery drain at a maximum output distortion of 10 percent.

The writer acknowledges the importance of the contributions made toward the practical use of this circuit by R. L. Miller.

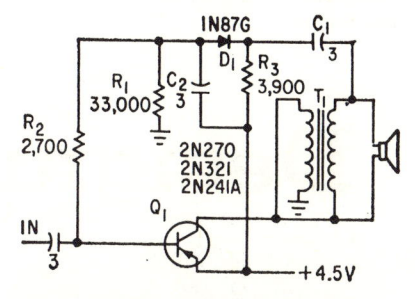

FIG. 2—Sliding class-A output has one transistor and no input transformer

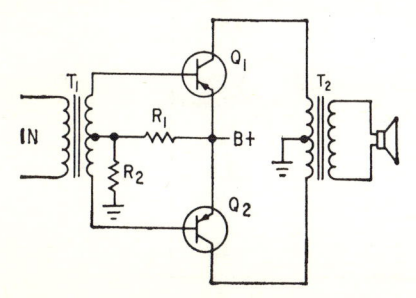

FIG. 1—Conventional push-pull class-B output uses two transistors

Fig. 1. Ribbed chassis provides both convection and radiation cooling

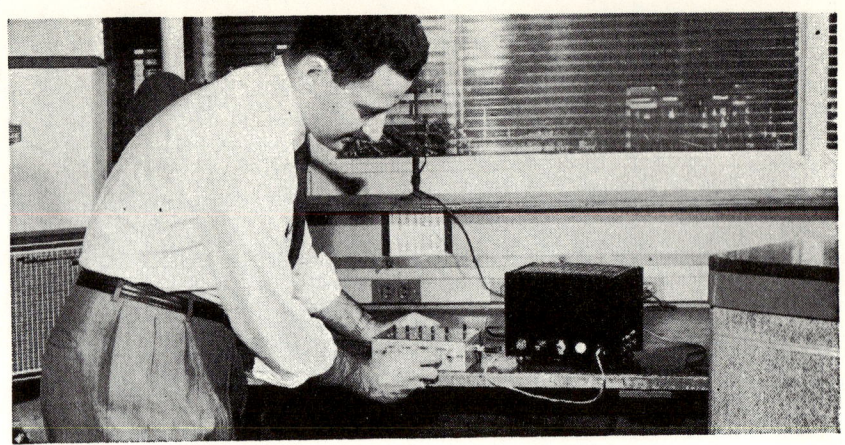

Preamplifier is adjusted in typical set up of record changer, preamplifier, power amplifier and speaker

Designing Audio

DESIGN techniques different from those used in low-power amplifiers are necessary for high-power transistor audio amplifiers. Both thermal and electrical limitations impose an upper limit on the power handling capabilities of the transistor. When transistors operate at high junction temperatures, the thermal considerations become important. For good circuit performance the chassis or other cooling facility employed must be capable of removing the heat generated by the transistor. Maintaining both a good thermal path from junction to air and a relatively stable d-c operating point over the intended temperature range will provide the necessary circuit stability.

Quasicomplementary Symmetry

The convection-cooled quasicomplementary symmetry amplifier[1] shown in Fig. 1 is designed to operate over a temperature range of -10 C to $+50$ C. The amplifier has outside dimensions of 8½ by 6½ by 6½ inches and weighs about 10 pounds. The circuit of Fig. 2, consists of three stages: a *pnp* class-A driver stage; a complementary transistor pair, which acts as a phase splitter; and a power output stage, consisting of two *pnp* transistors in single ended push-pull operation, capacitance-coupled to the load. The last two stages operate as class-B amplifiers.

Transistors Q_2 and Q_4 operate as common-collector amplifiers. When these transistors are conducting, the output current is $\beta_2\beta_4$ times the current supplied by the first stage, where β_2 and β_4 are the effective current gains of the phase splitter and output stages. Similarly, the output current when Q_3 and Q_5 conduct is $\beta_3\beta_5$ times the current supplied by the first stage, where β_3 and β_5 are the effective current gains of the phase splitter and output stages. If $\beta_2\beta_4 = \beta_3\beta_5$, the input resistance presented to the first stage is equal to $\beta_2\beta_4 R_L$, and the circuit is in balanced operation.

Output Stage

The output stage uses two 2N301A transistors connected in series with the d-c supply. For proper amplifier operation these transistors should possess certain characteristics.

One important property is that they have a thermal resistance from junction to case of less than 1.3 C/watt. Close thermal coupling to the chassis, achieved by mounting the transistor on a Mylar insulator coated with silicone oil, results in

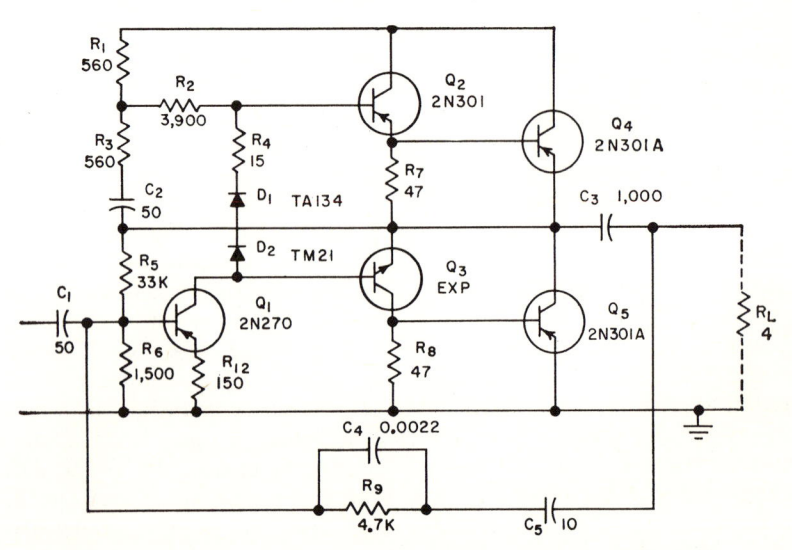

Fig. 2. Quasicomplementary symmetry amplifier delivers 45 w to 4-ohm load

High-power transistor audio amplifiers operate over ambient temperature range of −10 C to +50 C. Neither series type circuit nor quasicomplementary symmetry type circuit uses a driver or output transformer. Each amplifier can deliver 45w to a 4-ohm load. Output stages use *pnp* transistors exclusively

By MARVIN B. HERSCHER
Development Engineer, Radio Corporation of America, Camden, New Jersey

Power Amplifiers

a total thermal resistance from junction to air of about 3 C/watt. This resistance is sufficiently low so even the most severe condition of 10 to 12 watts dissipation from each transistor may be stably transferred to the air. The low chassis thermal resistance of about 1.6 C/watt is obtained both by convection and radiation cooling from the large surface area of the ribbed chassis configuration, as shown in Fig. 1. The amplifier chassis is painted a dull black to aid heat transfer by radiation.

Another important property of the output transistors is maximum collector-to-emitter breakdown voltage. This voltage depends on both the resistance connected between base and emitter and the junction temperature. It decreases when junction temperature is raised or emitter-base resistance increased (up to about 1,000 ohms).

Circuit Operation

The supply voltage is about 41 volts at full sine-wave power output. At peak signal swing the maximum inverse voltage applied to each output transistor is slightly less than the supply voltage. Under normal operating condition, a good d-c balance exists, and the center-point voltage at the collector of Q_5 is approximately equal to one-half the supply voltage.

The voltages across resistors R_7 and R_8, shown in Fig. 2 provide a small forward bias to the base-emitter junctions of Q_4 and Q_5. Forty-seven ohm resistors were chosen to minimize the nonlinear crossover region in the composite transfer characteristic of the class-B amplifier. They provide a bias of about 0.15 v at 25 C.

Bias voltage is ultimately determined by the voltage drops across R_4, D_1 and D_2. Diodes D_1 and D_2 are RCA TA-134 developmental temperature-compensating diodes. Forward voltage drop of the diodes decreases with increasing temperature and tends to hold the transistor emitter currents constant. The compensation is necessary since less forward base-emitter bias is required as junction temperature increases. For optimum temperature stability, three biasing diodes should be used.

A penalty must be paid to provide proper transistor bias and low distortion. The selection of small

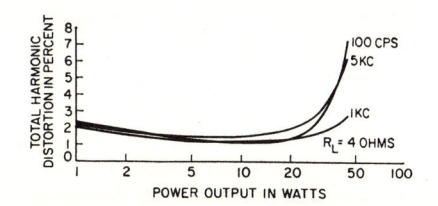

Fig. 3. Harmonic distortion increases at higher powers because beta falls off

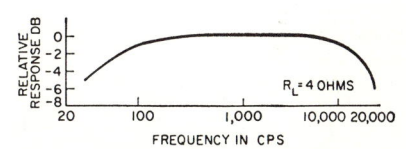

Fig. 4. Three db points are at 50 cps and 14 kc in amplifier response curve

value resistors for R_7 and R_8 results in a signal power loss of about 1 w at maximum power output. The loss occurs because the resistors are in shunt across the base-emitter junctions of the output transistors, which have a large-signal input impedance of approximately 20 ohms.

To minimize distortion the output transistors should possess a large-signal current gain of at least 25 at 4 amp of collector current. This is necessary since the peak collector current that flows in the transistors is between 4 and 5 amp. Because the output impedance of the driver stage is relatively low, variations in transconductance for the output stage are also important. Measurements show 2 to 1 variations may be tolerated.

Harmonic distortion of the amplifier at various power outputs is shown in Fig. 3. Distortion increases at higher power output because output transistor beta falls off at high collector currents. The reactance of C_3 gives an elliptical load line at low frequencies and as a result clipping occurs at high power outputs. At higher frequencies, transistor phase shift causes nonlinearities and increases the distortion.

Complementary Phase Splitter

Transistor Q_2 shown in Fig. 2, is a *pnp* 2N301A and Q_3 is an experimental *npn* transistor. Both operate class B and split the phase of the incoming signal from Q_1. To allow the amplitude of the peak signal swing to approach the supply voltage, bootstrapping is applied by way of C_2 and R_3 across R_1. Without the bootstrap action the voltage drop in Q_2 and Q_4 limits the output voltage swing to a value less than the supply voltage causing unsymmetrical clipping at high outputs. Resistor R_3 was selected to give symmetrical clipping at maximum sine-wave power output. Under this condition, the maximum collector efficiency of the output stage was increased to about 70 percent at room temperature.

Frequency response of the amplifier is shown in Fig. 4. The 3-db points are about 50 cps and 14 kc. The relatively low beta cutoff frequencies of the power transistors used in the circuit limit the high-frequency response of the amplifier. Low-frequency response is limited primarily by the capacitor C_3.

Driver Stage

Transistor Q_1 2N270, a medium power unit, acts as a class-A driver. Since the biasing of following stages depends heavily on the direct current flowing in the collector of this transistor, temperature stability of this stage is important. The transistor is biased through resistor R_5 which is connected to the midpoint of the output stage. Both d-c and a-c negative feedback are provided through the resistor. The temperature stability of this stage is further increased by emitter-current stabilization provided by the emitter resistance R_{12} in conjunction with R_0.

Source impedance driving the amplifier should nominally be 500 or 600 ohms for proper performance. About 9 db of negative feedback is applied through R_9 around the entire amplifier to the base of Q_1. Capacitor C_4 is connected in parallel with R_9 to give a step response in the feedback loop for stability.

At full power output, power gain of the amplifier is about 41.8 db. The input impedance is about 200 ohms at 1,000 cps and the output impedance is about 1.6 ohms.

Series Amplifier

The series amplifier, shown in Fig. 5, uses all transistors of like conductivity and requires no driver or output transformers. It consists of a split-load phase inverter, capacitance coupled to a class-B common-collector driver. The driver is direct coupled to a class-B common-emitter power output stage. Driver and output stages are each in series for the d-c collector supply.

The amplifier also weighs about 10 pounds and is similar in size and physical layout to the quasicomplementary amplifier. It is convection-cooled, uses 2N301 and 2N301A transistors throughout, and delivers 45 w to a 4-ohm load.

Power Output Stage

A pair of 2N301A output transistors Q_4 and Q_5, operate in the common-emitter mode and are capacitance coupled to the load. The output stage operates class B; therefore, with oppositely phased voltages applied to their bases the a-c collector currents add in the load. Hence Q_4 and Q_5 are in series for the d-c collector supply and in parallel for the a-c signals.

Output transistors were selected for the same properties as transistors in the quasicomplementary amplifier. The reasonable linearity requirements of the transconductance characteristic for these power transistors shows the advantage of driving large-signal amplifiers from a low-impedance source, that is, a voltage source. Distortion increases when a high impedance source is used because of the relative nonlinearity of the current transfer characteristic caused by beta fall off. In the series circuit, a low output impedance common-collector driver stage provides a relatively low source impedance for transistors Q_4 and Q_5.

The resultant distortion characteristics are shown in Fig. 6. Distortion can be reduced at low frequencies by increasing the values of C_2, C_3, and C_4. The 0.5-ohm resistors in series with the emitter of each output transistor improve d-c circuit stabilization for the output stage and reduce distortion, but at the expense of a decrease in power gain and power output. Overall collector conversion efficiency, including the power lost in these resistors, is still about 65 percent at maximum power output.

Frequency response of the amplifier is shown in Fig. 7. The 3 db points are about 40 cps and 30 kc. High-frequency response which is limited by the beta-cutoff frequency of the transistors depends on the source impedance and the r_{bb}' of the transistors. The best frequency response for a common-emitter stage is obtained when it is driven from a low source impedance. It is also desirable, therefore that r_{bb}' be as low as possible. The 2N301 and 2N301A transistors used in the cir-

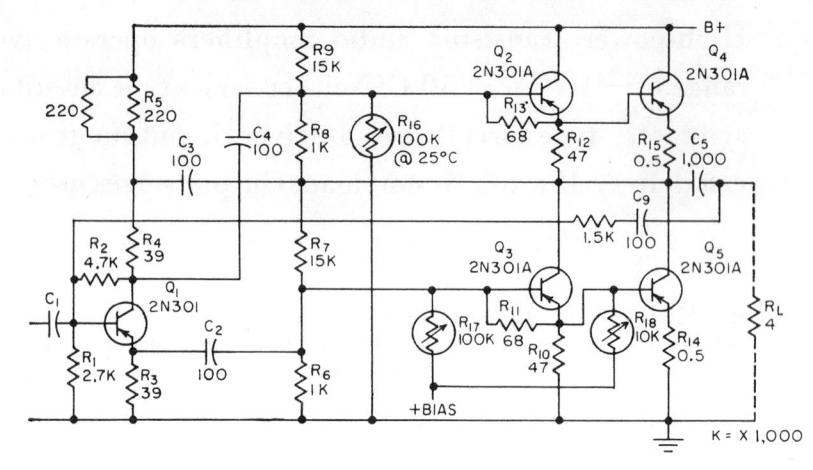

Fig. 5. Series amplifiers uses all transistors and requires no driver or output transformers

cuit have an r_{bb}' of about 20 or 30 ohms. The preceding stage is common collector and provides the desired low source impedance for the output stage. Low-frequency response is limited mainly by coupling capacitor C_5.

Series Driver Stage

The driver stage also uses a pair of 2N301A transistors connected as an emitter-follower. The stage operates class B and is direct-coupled to the output stage to eliminate a type of crossover distortion that occurs with capacitor coupling. Distortion arises because the coupling capacitor time constants for the charge and discharge paths are different during conduction and nonconduction. The difference causes a reverse d-c bias which is dependent upon the signal level, to be applied to the base-emitter junction. An increase in distortion results since the forward base-emitter bias applied to minimize crossover distortion is nullified.

In the driver stage, resistors R_{13} and R_{11} greatly reduced the crossover distortion caused by coupling capacitors C_2, C_3 and C_4. The resistors[2] linearize the input impedance presented to the phase inverter both during conduction and nonconduction. Tendency for a charge to develop on the capacitor and to produce crossover distortion is reduced by providing a low-impedance discharge path for the capacitors during nonconduction.

A 2N301 transistor Q_1 is used as a split-load phase inverter which feeds driver transistors Q_2 and Q_3. The stage operates class A and is biased at approximately 160 ma collector current with 35 v between the collector and emitter. This permits sufficient signal to be applied to the driver stage without introducing clipping at maximum power output. The transistor is closely coupled thermally to the ribbed chassis.

Resistors R_3 and R_4 provide a low source impedance for the driver transistors. The impedance presented by the upper half of the driver-output transistor combination is essentially equal to that presented by the lower half, since in each case there is a common-emitter stage preceded by a common-collector stage. Variations in the load impedance are reflected to the phase inverter in a similar manner for the upper and lower halves keeping the amplifier in balanced operation. A balanced output voltage is obtained by splitting the collector load and feeding the upper half of the amplifier from R_4.

Inherent negative feedback exists in the amplifier because of the phase splitter configuration, the common collector driver stage (100 percent negative voltage feedback)

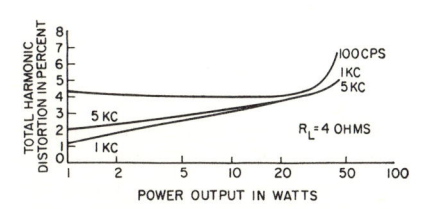

Fig. 6. Distortion can be reduced at low frequencies by increasing values of C_2, C_3 and C_4

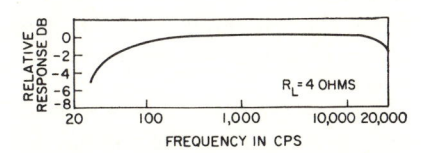

Fig. 7. Three db points are at 40 cps and 30 kc

and the unbypassed emitter resistors in the output stage. In addition about 8 db of negative feedback is applied through a 1,500-ohm resistor and capacitor C_9 around the entire amplifier. Bootstrapping similar to that used in the quasicomplementary amplifier is used to drive the upper half of the amplifier close to the collector supply voltage. The amplifier should be driven from a source having an impedance of about 500 to 600 ohms.

At full power output, the power gain of the series amplifier is about 30.6 db. Input impedance of the amplifier is about 113 ohms at 1,000 cps; output impedance, about 2.5 ohms.

Driver and Output Stage Bias

A small forward quiescent d-c base-emitter bias voltage is used to eliminate crossover distortion caused by the nonlinear transfer characteristics of transistors at low signal levels. Optimum bias voltage is obtained for the driver stages by the voltage divider action of resistors R_8 through R_{13}. The output transistor bias depends on the quiescent current of the driver stage and the voltage divider network. Quiescent current in the output stage is about 50 ma at room temperature.

As ambient temperature is increased, the operating points of the driver and output stage change, since both saturation current and input conductance increase.[3] The problem of d-c stabilization for the output stage is further complicated by the direct coupling of driver output stages. As temperature increases, the driver I_{co} increases, causing the forward voltage bias for the output stage to increase at the same time output transistor I_{co} is increasing and causes the quiescent current for the output stage to increase further. The one-half ohm resistors, R_{14} and R_{15} provide a negative current feedback that tends to keep output quiescent collector current constant.

In the common-collector driver stage, degenerative d-c feedback is obtained through resistors R_{10} and R_6 in the lower transistor, and resistors R_{12} and R_8 for the upper transistor. Thermistor compensation further stabilizes the d-c operating point of the output stage. The thermistors are mounted on the chassis near the output transistors to more closely compensate for the change in junction temperature rather than ambient temperature. Driver emitter current is kept constant by thermistors R_{16}, R_{17}, and R_{18} which provide a base-to-emitter voltage which decreases with temperature. A positive bias supply compensates the transistors at higher temperatures. It is actually necessary to apply a reverse bias to the base-emitter junction to maintain a relatively constant operating point at elevated temperatures.

This material was first presented at the Audio Engineering Society convention, New York, October, 1957.

REFERENCES

(1) H. C. Lin, Quasi-Complementary Transistor Amplifier, ELECTRONICS, p 173, Sept. 1956.
(2) A. I. Aronson, "Transistor Audio Amplifiers," *Transistors I*, RCA Laboratories, p 515.
(3) H. C. Lin and A. A. Barco, "Temperature Effects in Circuits Using Junction Transistors," *Transistors I*, RCA Laboratories, p 369.

Designing High-Quality

Completely transistorized audio amplifier uses junction transistors to deliver 25 watts output power. Frequency response is within ±1 db between 20 and 25,000 cps with low harmonic and intermodulation distortion

By ROBERT MINTON, Semiconductor and Materials Division, Radio Corporation of America, Somerville, N. J.

THIS TRANSISTORIZED AUDIO AMPLIFIER was designed to investigate the feasibility of using junction transistors in a high-quality audio amplifier. The seven-stage amplifier can deliver 25 watts when driven by a variable reluctance cartridge. It has low harmonic and intermodulation distortion and good frequency response, and may be used at ambient temperatures up to 55 C. Table I lists the amplifier characteristics.

Preamplifier

The preamplifier shown in Fig. 1 uses four alloy junction transistors and was designed to accept a signal from a variable reluctance cartridge having an output of approximately 10 millivolts at 1,000 cps with an inductance of 0.52 henries and a d-c resistance of 600 ohms. The preamplifier provides a frequency-corrective network for RIAA recording characteristics, variable bass and treble compensation, volume control and loudness control.

The recording frequency corrective network is composed of R_1, C_1 and C_2 connected as a feedback network between the collector of Q_2 and the emitter of Q_1. The feedback causes the overall frequency response of the first two stages to be the inverse of the recording characteristics and raises the effective input impedance of the input stage so that at any frequency the input impedance is much higher than the impedance of the inductive pickup. As a result, the equalization circuit provides flat response independent of the source impedance.

The treble control circuit consists of R_2, R_3, R_4, C_3, and C_4 and provides approximately +12-db boost and −17-db cut at the higher frequencies. The bass control circuit consists of R_5, R_6, C_5, C_6 and the input resistance of Q_3. This circuit provides approximately +15-db boost and −12-db cut at the lower frequencies.

Volume control R_7 is a current divider controlling the current fed to the loudness control and the input to Q_4. The loudness control circuit consists of R_8, R_9, R_{10}, C_7, C_8 and the input resistance of Q_4. Variable resistors R_8 and R_9 are ganged to form a dual control. The loudness control is a frequency selective current divider providing substantial boost to the low frequencies and slight boost to the higher frequencies while attenuating the mid-frequencies as the intensity level is decreased.

The power amplifier input stage Q_5 shown in Fig. 2 is a class-A, common-emitter power amplifier. Application of balanced negative feedback from the collectors of output stage Q_7 and Q_8 to the base and emitter of input stage Q_5 reduces the harmonic distortion and extends the overall frequency response of the power amplifier. The use of balanced feedback also effectively minimizes any residual hum appearing in the feedback loops because the voltages are opposite in phase and are cancelled.

Driver stage Q_6 operates class-A in a common-emitter circuit. The output is transformer-coupled to the class-B output stage. The secondary of driver transformer T_1 is bifilar wound to provide tight coupling, thus minimizing transient voltages when the current shifts from the base of one output transistor to the base of the other. The source impedance to the input of the class-B output stage has a pronounced effect on the total harmonic distortion in the output stage. As

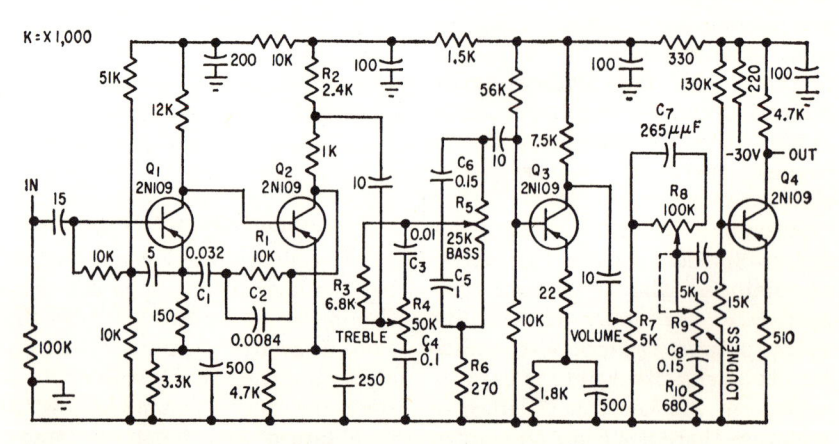

FIG. 1—Preamplifier accepts variable reluctance cartridge signal and contains bass, treble, volume and loudness controls. Four alloy junction transistors are used

Audio Amplifiers

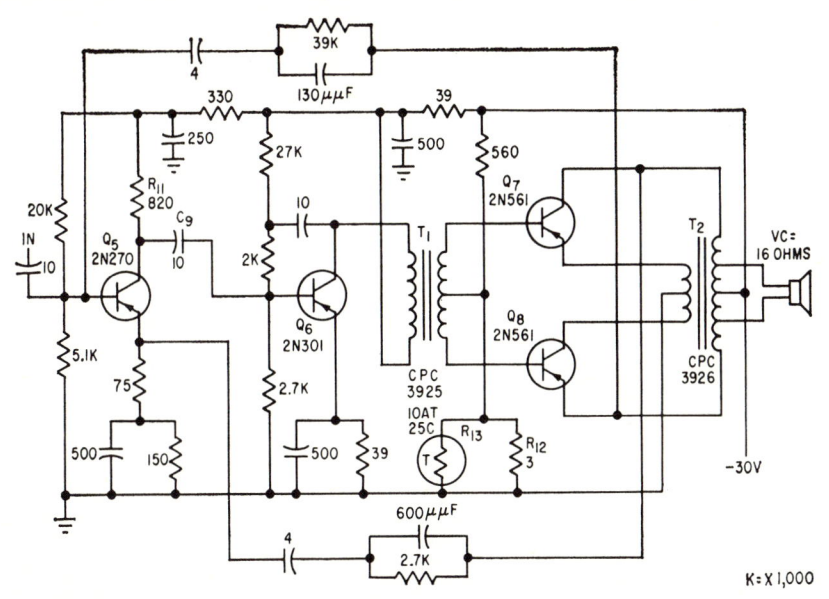

FIG. 2—Power amplifier uses balanced negative feedback and operates class-B. Output stage has emitter degeneration as further improvement

Table I—Amplifier Characteristics

Frequency Response	Within 1 db from 20 to 25,000 cps. 3 db down at 25-watt level
Power Output	25 watts
Harmonic Distortion	1.25 percent at 25 watts and 400 cps
Intermodulation Distortion	3.5 percent at 25 watts, 60/3,000 cps at 4:1
Hum Level	−75 db below 25-watt level
Maximum Ambient Temp.	55 C

the output impedance of the driver stage is too high, it is reduced by shunt feedback from the collector to the base. This also causes a reduction in the effective input impedance of the driving device and requires that the transistor used in the input stage operate as a power amplifier.

The power amplifier output stage consists of two pnp germanium alloy junction transistors operated in a class-B, push-pull circuit. Class-B operation is used because it provides maximum efficiency. The stage can be biased essentially at cutoff thus reducing standby power and minimizing zero-signal dissipation. A common-emitter type circuit provides the highest power gain but suffers from total harmonic distortion which is usually higher than acceptable; a common-collector circuit has lower distortion due to its inherent feedback but requires a considerable amount of driving power. The output stage used is a combination of both circuit types and has a power gain intermediate between the two.

Output transformer T_2 has a winding connected in the output stage emitter circuit to provide emitter degeneration and low harmonic distortion. Because the transformer d-c resistance is low, d-c power loss is minimized and higher operating efficiency results than would be obtained if resistors were inserted in the emitter circuit to provide the same amount of a-c degeneration. The output transformer has bifilar windings to reduce leakage inductance and winding capacitance. This method of winding provides nearly unity coupling between both halves of the windings reducing transient voltages.

Temperature stabilization of the output stage is provided by R_{12} and thermistor R_{13} in the bias circuit. The network provides a compensating shift in bias voltage with temperature variations which causes the d-c collector current to remain essentially constant.

The collector junctions of the power transistors in the driver and output stages are connected to the mounting flange and must be electrically insulated from the chassis which is at ground potential. At the same time, the mounting flange must make good thermal contact with the chassis. Mica insulating washers having a thickness of 0.002 in., are used to insulate the mounting flange electrically from the chassis. The mica washers have high thermal conductance and do not increase the total thermal resistance of the circuit appreciably.

Power Supply

The power supply consists of a transformer, two silicon diodes and a choke-input filter. The power supply should deliver 30 v d-c. The power transformer should deliver a maximum load current of 2 amps. Because the power supply works into a varying load impedance, a choke-input filter is used in conjunction with silicon diodes to provide good regulation. The choke has an inductance of 0.075 H at a current of 2 amps and a d-c resistance of approximately 1 ohm.

BIBLIOGRAPHY

H. Fletcher and W. A. Munson, Relation Between Loudness and Masking, *J. Acoust Soc Amer* 9, No. 1, p 1.

N. H. Crowhurst, Why Loudness Control, *Radio and TV News*, p 42, Apr., 1957.

N. H. Crowhurst, Loudness and Volume Controls, "Understanding Hi-Fi Circuits", 64, p 175, Gernsback Library Inc.

R. Minton, Class B Transistor Power Amplifier Design, *Electronic Design*, p 24, Sept. 1957.

F. Langford-Smith, McIntosh Amplifier, "Radiotron Designer's Handbook", fourth edition, p 594, RCA.

Feedback Design for

Negative feedback lessens the effects of temperature and transistor variations. For a specified current gain, the design method produces maximum available feedback. Alternatively, the equations and the approach may be used to obtain a specified input impedance

By THOMAS R. HOFFMAN, Professor of Electrical Engineering, Union College, Schenectady, New York

NEGATIVE FEEDBACK is known to contribute to amplifier stability and to lessen the effects of transistor variations. But the problem of considering all possible parameter variations over a given temperature range is complex and is best attacked with the aid of a fairly good-sized computer.

Since in many cases this procedure is not possible or economical, the usual design objective is to use as much negative feedback as possible and still obtain the required overall gain. The basic idea is to trade excess gain for stability.

Testing the completed circuit will reveal the extent to which the stability requirements have been met. If the maximum available amount of negative feedback has already been used, and further stability is required, another stage of amplification with feedback may be the solution.

All semiconductor devices, transistors included, are by nature temperature sensitive. Variations of the operating point also affect performance, often appreciably. The variation of parameters in production lot transistors is well known and again, negative feedback helps to minimize the effects of these variations.

Only the grounded emitter stage is considered here since it is the configuration with the highest gain and is the most often used. Negative feedback may be introduced into a grounded emitter stage by adding an unbypassed emitter resistor (R_E) or a feedback resistor (R_F) from collector to base. Although it might be thought that the results would be similar regardless of how the feedback was derived, the two cases produce greatly different input impedances. Thus the unbypassed emitter resistor makes possible a substantial increase in input impedance while the addition of feedback resistor R_F decreases the input impedance. The two cases will be considered separately.

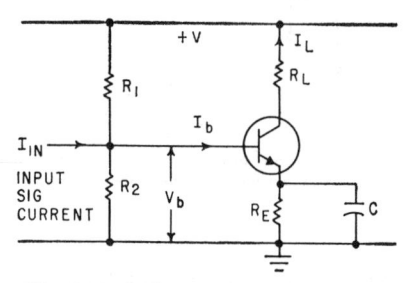

FIG. 1—Typical grounded emitter stage without feedback. The capacitor brings the emitter to a-c signal ground

Table I—Formulas for Feedback Amplifier Design

	Unbypassed R_E (Fig. 2 and 3)	Collector to Base R_F (Fig. 4 and 5)
Current Gain	$K_i = \dfrac{I_L}{I_b}$ $= \dfrac{R_E + r_e - r_m}{R_E + R_L + r_e + r_c(1-a)}$ $= \dfrac{R_E - 2{,}310{,}000}{R_E + 197{,}500}$	$K_i = \dfrac{I_L}{I_1}$ $= \dfrac{-aR_F}{R_L + R_F(1 - a + R_L/r_c)}$ $= \dfrac{-0.925 R_F}{0.079 R_F + 10{,}000}$
Input Impedance	$R_i = \dfrac{V_b}{I_b}$ $= r_b + (r_e + R_E)(1 - K_i)$ $= 300 + (R_E + 20)(1 - K_i)$	$R_i = \dfrac{V_b}{I_1}$ $= [r_e + r_b(1-a)](1 - K_i)$ $- \dfrac{R_L r_b}{r_c} K_i$ $= 42.5 - 43.7 K_i$
Overall Current Gain	$\left\vert \dfrac{I_L}{I_S} \right\vert = \left\vert K_i \right\vert \left[\dfrac{R_S R_B}{R_i(R_S + R_B) + R_S R_B} \right]$	
Overall Voltage Gain	$\left\vert K_V \right\vert = \left\vert \dfrac{I_L R_L}{E_S} \right\vert = \left\vert K_i \right\vert \left[\dfrac{R_L R_B}{R_i(R_S + R_B) + R_S R_B} \right]$	

Amplifier Stages

Figure 1 shows a conventional arrangement for an *npn* grounded emitter amplifier without feedback. Resistors R_1 and R_2 are chosen in conjunction with R_E and V to give the desired d-c emitter current. Capacitor C bypasses R_E so that the emitter is grounded for a-c. The quantities of interest are current gain, $K_i = I_L/I_b$, and input resistance, $R_i = V_b/I_b$. Current I_b is the portion of the input signal current I_{in} that actually reaches the transistor. Note that the overall current gain I_L/I_{in} will depend on the division of I_{in} between the bias resistance (R_1 and R_2 in parallel) and the transistor base lead. If K_i and R_i are known, this ratio may be readily determined.

If C in Fig. 1 is omitted, R_E becomes a factor in the signal behavior of the circuit. This circuit and its a-c equivalent are shown in Fig. 2A and 2B.

The current gain K_i may be expressed in terms of the device parameters. Input resistance R_i can be concisely expressed in terms of K_i. The equations for these two quantities are given in Table I.

Evaluation of the expressions for K_i and R_i is generally difficult because of the large number of parameters involved. At this point the designer must substitute average values of parameters for the transistor he plans to use. To illustrate, typical values for a low-level, silicon junction transistor have been chosen:

$r_b = 300$ ohms
$r_e = 20$ ohms
$r_c = 2.5$ megohms
$a = 0.925$
$r_m = ar_c = 2.31$ megohms

In addition, it will be assumed that R_L is 10,000 ohms. This, too, will be a known quantity in a particular design. The expressions for K_i and R_i with the above values substituted are included in the table.

Figure 3 is a plot of current gain and input impedance as a function of unbypassed R_E. Note that R_i starts to increase almost immediately as R_E exceeds zero, while K_i is not affected appreciably until R_E exceeds 10,000 ohms. From a practical viewpoint, it may be concluded that variation of R_E over a range that would still permit proper biasing will have no noticeable effect on K_i but will increase R_i drastically. However, if the input signal source is anything other than an ideal constant current generator, increased R_i will lower the overall current gain by reducing the amount of signal current that reaches the transistor base terminal.

For the transistor used, a 10,000-ohm load resistor is small with respect to $r_c(1-a)$. Current gain K_i and R_i are therefore relatively

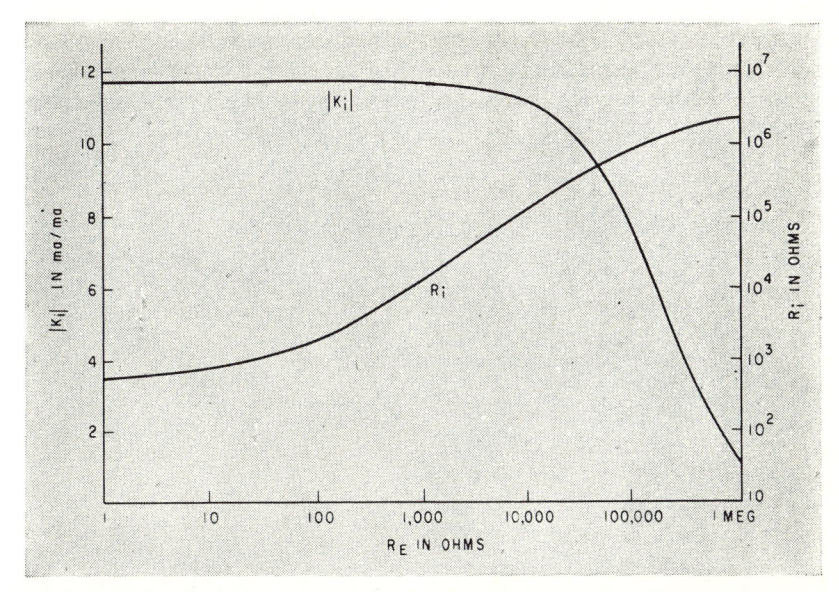

FIG. 3—Current gain K_i and input impedance R_i functions for feedback obtained from an unbypassed emitter resistor. Current gain is little affected until R_E exceeds 10,000 ohms. Curves are for small signals

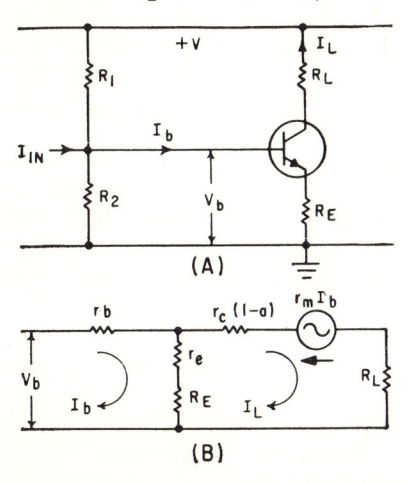

FIG. 2—Removing the emitter bypass capacitor produces negative feedback in transistor stage (A); equivalent circuit for small signals only (B)

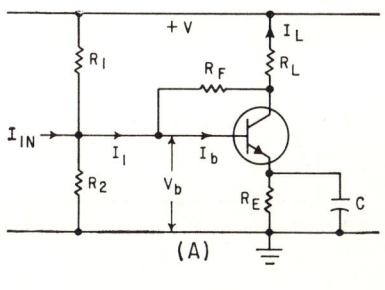

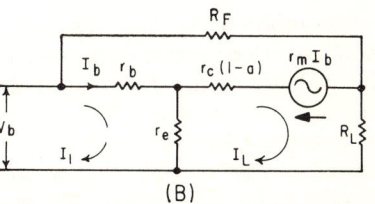

FIG. 4—Negative feedback from collector to base is obtained through resistor R_F (A). Bypassed R_E is not effective in small signal a-c equivalent circuit (B)

insensitive to changes in R_L.

Figure 4A shows the circuit for collector to base feedback and Fig. 4B the equivalent circuit. With the approximation that r_e and r_b are much less than $r_c (1 - a)$, normally true for transistors of all types, equations for K_i and R_i are developed and listed in the table.

be readily obtained from the relationships just presented by including the effect of bias and source impedances. An equivalent circuit which would apply for either feedback arrangement is shown in Fig. 6. The source has been represented by a constant current generator I_s paralleled by the source impedance

is also obtainable. The source is represented by its Thevenin equivalent (a constant voltage generator E_s in series with the source impedance R_s), and voltage gain can then be expressed in terms of R_i, K_i, R_s and R_B as follows: $K_v = I_L R_L / E_s = K_i \{ R_L R_B / [R_i (R_s + R_B) + R_s R_B] \}$. Again, this is readily evaluated for any R_E or R_F with the aid of the curves.

Design Examples

Use of the curves is best illustrated by examples. Assume a transistor with the average small-signal parameters of Figs. 3 and 5 is to be used in an application requiring an overall current gain of 8. The source impedance is 10 K, and the bias circuit impedance (R_1 and R_2 in parallel) is 20 K. Load resistance is 10 K.

For unbypassed R_E feedback, determine R_E to satisfy the requirement. The overall gain is: $|I_L / I_s| = |K_i| \{ 10^4 (2 \times 10^4) / [30,000 R_i + 2 \times 10^8] \} = 8$. From Fig. 3 it is evident that K_i is independent of R_E over a wide range. Thus, assuming K_i to be 11.7 (the no-feedback value), we may solve for R_i equal to 2,107 ohms. This requires an R_E of 122 ohms. (Note that if a d-c R_E larger than 122 ohms is desired for bias stability, the additional resistance can be bypassed by a capacitor. The 122 ohms represents the value of R_E that should be effective in the signal circuit.)

For R_F feedback, determine R_F to satisfy the requirement. The overall gain is: $|I_L / I_s| = |K_i| \{ 2 \times 10^8 / [30,000 R_i + 2 \times 10^8] \} = 8$. Using Fig. 5, a cut and try method is used. When R_F equals 400 K, the curve yields a K_i of 8.88 and an R_i of 430 ohms, which satisfies the above relationship.

For actual design work, the curves for K_i and R_i must be plotted accurately. Furthermore, it must be realized that the curves are for a particular type transistor and normally cannot be used in a design problem requiring other types. In many cases the same curves can be used, with little error, for several types of transistors.

Note that resistance, except for one case, is plotted to a log scale.

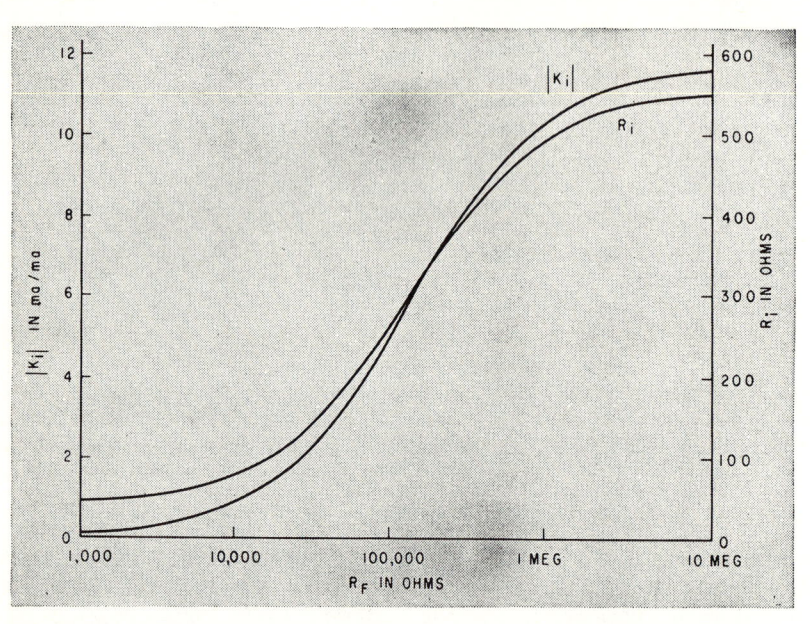

FIG. 5—Current gain K_i and input impedance R_i are shown for negative feedback from collector to base via resistor R_F. The gain and input impedance vary in nearly the same ratio. Curves are for small signals only and should not be extrapolated.

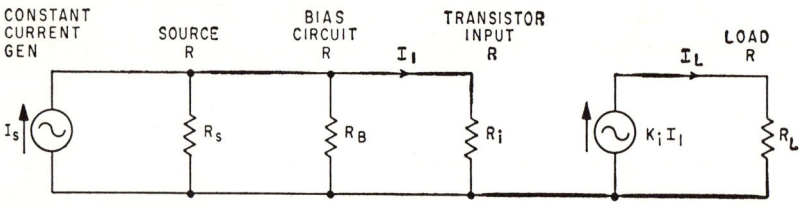

FIG. 6—Equivalent circuit for grounded emitter stage

Presentation of the results of these equations again requires specific values. Using the typical device parameters of the R_E case, the expressions for K_i and R_i are plotted in Fig. 5. Note that K_i and R_i vary together in much the same proportion as R_F decreases. For this case, the magnitude of the load resistance has a greater effect on gain than in the R_E case.

Circuit Design

In a practical problem, the parameter of interest is usually the overall current gain. This may

R_s. This representation is perfectly general since any source can be so shown by applying Norton's theorem.

Since R_i and K_i are known for any value of the feedback resistor (R_E or R_F), overall current gain can be readily predicted with the help of the curves (Fig. 3 or Fig. 5). The relationship is: $I_L / I_s = K_i \{ R_s R_B / [R_i (R_s + R_B) + R_s R_B] \}$. The factor in brackets is merely the current division ratio between R_i and the parallel combination of source and bias resistances.

If desired, overall voltage gain

Amplifier with 100-Megacycle Bandwidth

Shunt feedback networks around each stage reduce overall gain at low frequencies, trading gain for bandwidth. Circuit can be adjusted for use as pulse amplifier

By J. C. de BROEKERT and R. M. SCARLETT,
Stanford Electronics Laboratories, Stanford University, Stanford, California

PRESENTLY available drift transistors yield much greater video amplifier bandwidths than those easily attainable with conventional vacuum-tube pentodes, although with less output power. In this amplifier, shunt resistive feedback around each common-emitter stage exchanges gain for bandwidth, with a series inductance to peak the high frequency response. This arrangement results in an amplifier which is particularly simple by comparison with its vacuum-tube counterpart (often a distributed amplifier).

Simple design equations included relate gain and bandwidth per stage to both circuit and transistor parameters. A design example of an amplifier with 50 db gain and over 100 Mc bandwidth is presented. The construction of this amplifier is described, and choice of suitable components discussed in some detail. Performance measurements agreed satisfactorily with the design theory.

Other simple transistor video amplifier configurations have been discussed by Bruun[1] (simple resistive loading), Spilker[2] (emitter degeneration) and Scarlett[3] (common-collector common-emitter pairs). The configuration used here, with $L_f = 0$, has been discussed by F. H. Blecher.[4] He used it to shape loop gain in an amplifier which was gain-stabilized by a large overall feedback.

Design Procedure

A single stage is shown in Fig. 1, where R_L represents either the output load resistance, or the input resistance R_{in} of the next stage. The transistor is represented by the equivalent circuit model of Fig. 2, which is characterized by the five parameters shown. Element $1/\omega_t r_e'$ is made up by the capacitance of the emitter depletion layer and the storage capacitance. The frequency f_t is that frequency at which the common emitter short-circuit current-gain becomes unity. The d-c operating point will ordinarily be chosen to maximize f_t, subject to power dissipation limitations. Collector capacitance C_c should include

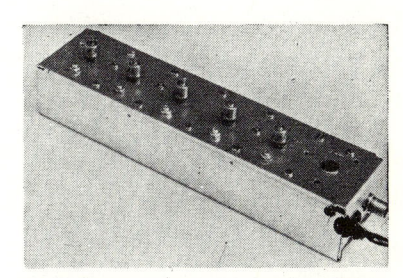

Neat surface appearance (A) and straightforward wiring and coupling networks (B)

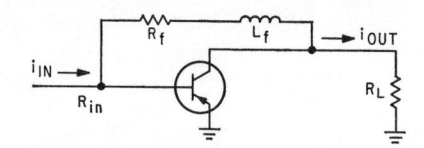

FIG. 1—Shunt feedback is applied to a single amplifier stage

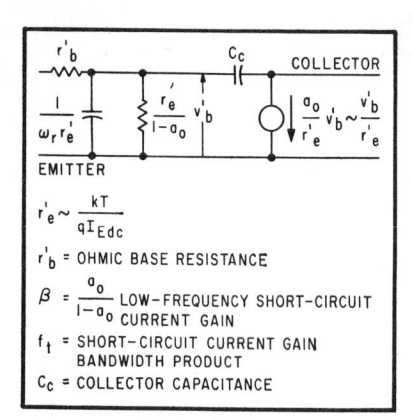

FIG. 2—Equivalent circuit of the transistor is valid over operating range

$r'_e \sim \dfrac{kT}{qI_{Edc}}$

r'_b = OHMIC BASE RESISTANCE

$\beta = \dfrac{a_0}{1-a_0}$ LOW-FREQUENCY SHORT-CIRCUIT CURRENT GAIN

f_t = SHORT-CIRCUIT CURRENT GAIN BANDWIDTH PRODUCT

C_c = COLLECTOR CAPACITANCE

all direct collector-base capacitance due to the leads and header. In the analysis, the resistance r_e' appears only in combination with r_b'/β; for convenience, this sum is denoted by r_1. Resistor r_1 is actually the common base input-resistance (at low frequencies) h_{ib}, and should be directly measured as such.

Design equations are derived from an exact analysis by the liberal use of simplifying approximations. A useful degree of accuracy is still retained, as the example demonstrates.

Low-frequency current-gain magnitude of one stage (see Fig. 1) is

$A_i = |i_{out}/i_{in}| = R_f/(R_L + r1)$ if $\beta \gg A_i$ (1)

Input resistance is

$R_{in} = A_i\, r1\, (1 + R_L/R_f)$ (2)

In an iterative situation, one has

$R_{in} = R_L = r1\,(1 + A_i)$ if $A_i > 3$ (3)

If $L_f = 0$, the 3 db bandwidth is, quite closely

$B_o = \dfrac{f_t}{A_i} \times \dfrac{1}{1 + (r_b' + R_L)/R_f + \omega_t R_L C_c (1 + r_b'/R_f)}$ (4)

For $L_f = 0$, the selectivity function is of the form $1/(1 + jx)$, and the bandwidth shrinkage as identical stages are cascaded is readily found.[5] The load on a stage other than the last (the input impedance of the stage following) is not in general a constant resistance. However, if R_{in} as given by (3) is within a factor of about 2 of r_b', Z_{in} will be sufficiently constant with frequency. Thus, (4) will give a good estimate of the bandwidth if R_{in} is substituted for R_L.

The peaking inductance L_f extends the bandwidth somewhat, and modifies the response shape to lessen the shrinking of bandwidth which normally occurs when stages are cascaded. If $L_f = R_f/4\pi B_o$, the frequency response is approximately flat (no ripple) with a bandwidth of about $(2)^{\frac{1}{2}}(1 + (r_b' + R_L)/R_f)^{\frac{1}{2}}$ times B_o. If $L_f = R_f/2\pi B_o$ some ripple occurs, the gain at B_o is nearly A_i, and the 3 db point is now typically 2 to 2.5 times B_o. If low-overshoot transient response is important, the value of L_f should probably be somewhat less than $R_f/4\pi B_o$.

Consider this example. For a large gain-bandwidth product a transistor with a large f_t and small C_c is required, for example, Western Electric M 2039. It was originally intended to be a high-frequency oscillator transistor; but it is also an excellent amplifier.[6] Five of these transistors were measured. Their average parameters were: $\beta = 30$, $r_b' = 50$ ohms, $r_1 = 10$ ohms, $f_t = 400$ Mc, $C_c = 2$ pf. The d-c operating point was $I_E = 5$ ma, $V_{CE} = 9$ v.

An amplifier of specified gain and bandwidth may readily be designed by trial and error with Eqs. (1-4). Because of the approximate nature of these equations, a precise design based on them is not justified. Some experimental adjustment of R_f and L_f may be necessary.

Suppose an amplifier of at least 50 db gain and at least 100 Mc bandwidth is required, and that some ripple is tolerable in the frequency response. Hence $L_f = R_f/2\pi B_o$ can be used to reduce bandwidth shrinkage, and an assumption of 120 Mc bandwidth for one stage should be safe. This means that B_o is about 60 Mc. Trial and error with Eqs. (1), (3), and (4) shows that A_i is between 3 and 4, or about 10-12 db for the above transistor parameters. Thus 5 stages are probably necessary for at least 50 db gain. From Eq. (3), the input resistance is about 50 ohms, which is also convenient as a load resistance for the last stage. Thus, all stages may be considered identical. Equation (1) gives 220 ohms as a convenient value for R_f. With these values a more exact B_o from Eq. (4) is 75 Mc. Hence $L_f = R_f/2\pi B_o = 0.47\,\mu h$.

Construction

As with most high frequency amplifiers, careful mechanical layout is necessary to minimize parasitic inductance and capacitance. With the shunt feedback used around each stage, the input and output impedances of each stage are quite low, requiring very short leads to minimize stray inductance. Transistor sockets can be used even at these frequencies because small capacitances are not as important as low lead-inductance.

Bypassing and decoupling networks must be arranged to present an exceptionally low impedance which must be maintained over the entire pass band of the amplifier. The construction necessary to achieve the foregoing requirements is pictured in the photographs showing the underside of the completed five stage amplifier.

Western Electric M 2039 transistors were used. Their short-

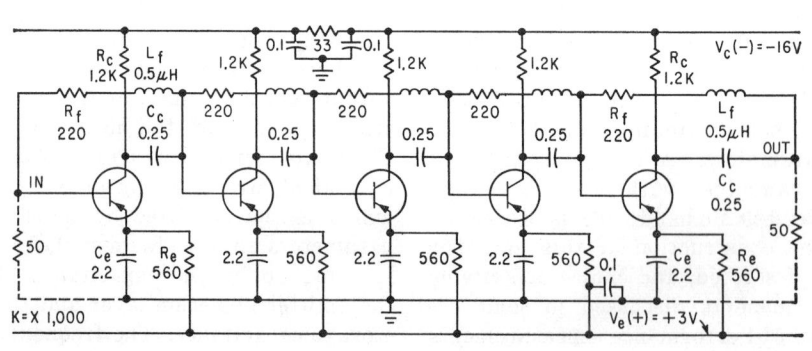

FIG. 3—Five cascaded stages produce overall gain exceeding 50 db ± 2 db over range 8 Kc to 130 Mc

circuit current gain-bandwidth product is 400 Mc at 5 ma emitter current and 9 v collector voltage. Since their gain-bandwidth product increases slightly at higher emitter currents and larger collector voltages, some increase in performance could be achieved by operating the transistors at a higher overall power dissipation. The operating points selected were chosen to keep the input impedance of each stage near 50 ohms.

Emitter Bypassing and Coupling

The emitter bypassing and coupling networks must have extremely low impedance over this exceptionally wide pass band.

The emitter bypass impedance must be small with respect to the 10 ohm impedance seen looking into an emitter terminal. The capacitor used was a disk-ceramic-type, 2.2 μf at 3 v. With the leads trimmed to the minimum, the series inductance—effective at high frequencies—is 9 nanohenrys. Series resonance of the capacitor and its lead inductance, and hence the minimum effective impedance, occurs at about 1 Mc. The magnitude of the impedance across the capacitor terminals is less than 5 ohms from about 10 kc to over 100 Mc.

The impedance requirement for the interstage coupling capacitors C_c is less stringent. This impedance must be small over the pass band as compared to the 50 ohm input impedance of the following stage. Metalized paper coupling capacitors of 0.25 μf, 200 v are satisfactory. An equivalent lead inductance of 10 nh can be obtained by shortening the leads to $\frac{1}{16}$ in. Lead lengths in the constructed amplifier were longer than this.

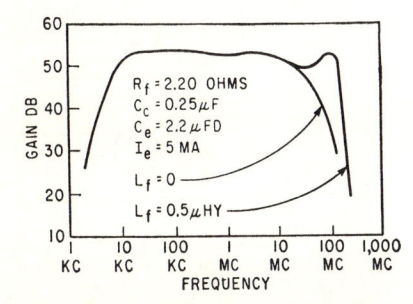

FIG. 4—Performance curve of amplifier shows improved frequency response contributed by inductive feedback

The feedback network around each stage consists of a series resistor R_f and inductor L_f connected from the output of the coupling capacitor back to the input base of each stage. R_f must remain a constant resistance up to the upper cutoff frequency. A low capacitance resistor such as the IRC type HFR is excellent. The feedback inductor L_f must have low distributed capacitance to keep it from resonating in the pass band and disrupting the high frequency response. It should therefore be an air-core single-layer solenoid on a low capacitance form.

Decoupling and Shielding

Decoupling is simple because the shunt feedback around each stage keeps its input and output impedance quite low. Therefore very little signal current flows through collector supply resistors R_c to be coupled back into the collector supply voltage lead. Sufficient decoupling is achieved with the network shown in Fig. 4. Because each emitter is bypassed to ground by C_e, very little signal current flows through emitter bias resistor R_e, and emitter positive supply lead can be decoupled by a single capacitor.

The amplifier is constructed on a conventional i-f strip chassis with all components inside the box formed by the chassis and its cover. The transistors are mounted in sockets through the outside of the box. Since the case of each transistor is internally connected to the collector, signal does appear outside the shielded box on the case of each transistor. This causes no great difficulty although there is some slight tendency toward external signal pick-up and regeneration. A similar transistor may become available with the case connected to the emitter for amplifier applications.

Performance

Measured performance is satisfactorily close to that predicted. In the response curve of Fig. 5, this performance is shown as a function of frequency. These gain measurements were based on the insertion power gain between a 50-ohm source and a 50-ohm load. In this example, where the source, load, and input resistances are all nearly equal, the insertion power gain in db is equal to the current gain i_{out}/i_{in} in db. The gain is plotted as a function of frequency both when $L_f = 0$ and when L_f is adjusted to provide reasonably uniform gain out to past 100 Mc.

Measurements of input impedance as a function of frequency are not conclusive. The input impedance appears to be resistive, and at low frequencies is about 45 ohms. At higher frequencies the impedance rises, until at 100 Mc it is about 100 ohms. The reactive portion of the input impedance is influenced by variations in the inductance of the measuring leads; the reading is therefore unreliable.

The maximum output power level is limited by the current the last stage can deliver into the 50 ohm termination. This stage is biased at 5 ma, and in the sinusoidal steady state can deliver less than 5 ma peak load current. Observable sine wave distortion at the output occurs at about 0.1 v rms across the 50 ohm load. This corresponds to a maximum undistorted output level of about −7 dbm. An increase in the maximum output power could be achieved by both increasing the collector current of the last stage and by operating into a higher value of load resistance.

If a higher gain than 50-db is required, it could be obtained by shielding between cases or collectors of the transistors. Owing to the peaking inductances, the circuit would have a poor response as a pulse amplifier. For good transient response, some compromise arrangement of peaking should be provided.

REFERENCES

(1) G. Bruun, Common Emitter Video Amplifiers, *Proc IRE*, **44**, p 1,561, Nov. 1956.
(2) J. J. Spilker, Jr., A Multistage Videoamplifier Design Method, *IRE Wescon Convention Record*, **2**, p 54, 1957.
(3) R. M. Scarlett, Some New High-Frequency Equivalent Circuits for Junction Transistors, *Stanford Electronics Laboratories*, **TR 103**, p 143, Mar. 20, 1956.
(4) F. H. Blecher, Design Principles For Single Loop Transistor Feedback Amplifiers, *IRE Transactions On Circuit Theory*, **CT-4**, p 145, Sept. 1957.
(5) Valley and Wallman, "Vacuum Tube Amplifiers", McGraw-Hill Book Company, New York, p 172, 1957.
(6) R. M. Warner, J. M. Early & G. T. Loman, Characteristics, Structure and Performance of a Diffused-Base Germanium Oscillator Transistor, *IRE Transactions on Electronic Devices*, **ED-5**, p 127, July 1958.

Single-Ended Amplifiers

Single-ended transistor amplifier uses capacitors and diodes to couple class-A driver to class-B output stage. Output circuit is discussed in detail and a practical high-fidelity amplifier circuit is shown. Here's design data, too

By H. C. LIN and B. H. WHITE,
Semiconductor Applications Laboratory, CBS-Hytron, Danvers, Mass.

SINGLE-ENDED OUTPUT circuits, using transistors operated class B, have advantages over the conventional push-pull output. One advantage is the elimination of the output transformer. The required load impedance of the single-ended output circuit is much lower than that of a push-pull output and can be directly coupled to commercial loudspeakers.

In the single-ended output the transistors are connected in series with the power supply, while in the push-pull circuit the transistors are connected in parallel with the power supply. Thus the maximum voltage and therefore the breakdown voltage requirements of the transistors are less in the single-ended circuit.

Table I—Amplifier Design Data

	Push-Pull	Single-Ended
Load Impedance	$\dfrac{2V^2}{P}$	$\dfrac{V^2}{8P}$
Peak V_{ce}	$2V$	V
Power Gain (Class B)	$\dfrac{h_{FE}^2 V^2}{2P r_{bb'}}$	$\dfrac{h_{FE}^2 V^2}{8P r_{bb'}}$
Peak I_c	$\dfrac{2P}{V}$	$\dfrac{4P}{V}$

V = supply voltage
P = max sine wave power output
$r_{bb'}$ = extrinsic base resistance
h_{FE} = collector to base current gain

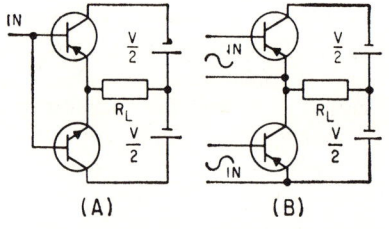

FIG. 1—Complementary (A) and push-pull (B) single-ended output circuits

Most of the known single-ended output circuits have some disadvantages. Use of the complementary circuit[1] (Fig. 1A) is limited due to the lack of commercially available *npn* power transistors.

In the single-ended output circuit using the same types of transistors (Fig. 1B), the inputs are of opposite or split phase. If the phase splitting is accomplished by an input transformer,[2] all the disadvantages of a transformer are present. Instead of a single driver, the quasicomplementary single-ended output circuit[3] requires two complementary transistors as a phase splitter.

Since the input circuit of the output stage is a series return path for the driver, the single-ended push-pull amplifier[4] shown in Fig. 2A can operate satisfactorily only in class A. How single-ended push-pull circuits can be adapted for class B operation to achieve high efficiency and high fidelity will be discussed.

In Fig. 2A, if the driver is to be operated class A and the output pair in class B, a return path for the driver other than the input of the nonconducting transistor must be furnished. If resistors R_1 and R_2 are used as a return path they will also shunt useful signal from the input and output stages. If capacitors C_1 and C_2 are used, they will be charged to a potential equal to the peak of the signal, reverse-biasing the transistors. The transistors will then be in class C operation, resulting in crossover distortion.

To couple a class A driver to a class B output stage the circuit shown in Fig. 2B is used. The driver and output stage are separated by blocking capacitors and diodes are connected between the base and emitter of the output transistors. The diodes are connected in a reverse direction to the base-emitter of the transistors.

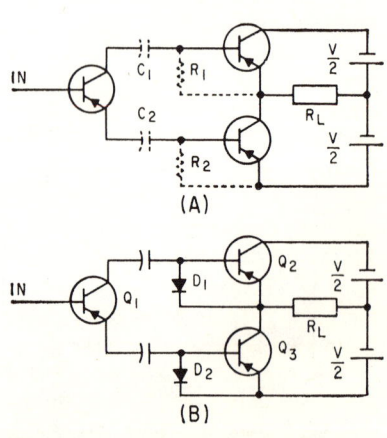

FIG. 2—Basic single-ended push-pull amplifiers for class A (A) and class B (B)

for Class-B Operation

Each diode provides a low impedance return path for the output current of the driver when its associated transistor is not conducting, but does not shunt the signal when its associated transistor is conducting. When the diodes are connected as shown in Fig. 2B no reverse bias is created at the base of the transistors to cause crossover distortion. Crossover distortion is completely eliminated by applying a small forward bias to the diodes and transistors.

Practical Circuit

A practical modification of Fig. 2B, using a single supply voltage, is shown in Fig. 3. The collector of the driver transistor is coupled to the base of Q_2 through R_3.

If C_1 were not used, the input impedance of the grounded collector of

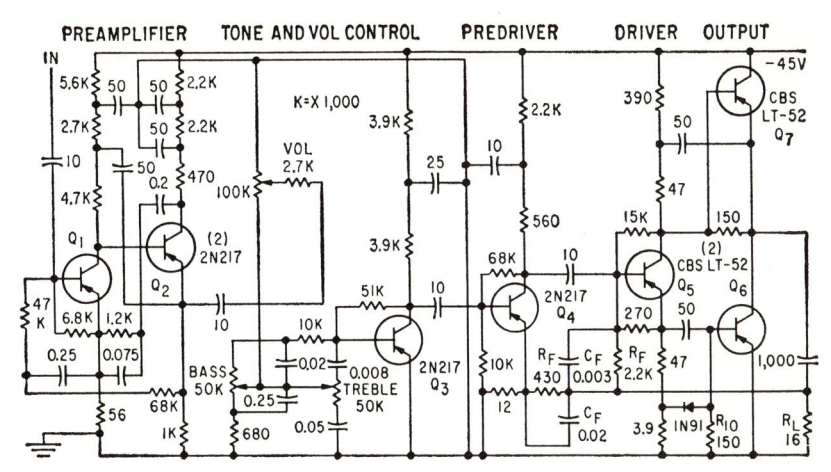

FIG. 4—Ten-watt high-fidelity amplifier employs class B single-ended push-pull output. Input stages are equalized for RIAA curve

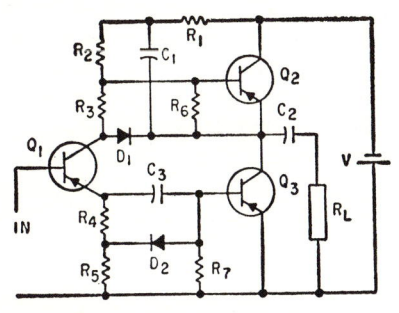

FIG. 3—Practical class B single-ended push-pull amplifier with single driver

Q_2 would be so high that a large amount of useful signal would be shunted by R_1 and R_2, and Q_2 would never saturate. When C_1 is used, Q_2 can be considered as in a common emitter configuration with R_1 connected in parallel with R_L in the collector circuit. The signal current flowing into the base of Q_2 returns through the parallel combination of R_1 and R_L. When $R_1 \gg R_L$ and $R_2 \gg r_{bb2}$ (where r_{bb2} is the extrinsic base resistance of Q_2), there is little signal degeneration.

The forward biases for Q_2 and Q_3 are supplied from the voltage drops across R_6 and R_7. The forward bias for D_2 is derived from a portion of the d-c voltage drop in R_4-R_5 in series with the emitter. The forward bias for D_1 is similarly derived from the d-c voltage drop in the collector's series resistance.

Actually, D_1 and R_3 can be eliminated without effecting performance appreciably. During the conduction of Q_2, since $R_1 \gg R_L$, negligible signal current flows through C_1 to reverse bias Q_2 and crossover distortion is negligible. During the cutoff period of Q_2, the signal current which would flow from D_1 to C_2 and R_L can detour, in the absence of D_1 and R_3, from R_2 to C_1 to C_2 to R_L. The bias on Q_2 due to this current through C_1 is in the forward direction and does not cause objectionable distortion. To compensate for the voltage drop due to the detoured signal current flowing through R_2 and to avoid collector saturation the supply voltage is increased by an amount equal to the peak driver current times R_2.

Complete Amplifier

Figure 4 is a schematic of a 10-watt high-fidelity amplifier employing the single-ended push-pull class B output circuit. The amplifier is designed to work between a variable reluctance pickup and a 16-ohm voice coil.

The only difference between the circuit shown in Fig. 3 and the last two stages of Fig. 4 is that negative feedback has been added to the complete amplifier to reduce distortion and to lower the output impedance for loudspeaker damping. Negative feedback is provided by R_f and C_f.

If the amplifier is to be operated over a wide temperature range, temperature compensation can be obtained by using a thermistor in place of R_{10}.

The total harmonic distortion is less than 1 percent at 10-watt output. Frequency response is flat within 1.5 db from 30 to 15,000 cps. Noise is more than 65 db below the maximum signal output.

Table I is used to design single-ended amplifiers with other output or load requirements than the amplifier just described. Information on the conventional push-pull amplifier is included for comparison.

Transistors used in the single-ended output circuit should have reasonably high gain at peak current to keep distortion low.

REFERENCES

(1) G. C. Szeklai, Symmetrical Properties of Transistors and Their Application, *Proc IRE*, June 1953.
(2) H. C. Lin, A 20-Watt Transistor Amplifier, "Transistors I," RCA Laboratories, 1956.
(3) H. C. Lin, Quasi-Complementary Transistor Amplifier, ELECTRONICS, Sept. 1956.
(4) A. Peterson and D. B. Sinclair, A Single-Ended Push-Pull Audio Amplifier, *Proc IRE*, Jan. 1952.

Amplifier Using

Versatility and reliability are gained in transistor a-c amplifier using multiple feedback loop. Shunt and series loop used in a single stage enable such circuit properties as voltage and current gain, input and output impedance to be preselected and accurately controlled independent of variable transistor parameters. Preselection of circuit properties permits amplifier to be adapted to fit a particular application

By HOWARD LEFKOWITZ, Electronic Engineer, U. S. Naval Ordnance Laboratory, Silver Spring, Maryland

ONE POSSIBLE METHOD used in designing reliably stable transistor circuits is to employ negative feedback in each stage. Since overall feedback loops around three or more stages present considerable stability problems, the use of negative feedback about each stage virtually eliminates problems of oscillation and substantially reduces design effort.

In addition, the use of both shunt and series feedback loops in each stage makes the amplifier versatile by enabling the circuit designer to preselect several circuit properties, such as voltage gain and input impedance or current gain and output impedance, to fit a particular application. With a large amount of negative feedback the amplifier properties may be made independent of the active device.

Circuits for Analysis

The steps in designing a transistor a-c amplifier are: analysis of a-c circuit (small-signal approximation), selection of operating point (large-signal considerations) and analysis of d-c circuit (bias-point stability).

Shown in Fig. 1 is the circuit to be analyzed. It uses both shunt and series feedback loops. Also illustrated is a possible method of obtaining bias-point stabilization through the use of both collector-voltage feedback, R_F, and collector-current feedback, R_E. For a-c analysis, if R_F is not much greater than R_f, then the effective shunt a-c voltage feedback loop resistance may be considered the equivalent resistance of resistors R_F and R_f in parallel.

Figure 2 illustrates the a-c equivalent circuit derived from Fig. 1 and is based on the assumption that $R_B \gg R_{in}$, $R_B \gg R_g$, and resistor $R_f \ll R_F$. Using this equivalent

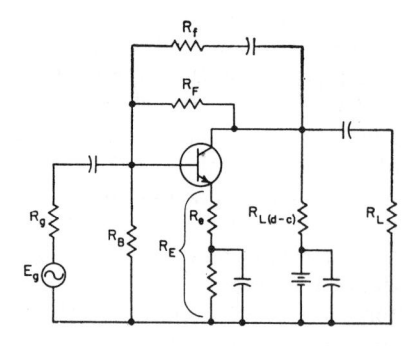

FIG. 1—Transistor a-c amplifier using both shunt and series feedback loops. Multiple loop enables circuit properties to be preselected and controlled independent of transistor parameters

circuit and reducing the transistor and its associated feedback loops to four-terminal networks valid in small-signal approximation, the circuit in Fig. 3 may be obtained. Matrix analysis techniques may be used for reducing the Fig. 3 circuit to one four-terminal network so that relations for the circuit properties may be found.

Using the assumptions of a good junction transistor, the small-signal voltage gain may be derived as follows

$$A_v \cong \frac{R_f R_L (R_e - \alpha_{fb} r_c) + r_c R_L (h_{11b} + R_e)}{R_f R_L (r_b + R_d) + r_c (h_{11b} + R_e)(R_L + R_f)} \quad (1)$$

where R_d is equal to $r_c + R_e$ and h_{11b} is equal to $r_e + r_b (1 - \alpha_{fb})$. Assuming $\alpha r_c \gg R_e$ the small-signal voltage gain may be simplified to

$$A_v \cong \frac{-\alpha_{fb} r_c R_f R_L + r_c R_L (h_{11b} + R_e)}{R_f R_L (r_b + R_d) + r_c (h_{11b} + R_e)(R_L + R_f)} \quad (2)$$

Assume $R_f \gg h_{11b} + R_e$ and $r_c \gg R_f$ the small-signal voltage gain becomes

$$A_v \cong -\frac{\alpha_{fb}}{h_{11b} + R_e} \times \frac{R_L R_f}{R_L + R_f} \quad (3)$$

If the two above assumptions hold and $R_L \gg R_f$ for example, large amount of shunt voltage feedback; and $R_e \gg h_{11b}$ for example, large amount of series current feedback, then

$$A_v \cong -\frac{\alpha_{fb} R_f}{R_e} \quad (4)$$

For good junction transistors alpha is about 0.95 or better. Therefore

$$A_v \cong -\frac{R_f}{R_e} \quad (5)$$

Small-Signal Current Gain

To compute the small-signal current gain additional assumptions

Multiple Feedback

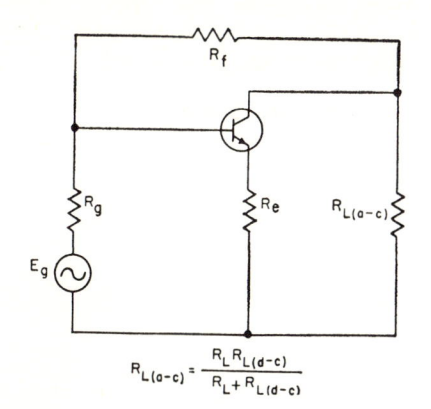

FIG. 2—Equivalent a-c circuit derived from network in Fig. 1

such as the following will have to be made:

$$r_c \gg R_f \text{ and } R_f \gg h_{11b} + R_e$$

Applying the above assumptions, the small-signal current gain may be determined as

$$A_i \cong \frac{\alpha_{fb} R_f}{R_L + (1-\alpha_{fb})R_f + (h_{11b}+R_e)} \quad (6)$$

For the assumptions of large series and shunt feedback and a good junction transistor: $R_e \gg h_{11b}$, $R_L \gg R_f$, $\alpha_{fb} = 1$

$$A_i \cong -\frac{R_f}{R_L + R_e} \quad (7)$$

The small-signal power gain may be computed using the following simple relation

$$A_p \cong A_v A_i \quad (8)$$

The small-signal input resistance may be found as

$$R_{in} \cong \frac{(R_L+R_f)(h_{11b}+R_e)}{R_L+(h_{11b}+R_e)+(1-\alpha_{fb})R_f} \quad (9)$$

For the following assumptions: $\alpha_{fb} = 1$, $R_e \gg h_{11b}$,

$$R_{in} \cong \frac{(R_L+R_f)R_e}{R_L+R_e} \quad (10)$$

For large shunt feedback $R_L \gg R_f$, the input resistance becomes

$$R_{in} \cong \frac{R_L R_e}{R_L+R_e} \quad (11)$$

and if $R_L \gg R_e$ the input resistance approaches the value of the series feedback resistor, R_e.

The small-signal output resistance may be determined as follows:

$$R_{out} \cong \frac{(R_f+R_g)(h_{11b}+R_e)+}{\frac{(1-\alpha_{fb})R_f R_g}{R_g+(h_{11b}+R_e)}} \quad (12)$$

With following assumptions: $R_f \gg R_g$, $\alpha_{fb} \cong 1$, $R_e \gg h_{11b}$, then

$$R_{out} \cong \frac{R_f R_e}{R_g + R_e} \quad (13)$$

If $R_e \gg R_g$, then the output resistance approaches the value of the shunt feedback resistor, R_f.

Practical Amplifier

Given the following specifications for the design of a reliable transistor amplifier: Minimum transistor short-circuit current gain $\alpha_{fb} = 0.97$, transistor short-circuit input impedance, $h_{11b} = 30$ ohms, generator impedance = 1,000 ohms, load impedance = 3,000 ohms, desired voltage gain = 15 and desired input impedance = 1,000 ohms.

Using Eq. 3 and assuming that $R_f \gg R_L$:

$$A_v \cong -\frac{\alpha_{fb} R_L}{h_{11b}+R_e}$$

Substituting and solving for R_e: R_e equals 164 ohms.

Since nearest standard value is 160 ohms, from Eq. 10

$$R_{in} \cong \frac{(3,000+R_f)(30+130)}{3,000+30+130}$$

With R_{in} equal to 1,000 ohms, R_f equals 16,700 ohms. Since nearest standard value is 16,000 ohms, from

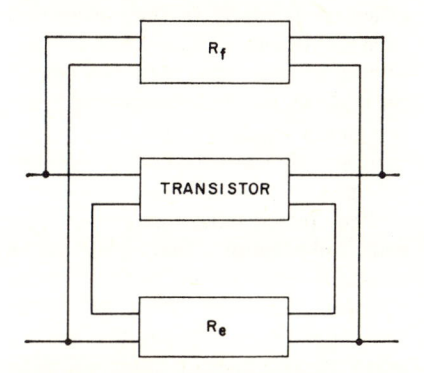

FIG. 3—Circuit shown can be used for matrix analysis technique. Matrix is derived by using equivalent a-c circuit shown in Fig. 2 and reducing transistor and feedback loops to 4-terminal networks

Eq. 3, quantity A_v is equal to -12.9. This is slightly less than the desired 15. Selecting a smaller series feedback resistor, R_e, will tend to increase the voltage amplification. Therefore, if R_e is selected to be 130 ohms, Eq. 9 becomes

$$R_{in} \cong \frac{(3,000+R_f)(160)}{3,000+160+.03R_f}$$

With R_{in} equal to 1,000 ohms, R_f equals 20,600 ohms. Let R_f equal 20,000 ohms, then A_v equals 15.8. From Eq. 9, R_{in} equals 980 ohms. Using R_f equal to 20,000 ohms and

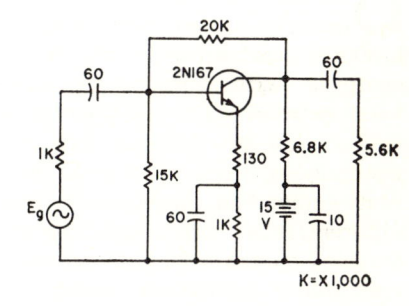

FIG. 4—Practical transistor a-c amplifier with selected component values. Use of 20,000-ohm shunt feedback resistor provides d-c and a-c voltage feedback and eliminates two components

R_e equal to 130 ohms, the desired voltage gain and input impedance may be obtained within 5 percent and 2 percent respectively. These may be designed closer to desired values using non-standard resistors.

Using the above determined values, the output impedance R_{out} from Eq. 12 becomes 3,280 ohms. The current gain A_i from Eq. 6 becomes 5.16. Therefore, the power gain A_p equals 81.5 or 19.1 db.

The design may now be completed by a selection of the quiescent operating point and the application of some type of d-c feedback to stabilize the operating point over the desired temperature range. Figure 4 illustrates a completed design having the above a-c characteristics. The 20,000-ohm shunt feedback resistor is used to provide d-c as well as a-c voltage feedback, resulting in a saving of two components.

D-C Operational Amplifier

Transistor operational d-c amplifier meeting military requirements for airborne computer applications is described. Circuit includes a low-level silicon transistor chopper with input impedance of approximately 1 megohm and a highly efficient output stage

By W. HOCHWALD and F. H. GERHARD,
Autonetics, A Division of North American Aviation, Inc., Downey, California

OPERATIONAL amplifiers are the important component building blocks in analog computers. When equipped with appropriate feedback networks and drift-compensating circuits, these devices are generally capable of voltage or current summation, integration and differentiation with computing accuracies on the order of 0.01 percent under laboratory conditions.

Development of a completely transistorized amplifier to meet equivalent accuracy requirements under airborne environmental conditions presents a number of problems. Foremost of these are the provision of input current sensitivity in the region of 100 $\mu\mu$a and a dynamic output range on the order of ± 50 v d-c. Furthermore, a filter with a time constant of about 10 seconds is normally required in this type of amplifier. This filter is difficult to obtain with inherently low transistor circuit inpedance levels. The amplifier to be described here uses silicon transistors throughout, including the chopper circuit, to achieve computing accuracies of 0.005 percent over an ambient temperature range of -55 C to $+85$ C.

Summing Amplifier

Figure 1 is a simplified schematic diagram of the amplifier connected as a summing amplifier with two inputs.[1] The input-output relationship will be independent of amplifier characteristics and approach the ideal relationship $e_o = -e_1 Z_f/Z_1 - e_2 Z_f/Z_2$ if the gain of the amplifier is sufficiently high and the effects of d-c drift in the transistor circuits are made sufficiently small. This is accomplished by a chopper amplifier, integrating amplifier, a-c preamplifier and output power amplifier forming the overall d-c operational amplifier.

The amplifier sections are connected to provide two signal paths for the small input error signal e_e. The d-c and low-frequency components are amplified in the chopper amplifier-integrating amplifier combination while the higher frequency a-c components are amplified in the preamplifier. The two signals are combined at the input of the output power amplifier and further amplified in this section. Effective preamplification of d-c input signal components in the essentially drift-free chopper section minimizes the effect of d-c drift in the succeeding direct-coupled circuits, while provision of the alternate a-c signal path achieves amplification of high-frequency signals beyond the capabilities of the chopper amplifier-integrator combination.

The integrating amplifier permits obtaining the long time constant filter at the output of the chopper amplifier with relatively low resistor and capacitor values, since time constant multiplication by the gain of the integrating amplifier is effectively achieved. The filter thus formed is required to remove chopper frequency signals originating in the chopper amplifier.

Chopper Amplifier

The chopper amplifier is the most critical portion of the entire amplifier and its accuracy is largely determined by the characteristics of the transistor chopper. A *pnp* microalloy silicon transistor was chosen for use as the chopper (Q_3 of Fig. 1) because of its excellent high-frequency characteristics and the inherently small magnitude and stable nature of the internal sources of offset errors. A 350-cps chopper was selected as a compromise between frequency-dependent offset errors in the transistor and time constant considerations in the amplifier.

Transistor chopper Q_3 is operated in the inverted connection to minimize internally generated offset voltages. Offset effects are further reduced by actuating the collector-base control junction with a half-wave rectified square wave which alternately provides sufficient base current for the closed-switch state or zero voltage for the open-switch state. This rectified square wave is provided by the flux oscillator circuit Q_1 and Q_2 and diode D_1. Operation with zero voltage across the control junction of the transistor chopper for the open-switch state is possible since the transistor open-switch resistance is greater than 50 megohms for the normally small d-c input signals to the chop-

With Transistor Chopper

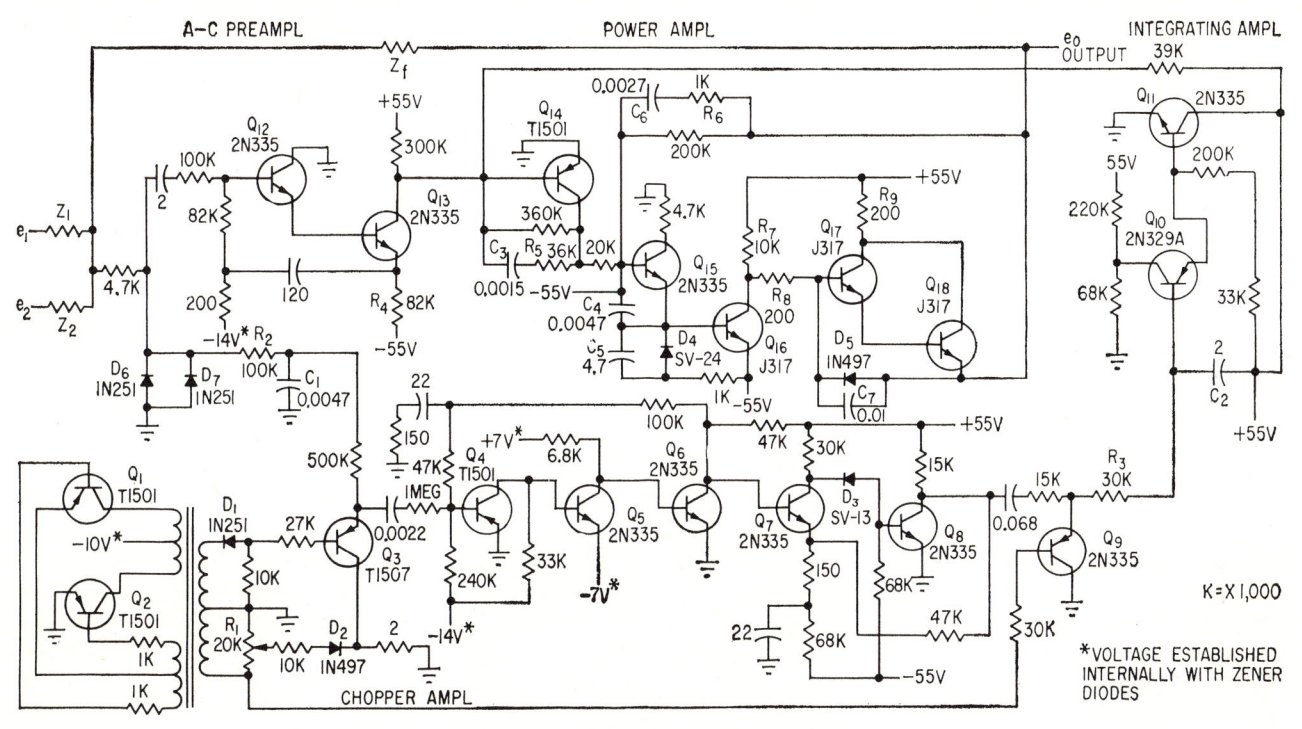

FIG. 1—Schematic of the operational amplifier. Drift compensation achieves accuracy of 0.005 percent or less in analog computers

per. This mode of operation avoids temperature dependent leakage current errors normally observed in back-biased chopper circuits.

Offset Effects

The zero control R_1 for nulling the nominal effects of offset in the modulator circuit inserts a variable square wave into the collector circuit of the transistor chopper. This method corrects the voltage offset directly at the source. The zeroing voltage is applied only during the instant when the transistor chopper is in the conducting state, thus again eliminating errors from leakage variations which would otherwise occur during the nonconducting state of the chopper. The low-pass filter R_2C_1 is incorporated to prevent a-c components of the input signal from overdriving the high-gain a-c amplifier.

The effects of switching transients or spikes from junction capacitance and carrier storage effects in the chopper transistor are reduced in the design of the a-c amplifier. Accurate clipping of the spike is provided at the output of the amplifier by saturation of Q_8 and control of the amplifier a-c input impedance and gain. This reduces the offset caused by the switching transient by an approximate factor of 3.

Bias stability in the a-c amplifier is provided using a minimum of capacitance. Two feedback networks are incorporated—a low-pass network from the collector of Q_6 to the base of Q_4, and another from the collector of Q_8 to the emitter of Q_7. This arrangement also provides the stable input impedance and gain essential for setting the accurate clipping level of the switching transient mentioned previously.

Demodulator Q_9 is used to recover the d-c component of the signal. Blocking of large positive voltages is accomplished by providing a large switching voltage at the base of Q_9. As mentioned previously, the amplifier has an internal chopper power supply, consisting of stages Q_1 and Q_2 and a saturable transformer in a flux oscillator configuration. However, the square-wave voltages required to actuate the chopper, demodulator and zero control circuits are all referenced to ground. It is therefore possible if desired to provide a single chopper supply for multiple installations without a separate isolation transformer in each d-c amplifier.

Integrating Amplifier

Figure 1 includes a schematic of the integrating amplifier. A time constant of approximately 12 seconds is obtained by use of a two-transistor circuit and a feedback network comprising a 2-μf capacitor C_2 and a 30,000-ohm resistor R_3. The amplifier employs an *npn* and *pnp* transistor circuit, which provides cancellation of the base-to-emitter voltage variations which are functions of ambient temperature. Base current level of the input stage of the amplifier is minimized by employing a *pnp* silicon transistor which has high d-c β at low collector currents.

Minimization of base-to-emitter voltage variations and reduction of the base current with resultant smaller magnitude variations are essential, since these variations effectively represent offset errors. These can be appreciable in spite of the fact that this portion of the circuit is preceded by the d-c gain of the chopper amplifier. The direct-coupled amplifier is bias-stabilized automatically since it is in the d-c signal path of the overall amplifier, which is normally stabilized by external feedback networks.

A-C Preamplifier

Again referring to Fig. 1, the a-c preamplifier consists of an emitter follower Q_{12} driving a common-emitter amplifier Q_{13}. This arrangement provides an a-c input impedance of approximately 300,000 ohms. The large emitter resistor R_4 provides sufficient bias stability to reduce offset variations referred to the input of the overall amplifier to a negligible value. Because of their inherently high frequency response characteristics, npn silicon transistors are used which permit extension of useful overall amplifier gain to frequencies above 100 kc.

Output Power Amplifier

Essential feature of the output power amplifier is efficient provision of a large dynamic voltage range and sufficient current capacity to drive a number of computing networks or other useful loads. Utilization of large dynamic voltage range effectively minimizes percentage errors from offset and drift in an operational amplifier.

To achieve these objectives the output power amplifier included in Fig. 1 was selected. This circuit consists of composite emitter-follower Q_{17} and Q_{18}, common-emitter amplifier Q_{16} and diode D_5. For positive output, the diode is back-biased and may be considered an open circuit. The circuit then consists of the composite emitter-follower stage Q_{17} and Q_{18} driven by common emitter stage Q_{16}.

For negative outputs, the diode is forward-biased providing a low resistance path from the collector of Q_{16} to the load. The base-emitter junctions of Q_{17} and Q_{18} are back-biased and are therefore nonconducting. The circuit then functions as a common-emitter output stage. This arrangement results in a great improvement in efficiency compared to conventional single transistor common-emitter or common-collector power output stages.

Other salient features of the output power amplifier are: two power resistors R_8 and R_9 limit the output current and thereby provide protection of circuit components in the event of accidental shorting of the output; two additional gain stages Q_{14} and Q_{15} are included to perform voltage level shifting operations and achieve the required power gain; negative feedback is used in amplifier stage Q_{14} and also from the emitter of Q_{18} to the base of Q_{15}.

This arrangement provides gain stability over the entire passband of the amplifier, which is essential for proper high-frequency cutoff control. Cutoff of the entire amplifier at 100 kc is accomplished by networks R_5-C_3 and R_6-C_6. These phase-control networks are designed to reduce the effects on frequency response of beta-cutoff and collector capacitance in the transistors of the a-c preamplifier and output power amplifier, as well as parasitic capacitances in the layout. Capacitors C_4, C_5 and C_7 stabilize the local feedback loop around Q_{15}, Q_{16}, Q_{17} and Q_{18}.

Connection of Q_{14} as a common-emitter amplifier with negative feedback provides the low input impedance required for efficient summation of the two signals from the integrating amplifier and the a-c preamplifier, respectively.

Power Supply

The amplifier circuit requires only two supply voltages, namely, ±55 v, although several convenient lower voltage levels are derived internally by Zener diode voltage-divider networks which are not shown in Fig. 1. The d-c input resistance of the amplifier is approximately 1 megohm. The nominal d-c gain of the chopper amplifier is 1,300, the d-c gain of the output power amplifier is 50 and that of the integrating amplifier, 200; this results in a d-c open-loop gain of approximately 13×10^6. The voltage gain of the a-c preamplifier and power output amplifier combined is approximately 3,000 at 100 cps, and 2,500 at 400 cps. The output voltage range is +45 v to −45 v into a 1,500-ohm load.

Extensive tests of the amplifier have shown that the offset remains consistently less than 300 μv referred to the input when the amplifier is operated over an ambient temperature range of −55 C to +85 C. The offset has been found to remain less than 50 μv over periods of several hundred hours under laboratory conditions. Care in mechanical layout and guarding of the critical input circuit portions by grounded circuit lines have resulted in leakage current errors of less than 10^{-9} amp, referred to the input under conditions of high humidity.

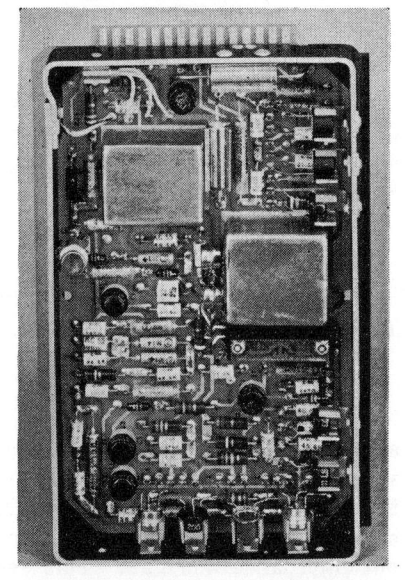

Layout of operational amplifier avoids closing of feedback loops through stray capacitances

REFERENCE
(1) Patent pending on this circuit.

BIBLIOGRAPHY
E. A. Goldberg, Stabilization of Wide-Band Direct-Current Amplifiers for Zero and Gain, RCA Review, 11, p 29, Ju 1950.

R. L. Bright, Junction Transistors Used as Switches, submitted as a Transaction paper, AIEE Winter General Meeting, 1955.

A. P. Kruper, Switching Transistors Used as a Substitute for Low Level Choppers, Submitted as a Transactions paper, AIEE Winter General Meeting, 1955.

W. Hochwald and F. H. Gerhard, A Drift Compensated Operational D-c Amplifier Employing a Low Level Silicon Transistor Chopper, paper presented at the NEC, Oct 1958.

D-C Amplifier
For Rugged Use in Field

Measurement of low-amplitude long-period surface waves from small islands in midocean led to the development of this simple, low-cost device

By WM. G. VAN DORN, University of California, The Scripps Institution of Oceanography, La Jolla, California

EXTENSIVE OCEANOGRAPHIC operations carried out on Pacific islands over the past six years by the Scripps Institution of Oceanography has led to the development of the battery-operated d-c transistor amplifier shown in Fig. 1. Designed for simplicity and low cost, this small, rugged amplifier is capable of driving a low-impedance recording galvanometer for extended periods in regions where no auxiliary power is available, and where a linearity and stability commensurate with that of other system components (1 to 2 percent) is sufficient. More than 100,000 total hours of continuous use have demonstrated its suitability and dependability in the field.

Design Features

The basic amplifier (Fig. 2) drives an Esterline Angus model AW recording milliammeter to full scale (± 0.5 ma) with an input signal of ± 75 mv ($\pm 15\mu a$) from a Statham model PM5-0.2d unbonded strain-gage pressure transducer. Design specifications and performance data are given in Table I.

The bilateral symmetry of the push-pull circuit using matched

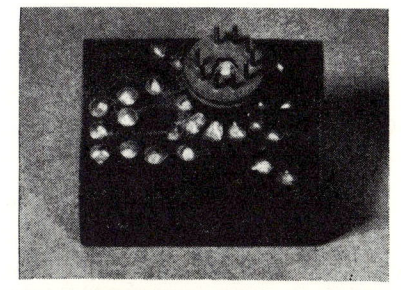

FIG. 1—Fixed components of amplifier are encapsulated in plastic on a 9-pin miniature tube base

2N65 transistors optimizes linearity and thermal stability. Stability is further enhanced by physically bonding the corresponding left- and right-hand circuit components together by a wrapping of copper wire and encapsulating the fixed components in plastic on a 9-pin miniature tube base.

Input and output circuit impedances are matched to those of the transducer and recorder, respectively, for maximum power transfer, and the T-pad gain control in the output circuit provides virtually constant amplifier loading at high-enough impedance to give adequate galvanometer response (2 sec) in the recorder. Separate banks of ten 1.3-v mercury cells, rated at 14,000 milliamp hr, provide power for the transducer, the accessory control panel (containing the gain and balancing potentiometers and calibrating network) and the recording milliammeter. Although designed for d-c operation, the a-c amplitude response of this amplifier was tested and found to be flat within 2 db to 50 kc, with negligible harmonic distortion to 30 kc.

Field Service Life

Field records show that the service life of these amplifiers averages about one year of continuous operation, failure usually manifesting itself by a progressively increasing drift to one side or the other as one of the transistors deteriorates. About one-third of the new amplifiers drifted initially after potting or failed to balance at all. This rejection rate was greatly reduced by refrigerating the units during curing of the resin. Although not tried, silicon transistors would probably insure a higher degree of stability than the germanium. This article is based on research carried out by the University of California under contract with the Office of Naval Research.

Table I—Design and Performance Data

Input impedance....	5k
Output impedance..	1.4k
Thermal stability	
Zero drift........	0.4%/deg C
Gain............	0.7%/deg C
Input voltage......	± 75 mv
Input current......	± 15 μa
Output voltage.....	± 0.7 v
Output current.....	± 0.5 ma
Power gain........	± 25 db
Linearity.........	1% of full scale
Supply voltage.....	12 v d-c
Supply current.....	3.4 ma

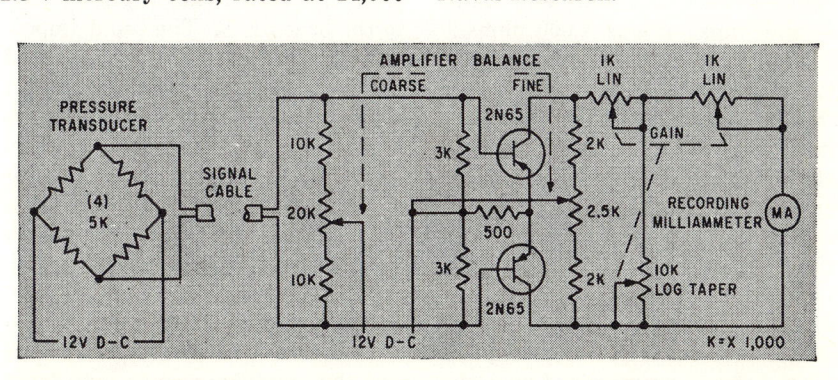

FIG. 2—Basic amplifier circuit. Calibrating network and auxiliary power-supply networks are omitted. Although designed for d-c operation, the a-c amplitude response is flat within 2 db to 50 kc

Low-Pass Filter for

Low-pass filter uses three transistors and solid tantalum capacitors for a flat frequency response from d-c to the 1 cps cutoff frequency, attenuation slope of 15 db per octave, near zero insertion loss and good temperature stability

By **R. C. ONSTAD,** Senior Electronics Engineer, Convair-Astronautics, San Diego, Calif.

DUE TO THE limited number of higher response channels available on a missileborne telemeter it is often necessary to commutate signals that have frequency components greater than one-half the sampling rate. According to sampling theory, in order that the multiplexer shall not superimpose interfering sidebands on the signal, it is necessary to insert a low-pass filter between the signal source and the multiplexer.[1]

Assume the signal requiring the filter is sampled at a 5-cps rate and has useful frequency components up to 1 cps and undesirable responses at 4.5 cps and higher. Because of the frequency characteristics of the signal, primary design objectives for the filter are for a 1 cps corner frequency and sufficient steepness of attenuation slope in the stop band so that the attenuation is 26 db or greater at 4.5 cps. Other criteria are for a flatness in pass band within 1 db, a d-c insertion loss not greater than 1 db and output impedance not greater than 10,000 ohms. The input signal voltage range is zero to 5 volts and signal source impedance is 4,000 ohms.

The required filter characteristics as defined by the above requirements are such that the filter should have a maximally flat frequency characteristic. Space and power supply limitations rule out a filter containing inductances or vacuum tubes. A passive RC filter would be unsatisfactory because of the insertion loss, output impedance and frequency characteristic requirement. The development of an active RC filter incorporating a highly stable simple transistor feedback amplifier and RC networks was the solution to the problem.

Amplifier Design

An active filter based on Fig. 3 (within the panel) has been developed for vacuum tubes using a gain stage and a cathode follower[2] A similar approach using a transistor in an emitter-follower configuration might be used. A d-c coupled emitter follower produces zero offset and introduces considerable drift with temperature. The amplifier incorporated in the filter has the desired characteristics of stable gain, zero d-c offset, low-temperature drift, high input impedance and low output impedance. The basic circuit diagram of this amplifier is shown in Fig. 4. In this circuit, drift is minimized by using a balanced input to transistors Q_1 and Q_2. The signal is direct coupled from Q_1 to Q_2 through the direct-connected emitters. The amplified signal on the collector of Q_2 is coupled to the output through emitter follower Q_3. Unity closed-loop gain and feedback ratio results from connecting the output directly to the base of Q_2. The input impedance of the amplifier is the input impedance of the common-collector input transistor and approaches as a limit the collector resistance of Q_1. The output impedance of this amplifier is essentially the output impedance of an emitter follower that has as its input a zero impedance generator. Due to the tight negative feedback loop the base of Q_3 sees essentially a zero impedance and the amplifier output impedance is less than 100 ohms.

There are three principle sources of zero drift due to temperature. Zero drift can result from variations in the d-c supply voltages. Maximum drift occurs when the plus and minus supply voltage are both changing in the positive direction or vice versa. Under this condition a change of 1 percent in each power supply causes a zero drift of 14 millivolts at the output.

Another source of zero drift is due to a shift in the emitter voltage—collector current transfer characteristics of transistors Q_1 and Q_2. This shift in parameters of Q_1 results in a zero drift that is cancelled by a corresponding shift in the parameters of Q_2 due to the rejection of common-mode signals that is characteristic of the differential circuit incorporated in this amplifier. The drift due to this source can be minimized if required, by using transistors that are matched with respect to temperature characteristics and are connected by a good thermal conductive path.

A third source of temperature drift is the variation with temperature of the input current of Q_1. This produces a zero drift at the input to the amplifier that is directly proportional to the source resistance of the input signal. If signal source resistance is of a value such that zero drift introduces significant error, compensation must be employed. If base current of Q_1 (Fig. 4) is plotted as a function of temperature it is found that a fairly linear plot is obtained over the temperature range of 0 to 100 C. The magnitude of current at a given temperature will vary for different transistors but the slope of the plot

Subaudio Frequencies

ACTIVE NETWORK THEORY

The filter network used is essentially a cascaded RC filter with feedback. Consider a two-section RC filter to give an attenuation slope of 12 db per octave. The two-section RC filter in cascade shown in Fig. 1 has a transfer function which equals

$$\frac{1}{\left[\frac{S}{\sqrt{\frac{1}{R_1C_1R_2C_2}}}\right]^2 + \left[\frac{S}{\sqrt{\frac{1}{R_1C_1R_2C_2}}}\right]\left[\frac{R_1C_1+R_2C_2+R_1C_2}{\sqrt{R_1C_1R_2C_2}}\right]+1} \quad (1)$$

which equals

$$1/[(S/\omega_0)^2 + (S/\omega_0)\,2\,\xi + 1] \quad (2)$$

where

$$\omega_0 = \sqrt{1/R_1C_1R_2C_2}$$

and

$$2\xi = R_1C_1+R_2C_2+R_1C_2/\sqrt{R_1C_1R_2C_2} \quad (3)$$

The shape of the frequency characteristic of $F(S)$ is dependent only on the value of the parameter ξ (see Fig. 2). The constant ω_o determines the position of the corner frequency in the frequency domain. From equation 3

$$\xi = R_1C_1+R_2C_2+R_1C_2/2\sqrt{R_1C_1R_2C_2} \quad (4)$$

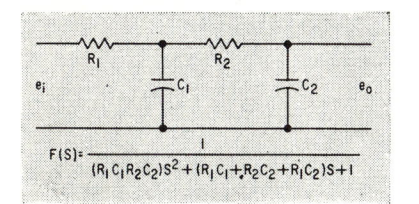

FIG. 1—Basic two-section RC filter with transfer function used in filter

Assuming

$$R_1C_1 = R_2C_2 = RC$$

Then

$$\xi = \frac{2\,RC+R_1C_2}{2\,RC} = 1 + \frac{R_1C_2}{2\,RC} \quad (5)$$

Equation 5 shows that the shape of the curves obtainable is restricted to those shapes corresponding to $\xi > 1$. These curves do not possess the more desirable frequency characteristics. When feedback is used, it is possible to obtain curves corresponding to $\xi < 1$.

Consider the original two-section RC filter inserted in the forward path of any amplifier with a gain K. The transfer function of a feedback amplifier of gain K and feedback factor β can be shown to be

$$1/(\beta + 1/K) \quad (6)$$

This transfer function is modified by the insertion of a two-section RC filter as shown in Fig. 3 so that

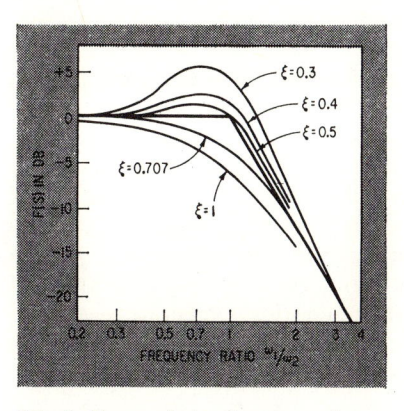

FIG. 2—Shape of the frequency characteristics is dependent on ξ

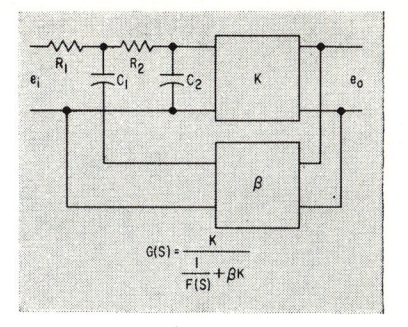

FIG. 3—Two-section RC filter is inserted in forward path of amplifier

$$G(S) = \frac{1}{\beta + 1/KF(S)} = \frac{K}{1/F(S) + \beta K} \quad (7)$$

Where $F(S)$ is the transfer function of the two-section RC filter. Substituting equation (1) for $F(S)$.

$$G(S) = K/(R_1C_1R_2C_2)S^2 + (R_1C_1+R_2C_2+R_1C_2)S+1+\beta K \quad (8)$$

The new values of ω_1 and ξ_1 are

$$\sqrt{\frac{1+\beta K}{R_1C_1R_2C_2}} \text{ and}$$

$$\frac{R_1C_1+R_2C_2+R_1C_2}{2\sqrt{(1+\beta K)(R_1C_1R_2C_2)}} \quad (9)$$

Assuming $R_1C_1 = R_2C_2 = RC$

$$\xi = \frac{2\,RC+R_1C_2}{2\,RC\sqrt{1+\beta K}} = \frac{1+[R_1C_2/2\,RC]}{\sqrt{1+\beta K}} \quad (10)$$

There is now means of adjusting ξ_1 to any value from $1 + [R_1C_2/2\,RC]$ corresponding to $\beta = 0$, see equation (5), to

$$\frac{1+[R_1C_2/2\,RC]}{\sqrt{1+\beta K}}$$

to obtain any set of curves within the limits of β and K.

---

for various transistors will agree quite closely. Compensation in the form of a resistance and thermistor network between the positive voltage source and the base of the transistor to be compensated provides a current of equal but negative slope to reduce the zero drift at the amplifier input from 1.6 millivolts/degree C to 0.4 millivolt/degree C over the range of —25 to +80 degrees C.

Active Filter Design

The 12 db per octave filter of Fig. 4 incorporates a two-section RC network and a temperature compensated amplifier. This feedback filter can be designed to have a response that very closely approximates the maximally flat frequency characteristic. The frequency characteristic is determined by the amount of feedback voltage as selected by the ratio of R_3 to R_4 and by the ratio of R_1C_1 to R_2C_2. A close approximation of corner frequency is $1/2\pi\sqrt{[1/R_1C_1R_2C_2]}$. The values of R_1 and R_2 should be made as small as possible to limit temperature

77

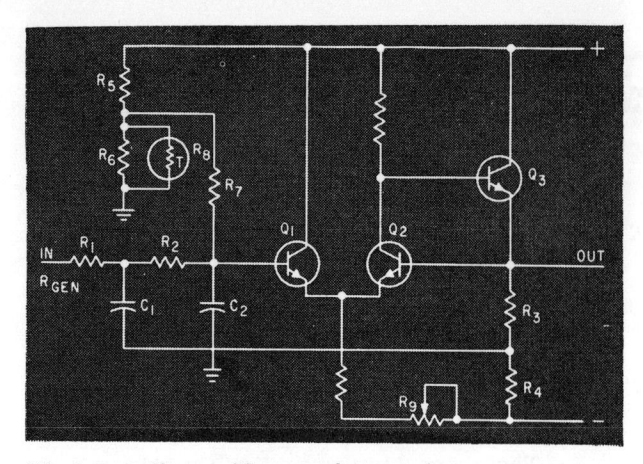

FIG. 4—Basic filter-amplifier uses three transistors

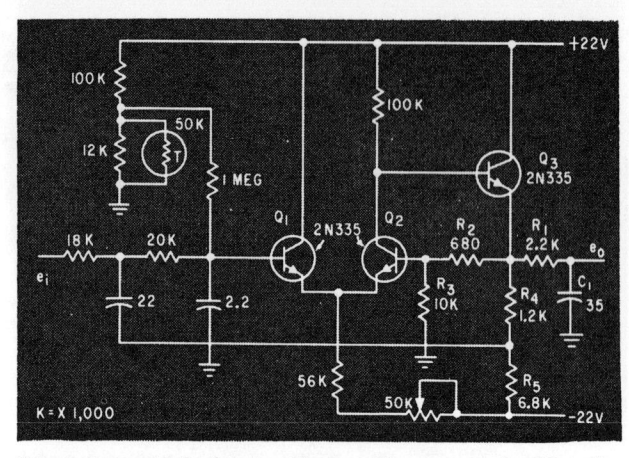

FIG. 5—Output of final design filter-amplifier uses passive RC section

drift. The temperature compensation network is made up of resistors R_5, R_6 and R_7 along with thermistor R_8. This network can be designed after the values of R_1, R_2 and R_{gen} are known. The filter is designed to accommodate an input signal that varies from 0 to + 5 v. Limiting occurs with negative signals that are greater than —1.5 v. Potentiometer R_9 is an adjustment for d-c zero set. This adjustment is set for zero d-c at the output with the input grounded.

The ultimate filter design which meets the particular requirements for a filter to precede a 5-rps commutator consists of an active filter cascaded with a passive RC network. Although the passive network may be placed either ahead of or after the active filter, it follows the active filter because input conditions are unchanged and output impedance remains sufficiently low. This design maintains the simplicity of the one active stage, requiring only the addition of a resistor and a capacitor, yet provides sufficient attenuation slope.

The active section of this filter shown in Fig. 5 is designed to have a frequency characteristic such that when its response is combined with the response of a passive section, R_1 and C_1, the resultant frequency characteristic closely resembles that desired. In order to achieve this type of frequency response in the active section of the filter, the factor βK must be greater than for an active filter without a passive section. The ratio of resistors R_2 and R_3 of Fig. 5 determines the gain K of the amplifier.

The ratio of R_4 to R_5 determines the magnitude of β. The product βK greatly controls the shape of the frequency curve. Proper choice of the two resistance ratios can provide an optimum overall frequency characteristic and zero overall d-c insertion loss. Figure 6 shows the nominal overall frequency characteristic of the cascaded active and passive filter. This final design of the filter has a corner frequency of not less than one cps (3 db point); insertion loss, adjustable to zero if desired, less than one db with no adjustment; flat passband, within ±1 db from d-c to 0.7 cps; attenuation slope of 15 db per octave; input signal voltage range of 1.5 to +7.5 v peak; zero drift of less than 0.4 millivolt per degree C; d-c input impedance of 0.5 megohm; and an output impedance of 2.5 kilohms.

The filter is packaged in a modular plug-in subassembly. Silicon transistors are used because of their superior temperature characteristics. Supply voltages are regulated to minimize zero drift.

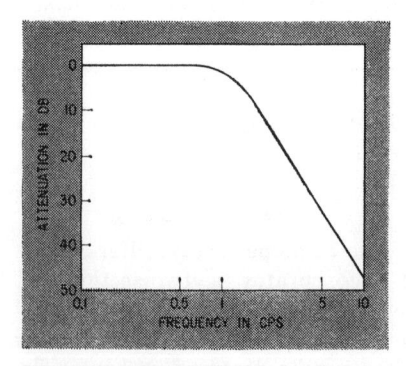

FIG. 6—Nominal overall frequency characteristic of filter-amplifier

In order to maintain the same degree of flatness in the pass band, a closer tolerance on capacitance is required for a cascaded active and passive filter than for a filter consisting of only active stages since a fall off in the pass band of the passive filter stage must be compensated by a corresponding peaking in the active stage to produce a maximally flat response. For this condition a given change in the capacitance of C_1 in Fig. 4 results in a greater change in the amplitude of the overshoot than when the frequency characteristic of the active stage is flat in the pass band. Using capacitors with a tolerance of ±20 percent, it was found impractical to combine the active filter stage with a passive stage of more than one section, and still maintain a flatness of ±1 db as necessitated by design requirements unless an additional adjustable resistance network were inserted in the feedback loop of the filter to vary β. The frequency characteristics of the filter might be maintained precisely within the limits imposed by the temperature coefficient of the capacitors by trimming the filter networks to keep constant the RC products of each section. In assembling the filter the capacitors can be measured and values of resistance selected to provide the required RC product.

REFERENCES

(1) Nichols and Rausch, "Radio Telemetry", p 37, John Wiley and Sons, Inc., New York, 1956.
(2) A. N. Thiele, Design of Filters Using Only RC Sections and Gain Stages, *Elect Eng*, 28, p 31, Jan. 1956.
(3) R. J. Millard, Reliability Data on Solid Electrolyte Capacitors, *Proc 1958 Elect Comp Conf*, p 21, Engineering Publishers, New York, 1958.

Audio Volume Compressor

Transistorized audio compressor has unity gain with expansion of 3 db, compression of 12 db. Gain adjustments are automatic

By E. C. MILLER, Technical Director, Inland Broadcasting Co., Weiser, Idaho

TO MAINTAIN an even recording level during tape-recorded interview sessions, the tape recorder operator may have to make gain adjustments to compensate for variations in speech levels of different speakers and for level changes due to changes in distance between the microphone and the person being interviewed.

At most radio stations, two volume compressors are used between the studio and transmitter. The first operates as an average program control and the second operates as a peak limiting amplifier to maintain a constant peak percentage of modulation. At some stations, inputs from tape recorders and remote lines have individual automatic gain devices. These are of little help if the signal from the remote line or tape recording is of insufficient level in relation to the line noise.

When interviews are being conducted and recorded by one man, the necessity of making gain adjustments sometimes breaks the reporter's chain of thought during the interview. The transistorized audio compressor is automatic and requires no operating controls. It has the added psychological effect of assuring the reporter of a good recording which allows him to concentrate on the event he is reporting without concern for precise microphone distances or gain control settings.

The audio compressor shown in Fig. 1 is a transistorized unity-gain amplifier having an expansion of 3 db and a compression of 12 db around an average level of approximately 45 db below 1 v at an impedance of 10,000 ohms. The compressor is inserted after the microphone amplifier.

Compressor Operation

The incoming audio signal is amplified by Q_1 and applied to the diode compressor through C_1. The diode compressor is essentially an L-pad attenuator with R_1 forming one leg and diode D_3 the other leg.

Diode D_1 rectifies the audio signal and applies the resultant d-c voltage through R_2 to diode D_3. The impedance of D_3 varies almost logarithmically in inverse proportion to the d-c voltage across it. Diode D_2 protects filter capacitor C_2 from any reverse polarity switching transients that may be applied from the output.

The curves of Fig. 2 show that about 15 db attenuation is available. A portion of this loss is made up in the transistor amplifier thus permitting installation of the unit as a unity-gain amplifier with an expansion of 3 db and compression of 12 db. Total consumption is approximately 0.6 ma.

The frequency response of the compressor for various input levels is shown in Fig. 3.

A tape recorder using this automatic gain compressor has been successfully used to report events from the quiet of an empty hall to the noisy cockpit of a small private plane.

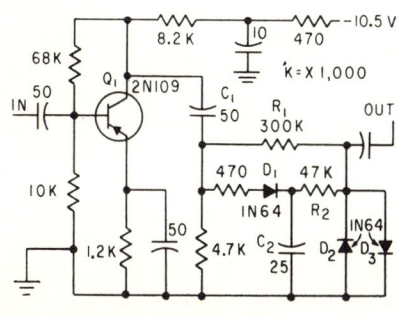

FIG. 1—Audio level compressor is basically a variable L-pad attenuator

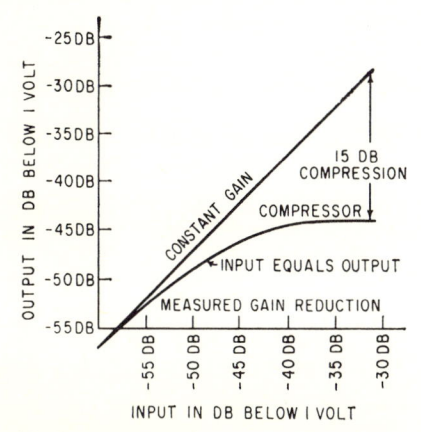

FIG. 2—Approximately 15-db compression is available from the device

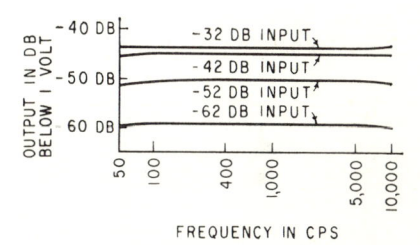

FIG. 3—Frequency response of compressor at various input levels

Amplifier Design Method

By **VICTOR R. LATORRE** Applied Research Lab, University of Arizona, Tucson, Arizona

SIMPLIFIED procedure for designing bandpass transistor amplifiers operating up to 50 mc uses an effective equivalent circuit. Design is exactly that used for vacuum-tube amplifiers.

Common-emitter hybrid-parameter equivalent circuit of a junction transistor is shown in Fig. 1.

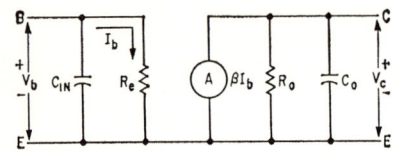

FIG. 1—Equivalent circuit of junction transistor

The hybrid parameters are: $r_s + a_{11}$ is input resistance with collector shorted to emitter; a_{22} is output admittance with base open; a_{12} is voltage feedback factor with base open; β is ratio of collector to base current with collector shorted to emitter; r_s is base-spreading resistance; C_{in} is input capacitance; and C_o is output capacitance.

C_{in} is equal to $C_{be} + C_{bc} + (1 + A_{bc})$.

For an amplifier with single-tuned output, an inductance is placed in parallel with the collector and emitter terminals. Since a_{12} may be neglected for small signals, expressions for gain and center frequency are obvious.

Multiple Stages

For more than one stage, accounting for the input circuit of the following stage greatly complicates the above expressions. The pole-zero diagram of a one-stage amplifier (Fig. 2) shows that design procedure would be greatly simplified if the real pole could be neglected. An effective equivalent circuit makes this possible.

All circuit impedances are assumed to be in parallel. Since the real pole is no longer present, the circuit is analogous to that for vacuum tubes. Base-spreading resistance r_s is not being neglected in the equivalent cicruit in Fig. 3.

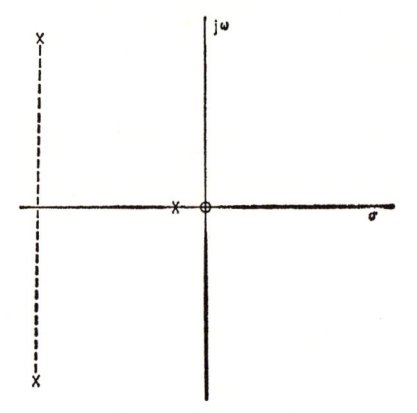

FIG. 2—Pole-zero diagram of one-stage amplifier using 2N384 transistor

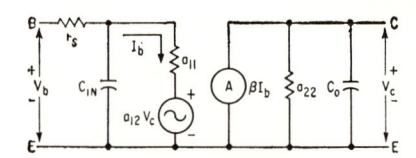

FIG. 3—Effective equivalent circuit of grounded-emitter amplifier

Parameters of the effective equivalent circuit were determined in the following manner. The output circuit was broadbanded (R_L very small), and a coil whose inductance and resistance are accurately known shunt the transistor input circuit.

By varying signal frequency, maximum voltage across the input terminals is found. The maximum occurs at input circuit resonance. Bandwidth of the input circuit is found by varying frequency on either side of resonance.

Input capacitance is then given by $C_{in} = 1/(\omega_o^2 L)$. Input resistance is calculated from $R = 1/(2\pi BC_t)$, where $C_t = C_{in}$ and $R = (R_{11} R_{ar})/(R_{11} + R_{ar})$. R_{ar} is the parallel resistance of the coil at ω_o.

This method was used and actual characteristics of the amplifiers were within 5 percent of theoretical values.

Chapter 5
OSCILLATORS

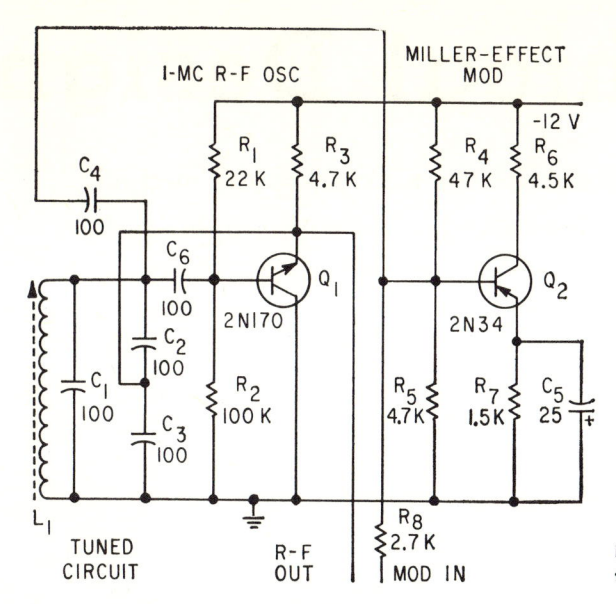

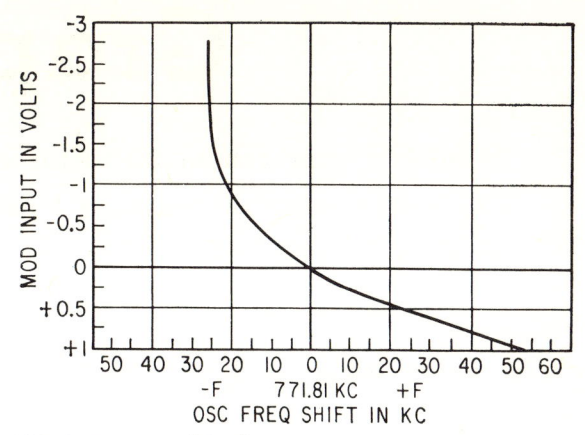

FIG. 2—Frequency shift plotted as a function of the d-c input voltage to the modulator

FIG. 1—A frequency sweep generator was constructed using two of these oscillator circuits and a diode mixer

Transistorized F-M Oscillator

Two-transistor circuit combines a Q multiplier with the Miller effect to produce a relatively stable but simple f-m oscillator and modulator

By **PAUL W. WOOD**, General Motors Corp., Detroit, Michigan.

A SENSITIVE MEANS of controlling or modulating frequency with a low control signal is provided by this transistorized f-m oscillator, Fig. 1. Operating at 1 mc, the configuration is similar to the Harris Q multiplier[1,2], with sufficient feedback to oscillate.

A low-impedance tap on the tuning coil is unnecessary because of the high input impedance of the emitter follower, Q_1. Negative feedback of Q_1 makes the circuit relatively stable.

The tuned circuit, L_1 and C_1, is center tapped by C_2 and C_3. The signal developed across the emitter resistor R_3 is fed back to the junction of C_2, C_3, sustaining oscillation at the frequency determined by L_1-C_1, C_2-C_3.

Resistors R_1, R_2 and R_3 determine the quiescent operating point of the transistor. The oscillator output signal is developed across the emitter load resistor.

The modulator is a conventional audio amplifier using Miller effect[3] to produce a change in the oscillator frequency.

The input capacitance of the common emitter circuit, Q_2, is approximately equal to the effective emitter capacitance times the transistor current gain. The modulation signal, applied to the base of Q_2, varies the current gain and thereby varies the input capacitance.

Frequency Shift

The input capacitance of Q_2 is coupled to the oscillator tank circuit through a small capacitor C_4. Thus as the gain of the transistor is varied, its input capacitance varies and is coupled to L_1-C_1, producing a corresponding shift in oscillator frequency. The values of R_4, R_5 and R_7 produce class-A operation of Q_2. Capacitor C_5 bypasses R_7 for audio frequencies. Resistor R_8 isolates the input signal source from the base of Q_2.

Figure 2 shows the frequency shift as a function of the d-c input.

Tests of this circuit, made in a variable temperature chamber, show a 4-kc decrease in frequency as a result of a temperature change from 40 F to 100 F. This temperature stability is good, being about 0.5 percent at 700 kc. Stability could be increased by use of negative temperature coefficient capacitors and thermistors.

A frequency sweep generator was constructed using two of the described oscillators and a diode mixer. One of the oscillators was modulated by a low-frequency sawtooth signal, while the other was operated at a fixed frequency.

REFERENCES

(1) H. E. Harris, Simplified Q Multiplier, ELECTRONICS, p 130, May, 1951
(2) G. B. Miller, Transistor Q Multiplier, *Electronic Eng*, p 79, May 1958
(3) Miller Effect, Radiotron Designer's Handbook, 3rd ed. p 182

Graphical Design

Simplified approach to transistor oscillator design uses graphical method. Universal method of designing feedback networks is used with graphic technique to design a highly stable transistor crystal oscillator operating at 1-mc.

By W. R. McSPADDEN* and E. EBERHARD, Motorola, Inc., Western Military Electronics Center, Phoenix, Arizona

A NEW METHOD of designing transistor oscillators described in this article is versatile in that it can be applied with little modification to many different circuits including crystal oscillators. It is comparatively simple, yet yields results accurate enough for most engineering design calculations.

Conditions

The design method is based on Barkhausen's criteria for oscillation which state that: while in a steady-state condition, the gain around the closed-loop portion of any oscillator must be unity, and the phase shift must be zero.

To simplify the problem, these two conditions will be considered separately. The first condition determines whether or not the circuit will oscillate. If the loop gain is lower than unity, the oscillator will not start, and if it is higher than unity, the oscillator will build up in amplitude until limiting occurs somewhere within the loop and the gain is thereby reduced. The second condition determines the frequency of the oscillator. If the phase shift is not zero, the frequency shifts in a direction to make it zero.

*Now at Univ. of Arizona

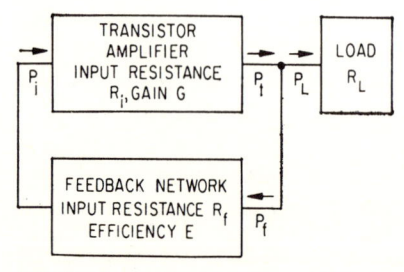

FIG. 1—Generalized oscillator

Design of transistor oscillators involves several problems that are not usually present for vacuum-tube oscillators. First, the ratio of output impedance to input impedance of a transistor is usually high, making it necessary to use an impedance transformer in the feedback network. Second, a transistor stage may have appreciable phase shift between its input and output. In addition, a transistor amplifier may have an input impedance dependant on the load impedance. Furthermore, variations of transistor parameters for different transistors of the same type may be considerable.

Loop Power Gain

Analysis of the power relationships within a generalized oscillator circuit uses an approach that is kept quite general so it may be applied to any type of feedback oscillator.[1] The oscillator is first divided into three blocks as indicated in Fig. 1. The transistor amplifier has a power gain G, a resistive component of input impedance R_i and a power output of P_t. The oscillator is shown supplying a useful load R_L with a power P_L. The coupling from the output of the amplifier back to the input is through the feedback network.

The feedback network has an input resistive component R_f which absorbs power P_f, and delivers a power P_i into the input resistance R_i of the amplifier. From Fig. 1 and the above definitions, the following relationships are apparent:

$$E = P_i/P_f \quad (1)$$
$$P_t = GP_i \quad (2)$$
$$P_L = P_t - P_f \quad (3)$$

Simple manipulation of these re-

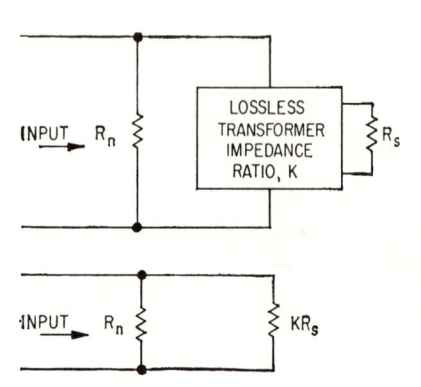

FIG. 2—Simplified representation of a transformer feedback network

lationships yields the following equation:

$$R_L = R_f/(EG - 1) \quad (4)$$

This equation is important in that it shows the relationship that must exist for unity loop gain between the input resistance of the feedback network and the load resistance.

Considering some of the general relationships that hold for the feedback network, it is assumed that this network is a lossless impedance transformer paralleled by a resistance R_n representing the open-circuit transformer loss. This simplification assumes that the short-circuit loss of the transformer is negligible. This has been found to be a good assumption in most practical cases. Considering Fig. 2, R_f is the input impedance, A is the actual impedance ratio, K is the theoretical impedance ratio, E is the transformer efficiency, and R_s is the load on the secondary. The following relationships hold:

$$A = R_f/R_s \quad (5)$$
$$A = K/(1 + KR_s/R_n) \quad (6)$$
$$E = A/K \quad (7)$$

The simple relationships ex-

# of Oscillators

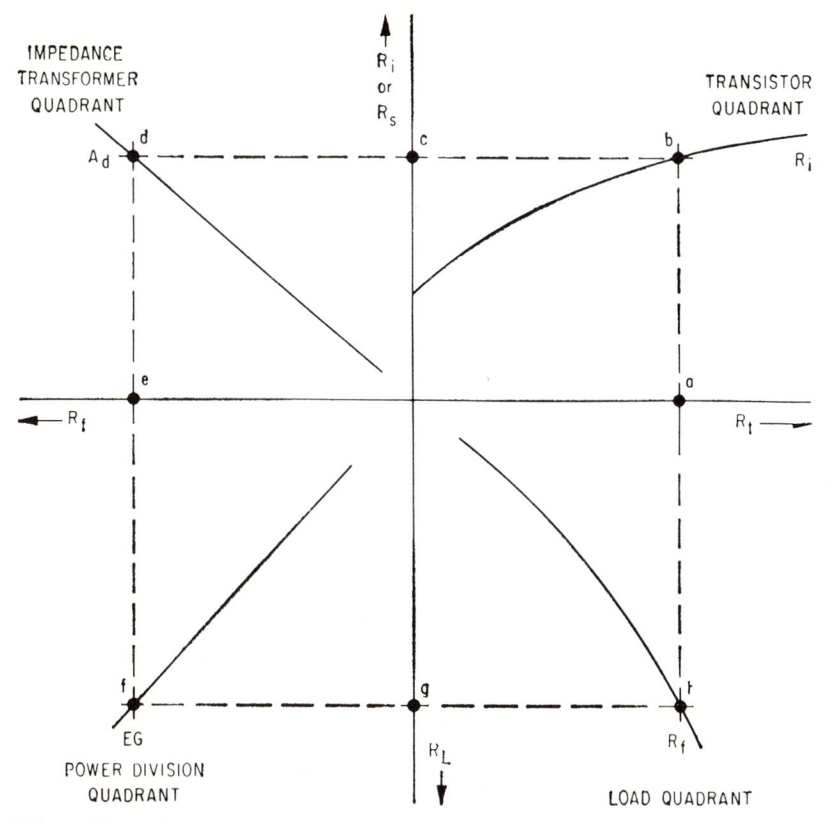

FIG. 3—Example of graphical solution for oscillator loop gain. Depending upon the known parameters, this procedure may be used to solve for other design factors

pressed in Eq. 6 and 7 may be used with good accuracy as long as R_s is not too small. Since it is comparatively easy to obtain R_n by measurement or by using the value of the circuit Q and since K is usually known, Eq. 6 and 7 can be used to obtain the actual impedance ratio and the transformer efficiency when loaded by R_s. These quantities can also be measured experimentally as functions of R, if a higher degree of accuracy is desired.

Excess Gain

Since it is not feasible to design an oscillator with exactly unity loop gain, any practical oscillator has excess loop gain. This fact simply causes the amplitude of oscillation to increase until limiting of some sort occurs and the actual gain exactly equals the minimum loop gain required to sustain oscillation. A factor τ is defined as the ratio of the actual class-A amplifier gain G_p, to the minimum amplifier gain G necessary to permit oscillation.

$$\tau = G_p/G \qquad (8)$$

When $\tau = 1$, oscillation is just possible.

For some cases a mathematical analysis of loop gain may be simple, but the problem is often complicated by the interdependence of several parts of the oscillator loop.

Graphic Method

The graphical method of approach used here was evolved with the idea of aiding the designer to solve this problem. It not only makes the solution easier, but also presents clearly the effect of varying the several circuit parameters.

Consider in detail the diagram shown in Fig. 3. Four resistance axes have been plotted. The upper right-hand quadrant is bounded by R_t, the total collector load on the transistor, and R_i, the input resistance of the transistor amplifier. In most simple oscillator circuits $R_i = R_s$; however, this is not the case for certain types of crystal oscillators. This quadrant is called the transistor quadrant in which a curve of R_i against R_t is plotted for the transistor to be used. The upper left-hand quadrant is bounded by the axes R_i and R_f. This quadrant is

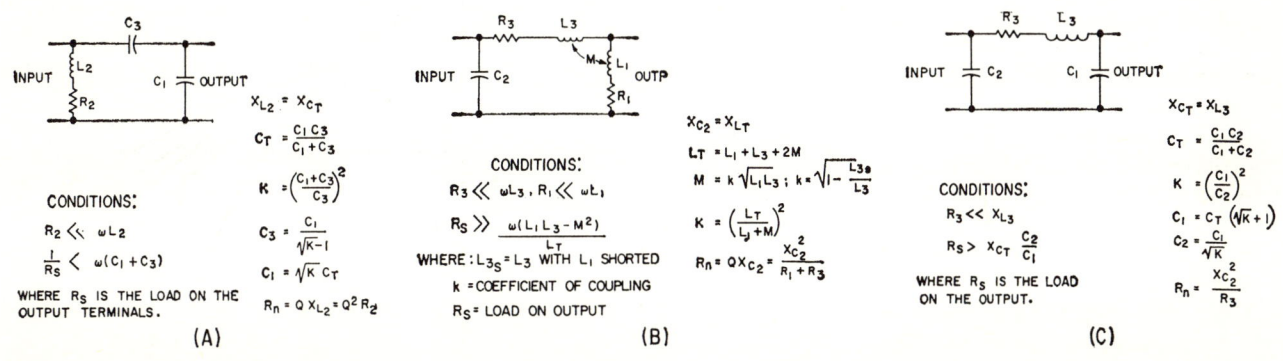

FIG. 4—Circuit and simplified design equations for three common types of feedback network. (A) L-C-C type (B) C-L-L type (C) C-L-C type. Type (C) is for a Colpitts oscillator

the impedance transformer quadrant and in it a straight line representing the actual transformation ratio A of the impedance transformer is drawn based on Eq. 5.

The lower left hand quadrant, bounded by the axes R_f and R_L, is known as the power division quadrant and is related to Eq. 4. This quadrant contains a straight line relating R_f and R_L to a given efficiency-gain product EG. In this quadrant it is possible to design for a given value of τ by simply making G equal to G_p/τ.

The lower right hand quadrant is termed the load quadrant. It contains a curve showing the relationship between R_t and R_L for a fixed value of R_f.

Using Diagram

To use this diagram, assume it is desired to operate with a total transistor load represented by point a on the R_t axis using a known impedance transformer having an actual transformation ratio A_d and an efficiency E. This value of A_d corresponds to a particular curve in the second quadrant. From point a project a vertical line to b, a horizontal line to c and d, and another vertical line downward to point e. Also project downward from a to h and horizontally to g. The particular R_f curve used in the fourth quadrant must agree with the value of R_f from point e. Point e represents the value of the input impedance of the feedback network and point g represents the value of R_L which in combination with R_f gives the original assumed value of R_t.

By simply extending lines g-h and d-e until they intersect at f, the efficiency-gain product necessary to just produce oscillation can be found. Since E was given and the EG product is known from point f, the minimum required gain G can be obtained. If this gain is larger than G_p, the class-A gain of the transistor, the circuit will not oscillate. If G is less than G_p oscillation will occur and τ (which is now greater than unity) may be evaluated from Eq. 8.

Many variations in this procedure will immediately be apparent. For instance, the design might be started with R_L, R_t, τ and E known and the impedance matching network to be determined.

Advantages

One of the advantages of this design procedure is its versatility. A simple modification of the basic plot of Fig. 3 permits this procedure to be applied to a crystal oscillator using the crystal in its series mode. This modification is accomplished by adding the transistor input resistance R_1 to the equivalent series resistance R_t of the crystal at resonance to obtain a new value of R_s. The efficiency E is then reevaluated to account for the power loss in the crystal. A later design example will illustrate this case.

To simplify the graphical process, an additional curve can be included in the first quadrant. This curve is a plot of transistor power gain P_t against the total load resistance R_t and facilitates the design by making it easy to evaluate the effect of operation at different values of R_t.

The design process can be speeded by starting with families of curves in all but the transistor quadrant. These can be plotted to cover a wide variation of impedance ratios, efficiency gain products, and values of R_t, R_L and R_f. If curves of power gain and R_i are known for limit transistors, these can also be readily plotted in the transistor quadrant.

Feedback Network

As previously explained, the feedback network has two parts: a shunt resistance R_n, and a lossless transformer with an impedance ratio of K, as shown in Fig. 2. Theoretically the study of any impedance transformer could be reduced to a study of the two factors R_n and K. Once these two factors are known for a given network, Eq. 5, 6 and 7 apply. In general, the feedback network is a band-pass filter tuned to the desired frequency of operation. Assuming that high-Q coils ($Q > 10$) are used in the network, then R_n is obtained simply by unloading the secondary ($R_s = \infty$) and calculating the input resistance of the network at resonance. An expression for K is obtained by assum-

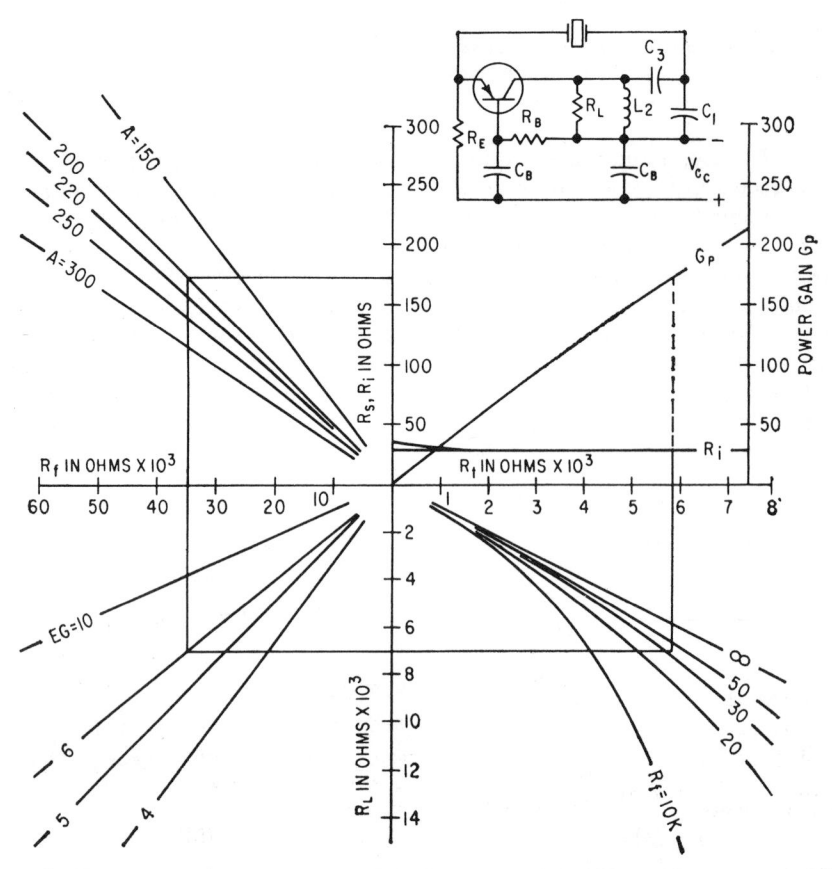

FIG. 5—An example of crystal oscillator design with an L-C-L feedback network. Transistor type is a 2N247 and the desired frequency of oscillation is 1 mc

ing lossless components of the filter and calculating R_t as a function of R_s.

Because of the assumption that the short-circuit losses are zero, there will always be some minimum value of R_s for which the design equations are valid. This relationship should be stated for each network since the errors can be serious if ignored.

Using the principles outlined in the above two paragraphs, the design equations for three common networks have been developed. The basic relationships and limitations are shown in Fig. 4.

An oscillator using the network of Fig. 4C is a Colpitts oscillator, and if L_3 and R_3 of this network are replaced by a parallel mode crystal, it becomes the familiar Pierce oscillator. The assumptions of Fig. 4C do not apply in the latter case; modifications may be found elsewhere.[3]

Design Example

A crystal oscillator with an L-C-C network has been chosen to illustrate the design procedures. It is assumed that the transistor, crystal and load resistance are specified. The frequency of operation is to be 1 mc, the transistor is to be a type 2N247 with a 6,900-ohm useful load and the crystal is to be a CR-19/U. Referring to Fig. 5, the design then proceeds as follows:

(1) Design the transistor bias network and evaluate the input resistance and power gain at 1 mc for several values of R_t. See any standard transistor text for the bias network design. Plot R_t and the power gain G_p against R_t as in Fig. 5. For this example $R_E = 4,700$ ohms; $R_B = 180,000$ ohms; $I_c = 1$ ma; $V_{cc} = 3$ v; R_1 of the crystal = 140 ohms.

(2) Locate all known points or curves on Fig. 5. $R_L = 6,900$ ohms; $R_i = 34$ ohms (approximately constant); $R_s = R_i + R_1 = 174$ ohms.

(3) Choose a value of R_t that is high compared to R_L. Let $R_t = 35,000$ ohms. Locate this point on the graph.

(4) Close the loop and record the following parameters: $R_L = 6,900$ ohms; $A = 200$; $R_s = 174$ ohms; $EG = 6.1$; $R_t = 35,000$ ohms; and $G_p = 170$.

(5) Now choose L_2. For this example a small ferrite toroidal inductor was available having the following constants: $L_2 = 286$ μh; $Q = 147$; $R_s = 12.2$ ohms; the tuning capacitor C_t required to resonate with this coil is 90 $\mu\mu f$.

(6) Calculate R_n and K: $R_n = QX_{L2} = 265,000$ ohms; $K = AR_n/(R_n - AR_s) = 231$.

(7) Calculate efficiencies and loop gain: $E_1 = A/K = 0.867 = $

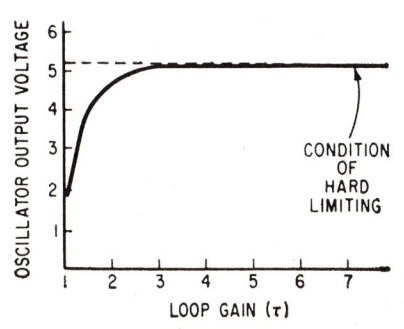

FIG. 6—Output voltage as a function of loop gain for an oscillator biased near center of class-A region

transformer efficiency; $E_x = R_t/R_s = 0.195 = $ crystal efficiency; $\tau = E_1 E_x G_p/EG = 4.7$. With this value of τ oscillation is assured.

(8) Calculate network capacitors: $C_1 = \sqrt{K C_t} = 1,370$ $\mu\mu f$; $C_3 = C_1/(\sqrt{K} - 1) = 96.5$ $\mu\mu f$.

The above oscillator was constructed and oscillated strongly as would be expected with the chosen value of τ. When tested for stability with respect to supply voltage variations it showed a frequency shift of 0.03 ppm for a 1 percent change.

Output Voltage

The determination of oscillator output voltage is often an important consideration in the design. The prediction is tied closely to the loop-gain analysis. In fact, the problem can be solved completely if the following two factors are known: the curve of power gain against output voltage with the desired load and at the desired operating point and the minimum gain G to permit oscillation. With these factors at hand, the resulting amplitude of the oscillator can be predicted with good accuracy.

In practice, the above data is often not known and the designer may want to make a rough estimate of the output voltage without obtaining the power gain versus output voltage curve. If the oscillator is biased such that its operating point is fixed and is near the center of its class-A region, the output voltage can be estimated from the amount of excess loop gain. This estimate makes use of the fact that limiting will occur sharply and gain will fall off rapidly when the a-c peak voltage at the collector equals the d-c collector voltage.

Experimental data shows (see Fig. 6) that if the loop gain, τ, is greater than about 1.8, the output voltage will be within about 15 percent of this maximum theoretical value. If τ is greater than about 2.5 the output voltage will be within about 5 percent of its maximum theoretical value. This relationship holds only if the amplitude limiting is quite sharp. For instance, if the transistor is biased near cutoff, limiting will start at much lower values of collector peak voltage, but the limiting action will not be as sharp and the output amplitude will continue to increase, even though the gain decreases, until the peak a-c voltage equals the d-c voltage. In this case it would be necessary to know the gain characteristic to predict the output voltage.

Oscillator Frequency

To investigate the factors influencing the frequency of the oscillator consider the Barkhausen criterion that the phase shift around the loop must be zero for any oscillator in a steady-state condition.

When the amplifier portion of the circuit is a transistor the assumption that the phase shift equals 180 deg through the device is not nearly good enough and it usually becomes necessary to consider the phase shift within the transistor to predict the operating frequency with accuracy. For best frequency stability the amplifier should have little phase shift. The factor of primary importance here is the magnitude of the phase angle associated with the transistor transfer characteristics h_{fe} or h_{fb}.

REFERENCES

(1) E. Eberhard, Chapter 14, "Handbook of Semiconductor Electronics", McGraw-Hill Book Co., Inc., New York, 1956.
(2) Final Report, Army Contract DA-36-039-SC-72837, Signal Corps Engineering Laboratories.

Solid-State Generator

Here is a tubeless microwave generator which produces 10 mw at 2,000 mc. Applications are in fields where light weight and high efficiency are musts

By **M. M. FORTINI,** Research Div., Philco Corp., Phila., Pa.

J. VILMS, Drexel Inst. of Technology and Philco Corp., Phila., Pa.

CONVERSION OF D-C POWER to a-c power can be accomplished in one step by an oscillator; it can be done in two or more steps by letting an oscillator operate at a frequency lower than the desired one and then converting the low frequency to the desired frequency with one or more harmonic-frequency converters. The two-step method to be described produces microwave power from a semiconductor source. This source uses a transistorized oscillator and amplifier which run at frequencies in the low-hundreds of mc and a diode harmonic multiplier.

Diode Harmonic Generator

Harmonic frequency conversion depends on the fact that a nonlinear element driven with a sinusoidal voltage or current gives a non-sinusoidal response. A semiconductor diode exhibits resistive, hole-storage and capacitive types of nonlinearity. The resistive mode suffers appreciable power loss in the nonlinear resistance due to average conduction current.[1] The hole-storage mode is based on the recovery time of a diode, the time required to remove the stored minority carriers when attempting to turn off the diode. For some diodes this recovery time appears as a narrow spike of reverse current which is rich in harmonic content. The two major limitations in using this method are high average-conduction-current losses and the requirement that the period of the applied signal be more than twice the recovery time of the diode.

The capacitive mode is far more suitable for harmonic generation because conduction current causes no power loss, and the frequency of operation is solely limited by the Q of the diode. Therefore this mode was chosen for the all-semiconductor system.

An equivalent circuit for a diode used as a nonlinear capacitor is shown in Fig. 1A. Diode dissipation factor $1/Q = \omega C_d r_s + 1/\omega C_d r_d + r_s/\omega C_d (r_d)^2$ and diode capacitance $C_d = K(V + \phi)^{-\frac{1}{2}}$. For good harmonic-generator performance, diode requirements are low spreading resistance (r_s), low dissipation factor and high reverse-breakdown voltage. Figures 1B and 1C show the nonlinearity characteristics of a good microwave harmonic-generator diode. The r_s and C_d were measured at 3.5 kmc. Diode Q was measured at 3 kmc, with Q ranging from 10 to 15 at a reverse voltage of 0.5 v to 20 v, respectively. The I-V characteristic indicates the voltage range of the capacitive-nonlinearity mode. Variation in r_s at low voltages may be explained by space-charge widening due to the reverse voltage. In a thin diode this widening is an appreciable portion of the diode thickness.

To achieve minimum conversion loss, the harmonic-generator circuit must provide the proper impedance transformation between source and diode and between diode and load. The circuit must also prevent power dissipation at other harmonic frequencies of the fundamental. Eliminating power dissipation at unwanted harmonics raises efficiency. Either short or open circuits for the unwanted harmonics eliminate this dissipation. To minimize losses in the nonlinear diode, which are primarily in its r_s, the matching sections must provide high source and load impedances looking in either direction from the diode.

Shown in Fig. 2A and B are two circuits suitable for harmonic generation in the nonlinear-capacitance mode. Circuit 2A short-circuits and circuit 2B open-circuits the undesired harmonics. Inductance L_f and capacitance C_f are resonant at the input frequency, L_h and C_h are resonant at the desired harmonic frequency, and C_1, C_2, L_1 and L_2 are for impedance matching.

The open-circuit arrangement appears to be more desirable because it eliminates device dissipation caused by the unwanted harmonic currents. However, analysis shows

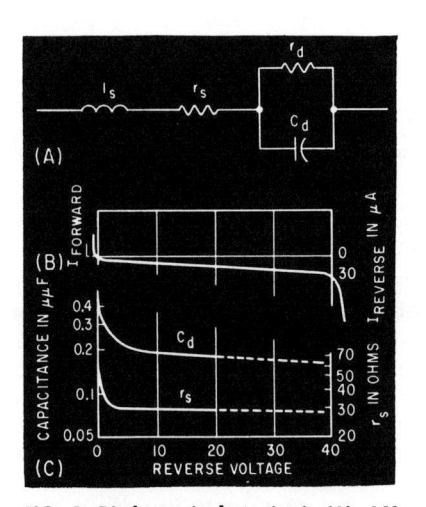

FIG. 1—Diode equivalent circuit (A); I-V characteristic (B); diode capacitance versus reverse voltage (C)

# of Microwave Power

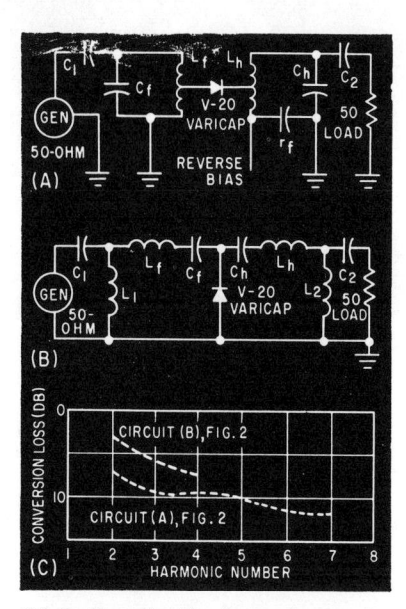

FIG. 2—Short-circuiting harmonic generator (A) and open-circuiting harmonic generator (B); their frequency response (C)

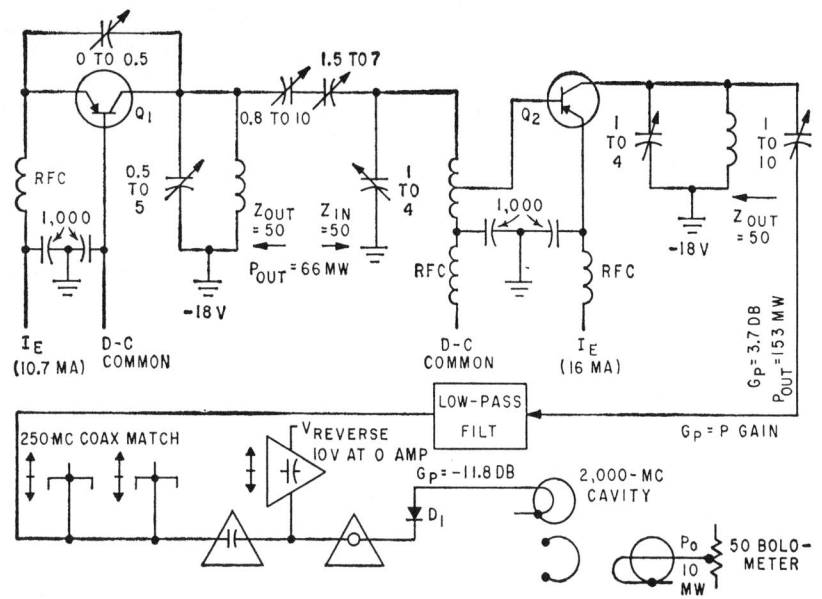

FIG. 3—Oscillator Q_1 and amplifier Q_2 deliver 153 mw at 250 mc to the coaxial matching section. Despite a conversion loss of 11.8 db in diode D_1, a 10-mw output at 2,000 mc appears across the 50-ohm bolometer

that one can open-circuit all undesired harmonics only for second-harmonic generation, due to the form of the diode C-V characteristic.

Another difficulty encountered with the circuit of Fig. 2B is the practicability of obtaining high Q's in the series-resonant branches. In general, the short-circuit arrangement is more suitable for high-order harmonic generation at high frequencies. Here the load and the source are coupled to the diode by parallel-resonant tanks, which short-circuit the undesired harmonics.

Figure 2C compares frequency-conversion loss versus harmonic number obtained, using a Varicap diode type V-20 as a nonlinear capacitor for the circuits of Fig. 2A and B. This diode has a capacitance of 10 to 60 $\mu\mu f$, an r_s of 15 ohms, and reverse breakdown of 45 v. Input frequency and power were held constant at 20 mc and 100 mw respectively, with impedance matching (and diode bias for Fig. 2A) optimized for each harmonic.

All-Semiconductor Source

Since the efficiency of a transistor is inversely proportional to the frequency of operation, it is wise to operate the transistorized section of a 2-kmc power source in the low-hundred-mc frequency range. Thus, at 250 mc the diode harmonic multiplier operates as an eighth-harmonic generator. Figure 3 shows the all-semiconductor 2-kmc power source. Two Philco high-frequency power transistors[2] are used as oscillator and power amplifier at 250 mc and a Transitron S-555G diode (D_1) operates in the capacitive mode as an eighth-harmonic generator. This system provides 10 mw at 2 kmc for a d-c power input of 423 mw; thus overall efficiency is 2.3 percent.

More recent data obtained using Philco experimental diodes as harmonic generators in place of the S-555G indicate the possibility of obtaining 10 mw at 2 kmc from only 300 mw of d-c power input to the entire system. Efficiency is then 3.3 percent. This increase is due mainly to the better efficiency of the Philco experimental diode and to the higher gain that the transistor buffer amplifier has when it provides less output. As efficiencies of diodes and transistors improve, further increases in efficiency will be feasible.

A simple way to modulate a variable-capacitance harmonic generator was devised. Since it is possible to vary the harmonic-generation efficiency, for a given r-f drive and for best matching and tuning conditions, by varying the back-bias voltage on the diode, this variation can be used for modulation purpose.

Figure 4 shows the change in output voltage at 2 kmc as the biasing voltage is changed. Since the diode impedance at audio frequencies is high, little modulating power is required.

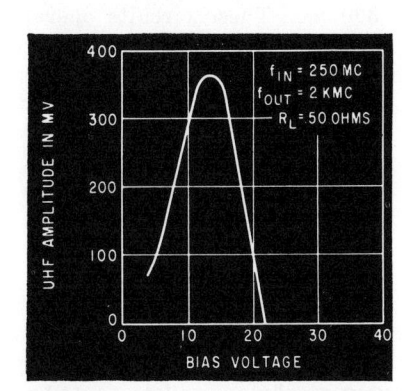

FIG. 4—Variation of uhf amplitude with diode bias

REFERENCES

(1) C. H. Page, Harmonic Generation with Ideal Rectifiers, *Proc IRE*, **46**, p 1,738, Oct. 1958.
(2) C. G. Thornton, J. B. Angell, Technology of Micro-Alloy Diffused Transistors, *Proc IRE*, **46**, p 1,166, June, 1958.

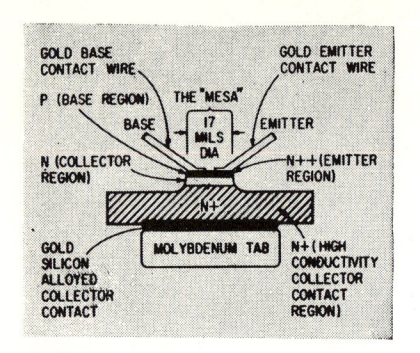

FIG. 1—Cross section of typical vhf silicon transistor

FIG. 2—Comb structure is used in high-power, high-frequency transistor

Design of High-Power Oscillators

New high-power transistors are usable at over 300 mc. Oscillator design is simplified with step-by-step procedure

By W. E. ROACH, Pacific Semiconductors, Inc., Culver City, California

HIGH-FREQUENCY, HIGH-POWER transistors require small structures and the best possible heat dissipation. Small size is necessary to meet the high-frequency requirement. Good heat dissipation is necessary since a large amount of heat is generated in a small volume.

The heat problem is solved by mounting the collector directly on the transistor case, which in turn is usually mounted on a metal plate or fin assembly. Stray capacitance, which must be dealt with in design, is thus introduced from collector to other parts of the circuit. Using the technique described in this article, and new transistors recently developed, oscillators can be built with over 100 watts output at 10 mc. Higher frequency oscillators are possible but at less power; for example, 0.2 watt at 300 mc.

The oscillator design method combines theoretical and experimental procedures to obtain an optimum circuit in minimum time. Amplifier and feedback sections of the oscillator are designed separately, then combined to produce the final circuit.

New Transistor

The cross section of a typical developmental transistor is shown in Fig. 1. It is a vhf power transistor that can dissipate approximately four watts at the collector with the proper external heat sink.[1,2]

A high-power transistor, which has a collector dissipation in excess of 100 watts with water cooling, is shown in Fig. 2. A special comb structure is used for the emitter to cut down current crowding which occurs in the emitter at high current levels.[3] The comb pattern gives the emitter a large edge-to-area ratio, thus minimizes capacitance for a given current rating. Silicon is used because of its good high temperature characteristics.

The transistors were designed for good performance below 30 volts but they can be operated at higher voltages. The low-voltage capability means, however, that in many cases d-c power can be taken directly from the primary power source of an aircraft, satellite or motor vehicle. High power output from a single stage permits substantial savings in power and components while the circuit simplification contributes to overall reliability.

Experimental Results

A series of similar oscillators was built for operation from 30 mc to 300 mc. These circuits operate with the collector tied directly to the chassis for maximum heat dis-

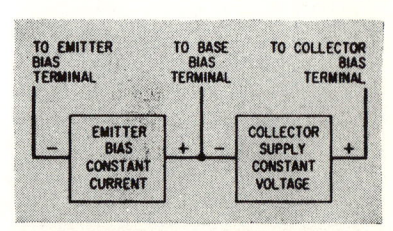

FIG. 3—Two power supplies are used for experimental work. Polarities are for npn transistors

90

Table I—Performance of Experimental Oscillators

Frequency in mc	V_{cc} in volts	I_c (d-c)	P_{in}, d-c, watts	P_{out}, a-c, watts	Eff. Percent
10	45	4.5 amp	202	110	54
30	67	150 ma	10	5.0	50
70	60	120 ma	7.2	3.1	43
200	50	80 ma	4.0	0.6	15
300	50	80 ma	4.0	0.2	5

Table III—Pi-T Transformation Formulas

$$Z_1 = \frac{Z_A Z_B}{Z_A + Z_B + Z_C} \qquad Z_A = \frac{Z_1 Z_2 + Z_2 Z_3 + Z_1 Z_3}{Z_2}$$

$$Z_2 = \frac{Z_B Z_C}{Z_A + Z_B + Z_C} \qquad Z_B = \frac{Z_1 Z_2 + Z_2 Z_3 + Z_1 Z_3}{Z_3}$$

$$Z_3 = \frac{Z_A Z_C}{Z_A + Z_B + Z_C} \qquad Z_C = \frac{Z_1 Z_2 + Z_2 Z_3 + Z_1 Z_3}{Z_1}$$

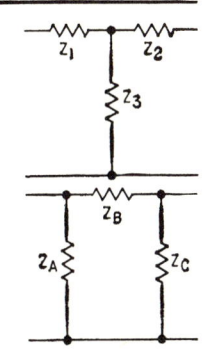

Table II — Pi-Network Design Formulas

The network input impedance is to look like R_1 at the input terminals with R_2 connected to the output terminals. Impedance X_B may be selected as desired, subject to a sufficient coupling restriction.

$$|X_B| \leq \sqrt{R_1 R_2}$$

$$X_A = \frac{-R_1 X_B}{R_1 \pm \sqrt{R_1 R_2 - X_B^2}}$$

$$X_C = \frac{-R_2 X_B}{R_2 \pm \sqrt{R_1 R_2 - X_B^2}}$$

Table IV—Complex Voltage Transfer Function for T and Pi Networks With Resistive Loads

Complex voltage gain, $V_2/V_1 = A + jB$

$$A + jB = \frac{jX_C R_L}{jX_B jX_C + jR_L(X_C + X_B)}$$

$$A + jB = \frac{jX_3 R_L}{(jX_1 jX_2 + jX_2 jX_3 + jX_1 jX_3) + jR_L(X_1 + X_3)}$$

Synthesis of network for specified complex voltage gain, $A + jB$

$$X_A = \frac{R_L B}{(A-1)(A^2 + B^2)} \qquad X_1 = \frac{R_L(A^2 + B^2 - A)}{B(A^2 + B^2)}$$

$$X_B = \frac{-R_L B}{A^2 + B^2} \qquad X_2 = \frac{R_L(A-1)}{B}$$

$$X_C = \frac{R_L B}{A^2 + B^2 - A} \qquad X_3 = \frac{R_L}{B}$$

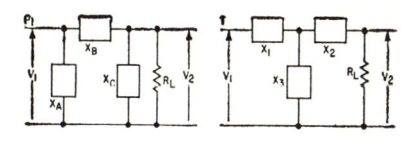

---

sipation. Table I summarizes the results. Transistors like those shown in Fig. 1 have produced results ranging from 5 watts output at 30 mc to 0.2 watt output at 300 mc. Efficiency decreases with increasing frequency as the efficiency of amplification drops, but appreciable power is obtained at the higher frequencies.

Transistors with the comb structure were tested in a modified circuit. Although tests are not completed, preliminary designs have given a power output of 110 watts at 10 mc, with a collector circuit efficiency of more than 50 percent.

Circuit Considerations

A good high-frequency unit of the type shown in Fig. 1 may typically have an alpha cut-off frequency between 100 and 200 mc. The maximum frequency of oscillation will be several hundred mc, and appreciable output power may be obtained at high frequencies even though efficiencies will drop as the frequency of operation is raised. In oscillator operation, the requirement for high gain through the transistor is not as important as in amplifier applications, since the oscillator circuit provides its own drive. A power gain greater than one is needed but, in many cases, it is more important to be able to operate the transistor at high d-c input power to get appreciable power output.

In the design of low-level linear amplifiers, small signal theory and a set of parameters such as h parameters permit straightforward design. For high level oscillators, which may operate class B or class C, such procedures are not straightforward. In addition, self-biasing d-c effects may exist in the transistor at high signal levels.

Mode of Operation

The output coupling circuit and load values required for an amplifier or oscillator are a function of the class of operation. In class A, the output impedance of the transistor should be matched. For class C operation, however, the requirements are different.

The load conditions for the circuit may be determined approximately by considering the tank circuit and load resistance only, assuming that the transistor itself is inactive over most of the cycle. Since the energy furnished to the load must come largely from the stored energy in the tank circuit, the amount of energy available in the L-C circuit must be appreciably larger than the amount furnished to the load between pulses of input current to the tank. This is equivalent to saying that the loaded Q of the tank circuit must be of a certain magnitude for sine-wave output. A minimum loaded Q of five is considered satisfactory.

Since the peak voltage swing across the tank should nearly equal

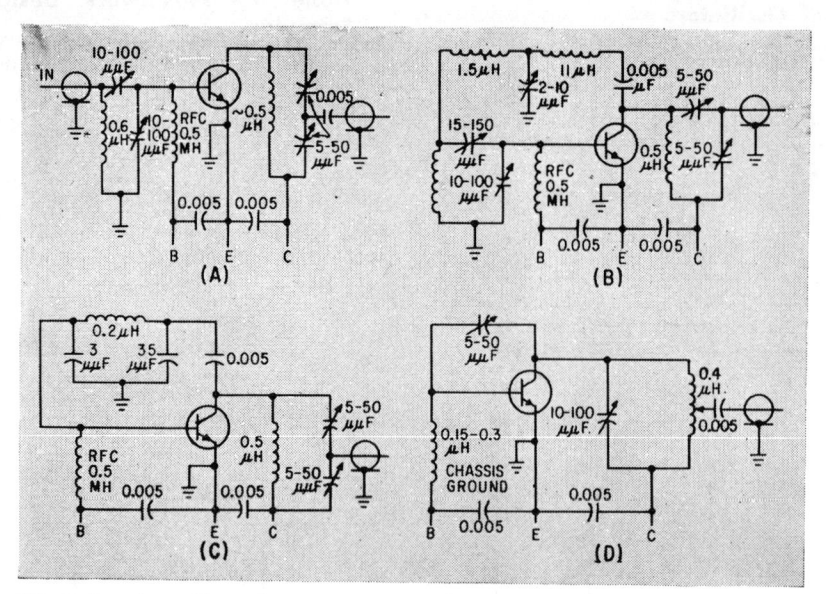

FIG. 4—Steps in oscillator design. First step is to build 70-mc amplifier (A); amplifier is converted to oscillator with separate input, output and feedback networks (B); feedback and input networks are replaced with pi network (C); simplified final circuit with collector conected directly to chassis and output from tapped tank (D)

the collector supply voltage (to provide the high collector efficiency mentioned previously), the load resistance for any given power is found from: peak-to-peak voltage swing across $R_L = 2V_{cc}$, where R_L = load resistance, V_{cc} = d-c collector supply voltage, and V_L = rms voltage swing across R_L; $V_L = 2 V_{cc}/2 \times 1.414 = 0.707 V_{cc}$.

The power delivered to the load is $P_L = V_L^2/R_L = (0.707 V_{cc})^2/R_L = 0.5 V_{cc}^2/R_L$; $R_L = 0.5 V_{cc}^2/P_L$.

For oscillator operation, there is also a starting requirement, which means that the oscillator must start in a condition approximating class A, though it may then change to a different type of operation. In many cases an internal d-c biasing effect will be observed in the input circuit of the transistor when an a-c signal is applied from a generator or fed back from the output circuit. This effect may shift the operating point.

Network Design

A detailed step-by-step procedure for oscillator design is given in the design box. In Step 1, networks are designed using standard formulas[4], shown in Table II, to match the input and output impedances of the transistor to the driving generator and to the load and/or power meter. Since these formulas are shown for resistive input and output, the transistor input and output should be made resistive by adding the necessary reactance in parallel with the input and output terminals. This added reactance may be combined with the matching network. These matching networks should have two or three of the elements variable to permit adjustment.

Two additional considerations may influence the choice of matching network configuration. One is the convenience of d-c bias introduction or isolation. The other is the network response at frequencies other than the operating frequency.

Conversion formulas for T-pi networks are shown in Table III.

The value of R_1 used for the output network may be made equal to the resistive output impedance of the network for class A operation, or may be calculated for class C operation by $R_L = 0.5V_{cc}^2/P_{load}$. For class C operation, the loaded Q of the matching circuit must also be considered.

These networks are checked by adding the proper terminating resistance to either the input or out-

---

Power Oscillator Design Procedure

Basic approach is to design and evaluate each network separately. Analysis or measurements of a complete complex network with several variables is usually difficult—if not impossible—but the problem is considerably simplified by this method.

Step 1—Design input and output matching networks for operation of the devices as an amplifier at the desired frequency

Step 2—Verify amplifier performance at the desired power level

Step 3—Estimate oscillator output by subtracting required drive from amplifier output power

Step 4—Convert the circuit to an oscillator by providing a feedback network

Step 5—Simplify the circuit by consolidating feedback and matching circuits

Design Example

Figure 4 through 4D illustrates the steps and procedures in the design of a 70-mc power oscillator. Typical characteristics of a transistor of the type shown in Fig. 1 are given in Table V.

Step 1—The small-signal common-emitter h-parameters shown in Table V were measured at 70 mc. Exact parameter values have limited meaning, since each impedance will be a function of the terminating impedance at the opposite terminals as well as of bias levels. In addition, at high signal levels the circuit operation is not linear. Nonetheless, such data serves as a useful starting point. Values measured for the unit used are

$h_{11e} \cong (20 + j2)$ ohms $\cong 20$ ohms

$h_{22e} \cong (5.6 + j4.6)$ millimhos

Calculation of the required matching networks (from Table II) resulted in the amplifier configuration shown in Fig. 4A. This circuit was set up and checked for power output as an amplifier with network elements adjusted for optimum performance.

Step 2—The results indicated that the unit could dissipate four watts with external heat sink provisions, and would provide an output as a 70-mc amplifier in excess of 3.2 watts with 7.2 watts d-c input. Input driving power was approximately 0.2 watt; power gain was 12 db.

Failure of oscillation will result in an approximately seven watts being dissipated in the unit, which will probably destroy the transistor unless protection is provided.

put and measuring the impedance at the opposite set of terminals. This procedure may also be used to determine the settings of variable elements in the network.

In Step 4, the feedback network characteristics to convert the amplifier to an oscillator are estimated. The impulses of collector current are to be 180 degrees out of phase with the collector voltage. Thus the feedback network must furnish a phase shift which, with the internal phase shift of the transistor, and that of the input pi network, will accomplish the desired result. The phase shift of the T or pi network is calculated from the formulas in Table IV.

If a specific value for the phase shift of the current gain (h_{21e} or $-\beta$) is not available, -90 degrees may be used, since the angle is near this value for frequencies between about 0.1 α-cutoff frequency to well above α cutoff.

With transistor and input network phase shifts known, the feedback phase shift may be calculated. The voltage gain desired may be approximated using the power levels and impedance levels at the input network and at the collector terminals. The complex voltage feedback is thus specified in polar coordinates; these may be converted to rectangular coordinates and the network calculated from the formulas in Table IV, using the generator impedance as the value of terminating resistance. The current at the transistor input may be assumed to be in phase with the voltage if the transistor input reactive component has been cancelled.

Step 5, simplification of the circuit, is carried out as shown in the example. In this procedure the internal feedback of the transistor is neglected. Depending on the particular device characteristics, this approximation may require further consideration.

Provision of separate d-c bias power supplies as shown in Fig. 3 has been found to be convenient for initial development and investigation of a-c circuit properties. Single supply biasing may be substituted later if required.

The transistors used were designed and fabricated by the vhf silicon transistor and the power transistor sections of the research and development department of Pacific Semiconductors. Some of the device development work was done under United States Army Signal Supply Agency Contract No. DA 36-039 SC-74887 and United States Air Force Contract No. AF 33-(600)-35088. The power output measurement of the 10-mc oscillator was performed by the Electronic Components Laboratory, Wright Air Development Center, Dayton, Ohio.

Table V—VHF Transistor Characteristics

Parameter	Measurement Conditions	Value
α_o	$I_c = 100$ ma	0.86
β_o	$I_c = 100$ ma, $V_c = 10$ v	6.5
C_c, $\mu\mu f$	$V_{CB} = 28$ v, reverse	12
V_{bc}, volts	$I_c = 1$ ma	150
V_{sat} (C. E.), volts	current gain = 5, $I_c = 100$ ma	0.45
h_{11e}, ohms	$V_c = 30$ v, $I_c = 30$ ma ⎫	$20 - j2$
h_{22e}, mhos	$V_c = 30$ v, $I_c = 30$ ma ⎬ $f = 70$ mc	$(5.6 + j4.6) \times 10^{-3}$
h_{21e} ($= -\beta$)	$V_c = 30$ v, $I_c = 30$ ma ⎭	$0.26 - j1.7$

References

(1) J. L. Buie, High Frequency, Silicon, NIPN, Oscillator Transistor, *IRE P.G.E.D. Conference*, Washington D. C., Oct. 1958.
(2) M. A. Clark, Power Transistors, *Proc IRE*, 46, p 1185, June 1958.
(3) N. H. Fletcher, Some Aspects of the Design of Power Transistors, *Proc IRE*, 43, p 551, May 1955.
(4) W. L. Everett and G. E. Anner, Communication Engineering, Third Edition, Chap. 11, McGraw-Hill Book Co., Inc., New York, 1956.

Step 3—Minimum power output capability as an oscillator was estimated by subtracting amplifier input from output, giving a result of 3.0 watts.

Step 4—A T-network was chosen for the feedback function.

If the network is connected from the collector to the input network as shown in Fig. 4B, Z_L may be assumed to be equal to the impedance of the generator used for the amplifier tests. The complex value of h_{21e}, ($-\beta$), the current gain of the transistor, was measured as $0.26 - j1.7$, as shown in Table V. The phase angle is thus -82 degrees. At the settings used to obtain the power shown in Step 2 above, calculation of the phase shift through the input matching network indicated a value of approximately $+40$ degrees. Assume that the transistor input current and voltages are in phase.

Thus there are -82 and $+40$ degrees of phase shift in the circuit, and a total of 180 is needed around the loop. Therefore an additional phase shift is needed of $180 + 42$, or approximately 220 degrees.

To determine the magnitude of the feedback function, calculate the voltage levels at the input and output of the feedback network, using the power levels and impedance values at these points. This calculation gives three volts rms at the input, 24 volts rms at the collector. The complex voltage gain of the feedback network is therefor $(3/24)\lfloor 220 = -0.11 - j0.08$.

Addition of the network resulted in the circuit of Fig. 4B. Output is 3.1 watts at 70 mc.

Step 5—The elements of the feedback and input networks may be consolidated to produce a less flexible but simpler circuit. This is done by reducing feedback and input networks with standard T-pi transformation and consolidation methods. A more direct method, however, is to calculate a new network having the voltage transfer characteristics of the feedback and input networks combined. In this example, the voltage transfer characteristics for the feedback and input networks are, respectively, 0.55 at an angle of 240 deg, and 0.67 at an angle of 40 deg. Multiplying these together gives 0.37 at 290 deg, or 0.37 at -70 deg; in rectangular coordinates, the value is $0.12 - j0.34$. Substitution of a pi network of this voltage transfer ratio, using a termination value of 20 ohms, yields the circuit in Fig. 4C. Substitution of this circuit, and use of a tapped tank for output power, resulted in the circuit of Fig. 4D, drawn as a grounded collector circuit. This circuit gives performance equal to that of the more complex circuit of Fig 4B. The transistor was mounted directly on the chassis to provide an adequate heat sink.

Transfluxor Oscillator

Magnetic-electronic oscillator retains last frequency setting for many hours after removal of control signal

By **RICHARD J. SHERIN**, Product Development Laboratory, IBM Corp., Poughkeepsie, N. Y.

DEMAND by industry for a simple device to replace the servomotor-controlled oscillator has existed for some time, particularly among manufacturers of military communications equipment, automobile radio receivers and frequency standards.

Previous continuously-variable voltage-controlled oscillators all drift cumulatively off frequency when the controlling voltage is removed. Although it is possible to approximate this type of oscillator to any desired degree of accuracy by using a digital register, such a system is not simple.

Basis of the transfluxor oscillator described is the circuit shown in Fig. 1. The small aperture of a transfluxor[1] is used as the magnetic core member in a Royer converter circuit[2]. A square-wave output is obtained having a period which is a linear function of the time integral of the voltage induced in the large aperture by the control signal.

Transfluxor Operation

To simplify the discussion, assume that the cross-sectional areas of transfluxor legs 2 and 3 are equal and that of leg 1 is twice this value as shown in Fig. 2A. Also, assume

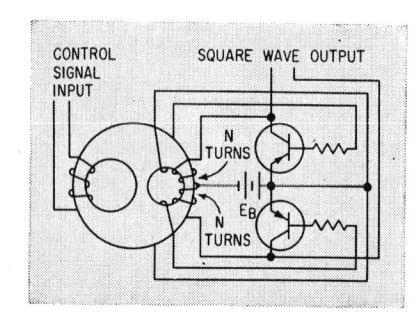

FIG. 1—Basic transfluxor oscillator circuit is essentially magnetic multivibrator with electronic frequency control

the material used in the transfluxor has a nearly rectangular hysteresis loop. Using the sign convention established in Fig. 2B, it is apparent that: $-2\phi_s \leq \phi_1 \leq 2\phi_s$; $-\phi_s \leq \phi_2 = \phi_3 - \phi_1 \leq \phi_s$; and $-\phi_s \leq \phi_3 = \phi_1 + \phi_2 \leq \phi_s$ where ϕ_s is the saturation flux for leg 2 and for leg 3.

Consider the magnetization of the small aperture when $\phi_1 \geq 0$. When the direction of magnetization is counterclockwise, leg 2 saturates at $\phi_2 = -\phi_s$ with leg 3 still unsaturated and $\phi_3 = \phi_1 + \phi_2 = \phi_1 - \phi_s$. When the direction of magnetization is clockwise, leg 3 saturates at $\phi_3 = \phi_s$ with leg 2 still unsaturated. Hence, the flux change, $\Delta\phi_3$, in leg 3 during clockwise magnetization starting with leg 2 saturated and ending with leg 3 saturated is given by $\Delta\phi_3 = 2\phi_s - \phi_1$. Similarly, for $\phi_1 \leq 0$, $\Delta\phi_3 = 2\phi_s + \phi_1$ and in either case $\Delta\phi_3 = 2\phi_s - |\phi_1|$.

Switching Ring C_1

Referring to Fig. 2C, when legs 2 and 3 are both unsaturated, the magnetomotive force (mmf) which must be applied to leg 3 to change ϕ_3 is at most equal to that value which will switch flux in the ring C_1 when it is unsaturated. In operation, when a sufficient mmf is applied to the small aperture, the flux density vector field pattern will be changed in such a way that the shortest closed path enclosing the small aperture and having no saturated segment is affected first. Paths of greater length are affected as the material closest in to the small aperture becomes saturated.

The change in the flux density field pattern proceeds from the inside to the outside until some path is reached which has sufficient length so that the applied mmf is less than the threshold mmf for this path. When the mmf is sufficient to overcome the threshold mmf for this path, it is great enough to saturate all the material inside the circle C_1 and, therefore, is an upper bound on the mmf re-

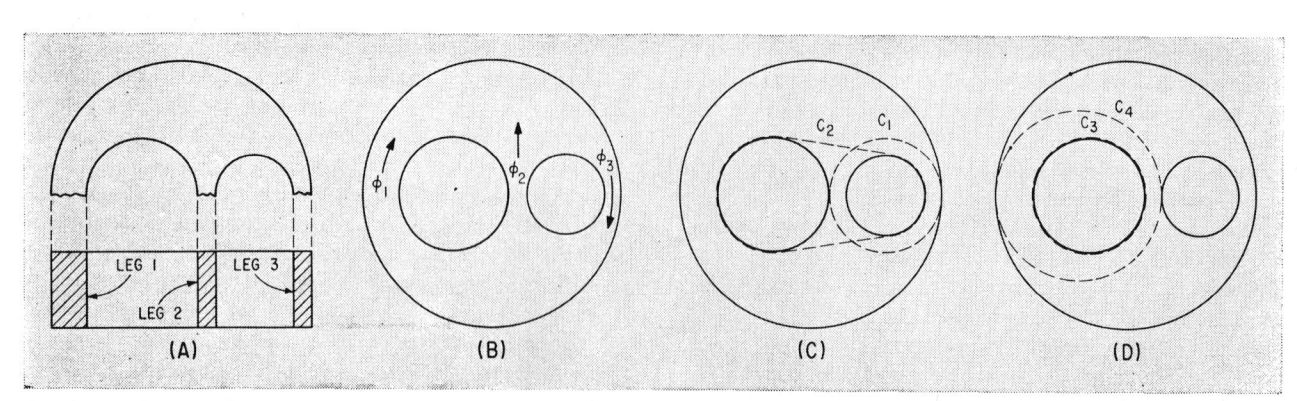

FIG. 2—Transfluxor construction and magnetization. Cross-sectional area of transfluxor legs (A); definition of fluxes used in description of transfluxor operation (B); paths determining the range of mmf which can be applied to leg 3 (C); and paths determining a maximum range for the mmf threshold presented to the control signal (D)

Gives Drift-Free Output

quired to switch the flux in a closed path through legs 2 and 3.

Switching Ring C_2

When leg 3 reaches saturation in the downward direction, an arbitrarily large downward mmf will cause no further flux change in the transfluxor. When leg 2 reaches saturation in the downward direction, a sufficiently large upward mmf on leg 3 can cause counterclockwise magnetization of a path which encloses both the small and large aperture thereby altering ϕ_1. However, if the counterclockwise mmf is limited to a value below that which will switch flux in the path C_2, then it is insured that the upward mmf on leg 3 will cause no further flux change after leg 2 has become saturated in the downward direction.

Circuit Parameters

The circuit of Fig. 1 is so designed that either transistor once ON is held in the saturated ON state for any collector current up to some maximum value I_o. Circuit parameters are chosen so that I_o is greater than that value which would just magnetize an unsaturated path C_1 but smaller than that value which would magnetize an unsaturated path C_2. Thus, the mmf is just greater than the threshold mmf for the length of the path referred to and will therefore alter the flux density direction along the path if the path is not already saturated in the direction of the mmf. These limit values are separated by a factor of two or more in a typical transfluxor, hence the adjustment is not at all critical.

Each time collector current I_o is reached in the ON transistor, the fed back voltage to the ON transistor decreases causing a corresponding decrease in the collector current. Because of the presence of a small amount of elastic flux excursion, this decrease in current causes the voltage to reverse across all the windings on the small aperture, the formerly ON and OFF transistors interchange states and the flux in leg 3 is then driven back toward the opposite saturation condition. The flux changed in leg 3 between the two saturation conditions is $|\Delta\phi_3| = 2\phi_s - |\phi_1|$, and the period of the oscillator is directly proportional to this value.

Frequency Control

When the transfluxor oscillator parameters are properly chosen, ϕ_1 is unaffected by the mmfs applied to the small aperture. Only when the mmf produced in leg 1 by the control signal exceeds a certain threshold will ϕ_1 begin to change. After this threshold is exceeded, the rate of change of ϕ_1 will be proportional to the voltage induced in an open circuited winding on leg 1, that is, the control voltage is a driving voltage minus an IR drop in the control winding.

This threshold mmf falls in the range bounded by the mmf values required to magnetize unsaturated paths C_3 and C_4 as shown in Fig. 2D, and corresponds to the static friction in the servomotor of a servomotor-controlled oscillator. It is the threshold effect which stabilizes the oscillator frequency against cumulative drift resulting from noise current in the control winding of the transfluxor oscillator in one case and from noise current in the servomotor drive voltage in the other case. The experimental oscillator (Fig. 3), uses an RCA XF3006 transfluxor.

REFERENCES

(1) J. A. Rajchman and A. W. Lo, The Transfluxor, *Proc IRE*, March 1956.
(2) G. H. Royer, A Switching Transistor D-C to A-C Converter Having an Output Frequency Proportional to the D-C Input Voltage, *Trans AIEE*, Part I, July 1955.

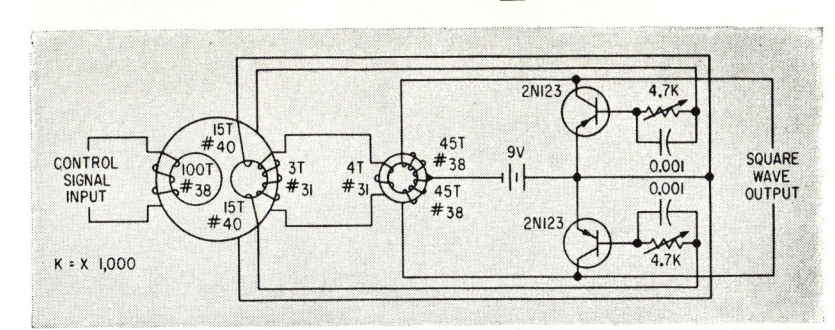

FIG. 3—Experimental transfluxor oscillator operates between 100 Kc and 1 Mc. In practical applications, a larger transfluxor would be used and the extra transformer eliminated

---

OSCILLATION TIME CALCULATIONS

Time in μsec for a half cycle of oscillation is that time required to switch the flux $\Delta\phi_3$ and is given by:

$$T = \frac{N}{E_B - \Delta E} \Delta\phi_3 = \frac{N}{E_B - \Delta E}(2\phi_s - |\phi_1|)$$

where all fluxes are measured in volt-μsec/turn and ΔE accounts for the sum of the voltage drop across an ON transistor and the IR loss in the conducting N turn winding. The relation between a corresponding flux and voltage is given by:

$$\phi_1(t) = \int_0^t \frac{e_c(t)}{N_c} dt$$

where e_c is the voltage induced in an open circuited winding of N_c turns on leg 1 and t is measured in μsec. Thus:

$$T = \frac{N}{E_B - \Delta E}\left(2\phi_s - \left|\int_0^t \frac{e_c(t)}{N_c} dt\right|\right).$$

Letting A be the cross-sectional area in cm^2 of leg 3 and B_s the saturation flux density in gauss for the core material, $\phi_s = B_s A/100$ volt-μsec/turn. The threshold mmf, F_c, in ampere-turns for the control signal falls in the range $H_T L_3/0.4\pi \leq F_c \leq H_T L_1/0.4\pi$ and the range of satisfactory adjustment for I_o is determined by $H_T L_1/0.4\pi < N I_o < H_T L_2/0.4\pi$, where H_T is the threshold mmf in oersteds for the core material and L_1, L_2, L_3 and L_4 are the lengths in cm of paths C_1, C_2, C_3 and C_4, respectively.

Generator uses keyboard of toy piano

Tone waveform is visually demonstrated by one of authors while output waveform is displayed on oscilloscope

Portable transistorized monophonic keyboard instrument permits demonstration of elementary principles of Fourier synthesis of musical tones. Circuits are simple, easily adjustable and unit may be constructed at moderate cost

Synthesizing Timbre

By W. S. PIKE and C. N. HOYLER, RCA Laboratories, Princeton, New Jersey

LECTURE DESCRIPTIONS on the synthesis of complex musical tones from series of harmonically related sinusoidal signals require visual and aural demonstration. Although at least one commercially available electronic organ could be used for this purpose, a lower cost and more readily portable instrument is desired.

The transistorized device to be described is a monophonic keyboard instrument having a compass of one octave, with its lowest fundamental frequency at about 250 cps. Electrical output consists of the fundamental and, within the limitations of the simple filters used, the second and third harmonics. The amplitudes of the three components of the output are individually adjustable and means for reversing the phase of the third harmonic is included.

Simple tunes may be played upon the instrument and the effect of changes in the three components of the output waveform can be demonstrated audibly and/or visually.

Tone Generation

The various tones are generated by a master oscillator and three aperiodic frequency dividers as shown in Fig. 1.

The master oscillator operates at six times the fundamental output frequency of the instrument, with a range of 1.5 to 3 kc. When any one of the playing keys is depressed its output drives two frequency divider chains. Chain A comprises a ternary divider followed by a binary divider, while chain B comprises a single binary divider.

Potentiometers across the filtered output of each divider are connected through mixing resistors to the output amplifier. Each filter substantially removes all harmonics above the fundamental frequency of its associated divider.

If the oscillator frequency is $6f$, the output at R_1 is $6f/(2 \times 3)$ or f; the output at R_2 is $6f/3$ or $2f$ and the output at R_3 is $6f/2$ or $3f$. Thus R_1 controls the amount of funda-

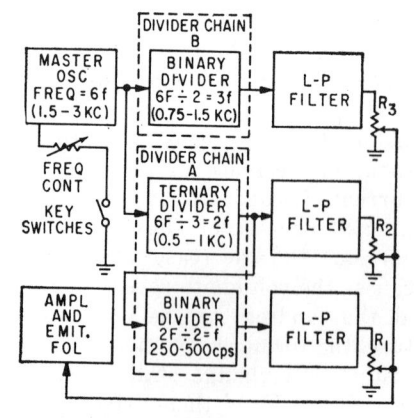

FIG. 1—System diagram of tone generator

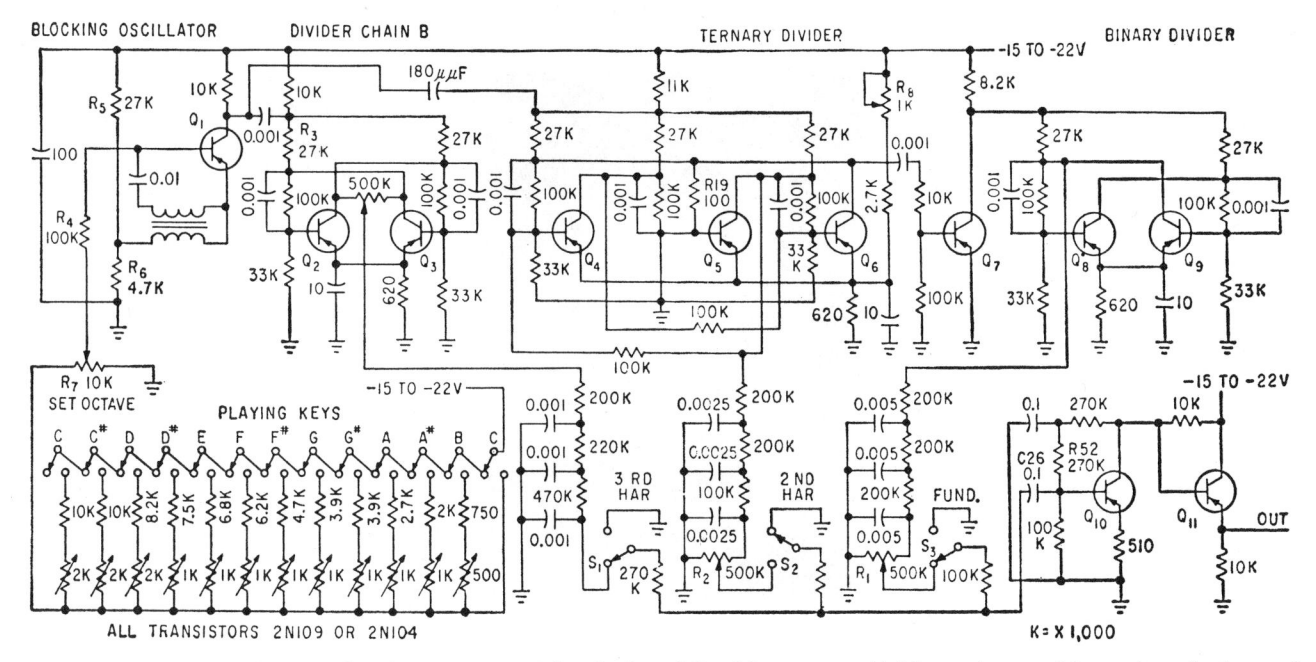

FIG. 2—Complete circuit of tone timbre demonstrator; switches S_1, S_2 and S_3 add or remove third harmonic, second harmonic or fundamental components from output signal to change tone quality

for Musical Tones

mental in the complex output tone while R_2 and R_3 control the second and third harmonics respectively. As the dividers are aperiodic, these relations will hold for any input frequency to the divider chain.

An inexpensive keyboard is provided by a toy piano. The original tone-producing portions of the piano were removed and suitable switches fitted to the key action.

Circuits

The complete circuit of the device is shown in Fig. 2. Blocking oscillator Q_1 is tuned over a range of one octave by varying the voltage to which base resistor R_4 is returned. This voltage is controlled by the resistors in series with the playing keys. The emitter of Q_1 is returned to a low negative voltage determined by R_5 and R_6. Unless a playing key is depressed Q_1 is biased off and will not oscillate as its base will be returned to ground through set-octave control R_7.

As chords cannot be played on the instrument, due to its monophonic nature, the key switches are so connected that if more than one key is depressed at the same time, only the highest note will sound. This scheme eliminates out-of-tune notes that would otherwise be produced.

Positive pulses from the collector of Q_1 drive the two frequency-divider chains. Chain B is a conventional binary divider comprised of Q_2 and Q_3. Divider chain A is similar except that it has two dividers, a ternary and a binary and Q_7 acts as buffer amplifier between ternary divider Q_4, Q_5, Q_6 and binary divider Q_8, Q_9. As the operating stability margin of the tenary divider is less than that of the binary, bias adjustment R_8 is provided.

The outputs of the two dividers of chain A are filtered by R-C filters to remove all but the fundamental frequencies of the dividers, which correspond to the fundamental and second harmonic of the output signal. The filters are a compromise between purity of waveform and uniformity of output over the range of the keyboard. They result in the upper tones of the keyboard being perceptibly weaker, but less filtering would not remove sufficient harmonics to produce a good oscilloscopic demonstration.

Phase Reversal

To demonstrate the ear's relative insensitivity to phase changes, the phase of the third harmonic has been made reversible. Potentiometer R_3, which controls the third harmonic amplitude, is not connected in the same way as the potentiometers controlling the fundamental and second harmonic. By connecting this potentiometer from the collector of Q_2 to the collector of Q_3, phase reversal of the third harmonic may be achieved. In this case the filter is placed after the potentiometer.

This method of phase switching avoids the generation of switching transients that might cause spurious triggering of the divider.

A disadvantage of this method is that due to the dissimilarity between the rise time and the fall time of the divider transistors, with R_3 set at midposition, the output is not zero but consists of a small pulse signal at the master oscillator frequency. When third harmonic in the output signal is not desired, shorting switch S_1 may be closed to eliminate this spurious signal.

Amplifier Q_{10} raises the output level of the instrument to compensate for the attenuation of the filters and mixing network; emitter follower Q_{11} drives a long cable to an external amplifier or drives the envelope control amplifier to be described.

Initial Adjustments

To set up the device, R_7 is adjusted to the middle of its range and, with a playing key held down, the outputs from the fundamental, second harmonic and third harmonic circuits are checked. The second harmonic should be an octave above the fundamental and the third harmonic should be an octave and a fifth above the fundamental.

The bias control on the ternary divider may require adjustment. Too little bias may result in continuous oscillation or triggered oscillation at $6f$ instead of $2f$. Too much bias will suppress all output from this stage and also remove the input signal from the following divider, resulting in zero output from the fundamental circuit as well.

If output is not obtained from any divider, the master oscillator is suspect and a different setting of R_7 should be tried.

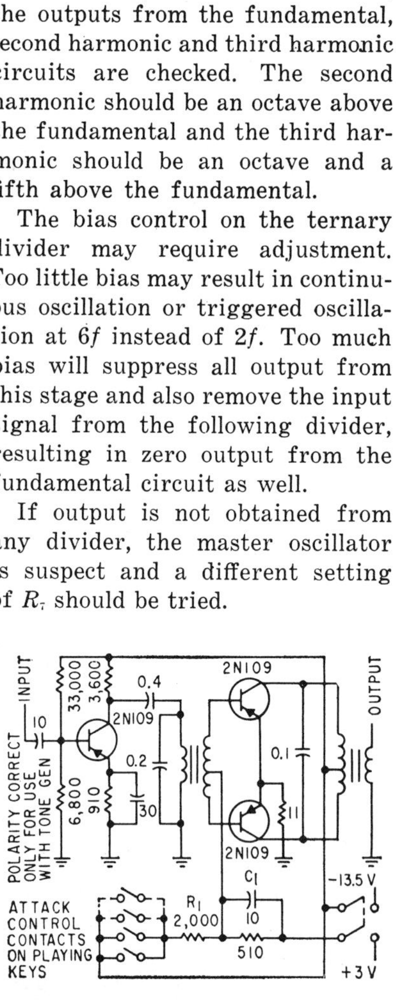

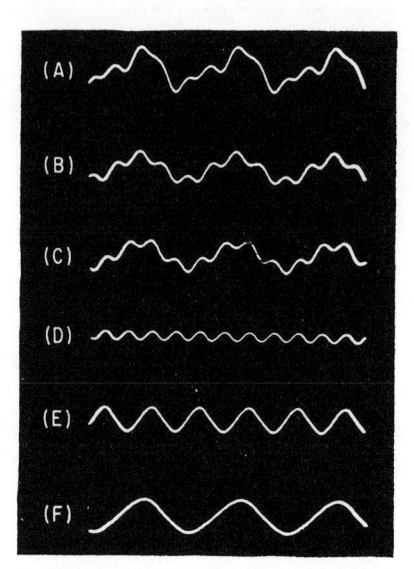

Typical output waveforms include fundamental, second and third harmonics (A), fundamental with third harmonic normal (B) and reversed (C), third harmonic (D), second harmonic (E) and fundamental (F)

Having obtained operation, the instrument must be tuned up. The adjustable resistor in series with the bottom note of the keyboard is set at its midpoint. This note and the note one octave higher are then alternately played while R_7 is adjusted to produce an octave relationship between these two notes. If the desired octave relationship cannot be achieved, the bottom note resistor is readjusted. The remaining notes of the scale may then be tuned in any convenient manner by adjusting their associated resistors.

Control R_7 may need occasional adjusting to compensate for battery aging.

Applications

The timbre demonstrator has been used by one of the authors in lectures and demonstrations to many groups of people. Usually an oscilloscope and a loudspeaker system are used simultaneously.

A note is held down on the keyboard and the various components of the output signal observed singly and in combination. Notes near the upper end of the keyboard are usually selected as better sinusoidal waveforms may be obtained for the individual harmonic components because of the filter characteristics.

The change of waveform due to change of phase of the third harmonic is best demonstrated by adjusting the fundamental control to about midposition and the second harmonic to zero. If the third-harmonic control is then rotated rapidly from one extreme to the other there will be a pronounced change in the appearance of the resulting waveform, but little or no change in the audible sound.

Musical Use

This device could be used for purely musical purposes if desired. For such use, its most serious limitations are its restricted range, marginal tuning stability, lack of control of the attack and decay of the tones produced. At the cost of slightly increased complexity this may be partially remedied by the attack control amplifier shown in Fig. 3, which requires an extra contact be added on each playing key.

The attack control amplifier has a push-pull output that is normally biased off. Depressing any playing key causes this bias to be removed at a rate determined by R_1 and C_1 thus producing a gradual attack. Each key contact for this amplifier should be adjusted to close slightly after the frequency control contact of the corresponding key.

The attack control amplifier will deliver about 250 mw to its load. In this form, the device could be used as a monophonic adjunct to another instrument such as a piano or, with readjustment of the filter and master oscillator frequencies, as an inexpensive pedal division of an electronic organ.

If more than one octave is to be covered, readjustment of the blocking oscillator frequency controlling circuit will be necessary. In addition, the filters should be simplified by removing at least one section from each to minimize the drop in output which occurs in the upper register of the instrument.

Other harmonics may be added by employing more dividers. An additional binary divider in each chain, for example, would permit the inclusion of the fourth and sixth harmonics. Alternatively, the sixth harmonic alone could be added by utilizing the direct output of the master oscillator.

FIG. 3—Attack control amplifier enables tone generator to be also used as musical instrument

Chapter 6
POWER SUPPLIES

Rectifier Gives D-C of Either Polarity

Diagonally symmetrical power transistor circuit permits smooth load current variation over range of several amperes at either polarity. Power into two-terminal load can be four times maximum allowable transistor power dissipation

By R. R. BOCKEMUEHL, General Motors Research Laboratories, Warren, Mich.

SMOOTHLY VARIABLE DIRECT-CURRENT SOURCES capable of supplying positive and negative currents in the ampere region are often needed for testing magnetic materials, solenoids, meters, reactors and other low impedance electrical devices. In many cases passive circuits are impractical for this purpose because of the poor resolution and contact noise of high-power, low-resistance rheostats.

Basic Circuit

In the basic diagonally symmetrical circuit shown in Fig. 1, power transistors Q_1 and Q_2 and voltage sources E_s form two loops in common with load R_L. Control voltage E_b is applied between the transistor bases resulting in a control current which flows through the two transistor base-emitter circuits with an opposite sense, thus promoting conduction of one transistor while cutting off the other. Reversal of control voltage polarity transfers conduction to the opposite transistor, thereby reversing the load current.

The conducting transistor loop forms a common-collector circuit having the emitter-base reverse resistance of the cut off transistor in series with the control voltage source. Voltage gain is approximately unity, current gain is approximately β and input resistance is approximately βR_L.

Characteristics of this diagonally symmetrical circuit are similar to those obtained more simply with complementary symmetry circuits[1]; however, complementary transistors which have high power capabilities are not commercially available.

Practical Circuit

A practical diagonally symmetrical circuit is shown in Fig. 2. Transistors Q_1 and Q_2 serve as drivers for the output transistors Q_3 and Q_4. The driver circuit increases the input resistance to $\beta^2 R_L$ which permits use of a 5,000-ohm, high-resolution control potentiometer and reduces battery drain. Greater resolution can be obtained using potentiometers with larger resistance, but a sacrifice in linearity resulting from potentiometer loading occurs.

Shunt Resistors

Emitter-base resistances are shunted by 220- and 20-ohm resistors which have a negligible effect on the input resistance of the conducting circuit. The 47,000-ohm resistors bias the transistors so that they conduct slightly with zero control voltage; hence, under any condition at least one transistor pair is conducting.

A maximum of 4 amp can be supplied to the 2.5-ohm load giving an output of 40 watts. Maximum collector dissipation is 14.5 watts. Similar circuits can be designed for different load resistance by suitable selection of supply and control voltages and base bias resistors.

Other Applications

Control voltage can be obtained from electronic circuits thereby permitting the general circuit configuration to be used as a relatively high-power output stage for d-c servo applications. Also, since the circuit responds to frequencies from d-c through the audio range, similar circuits have been designed for audio power amplification[2] with direct connection to the loudspeaker voice coil.

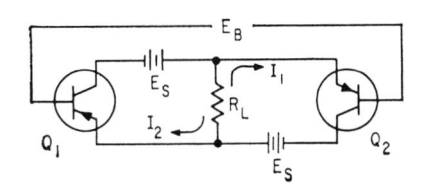

FIG. 1—Emitter currents in basic diagonally symmetrical circuit flow in opposite directions through load

FIG. 2—Automotive storage batteries are used for 12-v transistor supplies in practical circuit although rectifier supplies could be used. Maximum current drain from two 12-v dry cells in control voltage circuit is 7 ma

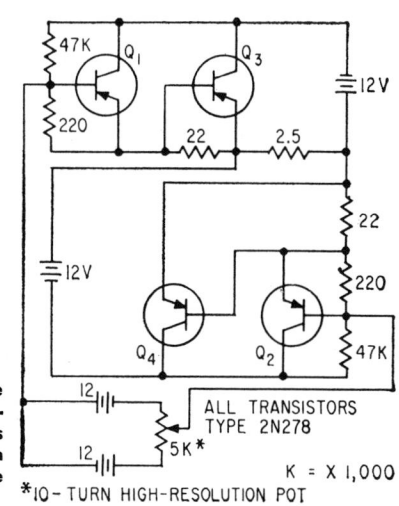

REFERENCES

(1) G. C. Sziklai, Symmetrical Properties of Transistors and Their Applications, *Proc IRE*, p 717, June 1953.
(2) Application Note 5-B, Delco Radio Division, General Motors Corp., Jan. 1958.

Design of

Design criteria for transistorized power converters are covered in this article. Examples cover silicon type delivering 15 w and a germanium type delivering over 100 w both from a 24-v source

By T. R. PYE, Applications Group, Texas Instruments Ltd., Bedford, England

ALTHOUGH TRANSISTORS are attractive for mobile equipment operated from low voltages, thermionic tubes may often be either indispensable or established in designs which it would be uneconomic to scrap. In these circumstances, the change to transistors must be gradual, and the need for a power unit to provide plate supplies from low voltage batteries remains.

If this power unit involves vibrators, fractured reeds and fused contacts must be expected. Rotary converter brushes and bearings will eventually wear out and sparking at contacts and brushes may cause serious interference. In many cases a transistor power converter will provide a more efficient and far more reliable replacement which can be readily incorporated into an existing system.

Additional applications of these versatile circuits include fluorescent lighting, electronic photoflash and servo system power supplies. Transistors now available allow output powers from a few milliwatts to several hundreds of watts. Efficiencies of 70-90 percent can be achieved, and input voltages may range from 1½ to 50 v. Series or parallel connection of transistors permits still higher output powers and input voltages.

Transistor Requirements

The maximum collector current of a transistor is decided on considerations of instantaneous dissipation and of linearity (fall of β at high collector currents).

During the half cycle when a converter transistor is cut-off, it will experience a collector voltage equal to twice that of the supply, and a small base voltage tending to cut it off still further. From this it would appear that a safe supply voltage would be half the open-circuit emitter breakdown voltage.

Unfortunately, at the end of the conduction cycle hole storage may allow current to continue when the transistor voltage has risen to the supply voltage or even higher. Under these conditions the transistor open-circuit base breakdown voltage is relevant, and should preferably also equal twice the supply voltage if avalanche breakdown is to be avoided.

Transistor dissipation is likely to limit power output only at high ambient temperatures. For example: the 2N389 silicon transistor has a maximum rated dissipation of 45 w at 100 C. With the maximum value of R_c, of 5 ohms and a total supply current of 1 ampere the mean dissipation will be only 2½ w in each transistor.

The corresponding figure for the 2N457 germanium transistor is 50 w at 25 C. For a typical R_c, of 0.05 ohm, 5 amperes will give a mean dissipation of less than 1 w.

These figures ignore leakage current, transient and input dissipations, as these will normally be small. They show that heat-sink requirements are usually modest.

It is important that the peak collector currents of the two transistors should be practically the same, otherwise the lower current will limit the output prematurely. The fall of β at high currents tends to equalize these peaks but transistors with widely differing β should not be used.

Since the feedback winding pro-

Transistor Converter Design Equations

Maximum Load Power Output

$$P_{Lmax} = \frac{V_s^2}{4 R_{cs}}$$

Current for Output Power P_L

$$I_c = \frac{V_s - \sqrt{V_s^2 - 4 P_L R_{cs}}}{2 R_{cs}}$$

Oscillation Frequency

$$f = \frac{V_s}{4 B_{max} n_p a} \times 10^8 \text{ cps}$$

Minimum Emitter Current for Oscillation

$$I_e = \frac{25}{\dfrac{R_L}{N} - \dfrac{r_b}{\beta}} \text{ ma}$$

P_{Lmax} = max load power output in watts
V_s = supply voltage
R_{cs} = transistor saturation resistance
I_c = current for output power P_L
f = oscillation frequency
n_p = number of primary turns
B_{max} = saturation flux density of core in gauss
a = core cross section area in cm^2
r_b = transistor internal base resistance
N = primary to feedback windings turns ratio
R_L = secondary load reflected across primary
I_e = minimum emitter current to insure oscillation
β = transistor beta

Power Converters

vides an almost constant voltage, variations of transistor input impedance will affect base (and hence collector) currents. The addition of a small base series resistor, of the same order as the input impedance, will tend to give constant-current drive. The minimum resistance possible must be used otherwise drive power will be lost and carrier extraction efficiency on switch-off impaired (although this may be restored with a bypass capacitor). The base resistor can be common to the transistors and in series with the starting diode.

As transistor and transformer parameters make exact behavior hard to predict, voltage and current waveforms of a converter design should be observed on an oscilloscope.

Transformer Design

The transformer core material must have low hysteresis loss when taken to complete saturation and, preferably, a high saturation flux density. A high permeability should be maintained until saturation, in order that the inductive current shall be small compared with the load current. All this suggests square-hysteresis-loop materials such as ferrites. However, ferrites show a comparatively high hysteresis loss when taken to complete saturation and have a low saturating flux density (about 3,000 gauss) and are therefore suitable only for low output powers. Certain nickel-iron materials also show this square-loop characteristic, and have a high saturating flux density (typically 15,000 gauss) with a low saturation hysteresis loss (around 750 ergs/cycle/cm³).

The choice of core material and of operating frequency (unless this is decided by other considerations, such as ripple frequency) will depend on the power output required.

At power levels of less than a watt, the maximum nickel-iron core volume for reasonable efficiency will become very small. A ferrite core may then be preferable, and high-frequency operation (at several kilocycles) will counteract the low saturation flux density and give reasonable copper losses.

For higher powers a nickel-iron core is essential, and a comparatively low operating frequency will be necessary. Core hysteresis losses are directly proportional to frequency and eddy-current and residual losses increase according to a square law. The optimum frequency will generally be from 200-800 cps.

A possible core is chosen for the power to be handled. As a starting point, about 1 cm³ of core volume should be allowed for each watt of output power required. On this basis a material having a loss of 700 ergs/cm³/cycle would give a core loss of 3 percent at the typical frequency of 400 cps. An operating frequency is chosen, and the number of primary turns is calculated. By allocating roughly half the winding space to the primary, the primary resistance, and that of the secondary reflected in the primary, may be estimated. The copper loss can then be compared with the core loss obtained from the core volume and the operating frequency. A high proportion of copper loss will suggest raising the frequency or increasing the core volume (and vice versa). The effect on the total loss of these alternative measures can then be compared.

The number of feedback turns are chosen to give the full load collector current, allowing for the starting diode and added base resistance voltage drop.

The secondary winding is chosen to give the required output voltage, bearing in mind that the primary voltage is less than the supply as a result of R_{cs}. Slight readjustment of the secondary turns may be necessary to cover the resistive drop in the winding.

Typical Designs

The silicon transistor converter shown in Fig. 1 was designed to give 15 w output from a 24-v

Placement of components in 120-watt power inverter

supply. It uses two 2N389 silicon power transistors and can operate in ambient temperatures up to about 130 C.

The 2N389 has a maximum collector-to-emitter voltage of 60 v, so that operation will be safe with supply voltages up to 30 v.

The maximum R_{cs} for transistors currently available is 5 ohms. Taking the worst possible case, if other losses are ignored, the maximum output is $V_s^2/4 R_{cs} = 28$ w for 50 percent efficiency. For 70 percent actual efficiency, which might be the minimum tolerable, the maximum output will be at least 15 watts.

In this case a square stack of laminations was chosen having a volume of 19.8 cm³ and a core area a of 1.61 cm².

The hysteresis loss of the material to the saturation flux density of 15,000 gauss was stated by the makers as 650 ergs/cm³/cycle. For a typical operating frequency of 400 cps, loss = 650 × 400 × 19.8 ergs/second = 0.515 joule/second or 515 mw. For an output of 15 w this corresponds to a loss of 3½ percent, and is therefore acceptable.

The collector current for $P_L = 15$ w is 0.76 ampere. If all losses apart from R_{cs} are assessed at 3 w, the actual I_c will be 0.88 ampere, and the voltage appearing across the primary will be 24 − (0.88 × 5) = 19.6 v. The primary will have 50.5 turns.

The area of the winding space

103

is approximately 0.4 sq in. Allowing 0.1 sq in. for each primary and a suitable space factor, 20-gauge wire is indicated. Two layers will give 45 turns, involving a change of frequency to 450 cps.

For a full load output voltage of 200 (rising to about 240 v on no load), allowing for secondary and rectifier voltage drop, the secondary $= 204/19.6 \times 45 = 470$ turns. This should be of 28 gauge in order to fill the remainder of the bobbin. The copper losses in primary and secondary are roughly the same as the core hysteresis loss and both are small compared with R_{cs} loss. The lamination pattern and operating frequency are therefore suitable.

The feedback turns can be calculated from the V_{bc}/I_c characteristic shown in the 2N389 data; 3.8 v will be required for an I_c of 0.9 ampere. Allowing 1 v drop across the starting diode and 0.7 v across the base resistor, 5.5 v will be required, giving 13 turns for each base winding. Wire of 28 gauge is suitable.

To give low leakage inductance it is preferable to sandwich the feedback windings between the primaries, which in turn should be placed in the middle of the secondary, wound in two equal sections.

The input resistance of the 2N389 at high current is about 30 ohms. Series base resistors of 10 ohms will drop just less than the 0.7 v allowed and give sufficient current equalization throughout the cycle.

The emitter current required for starting may now be calculated. Here, $R_L = 24/0.9 = 26.6$ ohms and $N = 45/13 = 3.5$. For low currents, r_b will be higher than usual, about 40 ohms, and the effective value will be increased by the added base resistors, giving a total of 50 ohms; β will also be lower, approximately 10. The minimum emitter current to insure oscillation will be 10 ma.

The total I_b required will be 2.0 ma so that $R = 12,000$ ohms. In fact, 5,600 ohms were found to be necessary. This difference can be explained by the rather arbitrary choice of r_b and β, both of which affect the result considerably, and the fact that the transistors oppose each other until a higher collector current is established.

At the full load of 15 w, 70 percent efficiency was obtained with a dissipation of less than 4 w per transistor. At temperatures up to 100 C, a copper or aluminum heat sink of about 3 square inches will be sufficient; at higher temperatures a larger one (preferably of copper) may be necessary.

Germanium Transistor Converter

The circuit of this converter is exactly the same as that of the silicon version. The component values are also in Fig. 1. Two 2N457 germanium power transistors are used, and over 100 w may be obtained from a 24-v supply at about 90 percent efficiency.

The R_{cs} of the 2N457 is typically

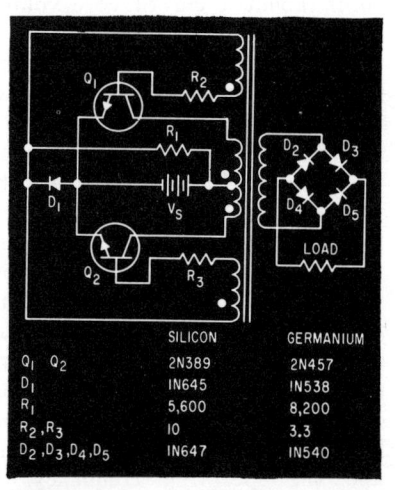

FIG. 1—Practical converter circuit with full-wave rectifiers. In the germanium unit, diode D_1 and the supply voltage must be reversed

	SILICON	GERMANIUM
Q_1, Q_2	2N389	2N457
D_1	1N645	1N538
R_1	5,600	8,200
R_2, R_3	10	3.3
D_2, D_3, D_4, D_5	1N647	1N540

0.05 ohm, with a maximum of 0.2 ohm. The maximum current of 5 amperes will give a maximum saturation voltage of 1 v. In this case, I_{cmax} and not R_{cs} limits the output; the arguments above concerning R_{cs} may be ignored, although an allowance must be made for the saturation voltage when considering the actual transformer primary voltage. A laminated core with a volume of about 40 cm³ was used and to reduce hysteresis losses, a frequency of 300 cps chosen.

By a similar process to that used for the silicon converter, the windings are calculated as follows: primaries, each 54 turns 18 gage; secondary, 588 turns 26 gage (for 250 v); and feedback windings, 10 turns each, 26 gage.

The input resistance of the 2N457 is about 10 ohms, and base resistors of 3.3 ohms give adequate current sharing.

The total effective r_b is now 13.3 ohms, $N = 5.4$, $R_L = 4.8$ ohms and β at least 30. This requires I_c of 55.5 ma, and $R = 6,500$ ohms. The experimental converter actually started with 8,200 ohms, showing that the β of one transistor was higher than supposed.

The mean R_{cs} dissipation at full load will be less than 2½ w per transistor, and a small heat sink about 4 or 5 square inches will allow operation in ambient temperatures up to about 65 C assuming a junction temperature of 80 C.

In neither design is smoothing shown, as ripple specifications may vary widely. As a rough guide, however, a single capacitor of 2 μf gave 4 v peak ripple (2 percent) on full load with the silicon converter, and necessitated changing the starting resistor to 3,300 ohms in order to start on full load. For the same percentage ripple, the germanium converter required 4 μf.

Although considerably higher power and efficiency can be obtained with germanium transistors than with silicon transistors at present available (at much lower ambient temperatures), it is anticipated that silicon transistors will soon be available with lower R_{cs}. The silicon transistor will then rival the germanium type.

Using selected 2N389 transistors having an R_{cs} of 2½ ohms, the circuit of Fig. 1 has given 30 w at 75 percent efficiency when operated from a 24-v supply.

Two 2N514B germanium transistors (25 amps, 80 v) have given 630 w at 90 percent efficiency. Thus the d-c converter can now meet any specification that previously required a vibrator power unit or rotary transformer.

Acknowledgement is made to the editor of Electronic and Radio Engineer, Iliffe and Sons, Ltd., London, England, for permission to reproduce some material.

BIBLIOGRAPHY
L. H. Light and P. M. Hooker, Transistor D-C Converters, *Proc IEE*, **102**, Part B, p 775, Nov. 1955.
J. L. Jensen, An Improved Square Wave Oscillator, *Trans IRE*, **CT-4**, 3, p 276, Sept. 1957.

Line Voltage Control Uses Zener Diodes

Compact five-transistor circuit uses breakdown diodes to regulate voltage inputs between 140 v and 113 v to within 0.5 volt of 110 v. Waveform distortion is no problem

By R. A. GREINER, Assistant Professor, College of Engineering, University of Wisconsin, Madison, Wis.

A SMALL, inexpensive, and light weight line voltage controller is described here. This controller is designed for use where the line voltage is always somewhat higher than the desired value. It provides regulation with some loss in voltage. In cases where the line voltage must be boosted, it is possible to use an autotransformer ahead of the regulator to increase the normal line voltage above the required regulation point at its lowest expected normal fluctuation.

Regulation

The regulator, shown in Fig. 1, was designed to deliver two amperes to the load. This capacity can be easily increased by the use of higher power transistors or by paralleling the output transistors.

The regulation for an input voltage range of from 140 v to 113 v is ± 0.5 volt at 110 volts. As can be seen from the circuit, the operation is relatively straight-forward. The output voltage is detected by the filament transformer and half wave rectifier combination. This signal is filtered and sent to a comparison amplifier where it is compared to a reference voltage established by the breakdown or Zener diode.

The signal from the comparison amplifier is then amplified by the 2N250 and sent to the control transistors. The impedance of the transistors is controlled by the current in the emitter-base loop which contains the 2N250 and the power supply.

Diode Protection

The diodes protect the transistors from large reverse bias on the emitters and forward bias on the collectors. A half-wave capacitor-filtered supply supplies the control currents. Again, breakdown diodes are used to regulate the supply voltage. Despite the simplicity of this regulator, it gives good results.

Waveform distortion was expected to be a problem with this type of regulator. It was found, however, that no distortion was observed on the oscilloscope. Since any small residual distortion was of no concern in the proposed application, this matter was not pursued further.

Filtering

It may be noted that in order to make a comparison between the alternating output voltage and a more convenient d-c reference voltage, it is necessary to rectify and filter the output signal. There are two limits to the amount of filtering which must be observed. In one case, that of inadequate filtering, ripple at the difference amplifier will overdrive the amplifier, reduce the effective gain, and give distortion in the regulated signal. In the second case, if filtering is too great, the response to sudden line voltage changes will be poor.

The 8-ohm shunt resistor may be adjusted to increase or decrease the range of regulation and the current capacity of the regulator over a considerable range. The resistor should be physically isolated from power transistors when line voltage is high.

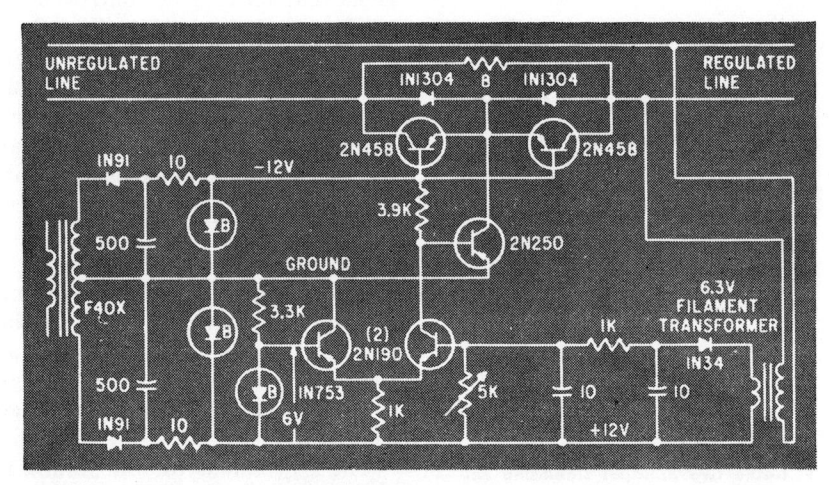

FIG. 1—Transistorized line voltage regulator can be modified for increased capacity

Designing Highly Stable Power Supplies

Summary of design techniques used in making extremely stable low-voltage power supplies for transistor circuits. Final circuit described has overall stability of plus or minus 250 microvolts and overall drift of less than 40 microvolts per hour

By E. BALDINGER and W. CZAJA, Institute of Applied Physics, University of Basle, Switzerland

CONVENTIONAL TRANSISTOR-STABILIZED power supplies use either shunt or series regulation. If the rectifiers and line transformer are suitably designed, the shunt-type circuit cannot be damaged by short circuits at the output terminals. The rectifier of the shunt-type circuit may be designed to have high efficiency and low cost since it operates at constant load. With no load, a high collector dissipation may occur within the shunt transistor. This may be prevented by using a suitable collector resistor which in no way influences the properties of the device. The low overall efficiency (approximately 40-percent) is a disadvantage.

In the series regulator, additional circuits must be used to prevent overload or damage to the series transistor. The control range at high voltage is limited by the maximum ratings of the series transistor as the absolute values of voltage variations increase with increasing input voltages. For a variable load, the rectifier internal resistance makes an important contribution to the input voltage variations that have to be controlled.

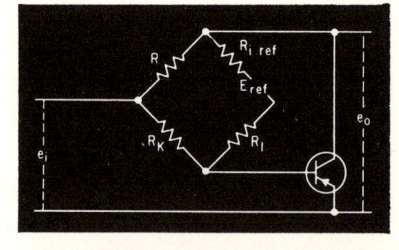

The circuits to be described show methods of improving circuit performance so that highly stable power supplies having an overall stability of ± 250 μv and overall drift of less than 40 μv/hour at outputs up to 17 volts can be designed.

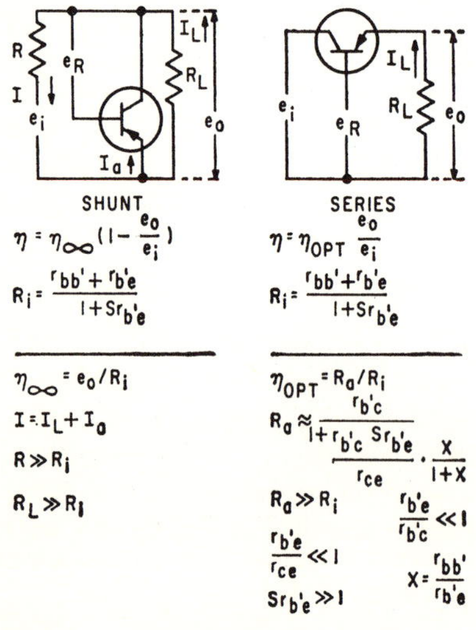

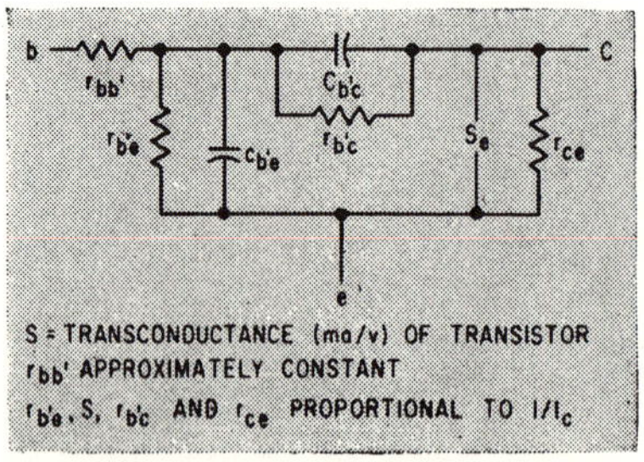

Basic Circuits

Basic series and shunt regulating circuits are shown in Fig. 1. The stabilization factor η is defined as the relative change in input voltage e_i to the change in output voltage e_o produced.

The expressions for the stabilization factor η and the internal re-

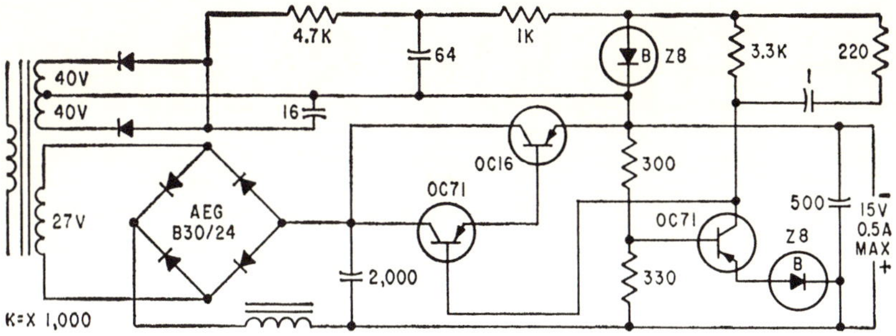

FIG. 4—Series stabilized power supply

FIG. 5—Shunt-type, short-circuit-proof, general-purpose power supply

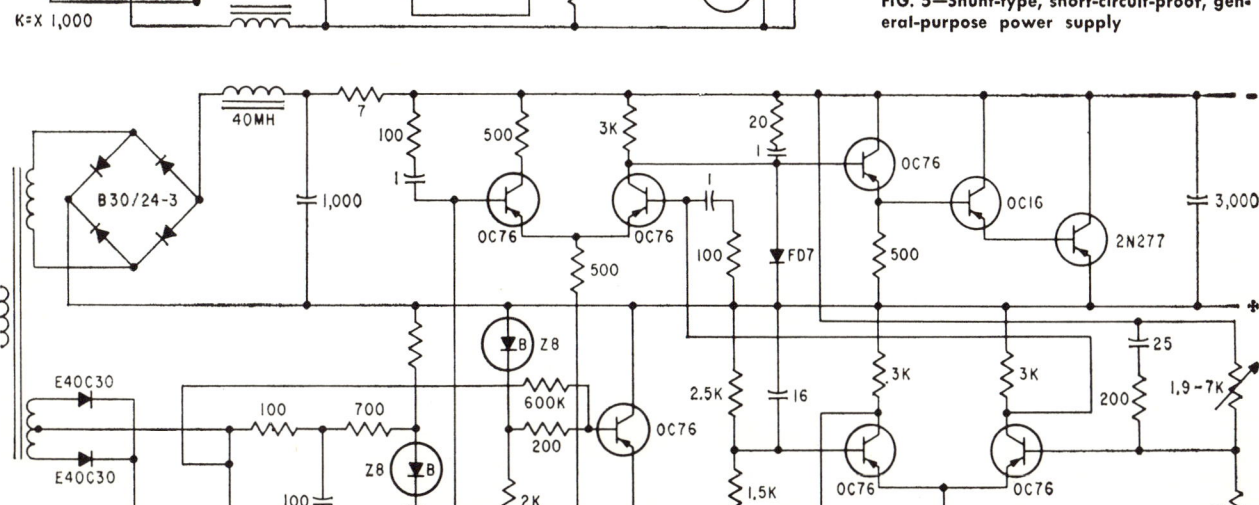

sistance R_i using small-signal parameters[1] are also shown in Fig. 1. The expressions are accurate under the following assumptions: small signal behavior of circuit; slow variations from equilibrium; and $r_{bb'}$ includes the internal resistance of the reference voltage source. Improved accuracy can be obtained by taking into consideration the approximate working point (collector current) dependence of the small signal parameters (see Fig. 2) and the internal resistance of the supply source. Modifications of the above expressions are published elsewhere[2].

Zener diodes may be used as reference elements instead of dry cells or similar devices. Cascading of an unstabilized power supply with a stabilized one may be used when high output voltages at high load currents are to be stabilized. The stability of cascaded supplies can be improved by driving the transformer of the unstabilized source with a servo motor to achieve optimum operating conditions for the stabilized supply.

The two basic means of improving the stabilization factor η or lowering the internal resistance R_i, or both, are by compensation or by preamplification.

Compensation

Figure 3 shows a simple method of improving the stabilization factor η of a shunt-type power supply. For exact compensation, the condition $R_{i\ ref} + R_1/R_i > 1$ must hold[2]. Exact compensation ($e_o = 0$, or $\eta = \infty$) is possible for only one load current value. The load current dependence of the output voltage may also be compensated, but this may be considered as compensation of R_i. Devices with a negative R_i can be realized this way.

Preamplification

By inserting a preamplifier with an amplification factor of A between the base of the transistor and the reference voltage shown in Fig. 1, both η and R_i can be improved. The stabilization factor η is multiplied by A and the internal resistance R_i is divided by A.

Stabilized power supplies with preamplifiers are complicated feedback systems and means must be provided to insure system stability for all load conditions. In most cases it is necessary to limit the frequency response by shunting one or more of the amplifier stages collector resistances with series R-C circuits and by employing a large output capacitor. Electrolytic capacitors are used because their high losses at high frequencies produce a suitable damping of unwanted oscillations produced by abrupt changes in the load.

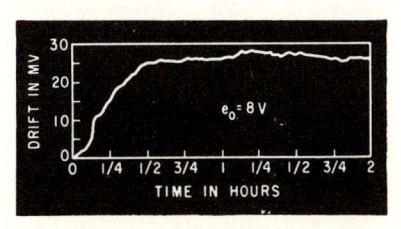

Comparison

The circuit with compensation is assumed to have a stabilization factor η_1 and the one with preamplification is assumed to have a stabilization factor η_2 equal to η_1. With $R_i/R \ll 1$ and $R_i/R_k \ll 1$, then the variations with load current of η_1 and η_2 are expressed as:

FIG. 6—Output voltage drift for the first two hours of operation for circuit of Fig. 5

$$\left.\frac{\delta \eta_1}{\delta I_L} \right| \frac{\delta \eta_2}{\delta I_L} = A\left(1 + \frac{R}{R_K}\right)$$

This equation shows that η_2 is about a factor A more insensitive to load current variations than η_1. This, and the fact that in compensated stabilizing circuits η is sensitive to transistor aging, show that a circuit should only be compensated in such a way as to improve η by about a factor of five.

Temperature

The temperature coefficient of the output voltage of the two circuits shown in Fig. 1 is small. The emitter junction voltage for constant collector current varies linearly with the absolute temperature with a slope of about −2 to −3 mv/degree C. This slope is a property only of the bulk material and depends strongly on the doping ratio. In the case where preamplification is used to improve η and R_i, the temperature coefficient is mainly determined by the temperature coefficient of the amplifier and of the reference voltage. The contribution of the amplifier is kept small, especially when symmetrical amplifier stages are used.

Stabilized Supplies

The series stabilizing circuit shown in Fig. 4 is of the Middlebroock type[5]. This supply is designed for constant output voltage

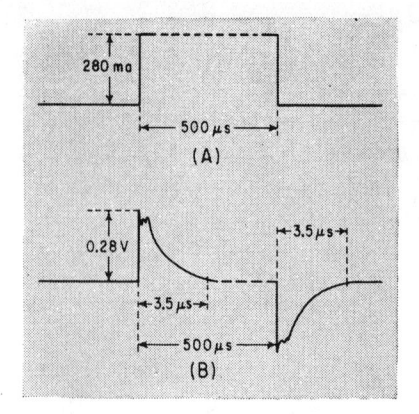

FIG. 7—When a current step (A) is applied to the output terminals of circuit of Fig. 5, the response is as shown in (B)

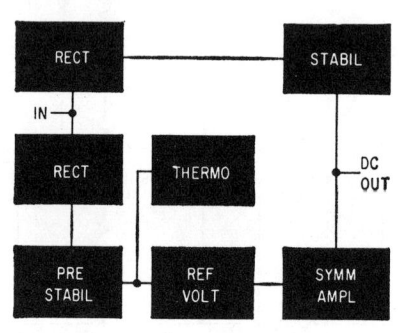

FIG. 8—Improved power supply showing temperature-controller coupling

and moderately variable load. This supply has an output of 15 v, an internal resistance of 0.1 ohm, a temperature coefficient of 1 mv/degree C, a ripple voltage of 4 mv peak to peak and a maximum current of 0.5 amp.

A short-circuit-proof, general-purpose power supply is shown in Fig. 5. This supply has an output variable from 1 to 17 v, an internal resistance of 0.0005 ohm, a ripple voltage of 1 mv peak to peak and a maximum current of 2.5 amp at 1 v and 0.8 amp at 17 v. After the first two hours of operation the output drift is negligible as shown in Fig. 6. One unit was initially set at 10 v and after one year of continuous operation had a drift of 130 mv. Figure 7 shows the output after the application of a square wave.

The limiting factors for η and R_i are variation of the reference voltage and temperature variations within the device. The circuit of Fig. 5 can be improved by improving the reference source and using a transistorized thermostat for controlling the reference voltage source and preamplifier. A block diagram is shown in Fig. 8.

Temperature Control

The temperature dependence of the operating point of a low-power transistor is used to drive a power transistor which in turn controls the temperature controller[4] shown in the improved stabilization circuit of Fig. 9. Fixing the base voltage of temperature-sensing transistor Q_1 produces a more stable circuit over long periods of time than fixing the base current. A low-imped-

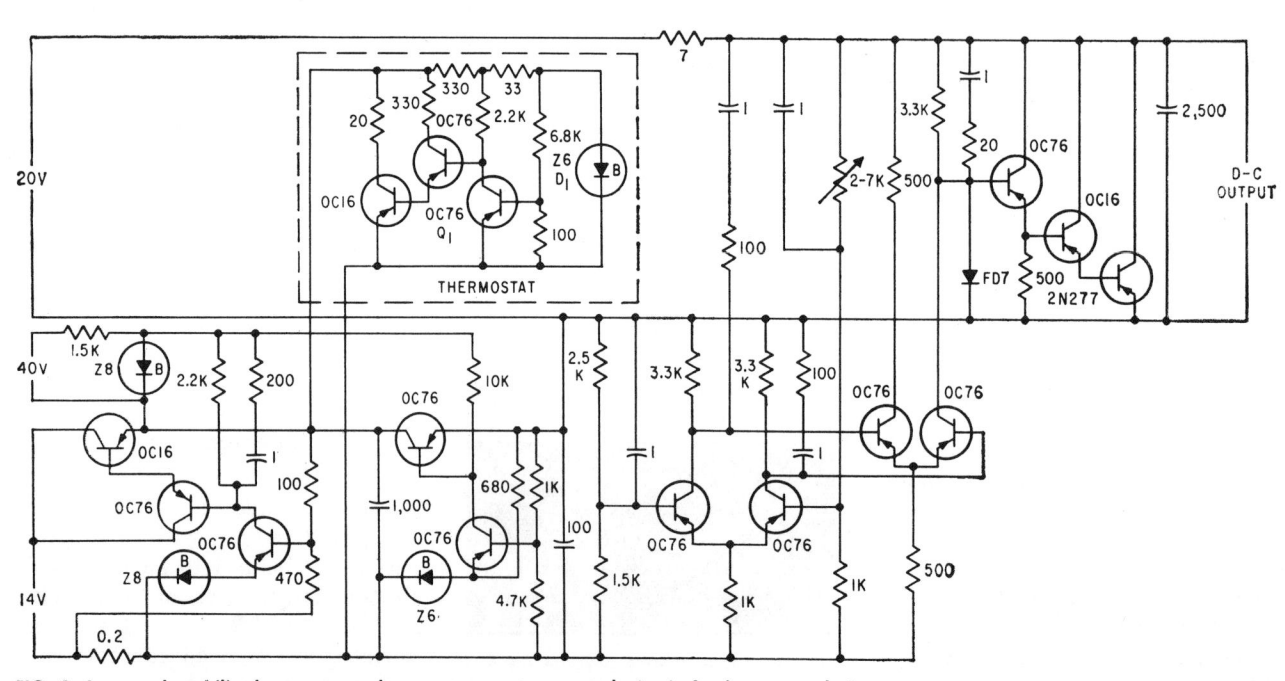

FIG. 9—Improved stabilized power supply uses temperature-control circuit for better regulation

ance input was provided for the first stage. Zener diode D_1, together with compensation, prevents variation of the controlled chamber temperature due to variations of supply voltages. The internal temperature may be varied by changing the operating point of the temperature-sensing transistor. The temperature control device used in Fig. 9 has a controlled temperature of approximately 40 degrees C. A temperature stabilization ratio of $\Delta T_{ambient}/\Delta T_{inside} \approx 40$ to 50 has been achieved. The overall drift of the controlled temperature of 40 C over a period of one year is 0.4 C. Conventional carbon resistors were used and no selection was made of the transistor. The circuit shown in Fig. 9 has a variable output between 1 and 17 v, an internal resistance of 1.7×10^{-1} ohm at d-c and less than 0.05 ohm at 500 kc, a temperature coefficient of 0.1 mv/degree C and a ripple voltage of 0.1 mv peak to peak maximum.

After turning on, 4 to 5 hours are required until equilibrium temperature is reached. From that time, the overall drift over a 15 hour period was 36 μv/hour at an output of 6 v.

The circuit shown in Fig. 9 is basically the same as that shown in Fig. 5 with a different reference source. A prestabilizer controls the temperature controller as well as a stabilizing circuit which serves as the voltage reference source.

Characteristics

Figure 10 shows the power supply output voltage plotted against input line voltage changes and loading. The major part of the short-term fluctuations is due to small temperature differences (0.05 C or less) within the thermostat.

Figure 11 shows the stabilized power supply compared with a lead-acid storage battery. Both the power supply and the battery were loaded. The long time period which the battery needs to establish equilibrium after load removal is typical.

The lead-acid storage battery has a time contant of about $\frac{1}{4}$ to $\frac{1}{2}$ hour resulting from the time required for diffusion of the chemical reaction products between the plates until equilibrium is reached.

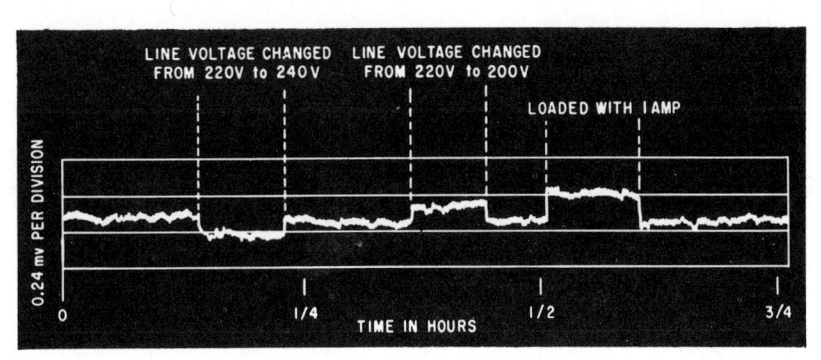

FIG. 10—Output characteristics of improved stabilized power supply

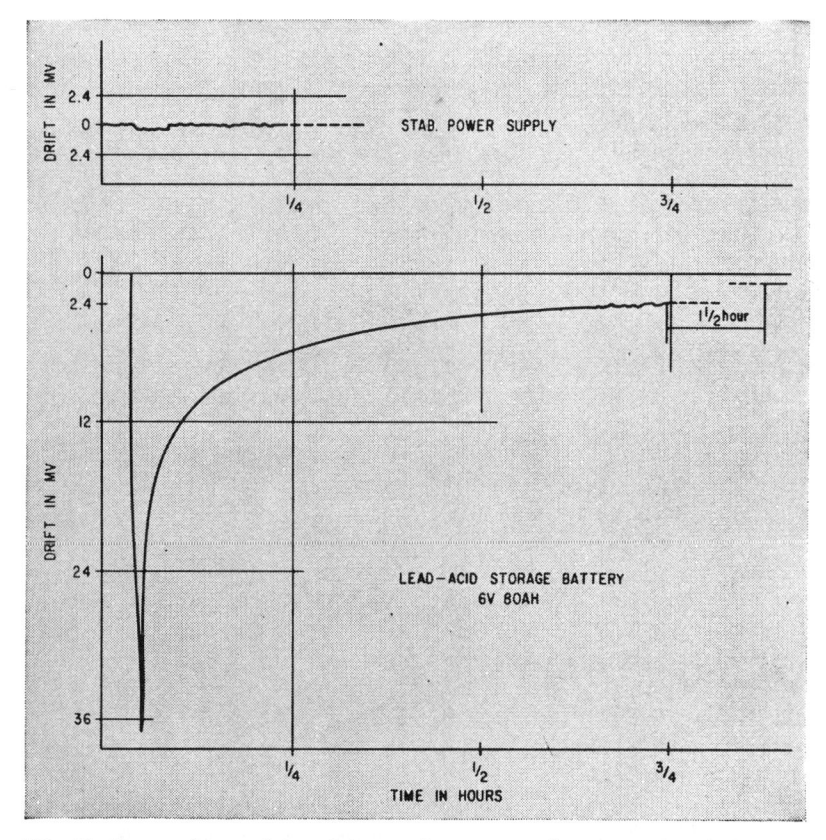

FIG. 11—Output characteristics of improved power supply compared with lead-acid battery (both 6 v)

During the preparation of this article, an American voltage reference (see last entry in Bibliography) was called to the authors attention.

This device has excellent stability and temperature coefficient, the output voltage is practically constant but the output current is much smaller having a maximum of 100 milliamperes.

There are a number of applications for this highly-stable power supply where up to the present, only lead-acid batteries connected in parallel with a power supply of poor regulation have been used.

Further consideration of the above described circuits show that this highly-stable power supply can be improved still further.

REFERENCES
(1) L. J. Giacoletto, RCA Rev. 15, p 506, 1954; J. Zawels, J. Appl. Phys. 25, p 976, 1954; E. Baldinger, W. Czaja and M. A. Nicolet, Zeitschrift fur Angewandte Mathematik und Physik (ZAMP), 7, p 357, 1956.
(2) E. Baldinger and W. Czaja, ZAMP, 9, p 1, 1958.
(3) R. D. Middlebroock, Proc IRE, 45, p 1502, 1957.
(4) E. Baldinger and A. Maier, ZAMP, 9, p 289, 1958.

BIBLIOGRAPHY
M. E. Pembeton, Marconi Rev. 20, 125, p 39, 1957.
W. W. Gartner, Proc IRE, 45, p 665, 1957.
W. Guggenbuel and B. Schnieder, Archiv fur Elektishe Uebertragung, 10, p 361, 1956.
Rev. Scientific Inst., 30, p 211, 1959.

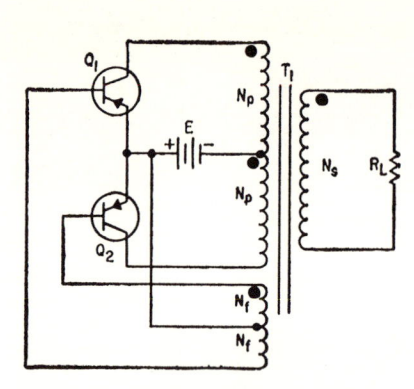

FIG. 1—Simplified schematic of oscillator-type power supply

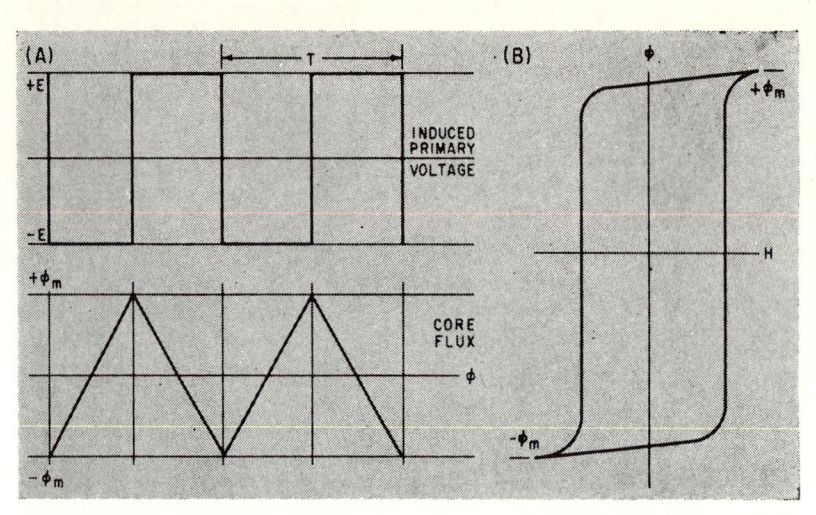

FIG. 2—Idealized waveshapes of primary flux and voltage in the transformer (A) and dynamic B-H hysteresis loop of transformer core (B)

Equations for

WITH THE DEVELOPMENT of high-current switching transistors, the d-c to d-c power supply has a new look. Transistorized power supplies are being utilized in applications that require small size, light weight and high efficiencies. Efficiencies ranging from 80 to 90 percent can be obtained. In addition, transistor power supplies contain no moving parts. A design procedure for engineering such a d-c to d-c power supply to meet specified operating characteristics is described in this article.

Operation

Figure 1 shows a simplified circuit of the oscillator-type power supply to be discussed. The transistors act as an ON-OFF switch to obtain a square wave a-c output voltage from a d-c source. This square wave can be stepped up or down by transformer T_1 for a desired output level and rectified to give a d-c output voltage. The transistors are switched ON or OFF by a feedback winding on the transformer, one transistor going ON (conducting) while the other transistor goes OFF.

Assume that the windings are phased as shown. If Q_1 is switched ON (conducting), Q_2 is OFF and the battery voltage E is applied to the top half of the primary, inducing a constant $d\phi/dt$ in the core (Fig. 2). When the core saturates, $d\phi/dt$ falls to zero. From Faraday's Law, $e = -N(d\phi/dt)$. Therefore, with $d\phi/dt = 0$, the induced feedback voltage must also be zero. At this point, with no base drive on either transistor, both transistors are OFF. There is no flux-producing current in either primary and the field now collapses, changing the direction of $d\phi/dt$.

This reversal of $d\phi/dt$ induces a voltage of opposite polarity in the feedback windings of the transformer, thus turning Q_2 ON and keeping Q_1 OFF. Now $d\phi/dt$ is again constant in the negative direction until the core saturates, thus giving one cycle of operation.

From the plot shown in Fig. 2A it can be seen that the flux ϕ has an excursion of $+\phi_m$ to $-\phi_m$, where ϕ_m is the saturation flux. This change of flux occurs in a half cycle or $T/2$ seconds, where T is equal to the period of one cycle. In $T/2$ seconds, the flux changes from $+\phi_m$ to $-\phi_m$ or a total change of $2\phi_m$. The average induced voltage is equal to the total change of flux divided by the time of the flux change. That is:

$$E = \frac{N_p 2\phi_m}{T/2} \times 10^{-8} \qquad (1)$$

also

$$T = 1/f \qquad (2)$$

and

$$\phi = BA. \qquad (3)$$

Combining these equations,

$$E = 4fN_p A B_m \times 10^{-8} \qquad (4)$$

where E is in volts, f is frequency in cps, N_p is number of primary turns or $\frac{1}{2}$ the total primary-winding turns, A is area of core in in.^{2}, B_m is saturation flux density in lines/in.^{2}

Core Selection

A core that has a high saturation flux and a low magnetization current has the ideal characteristics for transformer T_1. A square hysteresis loop is not required. There are many types of core materials on the market that will satisfy these requirements.

If the exact B-H hysteresis loop is not given for the core by the manufacturer, one can easily obtain this characteristic of the core material. To do so, one must measure and indicate both the flux and the magnetizing force simultaneously (see Fig. 2B) and record both at once on an cro screen. This can be done by the test circuit shown in Fig. 3, using the following relations: magnetizing force, H, in oersteds, is

$$H = \partial F/\partial L_f \cong F/L_f, \qquad (5)$$

where L_f is the mean length of magnetic path; magnetomotive force, F, in gilberts is

110

Here is a lucid development of basic design equations for a transistor d-c supply. Using power transistors, this type of supply can result in efficiencies from 80 to 90 percent. The design procedure is explained in detail

By **THEODORE HAMM, JR.**, Lawrence Radiation Laboratory, University of California, Livermore, California

Power-Supply Design

$$F = 0.4\pi i N. \quad (6)$$

Combining these equations,

$$H = 0.4\pi N i / L_f.$$

From Fig. 3, $i_p = e_1/R_1$. Instantaneous magnetizing force

$$H = 0.4\pi N_p e_1 / R_1 L_f.$$

Therefore, a horizontal deflection proportional to H can be presented on a cro screen by applying e_1 to its horizontal deflection plates.

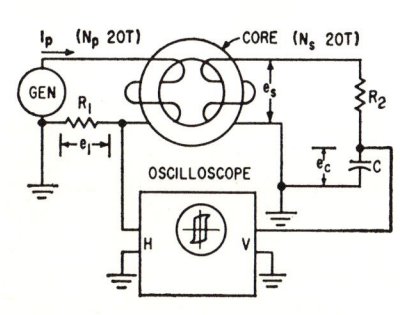

FIG. 3—Test circuit for measuring core flux density B against magnetizing force

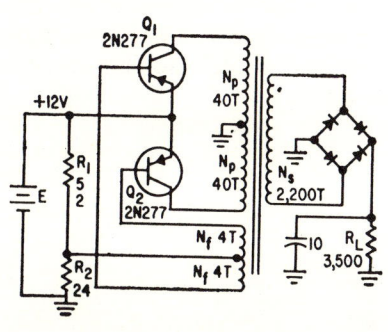

FIG. 4—Oscillator-type power supply with all circuit values shown

Since ϕ can be obtained as a voltage, it can be viewed vertically on the cro by using an RC integrating circuit with $R_2 \gg 1/\omega C$. This voltage can be applied to the vertical deflection plates of a cro and the voltage corresponding to $H(e_1)$ can be applied simultaneously to the horizontal plates to produce the B-H curve for the core material on the cro screen.

To obtain a voltage proportional to ϕ we integrate Faraday's Law:

$$\phi = (1/N) \int e_s \, dt.$$

The voltage across the capacitor C in Fig. 3 is

$$e_c = (1/C) \int i_s \, dt.$$

As the ratio of R_2 to $1/\omega C$ is 200:1 at the test frequency,

$$e_s \approx i_s R_2.$$

Substituting,

$$e_c = 1/R_2 C \int e_s \, dt. \quad (7)$$

From Faraday's Law,

$$e_s = (N_s/10^8) \times d\phi/dt, \quad (8)$$

where N_s is the number of secondary terms.

Combining the last two equations and integrating over one cycle,

$$2e_c = (N_s/10^8 R_2 C) \int_{-\phi_m}^{+\phi_m} d\phi,$$

and

$$2e_c = (N_s/10^8 R_2 C) \times 2\phi_m.$$

Substituting for ϕ_m the peak-to-peak voltage $2e_c$ is

$$2e_c = 2N_s B_m A_e / 10^8 R_2 C,$$

where A_e = effective area of core in in^2.

And

$$A_e = A \times S, \quad (9)$$

where S is the stacking factor.

The instantaneous voltage drop e_c across capacitor C is directly proportional to the instantaneous flux density in the core. Thus,

$$B_m = \frac{e_c R_2 C \times 10^8}{N_s A_e}$$

B_m can be measured at any frequency, but H_m must be measured at the frequency at which the core is to be operated. The higher the frequency, the wider the hysteresis loop and the higher the core losses. For testing, the core N_p and N_s can have a 1:1 ratio with about 20 turns. This is not critical.

Transformer Design

From Eq. 4 we have

$$N_p = E \times 10^8 / 4f A_e B_m, \quad (10)$$

where E is the battery voltage (Fig. 1). A good choice of frequency is 1,000 cps. The higher the frequency, the higher the core losses. The lower the frequency, the more turns required and the higher the copper losses. Saturation flux density is given by the manufacturer or by the method described above, using the oscilloscope B-H presentation.

Here are sample calculations on

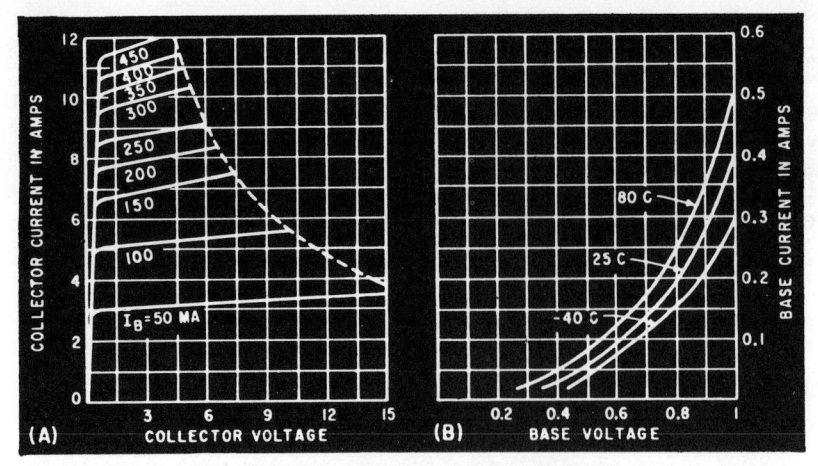

FIG. 5—These common-emitter characteristics of the 2N277 show output curves at 25 C (A) and input curves at several temperatures (B)

the transformer design using an Arnold Deltamax toroidal tape core 3T4178-D1. Let $B_m = 80,000$ lines/in^2 (measured by above method), $A_e = 0.125 \times 0.75 = 0.094$ in^2, $E = 12$ v, $E_s = 600$ v, $I_s = 170$ ma, $R_L = 3,500$ ohms, $P_o = 102$ w, and $f = 1,000$ cps. Assume a transformer efficiency of 90 percent, or Eff $= 0.9$. These values are substituted into the primary-turns equation

$$N_p = 12 \times 10^8 / 4 \times 10^3 \times 0.094 \times 8 \times 10^4$$
$$= 40 \text{ turns.}$$

N_s, the secondary turns, can be found by

$$N_s = N_p(E_s/E) \times 1/\text{Eff} = 2,200 \text{ turns.}$$

R_e, the equivalent transistor load resistance, is

$$R_e = (N_p/N_s)^2 R_L = 1.15 \text{ ohms.} \quad (11)$$

Transistor Selection

When the transistor is in saturated operation, parameter variations are held to a minimum and transistor heat dissipation is at a minimum.

Germanium power transistors are good for this application below temperatures of 80 C because of their low collector saturation resistance and low cost. Germanium power transistors have collector saturation resistances of less than 1 ohm, whereas typical silicon power transistors have collector saturation resistances values of 5 ohms.

I_p, the transistor switching current, is found by

$$I_p = E/R_e$$

to be equal to 10.5 amp.

A common emitter configuration is chosen because larger input powers are necessary for the common collector and common base configurations, thus giving lower transformer efficiencies. A 2N277 transistor, which is low priced, was chosen for this application (Fig. 4). The base current is determined by the collector EI characteristics shown in Fig. 5A. For the transistor to operate at saturation with a collector current of 10.5 amp, the base bias must be at least 400 ma. To insure saturation the base should be biased to 500 ma or more.

R_{cs}, the collector saturation resistance, is low, and R_1 should be $\gg R_{cs}$. Therefore, let R_1 be 5 ohms.

As

$$R_2 = E/I_b,$$

where I_b is the transistor base current,

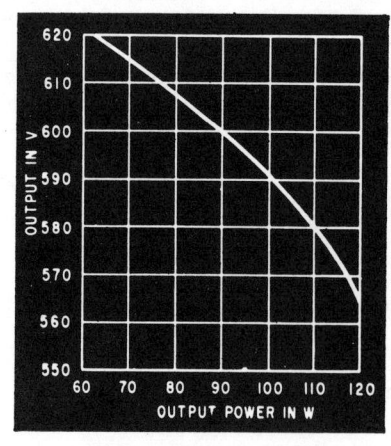

$$R_2 = 24 \text{ ohms.}$$

FIG. 6—Output characteristics of the supply for different loads

The base drive is taken from the base EI characteristics (Fig. 5B). For a base current of 500 ma, approximately 1 volt of base drive E_b is needed. Feedback turns N_f is found from

$$N_f = N_p(E_b/E) \times 1/\text{Eff}$$
$$= 40 \times 1/12 \times 1/0.9 = 3.7 \text{ turns};$$

4 turns were used.

Transformer Winding

The primary was wound on the core by hand, then the core was wrapped with glass tape, dipped in clear Glyptal and baked in an oven. The secondary was wound by a toroid-coil-winding machine. Then the core was again wrapped with glass tape, dipped in Glyptal and baked. Finally, feedback turns were wound, and the core wrapped with tape.

Wire sizes were based on 700 circular mils per amp. Winding wire sizes are: primary—no. 12, secondary—no. 29, feedback—no. 24.

Performance

If the power supply does not oscillate after construction, switch the feedback leads to the transistors. The measured d-c output voltage E_o was 590 v for $R_L = 3,500$ ohms. Due to the high frequency used, the d-c output voltage is easy to filter. With a 10-μf capacitor as a filter, the ripple was 0.2 percent.

Output power vs output voltage was measured for different loads (Fig. 6). For a ± 20-percent change in load there was only an 8-percent change in output voltage. Regulation can be improved by using different feedback turns ratios. This power supply has an over-all efficiency of 80 percent. If greater efficiency is needed, one should choose a core with smaller magnetizing force.

If the design of the power supply is to be held to close specifications, the saturation flux should be measured by the oscilloscope method as described above.

BIBLIOGRAPHY

E. Baldinger and W. Czafa, Designing Stable Power Supplies, ELECTRONICS, Sept. 1959.
Edward Gordy and Peter Haspenpusch, New-Design Transistor Regulated Power Supply, ELECTRONICS, Oct. 9, 1959.
L. P. Hunter, "Handbook of Semiconductor Electronics", p 13-26 to 13-28, McGraw-Hill Book Co., Inc., New York, 1956.
D. L. Waidelich, Diode Rectifying Circuits with Capacitance Filters. Trans AIEE, 60, p 1, 161 to 1,167, 1941.
O. H. Schade, Analysis of Rectifier Operation, Proc IRE, 31, p 341 to 361, July 1943.

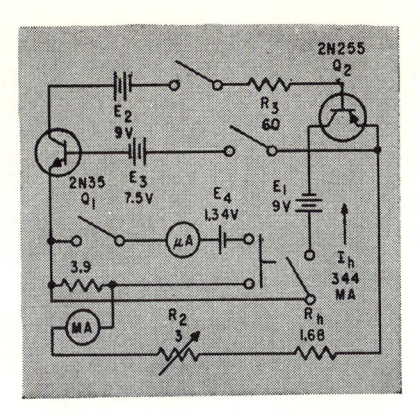

FIG. 1—Circuit used in moisture meter stabilizes heater current

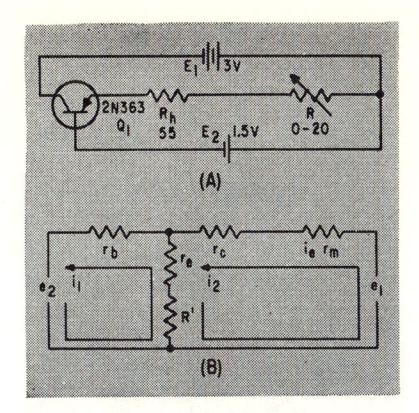

FIG. 2—Single transistor control circuit demonstrates circuit functions

Inverse Feedback Stabilizes Dry-Cell Current Sources

Errors in test instruments that draw heavy current from dry cells are reduced with transistor circuit that maintains constant current

By G. E. FASCHING, Bureau of Mines, U. S. Department of the Interior, Morgantown, West Virginia

HIGH PERCENTAGE of negative feedback is used in transistor circuit to offset effects on dry cells of heavy current loading and consequent voltage decay.

Moisture Meter

Meter for measuring moisture in pulverized coal or other powders required constant current supply from conventional zinc-carbon cells subjected to heavy loading. In the meter, six standard cells in series provide 344 ma to a 2-watt heater. Without control, current variation with time over the 20-minute test periods introduced errors.

The transistor version of a cathode follower in Fig. 1 provides direct compensation for limited battery voltage decay. A correction voltage, about equal to change in battery voltage but opposite in polarity, appears at the output of Q_2. The correction voltage is obtained by using inverse voltage feedback in a two-stage, direct-coupled amplifier. Voltages E_1, E_2 and E_3 are bias voltages; E_4 is a mercury cell used in a voltage null detector circuit for adjusting heater current I_h through heater R_h at the beginning of each test.

Adjustment of heater current at the start of each test is done with R_2, and R_3 provides enough voltage drop to prevent Q_1 from overheating.

For analysis, a simplified version of the circuit is shown in Fig. 2. It is accompanied by the a-c equivalent when small-signal T-parameters are used. Use of these is limited to operation within the linear region or a small region of the transistor characteristics.

Analysis

To determine heater current change, i_h, that would result from voltage decay of E_1 and E_2, it was found that for a voltage change e_1 in heater source E_1: $i_{h1} = i_2 - i_1 = e_1 r_b / [(R' + r_e + r_c - r_m) r_b + (R' + r_e) r_c]$. For a voltage change e_2 in base bias E_2: $i_{h2} = i_2 - i_1 = e_2 r_c / [(R' + r_e + r_c - r_m) r_b + (R' + r_e) r_c]$.

Upon inserting values of typical junction transistor parameters and external circuit constants into these equations, it was found that a unit change in E_1 causes only a small change in heater current because of the large dynamic resistance. This resistance, being in the order of megohms, permits heater-current deviation of only microampere magnitude coincident with a unit volt change in the heater supply.

However, for voltage changes in base bias E_2, dynamic resistance is relatively small and significant heater current deviation occurs for a unit decay of base bias. This shortcoming is partly offset by the lower discharge (hence decay rate) of the source. Bias current relative to load current is dictated by the inherent forward d-c transfer ratio of the transistor used (normally 40-150).

Performance

The circuit in Fig. 1 has high current stability in the 20-minute test period. Heater current varies less than 150 μa from the preset value of 344 ma. Without the control and under similar conditions, heater current deviation ranged from 8,000 to 15,000 μa. Stability improvement factor is over 50.

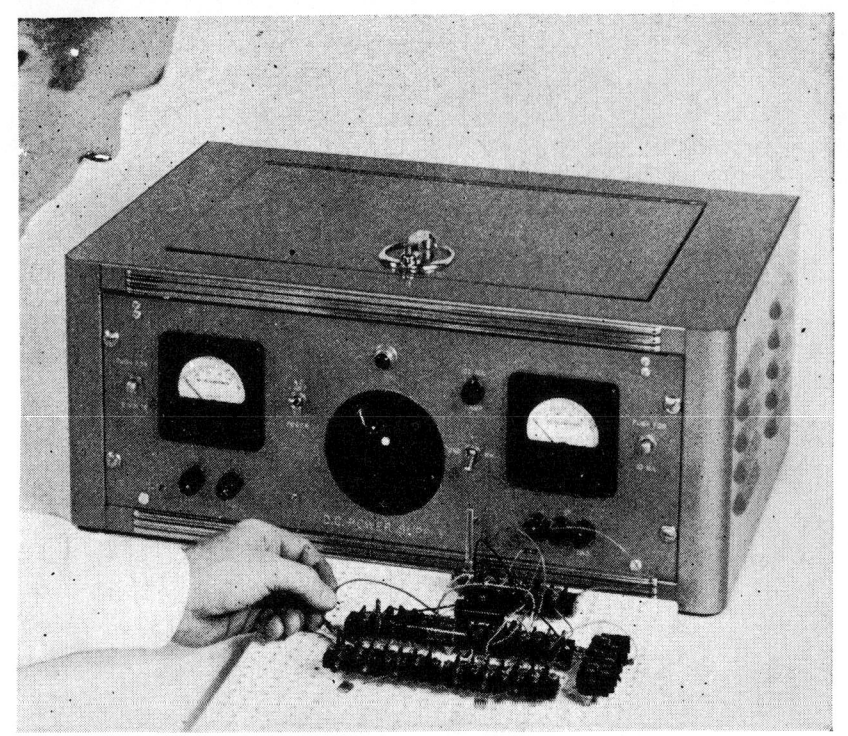

Experimental circuit is tested, using the closely regulated supply as a reliable source of constant voltage power

By EDWIN GORDY and PETER HASENPUSCH
Instrument Design and Development Dept.,
Roswell Park Memorial Institute, Buffalo, New York

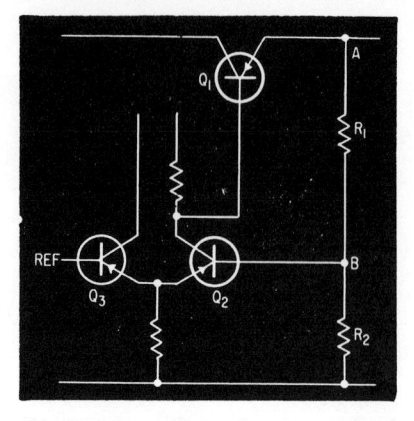

FIG. 1.—Error voltage of series-regulated power supply appears at point B attenuated by the R_1R_2 network

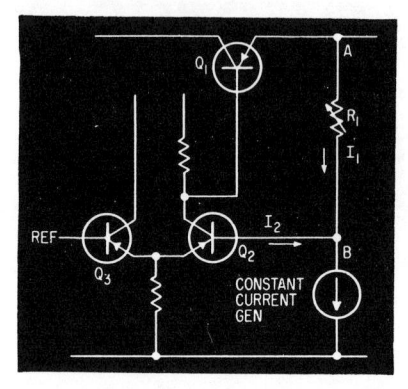

FIG. 2—Constant current circuit allows total error to appear at B

Constant-Current-Coupled

By driving a constant current through a fixed resistance, the total error voltage of a power supply can be fed into an error-correcting amplifier

SERIES REGULATED power supplies operate on the principle of using an error signal to control the series regulating device. If the gain of the control loop can be increased, the load regulating characteristics of the power supply can be improved. A similar improvement in regulation can also be obtained if the error voltage is sampled without attenuation by the customary resistive divider network.

The basic circuit of a conventional series regulated, transistor power supply is shown in Fig. 1. Transistor Q_1 is the series regulator and point B the input to the error correcting loop. The desired or reference voltage is set in at Q_3.

The error signal appears at point B and is the voltage at point A attenuated by the factor $R_2/(R_1 + R_2)$.

New Approach

Some factors which tend to lower the loop gain of the error correcting amplifier are reduced by a special circuit[1]. In addition, it is possible to lower the output impedance of the regulated supply by feeding the unattenuated output error voltage directly into a differential amplifier which is operated from two gas tube regulated auxiliary supplies[2].

A design using a new principle to obtain an unattenuated error signal is shown in Fig 2. A constant current generator feeds resistor R_1. For a fixed value of R_1, there will exist a constant voltage across this resistor equal to I_1R_1. The resistor with its constant current may then be considered the equivalent of a battery. Therefore, any error voltage existing at point A will be transferred, without attenuation, to point B, the input of the error-correcting amplifier. By making either I_1 or R_1 variable, the output voltage can be varied.

The ideal condition of constant current through R_1 is modified in practice by the finite current drawn by the base circuit of the input transistor of the error-correcting amplifier. The loading effect is

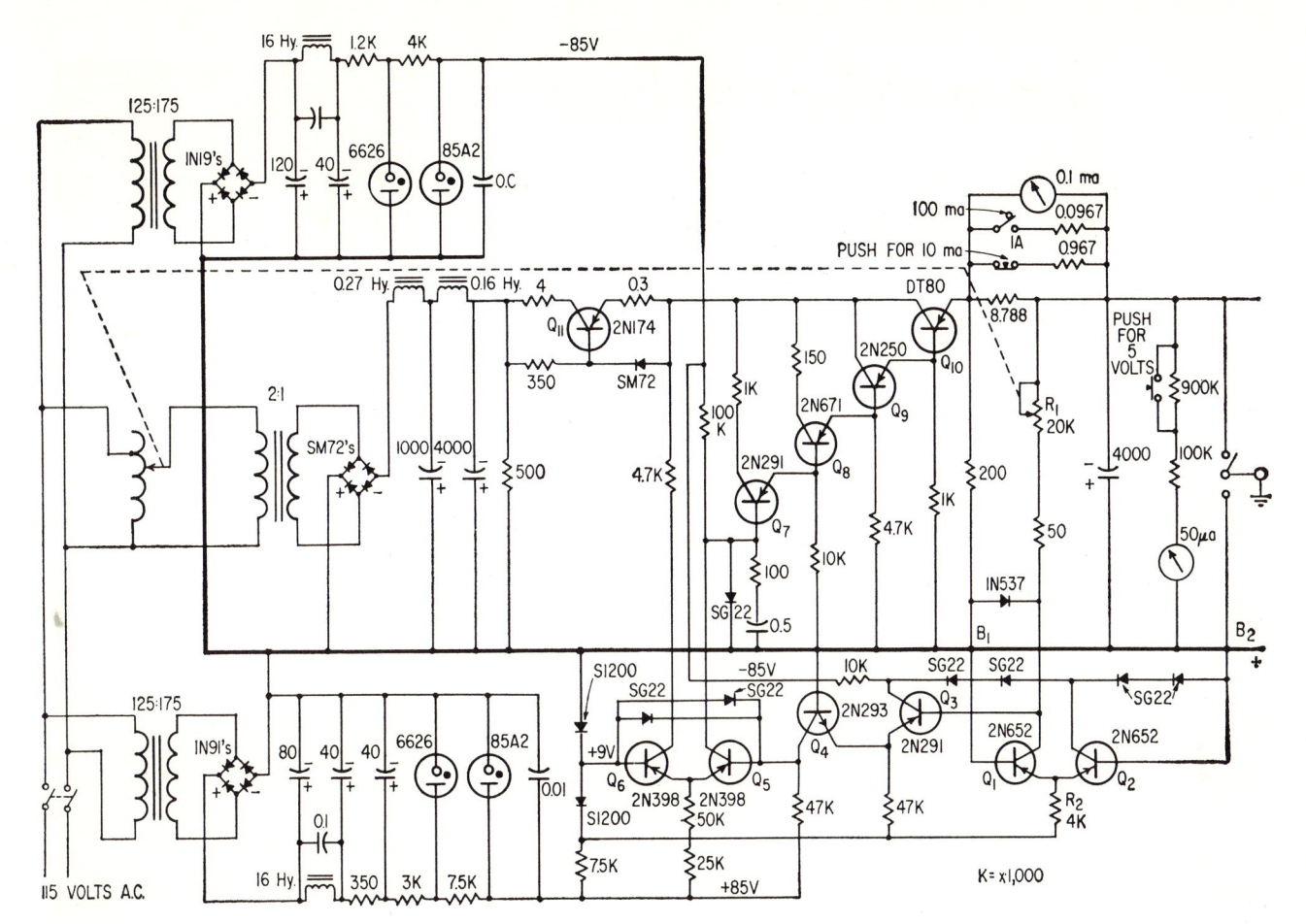

FIG. 3.—Regulated power supply uses two gas tube regulated auxiliary supplies to obtain performance. With a load current of 900 ma, drift in 30 minutes is about 5 mv

Power Supply

minimized by keeping constant current I_1 much larger than the base current of the input transistor.

Constant current coupling as used in this power supply is not unusual in d-c amplifiers designed for neurophysiologic use[3]. The application of this circuit to a regulated power supply appears to be a new approach to the problem.

A transistor operated in the common-base configuration has a high output resistance and can be used satisfactorily as a constant-current source. In the practical operational circuit of Fig. 3, Q_1 is the constant-current source and Q_2 is used as thermal compensation for Q_1. Both Q_1 and Q_2 are fastened to a common heat sink to keep them at the same temperature.

The points B_1 and B_2 on the B+ bus must be kept close together physically. Otherwise, the IR drop along as little as ¼ in. of solid number 12 copper wire is enough to cause a rise in output voltage as the load current increases.

The use of auxiliary positive and negative supplies follows from reference 2. The 6626 regulator tube is used in preference to an OA2 since it has better constant-voltage characteristics. The supply shown at the lower left of the schematic uses S1200 thermally compensated Zener diodes to obtain a drift-free reference voltage.

Performance

Regulation of the supply from zero load to 900 ma is 30 mv at 36 volts output. Ripple is less than 2.5 mv peak-to-peak. Line changes up to 10 percent do not affect the output.

Drift of the supply is less than 5 mv after 30 minutes operation at 750 ma. But temperature changes have a relatively large effect on drift. Without a one inch blanket of acoustic grade fiberglass over Q_1 through Q_6 transistors and circuits, drift was approximately 40 mv in 15 minutes. With the chassis removed from the cabinet and no thermal blanket, drift was 100 mv over a few minutes. Thus a 20 to one improvement in drift was obtained by relatively simple means.

The complete supply is shown in the photograph where it is being used to check an experimental circuit.

REFERENCES

(1) R. D. Middlebrook, Design of Transistor Regulated Power Supplies, *Proc IRE*, **45**, p 1502, Nov. 1957.
(2) T. H. Brown and W. L. Stephenson, A Stabilized D-C Power Supply Using Transistors, *Electronic Engineering*, **30**, p 425, Sept. 1957.
(3) J. Y. Lettvin, personal communication.

Inverters for Fluorescent Lamps

Inverter working at 1,250 cps permits 40-watt fluorescent light to operate from 24-volt d-c battery. Transistorized equipment has been installed in British railway coaches

By L. J. GARDNER, Lighting Equipment Engineer, A.E.I. Lamp and Lighting Co., Ltd., Leicester, England

CURRENT SOURCE for lighting British railway coaches is a 24-volt d-c battery. To provide fluorescent lighting from this supply, the current must be converted to a-c so that a step-up transformer can be used.

A small transistorized inverter, working at a frequency of 1,250 cps, has been devised for this purpose, and the first railcars embodying the new equipment have recently been completed. Either one four-foot 40-watt lamp or two two-foot 20-watt lamps can be run off each inverter.

Circuit Operation

Basic circuit for the inverter (shown in Fig. 1A) contains two *pnp* transistors (Q_1 and Q_2) and a transformer with three windings—primary W_1, feedback W_2 and secondary W_3. Capacitor C is included to allow self-starting and resistor R controls the value of feedback current. The circuit operates as follows. When the supply switch, S, is closed, the capacitor charges by way of the emitter and base of Q_1. This current, flowing through the base and emitter, reduces the emitter-collector impedance (which is normally very high) and current begins to flow into the lower half of the primary winding, W_1. Thus, in effect, the charging current into the capacitor switches on transistor Q_1.

While the current flowing in W_1 produces magnetic flux in the transformer core and voltages are induced in all the windings, the polarity of feedback winding W_2 is such that current passes through the emitter and base of Q_1, and Q_2 remains cut off since the voltage across its base and emitter is in the reverse direction. In these circumstances, the collector-emitter impedance is very low and the lower half of the primary winding is virtually connected directly across the supply voltage. The rise in current in this part of the winding is, therefore, rapid and is limited only by the inductance of the transformer. While the current is increasing, magnetic flux in the core is also increasing and the change of flux maintains the voltages on all windings.

Once the current exceeds the critical value, the collector-emitter impedance begins to rise rapidly and, consequently, the rate of increase of both the current and the flux is reduced, as are the voltage induced in the feedback winding and the current fed into the base of Q_1. With the decrease in the collector current brought about by the reduction in the current to the base of Q_1, the magnetic flux in the core begins to decline and the voltages across all the windings are reversed. This reversal of feedback voltage switches off Q_1 and switches

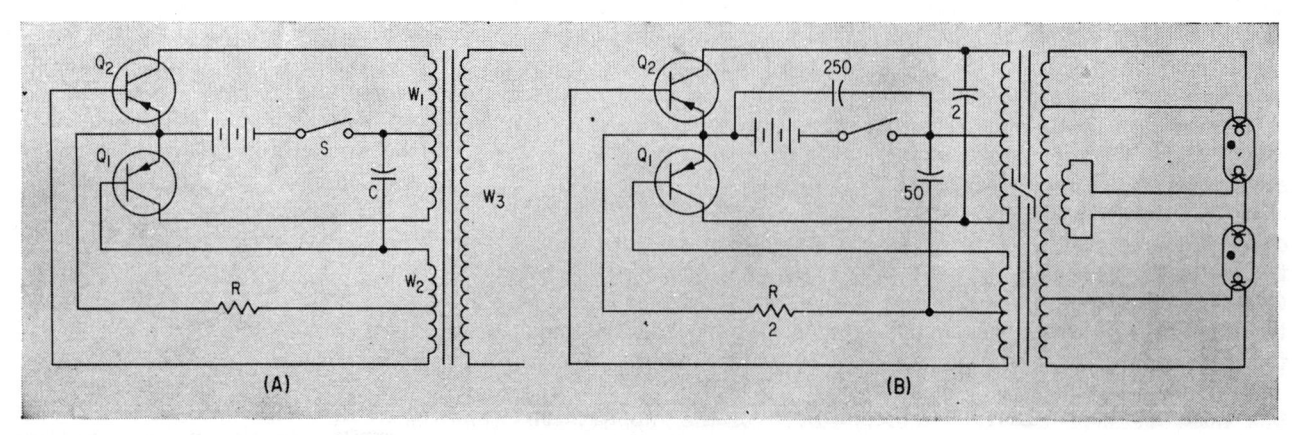

FIG. 1—Basic inverter circuit (A) is modified to operate with two 20-watt fluorescent lamps (B)

New British railway coach uses fluorescent lighting

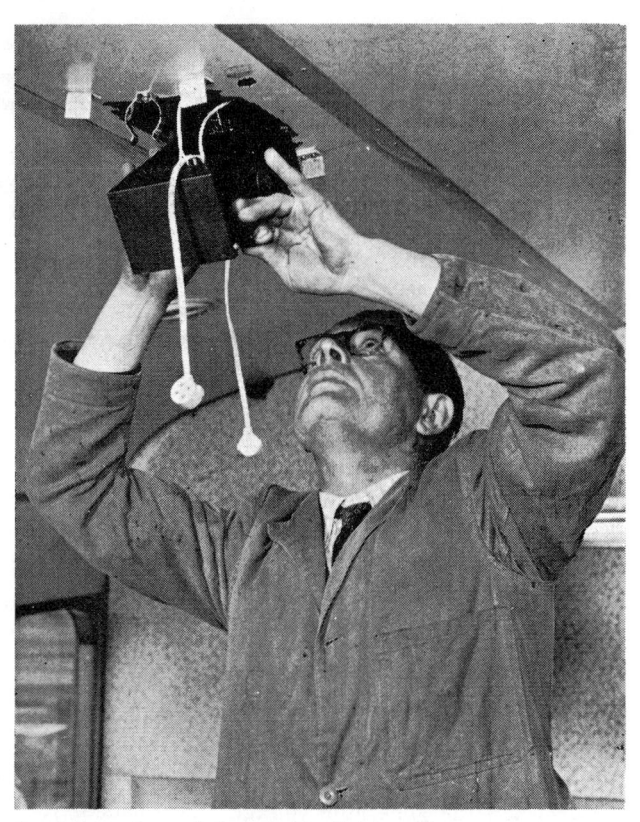

One inverter for each lamp is installed in ceiling

on Q_2, and the cycle begins again, but with current flowing into the upper half of W_1.

Actual Circuit

For the operation of a fluorescent lamp, some form of series impedance is required, owing to the negative volt-ampere characteristic of the discharge. If a separate inductance is used in series with the lamp, it will cause additional losses. Therefore, the transformer is designed as a high-reactance unit, with the secondary winding loosely coupled to the primary. The circuit for operating two 20-watt lamps is shown in Fig. 1B. It is similar, as far as the primary side is concerned, to that of Fig. 1A except that the starting capacitor is connected to the midpoint of the feedback winding instead of to the base of Q_1. The capacitor then charges up by way of the feedback winding, the base and emitter of Q_1 and Q_2 and the feedback resistor. There is sufficient asymmetry in the circuit to insure that either transistor is switched on by the current pulse. One other difference from the primary circuit of Fig. 1A is the addition of a smoothing capacitor across the supply and a surge-limiting capacitor across the full primary winding of the transformer.

Without the smoothing capacitor, the current taken from the supply is severely chopped. If several inverters are operated from a common supply line, the potential drop in the conductors will impose an a-c ripple on the d-c supply and, in some circumstances, may give rise to appreciable fluctuations in the light output.

Inverter has one transistor on outside of housing at left, the other on opposite side of housing

The surge-limiting capacitor reduces the magnitude of the surge voltages across the transistors. Each transistor operates as a switch, and as the circuit is inductive when the switch is opened (as at cutoff), a voltage surge is produced by the magnetic energy stored in the core of the transformer.

Except that the waveform of the output voltage and current is sinusoidal rather than square, the operation of the inverter with a lamp is similar to the process already described. The transformer is designed with leakage reactance between primary and secondary windings, so that the secondary voltage decreases rapidly as the current increases. In this way, stable operation of the lamp is automatically obtained.

One of the advantages of the quantity production of small transistors is that it allows the use of an inverter for each lamp, which means that the conventional low-voltage cable can be employed, and that consequently, tungsten lamps can easily be substituted for fluorescent, if required.

Solid-State Generator Regulator for Automobiles

Circuit using only semiconductors and resistors performs functions of conventional generator regulator. Two transistors and a diode regulate voltage; a transistor limits current; and a diode protects against reverse current

By **LEONARD D. CLEMENTS,** Consultants & Designers, Inc., New York, N. Y.

GENERATOR REGULATORS in contemporary automobiles are electromechanical devices designed to regulate generator voltage, limit maximum generator current to a safe value and prevent current flow from the battery through the generator when the generator voltage falls below the battery voltage. These functions are accomplished by voltage regulating, current limiting and cutout relays, respectively.

Severity of mobile service aggravates many of the problems inherent in relays. Pull-in of relays is affected by shock and the wide temperature extremes encountered. Armature springs used to calibrate relay pull-in change characteristics as the result of continual flexing, aging and temperature effects. Relay contacts stick because of mechanical defects or fusing and eventually deteriorate from mechanical wear and arcing.

To avoid these difficulties, an automobile generator regulator has been constructed in which the active elements are transistors and semiconductor diodes. Several regulator designs were built and tested. The circuit shown in Fig. 1 proved to be the best compromise based on performance, currently available semiconductor components and costs consistent with restrictions imposed by a price conscious market.

Voltage Regulator

Principal components associated with the voltage regulation portion of the circuit are transistors Q_1 and Q_2, Zener diode D_2, and a sampling network consisting of resistors R_8 and R_{10} and potentiometer R_9. Resistor R_5 limits maximum base current of Q_2 to a safe value during abnormal conditions, such as occur when the generator is overloaded. Resistor R_{11} provides a path for the collector leakage current of Q_1. Resistor R_4 reduces the collector dissipation in Q_2 but can be eliminated if a transistor having a higher collector dissipation rating is substituted for the 2N333.

Zener diode D_2 provides a reference voltage, e, against which the regulated output voltage is compared. Current is supplied to D_2 through resistor R_6.

Transistor Q_2 is connected as a difference detector with its base held at the constant voltage e and its emitter connected to sample a fraction of the regulated voltage, V_L, by means of the tap on potentiometer R_9. During operation, the emitter of Q_2 is more negative than its base resulting in a forward, or conducting, base bias current I_{b2}.

Regulated voltage V_L is related to the reference voltage e by the equation $V_L = e(r_1 + r_2)/r_2$ which is an approximation to the actual conditions existing. In this equation it is assumed that voltage drop from I_{b2} in the base resistance of Q_2 and R_5 is zero and that the emitter current I_{e2}, which must flow in the sampling network, does not alter the emitter voltage sampled by Q_2.

The extent to which these conditions can be approached is dependent upon achieving high current gain in Q_1 and Q_2, and choosing the sampling network resistors to have as low a resistance as practical considerations permit. Also, the generator field winding must be designed to best utilize the optimum dynamic range of Q_1 and Q_2.

Actual regulation obtained is shown in Fig. 2. Rapid decay of load voltage near 35 amps shown on Fig. 2A results from the current limiting feature of the circuit and should not be mistaken for the

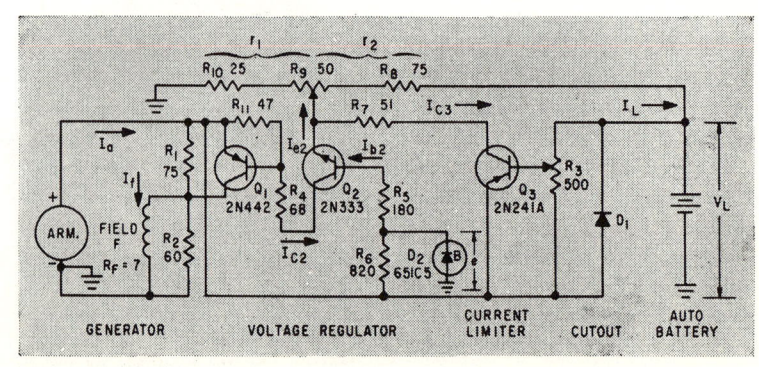

FIG. 1—Generator-regulator is not temperature compensated because different automobile electrical systems require individual selection of compensation parameters

normal voltage regulation curve. Further improvement of regulation can be obtained if positive feedback proportional to load current is introduced. This modification is not shown but would require that the generator operate with its low side ungrounded.

Temperature Compensation

Wide temperature extremes found in automotive applications require some form of thermally sensitive compensation for the effect of temperature upon transistors Q_1 and Q_2 and Zener diode D_2. If the voltage of D_2 is chosen near 5 volts, the voltage drift caused by D_2 will be negligible. As an alternate solution, D_2 can be chosen to have a temperature coefficient which will at least partly offset the voltage drift caused by transistors Q_1 and Q_2.

In addition to collector leakage current compensation, it is also necessary to compensate for the thermal characteristic of the base-emitter diode. Since both of these factors act to increase the output voltage, it is possible to compensate for them by using a single compensating method. For example, a negative temperature coefficient can be introduced into the r_1 portion of the sampling network by using a suitably chosen thermistor connected in series or in parallel with resistor R_8.

The nonlinear characteristic of a thermistor does not permit total compensation at all temperatures but this is tolerable since the required stability is not particularly severe. Since the desired characteristic of the incremental change of load voltage with respect to temperature is negative as shown in Fig. 3, it is necessary to overcompensate the regulator. This condition often proves easier to attain than a flat characteristic when using thermistors for temperature compensation.

Field Test Circuit

Because the circuit in Fig. 1 was designed for use in several different automobile electric systems which require individual selection of the compensation parameters, it does not include temperature compensation. Any compensation used must be adjusted to compensate for the characteristics of the entire automobile electrical system.

A silicon transistor was chosen for Q_2 to obtain the required collector dissipation rating of approximately 100 milliwatts at 80 C in a miniature package. Where space requirements permit the use of power transistor case sizes, it would be possible to substitute germanium units such as the 2N95, 2N326 or LT-5165L.

The only requirement for Q_2 is that it be an *npn* unit with the required beta and collector ratings. Beta values greater than 25 are suitable. Use of a germanium unit for Q_2 will require increased thermal compensation for collector leakage current but less compensation for base-emitter diode conductance temperature variations. Accordingly, the substitution of a germanium for a silicon transistor does not necessarily imply that better temperature compensation techniques are required.

Engine compartment heating effects on the circuit have not been conclusively evaluated. Two units were placed in field test and results in cool weather were satisfactory, although some thermally induced voltage drift was found. This drift could have been corrected by use of thermistors.

Current Limiter

Principal components associated with current limiting are transistor Q_3, resistor R_7 and potentiometer R_8. Diode D_1 is indirectly involved in current limiting since the voltage drop across it is a function of load current. Potentiometer R_3

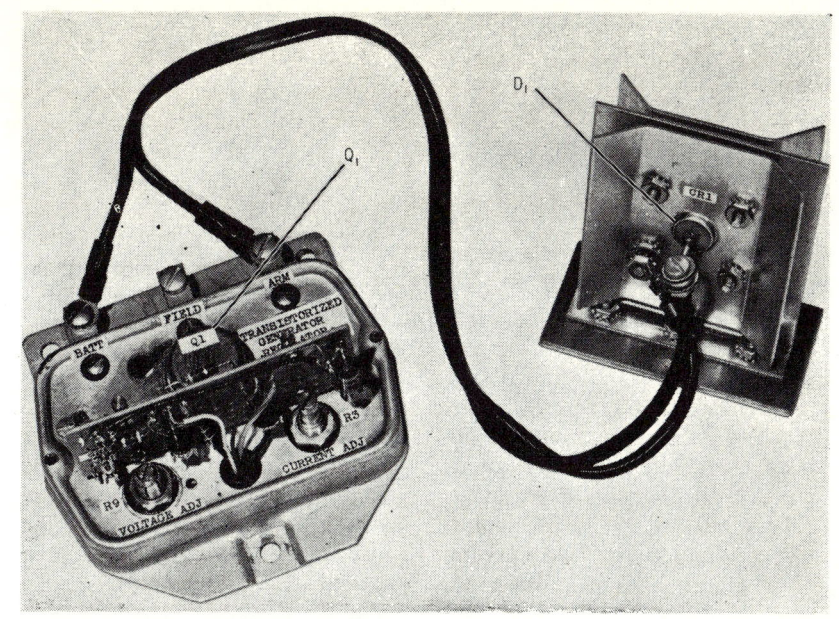

Field-test version of solid-state regulator. Power transistor Q_1 is electrically insulated from chassis by mica washer. Separate finned heat sink for cutout diode D_1 is electrically isolated by glass-epoxy mounting pad

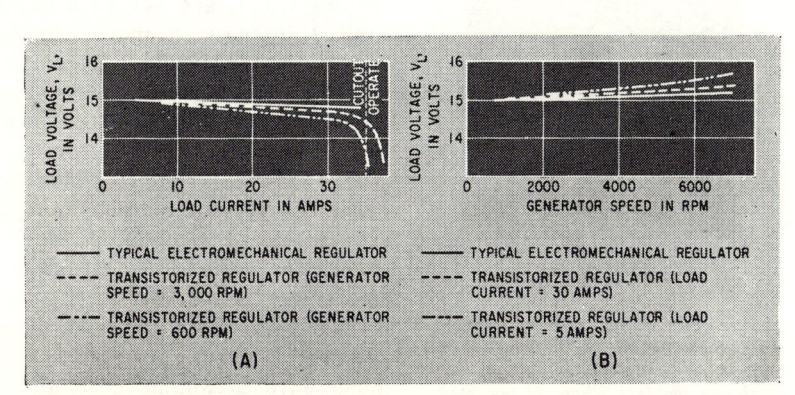

FIG. 2—Regulation afforded by regulator as a function of load current (A) and generator speed (B). Data were obtained using a standard 15-volt generator

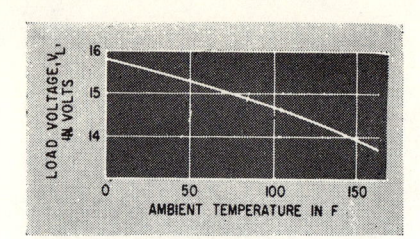

FIG. 3—Typical voltage versus temperature characteristics required by automobile electrical system

is connected across D_1, and a fraction of the voltage across D_1 and R_3 is applied to the base emitter-circuit of Q_3 by the wiper arm of R_3.

Proper adjustment of R_3 makes the base bias on Q_3 insufficient to cause conduction for load currents of less than 35 amps. When the load current reaches 35 amps, sufficient base bias is available and collector conduction starts.

Most of the collector current flows through r_2 by way of R_7. The increased voltage drop across r_2 attempts to cut off Q_2 thereby reducing the generator voltage and limiting the load current. Current limits at 35 amps and will not exceed 38 amps for further increases in speed or reductions of load resistance.

A test performed with the load short-circuited and the generator speed increased from 500 to 3,000 rpm showed that load current increased only 3 amps above the 35-amp value. To obtain these results the current gain of Q_3 should be at least 35 for the value of R_7 shown. Resistor R_7 limits the maximum collector dissipation in Q_3, but its value should be kept as low as possible to achieve good limiting.

It is not necessary to temperature compensate Q_3 to maintain calibration of the current limiting level since 5 or 10 percent variation is permissible in the average automobile electrical system. Although no separate compensation was attempted, it is worthwhile to note that utilization of the voltage drop across D_1 provides some measure of temperature compensation.

Figure 4 shows rectifier characteristics at several temperatures. The tendency of Q_3 to conduct at lower input voltage as temperature increases is offset by lower voltage across D_1. Unfortunately, successful utilization of this characteristic requires a large heat sink to insure that the temperature rise of D_1 is essentially caused by the ambient temperature rather than by the power it dissipates.

Cutout Diode

During periods when the generator voltage is lower than the battery voltage, reverse current is prevented by silicon rectifier D_1. Advantages of using a rectifier are that it prevents all reverse current (whereas a cutout relay typically requires several amps of reverse current before pull-in) and it does not require calibration, thus will never go out of adjustment. These advantages outweigh the disadvantage of the internal power dissipated in D_1 which may reach 40 watts when load current is at the maximum of 35 amps. Under most conditions the load current is considerably less than 35 amps; therefore, the use of a rectifier is not unreasonable.

Permissible reverse leakage current in D_1 can be as high as 150 ma without adversely discharging the battery during typical inoperative periods. Thus, the rectifier used can be a low cost unit or even a rectifier rejected for most other electronic applications if it is rated to carry the maximum load current (usually 35 amps). A unit rated at 25 piv installed in a 12-volt electrical system affords a satisfactory measure of protection against failure resulting from normal transients in the electrical system.

Heat developed in D_1 requires that it be mounted on a separate

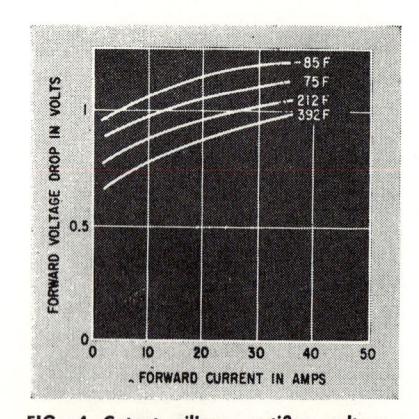

FIG. 4—Cutout silicon rectifier voltage-current characteristics as a function of temperature. Temperatures shown will approximate ambient when infinite heat sink is used

heat sink apart from the remainder of the circuit. This requirement is not a disadvantage since a production model would incorporate D_1 into the generator. Field tests were performed with D_1 mounted on a separate heat sink and located approximately one foot away from the rest of the circuit.

Possible Improvements

Although the circuit configuration shown in the photograph sufficed for field tests, it would be desirable to utilize printed circuit techniques and automated assembly for the production models. Except for diode D_1 and transistor Q_1, all components including the heavy gage metal terminal bars can be mounted on printed circuit boards.

Transistor Q_1 can be mounted on the metal regulator base using a mica washer insulator and then soldered to the printed circuit board during assembly. Diode D_1 must be mounted on a heat sink which is kept a foot or more away from the regulator when installed in the automobile. The heat sink can be made much smaller than that shown in the photograph if a good thermal contact with the automobile body or generator is provided.

Adjustment of the solid-state regulator can be automated since the two potentiometers need only be turned until the desired voltage or current limiting levels agree with predetermined references. This is simpler and more accurate than the corresponding operation on electromechanical regulators where mechanical stops and springs must be bent to obtain adjustments.

Applications

For industrial and military applications which are not so severely limited by cost, a more complex circuit is justifiable. Regulation which is less than 0.25 percent and control of several kilowatts are relatively easy to achieve. Moreover, the solid-state regulator does not generate r-f interference and response speed is limited only by generator inductance. By suitably modifying the circuit in Fig. 1, it is also possible to regulate alternators; however, the alternator must have a wound field and a ripple filter must be added.

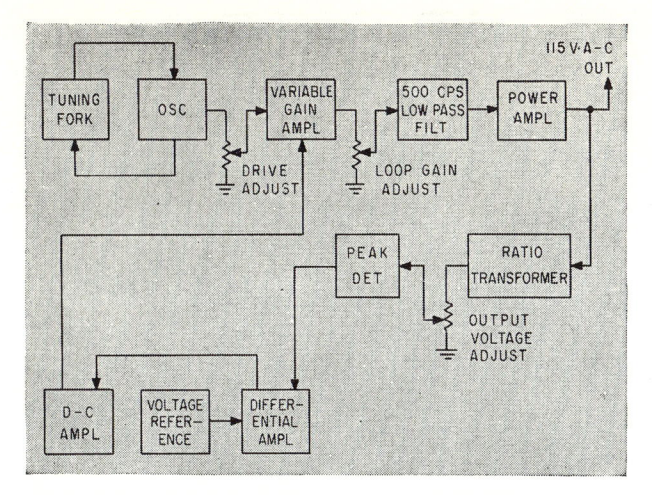

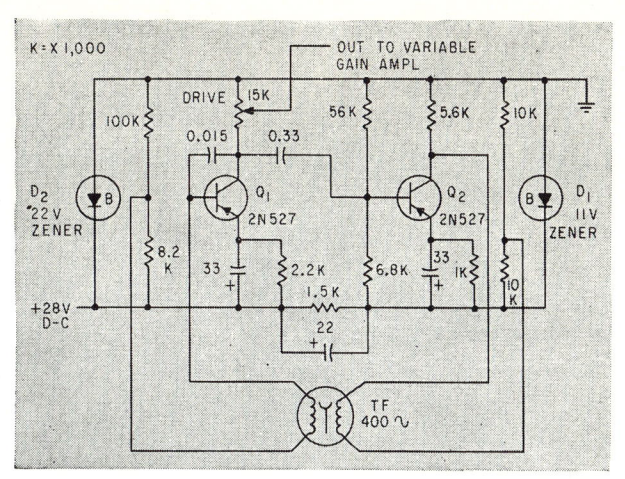

FIG. 1—Error correction mechanism of inverter is shown

FIG. 2—Accuracy of frequency is dependent upon tuning fork

Linear Circuits Regulate Solid-State Inverter

Transistorized d-c to a-c inverter uses linear circuits to obtain precise voltage regulation of output. Total harmonic distortion of 50-watt unit is 1 percent and 400-cps output is regulated to ± 0.2 percent. Tuning fork oscillator gives frequency accurate to 0.01 percent

By ROGER WILEMAN, Electronics Engineer, Packard-Bell Electronics Corp., Los Angeles, Calif.

RELATIVELY HIGH conversion efficiencies have been obtained by using transistors as saturating switching elements in d-c and a-c inverters. However, inverters of this type are severely limited with respect to distortion of the output waveform and accurate true rms regulation. If low distortion and precise voltage regulation are to be achieved with a transistor inverter, linear circuit techniques must be used, as the same properties which make a transistor efficient as a switch also contribute to its efficiency as a nonsaturating amplifier.

The requirements of low distortion and precise regulation have led to the development of a transistor static inverter using linear circuits having unique capabilities. A 50-watt unit capable of delivering a 115 volt rms 400 cps output regulated $\pm$ 0.2 percent is described here. Total harmonic distortion of this inverter is typically 1.0 percent. Output frequency of the inverter can be changed within the limit of 360 and 440 cps by means of plug-in tuning forks. Depending upon the type of tuning fork used, frequency accuracies up to 0.01 percent are obtainable.

Operation

Figure 1 is a block diagram of the inverter. The reference frequency generated in the tuning fork oscillator is coupled to the variable gain amplifier and then to the 500-cps low pass filter for complete removal of harmonics. Output of the filter is amplified to 50 watts and transformed to the output voltage of 115 volts. A precision ratio transformer divides down and isolates the a-c output voltage. The peak voltage from the secondary of the ratio transformer is detected and subsequently compared with a stable d-c voltage reference in a differential amplifier. The differential amplifier feeds an error voltage to a two stage d-c amplifier. Output of the d-c amplifier varies the output of the variable gain amplifier in inverse relationship to output voltage variations.

In this manner, regulation of the output against line and load variations is achieved.

Figure 2 is a schematic of the tuning fork oscillator. Essentially

FIG. 3—Tetrode transistor is used as variable gain device

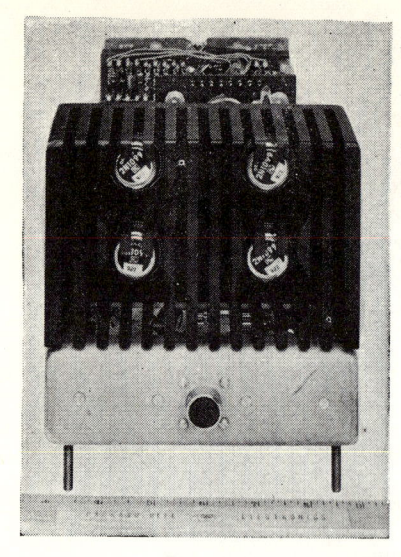

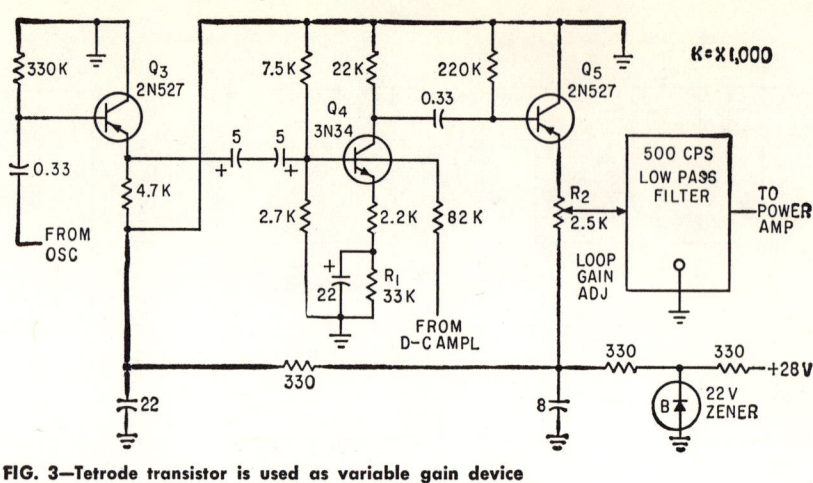

Photo at left shows packaging method employed

it consists of a two stage amplifier, the output of which is regeneratively coupled to the input through the tuning fork. The tuning fork acts as a frequency-selective network having a Q of 10,000 or more. A common problem encountered in the application of such an oscillator is the long build-up time after turn-on. In the circuit shown this problem is circumvented by making the open loop gain of the two stage amplifier high, such that the voltage build-up rate is high, saturating Q_2 soon after turn on. When saturation occurs, further build up ceases, as the fork drive voltage from the collector of Q_2 remains constant at a peak-to-peak value determined largely by Zener diode D_1. The voltage determined by D_1 is purposely made small to further cause early saturation of Q_2. In this manner build-up-to-stabilization times of 3 to 4 seconds have been achieved. As a means of keeping the oscillator output reasonably constant, D_2 is incorporated to regulate the supply voltage. The output taken from the collector of Q_1 is essentially sinusoidal due to the good harmonic rejection of the tuning fork.

Variable Gain Amplifier

The signal from the oscillator is applied to a 3N34 tetrode transistor through emitter follower Q_3 of Fig. 3. The 3N34 and 3N35 tetrode transistors have several unique properties which allow their use as variable gain devices.[1] If cross base current is applied to these transistors, a crosswise drift velocity is imparted to the minority carriers in the base region. The net effect of the crosswise drift is a removal of some of the carriers from the active base region and hence a decrease in base transport efficiency. Consequently, there is a significant reduction in h_{fe}, the common-emitter current gain.

The limits to which the reduction in h_{fe} can be applied are determined by the necessity for maintaining the operating point of transistor Q_4 within a given region. It is for this reason that the emitter resistor of this stage, R_1, is made very large. With the circuit shown in Fig. 3, it is possible to obtain gain variations of up to ten to one, maintaining less than 10 percent harmonic distortion at 1-volt output.

An error current from the d-c amplifier is applied to base two of Q_4 causing the a-c voltage gain of the stage to vary with the magnitude of the error current. The a-c output of Q_4 is then coupled into a 500-cps low-pass filter through emitter follower Q_5. Output of this filter is a sine wave at the reference frequency with less than 0.1-percent distortion. Potentiometer R_2 adjusts loop gain to compensate for variations in the nominal voltage gain and phase shift of the various stages in the inverter. When the loop gain exceeds a given value ringing of the output waveform occurs. By adjusting R_2 it is possible to find a condition which gives a good compromise between response time and overshoot under transient load conditions. Response time for this condition is typically 20 to 30 milliseconds.

The power amplifier used in the inverter is shown in Fig. 4. To provide good output voltage regulation, this amplifier must have characteristics of low internal impedance and low distortion. Low internal impedance keeps the loop error variations small with respect to load variations, thus reducing the regulator loop gain required for a given regulation tolerance. Because a peak detector is used in the inverter, harmonic distortion present in the output waveform affects the true rms regulation of the unit. If accurate regulation is to be achieved harmonic distortion must be minimized.

To achieve the above in a transistorized design, use is made of a compounded common collector output stage.[2] The application of this circuit is not without its problems however. Among the more severe problems are thermal runaway and parasitic oscillation.

Thermal runaway becomes a severe problem in this type of amplifier because of the extremely high effective alpha created by compounding. The problem is aggravated when the d-c resistance of the output transformer primary is reduced to a minimum. Two solutions were tried with a high degree of success. The first is to make the secondary winding resistance of T_2 as small as possible. In the present design it is 25 ohms base-to-base.

The second solution is to incorporate a high degree of d-c current feedback around the output transistors Q_{12}, Q_{13}, Q_{14} and Q_{15}. This was accomplished by incorporating

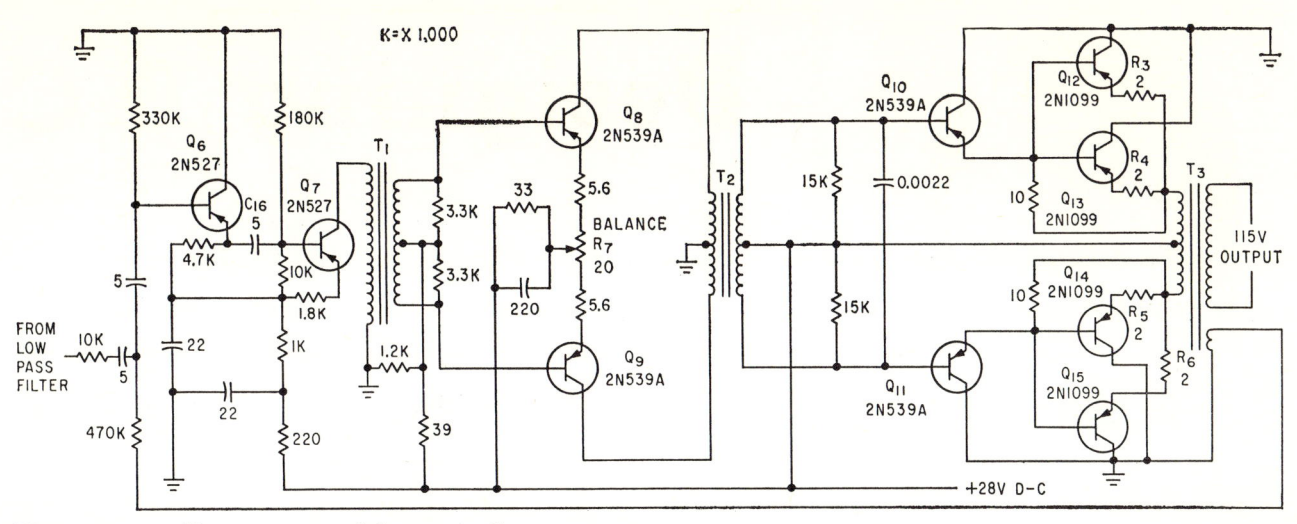

FIG. 4—Power amplifier uses compounded common collector output stage

R_3 to R_6 in series with the emitters of the four output transistors. Although this has a detrimental effect on the amplifier's internal impedance and efficiency, the stability resulting was well worth the compromise. When driven separately from the rest of the amplifier, the output stage delivers 55 watts at 400 cps with only 2.4 percent total harmonic distortion at 70 C ambient. (This corresponded roughly to a junction temperature of 100 C.) When the drive signal was removed, the d-c input current corresponding to leakage at the zero bias condition was less than 100 milliamperes. The unbalanced direct current in the output transformer was less than ten percent of this figure. For this reason an ungapped Silectron toroidal core was used in the output transformer with a consequent reduction in size and weight.

Parasitic oscillations are a problem with this type of amplifier due to the large reactances which exist in the circuit and the multiplicity of in-phase feedback paths. The power gain of the compounded stage can become quite high as in the present case where it was 26 db, and this in turn adds to the problem. The most successful suppression method used was capacitive loading of the secondary of transformer T_2.

The rest of the amplifier is of reasonably straightforward design. It might be mentioned that R_7, the balance control for Q_8 and Q_9 has another not-so-apparent use. This is as a load regulation adjustment and may be explained as follows. In any system which uses a peak detector to sense regulation to true rms the quantity of the harmonic content will cause a deviation from the desired output. Resistor R_7 by controlling to a certain extent the amount of second harmonic distortion in the output can affect the true rms regulation of the inverter. It is generally possible to find a setting of R_7 which gives optimum load regulation with only a slight increase in total harmonic distortion.

Another approach which yields good results is the use of a split load phase inverter as a driver for Q_8 and Q_9. When such a substitution is made performance at low levels is considerably improved.

Reference Element

The most critical area as regards the accuracy and stability of the inverter is in the reference and d-c amplifier circuit which is shown in Fig. 5. Careful attention was given to temperature compensation of the reference element and silicon transistors were used wherever d-c levels were to be handled. The reference element is a 1N2169A Zener reference which employs one reverse biased and two forward biased diode junctions. With a constant current input the voltage variation with temperature is less than 50 microvolts per degree C. A constant input current to this reference is achieved by matching the positive temperature coefficient of R_8 and TR_1 to the negative temperature coefficient of diode D_3.

As in the reference element, use is made of both positive and negative temperature coefficient characteristics of silicon diodes to set up a thermally stable collector voltage source for Q_{16}, Q_{17} and Q_{19}. A compensating control, R_9, is used to couple input voltage variations to the loop in opposition to supply voltage variations.

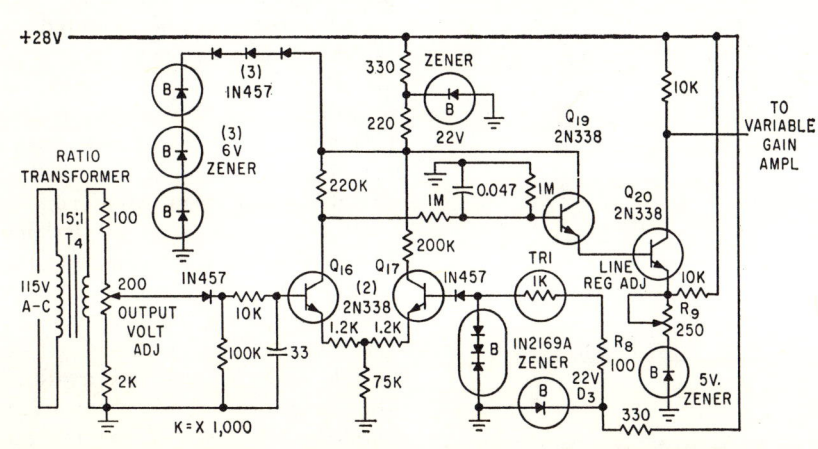

FIG. 5—Reference element is compensated to eliminate temperature variations

REFERENCES
(1) R. Hurley, "Junction Transistor Electronics", p 299, John Wiley and Sons, Inc., New York, 1958.
(2) R. F. Shea, "Transistor Circuit Engineering", p 130, John Wiley and Sons, Inc., New York, 1957.

Designing D-C

Transistorized, saturable-core relaxation oscillators are finding increased application as d-c to a-c converters. This article presents a rapid means for determination of necessary circuit parameters. The technique is applicable to many variations of the basic, two-transistor symmetrical circuit

By STANLEY SCHENKERMAN,
Senior Engineer, Missile Development Division, Ford Instrument Company, Long Island City, N. Y.

DETERMINATION of circuit parameters in transistorized d-c to a-c converters is simplified greatly by use of the accompanying nomographs.

A symmetrical form of converter[1] is shown in Fig. 1. It consists of transistors Q_1 and Q_2 operating as controlled switches, a battery E, and a core with square-loop characteristics.

Circuit Operation

Assume that Q_1 begins conducting. Current flowing into the top or dot end of L_1 causes all windings to be positive at the dotted end. The voltage across L_3 increases conduction still further, causing regeneration until Q_1 saturates. This switching occurs rapidly. The induced voltage across L_4 keeps Q_2 cut off.

When Q_1 saturates, almost all of the battery voltage appears across L_1. The core flux, ϕ, increases linearly with time until positive core saturation is reached. At core saturation, the transistor current increases rapidly in an attempt to maintain constant $d\phi/dt$, but the transistor current is limited by the base voltage developed by L_3.

When the current and the resultant flux can no longer increase, the induced voltages drop to zero and Q_1 is cut off. Cessation of current in L_1 allows the flux to decrease toward its remanance value, inducing voltages of the opposite polarity in all windings. This action holds Q_1 off and turns Q_2 on. The battery is now connected across L_2 and the flux builds up linearly to its negative saturation level.

The core oscillates between positive and negative saturation. If Q_1 and Q_2 are similar and $L_1 = L_2$ and $L_3 = L_4$, a square wave is induced in the output.

Design Procedure

Usual design procedures involve initial-core selection, completion of necessary design calculations, and a check to make certain that the required windings fit into the available area. The designer may then accept the initial core size or repeat the procedure with a larger or smaller core as influenced by his preceding calculations.

Core size may be dictated by factors such as window area available for windings, cost, weight, and overall size. Cores wound from thin tape and with high ratios of inside to outside diameter, give fast switching and steep waveforms.

Core materials such as Orthonol and Hymu "80", with narrow rectangular hysteresis loops, contribute to frequency stability and low core losses. Unfortunately, all desirable features are not available simultaneously. The narrower the loop, the lower the saturation flux density and the less rectangular the

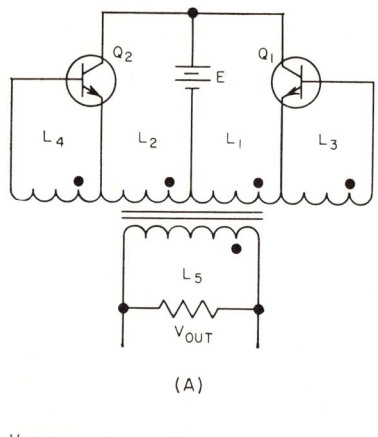

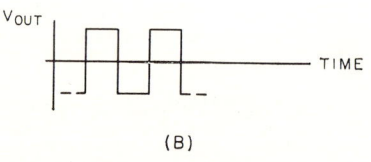

FIG. 1—Symmetrical converter circuit (A) and ideal output waveform (B).

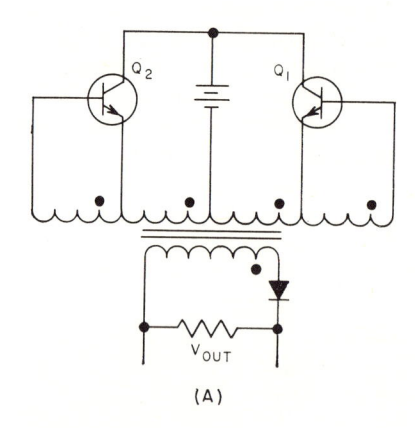

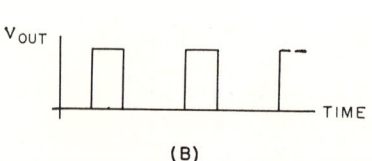

FIG. 2—Assymetrical circuit (A) and its ideal output waveform (B)

to A-C Converters

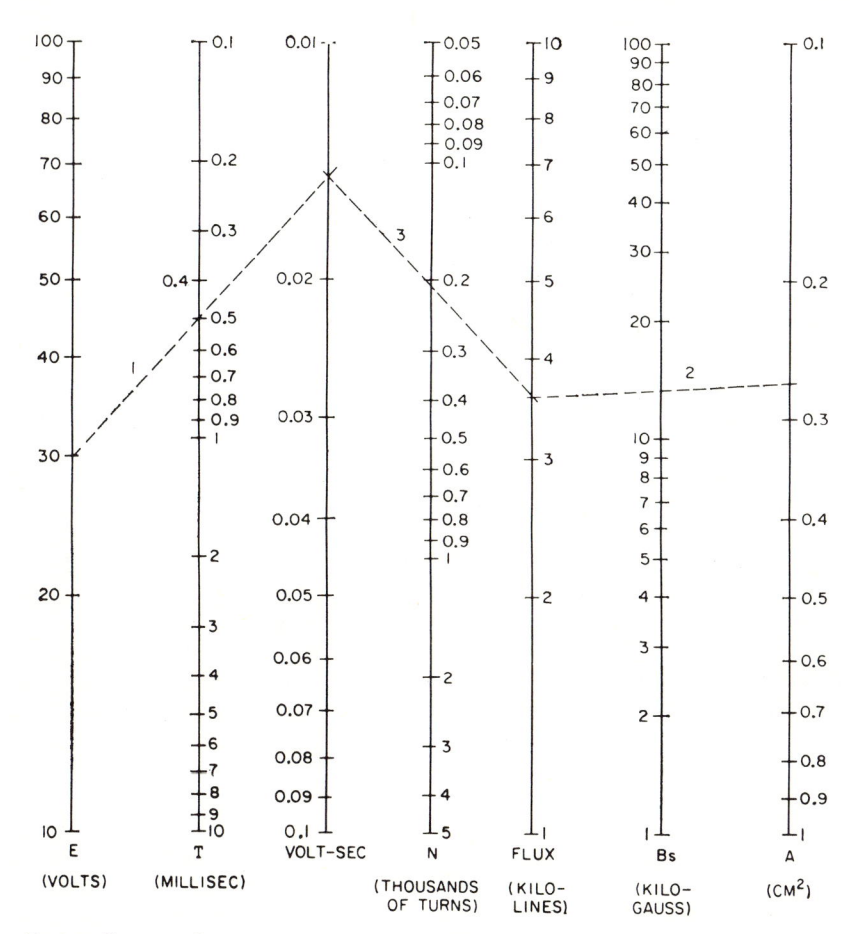

Timing Nomograph

dissipation region quickly. As long as switching time occupies a negligible portion of the cycle, the transistor may be assumed 95-percent efficient. Copper and core losses total about 15 percent. Overall conversion efficiency is 0.95 × 0.85 or about 81 percent.

The transistor must deliver current to a fictitious load whose resistance is about 85-percent of the reflected converter load. In addition, the transistor must deliver about 130 percent of the current required by the fictitious load. This value is necessary to insure that the core is driven well into saturation.

Peak voltage appears across the transistor when it is cut off. Assume zero saturation voltage for the transistors and ideal windings. Peak inverse voltage from collector to emitter of Q_1 is then the battery voltage plus the voltage induced in L_1 by current in L_2. In this case, peak inverse voltage is $E + E$ or $2E$ volts.

Feedback

Feedback turns of L_3 may be determined from the voltage nomograph once the required base voltage is found from the transistor output and input characteristics. The applicable formula is similar to Eq. 2 with the feedback voltage substituted for V_{out} and the number of turns on L_3 and L_4 for N_{out}.

Design now continues as in stand-

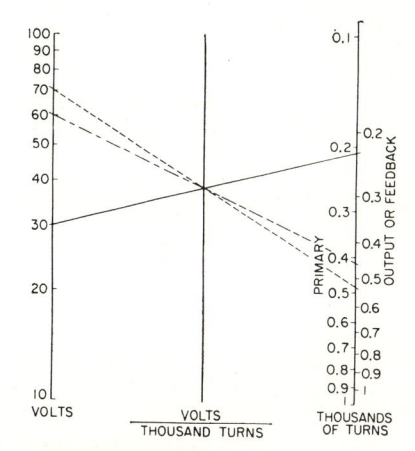

loop. The material selected must be a compromise of all factors.

Timing Nomograph

With a trial core in mind, the number of turns for L_1 and L_2 may be found from the timing nomograph. This is based on the relation governing the change in core flux:

$$N = \frac{ET}{2 B_s A} 10^8 \quad (1)$$

where T is time in seconds during which battery voltage is applied to the core winding; E is the battery voltage; N is the number of turns on the core; A is cross-sectional area of the core in sq cm; and B_s is saturation flux density in gauss. The factor of 2 is necessary because the core change from negative to positive saturation, or vice versa, is a change of $2B$.

For a specified load voltage, the number of output turns resulting from application of the ideal transformer equation must be increased by ten percent. This increase is necessary to compensate for leakage impedances and the transistor saturation voltage. The latter is normally less than one volt for germanium transistors but may be higher for silicon units. The voltage nomograph for output turns assumes that this ten percent factor is sufficient. The equation is

$$N_{out} = 1.1 N \frac{V_{out}}{E} \quad (2)$$

where V_{out} is the peak output voltage during one-half cycle.

At this point, the transistor may be selected. Although it must supply one-half the required load power and core and copper losses, its operating point passes through the high

Voltage Nomograph

125

ard transformer practice. Appropriate wire sizes are picked and the fit of the windings checked. It is important to employ a winding sequence that will result in close coupling between L_1 and L_2. Close coupling minimizes ringing caused by interprimary leakage reactance.

Regulation and internal heating are affected by the same factors that apply in standard transformer design. However the converter is one device that cannot be damaged by overload. Overloading causes the oscillation to cease and not resume until the overload is removed.

Example

As an example, assume a 1,000-cps symmetrical converter operating from a 30-v source is to deliver 120 v to a resistive load of 200 ohms.

Duration of each pulse is half the period of the 1,000-cps wave or 0.5 millisec. Using the timing nomograph, connect $E = 30$ and $T = 0.5$ with a straight line. Extend the line until it intersects the volt-sec axis.

As an initial core selection, try a 2-mil Orthonol core, Magnetics Inc., type 50018-2A. Cross-sectional area for this core is 0.257 sq cm. Saturation flux density is 14 kilogauss.

Connect $A = 0.257$ and $B_s = 14$ with a straight line that intersects the flux axis. Connect this intersection with that previously found on the volt-sec axis. This results in an intersection with the N axis at 0.21. Both L_1 and L_2 of Fig. 1, therefore, should have 210 turns.

To determine the number of turns for L_5 for an output of 60-v peak (120 v peak-to-peak), use the voltage nomograph. Connect battery voltage $E = 30$ with the primary turns, $N_1 = 210$, as read on the left scale. Note the intersection with the volts-per-thousand-turns axis. Connect this intersection with the peak output voltage (60 v) and extend the line until it intersects the turns axis. Read the value of N_5, the number of output turns, from the right scale as 460.

The fictitious reflected load to each transistor is $0.85 \times (210/460)^2 \times 200$ or 35.6 ohms. Transistor current during the pulse is 30/35.6 or 0.845 amp. Peak transistor current must be 1.3×0.845 or 1.1 amp.

Voltage and current requirements may be met by the type H-7 transistor. As determined from the transistor output characteristics, a base voltage of 0.7 v is necessary for an output current of 1.1 amp.

Bias Windings

Return to the voltage nomograph and connect the 70-v point with the intersection on the volts-per-thousand-turns axis found previously. Extend the line until it intersects the turns axis. The value 540 turns is read from the right scale. Since 0.7 v rather than 70 v is required, L_3 and L_4 should each have 5.4 or say 6 turns.

Wire sizes may now be selected and the winding fit checked. Table I shows the wire selection and required winding area for the coils.

Window area of the type 50018-2A core is 1.82 sq in. but space must be allowed for the winding-machine shuttle. The RW II toroidal-core winding machine uses a 13/16-in. diam shuttle for number 22 wire. Shuttle area is 0.518 sq in. Utilizing a 70-percent space factor, effective window area is 0.7 × (1.82 − 0.518) or 0.9114 sq in. This value leaves ample room for winding and insulation.

Admittedly, a generous amount of hindsight was used in formulating this example. In general, a few tries are necessary before a satisfactory design is achieved.

Circuit Variations

Two important variations of the basic circuit are shown in Figs. 2 and 3. In the circuit of Fig. 2,[2] power is delivered to the load only when the power transistor Q_1 conducts. Transistor Q_2 resets the core by driving it to negative saturation. During reset, the induced voltage in the output winding is such that the diode is cut off and no power is delivered to the load. Transistor Q_2 may be a low-power unit. It need only supply the core saturation current and any current required by the diode's finite back resistance.

Figure 3 shows the circuit for a single transistor converter.[3] Here, resetting action is caused by the capacitor. Again, the diode isolates the load during the reset interval and power is delivered to the load only when the transistor conducts. Value of the capacitor is chosen experimentally to provide a waveform with any degree of symmetry.

It has been assumed that the circuits under discussion are self-starting. At room temperature, germanium transistors usually have sufficient leakage current to start oscillation. This is not always true, however, and auxiliary starting circuits may be required.

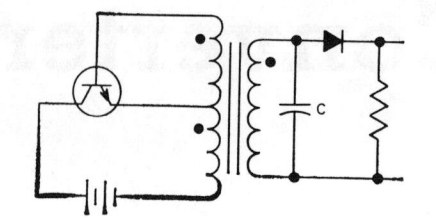

FIG. 3—Simple converter circuit using a single transistor

Table I—Wire Selection and Required Winding Area

Coil	Turns	RMS Current	Wire Size (Awg)	Coil Area (sq in.)
L_1	210	423	22	0.17
L_2	210	423	22	0.17
L_3	6	15	38	0.000156
L_4	6	15	38	0.000156
L_5	460	600	22	0.355

Winding area required = 0.6953 sq. in.

References

(1) Bright, Pittman, Royer, Transistors as On-Off Switches in Saturable-Core Circuits, *Electrical Manufacturing*, p 79, Dec. 1954.

(2) D. A. Paynter, An Unsymmetrical Square-Wave Power Oscillator, *IRE Trans*, CT-3, p 64, Mar. 1956.

(3) Chen, Schiewe, A Single Transistor Magnetic Coupled Oscillator, *AIEE Trans*, Pt. I, p 396, Sept. 1956.

Efficient Photoflash Power Converter

Single-transistor circuit has high efficiency and uses Zener diode to stabilize output voltage. Auxiliary transformer promises waveform improvements

By **RICHARD J. SHERIN**, Swampscott, Mass.*

IMPORTANT PHOTOFLASH converter characteristics are: high efficiency of energy conversion, rapid recycle, low standby current and regulated output voltage. The energy conversion efficiency in the converter to be described could exceed the 50-percent value which is the approximate theoretical upper limit on most conventional[1,2] circuits in this application. Converter action is halted when the desired output voltage is reached. The circuit then periodically replaces the charge lost by capacitor leakage.

Capacitor Charging

An elementary circuit is shown in Fig. 1. Basically the converter is a flyback (or ringing choke) oscillator which is freerunning when the voltage on the regulator capacitor C_2 is less than the Zener voltage for the reference diode D_2.

In operation the core is first charged through the conducting transistor Q and then, during a flyback interval in which Q is non-conductive, the core discharges its $LI^2/2$ energy through the diode D_3 into the storage capacitor C_3. The output winding N_3 is so poled that D_3 blocks during the core charging intervals and conducts during the flyback intervals. The storage capacitor is brought up to full voltage by many such consecutive on-off cycles of Q.

It will be seen later that the low voltage Zener diode D_1 is needed in the operation of the regulator.

*This work was done while the author was with the Boeing Airplane Company

FIG. 1—Elementary conveter circuit shows operating principle

The starting capacitor C_1 provides a path for the transistor base current at the start of each core charging interval before the regenerative action has made the hold voltage induced in N_2 sufficiently large to make D_1 conductive in the reverse direction.

Voltage Regulation

During each flyback interval, regulator capacitor C_2 is charged through D_1 and D_2 to a voltage approximately proportional to the voltage reached by the storage capacitor C_3. When full output is reached, the voltage on C_2 exceeds the Zener voltage for D_2 and biases the transistor base sufficiently positive with respect to the emitter to abruptly halt the oscillation. The low voltage Zener diode D_1 prevents C_2 from discharging through N_2, while still permitting the flow of base current during those intervals when Q is conductive. C_2 slowly discharges through resistor R.

When the off-bias on Q is sufficiently reduced, the on-off cycling of Q resumes and continues for several cycles until the storage capacitor has been restored to full voltage. The transistor is then cut off as before.

Improvements on Basic Circuit

Experiments have shown that the Zener diode D_1 of Fig. 1 could be replaced to advantage with parallel diodes as shown in Fig. 2. In the

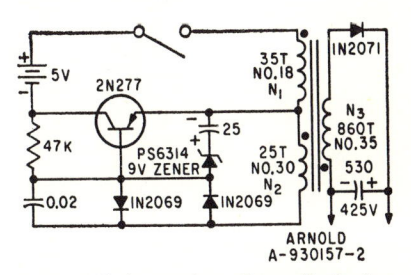

FIG. 2—Diodes replace Zener D_1 in improved version of photoflash circuit

elementary circuits the base current waveform is not optimum. Further experiments indicate that a current balancing transformer may be used to make the transistor base current increase with the collector current during the core charging cycle. This improvement will lead to a significant reduction in power wasted in overdriving Q at the beginning of core charging intervals.

REFERENCES

(1) G. H. Royer, A Switching Transistor D-C to A-C Converter Having an Output Frequency Proportional to the D-C Input Voltage, *AIEE Trans*, Part 1, July, 1955.
(2) H. A. Manoogian, Transistor Photoflash Power Converters, ELECTRONICS, August 29, 1958.

Fast Response Overload Protection

By F. W. KEAR
Integrated Dynamics Div.,
Globe Industries, Inc., Albuquerque, N. M.

THOUSANDS of dollars worth of transistors and other components have been saved with a transistorized overload circuit. Fuses were found to be unsatisfactory because of their time delay. The circuit switches power off much faster than conventional current protection devices.

The overload circuit in Fig. 1 is useful for production and maintenance testing with low d-c voltages. Current greater than 3 amp flowing through the 0.47-ohm resistor in the emitter circuit of current-switching transistor Q_1 drops voltage on the base of voltage-sensing transistor Q_2. This drop causes Q_2 to saturate, dropping bias voltage and causing Q_3 and Q_4 to saturate.

Transistor Q_3 opens the circuit immediately and keeps it open for the duration of the overload. For complete short circuits, Q_4 latches cut-off relay K_2, providing positive protection. Even if Q_3 were to fail, Q_4 would open the circuit, providing more reliable protection.

Transistor Q_3 also provides protection against fast-rising surges that would tend to damage Q_1 before relay K_2 could operate. The transistors are provided with heat sinks to protect them from heavy loads or extended periods of undetected overloads.

Resistors can be selected to protect circuits using greater load currents or supply or load voltages.

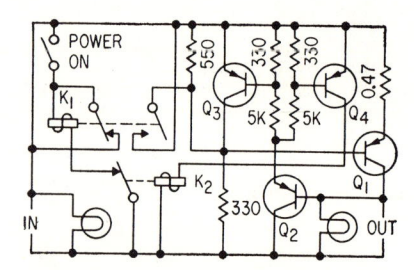

FIG. 1—Current greater than 3 amp drops base voltage of Q_2, rapidly cutting off output

Chapter 7
COUNTING CIRCUITS

Steering Circuits Control Reversible Counters

Transistorized reversible decade counter uses static voltage level on bases of steering transistors to determine direction of count

By ROBERT D. CARLSON, Electronics Division, Victor Adding Machine Co., Chicago, Ill.

DECADE COUNTERS which count both up an down are useful in control systems that measure temperature, pressure or other variables through combined analog and digital techniques. If the command and feedback of a servo are both expressed digitally, a reversible counter can be used to generate the error signal. Other uses for these counters are decimal arithmetic and in digital control systems.

In the counter, a static voltage level at the bases of steering transistors determines the direction of the count, whether up or down. Coding and feedback techniques minimize the number of components needed in gating and carry circuits when decades are cascaded. Pulses can be counted at rates up to 200,000 a second.

Logical Design

Many logical arrangements can be used to produce a radix-10 counter. Radix is used in the sense of base or root number and a radix-10 counter is the ten's system in general use. Four cascaded binary stages will produce 16 combinations of states (2^4) and this arrangement was chosen on the basis of cost.

Since 16 states or binary numbers are available, and only ten are needed, the excess six states must be nullified. The excess-three binary-coded system was chosen for several reasons. First, the nullification process of excluding the six unwanted states need not involve the first stage. Consequently, one pulse may be counted free before another pulse is passed to the stages in which cancellation feedback is used. Second, the code is represented by binary three to twelve in-

Plug-in decade modules being tested. Mounted in the rack are three readout modules

clusive, such that a unique code, binary two and thirteen, are available to activate the feedback system. Third and last, the code is complementary and uses the nines-complement arithmetic system.

Since complementary bistable memory elements are used, there are two outputs from each stage. One of the outputs will be referred to as the 8 4 2 1 set, and its complement as $\overline{8}\ \overline{4}\ \overline{2}\ \overline{1}$. Table I shows the relationships between the decimal and the binary code used in the counter. The 8 4 2 1 output increases as the $\overline{8}\ \overline{4}\ \overline{2}\ \overline{1}$ output decreases. The output of either set can thus be used to count up or down by changing the interstage coupling. The logic arrangement of the counter is shown in Fig. 1.

The current steering technique used in the interstage coupling requires using the previous binary-zero-to-one collector-transition as an input. When the coupling is derived from the $\overline{8}\ \overline{4}\ \overline{2}\ \overline{1}$ collectors, the 8 4 2 1 binary output increases in count. Conversely, if the interstage carries are derived from the 8 4 2 1 collectors, the 8 4 2 1 binary output decreases in count. When counting, the 8 4 2 1 binary coded decimal number is increased or decreased. Consequently, the decimal equivalent is counted up or down one count for each input transition.

When counting down, the process

---
* Now with Radiation Counter Labs, Skokie, Ill.

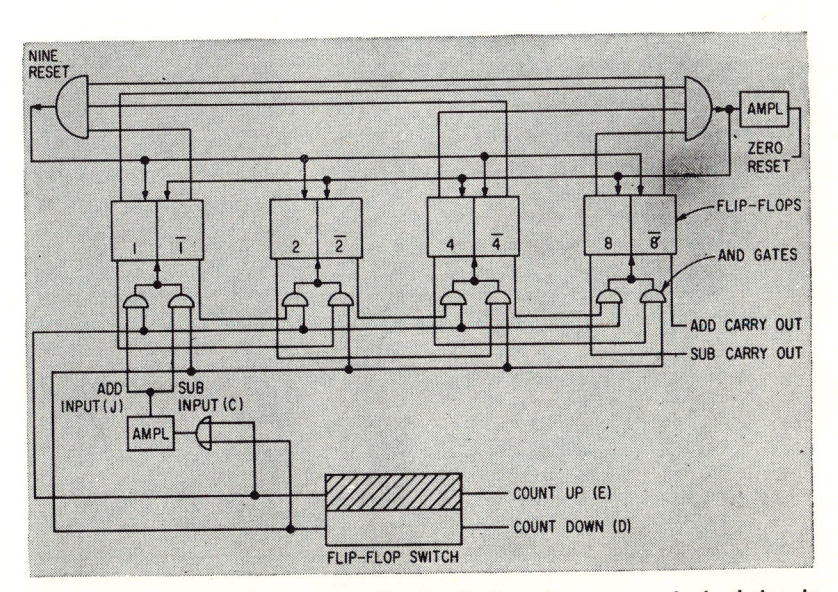

FIG. 1—Reversible counter uses four bistable flip-flops for storage. Logic design is based on nine's-complement arithmetic system

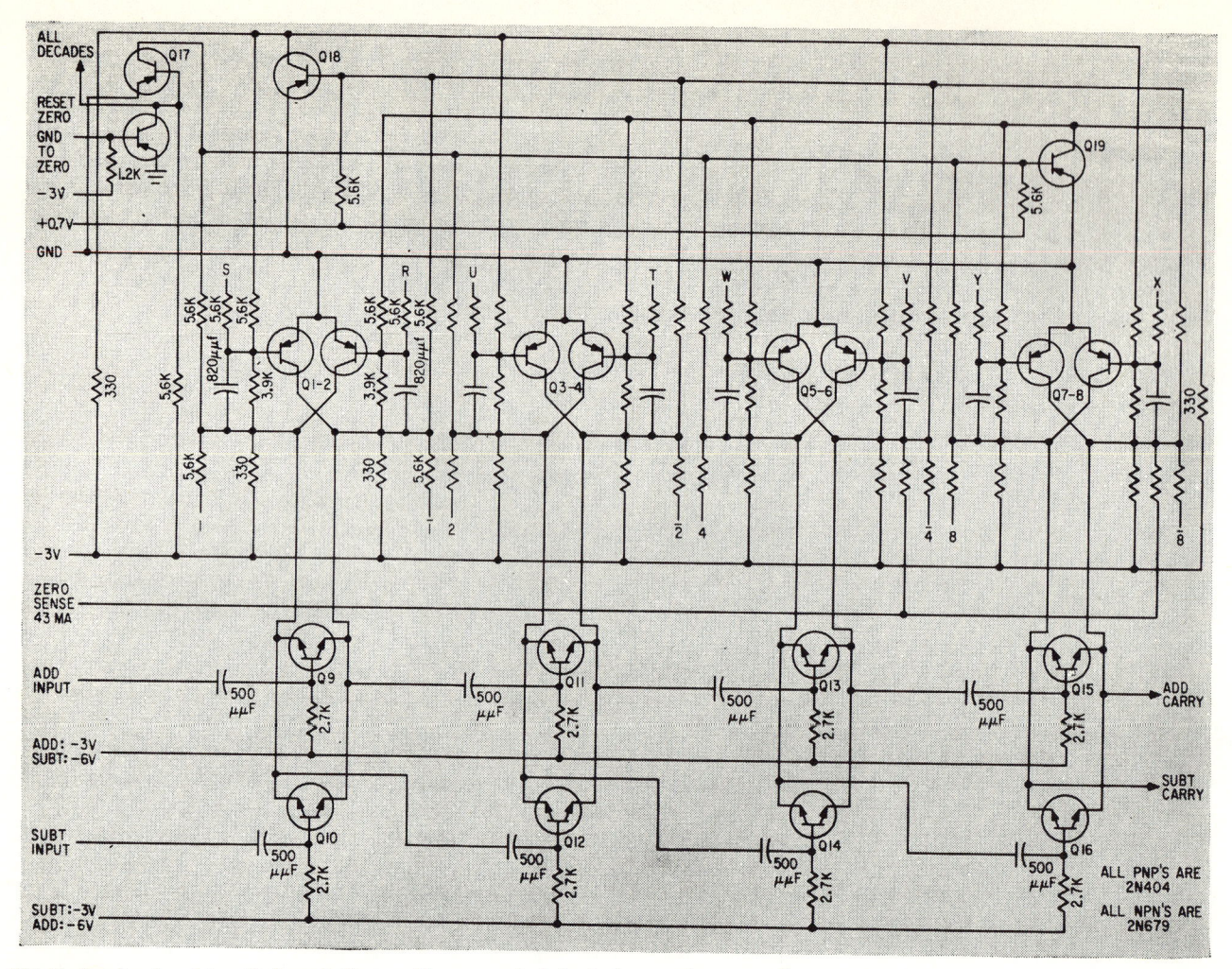

FIG. 2—Steering transistors Q_9 through Q_{16} provides economical circuit for counting up or down

continues until decimal zero is reached, 0 0 1 1 at which point the next state being 0 0 1 0 initiates a forced reset to nine, 1 1 0 0. Counting up, when decimal nine is reached, 1 1 0 0, the next state is 1 1 0 1, which initiates a forced reset to decimal zero, 0 0 1 1.

When counting down, it is not necessary to gate the forward forced reset line (9 to 0) since the combination 1 1 0 1 is never produced counting backwards. Similarly, it is not necessary to gate the reverse reset line when counting forward, as the combination 0 0 1 0 is never produced.

Circuit Considerations

The circuit, shown in Fig. 2, consists of four R-C coupled, complementary, saturated flip-flops. Symmetrical *npn* transistors, coupled between collectors of the flip-flops are used as trigger current amplifiers and for steering. The zero to one transition from the previous binary stage collector appears as a differentiated pulse at the base of the steering transistor (Q_9 through Q_{16}) by virtue of the coupling capacitor and base resistors. The steering transistors operate as emitter followers; the forward biased junction acts as the emitter-base junction; the junction with a reverse bias acts as the base-collector junction. The positive steered pulse appears at the OFF collector and at the ON base of

Table I—Code System

Decimal System	Binary Coded Decimal Excess Three	Binary Code				Binary Code Complement			
		8	4	2	1	$\bar{8}$	$\bar{4}$	$\bar{2}$	$\bar{1}$
0		0	0	0	0				
1		0	0	0	1				
2		0	0	1	0				
3	0	0	0	1	1	1	1	0	0
4	1	0	1	0	0	1	0	1	1
5	2	0	1	0	1	1	0	1	0
6	3	0	1	1	0	1	0	0	1
7	4	0	1	1	1	1	0	0	0
8	5	1	0	0	0	0	1	1	1
9	6	1	0	0	1	0	1	1	0
10	7	1	0	1	0	0	1	0	1
11	8	1	0	1	1	0	1	0	0
12	9	1	1	0	0	0	0	1	1
13		1	1	0	1				
14									

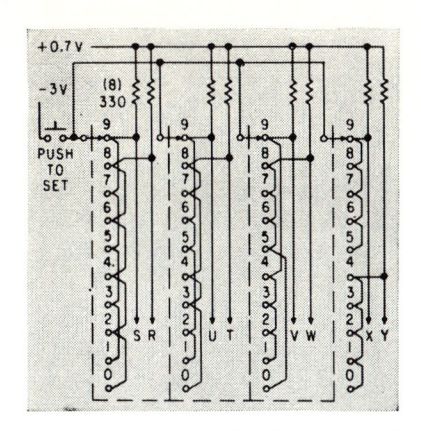

FIG. 3—Ganged switch allows any of ten counter states to be preset. Lines S, R, etc. connect to the circuit of Fig. 2

the flip-flop. An output appears even if the flip-flop is still in the transient condition. The only effect on the flip-flop is to speed up the transition. The *npn* transistors are not specified as bilateral units but at the current levels and speeds encountered their performance is adequate.

By changing the d-c return level of the steering transistor base resistors, the turn-on of the steering transistor can be controlled. Two steering stages are used for each flip-flop, the input of each connected to opposite sides of the previous flip-flop. Gating (count up or count down) is performed by controlling the base-resistor voltage return levels. The ON line is at a negative three volt level while the OFF line is at negative six volts.

Two resistor matrices sense that the count has exceeded the range of binary three through twelve. Their outputs operate either of the two forced-feedback amplifiers, Q_{18} and Q_{19}. Transistor Q_{19} is the forward forced-feedback amplifier which causes a forced zero reset by allowing the $\bar{8}\,\bar{4}\,2\,1$ base resistors to be returned to −3 volts, turning on those stages. Transistor Q_{18} functions as the reverse forced-feedback amplifier, controlling the $8\,4\,\bar{2}\,\bar{1}$ base resistor return level.

A separate isolation amplifier Q_{17} is provided for zero reset, 0 0 1 1, so that all counters can be connected to a common reset line controlled by one transistor amplifier. This base line must be at +0.7 v when counting.

Any positive input transition of 2 to 3 volts in amplitude, with a rise time of less than 1.0 microsecond or a pulse duration greater than 0.25 microsecond allows the counter to operate well above 100 kc.

Separate signal inputs and carry outputs for add and subtract are provided and are connected from the output of the previous decade to the input when cascading counters. This eliminates switching the carry outputs in each cascaded decade counter and allows isolated inputs for the first decade. First stage inputs may be connected together if only one input pulse source is used. When two separate inputs are used, they may be connected to provide control of the count up and count down lines directly. This is indicated in Fig. 1.

A pulse discriminator has been built which allows the counter to operate with random count up and count down inputs simultaneously without loss of information.

The number stored in the binaries may be changed to the nines complement by pulsing either control line to ground with a positive, 3-volt, 2-microsecond pulse. This enables the use of nines complement arithmetic when the counters are used in computers.

Eight resistor-isolated base lines, S, R, U, T, W, V, Y, X, shown in Fig. 3, are available for direct preset from any of the ten existing code combinations. These resistors, normally returned to a positive 0.7 v, are momentarily returned to the negative 3-volt level during reset. A four-pole ten-position switch, eight resistors, and a single pole pushbutton switch are required.

Eight inputs 8 4 2 1 and $\bar{8}\,\bar{4}\,\bar{2}\,\bar{1}$ each at 0.5 milliampere minimum are required for the indicator circuit of Fig. 4. The counter supplies these currents directly. Internal amplifiers provide the power to drive the resistor matrix which in turn drives voltage amplifiers controlling the Nixie tube.

The bias method allows the use of an unregulated power supply for the Nixie voltages, which may vary as much as ±20 percent. Reliable operation at high temperatures is maintained by reverse base-emitter bias during turn-off of the amplifiers.

The first binary stage was included in the feedback circuit to provide simple zero set. However, it need not be involved for higher counting speeds. Counters using regenerative type reset circuits and mesa or microalloy transistors are capable of reliable megacycle reversible counting.

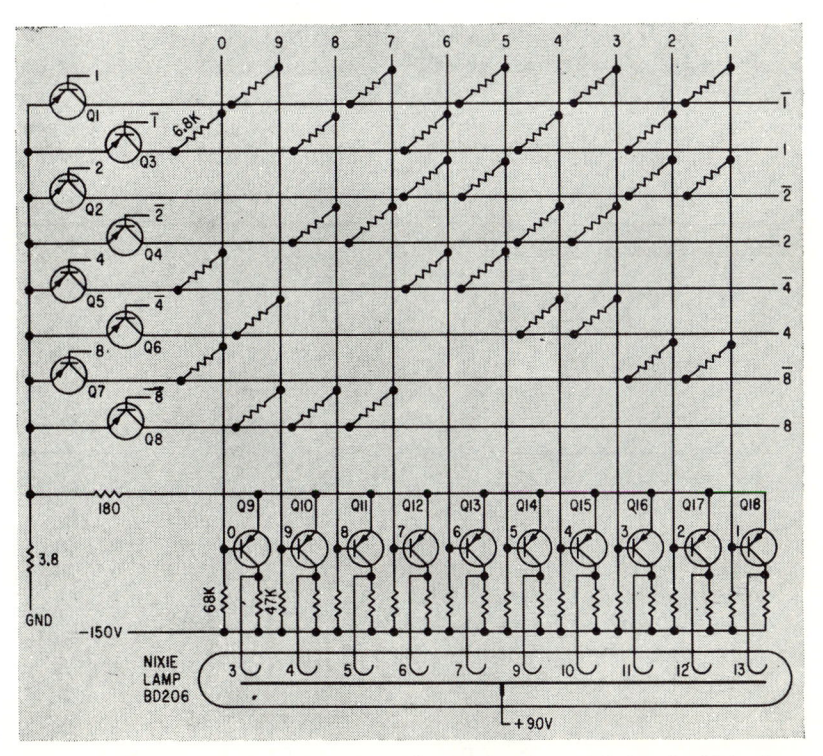

FIG. 4—Readout circuit has high gain, allows 20-percent variation in Nixie lamp supply voltage

Choosing Transistors for Monostable Multivibrators

By JOSEPH R. KOTLARSKI, Hughes Aircraft Co., Culver City, California

A SIMPLE DEVICE frequently used as a variable delay generator is the transistorized version of the monostable multivibrator (Fig. 1). When triggered, its output consists of a gate of either polarity. The natural period of this gate, determined by an R-C timing circuit, can be considerably shortened by a logical trigger.

Often the ratio of the natural period to recovery time, T/τ, designated σ, can be rather large (> 10). In such cases, particular care should be exercised in choosing the semiconductor unit.

This paper, which presents limitations encountered in design of such circuits, shows that the major limiting factor is the d-c gain, β, of the transistor; it shows further that the minimum gain necessary for predictable operation is a function of σ.

General Circuit

Figure 1 shows the general circuit. Two basic assumptions are made: voltage drop across the saturated transistor is zero; equal currents are switched from Q_2 to Q_1 during its quasistable state. Resistors R_1 and R_2 are chosen so that two states of the circuit are satisfied: Q_1 cut off during stable state, Q_2 saturated; and Q_2 saturated during the quasi-stable state, Q_1 cut off.

In the stable state, the current through Q_2 is determined by the external resistors R_L and R_E.

$$I_C = (V_{CC} - V_E)/R_L \cong V_{CC}/(R_L + R_E) \quad (1)$$

$$R_L I_C = V_{CC} - V_E \quad (1A)$$

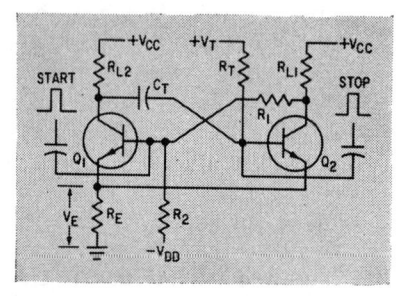

FIG. 1—A monostable multivibrator such as this is frequently used as a variable delay generator

Since there is a considerable variation in individual transistor β from a given family of units, the design must accommodate the lowest gain unit. Thus, for all units of a family to be in saturation in the Q_2 position, a base current of $I_{B\max}$ must flow:

$$I_{B\max} = I_C/\beta_{\min} = (V_T - V_E)/R_T \quad (2)$$

or substituting (1) into (2)

$$I_{B\max} = (V_{CC} - V_E)/\beta_{\min} R_L \quad (3)$$

Combining (3) and (2), and solving for R_T

$$R_T = [(V_T - V_E)/(V_{CC} - V_E)] \beta_{\min} R_L \quad (4)$$

Let $(V_T - V_E)/(V_{CC} - V_E) = \gamma$
Then Eq. (4) reduces to

$$R_T = \gamma \beta_{\min} R_L \quad (4A)$$

Since the delay of the circuit (used interchangeably with natural period of the blanking gate) is a function of $R_T C_T$, and since Eq. (4) shows that R_T is proportional to V_T, a smaller C_T can be used if R_T is returned to a high potential. This device may be employed when large delays are used and the physical size of the capacitor is limited.

As stated above, the delay is a function of $R_T C_T$. Explicitly, the wave form at the base of Q_2, upon the receipt of the initial trigger, can be expressed as

$$E_O(t) = [V_T + I_C R_L - V_E] [1 - exp(-t/R_T C_T)] \quad (5)$$

Delay of the circuit is uniquely defined because the base of Q_2 has regained its pre-trigger potential by rising $R_L I_C$ volts. At this point Q_2 fires and the circuit returns to its stable state. Therefore, when

$$I_C R_L = [V_T + I_C R_L - V_E] [1 - exp(-T_o/R_T C_T)] \quad (5A)$$

Eq. (5A) must satisfy the relation

$$T_O = R_T C_T \ln [1 + I_C R_L/(V_T - V_E)] \quad (6)$$

But using (1A) and (4A), (6) becomes

$$T_O = \gamma \beta_{\min} R_L C_T \ln [1 + 1/\gamma] \quad (6A)$$

After returning from its quasi-stable state, the circuit requires a finite time before it returns to its initial stable state. To assure predictable operation, the voltage at the collector of Q_1 must return to at least 95 percent of its initial value before the next trigger. The 95 percent

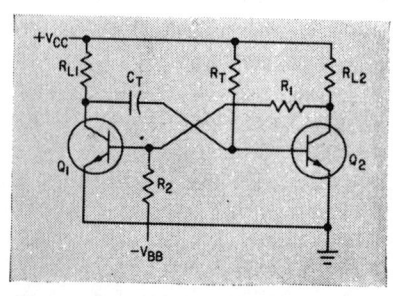

FIG. 2—In this circuit, charging network return is identical to the collector supply

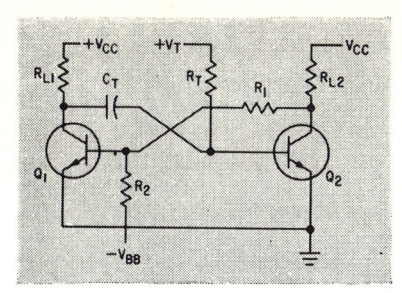

FIG. 3—Lower gain transistors may be used in this circuit if V_T is a higher source than V_{cc}

point, which corresponds to three recovery-path time constants, was arbitrarily chosen since an infinite time is required for the collector of Q_1 to completely recover. The recovery-path time constant is $(R_L + R_E)C_T$ and a duration equal to three time constants is considered sufficient for complete recovery of the collector of Q_1. This recovery time is designated as τ and during this time the collector of Q_1 will return to 95 percent of its initial value. Thus, the recovery time is

$$\tau = 3[R_L + R_E]C_T \quad (7)$$

and σ, which is T_o/τ, is

$$\sigma = [\gamma \beta_{min} R_L / 3 (R_E + R_L)] \ln(1 + 1/\gamma) \quad (8)$$

From Eq. (8) the relationship for minimum d-c gain, β_{min}, is readily obtained:

$$\beta_{min} = [3(1 + R_E/R_L) / \gamma \ln(1 + 1/\gamma)]\sigma \quad (9)$$

For cases where the charging network is identical to the collector supply (i.e., $V_T = V_{cc}$ and $\gamma = 1$)

$$\beta_{min} = 3\sigma(1 + R_E/R_L)/\ln 2 = 4.3(1 + R_E/R_L)\sigma \quad (9A)$$

If the circuit in Fig. 2 is used, where $R_{L1} = R_{L2} = R_L$ and $Q_1 = Q_2 = Q$, and R_1 and R_2 are determined as for Fig. 1, and since $R_E = 0$, then Eq. (9A) simplifies to

$$\beta_{min} = 4.3\sigma \quad (9B)$$

As a rule of thumb, Eq. (9B) can be used for choosing transistors, given σ and a single supply. When additional higher supplies are available, it is advantageous to employ them as returns for the timing resistor R_T. The advantages are twofold: smaller sized capacitors may be used and, more important from the designer's viewpoint, lower gain units may be used.

If $(V_T - V_E) \lessgtr 5(V_{cc} - V_E)$ or, equivalently, $1/\gamma \lessgtr 0.2$, the approximation $\ln(1 + 1/\gamma) \cong 1/\gamma$ may be used with less than a 10 percent error. Then Eq. (9) becomes

$$\beta_{min} = 3(1 + R_E/R_L)\sigma \quad (10)$$

If a circuit similar to Fig. 3 is used, where $R_{L1} = R_{L2} = R_L$, and R_1 and R_2 are determined as for Fig. 1, then, because $R_E = 0$,

$$\beta_{min} = 3\sigma \quad (11)$$

Thus, comparison of Eq. (9B) and (11) shows that lower gain units may be used by returning R_T to a higher source. The greater the ratio of V_T/V_{cc} the less error introduced by the logarithmic approximation.

Figure 4, a normalized plot of Eq. (9), shows the β_{min} necessary to obtain a given σ. Additional information needed for this graph is determined by other circuit considerations, such as load, stability, etc.

Examples

Example A—Given the circuit of Fig. 1, with the following values: $V_T = V_{cc} = 50$; $R_E/R_L = 0.1$; $R_{L2} = R_{L1} = 4,700$ ohms; $R_T = 36,000$ ohms; $R_E = 470$; $C_T = 0.015$; $Q_1 = Q_2 = 2N336$; and from Fig. 4, $\beta_{min}/\sigma = 4.8$. A σ of 13 is desired (where $T_o = 3,750$ μsec and $\tau = 280$ μsec.). Thus a β_{min} of 62 is required. The 2N336 can be used since its β_{min} is 76.

Example B—In systems with more than one supply, the same circuit requirements may be met by the same circuit (Fig. 1) using a lower gain transistor. Circuit values are: $V_{cc} = 50$; $V_T = 300$; $R_{L2} = R_{L1} = 4,700$ ohms; $R_T = 2.2$ megohms; $R_E = 470$; $C_T = 0.01$; $Q_1 = Q_2 = 2N338$. Ratio σ is 13; $V_T = 6V_{cc}$.

In this example, the numerical value of γ is required. To obtain γ, the voltage V_E, across the common emitter resistor, R_E, is required. Since $I_E = I_C + I_B$, and since usually $I_C \gg I_B$, $I_E \approx I_C$. Thus, $V_E = I_C R_E$. Value of I_C, calculated from Eq. (1), is 9.7 ma; thus, $V_E = 4.55$ v. Now γ is readily calculated since both V_{cc} and V_T are known.

For this example, $\gamma = 6.5$; and $R_E/R_C = 0.1$. By interpolation on Fig. 4, $\beta_{min}/\sigma = 3.4$, making $\beta_{min} = 44$. Therefore, a 2N338 can be used since its β_{min} is 45.

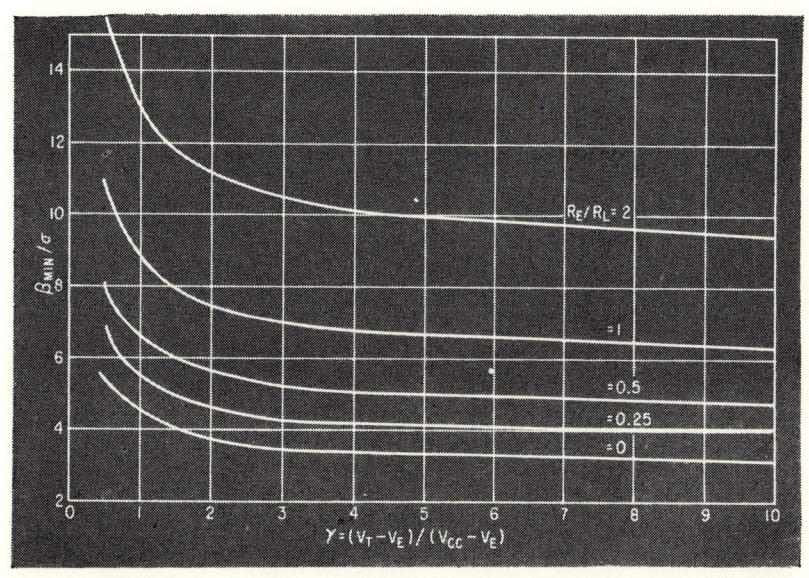

FIG. 4—The minimum d-c gain, β_{min}, needed to obtain a given σ, is determined from this normalized plot

Increasing Counting

Use of transistors in this glow-tube counter results in a cost reduction of one-half and an increase in reliability. Units can be cascaded to read as high as 10^5 and are used in nuclear instrumentation

By **HENRY A. KAMPF,** Consulting Engineer, Packard Instrument Co., Inc., La Grange, Illinois

USING A COMBINATION of transistors and glow-transfer counting tubes results in an inexpensive approach to reliability for counting systems. The absence of vacuum tubes and the use of transistors that are either cut off or saturated results in a nearly ideal counter which is unaffected by power supply variations as large as 20 percent and temperatures as high as 60 C. A single low-speed decade complete with decimal readout can be produced at less than one-half the cost of conventional decimal counting units.

Glow-transfer counting tubes perform the function of counting and simultaneously provide visual readout by the position of the glow of the tube. Each glow tube requires two negative pulses to advance the glow from one cathode to the next. One of these pulses is fed to the first guide which advances the glow one-third of the way. The second pulse is fed to the second guide which advances the glow the second one-third of the distance from cathode to cathode. The glow finally advances the last one-third of the distance as the pulse driving the second guide falls to zero.[1,2]

Pulse Timing

The negative pulses driving the guides are timed so that the second-guide pulse is nearly at full amplitude before the pulse at the first guide begins to fall as shown in Fig. 1A. Pulse amplitudes of at least 80 v are required to drive the tubes reliably.

Larger pulses produce faster glow transfer, giving faster counting rates. However, 80 v pulses will drive a GC10B glow tube at a rate of 1000 counts per second.

The pulse width also affects the counting rate. It is not possible to start glow transfer before the glow is resting on a cathode, and a finite time is required to transfer the glow from cathode to guide to guide to cathode; therefore a minimum period exists below which input pulses will not be resolved. Driving pulse width T_1 at one-half amplitude must not be wider than about 80 percent of T to allow for adequate glow-transfer time.

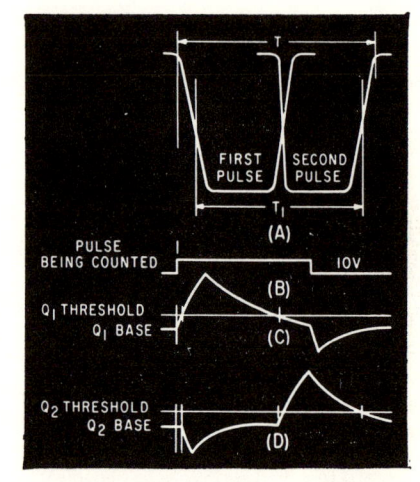

FIG. 1—Waveshape of pulses at the guides (A), input (B), transistor Q_1 (C) and transistor Q_2 (D) are shown

The circuit shown in Fig. 2 is a simple reliable circuit capable of 1-kc operation. It provides driving pulses of about 100 v. Since this driving-pulse amplitude is smaller than that usually used with the glow-transfer tube, it is necessary to make the output pulse widths wider than just described in order to accommodate the slower glow-transfer times. This circuit is essentially two amplifiers in cascade; both are saturated when no signals are present.

Circuit Operation

The positive input pulse is differentiated by the coupling capaci-

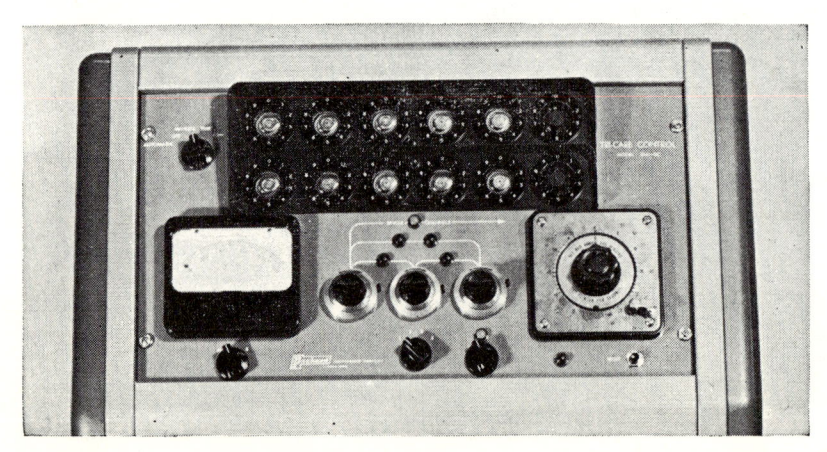

Reliable counter uses glow tubes and transistors

System Reliability

tor C_1 and R_1, the input resistor of transistor Q_1. The portion of this differentiated pulse that exceeds the cut-off threshold of Q_1, produces a large negative pulse at the collector of Q_1 as it is cut off. This pulse is fed to the first guide of the glow tube and also fed to transistor Q_2 to develop the second pulse. The pulse is differentiated by capacitor C_2 and R_2, the input resistor of Q_2.

Negative excursion of the signal at the base of Q_2 has no effect since Q_2 is already saturated. However, the positive part of this signal which exceeds the cut-off threshold of Q_2 causes the pulse output that is used to drive the second guide.

The time constant at the base of Q_1 controls the pulse width of the first pulse and the delay time of the second pulse. The time constant at the base of Q_2 controls the pulse width of the second pulse.

These 1-kc circuits are cascaded by connecting the input of one to the output of another. Registers as high as 10^5 have been obtained by using this method.

4-Kc Scaler

The circuit shown in Fig. 3 drives the glow tube at its maximum possible rate. It uses a single-shot multivibrator and step-up transformer T_1 to obtain the 300 v pulses necessary to drive the tube at its maximum rate of 4-kc. In this circuit the single driving pulse is fed to both guides at the same time.

The pulse arrives at the first guide after passing through differentiating network R_1 and C_1 while the pulse arriving at the second guide charges up capacitor C_2 of the pulse-stretching network. Therefore, as the pulse at the first guide is decaying the second guide-pulse voltage is still at a high value, and the glow is transferred as previously described.

The main-driving pulse width is determined by the multivibrator time constant and the pulse amplitude is determined by the loading on the step-up transformer. Pulse amplitudes as large as 300 v are obtained and, with a half-amplitude width of only 60 microseconds, are capable of driving the glow tube at a 4-kc rate. Some of the GC10B tubes have counted as fast as 6 kc with this circuit.

These circuits are quite tolerant of component variations in production and are used in the counting system of liquid scintillation spectrometers and other nuclear instrumentation.

Printed circuit layout of 1-kc and 4-kc scaler aids in assembly and packaging

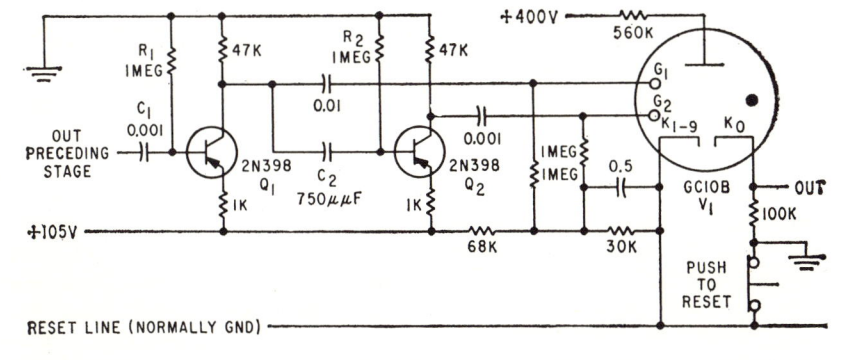

FIG. 2—Circuit for 1-kc scaler uses 5-percent tolerance resistors and 10-percent tolerance capacitors

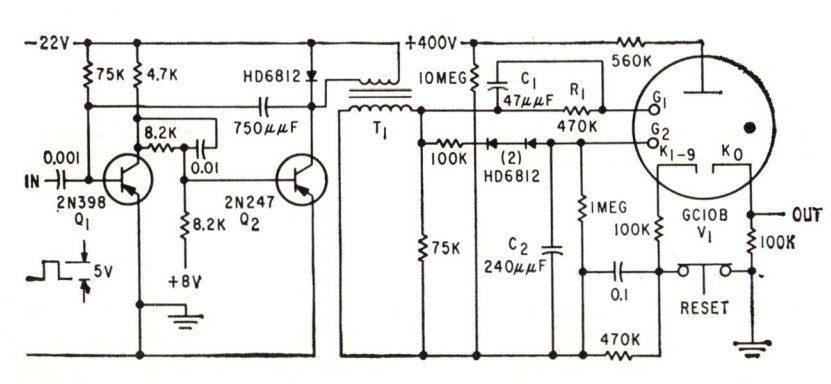

FIG. 3—All capacitors and resistors of 4-kc scaler have tolerance of 5-percent

REFERENCES

(1) J. H. L. McAsulan and K. J. Brimley, Polycathode Counter Tube Application, ELECTRONICS, p 138, Nov. 1953.

(2) R. C. Bacon and J. R. Pollard, The Dekatron, *Electronic Engineering*, p 173, May 1950.

Binary Circuits Count Backwards or Forwards

Transistorized binary counter uses small digital building blocks to perform adding and subtracting functions. Technique of using digital building blocks can be applied to other applications

By H. J. WEBER, Section Head, Servomechanisms, Inc., Goleta, California

BINARY COUNTERS generally consist of a number of stages of bistable storage elements whose purpose it is to count an incoming pulse train. This counting process is accomplished by cascading these devices so that two pulses applied to each stage result in a single output pulse which triggers the following stage. These bistable storage elements form a register; the number of pulses applied as an input appears as a binary representation in the states of the storage elements.

A counter of the type described above is capable of counting or adding a number of incoming pulses. This article describes the logic as well as the actual working circuits of a counter that is capable of adding and subtracting pulses at the control of the appropriate logic circuits.

This reversible binary counter[1] has been assembled using extremely small digital building blocks. The complete reversible counter fits on to a single plug-in unit 3.25 x 3.8 x 0.75 in. which includes the plug-in connector.

In explaining the theory of operation, the following definitions are used.

Input pulses applied to the forward and backward lines are of positive polarity. The ZERO state of a flip-flop is defined as *0 1*; the ONE state as *1 0*.

Transistors of the *pnp* type are used and the output from a flip-flop is as shown in Fig. 1.

ZERO state of a transistor is defined as the non-conducting state in which the collector is approximately $-V_{cc}$, the ONE state as the conducting state in which the collector is at approximately zero volts.

An output pulse of the proper polarity to trigger the following stage occurs only when the flip-flop changes from the ONE state to the ZERO state.

Finally, the flip-flops used in these circuits are operated in a saturation mode to provide maximum stability as well as independence from transistor parameters and selection.

Theory of Operation

The logic diagram for a four-stage reversible binary counter is shown in Fig. 1. It is understood that counters with any number of stages can be made; the method of assembly being simple repetitions of the four stages used here.

When this logic diagram is broken down to its submodule approach it becomes apparent that the reversible binary counter can be reduced to five flip-flops, six AND gates and four OR gates. Flip-flop 1 is the control or setting flip-flop which is operated in a set-reset manner. Flip-flops 2 to 5 comprise the storage register.

The reset switch shown in Fig. 2 is used to put the storage register in the ZERO state. Assume that flip-flops 2 through 5 are all in the ZERO state, *0 1*. Then, if an input pulse is applied to the forward bus, flip-flop 1 is cut off and the forward bus line is at $-V_{cc}$. The same input pulse is also routed through OR-1 to trigger flip-flop 2 and change its state to *1 0*.

Since the right-hand side of flip-flop 2 has changed state in the direction from *1* to *0*, an output pulse would have been routed through AND-1 but for the fact that the second input to AND-1 is tied to the backward bus which is at ground potential. One input to AND-2 is tied to the forward bus which is at $-V_{cc}$, but since the left hand side of flip-flop 2 changed state from *0* to *1* no output is forthcoming from this gate. The counter has received one input pulse and this information is stored as 0001 in flip-flops 2 through 5.

Indicator

The least significant digit is stored in flip-flop 2 and since this stage is in the ONE state the indicator lamp associated with it lights up to show its condition.

When the flip-flop output is zero Q_5 is nonconducting and the voltage at its collector is −45. Therefore, the voltage across the neon lamp is 40 v, which is insufficient to break

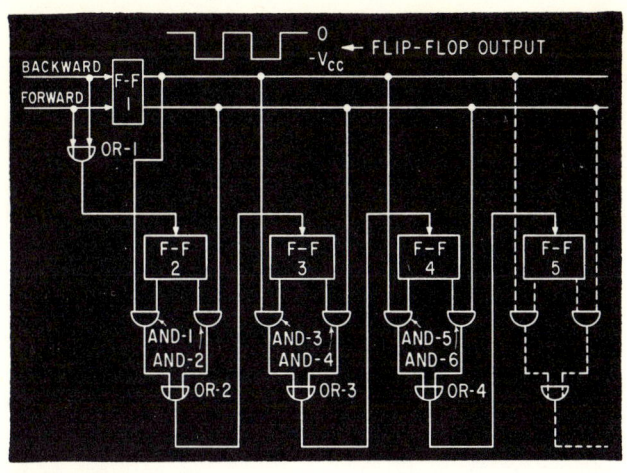

FIG. 1—Counter is broken down to basic building blocks. Dotted lines show how additional stages can be added

Use of submodules is shown in construction of counter. Larger printed boards are flip-flops, smaller boards are gates

down the neon. When the flip-flop is at -12 v, Q_5 conducts heavily and the voltage at its collector is essentially zero. The voltage across the lamp is thereby increased to 85 v which is sufficient to break down the neon and light the bulb.

Now assume that another input pulse is applied to the forward line. Although this pulse does not change the state of the control flip-flop, it serves to trigger flip-flop 2, which reverses its state to *0 1*.

This change of state applies two negative voltages to D_1 and D_2 of AND-2, cutting off transistor Q_7 and causing its emitter to drop to $-V_{cc}$. This negative voltage is transmitted through diode D_3 of OR-2, inverted and the resultant positive pulse used to trigger flip-flop 3. This changes the state of flip-flop 3 to *1 0*. The count is now stored as 0010 in the storage register.

Assume now that a pulse is applied to the backward bus line. This pulse causes the control flip-flop (flip-flop 1) to change state, causing the backward line to assume $-V_{cc}$. The same input pulse triggers flip-flop 2 reversing it to the *1 0* state. This change of state permits a pulse output to get through the AND-1 gate thereby triggering flip-flop 3 back to the *0 1* state.

The register now displays the total count as 0001 and has, therefore, shown that it can cycle in a forward direction as well as in a backward direction.

Modular Design

The plug-in module described is a separately packaged function which may be subdivided into individual submodules or digital building blocks. These fully engineered and tested building blocks can then be used to assemble the particular counter in a minimum of time and at the lowest possible cost.

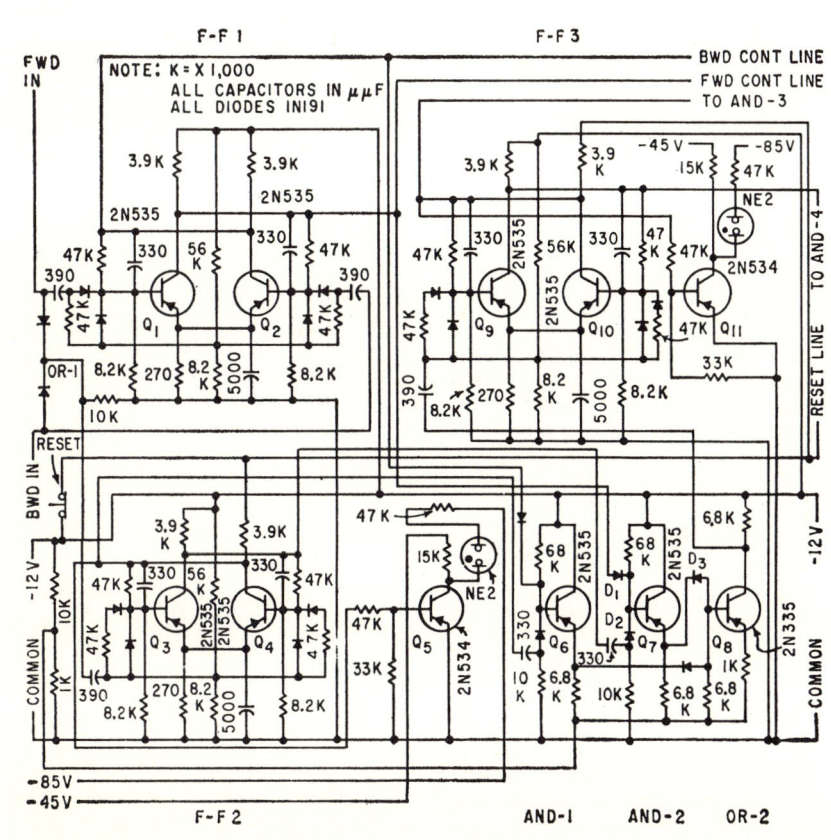

FIG. 2—Flip-flops used in counter are identical. Input voltages to flip-flops depend upon circuit function

Reference

(1) R. L. Trent, A Transistor Reversible Binary Counter, *Proc of NEC*, **8**, p 346.

Triggered Bistable Circuits

By J. B. HANGSTEFER and L. H. DIXON Jr., Solid State Products, Inc., Salem, Mass.

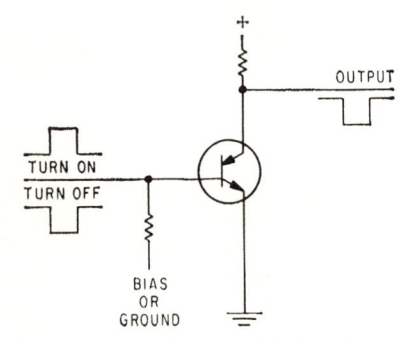

THIS *PNPN* semiconductor device is triggered on by a low-level positive pulse applied to its base. Once on, it will remain on without need for sustaining base current. A negative pulse applied to the base will turn it off where it will remain until triggered on again.

Present units are designed for operation in the range of 1 to 8 ma collector current. When on, the collector-to-emitter voltage drop is approximately 0.8 v. The dynamic resistance is in the region of 10 ohms. When off, the device has a high impedance with leakage current normally less than one microampere. Turn-on and turn-off times are approximately 0.4 μsec each and circuit repetition rates to 200 kc are possible.

The ohmic value of the collector resistor and the B+ voltage determine the on current level. For most circuits, the current level should be set between 3 and 5 ma to insure the best

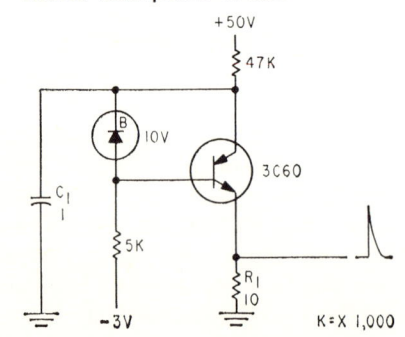

Turn on and off accomplished at base. Output pulse width determined by time between on and off input pulses is independent of pulse widths

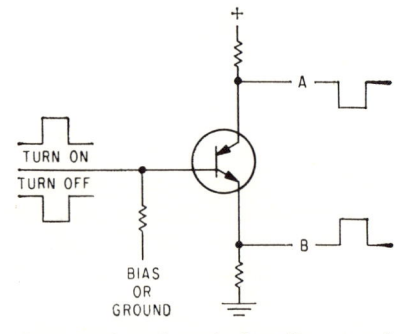

Output taken from both collector and emitter. Input voltage for both on and off increased by amount equal to output voltage at point B

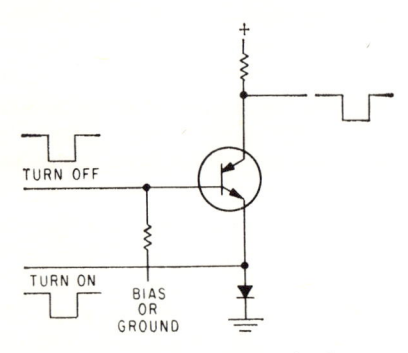

Negative trigger pulses used. Turn-on to emitter across silicon diode. Diode used because impedance is low with device on. Turn-off accomplished at base

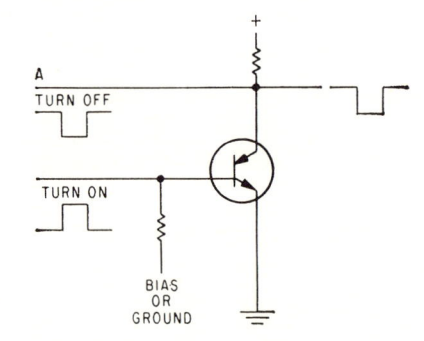

Turn off by driving collector negative. Essentially all collector current must be bypassed to point A for turn off to take place

Pulse generator delivers 1 ampere peak output current of 10 μsec duration. Higher outputs possible by reducing R_1 and increasing C_1

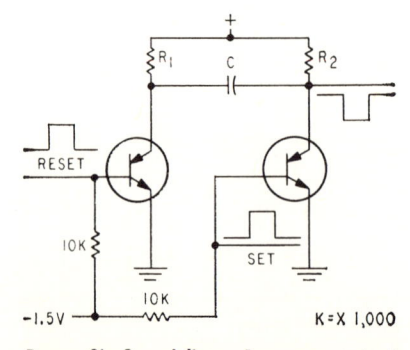

Power flip-flop delivers 1 ampere output current. On current determined by B+, R_1 and R_2. $C_{min} = 25I/E$ where I is on current, E is B+ voltage

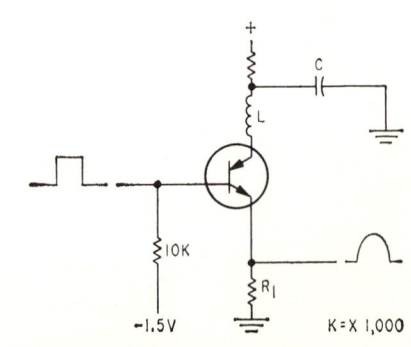

LC pulse generator has half-sinusoid output determined by L and C. R_1 should be less than $X_L/10$ to prevent excessive damping

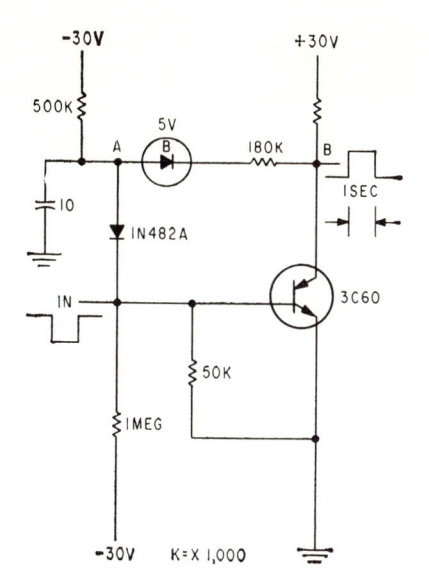

One-shot multivibrator provides up to 10 seconds delay. Circuit normally on. Point A at −11 v, point B at +1 v. Negative trigger operates circuit

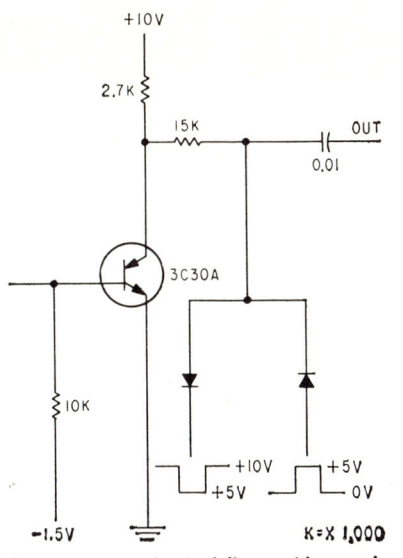

Basic memory circuit delivers either positive or negative pulse output accomplished by coupling diodes and transition memory capacitor

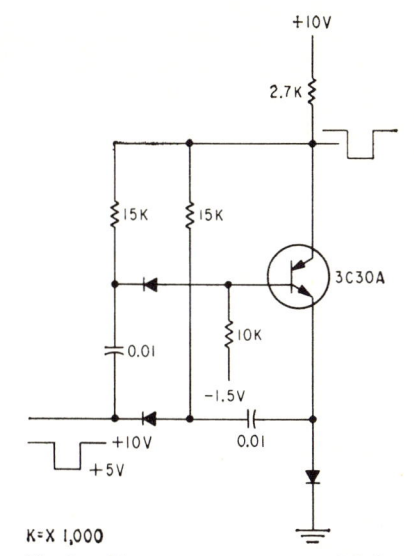

Flip-flop binary counter operates 2:1 by negative trigger pulses. Can be driven from identical flip-flop or from collector of npn silicon transistor

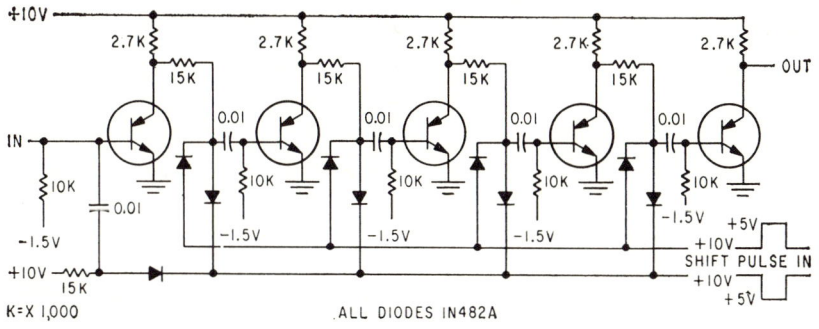

Shift register consists of five one-bit memory elements connected in cascade

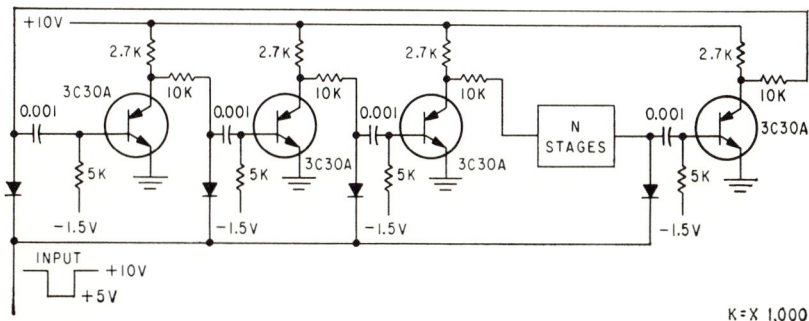

N-stage ring counter uses modified memory circuit. Input pulse turns off all stages except one following on stage

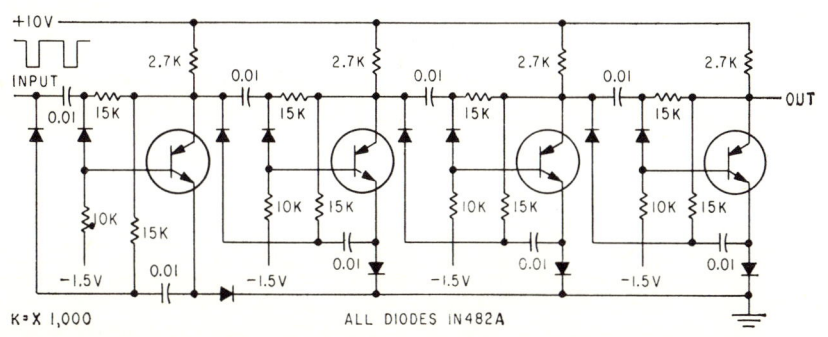

Four-stage binary counter has 16:1 division operating on negative trigger pulses

performance throughout the operating temperature range. The B+ voltage should be above +3 v to make the circuits insensitive to small voltage changes.

When the device is off, collector cutoff current can act as a positive gate current signal, tending to turn the device on. A base resistor is used to provide base bias current to insure stable off conditions throughout the operating temperature range. For operation up to 125 C, the base bias current should be a minimum of −150 μa. If a bias voltage of −1.5 v is used, the base resistor should be 10,000 ohms or less.

Smaller values of base resistor can be directly connected to ground or emitter at the expense of increased trigger current requirements and a reduced operating temperature limit. The base-to-emitter voltage must be less than +0.10 v at 100 C. Since collector cutoff current can approach 50μa at 100 C, the maximum value for base resistor at 100 C when directly grounded is 2,000 ohms. More base resistance can be used below 100 C. Above 100 C, negative bias is recommended.

The symbol used indicates the base contact is at the internal p region of the pnpn device.

Pulse Sorting With

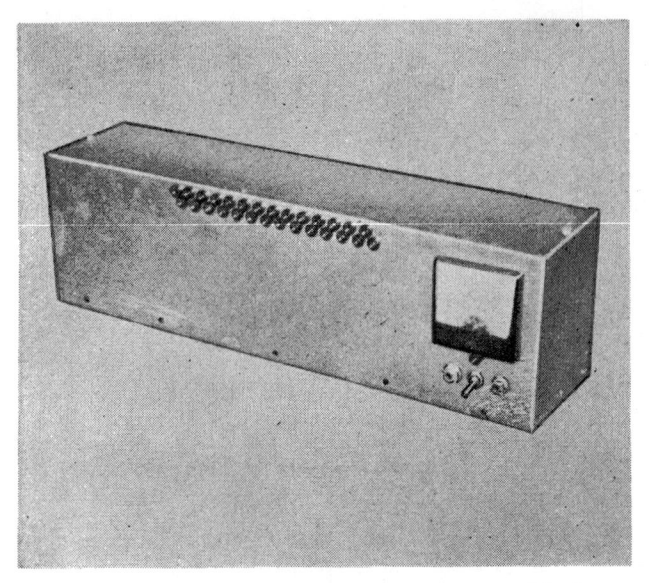

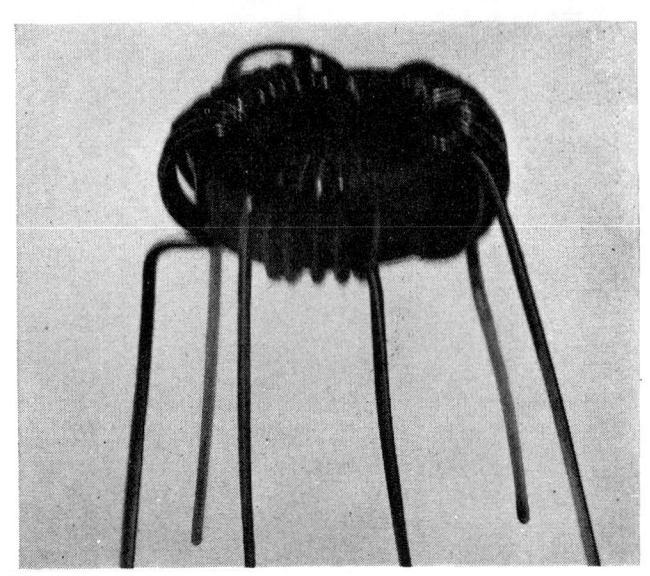

Pulse sorter is part of a battery-powered decommutator, left. Sorter uses triple-winding cores as shown at the right

When a pulse train is applied to the input of this solid-state pulse sorter, the sorter reproduces the width of each pulse at an output terminal that corresponds to the pulse's position in the train. Simple, reliable circuits handle rates greater than 1,000 pulses/sec

By **JOHN H. PORTER,** President, Portronics, Inc., Tarzana, Calif.

PULSE TRAINS sometimes convey information by presenting pulses of varying width. The pulse sorter to be described receives a train of pulses and presents each pulse at an output terminal which corresponds to the pulse's place in the train. After presenting the last pulse of the train, the sorter to be described readies itself for another pulse train or closes its input.

Figure 1 shows what happens when a pulse train preceded by a start pulse is applied to the sorter. The widths of the pulses that ap-

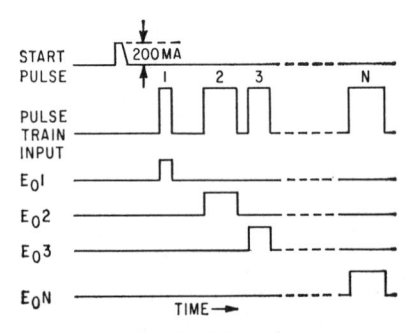

FIG. 1—Width of each pulse sent into sorter is reproduced at outputs E.

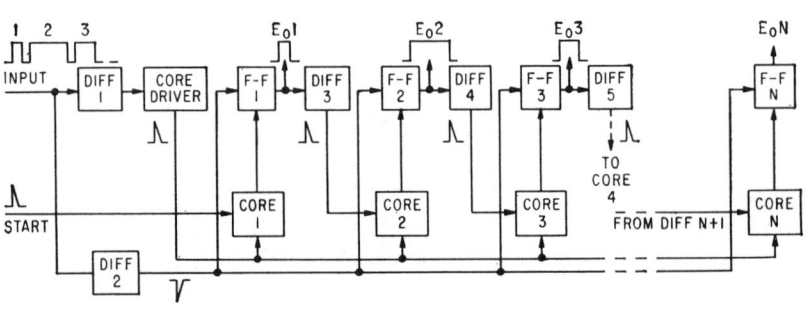

FIG. 2—Last output, E_oN, of sorter can be used to pulse core 1, thus setting up the sorter for next pulse train. Differentiators 3, 4 and 5 set up cores for pulse from core-driver

142

Transistors and Ferrites

pear at outputs E_o1, E_o2, E_o3 ... E_oN are equal to the widths of corresponding input pulses.

System Operation

A start pulse sets up the sorter (Fig. 2) to receive the pulse train. The start pulse switches off ferrite core No. 1, which turns off transistor flip-flop No. 1. The leading edge of the first input pulse is differentiated by differentiator 1, which pulses the core-driver. The core driver output switches on core No. 1, which delivers a positive pulse to flip-flop 1. Flip-flop 1, which sorts the first pulse of the train, goes on.

The trailing edge of the first input pulse is differentiated by differentiator 2 which applies a negative pulse to flip-flop 1. Flip-flop 1 goes off, thus duplicating the first pulse of the train.

When flip-flop 1 went on, it pulsed differentiator 3, setting it up for a negative pulse when flip-flop 1 goes off. Differentiator 3 goes on momentarily and pulses core 2. This pulse readies core 2 for the next input pulse of the train, in the same way as the start pulse set up core 1 for the first input pulse.

Flip-flop 2 and succeeding flip-flops duplicate succeeding input pulses at their outputs the same way as flip-flop 1.

Circuit Details

Transistor Q_1 and Q_2 (Fig. 3) drive the cores with 200-ma pulses. The cores require 2 amp-turns for switching. To gain a margin of safety, each core is wound with 12 turns. The core produces about 0.5 volt for each turn of output winding, and each flip-flop transistor, such as Q_3, triggers with 2-v pulses. For a margin of safety, the core output winding has 5 turns. Since the reset windings also require 2 amp-turns, they too have 12 turns.

The flip-flop output stages produce 200-ma reset pulses. As a 2N525 transistor has a nominal β of 40 at this current, the output stage requires 5 ma from its flip-flop. This requirement accounts for the large capacitance of C_1.

All transistors that drive cores, such as Q_2 and Q_5, are biased so that between pulses only the collector cutoff current, about 5μa, flows. Thus the drivers appear as impedances of at least 1 megohm.

The flip-flops are turned off and on by positive and negative triggers, respectively. Positive trigger amplitudes are low. Negative triggers turn on the flip-flops when the triggers rise above a well-defined threshold. These negative triggers are supplied by Q_9. Each flip-flop delivers a positive-going output pulse; as many as 88 flip-flops have been used in the pulse sorter.

Cores are scramble wound with No. 28 or 30 Formvex wire. The windings are in the same direction, and concentrated in three areas of the core. These ferrite cores will not function reliably above 60 C. Tape-wound cores could be used to get around this limitation.

Compared to the transistor-core sorter that has been described, an all flip-flop sorter would require four transistors for each input pulse. When a large number of pulses is sorted, the transistor-core sorter is considerably more economical, and is more reliable, than such an all-transistor sorter.

The author gratefully acknowledges the encouragement of N. Cushman.

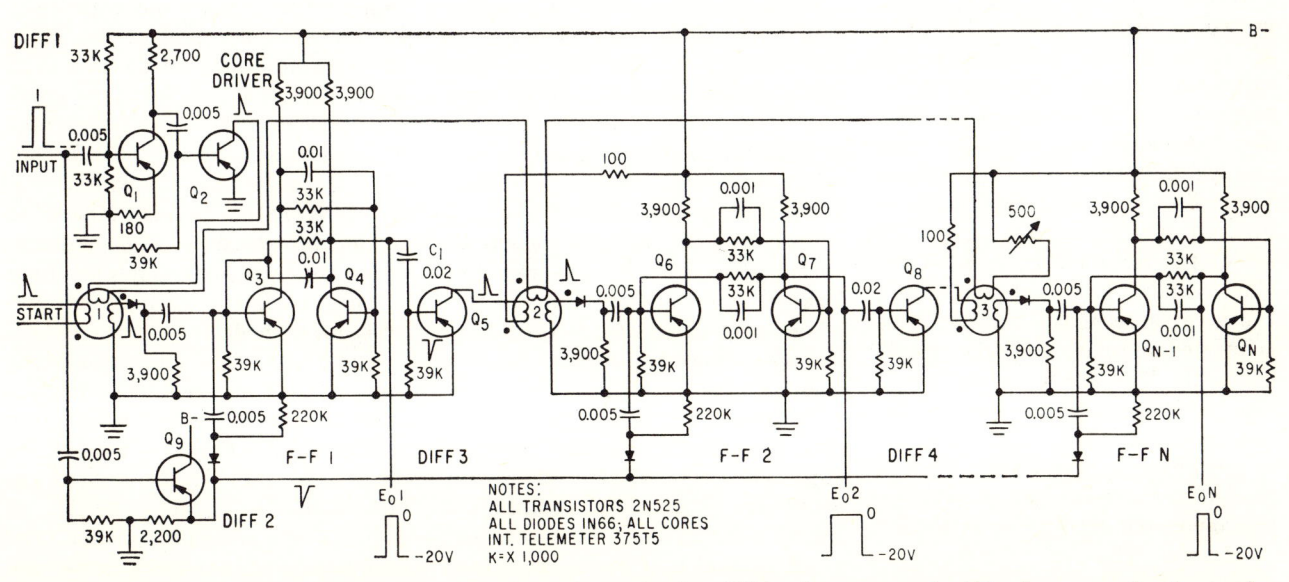

FIG. 3—Three transistors and one ferrite core are used for each sorted pulse. Widths of outputs equal widths of corresponding input pulses

Transistor Drives

One transistor blocking oscillator drives a cold-cathode counter tube to make a long-life decade counter having low power consumption. Waveform criteria for successful operation is discussed

By H. SADOWSKI* and M. E. CASSIDY,
U. S. Atomic Energy Commission, Health and Safety Laboratory, New York Operations Office, New York, N. Y.

THIS AUTOMATIC RECORDING system was developed to facilitate handling of data which are obtained from several hundred radioactive samples each day. The design of this system called for a large number of compact modular decade counters capable of storing data with visual and electronic readout. These units are connected in parallel to a 10-digit bus bar system for readout at a remote station. The maximum repetition rate of the system is slightly greater than 2,000 pulses per second.

Decade Counter Tube

The decade counter tube, commonly referred to as a glow transfer tube, used in this unit consists of a disk (anode) surrounded by 30 electrodes (cathodes) in the form of thin rods within a gas-filled envelope. The electrodes are connected as shown in Fig. 1A. Ten of these (every third one) are called cathodes and are brought out separately at the tube base. The electrodes adjacent on the same side of each cathode are tied in common internally and a single lead is brought out through the base. This set of ten is called guide 1 (G 1). The remaining ten electrodes, on the other side of the cathodes, are connected similarly and called guide 2 (G 2). The anode connection is also made through the base of the tube.

Initially, approximately 350 v is needed to cause conduction between the anode and cathode by a gaseous discharge. The tube voltage drop

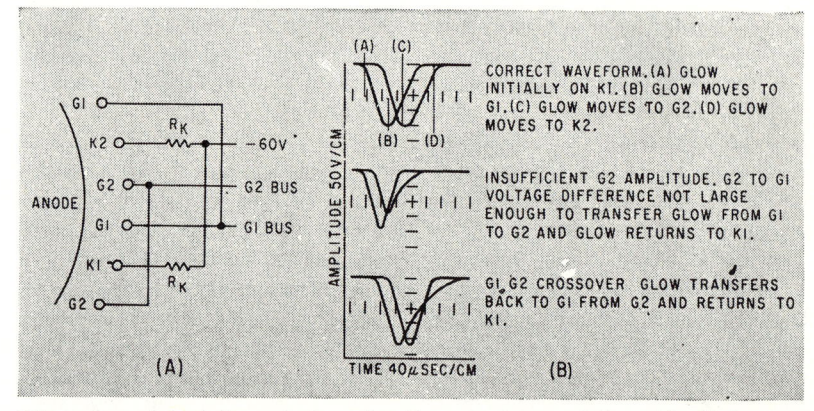

*Now with W. L. Maxson Corp., N. Y.

FIG. 1—A portion of the cathode and guide arrangement of the cold-cathode counter tube is shown in (A) and the waveform criteria for proper operation is shown in (B)

under these conditions is about 190 v, while the current is limited by an anode resistor to about 0.5 ma. A positive ion sheath is formed at the current carrying cathode, resulting in a characteristic glow of the gas in this region. This gives a visible indication of the information stored in the tube. The ions in the vicinity of this cathode also serve another purpose. If all other electrodes are tied in common and a negative voltage (greater than 50 v) is applied, these ions will allow an easy transfer of the discharge to an adjacent electrode rather than to one more remotely located. This preferential ignition of adjacent electrodes is the key to the operation of the glow transfer counter tube.

To transfer this glow in a particular direction, the guides must be supplied with pulses of proper waveform (Fig. 1B). Assume that initially the discharge is at cathode 1 (K 1) and that the cathode resistor, R_K has a voltage drop across it making this cathode a few volts negative with respect to ground. A 120 v negative pulse is now applied to the guide 1 bus and the discharge transfers to the nearest guide 1. The duration of this pulse must be sufficient to allow deionization of the gas around cathode 1. As the pulse on guide 1 begins to fall in amplitude, a large negative pulse (120 v) is applied to the guide 2 bus. When the negative voltage of the adjacent 2 guide exceeds that of 1 by more than 50 v, the discharge transfers to this guide. Had the gas at cathode 1 not been deionized, the discharge might have skipped back to the 2 guide that precedes cathode 1.

The full −60 v bias appears at the cathodes because all of the anode current is now flowing into guide 2, and there is no voltage drop across any of the cathode resistors R_K. As the potential at guide 2 rises toward ground, the −60 v bias at cathode 2 causes the discharge to transfer to this cath-

Cold-Cathode Counter

ode. The potential at cathode 2 will rise to a small negative voltage which is needed to insure a stable transfer of the discharge to this cathode. The duration of the second pulse must be sufficient to allow deionization of the gas around guide 1, to prevent skip-back to cathode 1 by this guide.

Figure 1B shows a comparison of waveforms at the guides. The top one is correct while the second illustrates insufficient amplitude of the guide 2 pulse. The discharge will not transfer but remains temporarily at guide 1 until the first pulse is over, and then returns to cathode 1. The bottom one illustrates a case where the second pulse has sufficient amplitude to transfer the discharge to guide 2 but decays more rapidly than the pulse on guide 1. Consequently, the discharge transfers back to guide 1 and finally to cathode 1.

Blocking Oscillator

The transistor blocking oscillator using emitter feedback shown in Fig. 2 was found to be a satisfactory driver. The transistor is biased near cutoff. When a negative-current pulse is applied to the base, the collector current through the primary of the transformer increases. Regeneration occurs since the tertiary winding is connected with a polarity that aids an increasing emitter current and the transistor goes rapidly into saturation. If the alpha cutoff frequency of the transistor is high enough, the rise time is determined by the primary to tertiary leakage inductance and their winding resistances in series with the tertiary shunt capacitance as well as the transformer loading. As the alpha cutoff frequency is reduced, the frequency response of the transistor ultimately limits the rise time

Operation

The upper waveform of Fig. 1B shows the waveform generated at the transformer secondary. The first part of the pulse charges capacitor C_1 through diode D_2 to produce a negative pulse at G_1. Because the voltage drop across D_2 is small, G_2 is held near ground potential. When the transformer reverses polarity, diode D_2 ceases to conduct and a negative pulse is applied to G_2. The voltage on capacitor C_1 adds to the transformer backswing voltage making the amplitude of the G_2 pulse comparable to that of G_1.

When capacitor C_1 discharges, diode D_1 conducts. This inhibits capacitor C_1 from charging in a reverse direction and clamps G_1 to ground. Resistor R_1 across D_2 helps discharge the circuit capacitance of G_2 when the second pulse is over.

To operate the circuit, a positive 5 v pulse of 100 μsec duration is required. Negative input signals are removed by diodes D_3 and D_4. Capacitor C_2 prevents d-c from reaching the transformer. Resistor R_2 converts the signal into a current pulse and reduces the amplitude of transients fed back to the blocking oscillator.

While receiving data, all cathodes are tied to the bias line through their load resistors. The current-carrying cathode is a few volts below ground due to the voltage drop across its load resistor. Electrical readout is accomplished by placing S_1 in the read position. This grounds the cathode bias line and drives the current-carrying cathode to positive 25 v while all others are held at ground. The positive signal is fed through a diode to the appropriate digit line. To reset the glow to a particular cathode, switch S_2 disconnects the cathodes from the bias. The elements then rise to a positive 100 v. The potential difference between the elements and the zero cathode is large enough to insure transfer to the zero cathode.

Acknowledgment is given to G. B. B. Chaplin of the Atomic Energy Research Establishment, Harwell, England, and H. J. Di-Giovanni of the Del Electronics Corp.

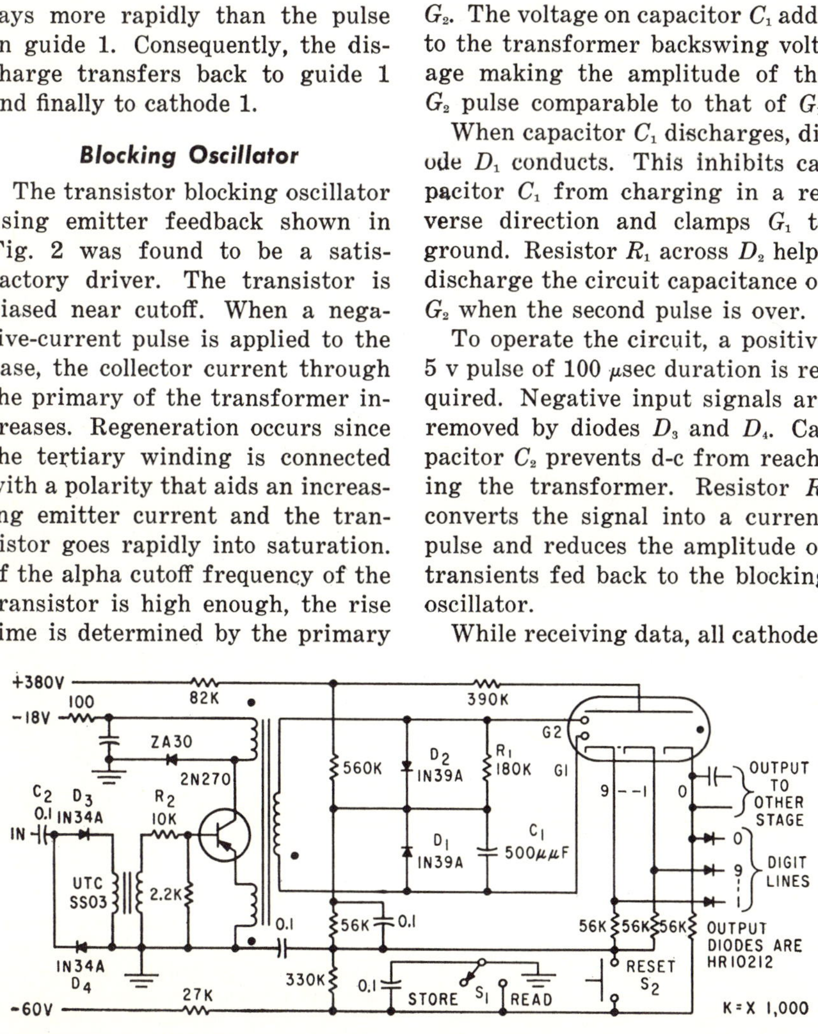

Typical installation shows mounting of glow decade modules

FIG. 2—A series of transistorized cold-cathode decade counters may be coupled together to increase the count range

Bibliography

R. Lee, *Electronic Transformers and Circuits*, John Wiley and Sons, Inc., New York, N. Y., 1947.

J. G. Linvall and R. H. Mattson, Junction Transistor Blocking Oscillator, *Proc IRE*, **43**, 11, p 1,632, Nov. 1955.

G. B. B. Chaplin and R. Williamson, Dekatrons and Electro-Mechanical Registers Operated by Transistors, *Proc IEE*, **105**, 21, Part B, p 231, May 1958.

IRE Conv. Record, March 23-26, 1959, part 9, PGNS.

Transistor Switch Design

Four significant parameters affecting transistor switches are tabulated for eight transistor types. Data presented permits designer to match transistor pairs for optimum operating characteristics

By ARTHUR GILL*, Research Division, Raytheon Manufacturing Co., Waltham, Mass.

TRANSISTORS can be used as sensitive spst switches by appropriately controlling base-to-collector voltage.[1] A pnp transistor switch and its emitter-collector characteristics are shown in Fig. 1.

When base voltage is positive, switch exhibits high resistance r_r; when negative, low resistance r_f. Ratio r_r/r_f is a measure of switch

---
* Now with the University of California

efficiency—the higher the ratio, the lower the average power dissipated in the switch as compared to that delivered to load resistor R_L.

An enlarged view of the v_e—i_e field near the origin is shown in Fig. 2. With the switch on, residual voltage V_p exists across the load; with the switch off, residual current I_p flows through the load. These two quantities impose a lower limit on signal voltage E.

Parameters r_r, r_f, V_p and I_p are significant when transistors are employed with low-level modulators. Mean value and deviation of the four parameters are given for various types of transistors in Table I. Deviation is defined as the amount by which 50 percent or more of the samples varied from the mean value. All transistors were switched on by a forward current of five milliamperes and off by a reverse voltage of 1.5 volts. It was observed that I_p and r_r were subjected to considerable drift while V_p and r_f remained constant.

Efficiency of Modulators

Sensitivity of modulators can be increased by connecting the transistor switches in opposition to cancel the effects of V_p and I_p. Since it is easier to match units with respect to V_p, the designer should strive to construct the modulating circuit so that the effect of I_p is negligible.

REFERENCE

(1) R. L. Bright, Transistor Chopper for Stable D-C Amplifiers, ELECTRONICS, p 135, Apr., 1955.

Table I—Switching Parameters for Various Types of Transistors

Type	No. of Samples	V_p in mv Mean	V_p in mv Dev.	r_f in ohms Mean	r_f in ohms Dev.	I_p in ua Mean	I_p in ua Dev.	r_r in meg Mean	r_r in meg Dev.
Audio 2N131	100	1.7	0.1	11	1.5	0.23	0.03	5.6	3
Bilateral CK870	50	5.6	1.5	2.3	0.2	3.1	0.08	17	5
CK871	50	4.3	0.9	1.7	0.2	3.3	0.09	14	6
Radio Frequency CK760	25	1.8	0.3	1.9	0.2	0.17	0.04	90	40
CK761	25	1.4	0.3	1.5	0.2	0.15	0.05	55	45
CK762	25	1.4	0.2	1	0.1	0.25	0.05	42	25
High Beta CK754	25	1.1	0.05	1.8	0.1	0.21	0.04	70	30
Silicon CK791	22	2.2	0.2	28	12	0.015	0.002	>1000	

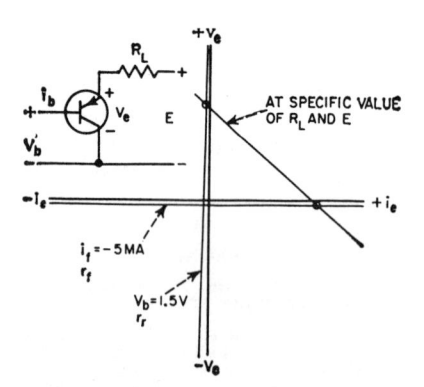

FIG. 1—Maximum power transfer occurs when R_L is geometric mean of r_r and r_f. Characteristics exist when $E < v_b$.

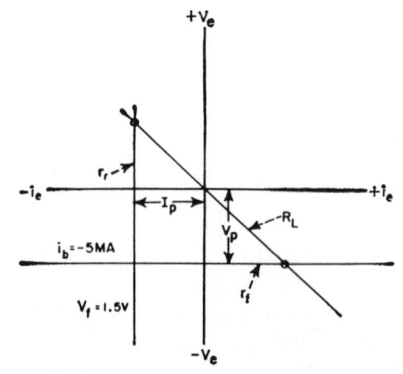

FIG. 2—Four parameters shown were evaluated for transistors connected in their inverted, or common-collector form

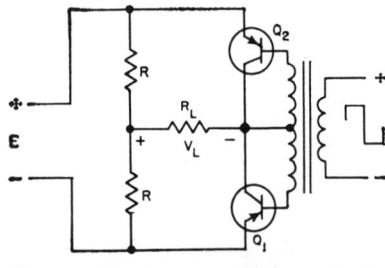

FIG. 3—Low-level modulating circuit. If V_p values are kept below 0.1 millivolt by using audio transistors, the output will contain a zero-signal a-c component whose maximum magnitude is 0.5 millivolt. Effect of mismatch in r_r and r_f is minimized if $R_L \gg r_f$

Insuring Reliability in Time-Delay Multivibrators

Silicon and Zener diodes improve performance of conventional germanium-transistor circuit. Timing network is isolated during timing interval

By PAUL E. HARRIS, Defense Systems Laboratory, Syracuse University Research Corporation, Syracuse, N. Y.

A PRECISION TIME DELAY is often required in radar and industrial electronics. Obvious applications of this multivibrator are the range gate delay in a doppler radar boxcar circuit and the time delay for an expanded range indicator sweep. An additional family of applications is the generation of gate waveforms.

A conventional transistor monostable circuit has been modified so that its performance is independent of transistor selection and its delay-stability does not vary appreciably with transistor temperature, trigger amplitude and power supply variations.

Tests show that the jitter of the output is less than four nanoseconds over a delay range of 3 to 35 microseconds. With input-trigger variations from 10 to 100 volts, variation in delay is 1.5 percent and a power supply change from 24 to 15 volts causes a delay change of 2 percent. A negative-going output pulse of nine-volt amplitude is obtained, and its steep edges (0.1 microseconds) are retained even when loaded with a 50-pf output capacitance.

Circuit Description

The multivibrator circuit is shown in Fig. 1. Reliability is achieved by introducing several modifications of the conventional monostable multivibrator[1]. Most significant of these modifications is the isolation of the timing network $R_4R_5C_1$ during the timing interval.

In the conventional multivibrator, the junction of C_1 and R_5 is directly connected to the base of Q_2. This connection leads to two difficulties. First, care must be taken not to exceed the base-to-emitter reverse voltage of Q_2. Second, and more disastrous, is the effect of i_{co}, the leakage current between collector and base of Q_2.

This leakage current flows into C_1 in parallel with the desired charging current furnished through R_4 and R_5. Since leakage current varies widely with temperature, relatively large variations in delay can be expected.

Diode D_3 is an extremely high back resistance silicon type. It serves as a voltage discriminator terminating the delay interval whenever C_1 has been charged back to ground potential. During the timing interval, D_3 is reverse biased and effectively isolates the timing circuit from the base of Q_2 and from all other components.

Circuit Operation

In the quiescent state Q_1 is cut off while Q_2 is conducting. The 9-volt Zener diode D_2 is open, allowing the base of Q_1 to remain at or near ground potential. The cathode of diode D_3 is at approximately -1 volt. A current flows through the forward biased diode D_3 to the base of Q_2 and to R_6 causing collector of Q_2 to remain bottomed.

Application of a satisfactory trigger causes conduction in Q_1. Voltage at the Q_1 collector falls toward ground potential. This voltage step is passed across capacitor C_1 to the cathode of D_3. The effect of this step is to reverse bias diode D_3 and reduce base and collector currents of Q_2 to near zero.

The collector of Q_2 rapidly approaches -24 volts. However, as soon as the collector voltage exceeds the 9-volt breakdown of Zener diode D_2, conduction occurs producing sufficient base current to bottom the collector of Q_1. The trigger is no longer needed to sustain circuit action.

The next phase of circuit operation is the sawtooth voltage runup at the cathode of D_3. That point was driven to a potential of approximately $+24$ volts by the collector of Q_1. Diode D_3 remains nonconducting and its cathode voltage moves back toward ground potential as C_1 is charged exponentially toward -24 volts by the current through R_4 and R_5.

The delay is terminated when C_1 has charged to a potential sufficient to first-produce forward conduction in D_3. Once conduction begins in D_3, turnoff is regenerative and quiescent conditions are rapidly re-established.

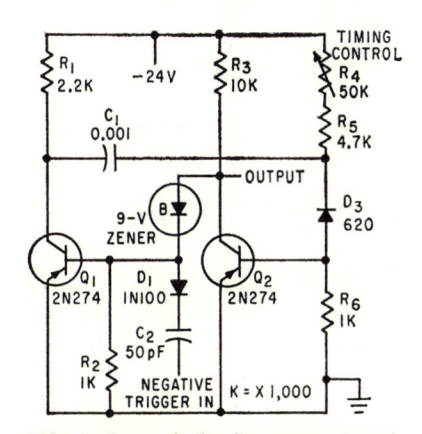

FIG. 1—Zener diode disconnects Q_2 collector from Q_1 base for rapid Q_2 cut-on. High back-resistance diode D_3 prevents collector-base leakage current from altering timing

References

(1) J. Millman and H. Taub, "Pulse and Digital Circuits", McGraw-Hill Book Co., New York, p 599.
(2) D. Sayre, "Generation of Fast Waveforms", 19, Radiation Laboratory Series, Mass. Inst. of Tech., p 187.

Feedback Stabilizes Flip-Flop

By PHILIP CHEILIK

Federal Telecommunication Laboratories, International Telephone and Telegraph, Nutley, N. J.

FEEDBACK enables a transistorized flip-flop to operate on pulses of 3 volts with 0.5 microsecond fall time. The flip-flop is very independent of changes in voltage and unbalance of transistors. The circuit was designed for use in a computer.

Operation

The common emitter resistor R_e in Fig. 1 provides d-c degeneration. For good trigger sensitivity, it is heavily bypassed in order to increase the gain around the regenerative loop. Resistor R_f connected between the bases of the two transistors provides negative feedback.

Assume that Q_1 is conducting and Q_2 is cutoff. A negative pulse is applied at the base of Q_1 cutting it off. Its collector rises to $+11$ v, and the rise is coupled to base B_2 through the cross-coupling network. As B_2 rises above B_1, a current flows in feedback resistor R_f. This current reduces the normal base current of Q_2 and prevents B_2 from rising too high. Base B_2, in turn, regulates the collector current drawn by transistor Q_2.

The flip-flop uses emitter followers in the cross-coupling network in order to match the high collector output impedance to the low base input impedance. These emitter followers also serve as low impedance output coupling to drive gating chains.

In the computer, the flip-flop is triggered with the differentiated trailing edge of a logic pulse to avoid the use of interstage delays for such circuits as shift registers. Capacitor coupling with a large time constant is used in case the logic pulse has poor fall time.

Attenuation is small even for slow fall time. The partially differentiated pulse is amplified, and the output of the amplifier is differentiated. Since phase reversal is undesirable, a grounded-base *npn* transistor provides a negative pulse to bring the amplifier into conduction from its normally cutoff state.

Design

For values of β between 20 and 100, transistors 2N124, 125, 126 and 167 may be used. The voltage swing required is six volts from $+5$ to $+11$. A self-biased multivibrator is used to improve d-c stability.

For a 2N125 at an ambient temperature of 55 C, 28.5 mw is maximum dissipation. With a collector current of about 1 ma for a transistor of minimum β, a 6,000-ohm resistor is needed for R_1 in order to get a 6-volt swing. A 6,800-ohm resistor, the nearest standard 10-percent value, is used.

When R_f is considered disconnected, 50 μa of base current is needed where $I_c = 1$ ma and $\beta = 20$. If the cross-coupling resistors are large compared to R_1, the equivalent Thevenin resistance $R = (R_2 + R_3)/R_2 R_3$. When $R = 50,000$ ohms, the base input resistance of a transistor is very low in comparison. The drop across the base-to-emitter junction can be ignored, and the total drop can be assumed to occur across R. For $I_b = 50$ μa, a positive voltage swing of 2.5 v across R is required. For a symmetrical peak-to-peak signal of 5 v at the base, the collector swing is 6 v so that $a = R_3/(R_2 + R_3) = 5/6$. Emitter follower gain is assumed to be one. Since R equals 50,000 ohms, R_2 and R_3 will be 60,000 and 300,000 ohms, respectively.

To obtain a base current margin, somewhat smaller resistors in the same ratio are used. Convenient values are $R_2 = 47,000$ and $R_3 = 220,000$ ohms.

When Q_2 conducts, $V_{c2} = 5$ v and $V_{b1} = 5/6 \times 19 - 14 = 1.8$ v. R_e must be large enough so that the voltage drop across it exceeds the 1.8 v necessary to cut Q_1 off. When $\beta = 20$, the base of Q_2 is at $(11 + 14) \times 5/6 - 14 = 6.8$ v.

Since $(V_{b2} - \beta I_{b2} R_e)/R = I_{b3}$, $I_{b2} = 89.5$ μa and $I_{c2} = 1.79$ ma.

For $\beta = 100$, the same procedure is used. However, since a clamp is present $\beta I_b R_e$ will represent the drop across R_e only if it is less than 5.

If V_e is assumed to be less than 5 v, the formula indicates that $I_{b2} = 31$ μa. Checking, $\beta I_b R_e =$

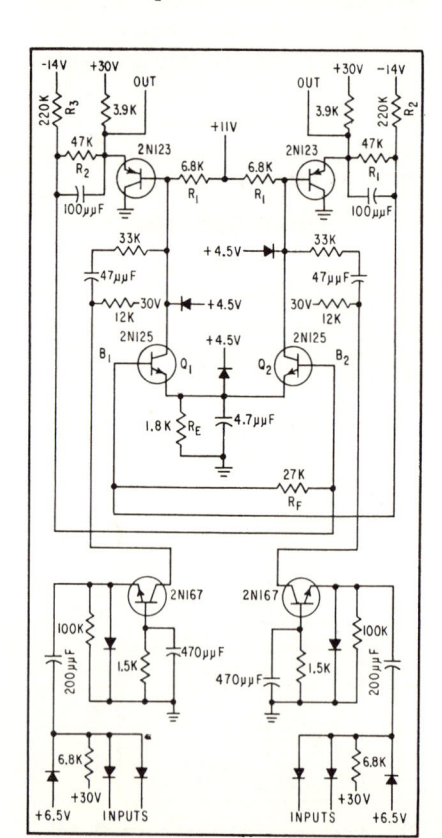

FIG. 1—Feedback resistor R_f in flip-flop makes it less sensitive to voltage variations and transistor unbalance

180,000 x 31 μa = 5.38 v. Therefore, the drop is not completely across R_e. Substituting 5 for $\beta I_b R_e$, I_{b2} = 45 μa. Collector current I_{c2} is therefore 4.5 ma.

The dissipation, $V_{c2} I_c$ = 22.5 mw, is within the allowable rating at 55 C.

Checking for stability for β = 100, α = 0.99. $S = (R_e + R_b)/[R_e + R_b (1-\alpha)]$ or 19.

The maximum I_{co} for any of the transistors previously mentioned is 2 μa. The stability against runaway is a function of S and of I_{co}. The maximum allowable dissipation of 23 mw is not exceeded for any value of β previously indicated. Clamp diodes used at the emitters and the collectors to prevent the transistors from saturating, speed the operation. The collector diode also serves to fix the lower level of the voltage swing.

The large emitter capacitor acts somewhat like a bias battery in conjunction with the emitter resistor. The cross-coupling capacitors were chosen so that the cross-coupling time constant is larger than the expected rise time but smaller than the period.

Assuming the base input resistance is small, the time constant is given by $R_1 C_o$ = 6,800 x 100 $\mu\mu$f = 0.68 μs.

The feedback resistor was picked experimentally. It is possible to analyze the circuit with the feedback resistor, to determine the operation with the chosen value of 27,000 ohms.

The effect of the feedback resistor R_f is that collector current varies by a smaller ratio, 2.08/1.08 or 1.93 as compared to 4.5/1.79 or 2.52. Also, the total collector current is much smaller so operation is within the maximum dissipation rating.

V_e never goes above 5 v so the diode at the emitter can be eliminated. Without the feedback resistor, it takes 2.5 to 4.5 volts to trigger the flip-flop, while with feedback, the range of voltage necessary is from 1.5 to 2 volts.

Magnetic Core Operates Counter

By E. H. SOMMERFIELD   Product Development Corp., Endicott, N. Y.

SINGLE-TRANSISTOR circuit generates trains of pulses in which each successive pulse is lower in amplitude than the preceding pulse. Number of pulses generated before pulse amplitude diminishes to some predetermined value is known.

The circuit is made into a counter by applying its output to the input of a voltage-sensing device that indicates when input voltage drops to a specified level. Thus its primary application is in timing circuits requiring a negative-going staircase input. Since a number of these circuits can be driven in series from a single constant-current driver, synchronization with different counting rates is possible.

Operation

A negative pulse applied to input terminal 1 in Fig. 1 drives the core of transformer T_1 in one direction of magnetic saturation. A subsequent positive pulse applied to input terminal 2 drives the core to magnetic saturation in the reverse direction. When the direction of magnetization within the core is reversed, a pulse is generated in the lower secondary winding that charges capacitor C_1 and also provides a pulse at the output terminals of the circuit. A positive pulse applied to input 3, the base of transistor Q_1, causes it to conduct, and C_1 discharges through Q_1 and the upper secondary winding, driving the core back to its first direction of magnetization. However, the level of magnetization is lower because of circuit losses.

A positive pulse is again applied to input 2, again reversing direction of magnetization of the core and generating a pulse in the lower secondary winding. This action again charges C_1 and delivers a pulse at the output terminals but of lower amplitude than the preceding output pulse. By alternating the application of pulses to inputs 2 and 3, a series of pulses diminishing in amplitude appears at the output terminals. A new train of pulses is generated by first applying another negative pulse to terminal 1 and the applying pulses alternately to terminals 2 and 3.

Conditions

Certain conditions must be met for proper circuit operation. With full drive at input 2, the value of capacitor C_1 and the number of turns of the two secondaries must be selected to ensure a net loop flux gain of less than one. Output must be sampled between the beginning of the input to terminal 2 and the beginning of the input to terminal 3. This is the time that C_1 is charged to different levels. At other times Q_1 discharges C_1.

Output must be buffered when

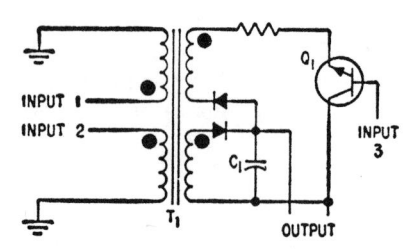

FIG. 1—Circuit delivers pulses of decreasing amplitude to operate counters

driving any load that would seriously disturb the energy stored on C_1 during the interval it remains charged. The time interval between the end of the pulse that is applied to input 2 and the beginning of the pulse that is applied to terminal 3 has a definite relationship to the value of C_1 and the number of turns in the two secondaries.

The transistor acts as a voltage switch to transfer energy stored on C_1 into the core. As such, collector voltage for Q_1 is supplied by C_1, which in turn is charged from energy supplied from the driving source via terminal 2.

Since flux in the core is decreasing, the number of counts is indirectly determined by the flux capacity of the core as well as the values of all of the previously mentioned components. Accuracy of the count is affected by all tolerances in the flux loop.

Chapter 8
RADIO RECEIVERS

Antenna occupies entire left temple piece while radio occupies other temple piece

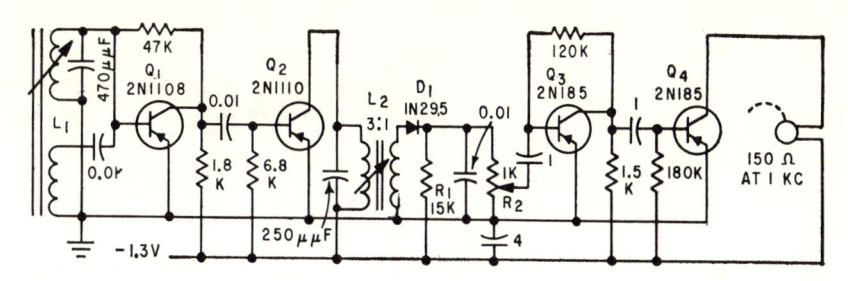

FIG. 1—Ferrite antenna has Q of 250. Base bias resistors in each stage are adjusted to give 0.5 ma collector current

Eyeglass Radio Receiver

Sensitive single-channel radio constructed in hearing aid eyeglass frame uses four transistors in tuned radio frequency circuit

By **HARRY F. COOKE,** Semiconductor Components Div., Texas Instruments Inc., Dallas, Texas

AUDIO MINIATURIZATION TECHNIQUES have been commercially used in the creation of hearing-aid eyeglasses. The eyeglass radio described in this article is a further extension of this idea.

Since a superheterodyne receiver is unnecessarily complex for single-channel operation and a regenerative detector is marginal on sensitivity and stability, a trf circuit, Fig. 1, was selected.

Circuit

The ferrite-core antenna has a high permeability which affords an effective area much greater than the actual size of the antenna. In the interests of sensitivity, the entire space within one temple piece is devoted to the antenna.

To obtain a reasonable r-f gain with a power supply consisting of a single 1.3 v mercury cell, grown-diffused transistors are used. These transistors have an h_{fe} (common emitter forward current gain) of 30 db at 455 kc. With fixed emitter current, the collector voltage may be dropped as low as 0.9 v without a serious loss in gain. The transistors were specially packaged in small diode cans.

For selectivity, a minimum of two tuned circuits are necessary with the antenna counting as one. A small ferrite E-transformer was selected for use as the second tuned circuit with the coils wound on the center leg. The transformer can be tuned by placement of the bar which closes the gap. The magnetic path is closed by the bar and there is minimum detuning once the bar has been cemented in place. The coil Q exceeds 100. The two tuned circuits are loaded to give a 10 kc at 6 db overall bandwidth.

Transistors Q_1 and Q_2 are resistance coupled to reduce complexity and give a degree of isolation between the two tuned circuits. The diode detector has a forward bias applied through R_1 to improve both detector sensitivity and linearity. The detected output is applied through volume control R_2 to audio stages Q_3 and Q_4.

The output load (earphone) has a 1,000 cps impedance of 150 ohms and requires 50 mv for a medium loud output.

Single-resistor constant-base biasing has been used. This type of biasing does not provide the best interchangeability and stability but is used because it requires the minimum number of parts. Base bias resistors are individually selected to produce 0.5 ma collector current in each stage.

The overall sensitivity of the receiver is 1,000 μv/meter for 50 mv across the earphone and the current drain is 2 ma giving a battery life of approximately 100 hours.

Antenna

The antenna primary is 79 turns of 10-44 Litz wire wound on a $\frac{1}{4} \times \frac{1}{8} \times 2\frac{1}{2}$ in. ferrite form. Adjust last ten turns for correct inductance. The secondary is 10 turns of the same wire wound at start of primary. Transformer L_2 primary is 65 turns 5-44 Litz wire and the secondary is 22 turns No. 40 enamel wire both wound on $\frac{1}{2} \times \frac{1}{8} \times \frac{9}{32}$ in. ferrite E-transformer core.

The eyeglass radio was constructed for use at a disk-jockey convention which was held at Miami, Florida.

Design of Reflexed

Reflex circuits in which i-f and a-f gain are achieved in the same transistor stage have recently been incorporated into economy broadcast receivers. Careful design is required to avoid motorboating

By JOHN WARING, Philco Corporation, Philadelphia, Pa.

CIRCUITS WHICH USE a single transistor for simultaneous amplification at intermediate and audio frequencies have been adopted recently for use in economy broadcast receivers. The reflexing circuit, normally used as a second i-f amplifier and first audio amplifier, can provide gain from a single transistor only a few db less than the gain obtained from two transistors in conventional circuits. It is not accomplished without sacrifice in distortion and power handling capabilities. Furthermore, care must be exercised in designing such circuits to avoid instabilities, particularly at high signal levels. The design of the reflex stage is also conditioned by system functions such as agc and overload level.

Fundamental Aim

In the design of a reflex stage the fundamental aim is to provide high gain and sufficient undistorted output power to drive the audio output transistor to its rated output level without permitting regeneration in the form of motorboating to occur. This can be accomplished by choosing the proper i-f and audio load resistances, selecting sufficiently high operating collector current and voltage to prevent audio and r-f clipping, and maintaining reasonable phase characteristics in the feedback loop.

A reflex circuit used in a portable broadcast receiver is shown in Fig. 1. In this circuit, transistor Q_1 amplifies a modulated i-f signal introduced at A and drives diode detector D_1 which produces an audio signal at the transistor base terminal. The transistor now amplifies at audio frequencies and delivers signal power to the audio load at B. The 450-ohm resistor across the output terminals represents the audio load presented by the audio output transistor and its biasing network.

Major Requirements

There are three major demands on the reflex stage during its operation. First, it must deliver sufficient undistorted audio power to drive the audio output stage to its rated level. Second, it must supply enough i-f power to the second detector to provide linear detection and agc voltage. A third requirement is stable operation throughout the range of signal levels which the stage is expected to handle.

Assume that for rated power output the audio output transistor requires a drive of 0.125 v peak across its input resistance of 450 ohms, that the d-c audio load is 560 ohms, and that the second detector requires a zero modulation i-f drive of 0.7 v peak across its effective resistance of 1,500 ohms to provide for linear operation and sufficient agc voltage. Assume that the reflex stage of Fig. 1 uses a linear transistor with an i-f output impedance of 30,000 ohms and a supply of 6 v.

If the transistor is assumed to be driven at audio frequencies only, the audio load line of 250 ohms (450 ohms in parallel with 560 ohms) can be drawn on the collector characteristic curves, as shown in Fig. 2, for a quiescent operating point of 4.9 v and 2 ma. The latter value is selected to provide adequate signal handling capabilities in the collector characteristic. The peak audio current swing on this line is $I = 0.125/250 = 0.5$ ma for rated audio drive.

Assume that the transistor is matched to the effective resistance of the detector. This matching requires an impedance transformation of 20/1 or a voltage transformation of 4.5/1. Thus, to obtain

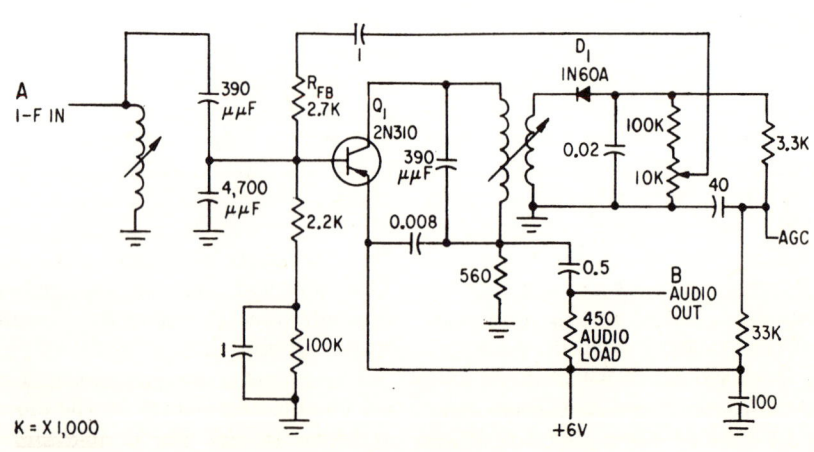

FIG. 1—Schematic of reflex circuit used in a portable broadcast receiver

Radio Receivers

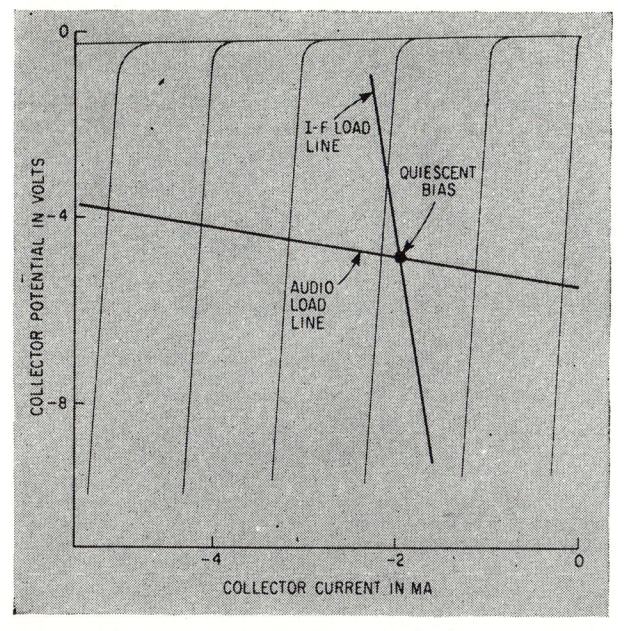

FIG. 2—Audio and i-f load lines drawn on collector characteristic

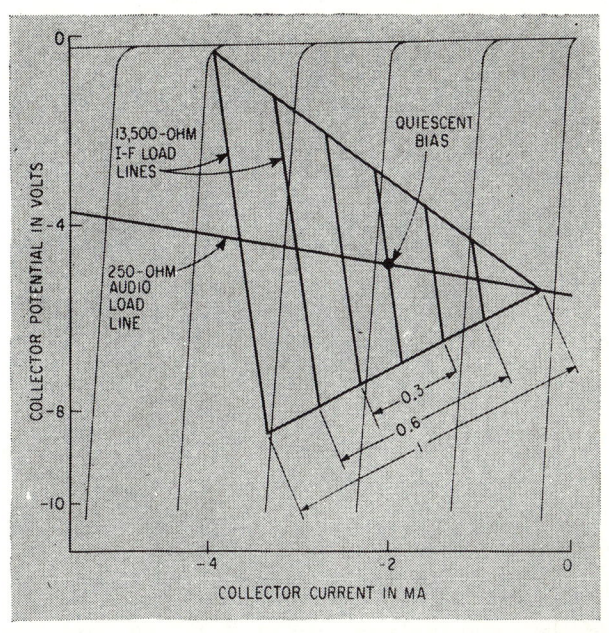

FIG. 3—Envelope of i-f carrier shown for various modulation indices

the 0.7-v i-f signal required to operate the detector, the zero-modulation peak collector swing would be 3.1 v, and at 100-percent modulation, peak collector swing would approach 6.2 v, causing clipping of the modulation. A choice of 3:1 as the voltage transformation ratio results in a peak collector swing of 4.2 v for a 100-percent modulated signal to give the desired detector drive. Assume that modulation is sinusoidal. The effective resistance of the detector is transformed to 13,500 ohms at the collector-emitter terminals of the transistor. If the transistor is assumed to be driven at intermediate frequencies only, the i-f load line can be drawn as shown in Fig. 2.

Combined Signals

In normal operation of the circuit, the transistor is driven through an appreciable portion of its collector characteristics by simultaneous audio and i-f signals. This operation can be described as a shift in the quiescent bias point of the i-f load line along the audio load line at an audio rate. Since the audio information is derived from the modulation on the i-f signal, the envelope of the excursions of voltage and current on the collector characteristics forms a definite pattern which aids in describing the behavior of the reflexed transistor.

Envelope Detector

It can be seen in Fig. 1 that the envelope detector produces a peak negative output at a time corresponding to the peak amplitude of the modulated i-f carrier. The detected envelope is fed to the base of the transistor, so the transistor is driven to maximum current on the audio load line at the same time. This assumes there is no phase shift in the feedback circuit. The transistor likewise is driven to minimum current on the audio load line at a time corresponding to minimum amplitude of the modulated i-f carrier. The envelope of the i-f carrier is shown in Fig. 3 for modulation indices of 0.3, 0.6 and 1, where it is assumed the i-f carrier drive to the reflex stage is held constant while the modulation index is varied. Since it is normal practice to specify the rated output power of a broadcast receiver for a 30-percent modulated signal, the 0.5-ma swing shown in Fig. 3 for a 30-percent modulated signal, is just sufficient for proper system operation. It is assumed here that the feedback resistance R_{FB} of Fig. 1 has been adjusted to provide this 0.5-ma audio swing with the volume control set at maximum output, when the second detector is driven with a 0.7-v, 30-percent modulated i-f signal.

This reflex stage used in a receiver capable of holding the i-f voltage at the detector to 0.7 peak by use of agc constitutes a working system. However, in an economy receiver the agc is seldom capable of holding the i-f level at the second detector constant, particularly at levels near overload. Figure 3 shows that failure to hold this level will cause clipping in saturation as well as cutoff. Clipping in saturation leads to the usual system faults near overload but clipping in cutoff can cause serious regeneration in the form of motorboating.

Motorboating Process

To visualize the regeneration process in the reflex amplifier, assume the i-f load line of the stage

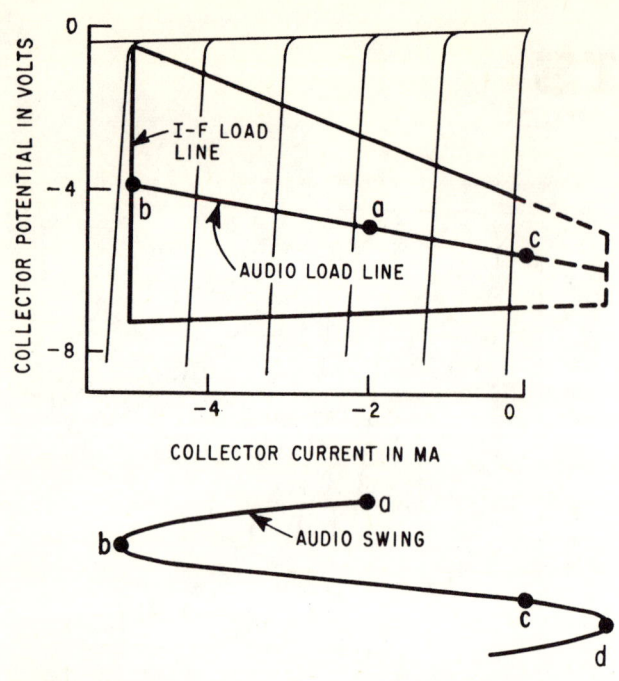

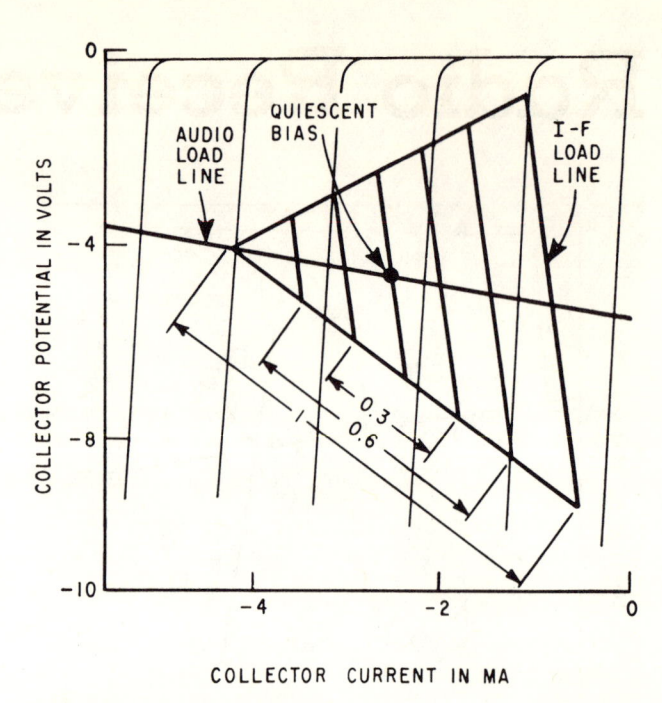

FIG. 4—Motorboating takes place as positive audio swing extends past point c

FIG. 5—Intermediate-frequency signal envelope when detector is polarized for positive output

is vertical and that both the value of the audio feedback resistor R_{Fn} and the volume control setting are selected to give a peak audio swing of 3 ma. Then the envelope of a 60-percent modulated i-f signal on the collector characteristic is as shown in Fig. 4, if no regeneration occurs. The behavior of the amplifier may be analyzed by starting from the quiescent point a and by assuming the amplitude of the i-f signal is increasing sinusoidally. Since the envelope detector is polarized for negative output, the transistor is driven further into conduction along the audio load line to point b which represents the peak envelope amplitude.

The amplitude of the i-f signal then drops to reduce the output from the detector driving the transistor toward cutoff at point c. The operation has been normal to this point and would proceed, as in normal overload, to clip off one third of the positive audio swing if it were not for the feedback characteristics of the reflex circuit. Actually, as the audio swing progresses more positive than c, the transistor is cut off and the i-f signal delivered to the detector becomes zero. Thus, the envelope detector develops a positive step which drives the transistor further into cutoff, and the latter remains cut off until the time constant of the detector and feedback circuit permits the transistor to return to conduction. At that time i-f signal abruptly appears at the detector which drives the transistor well into conduction.

When motorboating is severe, this drive into conduction is limited by driving into saturation and clipping of the i-f signal. The oscillation is similar to the behavior of a multivibrator which is stable only at saturation and cutoff. The motorboat waveform is largely a function of the modulation frequency, feedback circuit time constant and the strength of oscillation.

Positive Envelope Detection

In the circuit of Fig. 1 the detector diode is polarized to produce a negative output which will provide agc voltage for an *npn* transistor. When the polarity of this diode is reversed to provide agc voltage for a *pnp* transistor, the operation of the reflex stage is modified, because the maximum i-f signal is now associated with positive audio drive instead of negative audio drive. The envelope of the i-f signal on the collector characteristic is shown in Fig. 5 for modulation indices of 0.3, 0.6, and 1. Notice that the quiescent bias current has been increased from 2 ma to 2.5 ma to prevent clipping of the i-f signal in cutoff. The conditions which caused motorboating in the circuit of Fig. 1 are considerably alleviated, because regenerative clipping now occurs in saturation. When the transistor has good saturation characteristics, motorboating is improbable at high-modulation indices and is only moderately serious at low modulation indices. A disadvantage in this mode of operation is the cutoff clipping of the peak i-f signal which occurs at excessive audio drive. This condition can cause low overload level and loss of agc.

Other Considerations

The best transistor for reflex circuits is one which has collector characteristics showing excellent linearity and sharp knees. The transistor should have the qualifications normally expected of the i-f and audio amplifier including good gain and stability.

When the transistor is used in a low-voltage reflex circuit, its saturation voltage should not exceed a few tenths of a volt or system overload level may suffer. When I_{co} becomes an important transistor parameter, cold temperatures can cause motorboating while high temperatures can cause reduction in overload level, the sensitivity and the output power.

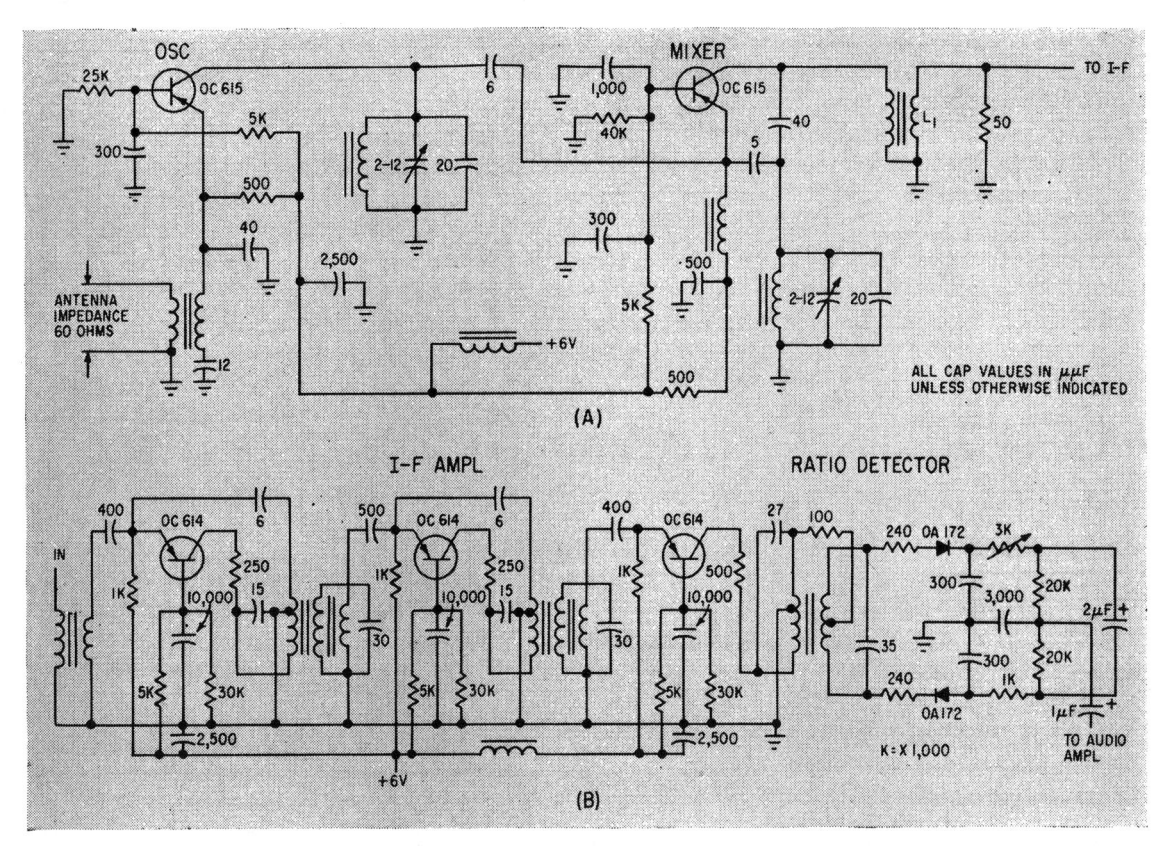

FIG. 1—Input stages, oscillator-mixer (A) and corresponding i-f amplifier (B) of vhf receiver developed by Telefunken, GmbH

European Designs for High-Frequency Radio Receivers

Schematics of recently-released vhf receiver circuits, coming out of Europe, point up some interesting design details

By RAYMOND SHAH, Engineer, Swiss Electrotechnical Institution, Zurich

COMMERCIAL AVAILABILITY of vhf transistors has focused interest in Europe on transistorized vhf receiver circuits. Characteristics of four European-made transistors are given in Table I. These are drift transistors with a metallic screening can and a corresponding fourth connection to the can.

CIRCUITRY. Figure 1A shows an input unit developed by Telefunken, using two OC 615 transistors. The oscillator plus mixer has a power amplification of 25 db. Voltage amplification is 16 db with an antenna impedance of 60 ohms and a load resistance of 50 ohms at the output of L_1. Noise factor is 10; image suppression is 1 to 20; parasitic radiation of the fundamental is 3.5 μv and of the first harmonic is 45 μv.

The corresponding i-f amplifier, Fig. 1B, uses a common-base configuration because the neutralization is simpler and variations of transistor parameters have less influence on circuit performance. Sensitivity of the complete vhf receiver is 0.85 μv for an

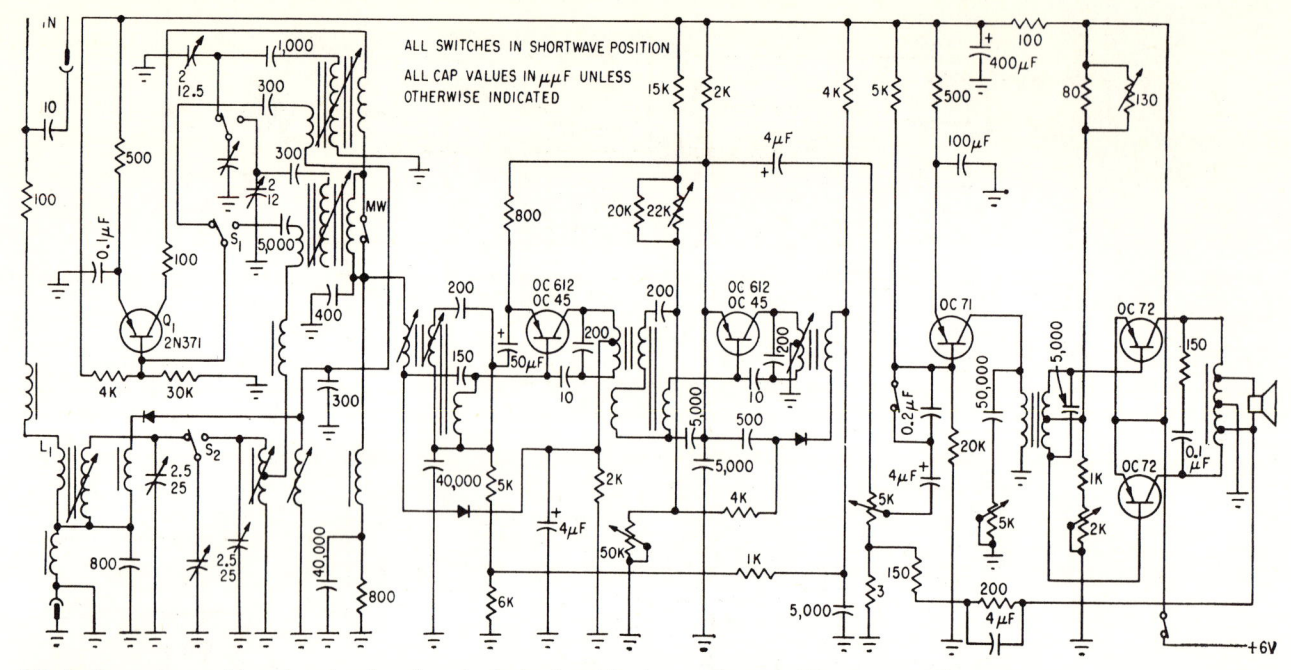

FIG. 2—Short-wave and broadcast-band receiver circuit developed by Loewe, Opta, GmbH uses RCA drift transistors

output of 50 milliwatts and 2 μv for 400 milliwatts. Eight transistors are used in all.

Some commercially available portables with shortwave and vhf bands use RCA drift-transistors 2N371 for the shortwave and 2N247 for the vhf band. One receiver, developed by Loewe Opta, GmbH, has two bands—medium and shortwave. The shortwave input circuit for 5.8 to 18.6 mc is shown in Fig. 2. Intermediate frequency and audio stages are common for medium-wave and short-wave bands. In the medium-wave position of bandswitches S_1 and S_2, the antenna signal is fed directly to transistor Q_1 which operates as a self-oscillating mixer stage. The receiver uses a total of six transistors. The shortwave sensitivity is 20 μv for 50 milliwatts output. Maximum output is 350 mw in a 7.5 by 4 inch oval speaker. Four 1.5-v cells are used. Total weight is 6 pounds.

VHF RECEIVER. Figure 3 shows the circuit diagram of a transistorized vhf receiver by Graetz KG.. The input stage, oscillator-mixer and the three i-f stages use RCA drift transistors 2N247. The receiver is designed for operation in the

Table I—European-Made VHF Transistors

Mfr	Type No.	Alpha Cutoff (mc)	Current Amp Factor	Collector-Emitter Capacitance
Telefunken	OC 614[a]	60	0.990	3.5 μμf at 10.7 mc
	OC 615[b]	90	($\alpha' = 100$)	2.5 μμf at 100 mc
Valvo, Mullard, Philips	OC 170[a]	70	0.987	1.6 μμf at 10.7 mc
	OC 171[b]	90	($\alpha' = 80$)	—

[a]—short-wave type, [b]—vhf type. Valves at $I_c = 1$ ma, $V_{ce} = 6$v

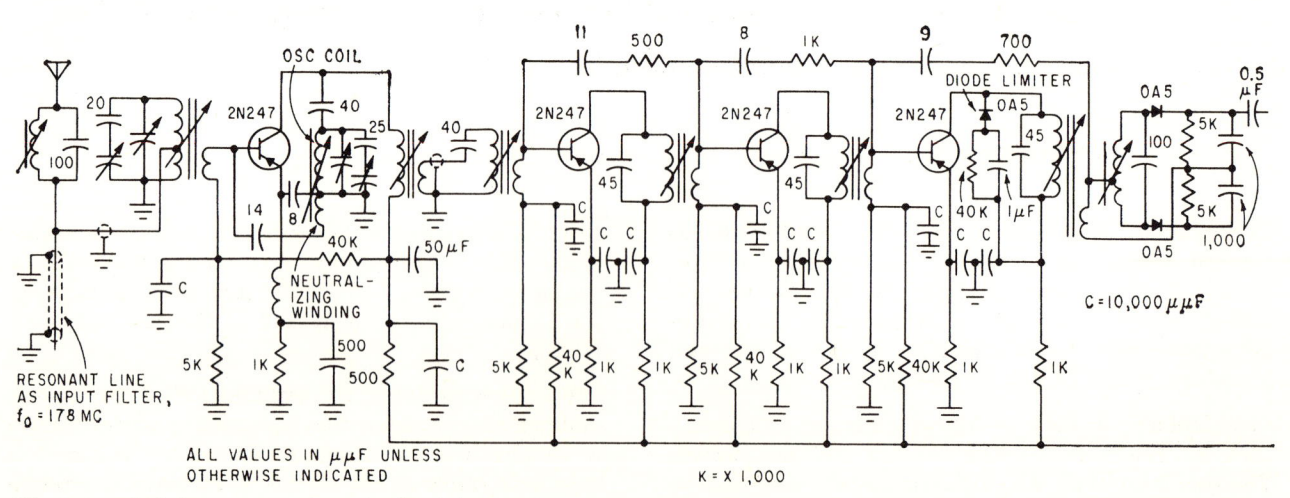

FIG. 3—Very-high-frequency receiver, developed by Graetz Kommanditgasellschaft, Altena (Westfallen). A low value i-f, 6.75 mc, provides

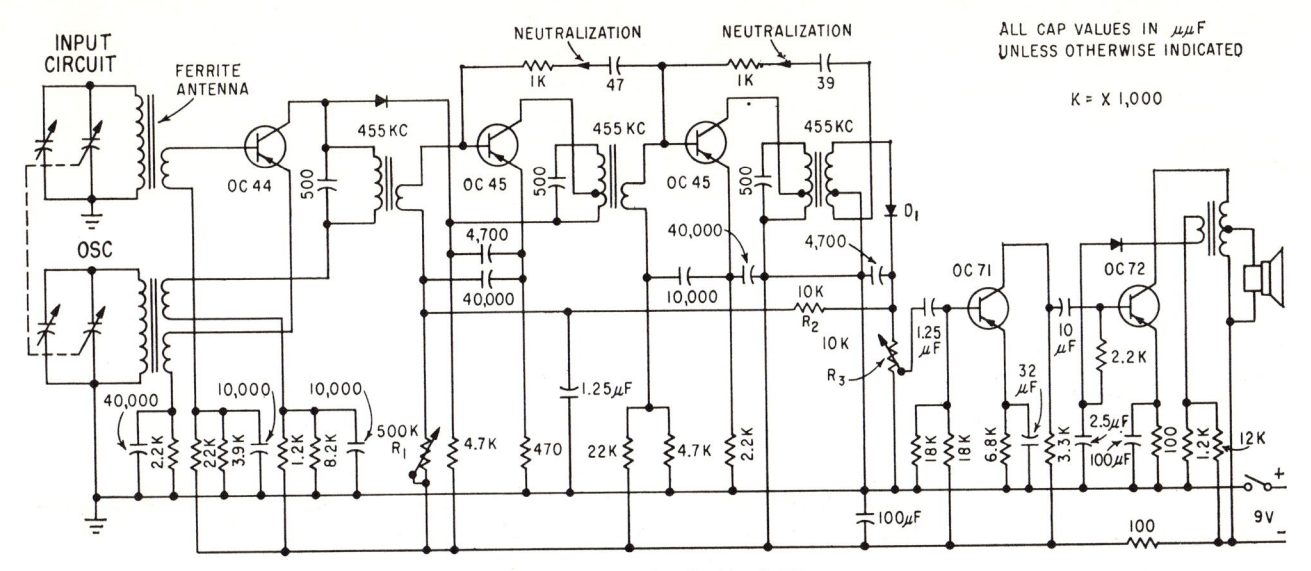

FIG. 4—Typical European broadcast-band receiver, the Peggie portable of Akkord AG

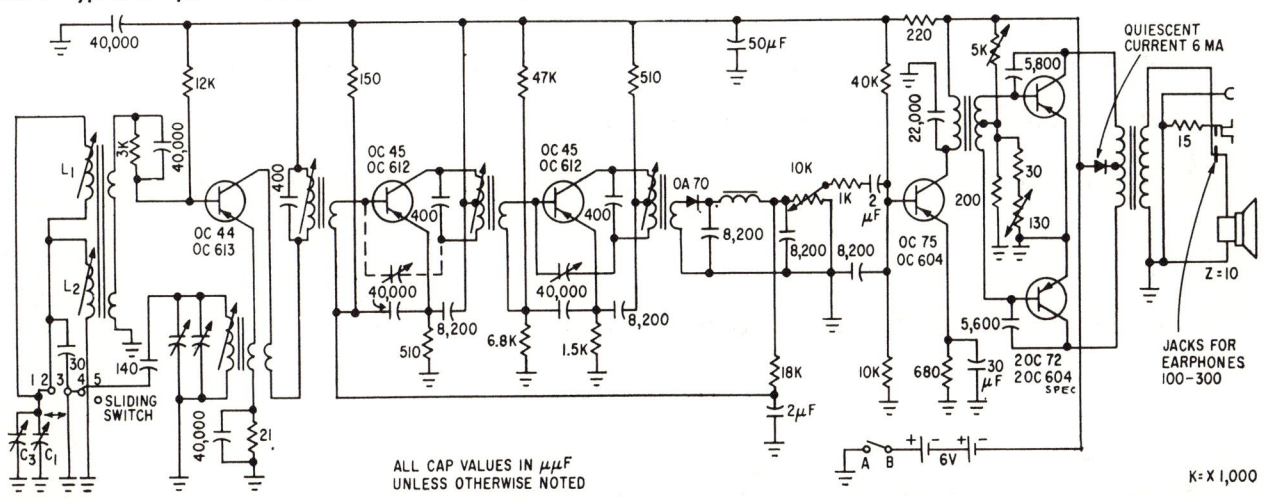

FIG. 5—Combined broadcast and long-wave receiver, developed by Braun GmbH. Sliding contacts are used for switching bands

87.5 to 101-mc band with an i-f of 6.75 mc.

Input sensitivity of the circuit for a signal-to-noise ratio of 30 db is 8 μv, with an antenna input of 240 ohms, and a signal having a frequency deviation of 22.5 kc. The receiver uses a 12-v battery. Current drain with full output is 55 ma.

Almost all circuits for transistorized portables having only the medium-wave broadcast band use five stages: an oscillator-mixer, two i-f's, an audio preamplifier and a power amplifier. Five or six transistors are used depending upon whether a push-pull or a single-ended output stage is used. The transistors are usually OC 44 or OC 613 for the oscillator-mixer stage; OC 45 or OC 612 for the i-f stages; OC 71 or OC 604 for the audio preamplifier stage and OC 72 or OC 604-special for the output stage.

Figure 4 shows a typical broadcast band receiver circuit as used in the Peggie portable of Akkord AG. Of special interest is the circuit consisting of R_1, R_2 and R_3 which establishes a fixed bias in the forward direction on demodulator diode D_1 for improving the detector efficiency at small signals.

One manufacturer, Braun GmbH has combined a broadcast and long-wave band receiver, Fig. 5.

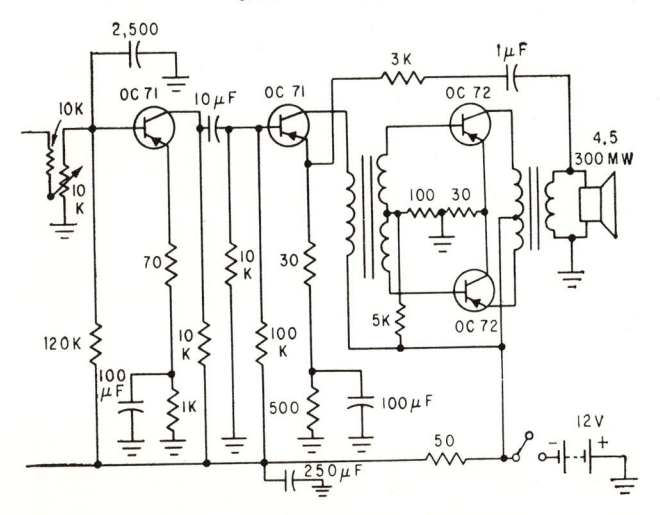

optimum amplification and effective a-m suppression

REFERENCE

(1) F. Mural, Dynamic Diode Limiter for F-M Demodulators, ELECTRONICS, p 146, Aug. 1955.

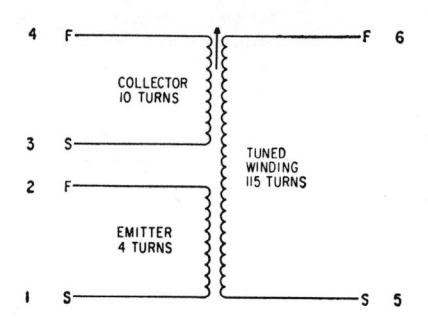

FIG. 1—Typical oscillator coil designed for emitter injection. Single pie gives less variation in coupling than double pie

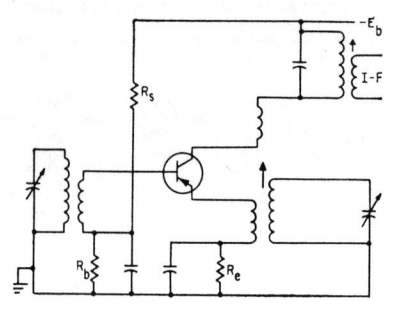

FIG. 2—Emitter-driven converter. Optimum injection voltage for direct emitter current of 0.5 ma is 0.1 to 0.15 v

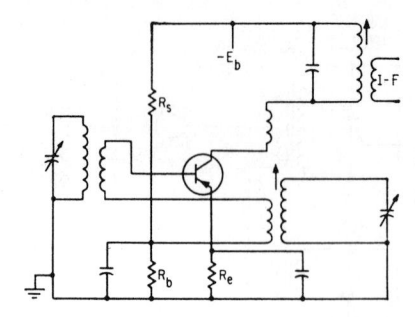

FIG. 3—Base-driven converter. Optimum injection voltage for direct emitter current of 0.5 ma is 0.2 to 0.3 v

Special Circuits for

Four portable transistor receiver circuits—autodyne converter, reflex circuit, avc overload diode and untuned r-f stage—are summarized in tabular form. Introductory text covers general design precedures for the entire receiver

By **WILLIAM E. SHEEHAN** and **WILLIAM H. RYER**,
Transistor Applications Laboratory, Semiconductor Div., Raytheon Manufacturing Co., Newton, Mass.

Circuit	Description and Design Hints	Problems and Solutions
Autodyne Converter (See Figs. 1, 2 and 3)	Used rather than separate mixer and oscillator. Basically an oscillator with special design considerations. Amplitude of oscillations must be limited by nonlinearity of emitter-base diode and not by any other form of nonlinearity such as collector bottoming. Impedance in collector circuit to oscillator frequency must be small enough so that peak-to-peak oscillator voltage does not approach battery supply voltage closely. Cost is about one-half that of separate mixer-oscillator circuit. Gain is about 30 db. Characteristics of transistor types generally used: input impedance to signal frequency, 500-1,500 ohms; conversion gain, 25-32 db; output impedance at 455 kc, 50,000-100,000 ohms. General design procedure: 1. Select oscillator coil design 2. Determine tuned-circuit parameters (oscillator and signal). Usually determined by convenient gang capacitor size 3. Select bias resistor values to give proper operating point and good temperature stability 4. Make necessary corrections to oscillator coil to give proper injection voltage across band. For more injection, increase number of turns on collector feedback winding by 10 to 20 percent and decrease turns similarly to lower injection voltage 5. Adjust number of turns on secondary of antenna to give fairly even sensitivity across band. Make set track by individual capacitor adjustments at each frequency 6. Adjust oscillator-coil and antenna inductances to make set track evenly across band	Greatest improvement in signal-to-noise ratio can be made by increasing size of antenna. Audio rate squegging may be eliminated by redesigning oscillator feedback circuit or changing values of coupling and/or blocking capacitors. Oscillation at i-f or some multiple may be corrected by relocating parts or revising tuned circuit to remove parasitic resonances. Spurious oscillation, apparent mistracking or oscillator pulling and worsened signal-to-noise ratio may be eliminated by removing oscillator feed-through in gang capacitor. Shield r-f section from oscillation section or reverse antenna secondary connections.

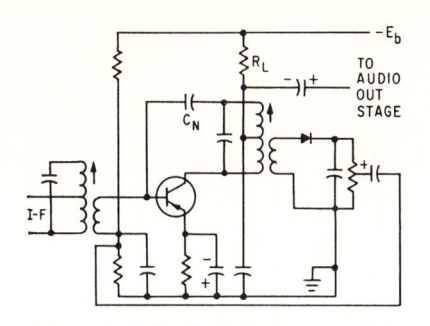

FIG. 4—Typical reflex circuit as used in portable transistor receivers. Transistor is used both as an i-f and audio amplifier

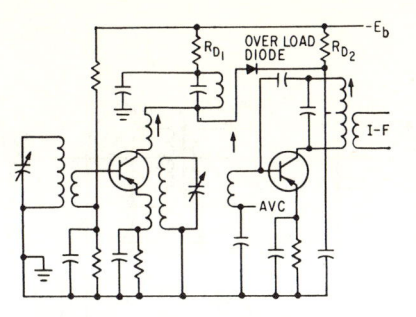

FIG. 5—Avc overload diode circuitry. Operates in conjunction with normal avc applied to the first i-f stage

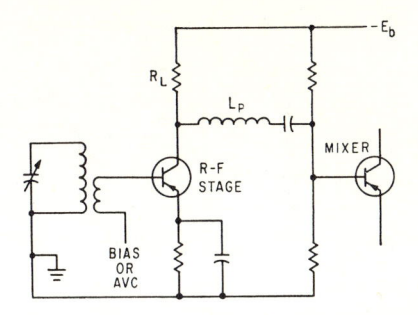

FIG. 6—Untuned r-f stage generally works into a mixer but could also work into a converter stage

Radio Receivers

OUTPUT STAGE gains of transistor superheterodyne receivers are about 25 to 40 db depending upon power level, supply voltage and circuitry.

The audio driver stage may be expected to contribute about 40 db of power gain. Overall gain should be about 60 to 80 db. For low distortion, negative feedback (3 to 7 db) may be added, reducing overall gain correspondingly. Knowing overall audio-system gain and input impedance, the second detector may be designed for optimum performance at its level. Diode detector losses are about 12 to 18 db. The avc circuits are usually designed after the i-f stages and make some adjustments in the detector necessary at a later time.

Preliminary design of the converter circuit is made to determine its output impedance. This, together with input impedance of the detector, is necessary for proper i-f system design. Generally, one or two i-f stages will be used. Selectivity requirements will determine whether single- or double-tuned i-f transformers must be used. The transistor type must be determined. Then, knowing transistor parameters and performance requirements, the i-f transformers can be designed by use of proper design equations If there is to be avc on the first i-f stage only, bias and avc circuits for the i-f stages can be designed and the second detector circuit optimized. Final step is design of the autodyne converter.

Circuit	Description and Design Hints	Problems and Solutions
Reflex (See Fig. 4)	Transistor is used as both an i-f amplifier and as an audio amplifier concurrently. Audio may be capacitance-coupled into next stage as shown or, if audio portion of circuit is used as driver for a class-B output stage, transformer coupling is used. Circuit can contribute 25-30 db i-f gain and 25-35 db audio gain. Its use enables elimination of an audio driver stage from overall circuit without decreasing performance.	Close attention must be paid to avc to prevent overload in the reflexed stage. Reflexed stage itself must be designed carefully to minimize overloading and cross-coupling. Special care must be taken to insure that operation is always in the linear region.
Avc Overload Diode (See Fig. 5)	Overload diode circuit operates in conjunction with normal avc applied to the first i-f stage. Its operation depends upon change in d-c conditions in this stage with changes in bias due to avc. Normally, an avc figure of merit of 30-35 db is obtainable in a set with two i-f stages with avc applied to the first i-f stage only. Addition of overload diode increases this to about 60 db and raises overload from 50,000 to 500,000 μv/m. Similar improvement can be noted in a set with only one i-f stage. Here, typical figure of merit without diode is 20-25 db; with diode, 50 db. Overload point is raised to about one v/m; without diode, about 200,000 μv/m. These figures are for 30-percent modulation.	Diode is connected for a-c across the converter to i-f transformer tuned circuit. D-c circuit is arranged so that diode is reverse-biased under weak signal conditions presenting a high impedance to i-f. Under these conditions, there is negligible shunting effect across the tuned circuit. Under strong signal conditions, d-c bias across the diode becomes a low impedance to i-f, loading down the converter to i-f transformer.
Untuned-R-f Stage (See Fig. 6)	Generally works into a mixer but probably could work into a converter as well. Inexpensive way of adding gain. Does not improve selectivity or image-rejection characteristics. Only makes slight improvement in noise characteristics. Adds 10-14 db more gain. Since circuit can use avc, avc figure of merit and overload point are improved considerably. Using untuned r-f stage, typical set with 300 μv/m sensitivity for 50-mw output can be improved to 50 μv/m at a 10-db signal-to-noise ratio with an avc figure of merit of 70 db.	Largest problem is to obtain a flat gain characteristic across the band. A small peaking coil may be added to the collector circuit to peak up at the high end of the band. Oscillator voltage must be prevented from feeding into the r-f stage by having a grounded shield plate between the oscillator and antenna sections of gang capacitor.

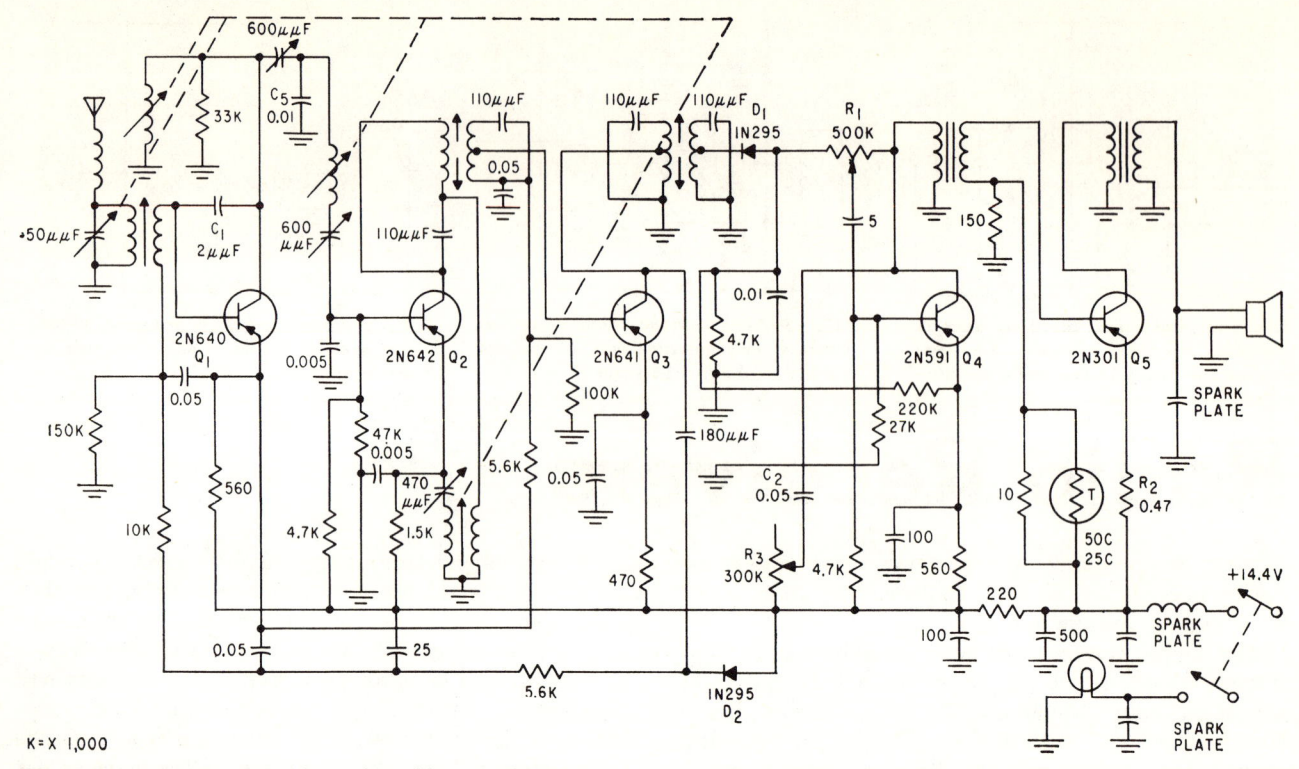

FIG. 1—Five transistors are used as r-f amplifier, autodyne converter, unneutralized 262-kc i-f amplifier, germanium-alloy audio driver and germanium power output stages

Design of Automobile

Automobile receiver using drift transistors has 2-microvolt sensitivity for one watt audio output. Single-ended output delivers 4 watts at less than 10-percent total distortion

TRANSISTORIZED AUTO RADIOS usually use seven or more transistors or are hybrids using both transistors and vacuum tubes. Drift transistors which inherently have a high maximum available gain and low feedback capacitance provide good performance with a minimum number of stages.

This five transistor auto radio has a sensitivity of 2 μv for 1 w of audio. The audio circuit can deliver 4 w at less than 10-percent distortion. The image rejection ratio varies from 85 db at the low end of the band to 78 db at the high end. The i-f rejection ratio varies from 89 db at the low end to 102 db at the high end. The receiver has a 60 db figure of merit using a 5,000 μv reference and a signal-to-noise ratio of 20 db at less than 5 μv.

No oscillator blocking was experienced with signal levels up to 2 v. Figure 1 shows the schematic of the radio.

Antenna Circuit

The whip antenna used for automobile radios may be represented as a voltage generator having a capacitive internal impedance. Although there is also a resistive component of impedance present, the value is so low that it may be safely neglected when a 6 to 8 foot whip antenna is used over the broadcast band. Analysis shows that when the loaded Q is fixed for bandwidth purposes, the maximum power transfer is obtained when the unloaded Q is as high as possible, trimmer and other shunt capacitors are kept to a minimum, and padders and other series capacitors are kept to a maximum. In addition, the power transferred decreases with decreasing frequency. These factors must be taken into consideration to obtain a good signal-to-noise ratio.

Another consideration is that of coupling the input impedance of the r-f transistor to obtain the proper antenna loading. Although capacitive division eliminates the need for an antenna coil secondary winding, this type of coupling requires the coil to be located between the antenna and the transistor base. This connection results in poor rejection of the high field strength 60 cps (power-line) interference occasionally encountered in auto radio applications. If the low end of the antenna coil is grounded selectivity

For automobile use, radio has manual and pushbutton controls

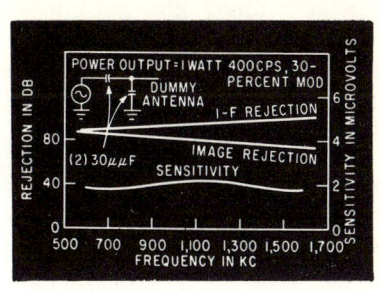

FIG. 2—Sensitivity and rejection characteristics of five-transistor receiver

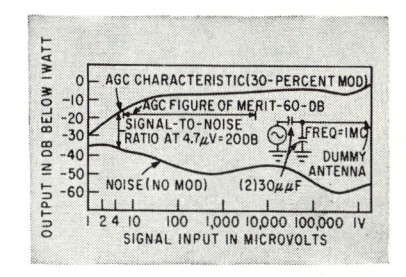

FIG. 3—Agc and noise characteristics

By R. A. SANTILLI and C. F. WHEATLEY, Semiconductor and Materials Division, RCA, Somerville, N. J.

Broadcast Receivers

falloff of 18 db/octave may be obtained. For reasons of 60-cycle rejection and of signal-to-noise ratio, an antenna coil having a secondary winding was chosen. Three r-f tuned circuits were required to obtain an image and i-f rejection in excess of 70 db across the band.

R-F Stage

To override converter noise adequately, r-f gain of the order of 15 db is required. Although neutralization may be used in a variable-frequency tuned amplifier, it was not considered necessary in this application.

Perhaps the most significant requirement for the r-f transistor is the agc requirement. The control of r-f input and output impedances and feedback capacitance is an obvious requirement. However, the range of signals encountered in an auto radio covers approximately 110 to 120 db. Consequently, the r-f transistor must provide 80 to 90 db of agc. As a result, the d-c beta (common-emitter current gain) and the I_{co} (cutoff collector current) must be controlled so that the agc system may supply sufficient power to utilize this cutoff range without amplification. This requirement becomes even more significant at higher ambient temperatures. The drift transistor is designed specifically to meet these stringent requirements.

Either top-side or bottom-side coupling may be used for the double-tuned r-f circuit, together with capacitive division on the secondary winding for impedance matching to the base of the converter transistor. The first r-f coil was designed to be tuned with a 600-$\mu\mu$f capacitor. It may be desirable to add a winding to the first coil and tune with a smaller trimmer. The value of 600 $\mu\mu$f was chosen to provide the correct collector loading as determined by dynamic stability considerations.

The r-f stage (Q_1) is operated at a collector voltage of -12 volts and a collector current of 0.7 ma and produces a power gain of 27 db at the low-frequency end and 20 db at the high-frequency end of the band. This gain is sufficiently below Q_1 maximum capabilities to assure excellent interchangeability and stability. At the low-frequency end of the band, the unloaded Q's of the first, second, and third tuned circuits are 65, 45, and 45, respectively, and the loaded Q's are 40, 30, and 50. At the high-frequency end, the unloaded Q's are 65, 65, and 65 and the loaded Q's 48, 40, and 55. The coefficient of coupling of the double-tuned circuit, which is 1.3 times the critical coefficient, provides a peak-to-valley ratio of approximately 0.2 db. This overcoupling is not sufficient to present alignment difficulties.

Converter Stage

The converter circuit is basically the autodyne type, in which emitter injection is obtained by capacitive division. The r-f signal is fed into the base of Q_2. The i-f output from the collector is fed through a double-tuned transformer to the base of i-f stage Q_3. Converter Q_2 operates at a collector voltage of -12 volts and a collector current of 0.6 ma, and produces a conversion gain of 37 db (262-kilocycle i-f). Again, this gain is below the transistor maximum capabilities and assures excellent interchangeability and stability.

The converter stage exhibits a strong tendency toward increased injection voltage with decreasing frequency. Conversion gain increases with injection up to a maximum point, and then starts decreasing with further increase of injection voltage. For the circuit shown in Fig. 1, conversion gain is approximately independent of injection voltage between 40 and 150 millivolts and decreases with injection voltage outside this range. Because the sensitivity is greater at the low end of the band than at the high end, the injection voltage was controlled with frequency to decrease the variation of receiver gain across the band. The emitter is not as well bypassed for low-frequency r-f signals as for high-frequency r-f signals and thus provides additional flattening of the gain/r-f characteristic.

If the tickler winding is between the collector of the converter transistor and the i-f winding, the capacitance of the tickler winding to ground shunts the i-f primary and changes the coefficient of coupling. When this arrangement is used, this shunting capacitance must be considered in the design of the i-f transformer.

The major problem in the design of the converter stage was that of blocking. When a very-high-level r-f signal is applied to the base of the converter transistor, the stage operates as a clamping circuit, thereby reverse biasing the transistor and preventing oscillation. Without oscillation there is no i-f output, and therefore no agc to reduce the high-level r-f input. If the incoming signal increases gradually, the agc has a chance to build up. However, it may come on abruptly if push-button tuning is used or if the radio is turned on in the presence of a strong signal. If the turn-on condition were the only one of concern, a suitable bypass capacitor could be incorporated so that the r-f stage would gradually obtain bias (a turn-on transit could be built into the circuit). However, this arrangement is no solution for push-button blocking, or blocking caused by tuning to a strong station.

The best method to handle blocking is to determine empirically the r-f collector signal level at which blocking occurs and then limit the collector signal below this level. The d-c collector-to-emitter voltage can be chosen so that collector limiting will result. The signal level can further be reduced approximately

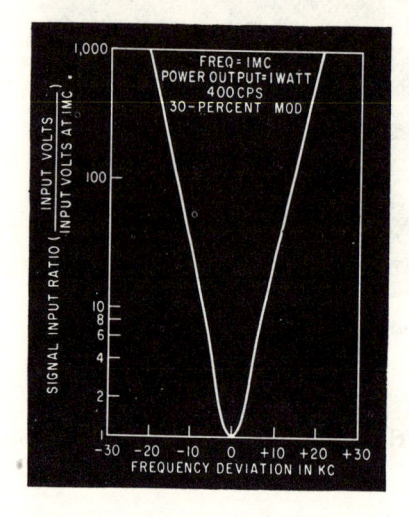

FIG. 4—Overall bandwidth of receiver at signal frequency of 1 mc

6 db by the use of a slightly back-biased diode which shunts the collector load. A much greater degree of limiting may be obtained by the use of two diodes appropriately biased in a conventional limiting arrangement.

Additional attenuation may be employed from the collector of the r-f transistor to the base of the converter transistor to eliminate blocking provided sufficient gain can be obtained elsewhere. Blocking is less severe when the converter current or injection voltage is increased.

I-F Stage

A drift transistor is used in the unneutralized 262-kilocycle i-f amplifier. Double-tuned input and output transformers are used to obtain the desired selectivity with coefficient of couplings set at 0.85 critical. Although more gain could be obtained with a higher collector load, and even more with neutralization, the i-f stage must deliver a high level of power. Consequently, the collector load is determined by large-signal class-A power amplifier criteria rather than dynamic stability alone. For the collector voltage of −12 v and the collector current of 2 ma, a collector load of 6,000 ohms is used. The i-f stage contributes 32 db of gain.

Audio detector D_1 is fed from a tap on the secondary winding of the i-f output transformer. The agc detector D_2 is fed by a capacitor from the collector of the i-f transistor. This arrangement provides a slightly wider bandwidth for the agc than for the audio, and also permits a high level of agc voltage. About 5 or 6 db of agc is obtained from the i-f stage.

AGC

The 110 to 120-db signal-handling requirement of this receiver makes agc a difficult problem. The germanium diode used for the agc detector develops approximately 2 v of agc. A tendency toward distortion at very high levels was corrected by the use of C_1 a 2 $\mu\mu f$ capacitor between the base and the collector of r-f transistor Q_1. This capacitor apparently extends the agc to some extent by introducing a feedthrough current which subtracts from the normal collector signal current. These currents are normally out of phase.

Another problem encountered in the agc system was that of spurious responses at very high levels. The agc bandwidth is fixed by the i-f bandwidth and is relatively narrow. When a strong signal is present and the receiver is tuned on carrier, the performance is as expected. As the receiver is tuned off carrier however, the agc is rapidly removed. This change permits much higher levels in all stages prior to the output of the i-f stage, and shows up first as distortion of the envelope (and detected audio distortion).

Further tuning off carrier removes the agc almost completely and permits a very high r-f signal on the converter. In fact, if r-f limiting is not employed, oscillator blocking may result. The wide agc system described previously reduces this effect.

It would also be possible to obtain freedom from blocking if additional agc voltage derived from the collector of the r-f transistor were added to the normal agc

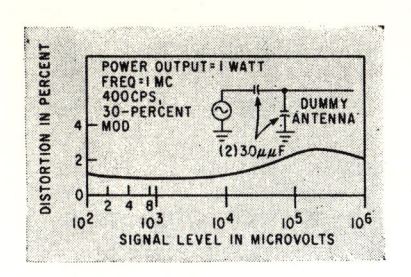

FIG. 5—Distortion of the receiver as a function of signal level

bias. This arrangement would not alter the receiver performance appreciably in normal reception areas. This arrangement was not incorporated in the receiver.

Detection

Point-contact germanium diode D_1 is used as the audio detector. For detection of small signals, maximum sensitivity is obtained by passing a certain value of forward d-c current through the diode. As the current is reduced from this value, a slight loss in sensitivity is encountered. For signals 5 or 10 db lower, however, a considerable loss of sensitivity is observed. A high degree of quieting in the absence of a signal can be obtained with only a slight reduction in normal sensitivity by utilization of this nonlinearity of the detector. This arrangement does not actually improve the signal-to-noise ratio at sensitivity, but it results in a subjective advantage similar to noise limiting. The compromise bias employed in this receiver causes a sharp curvature in the agc curve at sensitivity level. This curvature introduces detector distortion.

Maximum sensitivity is obtained when the input impedance of the audio transistor represents most of the load on the detector, that is, when the d-c load is small compared to the a-c load. Again, a high degree of distortion is introduced, particularly at high modulation levels.

For operation at sensitivity, however, the noise is sufficiently high to override the distortion. A signal-to-noise ratio of 20 db contributes as much undesired power as 10 percent distortion. At signal-to-noise ratios below 20 to 25 db distortion means little.

When the signal level is high enough to obtain a signal-to-noise ratio of 20 to 25 db (about 5 to 8 μv), the detector is fairly linear. The first 10 to 20 db of volume reduction inserts series resistance between the detector and transistor, unloading the detector to approach an a-c to d-c ratio of unity.

Audio Driver

Germanium alloy audio driver transistor Q_4 was specially designed for the high-temperature and high-voltage requirements of this application. The driver operates at a collector voltage of -12 v, a collector current of 3 ma and provides a power gain of approximately 44 db. Although the driver may contribute 3 or 4 percent distortion at very low r-f signal levels where the noise level is high, the volume control greatly reduces this distortion at normal r-f signal levels.

Volume Control

Volume control R_1 is a high-resistance potentiometer connected between the audio detector and the collector of the driver. When the variable arm (connected to the base of driver Q_4) is placed at the detector end of the potentiometer, maximum sensitivity is obtained. The resistance of the potentiometer is high enough so that appreciable collector-to-base feedback is not encountered. As the arm is moved from the detector to the transistor collector, series attenuation results, thereby unloading the audio detector. Further reduction of volume causes collector-to-base feedback, which not only reduces the driver distortion but also lowers the output impedance of the driver. When the output stage is driven by

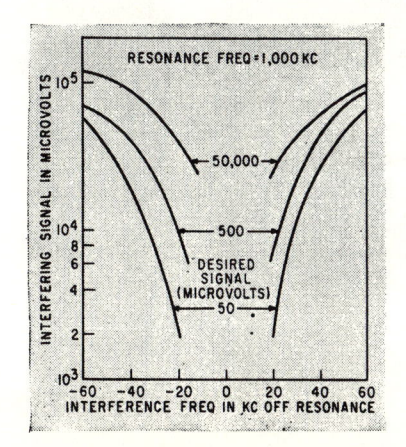

FIG. 6—Crosstalk characteristics for signal levels of 50, 500 and 50,000 microvolts

a high-impedance source, the presence of an unbypassed emitter resistor is not degenerative. Consequently, there is no loss of sensitivity although a greater dynamic range is required. When the output impedance of the driver is lowered by the feedback of the volume control, emitter resistor R_2 in output-stage Q_5 provides a significant amount of loop feedback. Further increase of the volume control setting results in additional feedback and attenuation produced by loading of the driver collector. Volume adjustment of about 100 db can be obtained by this method. For best effect, the volume control should have an S-taper with about 5-percent resistance at 35-percent rotation and 95-percent resistance at 65-percent rotation.

Audio output stage Q_5 consists of a germanium power transistor operated under class-A conditions. When this transistor is driven without regard to distortion, the power output level is 7 watts. The power transistor must be provided with an adequate heat sink to avoid exceeding a maximum junction temperature of 85 C.

Figure 2 shows the tracked sensitivity, image rejection and i-f rejection ratios as functions of frequency. With the dummy antenna shown in the insert, the sensitivity is 2 μv across the band.

The agc and noise characteristics of the receiver are shown in Fig. 3. The receiver has a 60 db figure of merit using 5,000 μv as a reference. A signal-to-noise ratio of 20 db occurs at less than 5 μv. No oscillator blocking was experienced with signal levels up to 2 v.

Figure 4 shows the overall bandwidth at a signal frequency of 1 mc, and Fig. 5 shows the distortion as a function of signal level. The crosstalk characteristic of Fig. 6 is given for signal levels of 50, 500 and 50,000 μv. This figure is a graphical presentation of antenna selectivity coupled with r-f distortion.

For a resonance frequency of 1 mc and three desired signal levels, the curves show the signal strength which an additional or interfering carrier off resonance would have to have to produce 3 percent crosstalk (1 mw of interfering audio for one watt of normal audio).

Single diffused-base transistor converter has only one variable tuning element in band-pass circuit. Last of three transistor i-f stages is reflexed as emitter-follower amplifier to provide audio gain and low output impedance. Ratio detector has 700-kc peak separation for low distortion and high a-m rejection

By **HARRY COOKE,** Circuit Development Branch, Texas Instruments Inc., Dallas, Texas

F-M Tuner Uses

FRONT END tracking and alignment is one of the more time consuming operations in the construction of a superhetrodyne tuner. The f-m tuner to be described incorporates a single transistor converter with only one variable tuning element to simplify tracking and alignment. The last of three i-f stages is reflexed as an emitter-follower audio amplifier with low output impedance. A 700-kc peak separation ratio detector provides relatively high a-m rejection and a low-distortion output.

The input circuit shown in Fig. 1 is a transistionally coupled series-tuned network. The low input resistance of the transistor with this type of tuning makes the required circuit values more convenient to work with. In addition, series tuning exhibits an impedance rise outside of resonance that is necessary to maintain oscillation over the required 113 to 133-mc range with a grounded-base oscillator. Since the f-m band covers approximately 20 mc, a 25-mc i-f frequency was selected as a compromise between i-f gain and primary-image rejection.

Feedback capacitor C_1 causes transistor Q_1 to operate as a two-terminal negative-resistance oscillator. The grounded-base configuration provides stability and uniform oscillator performance. Variable inductance L_1 is the tunable element, while C_2 is used only to align the oscillator to the tuning dial. Attempts to use C_2 as the tuning element will result in an undesirable shift in the tuning of i-f transformer primary L_2.

The combination of C_3 and L_3 is series tuned to the mean oscillator frequency to assist in reducing the voltage across L_2. Since, as far as the transistor is concerned, L_2 and C_3 could just as

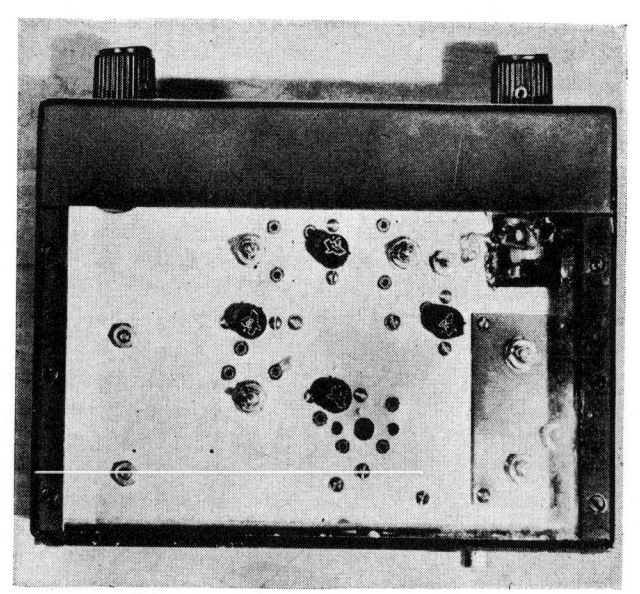

Bottom and top views of complete transistorized f-m tuner show component layout and integral power source

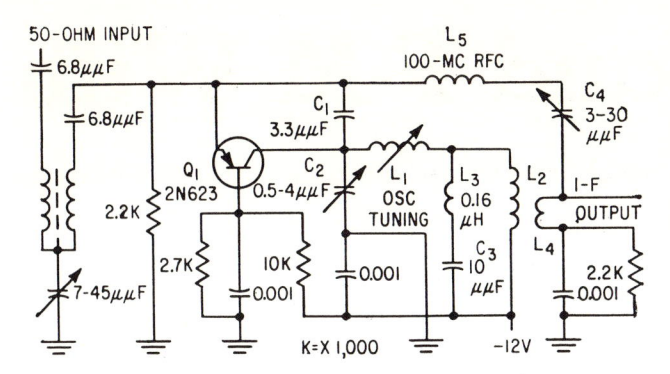

FIG. 1—Oscillator is tuned with brass-powdered iron hybrid tuning slug in L_1. Choke L_5 prevents C_4 from interacting with input

Tuner covers entire f-m broadcast band by varying single tunable element in local oscillator

Four Transistors

well comprise the oscillator tank circuit, admittance neutralization from secondary L_4 to the base of Q_1 is necessary. Loading L_2 at the oscillator frequency is only a partial solution as the mismatch required to prevent oscillation at the intermediate frequency also reduces the circuit gain to an impractical figure.

The neutralization shown is conventional except for r-f choke L_5, which prevents neutralizing capacitor C_4 from interacting with the input of the mixer Q_1 at the signal frequency. Since L_5 should be inductive at 25 mc, it is necessary to increase C_4 to keep a net capacitive reactance of the correct magnitude at the intermediate frequency.

At the oscillator frequency, the neutralization network is effectively not present because L_2 is a low impedance and there is little signal transfer from the transformer secondary back to the transistor input.

Conversion power gain of 10 to 12 db is obtained by operating Q_1 at 1-ma emitter current. Increasing this emitter current will increase the gain slightly, but neutralization and oscillator injection problems are increased.

In the preliminary design shown, the oscillator voltage appearing at the antenna terminals is marginal as far as the FCC regulations are concerned. Slight

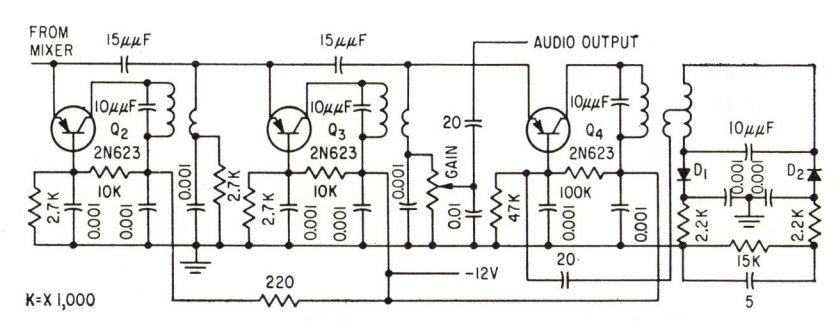

FIG. 2—Three-stage i-f amplifier has total gain of 60 db plus audio gain

changes in layout and circuitry should reduce this voltage to an acceptable value.

I-F Amplifier

The 25-mc i-f amplifer has three stages of 20-db gain per stage. The common-base configuration employed in Fig. 2 is quite stable and affords good interchangeability, even with fixed values of neutralizing capacitor C_4 (Fig. 1).

The transistors are operated at 1-ma emitter current. Increasing this value to 2 ma will raise the gain approximately 1 db per stage, to the deteriment of interchangeability with fixed neutralization.

The last i-f stage also operates as an audio-frequency emitter follower in a conventional reflex circuit that provides sufficient audio power gain and a low output impedance.

Design of the transistor-driven ratio detector, as compared to electron-tube versions, is modified by the fact that the loaded Q of the transformer primary is no longer determined solely by the transformer diode loading. In particular, the shunting effect of the output resistance of transistor Q_4 must also be considered. However, by adjusting the transformer's tertiary turns, it is possible to obtain a match between the collector output resistance and the transformed diode load.

The complete tuner has a sensitivity of 3 μv for a 10 db $(s+n)/n$ ratio with a signal deviation of 22 kc. Audio output is 10 mv for 1.5-μv input and 25-mv output for 10-μv input. The i-f amplifier has a 3-db bandwidth of 500 kc and detector peak separation is 700 kc.

The author acknowledges the help and suggestions of Roger Webster and Floyd Ducote.

167

Wideband F-M With

Here are two practical circuits which use voltage-variable capacitances as the means of modulating an oscillator tube

By **COLLINS ARSEM**, Diamond Ordnance Fuze Laboratories, Washington, D. C.

Schematic of this lab model of 100-mc oscillator is shown in Fig. 1

THERE ARE many useful systems for producing frequency modulation of vhf oscillators.[1,2,3,4] The system described here uses variable-capacitance diodes to produce frequency deviations as great as or greater than those obtainable with the other systems, without many of their shortcomings.

If a reverse voltage is applied to a semiconductor junction diode, the two regions adjacent to the junction are swept free of mobile carriers (holes and electrons) and assume opposite net charges due to the immobile impurity ions (donors and acceptors). The device thus forms a charged capacitor. Variation of the applied voltage will change the capacitance in a manner determined by the way in which the impurity-ion concentration changes with distance in the junction region. Capacitance variation is usually a function of the method of fabrication. For example, the capacitance of a fused-junction diode is proportional to $V^{-1/2}$, while the capacitance of a grown junction is proportional to $V^{-1/3}$.[5]

100-mc Oscillator

Choice of the type of variable capacitance diode is largely determined by oscillator frequency, desired deviation, maximum diode dissipation and the allowable load that can be presented to the oscillator by the modulator circuit. One complication is the fact that many circuit configurations which reduce the loading due to the diodes also tend to reduce the frequency deviation.

The oscillator tube should be chosen to produce an oscillator circuit of low internal impedance. This is especially important at high frequencies since the relatively low equivalent parallel resistance of the diodes might cause excessive loading of a high-impedance oscillator, particularly when the diode network shunts the tank directly.

One of the most satisfactory circuits of those investigated is the Hartley oscillator of Fig. 1 in which the plate current is limited by plate resistor R_1 instead of by grid-leak bias. The tube operates with the grid at essentially zero d-c potential, thus causing the values of transconductance (g_m) to be higher and the values of a-c plate resistance (r_p) to be lower than their negative-grid-region values. This type of operation thus tends to reduce the internal impedance of the oscillator tube and makes it better suited for this application.

Inductance L_1 is a ½-in. diameter air-core coil. The cathode is tapped to the 4½-turn coil at 2½ turns above ground.

The modulator circuit is a practical compromise between high deviation and low internal diode loss. This circuit consists of two similar variable-capacitance diodes (D_1, D_2) which are in series with respect to r-f and in parallel with respect to audio modulating signals and d-c bias. The net result is a variable capacitor having half the capacitance, but the same percent change in capacitance-per-volt as a single diode.

Oscillator frequency, which was about 100 mc, was deviated as much as 28 mc peak-to-peak (p-p) with modulating signals of less than 28 v p-p and negligible modulating power. Diode d-c bias is adjusted so that the peak modulating voltage does not cause diode conduction, and hence severe r-f losses.

Figure 2 shows frequency spectra obtained with a carrier frequency of 100 mc which is modulated by a 100-kc and a 10-kc sinewave. Only the envelopes of the spectra are actually observable with 10-kc modulation because the minimum resolution of the spectrum analyzer that was used in testing was 25 kc. Individual sidebands are discernible in the spectra for 100-kc modulation and deviations up to 4 mc p-p; but as the deviation is increased, the patterns tend to show only the envelopes of the spectra. Here once again the analyzer's resolution prevented the

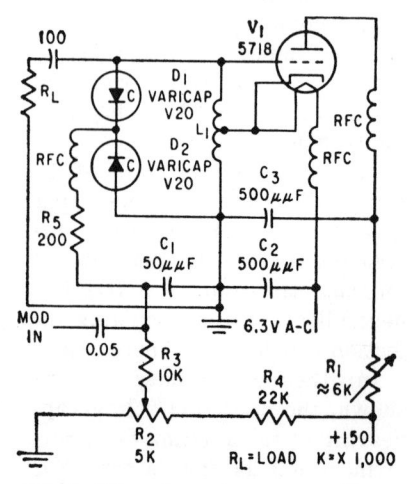

FIG. 1—100-mc lumped-constant oscillator

Capacitance Diodes

separation of the greatly increased number of sidebands.

Shown in Fig. 3A is a typical output curve of a lumped-constant oscillator similar to that shown in Fig. 1. The curve shows variation in the r-f voltage across the oscillator load for a frequency swing of 20 percent, peak-to-peak. Over all change in voltage is approximately 2.5 db. Figure 3B shows oscillator frequency as a function of diode bias. For all these measurements the d-c plate input power was about ½ W; the plate voltage and current were approximately 30 v and 17 ma, respectively.

400-mc Oscillator

Wide-band frequency modulation of a 400-mc distributed-parameter Colpitts oscillator (Fig. 4) was obtained with a bilateral (symmetrical) transistor in the modulator circuit. Transistor Q_1 is equivalent to two reverse-biased diodes connected in series with respect to r-f and in parallel with respect to modulating signals and d-c bias.

Inductances L_1 and L_2 are sections of transmission line; L_1 is 3¼-in. long and L_2 is 2 5/16-in. long. Both inductances are constructed of brass ⅛ in. diameter rod which is centered in a 1-in.² cross section channel. Collector and emitter of Q_1 are connected respectively to the rod (L_1) and channel wall; collector and emitter connections are approximately 1.8 in. away from the channel-wall connection of L_1.

Figure 5A shows the spectrum obtained for a frequency deviation of 15 mc peak-to-peak and Fig. 5B shows the nearly flat r-f output voltage obtained over this 15-mc band. The upper line in Fig. 5B is the voltage curve and the lower line is the base-line. Figure 5 was obtained while the oscillator was delivering 685 mw to a 50-ohm load with a plate efficiency of 22 percent.

The above results are typical examples of wide-deviation frequency modulated vhf oscillators. Useful frequency deviations of 10 percent or more are possible with the capacitance diodes mentioned. Operation at uhf should be possible if distributed-parameter type circuits are used.

Further increases in bandwidth and reduced a-m will be obtained by using improved variable-capacitance diodes now appearing on the commercial market.

Because of the high input impedance these variable-reactance diode modulators are ideally suited to wide-band high-frequency modulation with complex waveforms and little modulation power.

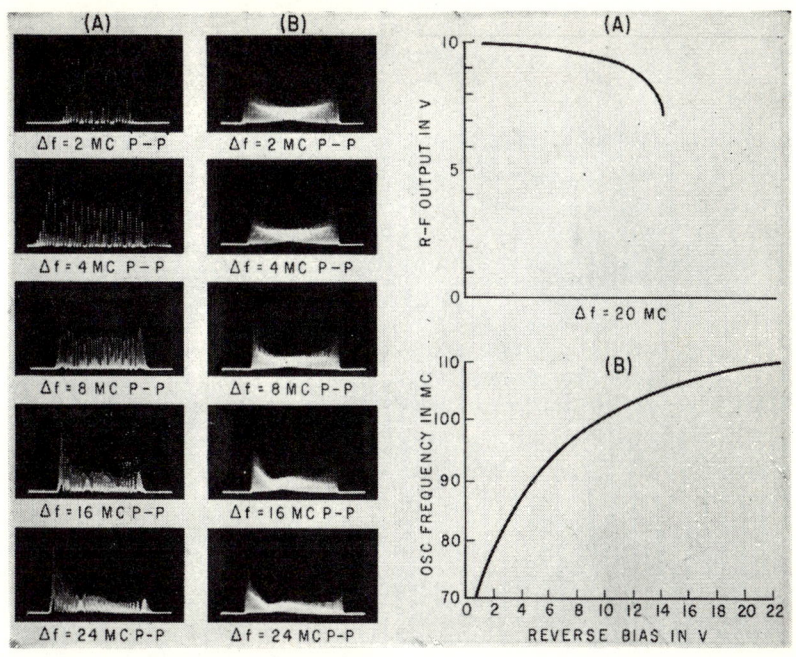

FIG. 2—Sinewave modulation of 100-mc oscillator at 100-kc (A) and 10-kc (B)

FIG. 3—Output for a 20-mc p-p deviation (A); f vs bias with R_L=2,000 ohms (B)

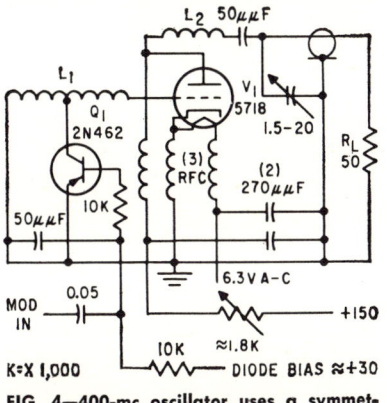

FIG. 4—400-mc oscillator uses a symmetrical transistor

FIG. 5—400-mc-osc. spectrum (A) and output voltage (B) for 15-mc p-p deviation

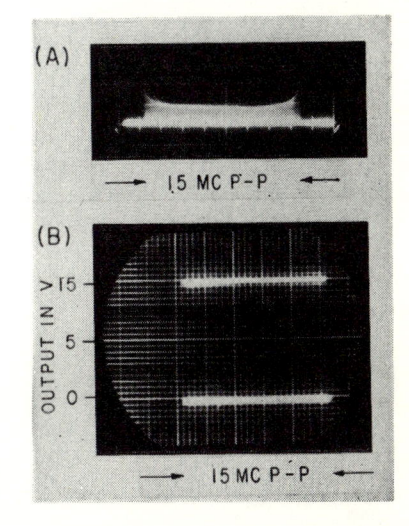

REFERENCES

(1) A. Hund, Reactance Tubes in Frequency Modulation Applications, ELECTRONICS, 15, Oct. 1942.
(2) M. Apstein and H. H. Weider, Capacitor-Modulated Wide-Range F-M System, ELECTRONICS, 26, Oct. 1953.
(3) C. G. Sontheimer, Applications of High Frequency Saturable Reactors, Proc of the National Electronics Conference, 9, Feb. 1954.
(4) W. A. Edson, "Vacuum Tube Oscillators", 16, John Wiley and Sons, Inc., New York, 1953.
(5) R. D. Middlebrook, "An Introduction to Junction Transistor Theory", 9, John Wiley and Sons, Inc., New York, 1957.

Chapter 9
TELEVISION RECEIVERS

Tuners for Portable Television Receivers

Microalloy diffused-base transistors applied to typical tv tuner design give 18-19 db power gain at 210 Mc with 12-db noise factor, characteristics sufficiently good for portable receiver applications

By **VICTOR MUKAI**, Senior Design Engineer, General Instrument Corporation, Newark, N. J.
P. V. SIMPSON, Group Engineer (Tv), Philco Corporation, Philadelphia, Penna.

A MAJOR PROBLEM in designing all-transistor battery-operated portable tv receivers has been application of the transistor to the tuning unit. One solution involves the use of microalloy diffused-base transistors which have sufficiently good characteristics in the 50 to 250-Mc region at cost low enough to warrant their use in home-entertainment equipment.

In the tuner discussed here, it was decided to adapt a commercially available tube tuner because of its form factor, relatively small cubic content, and reasonably good flexibility of switch design. As much external capacitance as possible was added to all tuned circuits to minimize transistor variations. To maintain low-impedance ground paths, a metal chassis was used instead of the printed panel specified in the original tube tuner.

Figure 1 shows the tuner's basic schematic. All transistors are *pnp* germanium MADT types. A single-tuned antenna preselector precedes the r-f stage, while a conventional double-tuned bandpass circuit is used between the r-f stage and the T-1600 mixer. Energy from the oscillator is fed into the mixer emitter. Total power drain is 130 to 140 mw as contrasted to approximately 10 w for the average tube tuner. Reduction in heat is an asset in controlling local-oscillator drift.

In Fig. 2, the tuner's average noise factor and power gain are compared to a typical commercial tetrode tuner. In this comparison, the tube-mixer i-f bandwidth and termination were adjusted to agree with that of the transistor tuner. Image and i-f rejections measure better than 55 to 60 db on the transistor unit, figures which are not inferior to those of tube tuners.

R-F Stage

Common-emitter configuration rather than common-base is used. Inherent degeneration of the common-emitter connection, if properly handled, provides greater production uniformity. Proper handling calls for neutralization of the collector-base feedback capacitance. This was arranged in the conventional manner shown in Fig. 1 where an adjustable capacitor feeds out-of-phase energy from the collector tank circuit back to the base.

This feedback control proved so effective that it was possible to over-neutralize to a predetermined point and improve high-channel gain by an additional 2 db, thus realizing an over-all average power gain of 18 to 19 db at channel 13 (210 to 216 Mc) with a 12-db noise factor. Stability of the r-f stage was excellent.

The collector output-resistance component of Q_1 under these operating conditions varies with frequency and is, of course, dependent on the degree of neutralization. Output capacitance is 1.5 pf. It is possible, with available parameters, to choose circuit constants for a double-tuned bandpass filter which present the sharp response curves obtained with tube tuners.

High-channel bandwidth can be varied by adjustment of the low-side mutual coupling derived from strategic placement of the low-potential return leads of the r-f stage and mixer tank circuits. Air mutual coupling is used for variation of coupling and bandwidth in the low channels, and is brought about by physical placement and adjustment of the associated inductances.

The input capacitance of the T-

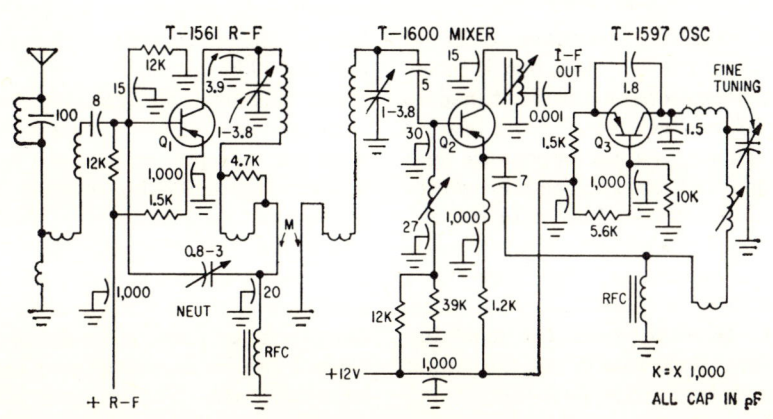

FIG. 1—Basic diagram of transistorized tv receiver tuner

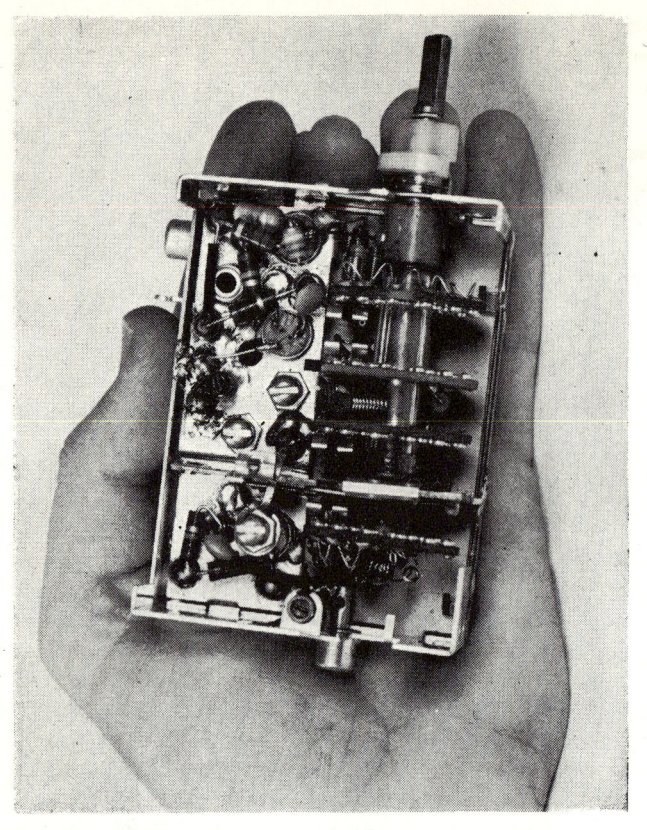

Dimensions of palm-sized tuner are 3¼ by 2½ by 1¾ inches

A typical commercial-type tuner adapted for transistors

1561 is about 10 pf. The resistance component is approximately 50 ohms at channel 13 and rises to 100 ohms at channel 2. A capacitively-tapped antenna resonant circuit was chosen for convenience in matching the transistor to the tank circuit and thence to a monopole rod antenna. The 70-ohm point for the monopole is inductively tapped on the antenna tank, and the tuner switching system readily adapts itself to this scheme. The shaft serves as a convenient tank ground point in this case. Use of this type input circuit also results in good low-frequency rejection, a significant consideration in cross-modulation.

The transistor experiences little noise-factor change when the source resistance is varied as much as 2 or 3 to 1 with respect to the input resistance. A grounded-cathode r-f amplifier tube, on the other hand, requires mismatching for best noise factor. Since such mismatch can cause transmission-line reflection problems, especially on channels 2 through 6, the advantage of the transistor is obvious.

Power gain of the T-1561 r-f stage closely follows a 6-db-per-octave curve at these frequencies in a matched, neutralized set-up. Typical single-stage gain figures are 9 to 13 db at channel 13 using low-loss tuned circuits.

Mixer Stage

In matching the mixer to its r-f circuit capacitive tapping is used. A series-resonant circuit, tuned to the i-f, and connected from base to ground forms a low-impedance path without which good mixer power gain cannot be achieved in the circuit used. A possible alternative is the connection of the base to a point on the tuned-circuit inductance. In either case, the most important termination of the mixer proved to be the base-circuit impedance at intermediate frequency. The series-resonant circuit obviated the necessity for i-f neutralization. Power-gain improvement was in the order of 5 db; noise figure was reduced about 6 db. Mixer power gain in the circuit used is somewhat less than the gain of the transistor as an r-f amplifier. Approximately 0.5 mw of oscillator power is required for good mixing.

The physical layout of the tuner is such that the bottom end of the oscillator tank coil can be conveniently returned directly to the mixer emitter. The ground path for the tank includes a 1,000-pf feed-through-type bypass capacitor and a small hairpin inductance connected between the bypass capacitor and the emitter pin. This in effect connects the emitter of Q_2 to a low-impedance point on the oscillator tank. By regulating the hairpin inductance, some degree of control can be exercised over oscillator injection. Voltages measured directly on the mixer emitter

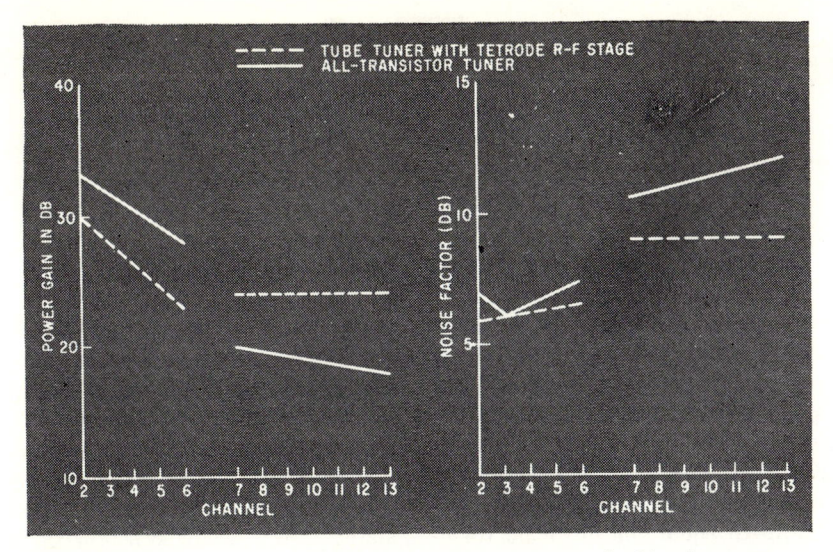

FIG. 2—Noise-figure and power of transistorized tuner compared with tube tuner

socket pin vary between 0.15 and 0.4 v rms.

The lead length for the 30-pf base-to-ground capacitance was selected to be series resonant in the high-channel oscillator range. This resonance keeps the base as close to ground as possible so far as oscillator energy is concerned, and has the added feature of improving mixer-diode efficiency; it also helps prevent oscillator energy from appearing in the interstage circuit, where it may be passed along to the antenna terminals and radiated. Thus, the resonance was a significant factor in keeping oscillator radiation within FCC limits. Another virtue of the injection system is freedom from interaction between the oscillator and r-f circuits.

The oscillator circuit shown in the basic diagram has proved capable of providing useful outputs at frequencies as high as 380 Mc. Feedback is controlled by the ratio of the collector-emitter and base-emitter capacitance. The oscillator functioned over a supply-voltage range of 8.5 to 14 v. Frequency drift characteristics with temperature had to allow for operation to 50 C.

The resultant change in oscillator frequency with change of supply voltage is shown in Fig. 3 for an average transistor. It was possible to keep short-term drift within ±300 Kc of the frequency reached after a one-minute warm-up time (25 C ambient). A fine-tuning range of 3 to 5 Mc cover these variables. Little change in inductance values was required in going from tube to transistor oscillator.

Of further interest is the necessity for keeping the d-c temperature stability factor as low as possible. Small changes in stabilization were found to have a decided influence on high-channel drift.

One of the properties of the MADT that lends itself to agc is dependence of its frequency characteristics on collector voltage. This is attributed to the presence of an

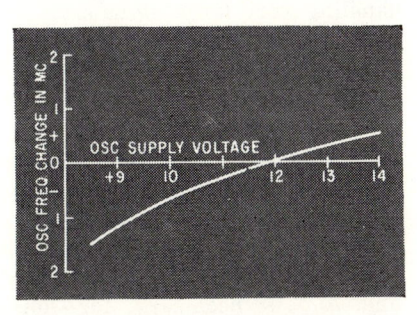

FIG. 3—Frequency stability characteristics of tuner oscillator set for channel 13

intrinsic region in the base. By inserting resistance in the collector circuit, a tuner gain reduction of about 30 db can be obtained with forward biasing. However, the disadvantage of the forward-bias technique with a 12-v supply is the limitation placed on achieving the maximum possible gain because of the lower initial collector voltage. Since measurements show this to be as much as 3 db on channel 13, this loss was considered important in terms of overall noise-figure performance.

It was, therefore, decided to control the r-f stage gain by the selection of specific supply-voltage points. A switch was used to change the r-f stage operating voltage in three steps corresponding to strong, normal and fringe operation. This allowed sufficient overlap of the r-f i-f system gain-control curves to insure good overload characteristics, medium signal snow and fringe performance.

Reverse biasing of the base-emitter diode by itself did not provide satisfactory gain control because of the poor overload capabilities encountered at low collector currents. With the tuner placed ahead of a conventional tube receiver, and the r-f gain varied in the manner described, no particular overload problem was found with levels as high as 0.4 v (300-ohm input).

Cross-Modulation

Cross-modulation characteristics were evaluated by field tests in which the transistor tuner was installed in a production-model tube receiver and compared to a similar model with its tube tuner. The method of field test was based on the vacuum-tube analysis which shows that cross-modulation percentage is proportional to the square of the interfering voltage and independent of desired signal voltage. Therefore, adjustment of the signal levels to each receiver can be made until the desired picture is free of interference.

Using this method, the transistor tuner was consistently within ±6 db of the tube tuner for various desired signal levels. Operating bias point of the transistor r-f stage was a negligible factor in the degree of cross-modulation with the gain-control system used.

Acknowledgment

Sincere thanks are extended to J. Waring, R. Booker and C. Simmons of Philco Corp. for their advice and help in the preparation of this article.

BIBLIOGRAPHY

"Transistors 1", p 422 to 430, RCA Laboratories.
L. P. Hunter, "Handbook of Semiconductor Electronics", section 12, McGraw-Hill Book Co., New York, N. Y., 1956.
D. DeWitt and A. L. Rossoff, "Transistor Electronics", McGraw-Hill Book Co., New York, N. Y., 1957.

Designing Television

Specially developed diffused-base mesa transistors permit design of tv tuners with noise performance equal to that obtained in tube tuners. Complete design procedure for r-f amplifier, mixer and oscillator stages is given

UNTIL RECENTLY, transistors were not used at very high frequencies due to high noise figure and low gain when compared with tubes. This is no longer true and, in fact, we may see the situation reversed in the next few years. The mesa transistors used in the tv tuner described here were developed for tv tuner and i-f applications.

In simplest terms, a tuner can be defined as that part of a receiver which selects the desired frequency from a spectrum and converts it to a fixed lower frequency with minimum degradation in signal-to-noise ratio and signal fidelity. Gain may be considered secondary if it is sufficient to keep the noise figure independent of the following stages. This is nearly always true in superheterodynes where gain is cheaper at intermediate frequencies.

Also of importance, but not necessarily tied to transistor utilization, are selectivity, oscillator radiation, swr (standing wave ratio) and agc (including overload). Automatic gain control is dependent on transistor parameters but can be enhanced considerably by the circuits. Actually, there is one more requirement which is not covered by this paper: suitable packaging for the tuner. Tuner packaging is an art in itself.

In actual numbers, present-day tube tuners provide the selectivity required with noise figures ranging from 6 to 12 db at 215 Mc. The gain of a tube tuner will vary considerably at 215 Mc, but in general lies between 25 and 40 db. Most tube tuners handle up to 1-v at the input.

The 2N1398 r-f amplifier described is tested to a maximum noise figure of 6 db at 200 Mc, while the mixer, 2N1399, is tested to a noise figure of 7 db maximum at 200 Mc. Typical net gain for the r-f amplifier in a practical circuit is 10 db at 200 Mc, the mixer 12 db at 200 Mc. The minimum noise figure of the r-f amplifier is 4 db at 200 Mc. Thus, with careful design, it is possible with mesa transistors to equal the noise performance of today's best tube tuners.

Automatic gain control can be made to equal or exceed present tube design, but overload still represents a problem. At a signal level approximately equal to the d-c base-to-emitter voltage (V_{BE}) transistor overload occurs. For germanium, V_{BE} is in the neighborhood of 0.2 volts. Forward, rather than reverse agc will sometimes enhance overload handling capability, but the order of improvement is small compared to what would be desirable. An input attenuator still remains the most effective method of handling very large signals.

H-F and Design Parameters

The parameters of interest can be divided into two categories: general high-frequency and design. General high-frequency parameters, such as $r_b'C_c$ product, f_α and f_{max} determine the gain capabilities of a transistor, while knowledge of r_b' and h_{fe}, is necessary for noise analysis. These parameters are useful to the designer in making a preliminary evaluation of the transistors. The measurement of $r_b'C_c$ is easy, and a simple test set is shown in Appendix 1.

The low-frequency common base forward current transfer ratio, a_o, can be measured by conventional test gear but f_α, the frequency at which $|\alpha|$ has decreased 3 db, is harder to ascertain. A high-frequency transfer bridge such as the GR1607A can be used to measure f_α directly. However, a somewhat simpler method is based on the fact that f_α is related to f_T (the frequency at which $|h_{fe}|$ becomes 1) by $f_\alpha = Kf_T$ where K is a factor which is usually between 1.2 and 2. The multiplying factor, K, depends upon the amount of drift field and, hence, the excess phase shift in the transistor. For transistors with no drift field (alloy transistors) the factor is 1.2. For the diffused-base mesa transistor K is approximately 1.9, so that $f_\alpha \cong 1.9\, f_T$.

The parameter f_T is determined by measuring $|h_{fe}|$ at 100 Mc and multiplying by 100, the frequency of measurement. Measurement of h_{fe} is covered in Appendix 2. From these parameters f_{max} and power gain at frequency f, is computed: $f_{max} = \sqrt{a_o\, f_\alpha/8\pi\, r_b'C_c}$ or PG = 20 log (f_{max}/f) db. Thus a transistor with an f_{max} of 1,000 Mc would have a gain of 14 db at 200 Mc and approximately 40 db at 10 Mc. The gain expression is approximate for frequencies much less than f_{max}.

The last high-frequency parameter that is of interest here is the noise figure. This may be computed from the following formula from Neilson[2]:

$$F = 1 + \frac{r_b'}{R_g} + \frac{r_e}{2R_g} + \left[1 + \frac{1}{1-\alpha_o}\left(\frac{f}{f_\alpha}\right)^2\right] \times [R_g + R_b' + r_e]^2/2\, h_{FEO} r_e R_g$$

where R_g is generator impedance; f = frequency under consideration; $r_e \approx 25/I_E = 25/$emitter current in ma; r_b' = high frequency base resistance, see Appendix 3; and h_{FEO} = d-c common emitter current gain (d-c β). The above formula shows reasonable agreement with actual noise measurements made with a noise diode from a 75 ohm source at 200 Mc.

Design Parameters

Once the short-circuit current gain of a particular transistor con-

176

Tuners With Mesa Transistors

By **HARRY F. COOKE**, Semiconductor-Components Div.,
Texas Instruments Incorporated, Dallas, Texas

figuration is known, the only additional design parameters necessary are the input and output impedances. These may also be used to compute the gain more accurately. (See Appendix 4.)

As a general rule, it is much easier to measure short-circuit impedances than open-circuit impedances at higher frequencies and, thus, the following will be used in design calculations:

r_{oep}, C_{oep}—Parallel equivalent output impedance, input short circuited.
r_{iep}, C_{iep}—Parallel equivalent input impedance, common emitter output short circuited.
r_{ies}, C_{ies}—Series equivalent input impedance, common emitter, output short circuited.
r_{ibp}, C_{ibp}—Parallel equivalent input impedance, common base, output short circuited.
r_{ibs}, C_{ibs}—Series equivalent input impedance, common base, output short circuited.

These parameters can be measured with such instruments as the Boonton RX Meter, Wayne-Kerr VHF Bridge, or General Radio Transfer Bridge. They closely approximate circuit conditions where the circuit is neutralized or where the input or output loads are such that the effect of transistor feedback is small. Above 100 Mc, it is recommended that input impedance be measured with the GR1607 Transfer Bridge, or an RX meter with coaxial adapter. Below 100 Mc, RX meter adapters can be used.

The critical design parameters used in the following are controlled by the transistor manufacturer. Methods for determining these parameters are given in Appendix 5. Design procedure for the three tuner stages—r-f amplifier, mixer, and oscillator—will be given now.

For an r-f amplifier, Fig. 1, two configurations are possible: common-emitter or common-base. The neutralized common-emitter stage will give less gain than the un-

neutralized common-base stage at channel 13. However, for a given set of parameter variations, the common-emitter stage will give less spread in total gain. The noise figure of both configurations is exactly the same.

Step 1—Select operating point: The current gain, h_{fe}, rises continuously up to about 7 ma, while the output impedance drops quickly. This gives a plateau of constant gain for a region lying between 0.5 and 2.0 ma emitter current (Fig. 1C). Noise figure is also minimum in this region. Therefore, an operating point of 1.5 ma will be selected. The collector supply voltage is usually determined by factors other than the tuner and will ordinarily be between 6 and 12 volts.

Biasing resistors are selected on the basis of the stability factor desired. A stability factor of 3 is satisfactory for operation to 60 C. The emitter resistor of a common base r-f stage should be large compared to the input impedance to avoid excessive signal loss. Base bias is not shown for this stage since it is derived from the agc system. Where agc is applied to the common emitter r-f stage, an r-f choke is used to connect to the agc source.

Step 2—Input Circuit: In the procedure which follows, and for the interstage and oscillator, a channel 13 (215 Mc) design will be shown since this is generally the most difficult. However, each channel is a separate design problem with different tuned circuits and transistor parameters. It is at this point that the transistor input and output impedances at 215 Mc are measured by the transfer bridge or RX meter. At 215 Mc, r_{ies} is about 40 ohms and r_{ibs} about 25 ohms. Typical values for c_{ies} and c_{ibs} are 35 pf and 0.012 μh, respectively.

There are a large number of in-

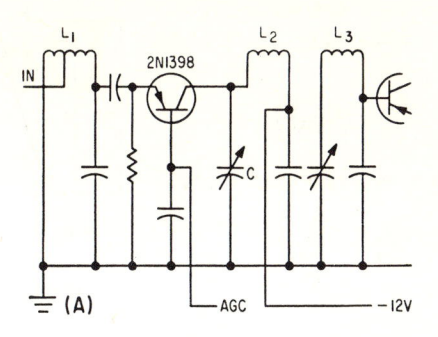

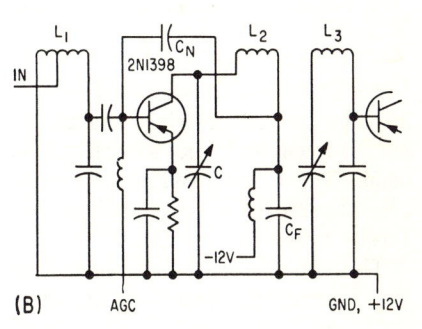

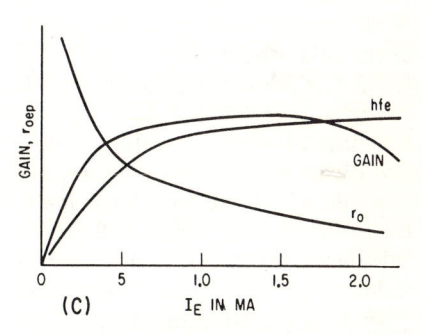

FIG. 1—Common base (A) and common emitter (B) connection for 215 Mc r-f amplifier. Transistor characteristics (C)

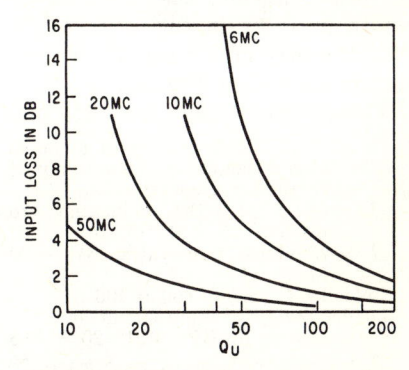

FIG. 2—Relation between input loss and unloaded Q for channel 13 varies with input bandwidth

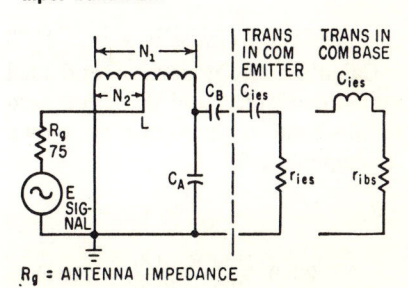

FIG. 3—Input network for r-f amplifier

$E_{PK} = E_0 + 1\,DB$

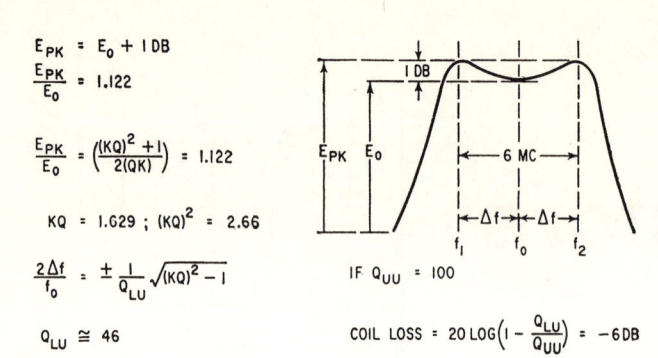

$\dfrac{E_{PK}}{E_0} = 1.122$

$\dfrac{E_{PK}}{E_0} = \dfrac{(KQ)^2 + 1}{2(QK)} = 1.122$

$KQ = 1.629\ ;\ (KQ)^2 = 2.66$

$\dfrac{2\Delta f}{f_o} = \pm \dfrac{1}{Q_{LU}}\sqrt{(KQ)^2 - 1}$

$Q_{LU} \cong 46$

IF $Q_{UU} = 100$

COIL LOSS $= 20\,LOG\left(1 - \dfrac{Q_{LU}}{Q_{UU}}\right) = -6\,DB$

FIG. 4—Interstage network bandpass suitable for channel 13

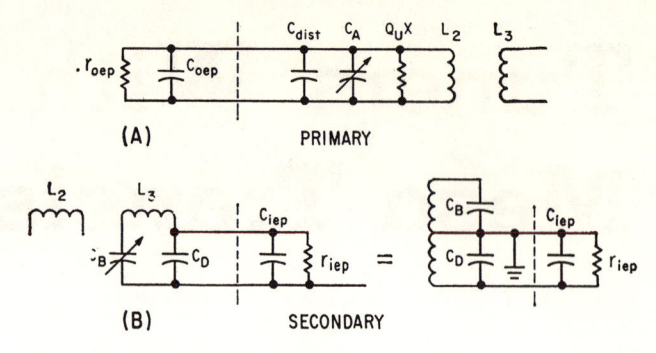

FIG. 5—Equivalent circuit of interstage network

put configurations possible, but usually the one which gives the greatest convenience in switching is used. Turret and switch type tuners, therefore, usually utilize different types of circuits.

The input bandwidth must first be determined. Considering selectivity alone an input bandwidth of 6 Mc would seem desirable; however, this would require a loaded Q (Q_L) of 36, and would necessitate using a high unloaded Q (Q_U) to keep input losses down. Input losses add directly to the noise figure, and it would be best to keep these losses to 1 db or less. Figure 2 shows calculated input circuit-loss versus unloaded Q for several bandwidths.

In a practical tuner it may not be possible to realize unloaded Q's of more than 100. If we use the 1 db loss criterion, the input bandwidth must be in the order of 20 Mc. We may now proceed with the design shown in Fig. 3.

Given: Frequency, $f_o = 215$ Mc.

$C_A = 25$ pf. This is selected to be as large as possible and is determined by the tuner construction.

$L = 0.002$ μh, the inductance to resonate 25 pf at 215 Mc

$R_g = 75$ ohms, Bandwidth, BW = 20 Mc

Unloaded Q (Q_U) of coil = 100

Loaded Q (Q_L) of input circuit = $f_o/\text{BW} = 215/20 = 10.8$

Calculate the circuit reactance, X

$X = \dfrac{1}{\omega C} = \dfrac{1}{2\pi(215 \times 10^6)25 \times 10^{-12}} \cong 30\ \text{ohms}$

Calculate the generator and load resistances (R_g' and R_L') referred to the total tank circuit as shown in Appendix 6.

$R_g' = R_L' = 2X\,\dfrac{Q_U Q_L}{Q_U - Q_L} = 2(30)\dfrac{(100 \times 10.8)}{(100 - 10.8)} = 720\ \text{ohms}$

Calculate tap point for antenna:

$N_1/N_2 = \sqrt{R_g'/R_g} = \sqrt{720/75} \cong 3$. This assumes $k = 1$. Tap can be moved up slightly to compensate for k being less than 1.

From GR-1607 transfer bridge: $r_{ies} = 40$ ohms; $C_{ies} = 35$ pf; $r_{ibs} = 25$ ohms; $C_{ibs} = 0.015$ μh, (input is inductive).

Calculate coupling capacitor, C_B

$X_{CB} = \sqrt{R_L'\,r_{ies} - (r_{ies})^2} = \sqrt{720(40) - (40)^2} = 168\ \text{ohms}$

$C_B = \dfrac{X_{CB}}{2\pi f_o} \cong 4.7$ pf, for common-emitter stage

$X_{CB} = \sqrt{R_L'\,(r_{ibs})^2 - (r_{ibs})^2} = \sqrt{720(25) - (25)^2} = 132\ \text{ohms}$

$C_B = \dfrac{X_{CB}}{2\pi f_o} \cong 5.6$ pf, for common-base stage.

Since $C_{ies} \gg C_B$, its effect can be ignored.

Where the input impedance is inductive, the procedure is the same except C_B must be increased slightly to give X_{CB} the correct magnitude.

The noise figure can be reduced about 0.5 db by mistuning the input circuit on the high side and by adjusting the input so as to allow the transistor to look back into an impedance slightly greater than its own input impedance. The transistor may also be operated directly from a 75 ohm antenna with negligible loss in gain, but a vswr of about 1.8 will result. Where the higher vswr is not objectionable, this method is satisfactory. For the diffused-base mesa transistor the best noise figure and gain each are obtained with almost exactly the same source impedance.

Step 3—The Interstage Coupling Network: The interstage coupling network is usually double-tuned, since it is here that the tuner must obtain most of its image rejection. A neutralized common-emitter stage requires some provision for neutralization from this network. Inductor L_2 (Fig. 1b) may be tapped, but a simpler neutralization method is to select a capacitor C_F for the low-impedance end of L_2 such that the series resonance frequency of L_2 and C_F is considerably lower than that of L_2 in parallel with C_total, where C_total is total effective parallel capacitance across the inductance. The voltage across C_F will then approach a 180 degree lag with respect to the collector voltage. If C_total is 5 pf, C_F should be about 25 pf. Then, if C_c, the capacitance to be neutralized, is 1 pf, $C_N = (25/5)(C_c) = 5(1) = 5$ pf.

Figure 4 shows a bandpass characteristic which could be considered suitable for channel 13. Using the equivalent circuit in Fig. 5, we may now compute the parameter values of the transformer for a common emitter (or common base) stage.

(a) Primary

Given: $Q_{LU} = 46 =$ Loaded, uncoupled Q, from Fig. 4; $Q_{UU} = 100 =$ Unloaded, uncoupled Q; $r_{oep} = 5000$ ohm (from RX meter); $c_{oep} = 1.5$ pf (from RX meter).

Using expression derived in Appendix 7, calculate

$C_\text{total} = \dfrac{Q_{UU}}{\omega r_o\left(\dfrac{Q_{UU}}{Q_{LU}} - 1\right)} =$

$\dfrac{100}{5000\left(\dfrac{100}{46} - 1\right)(2\pi)(215 \times 10^6)} \cong 13\,\text{pf}.$

Then $L = 0.04$ μh, the inductance to resonate with C_total at 215 Mc. $C_A = C_\text{total} - (C_\text{dist} + C_{oep}) = 13 - (2 + 1.5) = 9.5$ pf. C_A may then be a 4.7 pf fixed capacitor plus a 1 − 8 pf trimmer. Bandpass shape in Fig. 4 is used to determine $KQ_{LU} = 1.629$. Also, $K = KQ_{LU}/Q_{LU} = 1.629/46 = 0.035$. Letting L_3

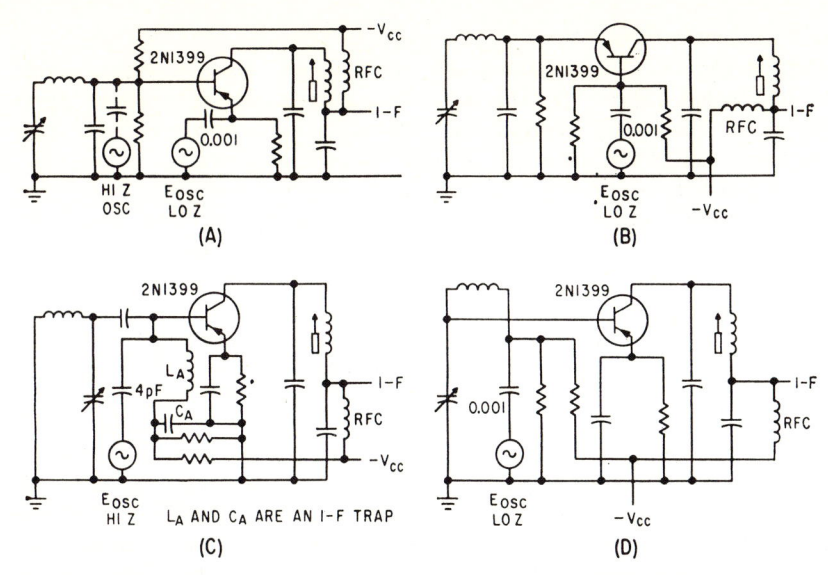

FIG. 6—Possible mixer connections include: signal in base, oscillator in emitter or base (A); signal in emitter, oscillator in base (B); signal and oscillator in base, emitter grounded (C); and signal and oscillator in base in series, emitter grounded (D)

$BW_{pk-pk} = 3.5$ Mc, $Q_{UU} = 70$
$f_o = 43$ Mc, Let $Q_1 = Q_2$
$KQ = 1.629$; same as interstage

$$Q_{LU} = \frac{f_o}{BW_{pk-pk}} \sqrt{(KQ)^2 - 1}$$

$$= \frac{43}{3.5} \sqrt{2.66 - 1} \cong 15.5$$

(a) Primary
$r_{oep} = 20,000$ ohms $\}$ RX Meter
$C_{oep} = 1.5$ pf $\}$ Data at 43 Mc

$$C_{total} = \frac{Q_{UU}}{r_o \left[\frac{Q_{UU}}{Q_{LU}} - 1\right] \omega} = \frac{70}{20,000 \left[\frac{70}{15.5} - 1\right] 2\pi \times 43 \times 10^6}$$

$= 3.7$ pf and $L_p = 3.7$ μh
$C_p = C_{total} - (C_{oep} + C_{dist}) = 3.7 - (1.5 + 2) \cong 0$
Omit C_p

(b) Secondary
$r_{iep} = 200$ ohms $Q_{input} = \omega C_{iep} r_{iep} = 0.55$
$C_{iep} = 10$ pf $(Q_{input})^2 = 0.3$

$$r_{ies} = \frac{r_{iep}}{1 + Q^2} = \frac{200}{1.3} \cong 153 \text{ ohms}$$

$$C_{ies} = C_{iep} \frac{(1 + Q^2)}{Q^2} = 10 \frac{(1.3)}{.3} = 43 \text{ pf}$$

$$C_{total} = \frac{1}{\omega r_{ies} Q_{LU}} = \frac{1}{2\pi 43 \times 10^6 \times 153 \times 15.5} \cong 1.6 \text{ pf}, \quad L_s = 8.8 \text{ μh}$$

$C_{total} \ll C_{ies}$; therefore, $C_s = 1.6$ pf
$KQ = 1.629$, $Q = 15.5$, $\therefore K = 0.105$

$$C_M \cong \frac{\sqrt{(C_{pri\,total})(C_{sec\,total})}}{K} \cong \frac{\sqrt{3.7(1.5)}}{0.105} \cong 20 \text{ pf}$$

$$\text{Efficiency} = \left(1 - \frac{Q_{LU}}{Q_{UU}}\right)^2 = \left(1 - \frac{15.5}{70}\right)^2 = -2.2 \text{ db}$$

FIG. 7—I-f transformer design equations used

$= L_2 = 0.042$ μh, then $M \cong K\sqrt{L_1 L_2} = 0.35 (0.042) \cong 0.00149$ μh. Adjust spacing between L_2 and L_3 for M and proper bandpass.

If we wish to have equal primary and secondary Q's, then r_{iep} of the mixer must be transformed to equal r_{oep} of the r-f amplifier. A tap on L_2 would accomplish this, but a pi section (or capacitance split) match is better since one less switch contact is required. Figure 5B shows such a network.

(b) Secondary, Fig. 5B

Let $C_{total} = 13$ pf (same as primary); $L_3 = 0.042$ μh (same as primary); $C_B = C_{total} = 13$ pf.

Assume common-emitter mixer and convert transistor input parameters from series equivalent to parallel equivalent.

$r_{iep} = r_{ies}(1 + Q_T^2) =$
$\quad r_{ies}[1 + (1/\omega C_{ies} r_{ies})^2]$
$= 40[1 + 0.25] \cong 50$ ohm
$C_{iep} = C_{ies} Q_T^2/(1 + Q_T^2) =$
$\quad 35 \times 0.25/1.25 \cong 7$ pf

Where Q_T is the Q of the transistor at its input. Calculate

$C_D = C_{total} \sqrt{r_{oep}/r_{iep}} =$
$\quad 15 \sqrt{5,000/50} = 150$ pf

The effect of C_{iep} can be ignored since C_{iep} is considerably less than C_D. To reduce oscillator transmission back through the network, C_D may be changed to a network to put a transmission zero near the oscillator frequency.

If the r-f stage is connected common-base, the output impedance for the r-f transistor is negative at 215 Mc. This impedance is not easy to measure since it depends not only on the internal gain of the transistor, but also upon the feedback capacitance. In practice a design based on a positive output impedance may be used if the secondary loading is increased by decreasing C_D. This adjustment can best be accomplished by empirical methods.

In theory, a double-tuned circuit driven by a negative resistance generator could be designed to have a large gain-bandwidth product. However, if a large increase in gain is realized, the circuit becomes critical to adjust, and small load changes may bring on instability. When the increase in gain is about equal to the coil loss (6 db), the circuit ordinarily will be unconditionally stable with the controls which are placed on transistor parameters.

Mixer

To perform the function of mixer efficiently, a transistor must have the following characteristics: good gain at the intermediate frequency; good emitter-base diode characteristics; and low emitter-base transition capacitance (C_{t_e}).

The diffused-base mesa transistor has all these characteristics. The gain at 43 mc is 25-30 db, and the diffused-base structure has an efficient emitter with good diode characteristics. Transition capacitance is also low making the transistor useful well into the uhf region.

When a transistor is to function as a mixer, the signal and oscillator voltages are introduced between the emitter and base. Whether the emitter or base is grounded is largely a matter of the designer's choice since, insofar as transistor mixing action is concerned, both connections are the same. Therefore, the oscillator and signal voltages may be both injected at the base or emitter or to either individually.

Maximum gain is obtained when the emitter-base diode is reverse biased by the oscillator voltage for a small fraction of a cycle. This corresponds to an injection level of 0.1 to 0.2 volts rms. Also, since the diode is reverse biased for only a short period, the input impedance of a mixer is close to that of an r-f amplifier.

D-c operating point for a mixer is a compromise between i-f gain and detection efficiency. The 2N1399 mesa transistor gives best gain with an emitter current of 1.0 to 1.5 ma. A few possible mixer configurations are shown in Fig. 6. Whichever configuration is used, it is always important to keep the i-f impedance low at the emitter and base terminals to reduce i-f degeneration. The necessity for neutralization is also removed. The tapped input coil and capacitive split accomplish this automatically, but where a small capacitor is used for impedance matching (Fig. 6C) an i-f trap ($L_A C_A$) must be used. In general, the Q of the trap should be less than that of the i-f transformer which follows it, so that the trap itself will not become a part of the bandpass circuit. The example in the final circuit (Fig. 9) is that of Fig. 6A with signal and oscillator both connected to the base.

Since the first i-f transformer is usually included in the tuner, one type will be shown here. It should be understood that the design is not intended to give a desirable total i-f response, since it comprises only two poles in the overall i-f system. The unloaded Q of the transformer is not critical as in the r-f stage, and a value of 70 is typical for an economical design. Figure 7 shows the computation for a transformer with combined parallel and series tuning. The parallel-series type is desirable since it uses low impedance coupling and is adaptable to situations where the tuner must be located remotely from the i-f amplifier.

Oscillator

The choice of oscillator circuit is considerably simplified by the fact that the common-base transistor connection is usually regenerative at high frequencies. With some added external feedback capacitance between emitter and collector, dependable oscillation is assured. The circuit shown in Fig. 8 is typical.

The most effective method of controlling the oscillator power in a circuit such as this is by varying the d-c emitter current. An oscillator with an emitter current of 1.5-2 ma can supply about twenty times the 100-300 microwatts required by the mixer. The emitter and base resistors shown give a stability factor of approximately 2 which is satisfactory for operation to 65 C. As in the common-base r-f amplifier, the emitter resistor is made large enough to prevent shunting of the signal path. The signal at the base of the oscillator should have a good low inductance path to ground such as provided by a feed-through type capacitor.

Mixer injection voltage may be taken from the oscillator in several ways. If the voltage is injected to the mixer at the same point as the signal, high-impedance feed from the oscillator collector via a small capacitor (C_a) is to be preferred. Where oscillator voltage injected into a separate element is desirable, the voltage may be taken from the bottom of L via C_b. Capacitor C_d determines the oscillator voltage and is also a low impedance path for the i-f voltage in the mixer.

Exact values for C_a and C_d must be selected by trial and error since residual circuit reactances play considerable part in determining the actual voltage at the mixer. The values shown in Fig. 8 are typical. Note that an r-f choke in series with the d-c supply is necessary where C_d is used to obtain the injection voltage. The trimmer capacitance C_e is used to compensate for the small variations in the total capacitance across L. In order that L may be a reasonable size C_e should be as small as possible. The circuit shown will have a total capacitance across L of about 6 pf, so that $L = 0.063$ μh at 257 mc.

Fine tuning may be obtained with either capacitance or inductance. Variable capacitance is, perhaps, easier to construct mechanically, but variable inductance gives a constant incremental frequency, Δf, for the tuning, which is desirable. For a 3 percent tuning range, $\Delta f = 6.75$ Mc. Then the required change in L would be $(1.03)^2 = 1.06$ or 6 percent.

A 2:1 change in inductance is obtained easily with a sliding core coil.

Let $L_{fT}/L_{tank} = N$; and $P =$ percent tuning ratio $= 1.06 = [(f + \Delta f)/f]^2$ when $L_{fT} =$ minimum inductance of fine tuning inductor. Then $N = (p - 2)/2(1 - p) = (1.06 - 2)/2(1 - 1.06) = 8$ and $L_{fT} = 8(0.063) = 0.504$ μh minimum and 1.008 μh maximum. The tank inductance must now be increased slightly to make up for the shunting effect of the fine tuning.

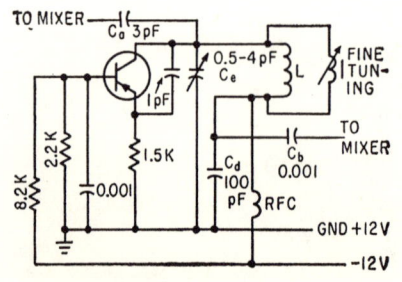

FIG. 8—Oscillator circuit shows typical component values

Conclusion

The complete tuner for channel 13 is shown in Fig. 9. In this article an attempt has been made to emphasize those aspects of design that are peculiar to transistor utilization. Areas of transistor utilization which have been described previously have been intentionally by-passed. Traps, baluns, and other refinements do not differ from tube counterparts and have been omitted. The tuner described has a noise figure of 6-8 db on channel 13 and a gain of 22 db. A similar design for channel 2 will have a noise figure of 5 db and a gain of 35 db.

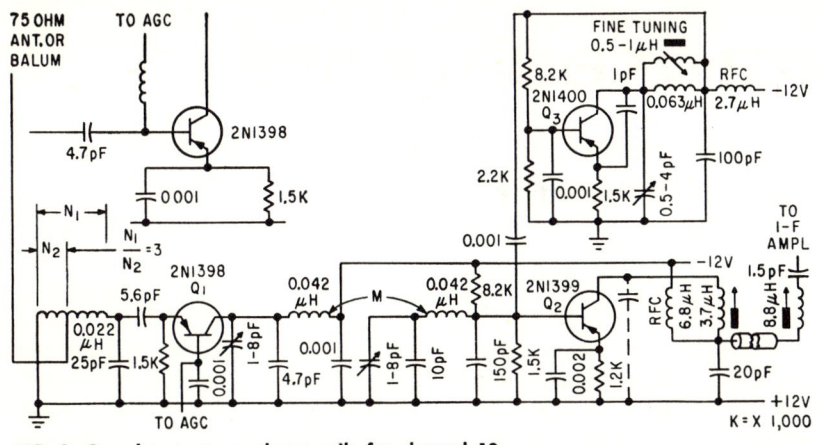

FIG. 9—Complete tv tuner shows coils for channel 13

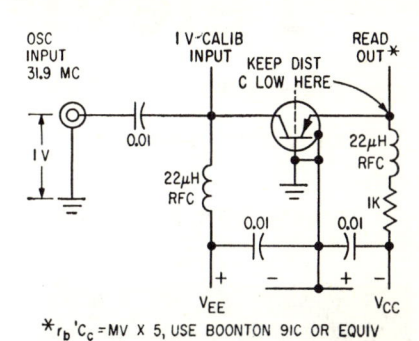

FIG. 10—Test set for $r_b'C_c$ uses calibrated input

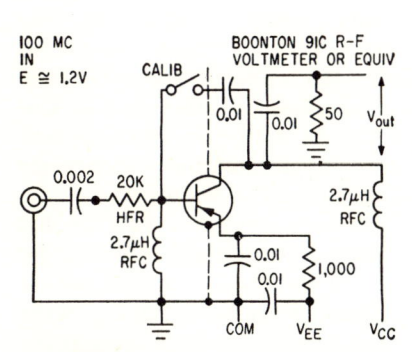

FIG. 11—Test set used to find h_{fe}

FIG. 12—Jigs and circuit used to measure short-circuit impedances below 100 Mc

APPENDIX

(1). $r_b'C_c$ Product; $h_{rb} \cong j\omega r_b'C_c$
Test set shown in Fig. 10. Separate 31.9 Mc oscillator and r-f voltmeter required. Oscillator may be transistorized and self-contained.

(2). h_{fe} at 100 Mc
Test set shown in Fig. 11. Separate 100 Mc oscillator and r-f voltmeter required. Oscillator may be transistorized and self-contained.

(3). r_b' referred to base input
(a) Measure $r_{ies} = R_e$ (h_{ie}) with General Radio 1607A transfer bridge at 100 Mc, or
(b) Measure r_{iep} at 100 Mc with RX Meter and jig shown in Fig. 12. Convert to series equivalent
$r_b' = r_{ies} = r_{iep}/[1 + (\omega C_{iep} r_{iep})^2]$

(4). MAG (maximum available gain)
$\cong h_f^2[r_{op}/4[r_{ip}/(1 + r_{ip}C_{ip})^2]]$

(5). (a) $r_{oep}, C_{oep}, r_{iep}, C_{iep}, r_{ibp}, C_{ibp}$
Use RX meter and jigs shown in Fig. 12.
GR-1607A bridge

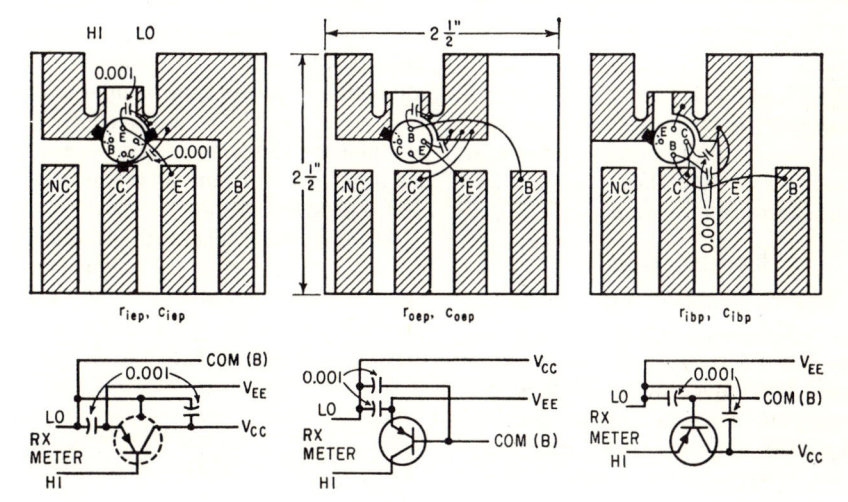

(b) C_c, the capacitive component of the reverse transfer admittance, common emitter.
Measure C_{oep} at 100 Mc or greater.
$C_c \approx C_{oep}$.

(6). Derivation of expression for generator and load impedances referred to the total tank circuit where $Q_{unloaded}, Q_{loaded}$, and capacity are given.

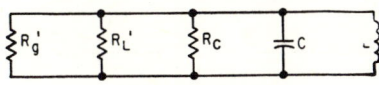

R_g' = Generator resistance referred to total tank circuit.
R_L' = Load resistance referred to total tank circuit.
R_c = Coil loss resistance referred to total tank circuit.
$Q_U = \omega C R_c, R_c = Q_U/\omega C = XQ_U$,
$Q_L = \omega C R_T$, $R_T = Q_L/\omega C = XQ_L$.
Where $R_T = R_{total} = 1/(1/R_c' + 1/R_g' + 1/R_L')$ and $X = 1/\omega C$.
Let the generator and load be matched ($R_g' = R_L'$). Then, $R_T = 1/(1/R_c + 2/R_g')$ and $R_g' = 2$
$R_T R_c/(R_c - R_T) = 2(Q_L X)(Q_U X)/(Q_U X - Q_L X) = 2 X Q_L Q_U/(Q_U - Q_L)$.

(7). Derivation of expression for total tuning capacity of a circuit where Q_U (Q unloaded), Q_L (Q loaded), and load resistance are known.

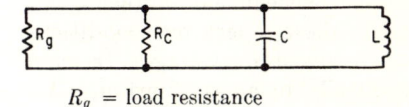

R_g = load resistance
R_c = coil loss resistance ($Q_U X$)
$Q_L = \omega C R_g R_c/(R_g + R_c)$,
$Q_U = \omega C R_c$,
Then, $Q_L/Q_U = R_g/(R_g + R_c)$ and $R_c = r_o (Q_U/Q_L - 1)$ also $R_c = Q_U/\omega C$
Therefore, $Q_U/\omega C = r_o(Q_U/Q_L - 1)$
and $C = Q_U/[\omega r_o(Q_U/Q_L - 1)]$

REFERENCES

(1) D. E. Thomas and J. L. Moll, Junction Transistor Short Circuit Gain and Phase Determination, *Proc IRE*, June, 1958.
(2) E. C. Neilson, Behavior of Noise Figure in Junction Transistors, *Proc IRE*, July, 1957.

Horizontal Deflection Circuits for Television Receivers

New horizontal-deflection and high-voltage circuits are designed around only two transistors. Circuits are stable and efficient

By **MARTIN FISCHMAN**, Sylvania Research Laboratories, Bayside, New York

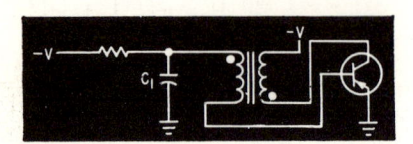

FIG. 1—Conventional transistor blocking oscillator

EFFICIENT AND STABLE, the 90-deg horizontal-deflection circuit and high-voltage generator that will be described uses only two transistors and a diode.

Blocking Oscillator

Before getting into the actual circuit design, a brief description of the operation of conventional transistor blocking oscillators will be given. Figure 1 shows that the blocking oscillator transformer provides regenerative feedback from the collector to the base of the transistor as soon as current starts to flow. The gain and feedback of the circuit cause the current to build up so that the transistor operates in the saturation region of its characteristic. In this region the collector current is substantially independent of the base current. Due to the absence of gain in the circuit under these operating conditions the transistor voltages remain practically in an equilibrium state for a period of time. This voltage-equilibrium state corresponds to the turned-on period or pulse width of the blocking oscillator. The equilibrium condition continues until the transistor operating point moves out of the saturation region into a nonsaturation region. A regenerative process then begins which rapidly turns off the transistor. Termination of the equilibrium condition may come about as a result of an increasing collector current, a decreasing base current or a combination of both; the cause depends on the relative values of capacitor C_1 transformer inductance and external resistive loading—if any exists. In this type of operation the pulse width is influenced by transistor characteristics, circuit loading, transformer characteristics and operating voltages.

An improved blocking oscillator circuit in which the pulse duration is accurately controlled by the transient response of a series L-C circuit is shown in Fig. 2A. Figure 2B shows the transient current and voltage waveforms of a series L-C circuit driven from a step voltage source. Current through the circuit is a damped sine wave oscillating about the zero current axis. Voltage across the capacitor is a damped cosine wave oscillating about the value of the step input voltage. The duration of the first half cycle of the sine wave is to a first approximation dependent solely on the L-C values.

The base circuit of Fig. 2A has a similar oscillatory response for the first half cycle. A step voltage provided by the base winding of the

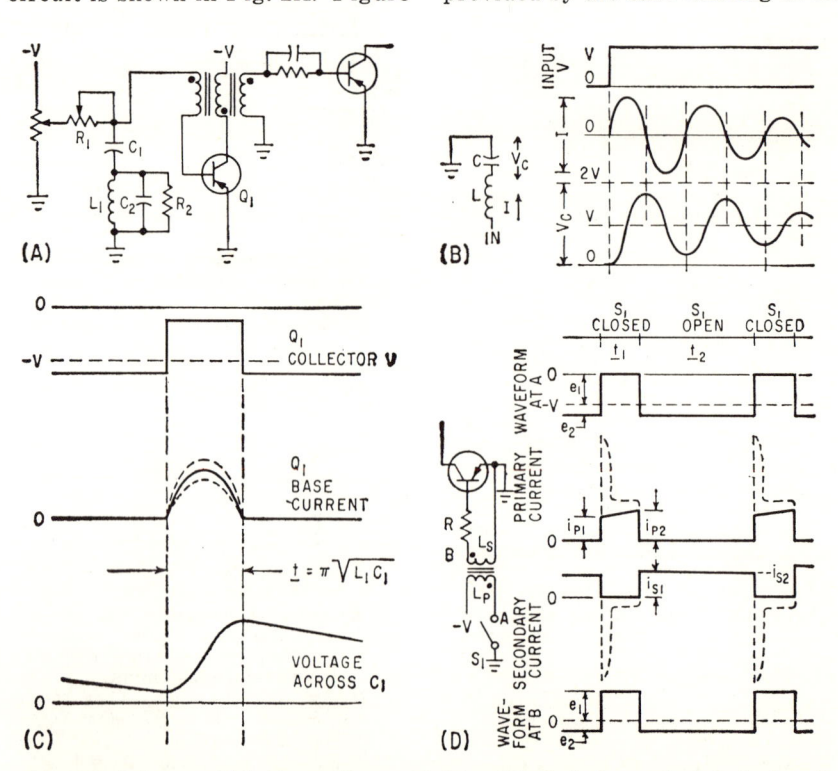

FIG. 2—Basic improved blocking-oscillator and driver (A). Equivalent series L-C circuit and its transient response to step input (B). Blocking-oscillator waveforms (C). Idealized waveforms of driver (D)

182

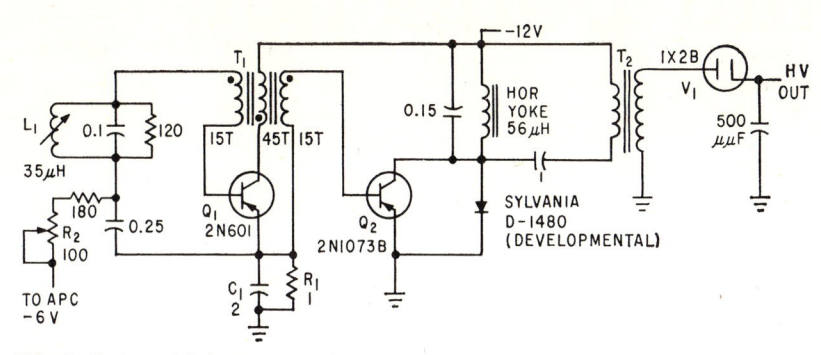

FIG. 3—Horizontal-deflection and high-voltage circuits. In typical operation, oscillator current is 0.12 amp, output-stage current 0.72 amp and p-p yoke current is 11 amp

Spacing between the high-voltage-transformer primary and secondary is about ¼ in.

transformer drives the series circuit consisting of L_1C_1 and the low base resistance of the forward-biased base-emitter diode. During the first half cycle of oscillation the transistor is operated in the saturation region; thus the collector current is independent of the base current drive for practically all of the base-current waveform. When the base current goes through zero after the first half cycle of oscillation, the transistor falls out of saturation and a regenerative turn-off process begins.

The base-emitter diode is now reverse biased and oscillation of the L-C circuit terminates. Capacitor C_1 remains charged with the proper polarity to maintain the transistor cut-off. Idealized waveforms at various points in the circuit of Fig. 2A are shown in Fig. 2C. The base-current waveform of Fig. 2C shows that the critical crossing of the zero current intercept in the base circuit occurs after a time, t, equal to $\pi\sqrt{L_1C_1}$, and is practically independent of the peak current amplitude. Due to storage effects a reverse current flows for a short interval before the transistor is turned off.

The frequency, or repetition rate, of the oscillator is determined mainly by the time constant C_1R_1 and the voltage to which R_1 is connected. Capacitor C_2 provides a high frequency by-pass across L_1 thereby increasing the initial rate of rise of base current during the edges or regenerative intervals of the pulse. Resistor R_2 provides damping of the L_1C_2 parallel-resonant circuit.

Driver Operation

The base-emitter diode of the output stage requires forward-current drive during the scan interval and reverse drive during the retrace interval. A simplified diagram of the circuit that accomplishes this is shown in Fig. 2D. Switch S_1 is closed during the retrace interval, t_1, and open during scan interval t_2. After steady-state conditions have been established the waveforms across the similar windings L_p and L_s have zero average values over a complete cycle; thus $e_1t_1 \approx e_2t_2$.

During interval t_2, forward-drive current approximately equal to e_2/R flows in the base of the output stage. Resistor R represents the total series resistance in the base circuit. Time constant L_s/R is assumed to be large in comparison to t_2. During interval t_2, energy is dissipated in the resistive elements. This energy of approximately $e_2I_2t_2$ is replaced during retrace interval t_1 by energy from the power supply equal to $e_1I_1t_1$; I_1 and I_2 are average values of the currents during the intervals t_1 and t_2 respectively. The energy replaced by the power supply during t_1 is stored in inductance L_p. The energy level at the beginning of the interval t_1, $L_p(i_{p1})^2/2$, is increased to $L_p(i_{p2})^2/2$ at the end of t_1. Upon opening switch S_1 the energy is transferred from the primary to the secondary circuit. During the next scan interval, t_2, energy level $L_s(i_{s1})^2/2$ is reduced a small amount to $L_s(i_{s2})^2/2$. The energy reduction is equal to the energy dissipated in the resistive elements in the secondary circuit during t_2.

Efficient output-stage operation requires rapid turn-off of the transistor at the end of the forward-drive interval. In order to accomplish this the ratio of reverse to forward drive current must be large. The ratio of reverse voltage to forward voltage across L_s is equal to t_2/t_1. If it is assumed that the forward and reverse base resistances are equal, the ratio of initial reverse base current to forward base current will be approximately equal to t_2/t_1. It is desirable for this ratio to be larger and means for increasing it will be discussed below. Secondary and primary currents flow simultaneously during the reverse-drive interval, unlike the forward drive interval in which secondary and primary currents flow alternately. The dashed lines of Fig. 2D show idealized waveforms of the relatively large-amplitude, short-duration drive pulses that flow in the circuits during the initial part of the retrace interval. Throughout the

Table I—Transformer Design Data

Blocking-Osc Transformer

Winding	Turns	Inductance in millihenries
Collector	45	4.5
Output	15	0.5
Base	15	0.5

Wire: No. 30 Formvar
Core: Allen-Bradley No. 1,620–160A Ferrite WO-3

High-V Transformer

Pri.: 20 t No. 28 Formvar; single layer, close wound
Sec.: 2,200 t No. 38 Bondeze-2, pie-wound
Core: Allen-Bradley No. 1,620–160B, Ferrite WO-3

---

above discussion a transformer ratio of 1 to 1 has been assumed. In the actual circuit a different ratio is used in order to match the driver and output-base circuits.

To obtain an efficient, fast turn-off of the output transistor the forward base drive at the end of the scan interval should be just enough to maintain the transistor in saturation. This operating condition produces a minimum of stored base charge and reduces the reverse base-driving power requirement.

As previously mentioned, it is desirable to provide a high ratio of reverse to forward base drive. In Fig. 3, which shows the actual circuits developed, the circuit of R_1C_1 helps to accomplish this condition. This RC circuit also tends to reduce the steady reverse base-emitter current of Q_2 that normally flows when operating the emitter junction of Q_2 in the breakdown region. This current reduction decreases blocking-oscillator input power since less average power flows during reverse drive. Breakdown current is reduced because of the voltage drop across R_1 produced by the oscillator's collector current. Capacitor C_1 provides by-pass action during the initial interval of reverse base current flow when the stored base charge is swept out.

Blocking-oscillator transformers usually require damping after turn-off to prevent a large oscillatory voltage from appearing across the windings and retriggering the oscillator after a half cycle of the output wave. An efficient method of damping may be effected by a diode and series resistor placed across one of the transformer windings. In the circuit of Fig. 3 the emitter diode of transistor Q_2 damps the transformer when it is driven in the forward direction. If the blocking oscillator is to operate properly the emitter diode of the output stage must always be connected.

Automatic phase control of the oscillator may be obtained by connecting a control voltage to variable resistor R_2. The frequency sensitivity at this point is approximately one kc/v.

Deflection System

A basic energy-recovery deflection circuit is shown in Fig. 4[1]. This circuit consists of a voltage source E, a low-loss bi-directional switch and a deflection inductor shunted by a capacitor. Switch S_1 is periodically closed during the scanning interval and opened during the retrace interval. During the scanning interval the deflection current

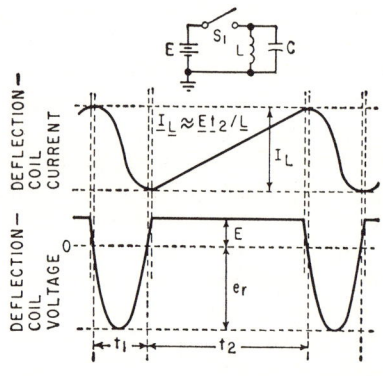

FIG. 4—Waveforms for equivalent deflection circuit indicate energy-recovery process

builds up at a rate $di/dt \approx E/L$. The switch opens during retrace interval t_1 ($t_1 = \pi\sqrt{LC}$) and the inductor current starting at its maximum value oscillates as a cosine wave for a half cycle, thus reversing its original polarity. In the low-loss case the reversed current that flows back into the power supply approaches the value of the previous forward current with the result that the net power taken from the supply approaches zero. Inductor voltage and current waveforms shown in Fig. 4 are for ideal conditions.

Peak Transistor Voltage

For a given retrace time and waveform, the peak collector voltage is determined solely by the power supply voltage. In the simplified circuit of Fig. 4 the average value of the voltage across the pure inductance L is zero over the cycle. It follows that the product $E \times t_2$ is approximately equal to the product of the average voltage during t_1 multiplied by t_1. Since the average value of the sine-wave retrace voltage over the interval t_1 is $e_r \times 2/\pi$ the ratio of peak retrace voltage to supply voltage is

$$\frac{e_r}{E} \approx \frac{t_2}{t} \times \frac{\pi}{2}$$

Supply voltage E appears across the transistor in addition to the retrace voltage. The ratio of peak collector-emitter voltage to supply voltage is therefore

$$\frac{e_r + E}{E} \approx \frac{t_2}{t_1} \times \frac{\pi}{2} + 1$$

For a given supply voltage the retrace interval is adjusted to be of long enough duration to limit the transistor collector voltage to its maximum safe value.

Additional factors to be considered in determining the retrace time are:

1. The ratio of peak retrace voltage to supply voltage may be reduced in the case of non-sinusoidal retrace voltage waveforms. The harmonic waveforms introduced by a combined high voltage and scanner arrangement may reduce the above ratio by about 20 percent.

2. Allowance should be made for the reverse base voltage pulse during retrace. The peak reverse collector-to-base voltage depends on the type of drive circuit employed.

3. Some allowance is required for possible low-frequency operation during oscillator frequency adjustments and for out-of-sync operation due to other causes. A reasonable estimate for increased collector voltage due to low-frequency operation might be about 5 to 10 percent.

4. Power-supply variations above the nominal voltage result in pro-

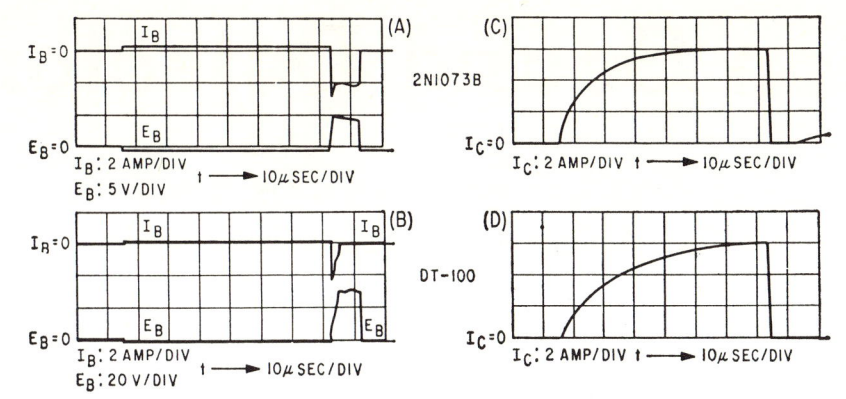

FIG. 5—Reverse base current in the 2N1073B flows during a positive pulse into its base (A) but in the DT-100, reverse current stops shortly after a positive base pulse appears

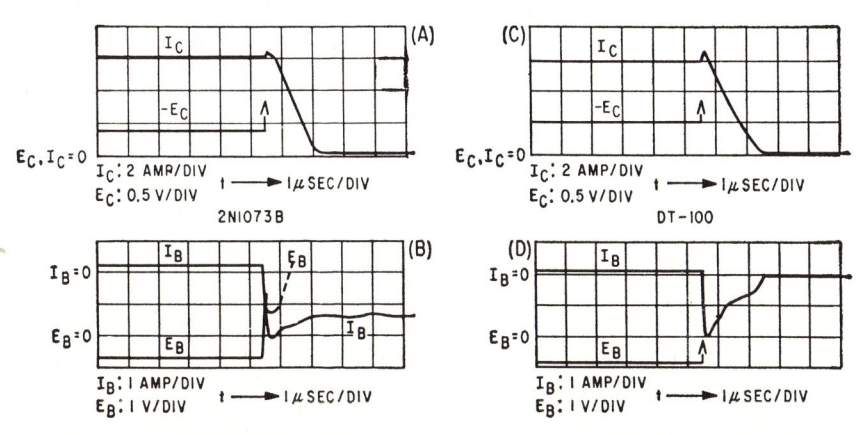

FIG. 6—For the 2N1073B, a reverse base current of 2 amp turns off a collector current of 6 amp in about 1.5 μsec. Waveshapes obtained with test circuits, as were waves of Fig. 5

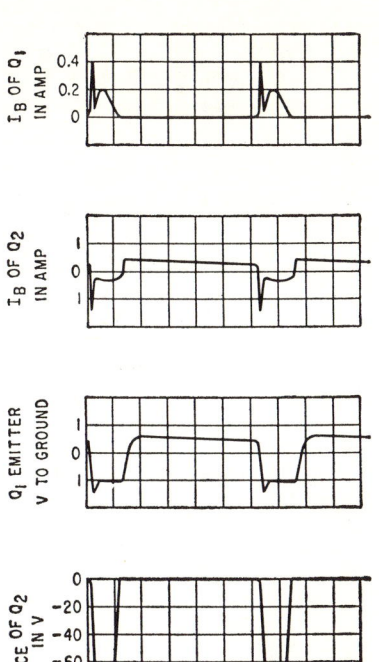

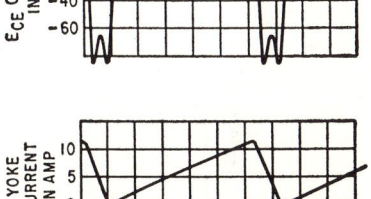

FIG. 7—These operating waveforms for the circuit of Fig. 3 are drawn to a common time scale

portionate increases in peak collector voltage.

The above considerations lead to a retrace time of not less than 12 μsec for a single transistor output stage having a maximum collector rating of 120 v and using a nominal 12-v supply.

Deflection and High Voltage

As mentioned previously, efficient operation of the output stage requires fast turn-off of collector current. This can be accomplished with alloy power transistors such as the DT-100 by providing large peak reverse drive current. However, as the voltage drop across the emitter junction during sweep-out of the base charge is considerable, large peak power is required from the driver stage. The diffused-alloy power transistor 2N1073B has the advantage of a much smaller voltage drop across the emitter junction during reverse current flow. Therefore, the required peak driving power is much reduced. To achieve the desired collector current cut-off time the 2N1073B is driven into the region of emitter-junction breakdown.

Some test-circuit comparisons (with resistive load) of the 2N-1073B and DT-100 power transistors are shown in Fig. 5 and 6. Fig. 5A and B show the difference in emitter junction voltage drop for similar reverse current between the two transistors. As indicated by Fig. 5C and 5D, forward-base drive performances are similar. Figure 6 compares transient responses during reverse drive.

Waveforms at various points in the oscillator-driver and output circuit are shown in Fig. 7 for typical operating conditions.

The deflection yoke is a Sickles

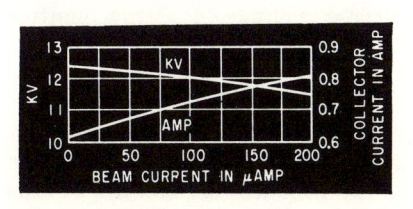

FIG. 8—Regulation of high-voltage output

17496-11 90-deg model with an inductance of 56 μh. The yoke drives a ST2587A (Sylvania) 9/8-in.-neck tube. A step-up transformer with a turns ratio of 110 to 1 is driven from the deflection circuit and provides sufficient fly-back voltage for tube V_1 of Fig. 3. Leakage inductance of the secondary of transformer T_2 is adjusted by varying the lateral spacing between primary and secondary. This adjustment, which is normally about ¼ in, increases secondary voltage peaking, reduces peak collector and yoke voltages. Transformer data is given in Table 1.

Figure 8 shows the high-voltage rectifier output as a function of beam current.

Future improvements of reduced retrace time and less input power may be expected when power transistors having higher voltage-breakdown ratings and improved switching characteristics become available.

REFERENCE
(1) A. D. Blumlein, U. S. Patent No. 2,063,025, Dec. 8, 1936.

Television Sound Detector

Drift-transistor slope detector operating in an oscillating mode gives superior performance compared with passive detector in a-m rejection, audio recovery and linearity at low signal levels. At larger signal levels, performance equivalent to a passive detector is obtained

By MARVIN METH[*], Member Technical Staff, RCA Laboratories Div., Industry Service Laboratory, New York, N. Y.

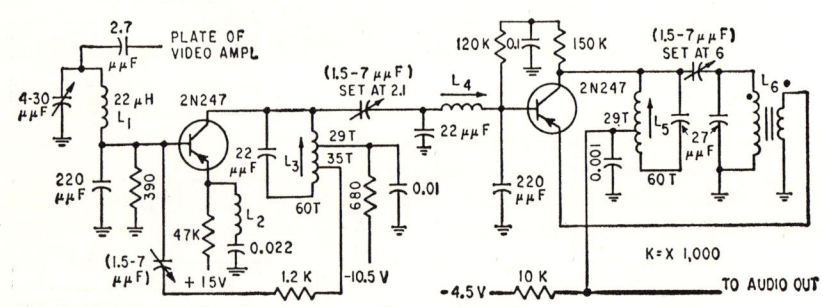

FIG. 1—Experimental f-m detector intercarrier-sound circuits

DESIGN of a transistorized tv receiver requires an efficient, low-cost sound strip. The circuitry described uses an efficient, highly sensitive oscillating linear-slope detector, injection locked by a one-stage sound driver. The combination of sound driver and detector is capable of overdriving the output audio amplifier when driven from the first video amplifier.

By operating in an oscillating mode, detector threshold level is reduced, a-m rejection is uniformly high over the full detector bandwidth and audio output is maintained at a constant level independent of carrier strength.

Detector

A 2N247 drift transistor functions as a slope detector in an oscillating mode. Oscillations are maintained by collector-to-emitter feedback through an overcoupled double-tuned circuit as shown in the right half of Fig. 1. The oscillator is injection locked to the sound signal which is applied to the base electrode by the driver stage through an impedance-matching network. Detected audio appears across the 10,000-ohm collector resistor and is obtained at r-f ground of the primary winding. Required forward bias is obtained by bleeding current from the collector into the base through two series-connected resistors bypassed at their junction to prevent audio and r-f degeneration.

Collector characteristics of drift transistors differ from conventional bipolar transistors, as shown in Fig. 2. Note that for negative collector voltages, the zero-bias characteristic of the drift transistor is similar to that of conventional units. For positive voltages applied to the collector of a *pnp* transistor, the collector acts as an emitter and the emitter as a collector. The applied voltage is a reverse voltage for the emitter junction.

Symmetrical Breakdown

Grading of the base layer of drift transistors is in such a sense as to produce breakdown of the emitter junction for relatively low voltages. Three to five volts is a typical value. Breakdown of the emitter junction is reflected in the collector as a large increase in current. The low breakdown voltage shown in the first quadrant of Fig. 2 will be referred to as symmetrical breakdown.

Positive peaks of the collector voltage are clamped at the symmetrical breakdown level. Time constant of the clamp, which is in the decoupling network, corresponds to at least 20 r-f cycles. Collector current adjusts to maintain a sinusoidal voltage waveform at the collector electrode. Positive excursions of the collector waveform are held at the symmetrical breakdown level. Negative excursions are limited by the available collector voltage.

Driving Impedance

The average collector current and oscillator amplitude are a function of the collector-circuit impedance and the coupling coefficient of the transformer. The link is a convenient means for obtaining a sufficiently low driving impedance for efficient coupling between the collector and the emitter.

The double-tuned coupling arrangement restricts oscillator operation to either the positive or negative slope of the impedance characteristics of the collector tank circuit. This action prevents excessive distortion which would arise from a slope reversal. Assume that the collector and emitter

[*] Now with Department of Electrical Engineering, City College of New York

Uses Drift Transistor

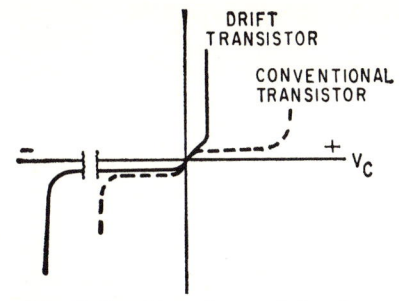

FIG. 2—Zero-bias characteristics for a drift and conventional pnp transistor

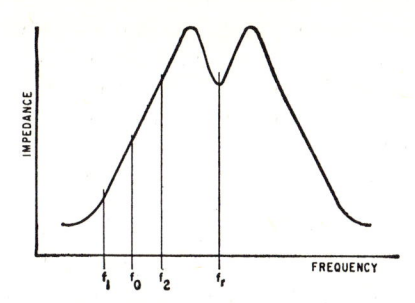

FIG. 3—Impedance of the collector tank where f_1 and f_2 are synchronizing range

tank circuits are resonant at the same frequency f_r. A signal of frequency f_r would experience a 90-deg phase shift in passing through the coupling network. The direction of phase shift, either lead or lag, is dependent on the link polarity.

Injection Locking

The oscillator is injection locked to the sound signal which is applied to the base electrode. Natural frequency of the oscillator f_o is set equal to the carrier frequency. Deviations of the signal about the carrier frequency are followed by the oscillator.

The oscillator will assume the frequency of the injected signal if two conditions are met. First, the circuital phase requirement for self-oscillation at the injected frequency must be satisfied. Second, the injected power level must be greater than a minimum. The synchronizing range is restricted to those frequencies for which the coupling network introduces less than 90 deg of additional phase shift relative to the shift at the natural oscillating frequency. The synchronizing range is restricted to either the positive or the negative slope.

Figure 3 shows the collector tank impedance for a circuit where the link polarity corresponds to synchronization on the positive slope. Frequencies f_1 and f_2 are the limits of the synchronizing range for a given injected power level.

Two restrictions are imposed on the collector voltage swing. Positive peaks are clamped at the symmetrical breakdown level. The average value is set by the effective collector voltage. The clamping time constant—approximately 10 μsec—is made fast enough to follow the amplitude variations encountered in the application. As a result, the dynamic and static limiting characteristics can be considered similar.

Ability of the detector to reject amplitude variations is not dependent on the injected signal providing the injection is of sufficient level to lock the oscillator for full carrier deviation. If the injection is below this threshold, the a-m rejection will be maintained only over the synchronizing range. Beyond this range, the output contains the beat between the injected and the oscillator signal. Also, for signals in excess of the threshold level, the audio output is maintained constant independent of carrier level.

Driver Stage

Maximum power transfer between the sound takeoff and the detector stage is provided by the driver for signals below the limiting threshold. Signals that would otherwise overload the detector are limited symmetrically to maintain a constant injection into the detector. Limiting action is rapid enough

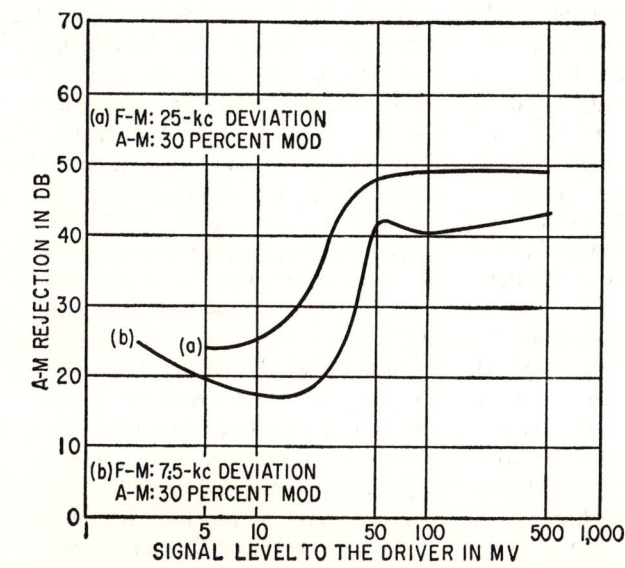

FIG. 4—A-m rejection as a function of signal level to the driver

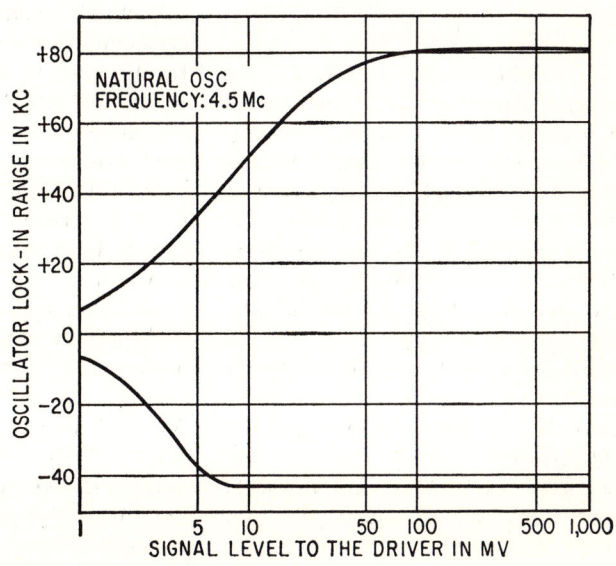

FIG. 5—Oscillator pull-in characteristics

187

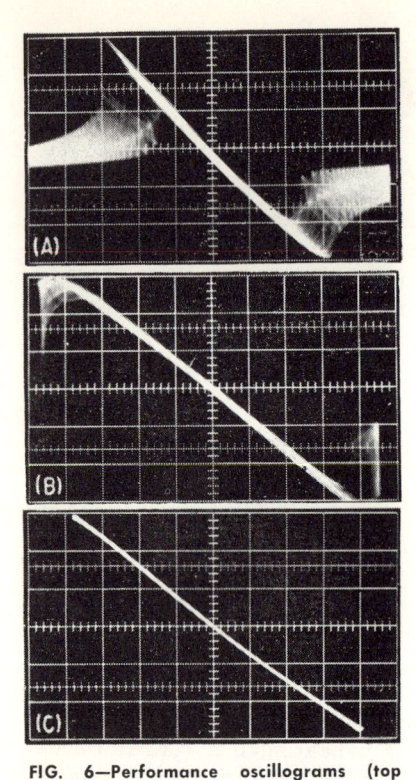

FIG. 6—Performance oscillograms (top to bottom)—signal level to driver, 1.5 mv, 10 mv and 30 mv; f-m, 25 kc at 50 cps, 50 kc at 50 cps and 50 kc at 50 cps; a-m, 30 percent at 400 cps for all three

to respond to the video modulation of the sound carrier.

A double-tuned critically coupled circuit matches the driver and detector stages. The primary is tapped to provide an impedance match for the neutralized driver stage. A signal of opposite phase to the collector signal is available at the second tap. It is used for neutralization of the transition capacity in a conventional feedback arrangement.

Biasing

A highly degenerative biasing scheme fix-biases the transistor at a high g_m point. The emitter current is set predominantly by the emitter resistor, bypassed to ground for signal frequencies, and by the positive supply voltage. Collector voltage is determined by the negative supply voltage and the decoupling resistor.

For large positive swings at the base of the driver, the transistor cuts off. For large negative swings, the transistor is driven into collector-voltage saturation. Quiescent conditions are proportioned for these effects to occur for equal positive and negative swings. This arrangement provides the desired symmetrical clipping of the input waveform.

The transistor has an exponential transfer characteristic and produces a rectified component of emitter current, reducing the forward bias of the emitter junction. For large input signal amplitudes, highly modulated by the video information, the driver stage can be cut off during the sync interval. This condition will occur only if the emitter circuit cannot respond rapidly to the required shift of the quiescent operating point. This effect may be reduced greatly by using a larger emitter resistance. As a result, rectification efficiency of the emitter junction is reduced.

Another technique is to limit the maximum emitter time constant to a value that permits bias adjustments at a horizontal line rate. A 4,700-ohm emitter resistor bypassed with a 0.022-μf disk capacitor with leads cut and coiled for resonance at 4.3 mc is a satisfactory compromise. Incomplete emitter bypassing introduces 0.6 db of negative feedback at 4.5 mc. The driver stage can follow deep modulation of the sound signal.

Circuit Details

The schematic diagram of the sound strip installed in a commercial chassis is shown in Fig. 1. This circuit shows high-side capacitance coupling. But mutual, or a combination of mutual and high-side coupling, may be substituted.

Detection linearity and a-m rejection, Fig. 4, are dependent on the coefficient of coupling in the double-tuned coupling arrangement of the oscillator loop. A coupling factor of 1.5 times critical coupling gives good a-m rejection and reasonably good linearity. Larger values of coupling distort the detection characteristic S curve; smaller values of coupling reduce the amplitude-modulation rejection.

Base and emitter networks are designed to suppress spurious oscillations arising from input-circuit feedback due to high input capacitance of the transistor. Driving-point impedances of these networks have been designed so that for all frequencies for which the base reactance is positive, the emitter reactance is also positive. Consequently, feedback from emitter to base is degenerative for all frequencies and stable operation results.

The detector operates in an oscillating mode over the full range of input signals provided by the driver. The holding, or lock-in, range is a function of signal level. Lower limit of the holding range, Fig. 5, corresponds to frequencies for which the tank impedance is too low to support oscillations. Theoretically, the oscillator can follow positive deviations up to the tank resonant frequency. Practically, these limits are dependent on the maximum power capability of the driver.

Performance Results

Overall performance is illustrated best with the oscillograms of Fig. 6. They were obtained by passing a carrier that is simultaneously amplitude- and frequency-modulated through the sound strip and displaying the detected output with the f-m as a time base. These oscillograms indicate that a-m rejection is not critical to center tuning. A-m rejection characteristics are given for 30- and 100-percent f-m where 100-percent modulation corresponds to 25-kc deviation.

At low signal levels the a-m rejection is 20 db, which is the inherent a-m rejection of the detector. This rejection increases as the driver limits. For input signals of less than two mv, a-m rejection is still high.

Audio recovery is 75 mv per kc of deviation, independent of carrier level. For 100-percent modulation, the detector develops an open-circuit output of 1.3-v rms. Maximum power transferred by the detector into an audio load is 10 dbm and is maintained over a wide range of loading centered about 4,000 ohms.

As loading varies from 500 to 25,000 ohms, output is maintained within three db of the maximum level.

For 25-kc deviation, the audio output contained from 2.5 to 3.5 percent of rms harmonic distortion depending on center tuning and on signal level. As the deviation was increased to 50 kc, rms harmonic distortion increased in the range of four to five percent.

Low-Distortion Television Monitor Amplifier

Monitor amplifier for broadcast duty uses production lot power transistors for high fidelity and low distortion. Technique has other applications

By HAROLD J. PAZ, Radio Corporation of America, Camden, New Jersey

POWER TRANSISTORS are used in this monitor amplifier to eliminate the problems of hum, microphonics and heat. In addition, a reduction of size is obtained which is of value in many presently cramped studios.

The design objective was to produce a low-distortion, high-fidelity all-transistor power amplifier that does not require laboratory adjustment and selection of power transistors. Most present amplifiers require a power transistor with a beta cutoff of 30 kc for low distortion at 15 kc. But presently available power transistors have beta cutoffs of about 6 to 9 kc.

Distortion

There are a number of reasons for distortion at midband frequencies with presently available power transistors. In a class-B amplifier, beta mismatch distortion will cause one-half of the output signal to be larger than the other. There is, of course, a wide variation in the current gain, beta, of most types of power transistors.

The input impedance of a transistor has an effect on the gain and distortion of the driver and output stage. The common-emitter input impedance is $z_{in} = r_b + (\beta + 1) r_e$, where r_b is the base resistance, r_e is the emitter resistance and β is the current gain. But r_b, r_e and β are all inversely proportional to the current flow in the emitter. Typically, the input impedance can change from 2,000 ohms at an emitter current of 1 ma to 15 ohms at 1 amp.

Changes in the large signal current transfer ratio with the input base current drive is another reason for distortion in a high-power transistor amplifier. Figure 1 shows how rapidly the current transfer ratio of a power transistor decreases with collector current. The

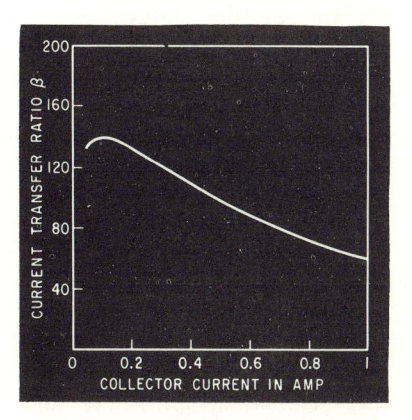

FIG. 1—Current transfer ratio, beta, of a 2N301 or 2N301A power transistor varies with collector current

current transfer ratio is 140 at 100 ma but decreases to 60 at 1 amp.

Local Negative Feedback

A new approach to the problem uses negative feedback to shift the dependence for ultralinear amplification from the critical selection of the power transistor to the circuit design. Negative feedback may be used to improve frequency response and reduce distortion. However, in a transistor amplifier using four or five stages, too much loop feedback can result in oscillation. The phase shift in each stage of a transistor amplifier is great. Therefore, the total phase shift of the amplifier greatly limits the maximum amount of loop feedback. Hence, loop feedback is not as successful in reducing distortion in a transistor power amplifier as local feedback.

An unbypassed emitter resistor provides local negative feedback. This external emitter resistor must be larger than the internal transistor emitter resistance. Low dis-

Transistorized monitor amplifier is 5 in. wide, 4¾ in. high and 12 in. long

tortion in a transistor amplifier is obtained when the transistor is driven by a low impedance source. Since the input impedance of a transistor operating in class B varies considerably with emitter current, a shunt resistor at the base input terminal will control this impedance variation. When using the 2N301A power transistor, a shunt base resistor 15 to 18 times the unbypassed emitter resistor will provide enough local feedback to control the effective current gain as well as the input impedance. This local feedback also compensates for beta mismatch, beta variation with base current drive and phase shift, and boosts the beta cutoff frequency from 9,000 to 30,000 cps.

Hybrid Transistor Amplifier

The new approach is a hybrid of the series[1] and the quasicomplementary[2] transistor amplifiers and is shown in Fig. 2. The input stage, Q_1, is a low-level class-A stage.

The complementary symmetry pair of transistors, Q_2 and Q_3, is used as a class-B direct-coupled phase inverter. Transistors Q_2 and Q_3 are directly connected to Q_4 and Q_5 which are two *pnp* transistors operating in class B and are used to drive the output pair of transistors, Q_6 and Q_7. Transistors Q_4 and Q_5 are coupled to Q_6 and Q_7 in a way similar to that used in the series power amplifier.

Local feedback has an important effect on the high-frequency response and distortion. By careful selection of resistors R_2 through R_7, local negative feedback is introduced. Surprisingly low distortion is measured when the ratio of R_4 to R_6 and R_5 to R_7 is 15 and the ratio R_2 to R_4 and R_3 to R_5 is 18.

Circuit Operation

In Fig. 2A, the collector voltage of Q_1 is one-half the supply voltage, V_{cc}, at zero input signal. The complementary symmetry phase inverter, Q_2 and Q_3, does not have a potential across the base to emitter junction because the output capac-

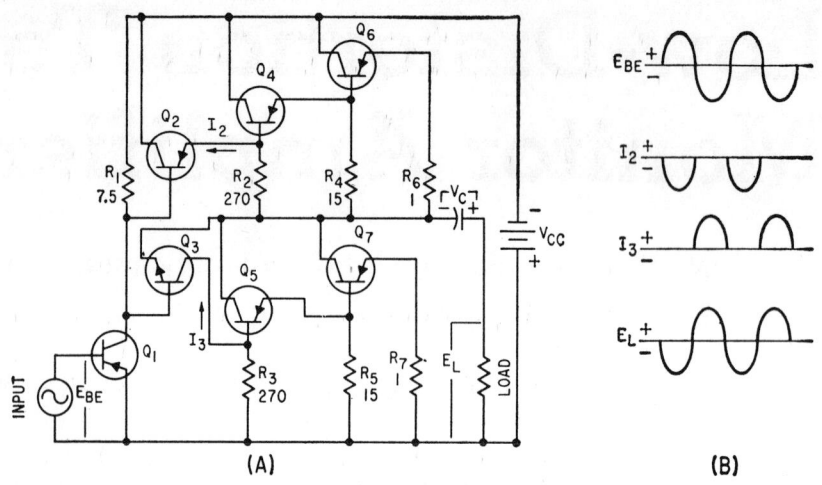

FIG. 2—The basic circuit of the hybrid power amplifier (A); and circuit waveforms (B)

Obsolete monitor amplifier is shown at left. New transistorized unit is shown at lower right and unit it replaces is in the center

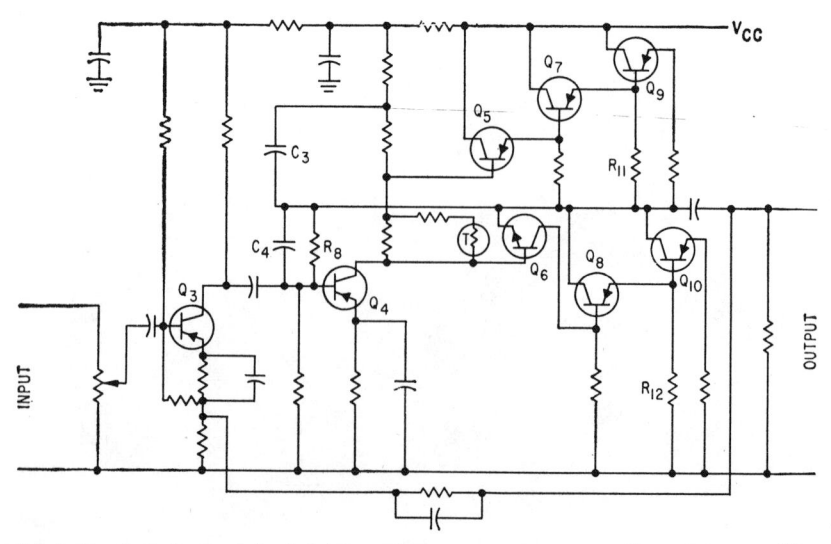

FIG. 3—Practical circuit of the hybrid complementary symmetry amplifier. Output at 10 w has less than 0.25-percent distortion

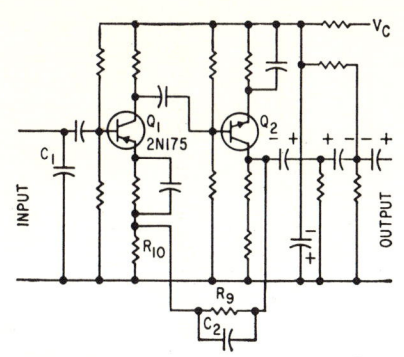

FIG. 4—Low-noise preamplifier has loop feedback and low-frequency noise filter at output

itor voltage V_c is equal to the collector voltage of Q_1.

If the collector voltage of Q_1 becomes more negative, then the base of Q_2 will be more negative than the emitter. This will turn on Q_2 and a voltage drop will appear across R_2. This voltage drop causes Q_4 and Q_6 to conduct. The base voltage that causes Q_2 to conduct is also applied to Q_3. This puts a reverse bias on Q_3 which prevents conduction. Since Q_3 is cutoff, Q_5 and Q_7 are also cutoff.

Thus the top half of the circuit, Q_2, Q_4 and Q_6, conducts only when the collector voltage of Q_1 is greater than $V_{cc}/2$, or when the base to emitter signal voltage E_{BE} of Q_1 is positive. When E_{BE} is negative, the collector voltage of Q_1 becomes less negative than the capacitor voltage V_c. This makes the emitter of the npn transistor Q_3 more negative than the base, and starts conduction in the transistor. However, this potential prevents conduction in pnp transistor Q_2.

As the npn transistor conducts, it produces a voltage drop across R_3 which puts Q_5 and Q_7 into conduction. Fig. 2B shows the relationship between input voltage E_{BE} and the current flow in Q_2 and Q_3. Notice

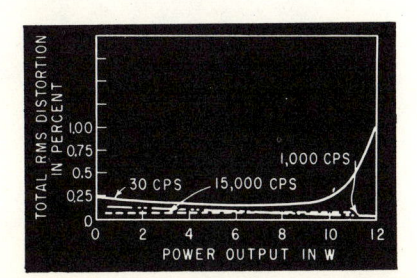

FIG. 5—Distortion of hybrid amplifier is less than 0.25 percent over frequency range from 30 to 15,000 cps

that the voltage across the load, E_L, is 180 deg out of phase with the input signal E_{BE}.

The practical hybrid complementary symmetry circuit shown in Fig. 3 is part of a transistor monitor amplifier which is designed for use in broadcast studios. Used with a two transistor preamp, the circuit provides 104 db of gain. This gain is enough to permit a microphone to drive 10 w of power into a loud speaker.

The input impedance of Q_3 should be high to prevent loading of the transistor noise filter used in the preamplifier. Capacitor C_3 provides positive feedback to the driver stage Q_4. The output stage operates at full power efficiently because the positive feedback reduces the drive

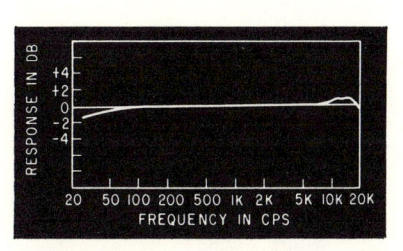

FIG. 6—Frequency response is within ±2db from 30 to 20,000 cps

voltage requirement. To reduce phase shift in Q_4, a drift transistor, type 2N247 was selected. The network R_8C_4 provides a third internal feedback loop from the output to the base of Q_4. High-frequency stability is improved by this loop.

Preamplifier

A two stage preamplifier is shown in Fig. 4. Transistor Q_1 is a low-noise element which can handle a high-input level without distortion because of loop negative feedback. Resistor R_9 feeds back some of the collector current to the emitter of Q_1. To keep low-frequency phase shift at a minimum, a series capacitor is not used in the feedback loop. The emitter feedback increases the input impedance of Q_1 and decreases the loading on the signal source. For good low-frequency response from a ribbon type microphone, the input impedance of the transistor amplifier should be over 4,000 ohms. Resistors R_9 and R_{10} should be 1-percent carbon film resistors to control preamplifier noise and gain. The preamplifier does not like to see an inductive load at the input terminals so capacitor C_1 is used. Capacitor C_2 is used to limit the preamplifier bandwidth.

A three-section R-C filter is used to cutoff the low-frequency response of the preamplifier. This network will filter out the low-frenquency transistor noise content. Transistor noise is also called flicker noise and is quite large below 30 cps.

Temperature Stability

Since the hybrid complementary symmetry amplifier (Fig. 3) uses direct coupling, control of the d-c collector current of Q_4 is important for temperature stability. Any drift here is amplified by the three direct-coupled class-B stages. Since the transistors operate in the class-B mode, the resistance of R_{11} and R_{12} is kept low enough to shunt most of the temperature sensitive leakage current, I_{co}, thus preventing its amplification. The thermal change in collector current of the class-B stage is mostly caused by a change in the transconductance of the transistor. To compensate for this, a thermistor is used to stabilize the quiescent operating point.

Distortion of the 10-w amplifier is shown in Fig. 5. At 10-w output, distortion is less than 0.25 percent over the complete frequency range. At 30 cps, distortion begins to rise at 10 watts because the filtering action of the power supply decreases with signal frequency.

Frequency response of the monitor amplifier is shown in Fig. 6 for an output of 1 w. When terminated with the recommended impedance the response is within ±2 db from 20 to 20,000 cps.

A model of the hybrid power amplifier is shown in the photographs. A metering switch is provided to measure transistor voltages. Input and output transformers are used to provide circuit isolation, with input impedances of 37.5, 150 or 600 ohms. The output transformer can handle loads of 4, 8, 16, 150 or 600 ohms.

REFERENCES
(1) M. B. Herscher, Designing Transistor A-F Power Amplifiers, ELECTRONICS, p 96, April, 1958.
(2) H. C. Lin, Quasi-Complementary Transistor Amplifier, ELECTRONICS, p 173, Sept. 1956.

Chapter 10
COMMUNICATIONS APPLICATIONS

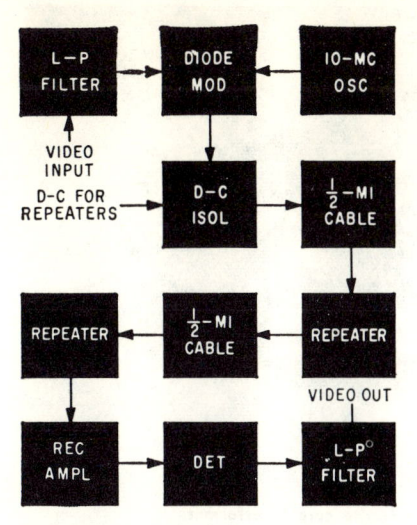

FIG. 1—Basic transmission system

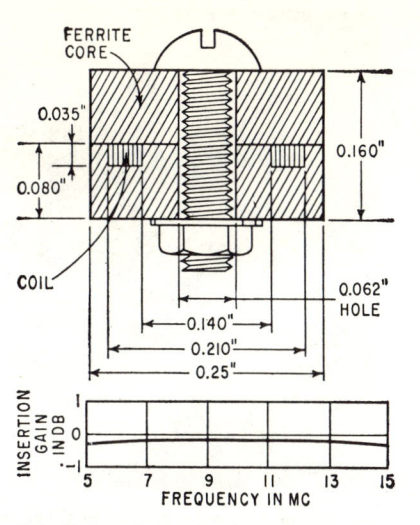

FIG. 2—Transformer and response

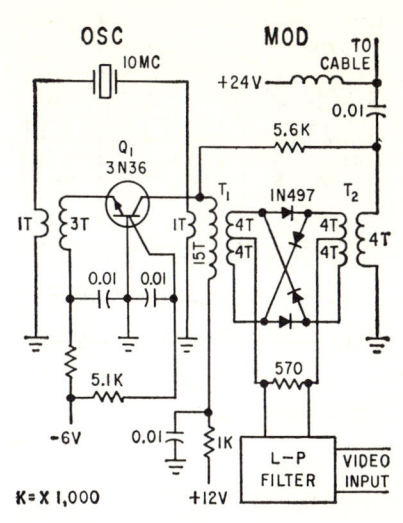

FIG. 3—Transmitting terminal

Carrier Transmission for Closed-Circuit Television

Simple and inexpensive coaxial-cable transmission system passes high-quality television signal with 4.5-mc bandwidth. Transistorized terminal and repeater circuits minimize space and power requirements. Unusual power supply sends d-c repeater power through signal cable

By **L. G. SCHIMPF,** Bell Telephone Laboratories, Murray Hill, N. J.

INEXPENSIVE TV TRANSMISSION SYSTEMS for 4.5-mc bandwidth high-quality signals would be useful in closed-circuit television where one common pickup is used to transmit a picture to one or more viewers.

In such service the transmission links are usually short (up to several miles) and it is seldom necessary to connect a number of systems in tandem. With these limitations, a system has been designed to meet the bandwidth and picture-quality requirements without meeting the exacting requirements of each link in a transcontinental transmission system.

To save power and keep the apparatus small, terminals and repeaters were transistorized. A double-sideband carrier system was used even though it requires about twice the bandwidth of the other coaxial-cable transmission systems. Transmitting twice the bandwidth results in a less complicated system since the terminal equipment for the double sideband system is simple.

System Considerations

A 10-mc carrier frequency was selected as a compromise between the various factors in favor of a higher or a lower frequency. The system is blocked out as shown in Fig. 1.

Carrier frequency is generated by a crystal-controlled transistor oscillator; the output is modulated by the video signal in a semiconductor modulator. Before being fed to the cable, the modulated signal passes through a d-c isolation circuit that permits d-c power to be fed to the repeaters through the cable.

After passing through a length of cable, the signal is amplified by a transistorized repeater, which has a gain characteristic designed to match the loss characteristic of the cable. Since the output level from a repeater is too low for good linear detection, an amplifier is used in the receiving terminal between the last repeater and the detector. A low-pass filter removes the carrier from the video output.

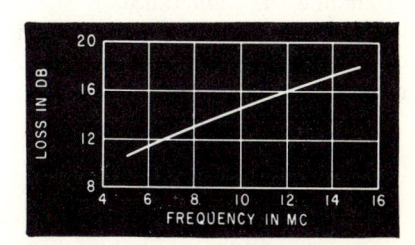

FIG. 4—Transmission characteristic of 0.5-mi cable

Transformers

The transformers are of unusual design. Their size and general method of construction are indicated in Fig. 2. The two sections of the ferrite core are held together

195

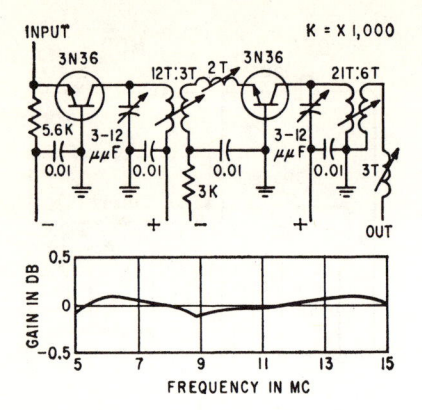

FIG. 5—Repeater and response

Closeup of terminal repeater amplifier shows ferrite core transformers

by a No. 0-80 machine screw, which also serves as the transformer mounting.

Some of the coils employed in the repeaters are adjustable. In these cases the same general construction is used, but a slot is milled in each half of the core.

The transformer is assembled so the faces with the slots are together. With this arrangement, the effective length of the air gap can be varied as one half the core is rotated with respect to the other to change the inductance.

The transmission characteristic of one of these transformers operating between 75-ohm impedance levels is also shown in Fig. 2; the loss is only a few tenths of a db and the characteristic is flat to within 0.1 db from 5 to 15 mc.

Transmitting Terminal

The crystal-controlled transistor oscillator for the 10-mc carrier is shown in Fig. 3. An experimental junction tetrode transistor, with an alpha cutoff of about 30 mc, was employed in a common-base connection; A 3N36 tetrode transistor should be a suitable commercial substitution, though undoubtedly a conventional triode transistor with a similar alpha cutoff would be satisfactory also.

Neither of the two miniature transformers, about ¼-in. dia., is tuned. Output transformer T_1 couples the high impedance of the collector into the relatively low impedance of the modulator.

A third winding on T_1 couples the collector circuit to a low-impedance feedback circuit, which is best traced by starting with the transistor emitter. At an emitter current of 1 to 1.5 ma, input impedance of Q_1 is 50 to 75 ohms. This impedance is stepped down by a 9:1 impedance ratio in the transformer connected to the emitter. At series resonance, the impedance of the crystal is 8 to 10 ohms,

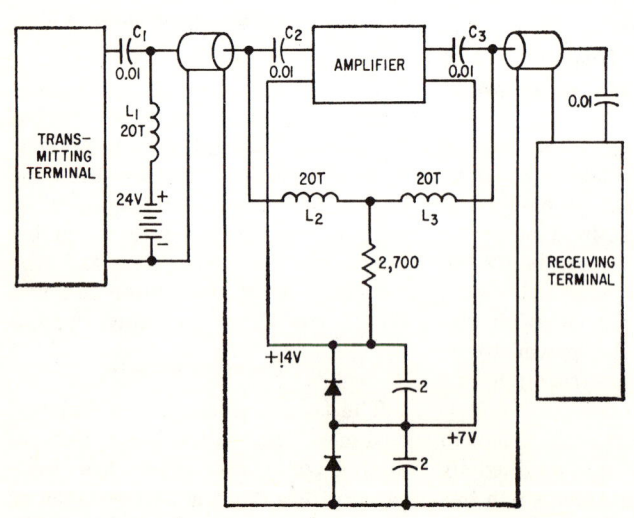

FIG. 6—Repeater d-c power supply

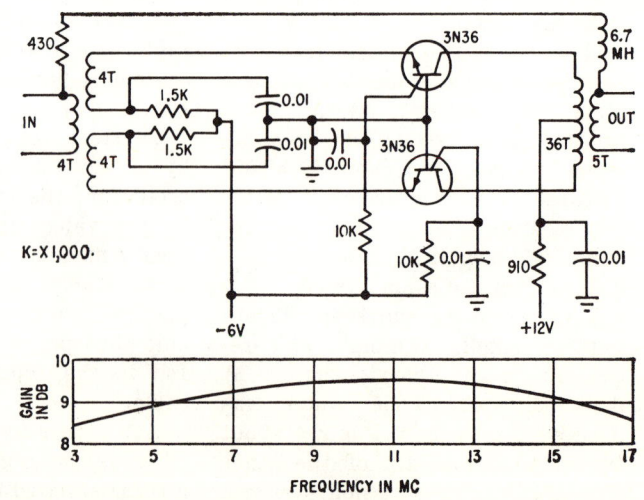

FIG. 7—Receiving amplifier and response

causing the feedback path to have a low impedance; thus sufficient power is fed back from the output to set up oscillations.

Operation at the fundamental mode of the crystal takes place for several reasons. At higher frequencies, the gain of the transistor decreases. Also, the losses in the transformer increase with frequency and the impedance of the crystal at series resonance increases for the higher modes.

The circuit has shown no tendency to operate at harmonic frequencies of the crystal. Power output is of the order of a few milliwatts.

Diode Modulator

Four gold-bonded diodes are employed in the modulator, which is also shown in Fig. 3.

The signal levels are adjusted to give a peak modulation of about 70 percent and polarization of the video signal is such that the peak signal occurs during the synchronizing pulses. The peak level at the output of the modulator is about 4 db below 1 mw.

The diodes used in this modulator are characterized by low capacitance, short storage time and a fairly high ratio of back-to-forward impedance. Ideally, a switch with zero impedance in the forward direction, infinite impedance in the reverse direction and an operate time that is short compared to 10 mc is desired. The IN497 diode approaches these characteristics.

Use of a balanced modulator eliminates a number of unwanted sidebands from the output. Since the carrier is also eliminated, it is necessary to introduce it into the output for double-sideband operation. This is done by connecting the collector circuit of the oscillator to the output through a 5,600-ohm resistor.

Low-frequency video signals cannot be transmitted to the line because of the low-frequency cut-off characteristic of T_2. This is a desirable effect. Any high-frequency video signal that appears on the line will be attenuated because the repeaters are not equalized to amplify these frequencies.

Design of a repeater for a system of this kind depends on the loss characteristics of the cable employed A $\frac{3}{16}$-in. dia. coaxial cable using partially blown-up polystyrene was used; transmission characteristics are shown in Fig. 4 for 0.5 mile of cable.

The experimental two-stage amplifier shown in Fig. 5 was designed to have a gain characteristic to match that of the cable loss. The Tetrode transistors used are shown in simplified form.

Referring to Fig. 4, a gain of about 18 db is required at 15 mc for 0.5-mile repeater spacing. As the amplifier has a gain capability in excess of this value, mismatching is employed at the input to the first transistor and between stages to stabilize the gain. This stabilization technique was chosen instead of negative feedback, because of its simplicity.

Six variable circuit elements in each repeater, make it possible to match the transmission loss of a section of cable closely, as well as to compensate for variations between transistors. Measurements on one repeater with a half mile of cable are also shown in Fig. 5.

Spacing between repeaters is not limited to 0.5 mile. With two repeaters connected in tandem and one-mile spacing, circuit noise was not sufficient to be a problem. Since the repeaters are small and fed with d-c power over the cable, they could be placed at locations other than manholes; this might make closer spacing of units having less gain and fewer components economically feasible.

D-C Power Supply

The d-c power supply system is shown in simplified form in Fig. 6. At the transmitting terminal, blocking capacitor C_1 isolates the d-c from the terminal equipment. Coil L_1 in series with the 24-v battery prevents shorting of the signal.

The center conductor of the cable

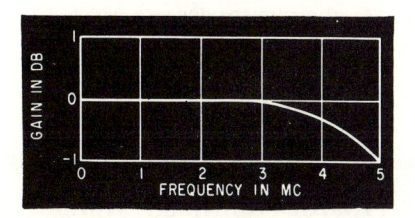

FIG. 8—Response of 1-mile system

is held at +24 v, which is isolated from the amplifier circuit by two blocking capacitors C_2 and C_3. Coils L_2 and L_3 prevent shunting of the signal through the d-c supply circuit.

Since the amplifier employs tetrode transistors, both a positive and a negative voltage are available from the bias source. Two 7-v avalanche diodes in series serve as voltage regulators.

If a number of repeaters, connected as shown, were employed in a system, they would all be in parallel and the cable voltage would be 24 v regardless of their number. Total line current is 5 ma times the number of repeaters used.

Receiving Terminal

To make up for the last section of cable, a repeater is used as the first element of the receiving terminal. Linear detection requires that a certain minimum amplitude of signal, greater than that provided by the repeater, be maintained. Therefore, the repeater is followed by an amplifier which increases the level by about 10 db.

This experimental amplifier and its response are shown in Fig. 7. Selective negative feedback improves the transmission characteristic.

The output of the amplifier is rectified by a full-wave circuit and passed through a low-pass filter to remove the 10-mc carrier frequency. If desired, this video output can then be amplified to a standard output level.

Results

A 1-mile experimental system has been set up for demonstration purposes. On an A-B test, no degradation could be detected by most observers when a standard television signal was transmitted over the system. This would indicate that a system several miles long could be employed without undue degradation. The video-frequency response of this one mile of system is shown in Fig. 8.

The author is indebted to R. L. Wallace, Jr. for many helpful suggestions. The transistor repeaters were designed by J. G. Linvill, who is at present a member of the staff of Stanford University.

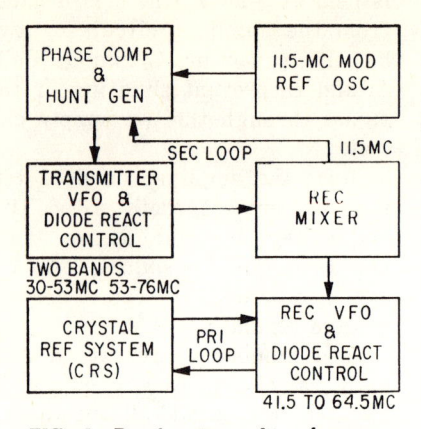

FIG. 1—Receiver-transmitter frequency control system, showing major frequency determining components

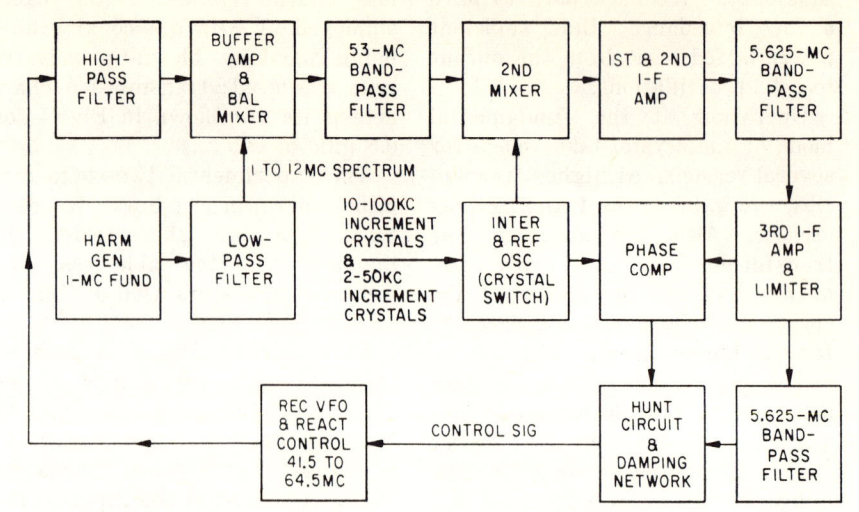

FIG. 2—Crystal reference system is used for channel derivation and controls the local oscillator frequency within 3 kc of nominal channel

Controlled-frequency transceiver, operating in two bands, uses improved bandpass filter techniques that double the number of channels per megacycle of spectrum. Oscillator-stabilized system, used in military vehicular communications, is designed for 50-kc channel spacing and selects any of 920 channels between 30 to 76 mc

Mobile Radio System

By FRANK BRAUER and DON KAMMER
Communications Equipment Group, Avco Manufacturing Corp., Crosley Division, Cincinnati, Ohio

SINCE THE FREQUENCY spectrum cannot be enlarged, the present urgent need for more reliable communications channels can be satisfied by decreasing channel spacing, thereby stretching the usable radio spectrum. To accomplish this, the frequency-determining devices in a communications system must be designed for increased stability.

This article explains an approach used to stabilize the receiver and transmitter oscillators in a transistorized vehicular communications system (AN/VRC-12) developed for the U. S. Army Signal Corps.

50-Kc Channel Spacing

The system furnishes 920 channels of communication between 30 to 76 mc and operates from a 24-v vehicular battery. This equipment is a transistorized military vehicular system designed completely for 50-kc channel spacing. It shows certain advantages over present military 100-kc channel-spacing equipment that now uses three different receivers to cover the 20- to 54-mc range.

The receiver has a sensitivity of 0.3 μv for a 10-db signal-plus-noise/noise ratio and spurious responses are down 85 db. Transmitter output is 30 watts and all transmitted spurious frequencies are also down 85 db. The transceiver has ten preset channels and can be tuned manually or automatically. For automatic tuning, a servo system provides 3-second tuning between any two preset channels. The system operates over an ambient temperature range of −40 to +65 C.

Stable frequencies are derived by free running oscillators regulated by special control systems. A frequency control system implies a reference or standard, along with conventional regulation, direction, or error correction. Figure 1 shows the major frequency determining components.

Two frequency control loops are shown. The primary loop, the crystal reference system (crs), is used for channel derivation and controls the frequency of the receiver local oscillator within 3 kc of nominal channel frequency over the temperature range of −55 to +85 C. Since reception and transmission occur on the same channel, the stabilized receiver local oscillator is used as one of the references in the transmitter frequency control loop.

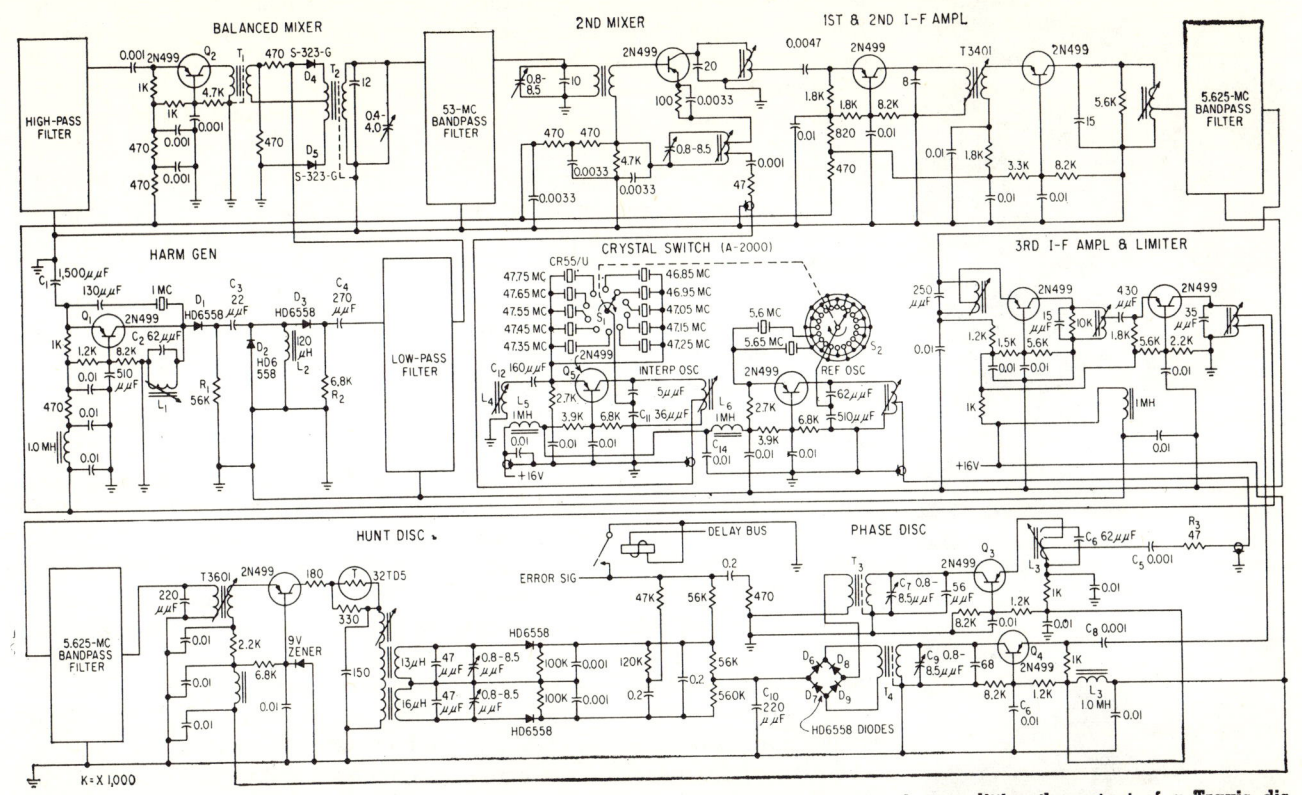

FIG. 3—Crystal reference system schematic. Hunting is achieved through instability, by phase splitting the output of a Travis discriminator, direct coupling one output and capacitance-coupling the other output to the reactance control circuit of the oscillator

Provides 920 Channels

The 11.5-mc modulated oscillator, controlled by a crystal discriminator afc loop, performs the sidestep function since it operates at the receiver intermediate frequency. The error signal in both controlled loops is derived by phase comparison of the controlled frequencies with crystal-controlled references. Frequency correction is obtained by using semiconductor reactance devices in the vfo circuits. Electronic hunting circuits in both loops bring the vfo within holding range of the phase comparators.

Two bands cover the 30 to 76-mc operating range of the receiver: 30 to 53 mc, and 53 to 76 mc. An 11.5-mc i-f was selected to obtain high- and low-side mixing of the receiver local oscillator. Operational range of the local oscillator is thus restricted to 41.5 to 64.5 mc.

Crystal Reference

A sampling of the local oscillator is fed to the crs (see Fig. 2). The 50-kc spacing is achieved by three successive interpolations.

The first interpolation occurs in the first or balanced mixer and divides the 23-mc band coverage into 1-mc increments. Since high and low mixing are employed in the balanced mixer, only the first 12 harmonics of a 1-mc crystal-controlled oscillator are required.

The second interpolation, which divides the 1-mc increments into 100-kc increments, takes place in the second mixer and is accomplished by a mechanically switched and synchronized crystal interpolation oscillator that employs 10 crystals in 100-kc steps between 46.85 and 47.75 mc.

Final interpolation is carried out in the phase comparator by synchronized selection of one of two crystals in the reference oscillator at 5.6 and 5.65 mc.

The switching frequency is the receiver local oscillator signal in the first or balanced mixer. Since this signal level is 25-db higher than the outputs of the harmonic generator, balancing of the first mixer provides nearly constant output at all receiver local oscillator frequencies. A buffer amplifier is used for isolation, since mixed products of the harmonic generator, present in the first mixer, fed back to the receiver would lead to spurious responses.

The crystal reference system filter complement keeps spurious responses down at least 85 db. The high-pass filter at the input of the crs further attenuates undesirable frequencies below 41.5 mc. The low-pass filter cuts off at 12 mc and attenuates undesirable frequencies, particularly those around the 53-mc first i-f of the crs. The 53-mc bandpass filter attenuates the image and half i-f responses. A bandwidth of 1,750 kc accommodates the 400-kc control range of the receiver local oscillator, the 900-kc range of the ten 100-kc increment crystals in the second crs oscillator, and the 50-kc

Design engineer checks crystal reference system for the 920 channel vhf f-m receiver-transmitter

Crystal reference system, covers removed, shows packaging techniques employed. Individually packaged, prealigned, plug-in assemblies facilitate field maintenance

displacement between the crs reference oscillator crystals.

The 5.625-mc bandpass filters limit action to the desired frequency signal in the crs second i-f. These filters prevent undesired signals, 1-mc either side of the desired frequency, from disturbing discriminator response necessary for automatic hunting.

Action of the harmonic generator causes first and second i-f responses every 1 mc. This desired interval response is predetermined by the receiver local oscillator, positioned within 400-kc of the required frequency. The nearest spurious signal is then 600-kc away.

The proper crystals in the interpolation reference oscillators are selected automatically by gear-coupling the crystal switches to the main tuning system. The three second tuning on the overall system has a switching speed of 460 revolutions per minute for the interpolation and reference oscillators. Each revolution of the switch shaft switches the control system over a 20-channel range.

An automatic hunting system brings the frequency within the holding range of the phase comparator. Hunting is accomplished by phase splitting the output of a Travis discriminator, direct coupling one output, and capacitance coupling the other output to the reactance control circuit of the oscillator. Hunting action is achieved through instability and automatically stops during phase lock.

Phase comparator output is applied directly to the reactance control circuit of the receiver local oscillator. The output overrides the hunting voltage and compensates for the frequency error of the receiver local oscillator. A damping network reduces control loop gain below unity at the critical frequency and prevents loop instability. The critical frequency is the error frequency at which the loop phase shift is 360 deg. And here instability occurs if the loop gain is greater than unity.

The maximum frequency error of the controlled oscillator is determined solely by the additive errors of the various crystal oscillators. A total of 13 crystals, used with 12 transistors, in the crs provides the 920 crystal controlled channels.

Harmonic Generator

As seen in Fig. 3, transistor Q_1 is connected in a common-base circuit. Oscillation is achieved with 100-percent feedback from the collector tank to the input circuit through the 1-mc series-mode crystal. Capacitor C_1 corrects the oscillator frequency to within plus or minus five cycles of the nominal 1-mc. The transistor is driven into saturation, and the waveshape across C_2 and L_1 is clipped on half the cycle.

This waveshape passes through a clipper network D_1 and R_1. This square wave is differentiated in network C_3 and D_2. Coil L_2 suppresses the lower order harmonics, and evens amplitude spread. Diode D_3 and R_2 act as a second clipper. Capacitor C_4 and input resistance of the low-pass filter form a second differentiating network. Thus the output is a pulse train with a 1-mc repetition rate and a pulse rise of 0.03 μsec.

Balanced Mixer

Buffer amplifier transistor Q_2 develops only 6-db gain and is broad-banded to accommodate the 41.5- to 64.5-mc input signals from the receiver local oscillator. It isolates the receiver tuner from balanced-mixer cross products which would cause spurious responses in the receiver mixer.

The receiver local oscillator signal, used to drive the mixer, is 30-db higher than the individual harmonics of the harmonic generator. Careful circuit layout, and shielding of T_1 and T_2, provide a 30-db rejection while permitting a small variation of mixer output amplitude and meeting the 85-db spurious-signal rejection demands of the receiver. Diodes D_4 and D_5 achieve mixer saturation at a low driving signal level and permit a reasonable variation in driving level without upsetting the balance. The tuned output is essentially an end section of the 53-mc filter. Net loss of the balanced mixer, referred to the power level of the harmonic generator individual harmonics, is 10 db.

Phase Discriminator

Capacitor C_5 couples the fixed input of either 5.6 or 5.65 mc to an input tank, C_6 and L_3, which is tuned to 5.625 mc. Resistor R_3 minimizes

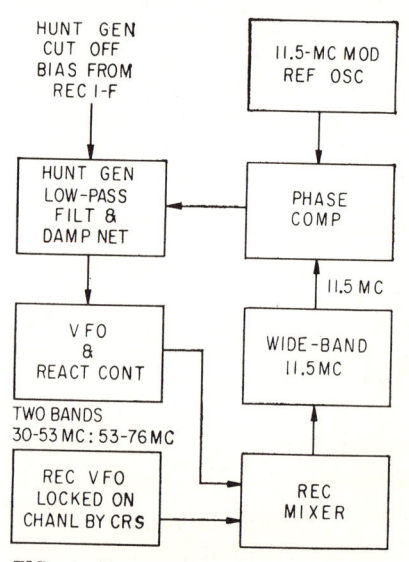

FIG. 4—Transmitter frequency control (secondary control loop) block diagram

effects of the wide impedance swing of the buffer input on the reference oscillator. Buffer amplifier Q_3 is a common base transistor whose collector output is applied to the tuned circuit of C_7 and the primary of T_3 which is tuned to 5.625 mc. Transformer T_3 secondary output is applied to the ring modulator D_6, D_7, D_8 and D_9. Capacitor C_8 couples an input from the crs limiter to driver amplifier Q_4, a common-base transistor.

The tank, primary of T_4 and C_9, is tuned to 5.625 mc. The secondary of T_4 couples the driver section output to the ring modulator. The diodes are matched impedance, low conductance pairs. They are matched to hold residual d-c output voltage as low as possible. A d-c voltage, resulting from a phase difference between the output of the buffer and driver sections of the phase discriminator when they are on the same frequency, is developed across C_{10}.

Crystal Switch

The crystal switch interpolation oscillator, Q_5, is a high-frequency transistor, connected in a Colpitts oscillator circuit. Oscillation is sustained by feeding back a portion of the energy from the collector tank circuit into the emitter circuit, through switch S_1 and its properly selected crystal. Feedback is controlled by the ratio of the 5-$\mu\mu$f capacitor and C_{11}. The C_{12} and L_4 network compensates for stray capacitance due to switch leads and operates on the crystal frequency.

The crystals themselves are series-mode third-overtone crystals (CR55/U) having a frequency stability of 50 ppm. Oscillator output is tapped off the collector tank circuit and fed through a 50-ohm coaxial cable into the crs second mixer through a connector. Decoupling network L_5 and the 0.01-μf capacitor prevent the oscillator output from getting back into the 16-v regulated bus.

The reference oscillator is basically the same circuit as the interpolation oscillator, operating at either 5.6 or 5.65 mc. Each reference-oscillator crystal, used once with each interpolation-oscillator crystal, achieves the final 50-kc spacing. Decoupling network L_6 and C_{14} keeps the 5.6 and 5.65-mc signal from leaking into the 5.625-mc i-f via the interpolation oscillator, as well as into the i-f itself.

High-speed operating ganged switches S_1 and S_2 are mechanically linked to the main tuning capacitor. These make-before-break switches always present two inputs at ring modulator D_6 thru D_9 of the phase detector and prevent a large undesired residual d-c output. One revolution of the switch for each megacycle represents a speed of nearly 500 rpm when a tuning requirement of three seconds is imposed on the servo operated receiver transmitter. A cam-operated switch withstands this speed.

Close tolerances achieve proper crystal selection with 20 different reference oscillator positions and 10 interpolation oscillator positions for each revolution of the shaft. Molding the cams maintains these close tolerances from unit to unit. The reference oscillator is well shielded from the interpolation oscillator by heavy metal housing.

Transmit Frequency

Transmit channels are derived from the receiver local oscillator, controlled by the crs (See Fig. 4). The transmit frequency is sampled by the receiver, and the receiver mixer output frequency of 11.5 mc is compared in a phase comparator with the output from a 11.5 mc modulated afc reference oscillator.

The error signal from this phase comparator is applied to a semiconductor reactance control device on the transmitter vfo to maintain channel frequency. Modulation is accomplished at a single frequency thus affording constant modulator sensitivity. Also, the modulator circuit parameters are easier to control at the single lower frequency.

The transmit frequency is fed directly into the receiver mixer from the transmitter vfo with the B+ removed from the receiver r-f amplifier.

A low-pass filter, inserted into the control line, attenuates the 11.5-mc signal leaking into the control from the phase comparator. The 75 db attenuation is obtained near i-f harmonic crossovers of 34.5, 46.0, 57.5 and 69.0 mc.

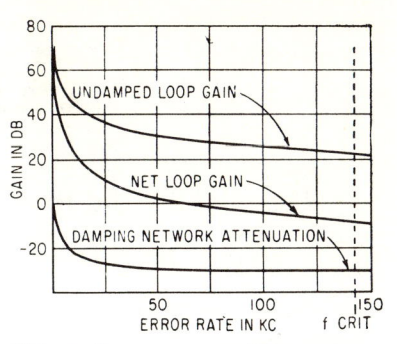

FIG. 5—Frequency-control-loop gain

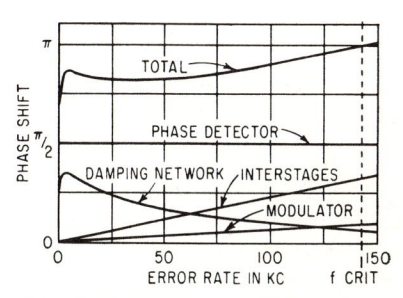

FIG. 6—Frequency-control-loop phase shift characteristics

Hunting is accomplished by a multivibrator operating into the reactance control circuit of the vfo. The multivibrator is biased with the rectified 11.5-mc signal derived from the receiver i-f after crystal frequency selection.

Figure 5 indicates gain characteristic of the transmitter frequency control loop vs sinusoidal error rate or modulation. Loop gain at the critical frequency is −10 db or less and the critical frequency occurs above 100 kc.

Frequency-control-loop phase shift is shown in Fig. 6.

The speech limiter in the transmitter prevents deviation in excess of 15-kc.

The control loop also overcomes the hunting voltage, compensates for vfo frequency error, and achieves maximum audio modulation because of the use of a voice-operated r-t relay.

The authors acknowledge assistance of engineers at the Signal Corps Engineering Labs., in particular A. C. Colaguori, A. Sills, and E. Jacubowics who formulated many concepts used in the crystal reference system. J. J. Lamplot, R. Midkiff and E. Wilson of the Communications Equipment Group of AVCO Crosley Division contributed significantly to the design of the frequency-control systems.

Amplifier for 16-MM

Use of 12-v d-c to 110-v a-c power supply and transistor sound recording amplifier permits 16-mm sound-on-film camera to be operated in field from 12-v nickel-cadmium cells. Transistor amplifier drives 10-ohm recording galvanometer for variable-area optical sound track

By **EDWARD M. TINK**, Assistant Chief Engineer, WLAC-TV Inc., Nashville, Tenn.

SIXTEEN-MILLIMETER cameras with optical sound recording have long been used in television broadcasting for on-the-spot news coverage of both picture and sound. One camera, having a film capacity of 100 ft, for 2¾ min, is housed in a relatively large portable case with a weight of 45 lb or in two smaller cases weighing a total of 60 lb. Though the recording amplifier has a self-contained battery supply, the camera drive motor requires an external source of 115-v 60-cps power.

A smaller and more desirable system results if a miniature transistor amplifier, mounted directly on the camera, is substituted for the vacuum-tube amplifier and a portable 115-v power pack is used to supply the motor. A commercially available power pack employing two 6-v nickel-cadmium cells driving a transformer and vibrator supplies the 115 v. The 12 v also furnishes collector voltage to the transistor amplifier and current to the sound exposure lamp in the camera, thus eliminating all additional batteries.

The variable-area recording galvanometer in the camera has an impedance of 10 ohms at 400 cps and is designed to be driven from a 50-ohm source through a 40-ohm series isolating resistor; 100-percent modulation is represented by an output of 1.6-v rms (52 mw) across the 50-ohm output. The equalization required for speech recording results in a frequency response 12 to 14 db down at 100 cps, rising to +3 db at 3 kc and returning to 0 db at 5 kc.

Amplifier Design

A class-A push-pull output circuit was decided on for low distortion. Due to its greater power sensitivity, the common-emitter configuration was used as can be seen in the circuit of Fig. 1.

A thorough search of available miniature transistor output transformers indicated that there were practically none designed to work into a 50-ohm load. It was also observed that the transformers had efficiencies varying from 15 to 50 percent. For these reasons a ferrite-core transformer having an efficiency greater than 70 percent was designed for this particular application. The design was simplified by the fact that the low-frequency requirements of the amplifier are not too severe.

One of the greatest objections to the use of the ferrite core is that it saturates quite easily and therefore the effective d-c current through the primary winding must be kept low. It was at first thought that the requirement of a low value of d-c unbalance in primary current would require the use of matched output transistors. This, however, was not a problem since all of the output transistors which were used at various times (2N43, 2N44, and 2N188A) were matched within less than 1 ma.

Collector Dissipation

Due to the low collector-to-emitter voltage available (approximately 10 v) it was necessary to operate the 2N43 collectors near their maximum dissipation rating to obtain the required output power with low distortion. The allowable transistor power dissipation is dependent upon the maximum ambient temperature which will be encountered and is given by the expression $P = (85 - T_A)/K$ where T_A is the maximum ambient and K is the thermal resistance in deg C per mw.

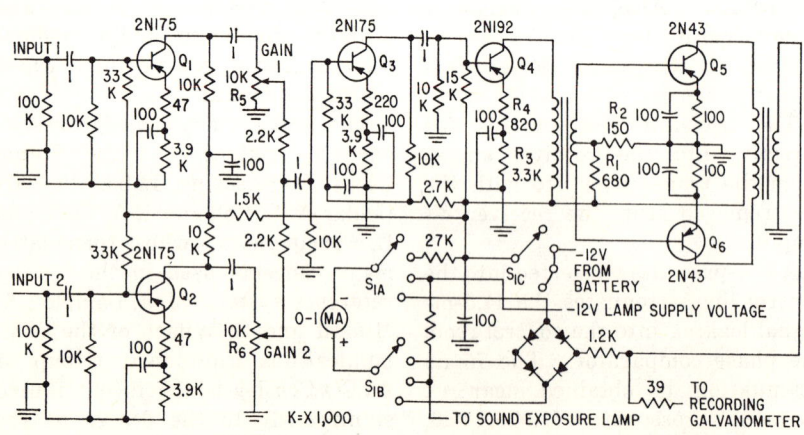

FIG. 1—Dual-input amplifier for Auricon CM-72A drives recording galvanometer

Sound Movie Camera

It was felt that the amplifier might be expected to operate in ambient temperatures as high as 110 F (43 C). Since the thermal resistance for the 2N43 is 0.2 C per mw (with a heat sink whose minimum area is 0.95 sq in.), the dissipation must be limited to 210 mw per transistor. This limits the collector current to a maximum of 21 ma with a collector-to-emitter voltage of 10 v.

The heat sink consists of a 1-in. sleeve of copper tubing. Good thermal contact can be maintained between the transistor body and the heat sink by removing the paint from the transistor body.

With a d-c collector-to-emitter voltage of 10 v, the maximum peak-to-peak excursion of voltage across the transformer is slightly less than 40 v. With a turns ratio of approximately 5.5 to 1, the maximum voltage across the 50-ohm secondary will be approximately 7 v. peak-to-peak. Thus the maximum power output before clipping occurs will be in the vicinity of 125 mw.

Bias Stabilization

The 100-ohm resistors to ground, in series with the emitter-to-base junction of the output transistors, help prevent thermal runaway. In addition, base voltage stabilization is obtained by the network consisting of R_1 and R_2. As this stabilization was adequate, no nonlinear stabilizing units were employed.

The driver stage employs a 2N192 transformer coupled to the output circuit. The relatively large value of R_3 in the emitter circuit, along with the base voltage stabilization, again contribute to excellent temperature stability. A smaller unbypassed resistor, R_4, also placed in series with the emitter provides approximately 16 db of gain reduction and results in excellent overload characteristics, with uniform gain and frequency response with variations in transistor parameters.

Input Stages

The two input stages and the following stage all employ 2N175 low-

Transistor amplifier is mounted on camera

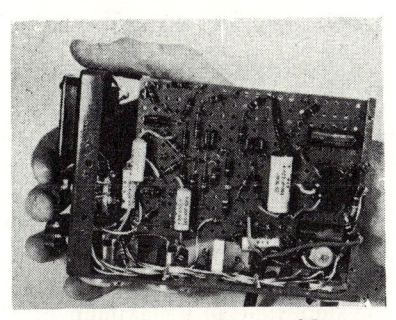

Inside view of recording amplifier

noise transistors. Good temperature stability is obtained by the methods used in the driver stage and in addition d-c collector-to-base feedback is employed. The unbypassed emitter resistors result in a gain reduction of 12 db in Q_3 and 5 db in Q_1 and Q_2.

There is a maximum interaction of 5 db between gain controls R_5 and R_6 under the worst possible condition—feeding a signal in channel 1 and changing gain control 2 from minimum to maximum. Measurements indicate that no frequency discrimination occurs at any combination of settings of the gain controls.

A miniature meter included in the circuit measures 12 volts d-c from the batteries, sound-exposure-lamp current and audio output level. The shunts and multipliers are so calculated that the proper values of voltage and current (12v, 140 ma and 100-percent modulation) in all three meter positions results in exactly half-scale deflection.

Amplifier Performance

A signal-to-noise ratio of 56 db was measured in both channels. Since each channel has a gain of approximately 81 db, this results in an equivalent noise input of 137 db. The harmonic distortion, measured at 1,000 cps across the 50-ohm secondary winding, was 1.9 percent with an output level of 1.6 v. rms (100-percent modulation).

The measured low-frequency response was down 14 db at 250 cps, but listening tests indicated that this slight deficiency in low-frequency response was not at all objectionable. As the low-frequency response fell, the distortion rose to a value of 3.2 percent at 250 cps. As the frequency increased beyond 1,000 cps, the distortion remained equal to or less than the value for 1,000 cps. High-frequency response was flat within ±1 db to 9,000 cps, which is beyond the upper limit of 16-mm sound-on-film recording capability.

It was necessary to make some wiring changes in the Dormitzer DY-1012 power pack to obtain the proper polarity for the 12-v d-c output. One of the parallel pilot lamps in the battery charging circuit was eliminated to bring the charging current down to 450 ma, which is within the rating of the new rectifier. Finally, a potentiometer was added to control the sound-exposure-lamp current. A large amount of filtering was required to remove the vibrator hash from the battery leads. Although this modified camera has been in use only a short time by the news department at WLAC-TV, early results indicate that its operation is reliable and that good quality sound-on-film recordings are readily obtained by personnel not particularly accustomed to operating electronic gear.

The author expresses his appreciation to Chief Engineer R. L. Hucaby and to the station engineers for their assistance.

Portable Multiplexer for

Stabilized transistor circuits enable four-channel ppm multiplex unit to operate from −54 to +65C. Amplitude modulation of a microwave radio system is pulse position modulated by the multiplexer. Circuit operation of the modulator sweep generator, video pulse shaper, demultiplexer synchronizer and the demodulator flip-flop are described

By PAUL W. KIESLING, JR.,
Communications Department, Raytheon Manufacturing Company, Wayland, Massachusetts

INVESTIGATION of the applicability of transistors to pulse-position communications equipment for military use resulted in the development of a portable four-channel all-transistor multiplexer. The equipment has circuits for four channels of modulation and demodulation, the multiplexing and demultiplexing of these channels, line terminal facilities for both two and four-wire operation, and low-frequency signaling facilities. It is packaged in a watertight aluminum case that can be carried on a packboard, and weighs 55 pounds.

Operation

The audio or ringing signal to a modulator is sampled at an 8-kc rate or every 125 μsec. Amplitude of the signal at the instant of sampling alters the position of the channel pulse from its normal position by an amount proportional to the amplitude, or a maximum of ±1 μsec which is 100-percent modulation. The sampling signal originates at an 8-kc oscillator but passes through a pulse shaper to a large delay line before sampling each channel. The delay line provides timing pulses for the four voice channels and the synchronizing channel.

After transmission by a microwave radio link, the video signal, consisting of pulses which have a rise and fall time of 0.1 μsec and a width of 0.5 μsec, is sliced of noise at the top and bottom, amplified and each channel individually demodulated. Synchronization of the receiving circuits to the video signal is provided by the output of a circuit that detects the presence of the closely spaced pair of synchronizing pulses. A large delay line similar to that used in the multiplexing circuits provides the timing pulse to gate the proper video pulse to each channel and to convert the pulse-position modulation to pulse-width modulation. Complete demodulation is obtained by passing the pwm signal to a low-pass, 3.5-kc filter and amplifying the signal output.

Additional circuits convert the four-wire unbalanced circuits to

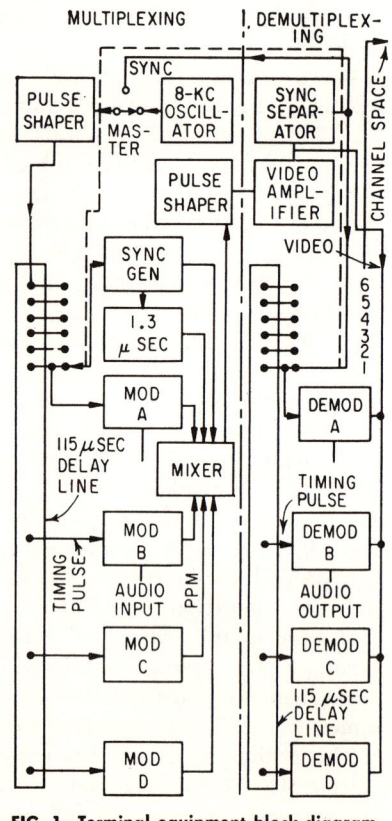

FIG. 1—Terminal equipment block diagram of multiplexer

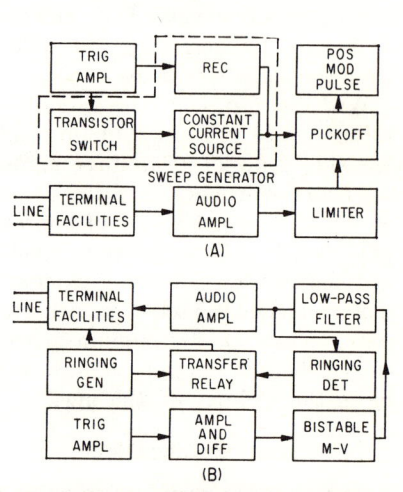

FIG. 2—Basic functions of the modulator (A) and demodulator (B)

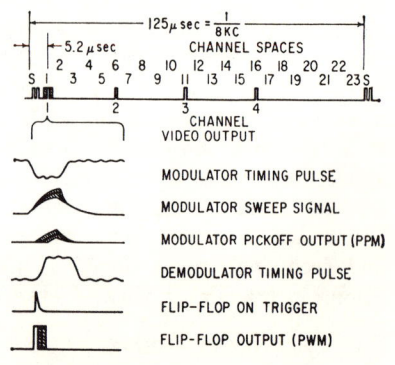

FIG. 3—Video output and modulator and demodulator waveforms for channel one

Telephone Communications

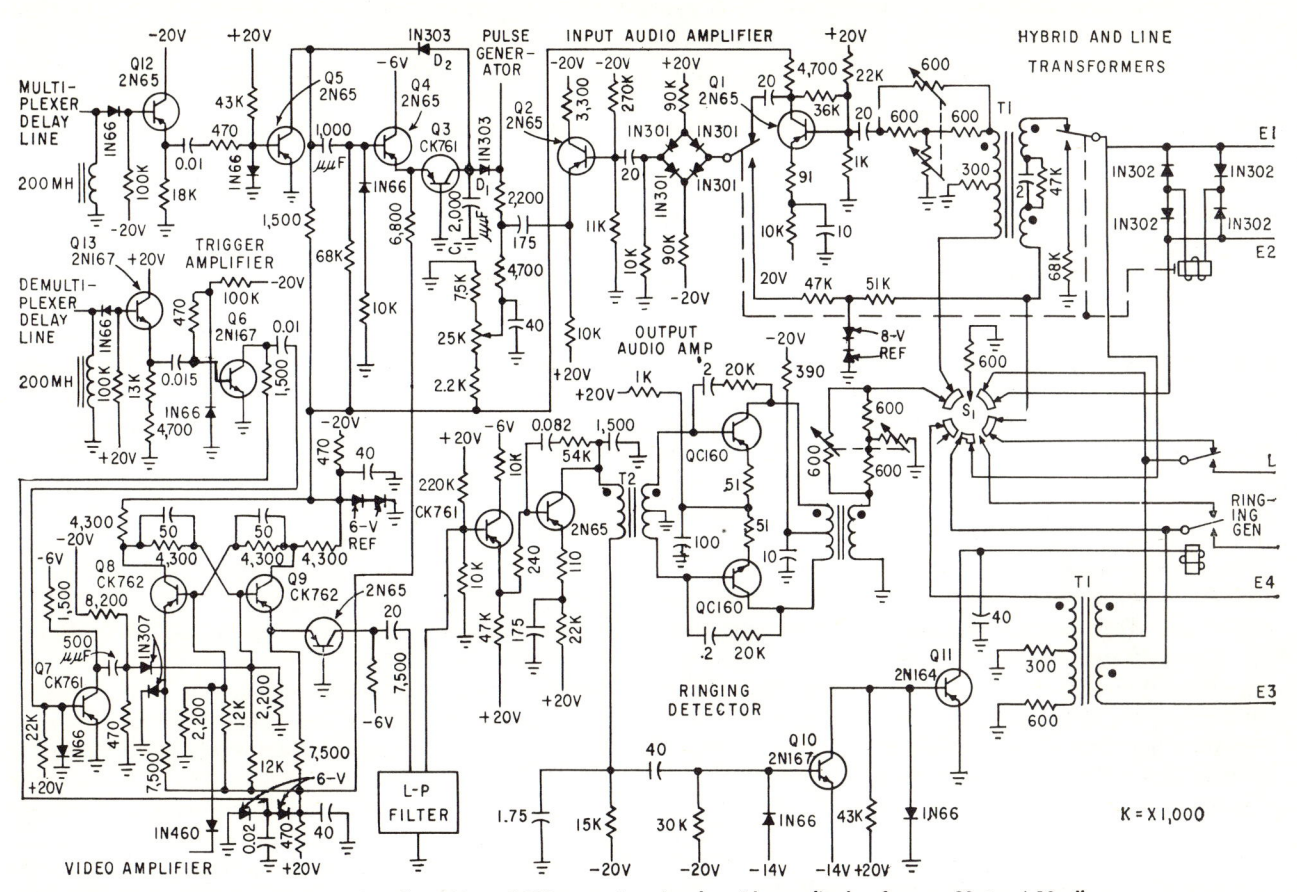

FIG. 4—Modulator-demodulator circuit handles 300 to 3,500-cps voice signals with amplitudes from −20 to +10 dbm

either two- or four-wire balanced circuits and detect low-frequency modulation for signaling. The block diagram of Fig. 1 shows the basic multiplexing and demultiplexing functions and the diagram of Fig. 2 shows the basic functions of a modulator-demodulator. The waveforms of Fig. 3 illustrate the video output and the more important modulator and demodulator waveforms for channel one.

Modulator

The modulator - demodulator shown in Fig. 4 handles 300- to 3,500-cps voice signals with amplitudes from − 20 to +10 dbm and provides output audio signals with the same range of level. In addition, a two-position rotary switch makes it possible to instantly convert from two- to four-wire operation. Additional circuits limit and transfer a high-voltage 20- to 30-cps ringing signal to the modulator and detect the same frequency signal in the demodulator to transfer the output of a common ringing generator to the line terminals.

The upper half of the schematic is the modulator which uses six transistors: two are used to isolate the sweep circuit from the delay line and to trigger the sweep; two are used in the sweep circuit; and two are used in the audio circuits to terminate the bridge-T attenuator, provide gain and isolate the bridge limiter from the pickoff.

Delay-line isolation is provided by transistor Q_{12} which is normally cut off. The R-C components in the emitter circuit were chosen to cause Q_{12} to conduct and Q to conduct and saturate when the negative delay-line trigger reached 50-percent amplitude.

The 10- to 15-μsec pulse at the collector of Q_5 cuts off sweep recovery diode D_2 and emitter follower Q_4. Grounded-base sweep transistor Q_3 acts as a constant-current source of 2 ma to charge capacitor C_1 at 1 v/μsec. The normal pickoff voltage at the cathode of diode D_1 is about 5 v so the normal unmodulated position of the channel pulse is about 5μsec after the sweep is triggered. Precise adjustment of the pulse position is made by a centering control that adjusts the d-c pickoff voltage. An audio signal of 1-v peak at the pickoff gives 100-percent modulation of ±1 μsec.

Amplifier Q_1 drives the limiter and terminates the bridge-T attenuator in 600 ohms. Proper operation of the limiter depends on the high back impedance of the four diodes. High-voltage ringing signals bypass the attenuator, are injected into a limiter and fed into the audio limiter. The ppm output of the modulator has poor rise time and large audio content.

The pulse shaper removes the

audio component from the modulator output to prevent crosstalk and shapes the pulse to the required rise and fall time and width. A typical pulse shaper is shown in Fig. 5. The first stage acts as a high-pass filter and second stage Q_{14} saturates providing a fast rise time pulse 5-μsec wide. This pulse is differentiated and triggers amplifier Q_{15}. The output of this stage triggers a blocking oscillator.

A shorted delay line of 0.5-μsec round-trip length in the base of the 6-0 transistor gives an inverted pulse to turn the oscillator off. This determines the pulse width. The oscillator transformer is wound on a ferrite core. Changes in permeability with temperature do not affect the pulse width since the natural width of the oscillator is always greater than that determined by the delay line. A third winding on the transformer provides drive for output transistor Q_{16}. The collector of this and similar transistors in the other four shaper circuits in a multiplexer are tied together and act as mixers to provide the video train shown in Fig. 3.

Synchronizer Detector

After slicing and amplification, the video pulses in the demultiplexer go to the synchronizer or sync separator shown in Fig. 6. The open-circuited 0.65-μsec delay line in the collector of Q_{17} causes a reinforcement of the second synchronizing pulse which is 1.3 μsec from the first synchronizing pulse.

A slicing circuit in the base of the second stage, Q_{19}, is adjusted to permit only the reinforced pulse to fire this stage. This action provides a pulse every 125 μsec that is synchronized to the incoming video signal.

Transistor Q_{18} keeps the slicing voltage constant by compensating for the saturation current of transistor Q_{19}. This enabled the sync separator to operate from -60 to $+70$ degrees centigrade. The output of Q_{19} drives a blocking oscillator that provides the drive for the demultiplexing delay line. This delay line is tapped at appropriate points to provide the demodulator with timing signals that gate the individual video pulses to each demodulator and convert to pulse-width modulation.

Demodulator

The demodulator (Fig. 4) uses 12 transistors: six in the demodulating circuits and six in the audio amplifier and ringing detector circuits. Flip-flop Q_8 and Q_9 converts the pulse-position signal to pulse-width modulation. The flip-flop is turned on by the delay-line timing trigger and turned off by the video pulse following the delay-line trigger. The delay line is isolated from the flip-flop by emitter follower Q_{13} which operates similarly to stage Q_{12} in the modulator. Transistor Q_6 provides gain and Q_7 provides a pulse with a sharp rise time that is differentiated before turning the flip-flop on.

The output of the nonsaturating flip-flop is width modulated on the trailing edge and has an unmodulated width of 2.5 μsec. When transistor Q_8 is turned on, its base is slightly negative, thus holding off the diode on its emitter. The collector is not quite at saturation voltage and the base of transistor Q_9 is positive by about 1 v. The diode in the emitter of Q_9 is conducting, clamping the emitter slightly above 0 v.

In operation, this second emitter diode was replaced by a grounded-base transistor stage to drive the low-pass filter. The collector voltage of an ON transistor is determined by its emitter bias circuit and by the voltage divider to the base of the OFF transistor. The voltage divider to the base of the OFF transistor is adjusted to keep that base more positive than the conducting diode in the emitter.

Filter

The low-pass filter, in addition to demodulating the pwm, has an 8-kc notch so that the sampling frequency level is at least 50 db below the audio level in the output. The audio amplifier consists of an emitter follower to properly terminate the filter, one single-ended amplifier and a class-A push-pull amplifier capable of delivering 30 mw of audio power with less than 3-percent harmonic distortion to 600 ohms. The first stage of amplification amplifies a ringing signal of 20 to 30 cps as well as audio. The coupling transformer acts as part of a low-pass filter to pass the ringing signal to ringing detector Q_{10} and Q_{11}. Transistor Q_{11} operates a double-pole double-throw relay to transfer the output of a local ringing generator to the line terminals.

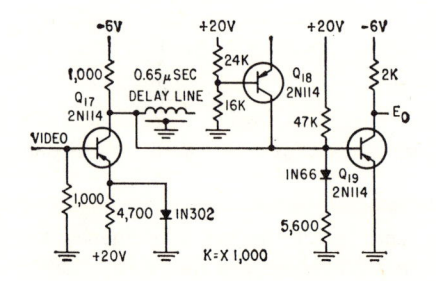

FIG. 6—Pulse synchronized to the incoming video is provided every 125 μsec. Transistor ppm equipment operates over a wide temperature range

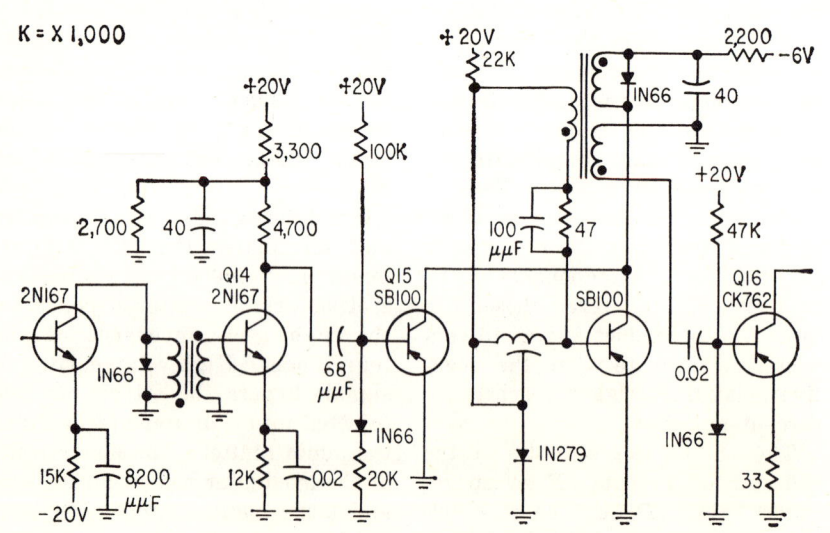

FIG. 5—Pulse shaper removes audio from modulator output to prevent crosstalk

Versatile F-M Transducer

Single-transistor oscillator on phonograph pickup arm is frequency modulated by stylus. Radiated signal goes to monitoring receiver

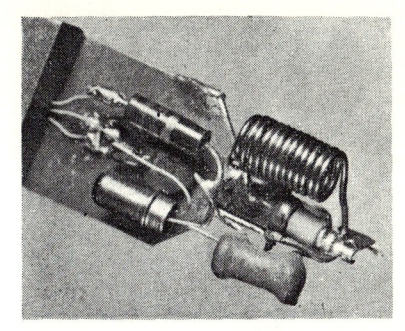

By C. S. BURRUS, E. E. Dept., Rice Institute, Houston, Texas

Oscillator is mounted on phono arm

SOME ADVANTAGES of the pickup system to be described are light weight, no need for connecting wires, excellent frequency response and insensitivity to hum. These and other qualities make this type of pickup useful in locations that are inherently difficult to monitor.

Success of this oscillator has pointed to several possible uses in measurement. Measurement of temperature using a bistable element, of pressure using a metal diaphragm, of voltages using a voltage-variable capacitor, or of mechanical vibrations in different locations are possible.

Circuit Description

Frequency of oscillation is determined by L_1, C_1, C_2, and the output capacitance of Q_1 (Fig. 1). Capacitor C_1 is formed by a stationary plate and the stylus arm, which are separated by approximately 0.025 in. They have a common area of approximately 0.0064 in^2 to give a static capacitance of approximately 0.06 $\mu\mu f$. The variation of this value produces the frequency modulation.

Since the emitter-to-base impedance is very low, the feedback element C_2 is also across the tank. Output capacitance of the 2N588 is a function of frequency and collector current; at 100 mc and 3 ma circuit output capacitance is about 2 $\mu\mu f$. This gives a total, including strays, of approximately 9 $\mu\mu f$ which requires L_1 to be 0.34 μh. Thirteen turns of No. 20 wire, 0.25 in. in diameter gives this value at a Q of 160.

The 500-ohm emitter resistor and the 1.5-v battery establish the 3-ma collector current; this combination and 6 v on the collector provide strong oscillation. A single-battery supply could be used with appropriate biasing resistors.

The common-base configuration was chosen because of economy of parts and ease of obtaining feedback. Feedback is through C_2 to the emitter. It would probably be better to bring C_2 to a tap down on L_1 and use a larger value of C_2.

Linearity

Two inherent sources of nonlinearity in this system must be minimized. First, frequency varies with the inverse square root of tank capacitance and second, capacitance varies with the inverse of the spacing of the plates. By padding the varying capacitor C_1 with C_2 and the transistor's output capacitance, the ratio of fixed to variable capacitance is large enough that frequency varies very nearly linearly with C_1. To reduce distortion from the relation of C_1 to spacing, the fixed spacing is made large compared to the expected deviations. With this arrangement only on extreme excursions that place the stylus arm near the fixed plate is there appreciable distortion.

The frequency deviation must be such that the receiver's band width

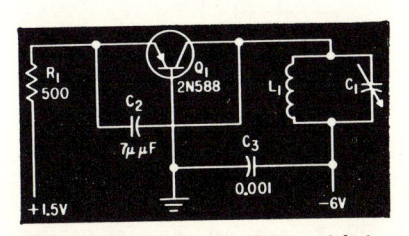

FIG. 1—Stylus varies C_1, thus modulating the oscillator

will accept all the side bands. This is controlled by the amount of padding of C_1. Fortunately, the padding necessary to reduce the frequency swing to approximately the same as that of broadcasting stations is also sufficient for the reduction of the nonlinearity mentioned above.

Construction

The entire assembly with the exception of the batteries is mounted at the head of the phono arm and is no larger than most magnetic pickups. The plate that supports the assembly is of 1/16 in. copper and is common to the 6-v battery lead. A small ceramic stand-off insulator supports the fixed plate of C_1. Also connected to this point is L_1, C_2 and the collector lead of Q_1. By-pass capacitor C_3 is positioned between R_1 and the transistor. The assembly is made rigid since the pickup tends to be microphonic. Cost of parts is less than 10 dollars.

Performance

Frequency response of the pickup system is determined by the stylus assembly, the arm, and the f-m receiver. The h-f limit is determined by the mass, compliance, and self resonance of the stylus assembly and the low frequency limit is determined by the tone arm's ability to remain stationary as the stylus moves. The constructed unit was flat from 30 to 15,000 cps. The range was probably greater but this was the limit of the test record. This device is displacement sensitive, not velocity sensitive as are magnetic pickups. It tracked reliably at less than one gram.

How to Construct a

Here is a portable transistor transmitter that is easy to build with standard commercial components at a cost of about $25. Range of 200 feet makes it a handy part of public-address system

By D. E. THOMAS and J. M. KLEIN, Bell Telephone Laboratories, Inc., Murray Hill, New Jersey

PUBLICATION of the description of an experimental transistor f-m transmitter[1] was followed by numerous requests for information on where the components to build a similar transmitter could be obtained. Unfortunately, the point-contact transistor used in the original transmitter was an experimental device that became obsolete even before it could be coded for manufacture. The experimental f-m transmitter to be described can be built with readily available commercial components at a material cost of approximately $25.

Circuit

The circuit of the new transmitter, which was designed as a cordless hand microphone operating in the commercial f-m band is shown in Figure 1.

The 2N499 transistor performs the three functions of r-f oscillation, frequency modulation, and audio amplification.

Since the frequency of oscillation of this transistor is somewhat closer to its cutoff frequency than the previously used point-contact transistor, less phase shift is required in the feedback coupling circuit to maintain oscillation. Feedback is therefore obtained from a tap on tank coil L_1 through the series combination of R_2 and C_1, rather than through a capacitance only. Coil L_1 consists of 6 turns No. 24 bare tinned copper wire wound on ¼-in. polystyrene form. Feedback tap is at exactly one full turn.

Frequency modulation is accomplished by the feedback-loop phase shift resulting from alpha-cutoff frequency shift under the control of an audio-frequency signal input. The high common-emitter current gain of the 2N499 at audio frequencies amplifies the audio output of the microphone to the level needed to produce adequate frequency deviation.

The audio signal is fed into the base of the transistor as shown in Fig. 2, the equivalent circuit of the transmitter at d-c and audio frequencies. As R_1 appears in series with the emitter of the transistor at audio frequencies, the maximum available common-emitter current gain is reduced.

More audio gain can be obtained by shunting R_1 with the LRC circuit shown dotted in Fig. 1. Inductance L_2, which can be any low-capacitance r-f choke with inductance greater than 1 μh, is needed to avoid

Communication is convenient with this miniature f-m transmitter

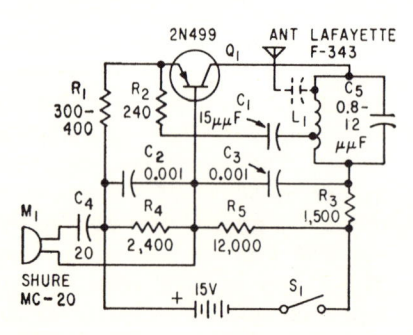

FIG. 1—Miniature transistor f-m transmitter

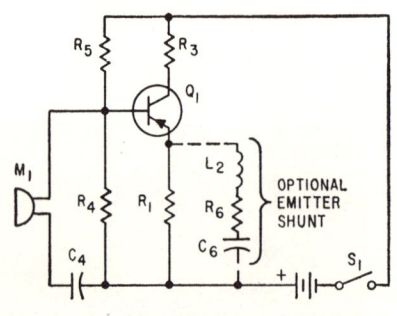

FIG. 2—Equivalent circuit of transmitter for audio and subaudio frequencies

Miniature F-M Transmitter

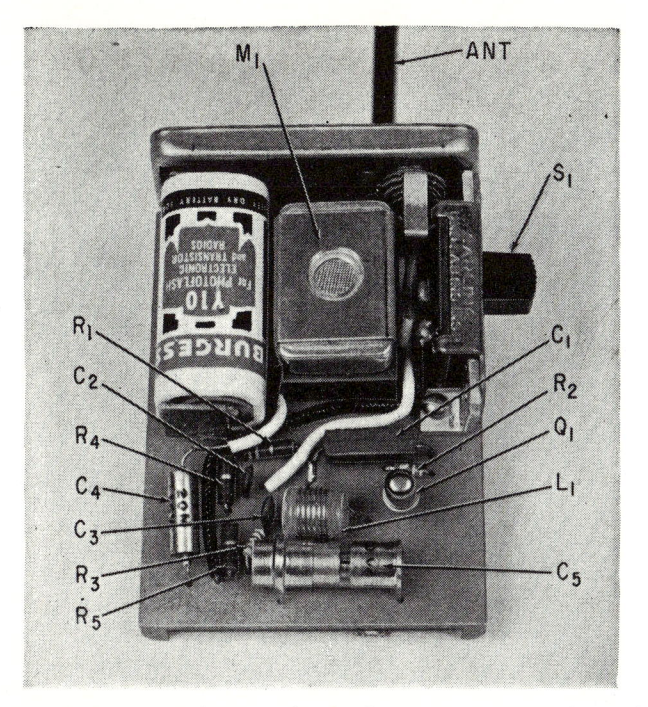

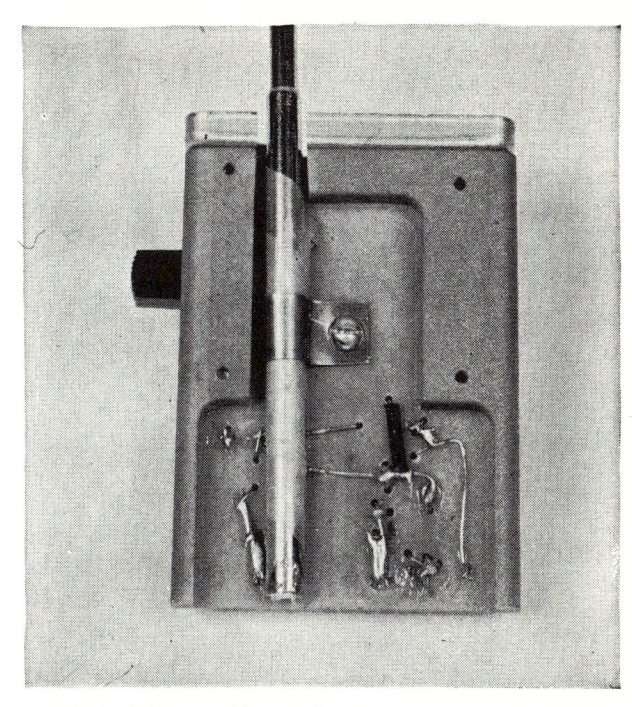

FIG. 3—Component layout, wiring and mounting. Antenna is coupled to tuned circuit by parasitic capacitance

shunting the emitter at r-f and interfering with r-f oscillation. Capacitance C_6, which should be of the same value and type as C_4, is needed to avoid disturbing the d-c biasing of the transistor. Audio gain can be adjusted by varying resistor R_6.

Component Layout

At vhf frequencies, physical position of components may play an important part in performance. Therefore, the physical layout of the components is shown in Fig. 3. All components except the antenna are mounted on one side of a 2½ by 1⅞ in. panel. Most of the parts are supported by their own leads, which are passed through holes in the panel and wired on the underside.

The bottom section of the antenna is wound in polyethylene sheet and clamped to the underside of the panel directly beneath the L_1 C_5 tank circuit. As the only r-f coupling between the antenna and the tank circuit is the parasitic capacitance between them, radiation is limited to that needed for a portable public address microphone with a range of about 200 ft; also, frequency pulling due to any change in radiation impedance of the antenna is reduced.

Adjustments

The value of resistor R_1 should be adjusted for the particular transistor used so that the d-c voltage drop across resistor R_3 is approximately 6 v when the battery is new. This adjustment, plus the adjustment of capacitor C_5 to give the desired carrier frequency, have been the only adjustments necessary. However, for those 2N499 transistors which happen to be on the edge of the distribution of production acceptability, the circuit may not oscillate. In such cases, a variation of R_2 and/or C_1 should produce oscillation.

Alternate Microphone

An external microphone of different impedance or larger size may be used. The space occupied by the small reluctance microphone could then be utilized for a suitable audio impedance matching transformer.

The alternate microphone selected should be connected to the transmitter by a coaxial microphone cable. In this case, the microphone cable will provide sufficient radiation for limited local transmission and the collapsible dipole antenna may be omitted.

The audio input impedance of the transmitter will vary widely depending upon the transistor used, the value of resistor R_1 and whether an emitter shunt is used to increase the audio gain. If an audio impedance of 1,000 ohms is assumed and an audio transformer is selected to match the microphone selected to 1,000 ohms, the actual audio impedance match will be satisfactory.

Considerably greater range could be obtained by use of an efficient radiator suitably coupled to the transmitter. However, FCC regulations with regard to transmitter frequency, stability of frequency and power must be complied with.

The concept of a frequency-modulated transistor transmitter originated with R. L. Wallace, Jr. The earliest transmitter was built by L. G. Schimpf using an early tetrode transistor modulated by a capacitor microphone across the tank circuit.

REFERENCE

(1) D. E. Thomas, Single Transistor F-M Transmitter, ELECTRONICS, p 130, Feb. 1954.

Chapter 11
TEST INSTRUMENTS

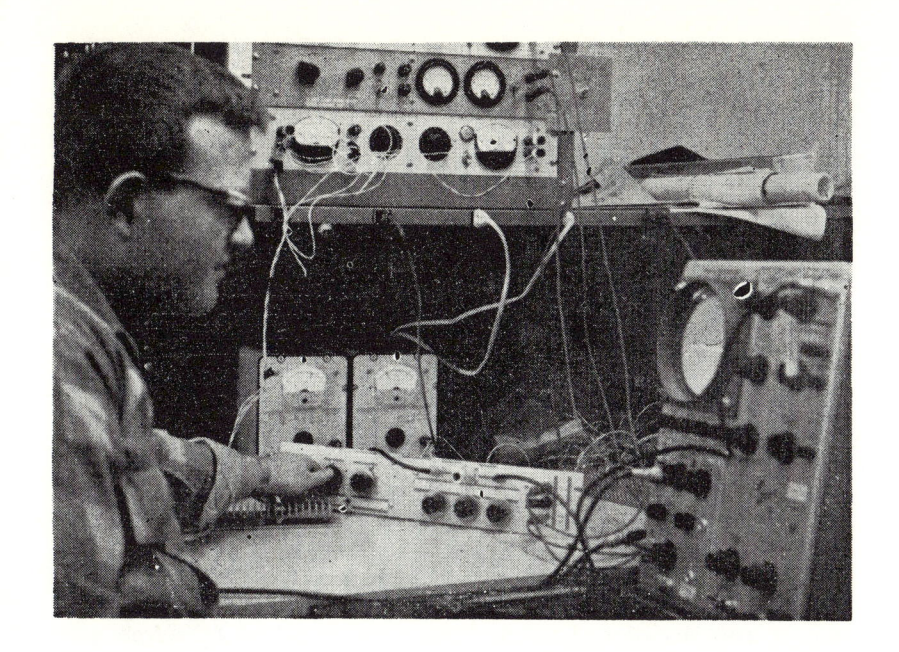

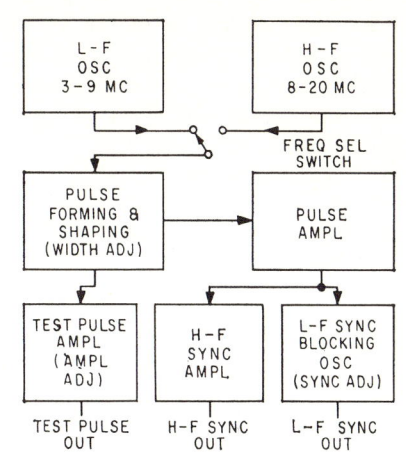

FIG. 1—Pulse generator uses two oscillators and has three outputs

Providing synchronizing signals by use of the sync blocking oscillator output

Generator for Pulse Circuit Design

Versatile instrument produces pulses with widths ranging from 25 to 35 millimicroseconds at repetition frequencies from 3 mc to 20 mc and amplitudes from zero to 2 v. High- and low-frequency Hartley oscillators drive novel pulse forming and shaping circuit

By LEOPOLD NEUMANN, Staff Member, M.I.T. Lincoln Laboratory, Lexington, Mass.

TESTING HIGH-SPEED digital computer circuits, using 30 millimicrosecond pulses, requires a versatile pulse source. This article describes a transistorized, variable-frequency pulse generator designed to test recently developed circuits.

The pulse generator has three outputs. One output provides the test pulse. The second and third outputs supply pulses for synchronizing oscilloscopes.

The test pulse is variable from 25 to 35 millimicrosec. This permits testing circuits with pulses both narrower and wider than the normal input pulses.

The pulse-repetition frequency of the test pulse can be varied from 3 mc to 20 mc. This allows testing the prf sensitivity and determining the h-f limits of the circuit being checked.

An internal amplitude control varies the test pulse amplitude from zero to 2 v. The amplitude sensitivity of a circuit can be determined by varying the amplitude of the test pulse.

The second output provides a h-f 50-ohm cable positive pulse. Traveling-wave tube oscilloscopes are synchronized by using this output. The third output is a l-f submultiple of the test pulse. This l-f pulse is used to synchronize conventional laboratory oscilloscopes.

As shown in Fig. 1, the pulse generator consists of two oscillators, a pulse forming and shaping circuit, three pulse amplifiers and a sync blocking oscillator.

Oscillators

When a switch is used to interchange coils in a single oscillator, poor performance results. This poor performance is caused by switch capacitive feed through. For this reason independent h-f and l-f oscillators are used.

The oscillators, Fig. 2, are Hartley type, sine-wave oscillators. Each is tunable over a 3 to 1 frequency range by capacitor C_4. Use of a 2N501 switching transistor in the oscillator circuits allows all of the transistors used in the pulse generator to be of the same type.

Resistors R_1, R_2 and R_3 control

the d-c bias of the oscillator and provide bias compensation for variation in I_{co} and β of individual transistors. Some degeneration for equalizing the gain of individual transistors is provided by R_4. The function of R_5 is to limit the maximum collector-to-base voltage to a safe value.

A spdt switch permits the output of either the h-f or l-f oscillator to be used to drive the pulse-forming circuit. Independent h-f and l-f oscillators make it possible to rapidly switch between two preset frequencies.

Basic Pulse Circuit

The basic pulse-forming and amplifying circuit[1] is shown in Fig. 3A. When a negative step is applied to the base of Q_1, the emitter tends to follow this input voltage and causes a current in the collector load. The collector voltage rises toward the emitter voltage and the transistor saturates.

When the transistor saturates the voltage across the collector load is $(V_s - V_c - V_{ce\,sat}) = (V_s - V_{in})$ = a constant. The current drawn by the load is $i_c = (V_s - V_{in})[(1/R_1) + (t/L)]$. The emitter resistance can supply only a $I_{e\,max} = (V_{in} - V_{be})/R_2$. If the input step is continued beyond the point where I_c equals $I_{e\,max}$, the transistor comes out of saturation and the collector falls to $-V_s$ with a time constant L/R_1.

However, C_1 transiently lowers the emitter impedance for changing inputs. Hence, C_1 improves gain to the leading edge of the input step by decreasing the degeneration caused by the emitter resistor.

Also, as seen in Fig. 3B, when V_c

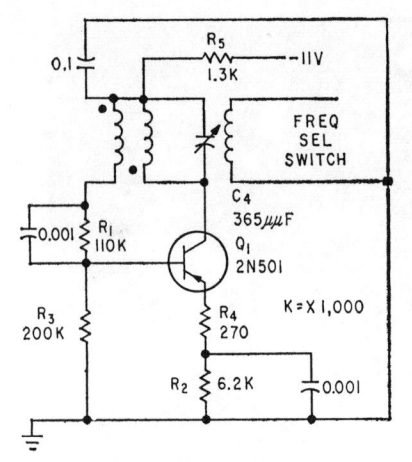

FIG. 2—Oscillator for h-f and l-f are identical with exception of coils used

falls due to the transistor coming out of saturation, C_c pulls the base transiently more negative. Because of C_1 in the emitter, the circuit gain to this input change is high. Thus a pulse of current is fed to the collector load, tending to square off and effectively decrease the damping of the decay. If C_1 is increased the decay becomes less damped and may overshoot. In the third pulse-forming stage and in the pulse amplifiers the effect of C_1 is used to speed up the pulse fall.

When the negative input pulse is removed and the base made positive, the transistor is quickly turned off, reducing the collector current to zero. The collector then overshoots by an amount $i_L R_1$. This overshoot decays with a time constant L/R_1.

Pulse-Forming Circuit

The pulse-forming portion of the generator is shown in Fig. 4. The pulse former consists of three stages, each of which is a variation of the basic circuit just described.

Stage one of the pulse former amplifies the input sine wave. This amplifier provides the current gain necessary to drive the second pulse-former stage. Stage one also squares off the output of the sine-wave oscillator.

The stage is designed so that it does not come out of saturation until the input sine wave starts to go positive. Increased stage gain is provided by emitter capacitor C_1.

The overshoot of this stage drives the second pulse-forming stage. This overshoot has a sharper leading edge than the pulse itself. The overshoot is controlled by the collector inductor rather than the input waveform. In Fig. 5B this sharp leading edge may be observed as the continuation of the turnoff fall. The overshoot falls from zero to -1 v in three millimicroseconds, then slows down its fall rate due to the loading of the next stage.

A 40-millimicrosecond pulse is formed in the second stage. This stage has a low-inductance collector load and a large capacitor across its emitter resistor as shown in Fig. 4. The high emitter capacitance boosts the high-frequency gain of this stage so that the sharp leading edge of the first stage overshoot quickly turns on the transistor and drives it into saturation. The collector rises until the transistor is in saturation and then follows the input waveform down. When the input waveform begins to rise again toward ground, the transistor comes out of saturation and the collector inductor pulse overshoots. This overshoot (Fig. 5C) is used to drive the third stage. The positive-going output cannot be used to drive the next stage because a second positive-going ring which may occur at

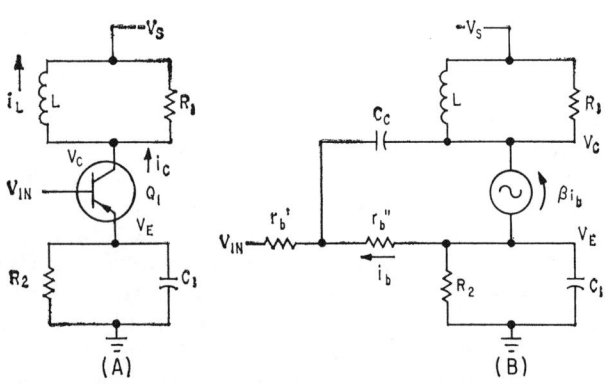

FIG. 3—Basic pulse circuit (A) and its simplified equivalent circuit (B)

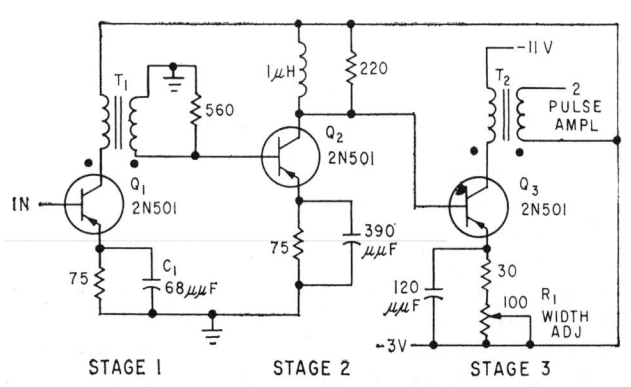

FIG. 4—Pulse former and shaper has three stages

FIG. 5—Oscilloscope photos of output of oscillator (A) first shaping stage (B), second shaping stage (C), third shaping stage (D), generator test pulse amplifier (E) and sync blocking oscillator (F)

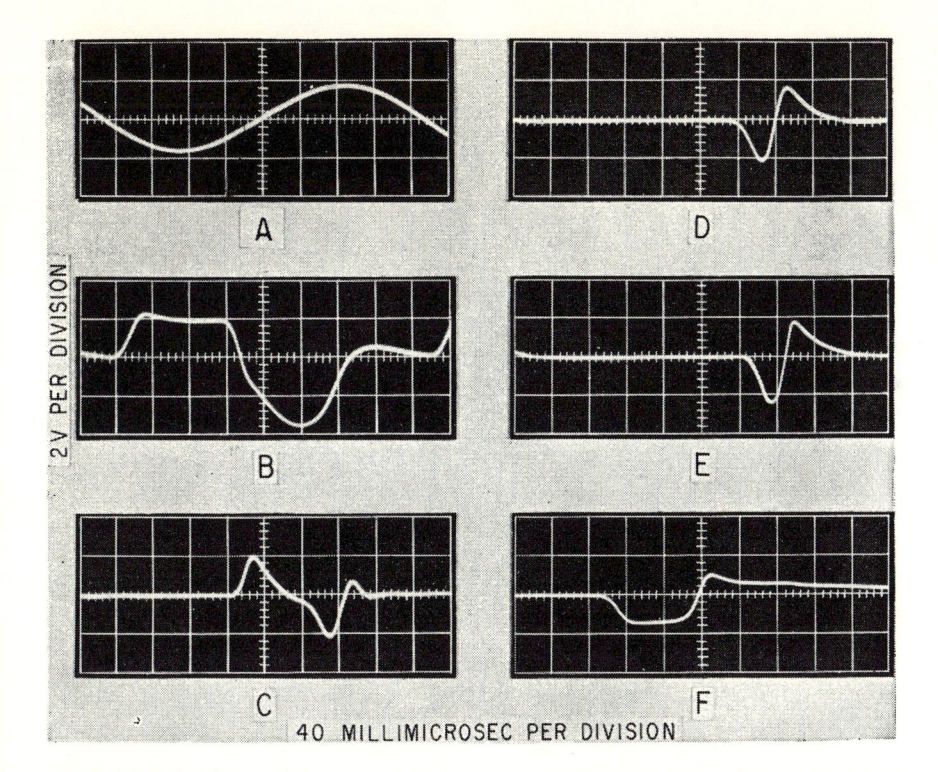

FIG. 7—Low-frequency sync blocking oscillator provides synchronizing signals for conventional oscilloscopes

low frequencies could cause false operation.

Since the collector supply voltage of the second stage (Fig. 4) is equal to the emitter return voltage of the next stage, a coupling transformer is not required between stages. Under these conditions no shift of d-c level is necessary.

The third pulse-forming stage narrows the pulse and changes it into the conventional pulse form (Fig. 5D). The variable resistor R_1 (Fig. 4) in the transistor emitter circuit provides pulse-width control. The resistor controls the maximum emitter current which determines the length of time the stage can remain saturated.

Pulse Amplifier

The pulse amplifiers amplify and reshape. In Fig. 6 a 2-v input pulse at the base of Q_1 produces 1.8 v at the emitter. The emitter will then supply 29 ma to the collector. When Q_1 is in saturation, approximately 6 v appears across the collector load. Since the collector current is of the form $i_c = V\,[(1/R_{\text{load}}) + (t/L)]$, the collector draws maximum emitter current when t equals 18 millimicrosec. With pulse rise and fall times included, a 25 to 35 millimicrosecond pulse is produced. The pulse overshoot time constant is 19 millimicroseconds.

Capacitor C_1 increases the circuit gain and its value is chosen to provide approximately critical damping to an emitter-controlled collector decay. The circuit provides a 2-v output pulse.

Three pulse amplifiers are used in the pulse generator. Two pulse amplifiers are driven by the third pulse-shaping stage. The first pulse amplifier driven by the third pulse-shaping stage provides 2 v across a 50-ohm amplitude-control potentiometer. The tap of the potentiometer is the pulse-generator output. The second pulse amplifier driven by the last pulse-shaping stage drives the sync blocking oscillator and the third pulse amplifier.

This third amplifier provides the h-f 50-ohm positive pulse for synchronizing a traveling-wave tube

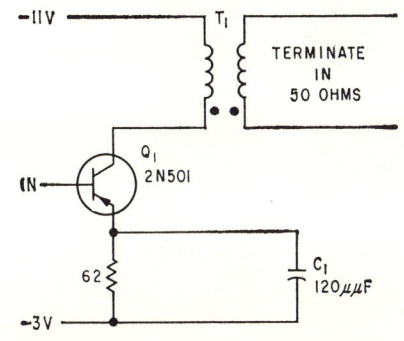

FIG. 6—Pulse amplifier uses circuit similar to last pulse former and shaper stage

oscilloscope. A 500-kc to 1-mc pulse is supplied by the sync blocking oscillator to synchronize conventional laboratory oscilloscopes.

Sync Blocking Oscillator

The l-f sync blocking oscillator (Fig. 7) consists of a grounded-emitter stage whose free-running period is made to lock in with the frequency of the pulse generator. This is accomplished by mixing the normal blocking oscillator feedback signal and the pulse generator output at the base of the transistor.

The free-running period is manually adjusted to a near submultiple of the pulse generator output by R_1 in the base circuit. The mixed input pulses to Q_1 cause the blocking oscillator to lock in with the pulse generator. This technique permits the single blocking oscillator to perform the large variable countdown ratio (up to 40 to 1) required to provide synchronizing signals over the entire range.

The work described here was performed under contract at Lincoln Laboratory, operated by MIT with the support of the Army, Navy and Air Force.

REFERENCE
(1) W. N. Carroll and R. A. Coopper, Ten Megapulse Transistorized Pulse Circuits for Computer Applications, Proc of 1958 Transistor And Solid-State Circuits Conf.

Battery-Operated Transistor Oscilloscope

Using 39 transistors and 3 vacuum tubes, this self-powered instrument has response from d-c to over 5 Mc. Deflection-blanked cathode-ray tube and automatic battery charger are among new circuits used

By OZ SVEHAUG and JOHN R. KOBBE, Engineering Division, Tektronix, Inc., Portland, Oregon

RAPID INCREASE in the variety and performance of transistors coupled with the growing need for a high-quality portable oscilloscope for use where conventional power is not readily available led to the design of this battery-operated transistorized oscilloscope.

With the exception of the charger circuit, power sources and the deflection-blanked cathode-ray tube, the block diagram shown in Fig. 1 follows well-established oscilloscope practices.

Vertical Amplifier

A conventional compensated input attenuator is used as shown in Fig. 2. The input amplifier is a vacuum tube because it has high input impedance, requires no temperature compensation, has negligible grid current, is economical and has good bandwidth. Regulated d-c is used for the filament supply to insure low drift.

A swing of better than 30 v is required at each cathode-ray tube deflection plate. Since drift transistors have good BV_{ce} (breakdown voltage collector to emitter), drift transistors tested to 50 v are used. The limitation is power.

To achieve the desired bandwidth, more power was required in the output stage. Checking the thermal resistance showed that with a proper heat sink, 50 percent additional power dissipation could be attained. This permits operating near the manufacturer's maximum ratings with a good margin of safety. A life-test initiated with 50 units being operated near the limit of power rating at 45 v collector to emitter revealed nearly 2,000 hours per unit with no failure.

Saturation of the output transistors had to be avoided as their recovery time is slow. One interesting effect is the offset or thermal time constant. A large low-frequency input signal causes internal heating of the transistors and an increase in low-frequency gain. The proper choice of RC time

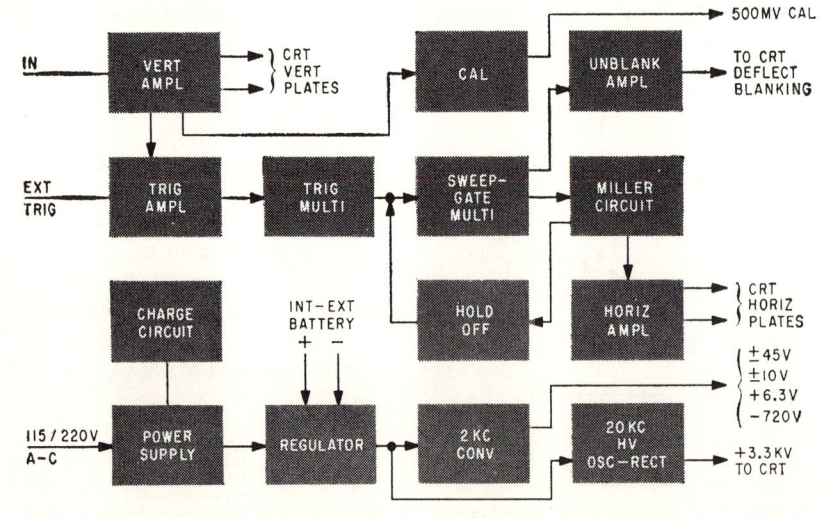

FIG. 1—With exception of power supply and associated circuits, compact oscilloscope follows established design practice

216

constant compensation increases the a-c gain enough to correct for this effect. A balanced circuit is used for temperature compensation. The bandwidth of sample amplifiers varies from 5.3 to 7 Mc.

Trigger takeoff Q_1 (Fig. 2) is an amplifier that reproduces a sample of the vertical signal for use by the sweep trigger circuit.

Calibrator Q_2 is an overdriven amplifier that gets an a-c input signal from the power supply. Its output calibrating signal is a square wave whose amplitude is determined by ground on one side and diode clipping on the other. The 2-kc output square wave has approximately a 1 μsec rise and fall time that is suitable for probe compensation and amplifier gain checking.

Trigger

The trigger input amplifier Q_1 and Q_2, shown in Fig. 3, is an emitter-coupled amplifier. Trigger multivibrator Q_3 and Q_4 is a conventional Schmitt circuit with one exception. The multivibrator is normally free running at about 50 cps to produce a trace on the cathode-ray tube. By adding R_1, R_2 and C_1, the multivibrator can be triggered to produce a stable presentation at 2 Mc and can be synchronized up to 4 Mc.

Gating Multivibrator

Sweep-gating multivibrator Q_1 and Q_2, shown in Fig. 4, controls the starting and termination of the sweep. It is a Schmitt multivibrator having large hysteresis.

The trigger and hold-off signals are mixed at the input to the sweep-gating multivibrator. A positive-going trigger signal starts the multivibrator to produce the sweep. When the sweep voltage reaches approximately 20 v, hold-off circuit Q_7 couples back a negative-going signal causing the sweep-gating multivibrator to revert to its normal state. This stops the sweep and causes it to retrace.

A portion of the sweep-gating multivibrator output signal is applied to unblanking amplifier Q_8 and Q_9 to turn on the cathode-ray tube during the sweep.

Sweep generator Q_3, Q_4, Q_5 and Q_6 (Fig. 4) is essentially a Miller cir-

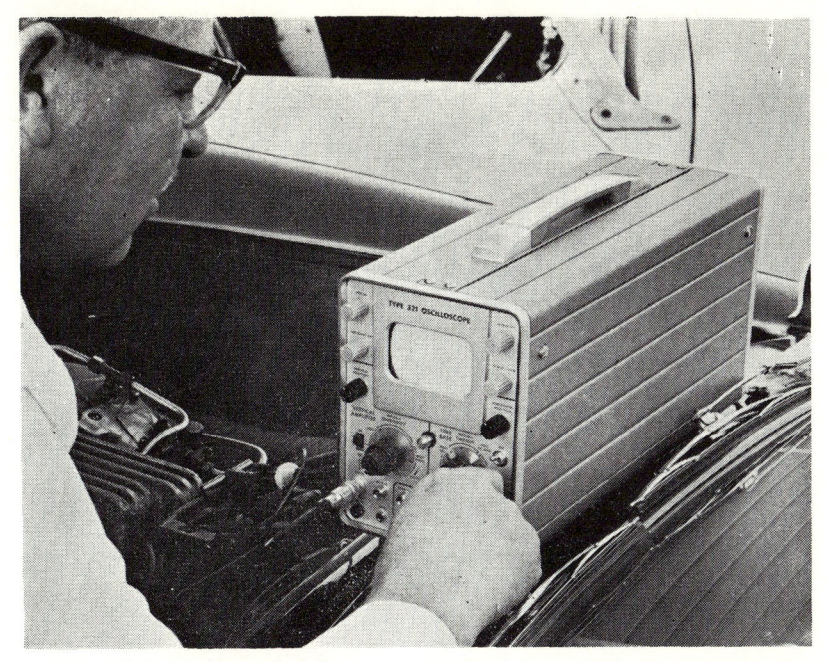

Compact self-powered oscilloscope can be used in areas remote from conventional power

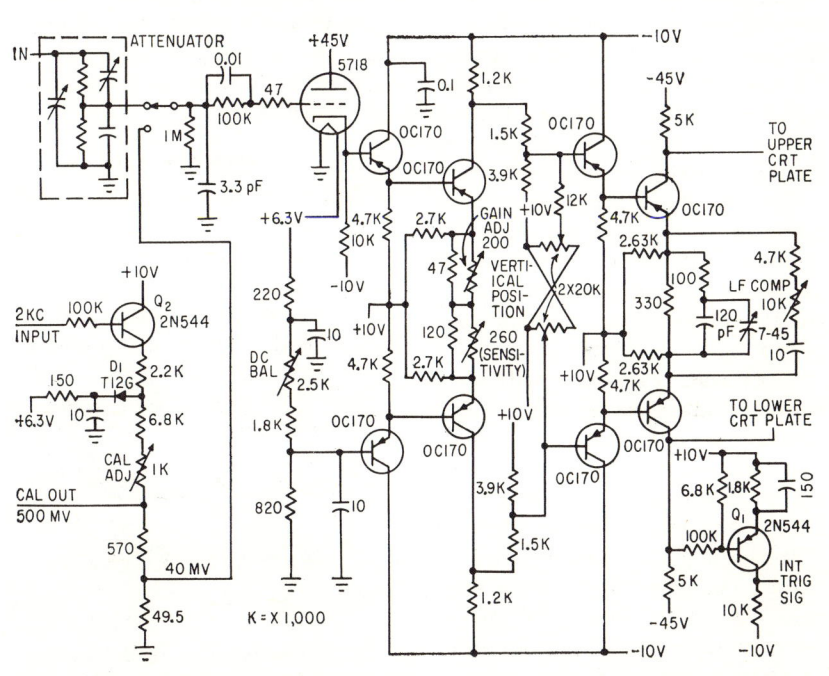

FIG. 2—Calibrator circuit generates 40-mv square wave at 2 Kc. Internal trigger is derived from vertical amplifier

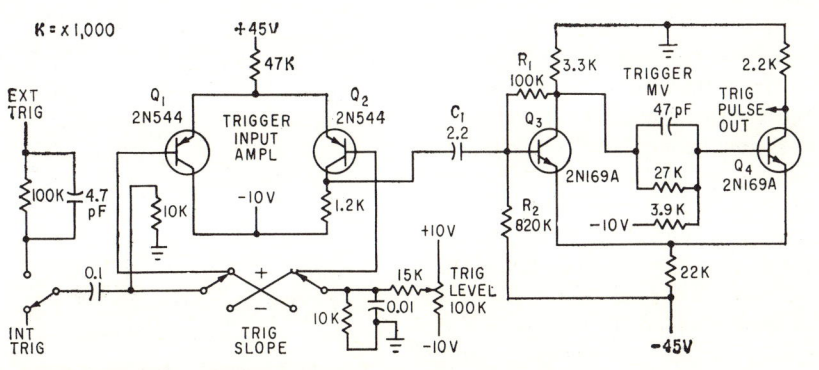

FIG. 3—Trigger multivibration can be synchronized up to 4 Mc

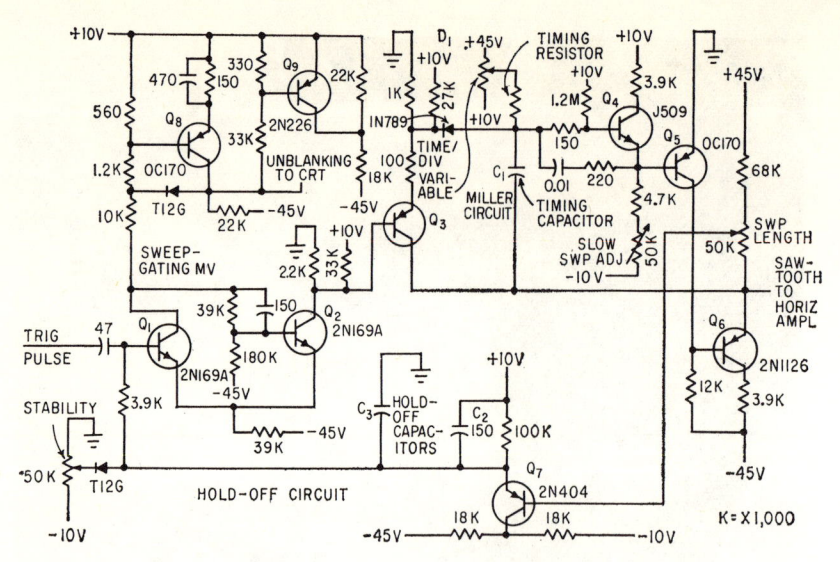

FIG. 4—Unblanking amplifier supplies signal to crt deflection blanking plates. Hold-off circuit insures trace starting from same point every sweep

cuit. Timing capacitor C_1 is initially discharged. When the sweep commences, the capacitor applies a positive voltage to the input of emitter follower Q_4. The change is amplified by sweep amplifier Q_5. Emitter follower Q_6 pulls the timing capacitor in a negative direction.

The action continues with the net result of a 20-v sweep signal at emitter follower Q_6. With a gain of about 400, the signal non-linearity at the input to Q_4 is about 50 mv.

A silicon transistor with high beta and low leakage is required in the sweep generator as any variation in leakage will cause a timing error. To reduce this effect, the timing capacitors are much larger than those used in vacuum-tube circuits. The charging currents are made larger so that accuracy can be maintained over a wide variation of ambient temperature.

The retrace is accomplished by applying a negative signal from the sweep-gating multivibrator to Q_3 which will discharge the timing capacitor through diode D_1.

Since the diode is connected to the timing elements, it must have very low leakage and since the fast sweeps are to be linear, the diode must have a fast recovery time.

It is essential that the trace always start from the same place on the cathode-ray tube screen. This is accomplished by the hold-off circuit. During the sweep, a charge is built up on hold-off capacitors C_2 and C_3. This charge is used to block the input to the sweep-gating multivibrator until the timing capacitor has fully discharged. The hold-off time is 20 to 30 percent of the sweep time.

Unblanking

As the cathode-ray tube grid is operating at approximately −670 v, it is difficult to obtain the required grid blanking signal. To get around this, deflection blanking is used. Additional deflection plates are

FIG. 5—Electrostatic crt has deflection plate use reversed (horizontal plates nearer gun) and has added deflection plate for beam blanking

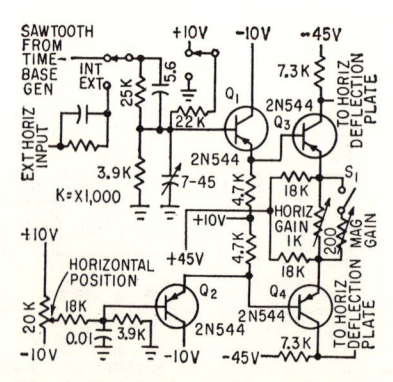

FIG. 6—Horizontal amplifier includes 5 times magnification

added to the electron gun of the cathode-ray tube as shown in Fig. 5.

The electron beam is deflected out of alignment with the exit aperture by less than 25 v applied to the deflection blanking plates. Power supply regulation is simplified as this method of blanking does not alter the cathode current.

The unblanking waveform is derived from overdriven amplifier Q_8 and Q_9 of Fig. 4. This yields a fast rise and fall time to turn the trace ON and OFF rapidly. Deflection blanking has the advantage that the electron gun control grid is available for Z-axis modulation. The control grid has a higher impedance than the cathode that is usually used.

The cathode-ray tube uses a 2-watt heater in place of the conventional 4-watt filament. A new cathode-ray tube is being developed with a 0.68-watt heater.

Due to the limited voltage swing that can be obtained from small high-frequency transistors, a limit of 30 v per deflection plate is a good compromise. The sweep voltage would then have to come from tubes or stacked transistors. Stacking transistors will work but component tolerance and cost is high.

To overcome this problem, the functions of the deflection plates are interchanged. The usual horizontal plates were brought closer together and used for the vertical sweep. This increased the sensitivity to where they could be driven by transistors.

The oscilloscope uses 39 transistors, 3 vacuum tubes (including the cathode-ray tube); nearly half the total power consumption of 9.2 w is used to heat the 3 vacuum tubes.

Horizontal Amplifier

A balanced circuit as shown in Fig. 6 is used for temperature compensation. The base-to-emitter 2 mv/degree C bias change will cancel as in the vertical amplifier.

Emitter followers Q_1 and Q_2 drive emitter-coupled amplifiers Q_3 and Q_4 respectively. Output stages are pretested on a curve tracer. Operation of switch S_1 permits 5 times magnification. The sensitivity of the horizontal amplifier is 1.5 v per division with approximately 1 Mc bandwidth.

The nickel-cadmium batteries used have a very shallow discharge slope, making it difficult to determine the state of charge of the batteries. Constant voltage charging is elaborate and expensive while constant current charging requires timers or supervision.

To make the charger trouble free and eliminate human errors, the circuit shown in Fig. 7 is used.

The charging circuit consists of thermistor R_1 as part of a bridge circuit that senses the battery temperature as the battery is charging. When the battery approaches full charge, more power goes into heat and less into storage. As the battery temperature rises, the bridge unbalances and turns off the charger. Since the instrument must operate in a variety of ambient temperatures, thermistor R_2 is added so that when battery temperature rises approximately 20 F above ambient, the charger will turn off.

Battery

Since the nominal current drain is 0.8 amperes, the internal battery will operate the instrument for approximately 5 hours. A 12-v car battery can be connected externally should longer continuous operation be desired.

Regulator

The regulator permits operation from 11.5 to 35-v d-c external sources, and maintains a continuous 10-v output. Zener diode D_1 is used as a reference.

At the higher d-c input voltages, series regulator transistor Q_1 has to dissipate more power. The thermal resistance should not exceed 4 degrees C per watt, as this amount would raise the internal temperature; therefore transistor Q_1 is mounted on a heat sink on the rear external wall of the instrument. The 10-v output of the regulator supplies the other power supplies with primary power.

With the exception of the cathode-ray tube voltages, all instrument voltages are derived from a two-transistor d-c to d-c converter operating at 2 Kc. The frequency is a compromise between filtering requirements and efficiency. Separate rectifiers and filters are used for the horizontal and vertical circuits to reduce crosstalk to a minimum.

The 3.3 Kv required for the cathode-ray tube post accelerating anode is derived from a separate 20-Kc high-voltage oscillator supply as shown in Fig. 8. The high-voltage rectifier is a vacuum tube; semiconductor high-voltage rectifiers are not commercially practical as their leakage current is equal to the normal load current required in this application.

The diodes located in the various power input leads to the regulator (Fig. 7) perform the switching of the power sources; the highest source of supply voltage is the one automatically selected.

Nickel-cadmium batteries with 4.3 ampere-hour capacity are used.

Since the nominal current drain of the instrument is 0.8 ampere, the internal battery will operate the instrument for approximately 5 hours. In normal use, 5 operating hours should enable the operator to complete an 8-hour day with some reserve power if the instrument is turned off when not in use.

Table I compares the specifications of this oscilloscope with those of a similar vacuum-tube version made by the same company.

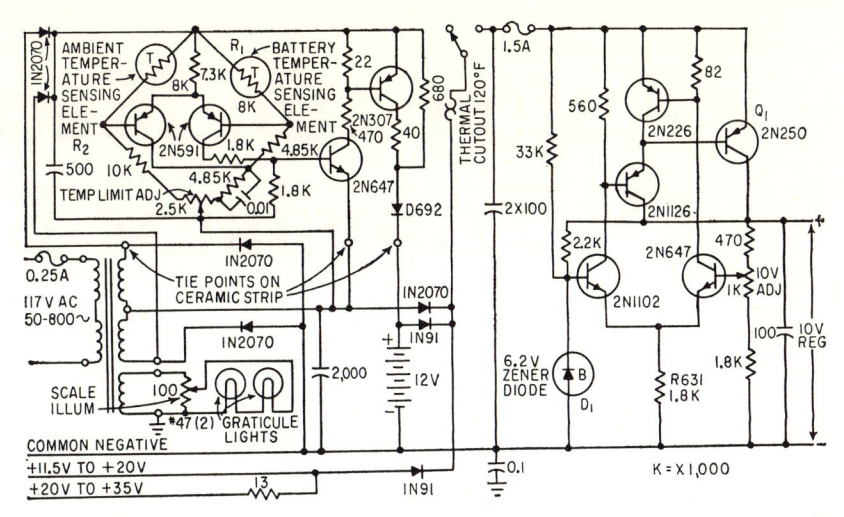

FIG. 7—Regulator circuit maintains constant 10-v output from internal 12 v, from external d-c voltages up to 35 v or from 117 v a-c line

Table 1—Instrument comparison

	Vacuum Tube	Transistor
Power Consumption (w)	175	12
Weight (lbs)	23½	13½
Bandwidth	d-c to 4 mc	d-c to 5 mc
Sensitivity	0.1 v/div	0.01 v/div
Size	10 × 6¾ × 17	8¾ × 5¾ × 16
Tube or Transistor Complement	30 tubes + crt	39 transistors, 2 tubes + crt
Power Source	105–125 v, 50–800 cps	105–125 v, 50–800 cps, 11.5–35 v ext. battery, 12 v internal battery
Reliability	—	better
Ruggedness	—	better

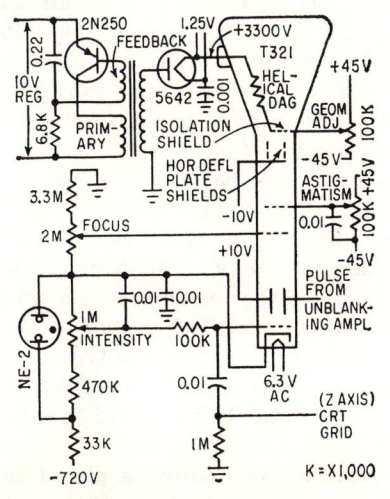

FIG. 8—High voltage circuit for crt uses vacuum-tube rectifier

Automatic Measurement Of Transistor Beta (h_{fe})

Transistorized checker sets and holds collector current to a predetermined value while base current is measured and beta determined. Accuracy and speed of testing are increased with consequent cost savings

By **E. P. HOJAK,** International Business Machines Corp., Federal Systems Division, Owego, New York

DIRECT CURRENT gain, or beta, is a reliable measure of the usefulness of most transistors. For production lot transistors, beta (also called h_{fe}) has wide variations and it is usual practice to test all units of a given batch.

But the circuits for measuring beta usually require a number of manual adjustments. Since beta varies widely from transistor to transistor, and since it also varies significantly with the magnitude of the collector current, it is typical to make the tests at a specified collector current. Typically, the base current is increased until the desired collector current is obtained. Beta is then the ratio I_c/I_b, where I_c is collector current and I_b is base current.

When a large volume of transistors must be tested, the manual operations are time consuming and subject to error by the operator. A circuit has therefore been developed which automatically adjusts the base current to give a preset collector current. The collector current is held constant while the reading of the base current is being made. Base current is thus a direct measure of beta and the output meter is calibrated in terms of this parameter.

Circuit Operation

Assume that beta, (h_{fe}), of a group of *npn* transistors is to be measured. Since beta varies with collector current, all measurements are to be made at a collector current of 10 ma and a collector voltage of 2 volts. The test circuit is shown in Fig. 1 where the supply voltage V_{cc} is set at 12 volts. With 10 ma flowing through R_1, producing a 10 volt drop, a 2-volt drop, V_{ce}, is available across the test transistor. The reference voltage is set to the desired value of V_{ce}, in this case to 2 volts. The battery voltages remain constant (1.5 and 22.5 volts as shown) for any bias values within the limits of the tester. The circuits of the tester are such that Q_1 and Q_2 must be type *pnp* if the transistor being tested is type *npn*. All polarities must be reversed if the transistor being tested is type *pnp*.

The test transistor is placed in the circuit and then the switch (not shown), that connects the circuit voltages is turned on. The operating voltage for relay K_1 is also applied with this same switch. But capacitor C_3 across the coil of K_1 acts to delay the closing of the relay. Thus the 22.5 volts available at the base of the transistor under test is shunted through C_2. The circuit protects the test transistor from a high current surge into the base.

The collector of Q_x, point A on the drawing, is momentarily at 12 volts when the bias is initially applied. Point B, the collector of Q_1, and the base of Q_2, is initially at -1.5 volts. Maximum current thus flows in the base circuit of Q_2 and produces maximum collector current in this transistor. The collector circuit of Q_2 includes the 22.5-volt battery, 10-k resistor R_2 and the base circuit of Q_x. Since the relay is not energized during the first part of the sequence, Q_2 collector current is shunted by C_2. As C_2 begins to charge, base current starts to flow in Q_x. With current flowing through Q_x and R_1, point A begins to drop from its initial value of 12 volts.

Relay K_1 now operates, removing C_2 from the base circuit, and test transistor Q_x can go into saturation. The collector current of Q_x will attempt to go beyond the desired current of 10 ma. At just 10 ma, point A is at 2 volts, with 10 volts being dropped across R_1. As the collector current tries to go above 10 ma, point A drops below 2 volts and the base of transistor Q_1 becomes more negative than its emitter,

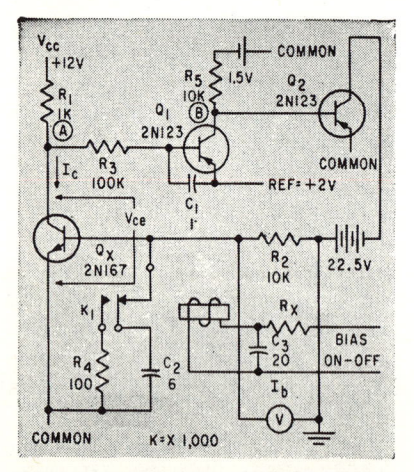

FIG. 1—Automatic beta checker holds collector current of test transistor Q_x at a preset value. Points marked common indicate bus bar connections

Automatic transistor measuring system has console, digital voltmeter and punch card readout. Beta is just one of the transistor parameters that the system can measure

which is at a +2-volt reference.

Because Q_1 is a *pnp* transistor, base and collector current will flow from these bias conditions and the voltage drop across R_5 will increase. The initial voltage drop across R_5 was small and was caused by the small base current of Q_2. Since point B is now less negative, the base current of Q_2 is reduced. Collector current of Q_2 and thus base current of Q_x are therefore reduced. Point A now becomes more positive again because of the decreasing collector current in the test transistor. As point A attempts to go above 2 volts, transistor Q_1 will cut off. The turning-off and turning-on action is filtered by R_3 and C_1, thereby presenting a smooth d-c bias to the circuit.

The actuation of K_1 removes C_2 and shunts it with a discharge resistor so the circuit will be ready for the next transistor. With the collector current of Q_x held at the desired value, the base current is measured and beta is determined. Base current is determined from the voltage drop across R_2.

Adjustments

Bias adjustments should be completed with a test transistor in the circuit because when transistor Q_1 is turned on it will have about 0.1-volt drop between emitter and base. The reference voltage, therefore, will always be about 0.1 volt more than the desired collector voltage. Supply voltage V_{cc} and the reference voltages may be adjusted while monitoring point A for collector voltage and voltage drop across R_1 for collector current.

For testing transistors at a low collector current (about 1 ma), the beta of Q_1 should be at least 60. Base current of Q_1 will then be negligible compared to the collector current of the test transistor.

The discussion has assumed that an *npn* transistor was being tested, with *pnp* units in the test circuits. For testing *pnp* elements, the test circuits use *npn* units and all voltage polarities are reversed. In the production model of the tester, these polarity changes are made with a switch.

For minimum noise pick-up at the readout equipment, a shielded cable is used as a return. Because this requirement means that one side of base-current-measuring resistor R_2 must be grounded, all other points in the circuit are left ungrounded. A bus bar wire is used to connect the points marked common in the drawing.

Accuracy and Limitations

With the type transistors shown in Fig. 1, h_{fe} values from 6.5 to 125 were measured. These values are well within the usual range for most transistors of this type. Collector currents from 1.0 ma to 100 ma were realized without difficulty. Using one-percent precision resistors for R_1 and R_2, and maintaining one-percent accuracy in the associated readout equipment, over-all system accuracy of the transistor tester is about two percent.

High power transistors can be measured with the same basic technique. Higher power transistors are used for Q_2 and the capacity of the bias and power supply are increased accordingly.

Because of its accuracy and speed of operation in measuring transistor parameters, the automatic beta tester is an improvement over other methods of measuring this parameter. The bias points are automatically established at the same levels each time, thus minimizing error and allowing a greater number of units to be tested in a given period of time.

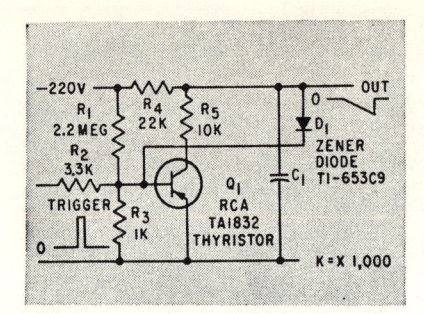

FIG. 1—Sawtooth pulse generator uses a semiconductor switch, Q_1, whose amplitude is controlled by Zener diode D_1

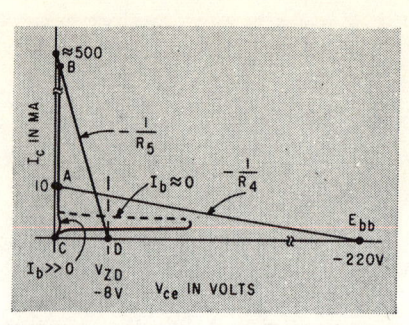

FIG. 2—Curve shows the operating characteristics of the high-speed switching transistor, TA 1832

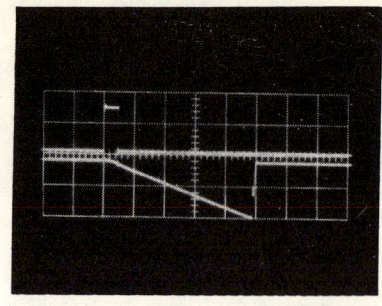

FIG. 3—Typical input trigger and output waveform obtained with the sawtooth generator

Generating Accurate Sawtooth

Designers present simple, reliable circuits that generate stable sawtooth and rectangular pulses. Output pulse widths, amplitudes and waveform timing are independent of the active elements in the circuit

By C. A. VON URFF and R. W. AHRONS,
Radio Corporation of America, RCA Laboratories, David Sarnoff Research Center, Princeton, New Jersey

SAWTOOTH and pulse generators are important building blocks in many electronic systems. In the past, complexity and increased power consumption has been the price paid to obtain desired reliability and stability.

Excellent switching characteristics of a high-speed switching transistor (Thyristor)[1], controlled by a Zener diode, now make it possible to construct two simple circuits that consume little power. Simplicity, low power consumption and small size are some of the desirable features which these circuits provide.

Sawtooth Generator

Linearity and amplitude stability of the generated sawtooth are usually of prime concern in the design of a sawtooth generating circuit. There are many ways of achieving these requirements. The method chosen is to use only a small part of the R-C charging curve and an amplitude controlled switch, shown in Fig. 1.

In this circuit R_1 and C_1 form the basic charging circuit. Transistor Q_1 is used as the switch and the amplitude is controlled by 8-v Zener diode D_1. Figure 2 shows the transistor characteristic curve with load lines and operating points.

At point A the transistor is conducting with current set by R_1 and E_{bb}. While the transistor conducts, the voltage across the capacitor is zero because of the low conducting resistance of the transistor and the small current flow. Therefore, the starting voltage of the sawtooth is always the same, with no voltage jitter. This forms a built-in clamp.

When a positive pulse is applied to the base, the transistor turns off (point A to point C, along the $I_b \gg 0$ curve Fig. 2) and the capacitor starts to charge (point C to D along $I_b \approx 0$ curve Fig. 2). As the voltage across the capacitor reaches the breakdown voltage of the Zener diode, the diode conducts, current flows through the diode into the base of the transistor and the transistor switches to high-conduction.

Since the voltage on C_1 can not change instantaneously, the voltage, V_{ZD}, appears across R_5 and the transistor (point D Fig. 2). Discharge of C_1 takes place through R_5 and the forward resistance of the transistor. This is a short time as $R_5 = 10$ ohms and $R_f = 3$ to 5 ohms. Resistor R_5 limits the peak current and prevents damage to the transistor.

Linearity

The degree of linearity achieved with this circuit depends on the portion of the R-C charging curve used. Here an 8-v diode was used and the supply was -220 v. Using the for-

222

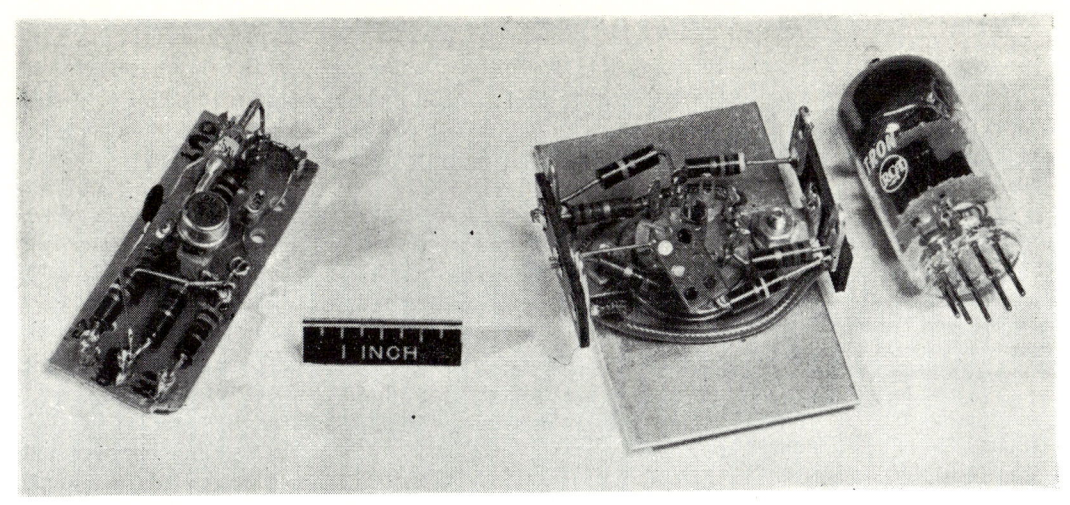

The entire pulse generator circuit, left, is shown alongside its equivalent tube version

and Pulse Waveforms

mulas given in the box, the linearity deviation, $(t_2 - t_1)/RC$ is 0.001. This is acceptable for all but the most stringent requirements.

Sawtooth Amplitude

The amplitude of the sawtooth, determined by the breakdown voltage of the Zener diode, is independent of the transistor characteristics. Thus the transistor characteristics can vary over wide limits without affecting circuit performance. Zener diode characteristics, which are important, are stable with life and temperature. A typical figure is 0.06 percent per deg C. The most important feature needed in the Zener diode is a sharp break at low values of operating current.

If the impedance of the trigger pulse source is too low, the feedback pulse from the Zener diode will be shorted, and the transistor will fail to switch states. Resistor R_2 isolates the base feedback circuit from the trigger generator and R_1 gives a slight forward bias.

The amplitude of the trigger needed to switch the transistor off is directly proportional to the magnitude of collector current at the time of switching. Since the required trigger increases with increasing collector current, the designer operates close to the sustaining current to minimize the required trigger power.

When trying to obtain narrow sawtooth outputs it is important to use a trigger pulse narrow with respect to the output pulse. Otherwise, interaction between the trigger pulse and output waveform results. However, a greater amplitude is required with a narrow trigger pulse than with a wide trigger. The narrowest sawtooth obtained was about 1 to 2 μsec. Typical waveforms obtained with this circuit are shown in Fig. 3.

Pulse Generator

A most desirable feature in pulse generator design is having the output pulse width and amplitude independent of the tubes or transistors. This independence of active elements is achieved in this pulse generator by using the transistor only as a switch, Fig. 4.

The transistor is normally conducting. Hence, voltage V_1, Fig. 4, is approximately zero. When a positive trigger pulse is applied to the base of the transistor, the transistor switches to low-conduction and V_1 rises to the supply voltage. Diode D_1 is back-biased and C_1 starts to charge through R_5 toward the supply voltage.

Capacitor C_1 continues to charge until the voltage across it is equal to the breakdown voltage of Zener

---

LINEARITY DEVIATION

The method chosen to define deviations from linearity first computes the time required to charge to a given voltage, based on exponential charging. The method then computes the time required for linear charging where the linear charge has a slope given by the slope of the exponential at the origin.

Using these definitions: t_1 = linear charge time to charge to V_{ZD}; t_2 = exponential charge time to charge to V_{ZD}; V_{ZD} = breakdown voltage of Zener diode; E_{bb} = supply voltage of circuit; E_{bb}/RC = slope of exponential at time $t = 0$.

And this formula:

$$\frac{t_2 - t_1}{RC} = \left\{ ln\left[\frac{1}{1 - \frac{V_{ZD}}{E_{bb}}}\right] - \frac{V_{ZD}}{E_{bb}} \right\}$$

The degree of linearity is defined by substituting known values

$$\frac{t_2 - t_1}{RC} = \left\{ ln\left[\frac{1}{1 - \frac{8}{220}}\right] - \frac{8}{220} \right\}$$

$$\frac{t_2 - t_1}{RC} = [0.001]$$

223

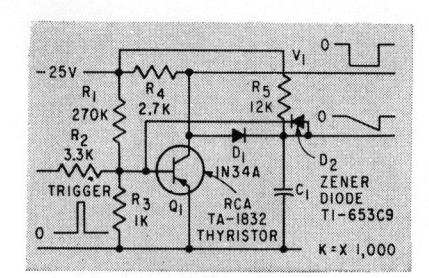

FIG. 4—Rectangular pulse generator operates in almost the same manner as the sawtooth circuit of Fig. 1. Independence of active elements is achieved by using the transistor only as a switch

diode D_2. At this point the Zener diode conducts, sending a negative pulse of current back into the base of the transistor, turning the transistor on. Voltage V_1 now drops to approximately zero. The collector waveform is therefore a rectangular pulse.

If this circuit is to operate at high repetition rates, C_1 must be discharged rapidly. This is done by diode D_1 and the conducting transistor. When the collector voltage, V_1, drops back to zero, the anode of D_1 is at ground potential. Due to the voltage on C_1, the cathode is at a negative potential. Hence, the diode is forward biased and C_1 rapidly discharges through D_1 and the conducting transistor.

As soon as C_1 discharges completely, D_1 stops conducting and the circuit is ready for another operation. This pulse generating circuit has as an ultimate limit in repetition rate—the output pulse width.

Resolution Checks

Double-pulse resolution checks showed that for reliable operation the circuit requires an extra 10 percent of output-pulse-width dead period between triggers. That is, if 10-μsec pulses are being generated, the triggers can be no closer than 11 μsec. This statement is also true for the sawtooth-generating circuit since the pulse and sawtooth circuits operate in almost the same manner. Feedback isolation, furnished by R_2 and R_1, provides a slight forward bias.

The excellent rise and decay times of the output pulse are shown in Fig. 5. The delay between the start of the trigger and the start of the output pulse is of interest. It is due to the fact that the circuit operates well above sustaining current; therefore, there is a delay while the trigger pulse forces the collector current to drop to the point where the unit can switch off. The effect is noticeable also in the output of the sawtooth generator circuit, Fig. 2.

By adding a high impedance audio generator across C_1, the sawtooth and pulse circuits not only generate an output pulse for every trigger but the output pulses have their trailing edges modulated at an audio rate. This results in a simple pulse-length-modulation scheme.

Circuit Stability

Twenty four unselected transistors were checked to see if their characteristic covered the expected production spread.

Tests using these transistors were conducted on both circuits to determine if changing the transistor affects the output pulse width. In this test, the limit transistors produced only a 1.6-percent deviation in pulse width, a tolerable error for most applications. Data taken to determine the effect of varying the trigger pulse width and amplitude showed that a change of 5 to 1 in pulse amplitude or 10 to 1 in width produced less than a 1-percent change in output pulse width.

Temperature Effects

Since semiconductors are used, changes in ambient temperature have an affect on circuit performance. The first test in this area determined the effect of temperature on overall circuit performance over a range of 10 to 80 C. This test evaluates the effect of varia-

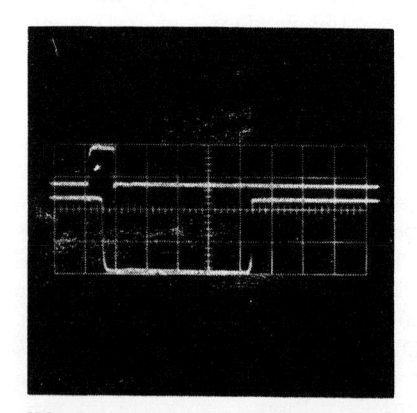

FIG. 5—Input trigger and output wave obtained with the rectangular pulse generator. Note the excellent rise and decay times of the output pulse

tions in back resistance of the transistor and the temperature coefficient of the Zener diode. A variation from 10 to 80 C resulted in only a 1-percent deviation in output pulse width.

Variation in required trigger amplitude with temperature was also checked. Any change here is due to changes in I_{co}. Over 20 to 80 C there is required a 4-to-1 increase in trigger pulse amplitude.

Output waveforms with durations as short as 1 to 2 μsec are possible with these circuits, and in all cases the required trigger power is small.

These circuits should find wide use in any field where small size, low power consumption, small trigger requirements and excellent timing stability are desired.

Thyristor

Operation of the Thyristor is similar to that of the thyratron hence the similarity in names. It is specially constructed so that at some critical value of collector current, called the breakover current, a_{ce} becomes greater than unity.

In the grounded-emitter configuration when a_{ce} becomes greater than one, the collector current increases and the voltage across the unit drops to a very low value. In the Thyristor, this action is regenerative and once the unit is switched, the base current can be discontinued and the thyristor will stay in the high-conduction mode. This feature of the device makes it especially suited to switching service.

The inherent switching time of the Thyristor used in these circuits is 0.1 μsecs when used in the high-conduction, low-conduction method of operation.

In contrast, when the Thyristor is used in the region below the breakover current and kept out of saturation, it is capable of switching times in the order of 20 milli-μsecs. In the high-conduction state the current is usually limited by the external circuit resistance since Thyristor resistance is 3 to 5 ohms.

BIBLIOGRAPHY
(1) Mueller, C. W., Hilibrand J., The Thyristor, A High Speed Switching Transistor, PGED of IRE, Jan. 1958.
(2) Sylvan, T. P., Solid State Thyratrons Available Today ELECTRONICS, Mar. 6, 1959.

Clock Source for Electronic Counters

Small, stable, fixed frequency, 100-kc oscillator is a transistorized version of the Pierce oscillator. When the crystal is replaced with a high-Q inductor-capacitor tuner, it becomes a transistorized Clapp oscillator

By **THOMAS F. MARKER,** Manager, Automated Data Development Dept., Sandia Corp., Albuquerque, New Mexico

APPLICATIONS of transistors to computer and communication equipment today requires small transistorized oscillators with frequency stabilities better than ±1 part in 10^6. Limitations of space and available power often prevent the use of elaborate circuits to control frequency and output voltage from source oscillators.

The circuit shown in Fig. 1 has been used for applications where a small, stable, fixed-frequency oscillator is needed. The circuit consumes 1.5 percent of the power and can be packaged into less than 10 percent of the space required for an electron-tube equivalent.

The circuit is a transistorized version of the Pierce crystal oscillator. When an E cut crystal plate is employed, the circuit constants are suitable for operation at 100 kc. For stable operation at 60 kc, C_1 and C_2 are increased to 0.01 μf. The tuned circuit consists of the mesh which includes the crystal and C_1, C_2. The capacitors produce an impedance transformation which aids in isolating transistor parameter variations from the crystal. The crystal oscillates in the positive reactance mode slightly above its natural series resonance frequency. To achieve maximum frequency stability, the oscillation frequency should approach the series resonant frequency of the crystal. As the two capacitors across the crystal are made larger, improvement in stability is achieved; however, this is at the expense of requiring more current gain from the transistors. When larger current gains are employed by using more favorable transistor types, the effects of transistor characteristic variations

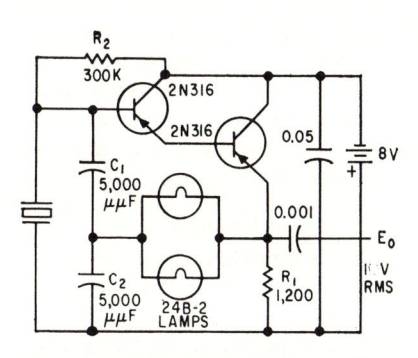

FIG. 1—Stabilized transistor oscillator circuit is used in computer equipment where the temperature is held close to room ambient

become more pronounced so that only moderate improvements in stability are achieved.

Gain Regulation

The circuit uses two 2N316 transistors connected in a composite common emitter configuration. Because of production variations in the current gain of transistors and crystal Q, the gain is automatically adjusted to the value which just sustains oscillation. The lamps, connected between C_1 and C_2 to the output emitter, are used for this function. Two type 24B-2 switchboard lamps connected in parallel between C_1 and C_2 provide sufficient regulation to limit the amplitude of oscillations so that the output is essentially sinusoidal. Lamps are selected so that their cold resistances do not exceed 40 ohms.

Bias current for the transistors is provided by R_2 connected from the collectors to the input base and R_1 connected from output emitter to ground. Since the oscillator is used in computer equipment where the temperature is held close to room ambient, there is no requirement for elaborate temperature-compensating bias circuits.

Oscillator stability varied less than ±1 part in 10^6 over periods of several hours in an environment where the temperature varied not more than ±2 F from ambient, and the 8-volt collector supply was constant to ±0.5 volt.

At room temperature, oscillator frequency closely follows the temperature characteristic of the crystal. With the E plate used, this amounts to a frequency shift of approximately 1 ppm/deg C about the crystal turnover temperature and becomes as large as 4 ppm/deg C at temperatures substantially above and below the turnover temperature. For applications with limited ambient variations, the crystal is enclosed in a small container lined with ¾ in. thick Dylite expanded polystyrene foam.

The circuit has been employed in electronic counters as a clock source.

Zero-Crossing Synchronizes

Sinusoidal wavetrain output starting at the zero crossing of a sine wave is produced by a gating circuit. This wavetrain generator is used to determine characteristics of ultrasonic delay lines and other ultrasonic equipment

By JOHN A. WEREB JR., *Staff Member, MIT, Lincoln Laboratory, Lexington, Mass.

EVALUATION and testing of ultrasonic equipment requires a variable frequency sinusoid wavetrain generator whose output is synchronized with the zero crossing of the sine wave. The generator is used to determine the attenuation and velocity characteristics of sound and ultrasound in wire type delay lines using magnetostriction transducers.

Gated oscillators, usually used in these applications, may be undesirable as the first few cycles of output may be unstable and the output wave may contain a d-c component. The unstable cycles cause trouble when a test requires only a few cycles. The d-c component generates unwanted transients in the system being tested.

Synchronizer

A continuous sine wave is fed into the wavetrain generator or synchronizer and the output shown in Fig. 1 is produced. The output wavetrain starts at the zero crossing of the sine wave. This gives a stable presentation on a cathode-ray oscilloscope and greatly reduces input harmonics.

Gate length t is variable from a fraction of a cycle to many cycles. Gate length is limited only by the relationship $(T-t) = 5t$, where T is the repetition period.

Two synchronizers are used to cover the frequency range from 20 cps to 2.5 mc. One operates from 20 cps to 300 kc while the second covers from 300 kc to 2.5 mc. Two units are required because the large components required at low fre-

* Now with Sanders Assoc., Nashua, N. H.

FIG. 1—Output wave starts at zero crossing of sine wave

quencies do not perform well at the higher frequencies.

Up to 120 kc the gating effectiveness is approximately 50 db and decreases with increasing frequency due to diode capacitance. Cascading of two or more switches gives larger gating values at upper and lower frequencies.

Circuits

A block diagram of the synchronizer is shown in Fig. 2. An oscillator capable of providing 4 mw at 2 v is connected to the electronic switch and the slicer. The slicer is a Schmitt trigger with a level control in front which causes ON action at the zero crossing.

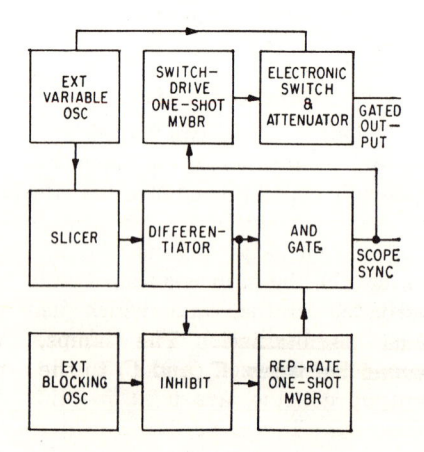

FIG. 2—Inhibit circuit eliminates jitter

The output of the slicer is fed to a differentiator then to an AND circuit. The AND circuit delivers one pulse per cycle. One or more pulses pass through the AND whenever the repetion-rate one-shot multivibrator is on. The first pulse triggers the switch-drive one-shot multivibrator and opens the electronic switch for the time t. This operation repeats at the time T.

By adjusting the period of the repetition-rate one-shot multivibrator to be greater than the period of the sine wave, T will hold to within one cycle of the sine wave.

Jitter

Without the inhibit circuit a wavetrain could, at high frequencies, be out of synchronism with the zero crossing. The number of degrees from zero at which synchronization occurs is called jitter.

The pulse from the AND gate is slightly wider than necessary to trigger the switch-drive one-shot multivibrator. If the repetition-rate one-shot multivibrator is turned on while the pulses from the differentiator are imposed on the AND gate, the rep-rate multivibrator output may be sufficient to turn on the switch-drive one-shot multivibrator. Since only the front end of the pulse is synchronous a slight jitter will occur.

Inhibit Circuit

The simple inhibit circuit eliminates jitter completely. For negative pulses coming from the blocking oscillator shown in Fig. 3 diode D_1 is an open circuit and the rep-rate one-shot multivibrator will trigger. If the positive pulse from

226

Wavetrain Outputs

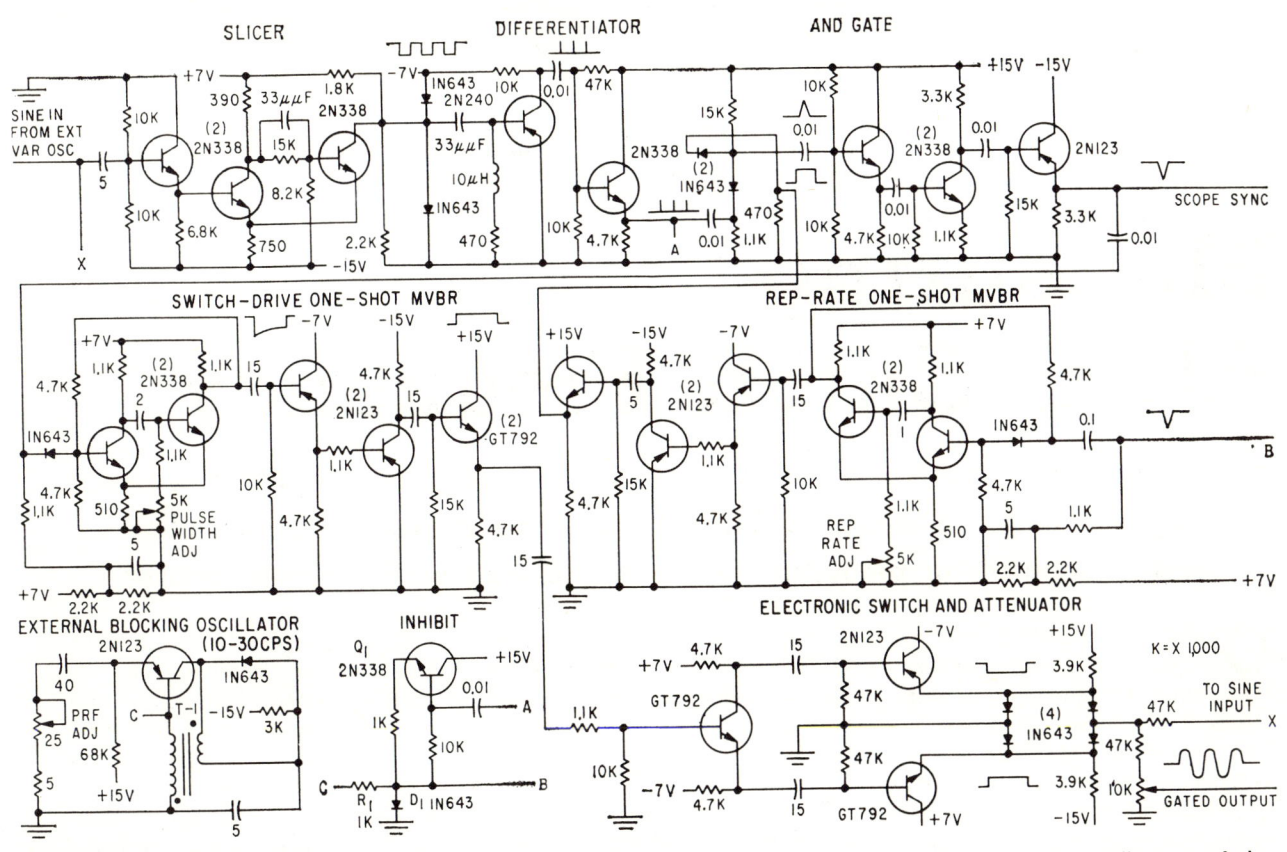

FIG. 3—Detailed circuit of a synchronizer covering the frequency range from 20 cps to 300 kc. External blocking oscillator allows use of alternative repetition-rate generator when required

the differentiator is imposed on the AND gate the pulse turns on the transistor Q_1 in the inhibit circuit, causing D_1 to conduct. Diode D_1 will then be a short circuit to the pulse from the blocking oscillator and the rep-rate one-shot multivibrator will not trigger, eliminating jitter. For this reason pulses of opposite polarity are chosen to drive the rep-rate one-shot multivibrator and the AND circuit.

Resistor R_1 produces an internal resistance in the blocking oscillator so that D_1 looks like a short circuit when conducting.

Germanium transistors type 2N388 can be directly substituted for the silicon transistors used in this instrument if temperature conditions allow.

Application

The synchronizer may be used to determine the characteristics of ultrasonic delay lines. Figure 4A is a block diagram of a system for locating inhomogeneities in a delay line. An artificial flaw is created on the line by clamping it at a point between the receiving transducer and the end, setting up a region of inhomogeneous stress which produces a reflection.

Details of the delay line are shown in Fig. 4B. For the input signal shown in Fig. 4C the output of Fig. 4D is produced. Location and extent of the flaw can be determined by examining the output waveform.

Acknowledgement is made to Vincent Sferrino, of Lincoln Laboratory, for his contribution to parts of the circuit design. The work reported here was supported jointly by the Army, the Navy and the Air Force under contract with the Massachusetts Institute of Technology.

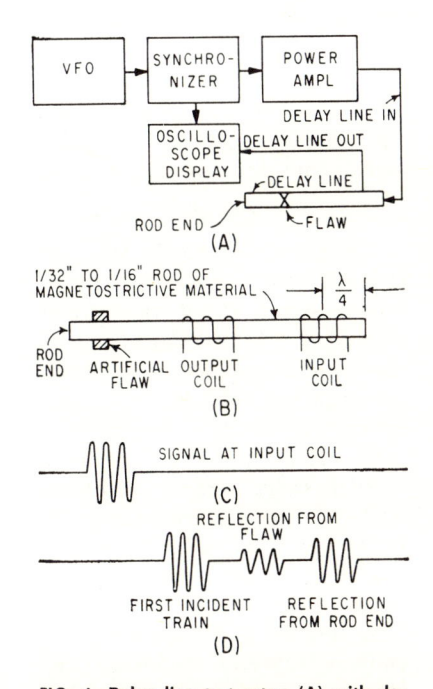

FIG. 4—Delay line test setup (A) with details of delay line (B) and input (C) and output (D) signals. Output wave shows location and extent of flaw

By G. I. TURNER
Semiconductor Measurements Project Leader, Standard Telecommunications Laboratories Ltd., Enfield, United Kingdom

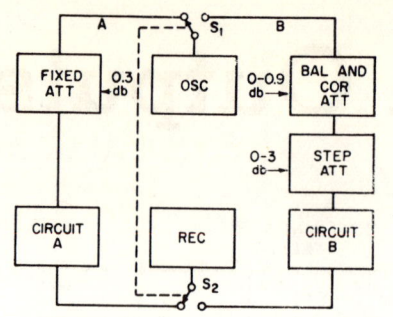

FIG. 1—Signal output of paths A and B must be equalized within 0.1 db in measuring technique described in text

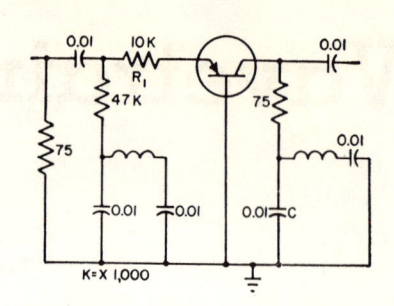

FIG. 2—Capacitor C of transistor circuit to be measured must be increased in value for frequencies below 1 mc

Measuring Alpha-Cutoff Frequency

Alpha-cutoff frequency of high-frequency transistors is measured with accuracy exceeding ± 3 percent up to 30 mc and ± 5 percent up to 100 mc. Method can be adapted for production testing by ganging oscillator and receiver

QUANTITY PRODUCTION of junction transistors whose alpha-cutoff frequencies are in range 5 to 30 mc or higher is now a reality. But the upper frequency limit of currently available high-gain wide-band amplifiers requires a fresh approach to measuring alpha-cutoff frequency f_{aco}.

The method to be described compares the transistor with a short circuit. At the alpha-cutoff frequency, the gain through the transistor path A in Fig. 1 equals the gain through path B.

Circuit

Constant current passes through the 10,000-ohm emitter feed resistor R_1 of the circuit shown in Fig. 2. Variations in the input resistance of the transistor should not affect the input current by more than one percent. The output circuit is a virtual short circuit compared with the output resistance up to at least 50 mc. Capacitors of 0.01 µf show no resonance effects up to 100 mc.

A duplicate circuit with the emitter-collector connections short circuited is in the other path. The signal generator supplies a modulated 0.5-v rms for normal low-power h-f transistors. Ability to switch off agc improves overall accuracy when comparing paths.

Operation

The transistor whose f_{aco} is to be measured is connected in path A and its operating point is set. Signal generator and receiver are tuned to a frequency one twentieth of the minimum estimated f_{aco} where the current gain may be assumed to have its maximum low-frequency value. Balance attenuator in path B is adjusted so that the output signal is the same for each path within 0.1 db. Neither transistor nor receiver must be overloaded.

The step attenuator is set from 0-to-3 db and the frequency where the outputs from each path are identical is found. Initially the procedure is repeated with the oscillator level halved. If the same value for f_{aco} is found, the input signal may be considered small and the transistor considered to be operating linearly.

For maximum accuracy, the transistor, short circuit and attenuators should then be interchanged, the process repeated and resultant mean taken as f_{aco}.

In production testing the oscillator and receiver can be ganged, fitted with click stop mechanisms or a switched crystal-controlled oscillator can be used with the same number of receivers or a special receiver with switched crystal-controlled local oscillator.

Calibration and Accuracy

Differences in the two paths and ability to set the indicating meter to the same reading for each path affect accuracy. Path differences are compensated for by short circuiting the transistor connections for both paths. After an initial balance setting at a low frequency by the balance-attenuator, the frequency can be increased to the required maximum and the differences noted. A correction graph is drawn and a level response held when testing transistors by inserting corrections at proper frequencies with a 0.1-db step attenuator.

By replacing the transistor circuits with transistor amplifiers, insertion gain measurements over a wide frequency range can be arranged. Power gain measurements can be made with allowances for transistor impedance changes.

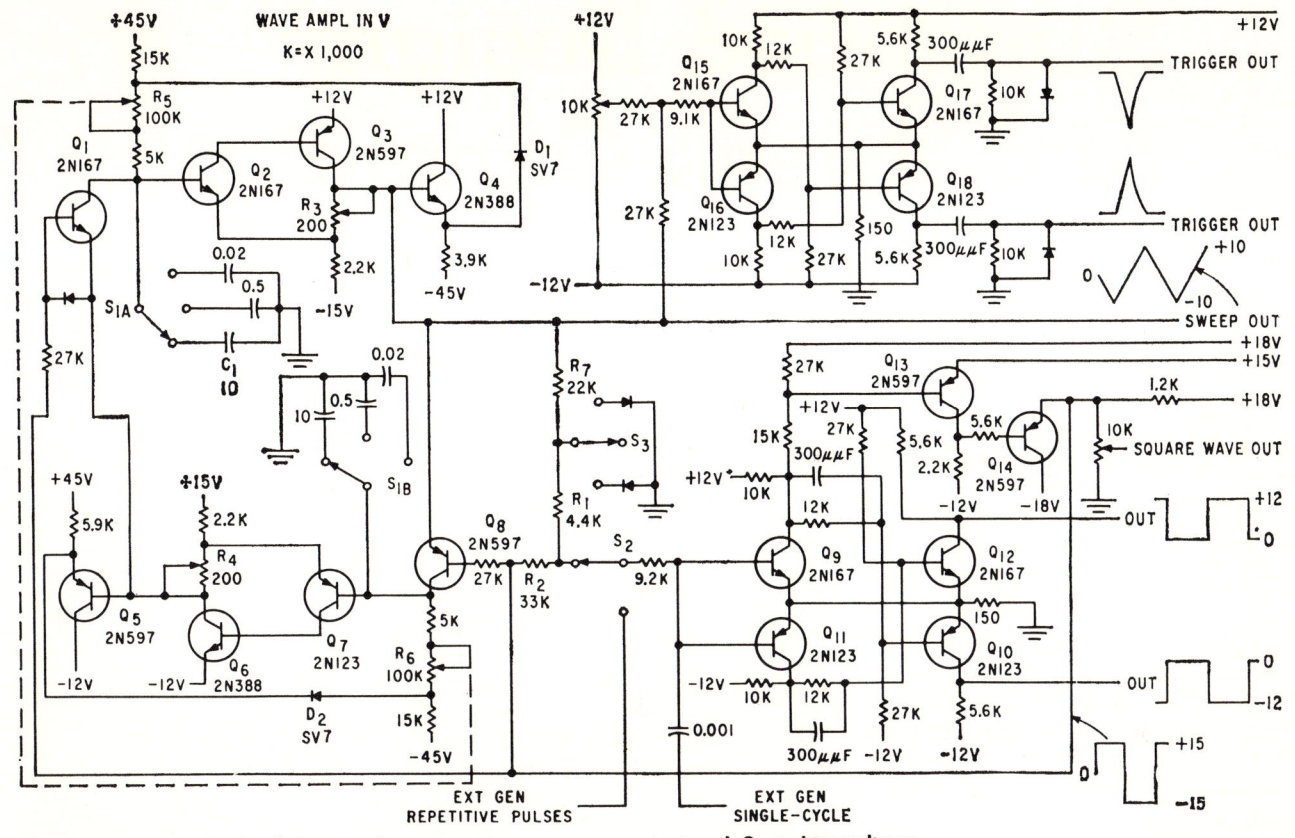

FIG. 1—Resistors R_7, R_1, R_2 of the waveform generator sum sweep-output and Q_{14}-emitter voltages

Multiple-Waveform Generator

Features double-bootstrap sweeps which generate a triangle-shaped wave. A polarity-sensitive trigger circuit controls these sweeps

By **JAMES E. CURRY**, Tasker Instruments Corp., Los Angeles, Cal.

WAVEFORMS GENERATED are triangle, rising and falling sawtooth, and square. In addition, there are two trigger outputs. The generator can be free running or driven by an external generator. The following circuit discussion describes the free-running mode.

Double-bootstrap sweeps Q_1 to Q_4 and Q_5 to Q_8 (Fig. 1) are controlled by the polarity-sensitive-trigger circuit Q_9 to Q_{12}. In turn, the sweep output controls this trigger circuit, switching trigger input transistors Q_9 and Q_{11} when the normally triangle-shaped output goes to its positive and negative apexes ($\pm$ 10 v).

The sweep circuits are complements of one another. Switching transistors Q_1 and Q_8 of the sweeps are controlled by Q_{14}, whose output depends on the state (ON or OFF) of Q_9 (and Q_{11}).

When either Q_1 or Q_8 is saturated, the input of one sweep circuit is tied to the output of the other. When, for example, Q_6 is tied to C_1, capacitor C_1 charges towards a voltage that corresponds to the triggering point. This point is reached when the collector of Q_3 (the sweep output) drops to -10 v.

The actual firing points of Q_9 and Q_{11} are + and -0.15 v, respectively (measured at junction of R_1 and R_2).

Zener diodes D_1 and D_2 simulate constant-voltage sources in the feedback paths of the sweeps.

Potentiometers R_3 and R_4 compensate for voltage drops between inputs and outputs of the bootstraps. Frequency on any range can be modified by a factor of 20 by R_5 and R_6. Switches S_{1a} and S_{1b} vary rising and falling portions of the triangle wave.

When using external repetitive pulses as triggers (with S_2 at *ext gen*), the frequency of the external generator must be greater than that of the waveform generator. When driven by a sine or triangle-wave generator, sweep output is 90 deg out of phase with the input.

When using an external trigger to gate one cycle of the waveform generator, S_3 is thrown to one of its diode positions.

Simplified Design of a Sweep Generator

Thoughtful selection of components takes care of both high and low temperature operation of four-transistor circuit. Precision multiple-range sweep generator is used in an airborne radar system

By H. P. BROCKMAN*, Air Arm Division, Westinghouse Electric Corp., Baltimore, Md.

THIS SIMPLE multiple-range sweep generator, designed for use in an airborne radar system, contains only four active elements and provides a constant amplitude output sawtooth for all ranges, plus a fast rising gate-pulse (or pedestal) equal to the duration of the sawtooth.

How It Works

The circuit diagram, Fig. 1, consists of a conventional flip-flop, Q_1 and Q_2; a transistor switch, Q_3; an R-C timing network; an emitter follower, Q_4; and a diode feedback circuit composed of R_8, R_9 and D_1. Initially, Q_1 is cut off and Q_2 is in saturation, so that switch Q_3 is held in saturation by base current supplied through R_5. A positive trigger on the base of Q_1 drives Q_1 to saturation and opens switch Q_3. At this time, the potential on the collector of Q_3 rises exponentially from zero toward +300 volts at a rate determined by the time constant $R_7 C_6$, assuming switch S_1 is in the position shown.

The sawtooth is coupled through emitter follower Q_4 and a portion is fed back to the base of Q_2 through D_1, so that at some predetermined bias-level the flip-flop is reset with Q_2 in saturation and Q_1 cut-off. Switch Q_3 is driven quickly into saturation by a positive pulse through C_4 to the base of Q_3, thereby providing fast recovery for timing capacitor C_6.

Range switch S_1 provides a range sweep of 15, 50, and 200 miles by selecting a single timing capacitor

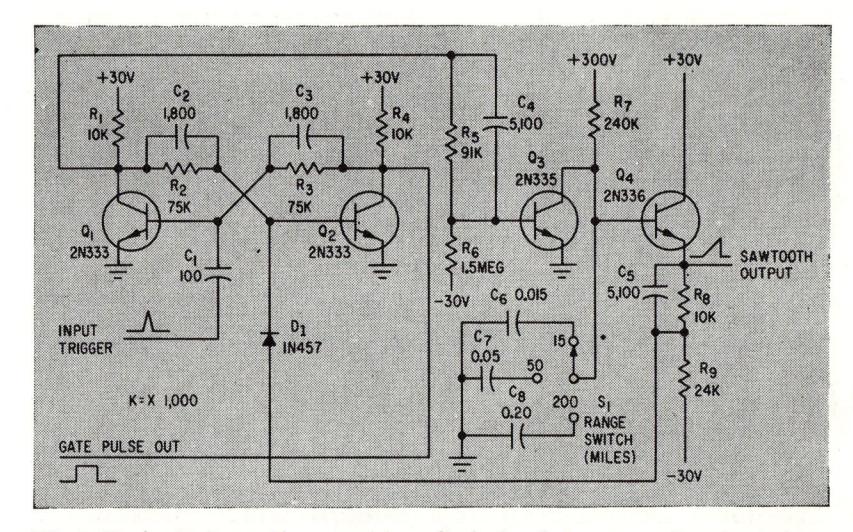

FIG. 1—Simple circuit provides constant amplitude for three ranges. Linearity is kept within one percent without use of a bootstrap circuit

for each range. The output sawtooth is obtained from the emitter of Q_4; a positive gate-pulse is obtained from the collector of Q_2. Precision timing of the sawtooth is obtained without adjustments by using precision parts in the R-C timing network and by making the input impedance to emitter follower Q_4 high compared to the value of R_7, so that the timing of the waveform is independent of the active elements of the circuit. The linearity of the sawtooth is held within one percent without the use of a bootstrap circuit by using a +300-v supply for the charging potential and by restricting the amplitude of the sawtooth to about 15 v.

Constant Amplitude

The amplitude of the sawtooth is determined by the bias level at the junction of R_8 and R_9 and by the energy required to trigger Q_2. Constant amplitude regardless of range selection is maintained by choosing the proper value of C_5. The trigger amplitude required for a slow rising sawtooth is less than for a fast rising sawtooth because the time duration, and thus the energy, of the former is greater. The value of C_5 is such that the feedback is increased for a fast rising sawtooth but it has very little affect on the feedback of a slow rising sawtooth.

The circuit is designed to operate over a temperature range of −55 to +85 degrees C. This required care in component selection.

---

* Now a member of the Electronic Development and Design Group, Missile and Surface Radar Division, Defense Electronic Products, Radio Corporation of America, Moorestown, N. J.

Chapter 12

SOLID-STATE SWITCHING AND CONTROL

Industrial Temperature Controller

By H. SUTCLIFFE Dept. of Electrical Engineering, University of Bristol, Bristol, England

A LOW-POWER transistorized thermostat takes advantage of reverse characteristics of *pnp* junctions to control temperature. Variation of junction conductance in reverse direction with change in temperature permits use of transistor as a temperature-sensing device. The thermostat described is more sensitive than a thermistor and provides continuous control. Additional features include quiet operation, remote resetting of temperature and small thermal time-constant. Principal disadvantage is high-impedance.

The change of reverse current in a germanium transistor junction is particularly spectacular[1]. The type studied was an OC71 *pnp* fused-junction low-power audio transistor. The reverse characteristics of this transistor are similar to those of single junctions[2].

To promote rapid heat flow the emitter and collector leads should be common, as shown in Fig. 1A. The output voltage V is obtained by graphical construction on the I-V characteristic curve. In the equivalent dynamic circuit of Fig 1B the parameters g and R_a are derived as follows:

$$\frac{dV}{d\theta} = -g\frac{R_a R_L}{R_a + R_L} \cong -5 \text{ volts per deg C}$$

Assuming V_B is 85 volts and R_L is 5 megohms the equivalent circuit shows that

where

$$g = \frac{\delta I}{\delta \theta} \ (\cong 1.5\ \mu \text{ per deg C when}$$
$$\theta = 40 \text{ C})$$

$$R_a = \frac{\delta V}{\delta I} \ (\cong 10 \text{ megohm when}$$
$$\theta = 40 \text{ C})$$

The source impedance is several megohms, so before an amplifier is selected for use in conjunction with the device this loading requirement should be considered. One possible choice is a 2D21 tetrode thyratron whose control grid current before firing is a small fraction of a microamp. The thermostat described and shown in Fig. 2 uses a 6AM6 vacuum radio-frequency pentode as a d-c amplifier. Although this amplifier satisfied the loading requirement, care had to be taken in circuit design to avoid unwanted effects due to grid current.[3]

The particular application considered was the continuous control of the cold-junction temperature of a thermocouple, using a 250-v d-c power supply. Ambient temperatures of the enclosure were expected to range from 5 to 30 C. Measurements on a d-c amplifier circuit using a small r-f pentode indicated that 100 to 600 mw of output power can be obtained for a grid-voltage swing of 1.5 volts, with negligible grid current. If the heat leakage from the enclosure is 15 mw per deg C (excess over ambient) and the temperature is to be stabilized at 40 C, then the heating power required can be expected to vary between 150 and 525 mw. If the control loop is stable the change of temperature of the enclosure, from the equation, should be less than 1.5/5 or 0.3 C for the full range of ambient temperature.

When the temperature is too low, which is the condition when the thermostat is placed in operation, the control voltage V rises until it is limited by grid current. The amplifier bottoms and about one watt is applied to the two 100K heating resistors in the enclosures. When the temperature is too high, the amplifier is cut off. In the region of correct temperature, proportional control of heating power exists.

The stability of the control loop is dependent on a number of thermal time-constants which are not easily calculable. A safe principle to follow in feedback circuits is to incorporate a relatively large time-constant and to keep the remaining time lags as small as possible. This principle was applied by mounting the enclosure components in a solid aluminum cylindrical block 1 in. in diameter and ⅝ in. long. The two 100,000-ohm heating resistors, ½ watt carbon type with insulating sleeves, were wedged in holes drilled longitudinally in the cylinder. The transistor was mounted in a similar manner mid-way between the resistors. Its collector and emitter leads were kept as short as possible and soldered to a lug screwed to the cylinder. The cylinder was fixed at the center of a 3 x 3 in. cylindrical can, supported by cork and surrounded by loosely-packed cotton wool. The thermojunction was wedged in a small hole drilled in the block. The four leads to the resistors and transistor were 0.01 in.-diameter resistance wire, used to restrict the loss of heat by conduction.

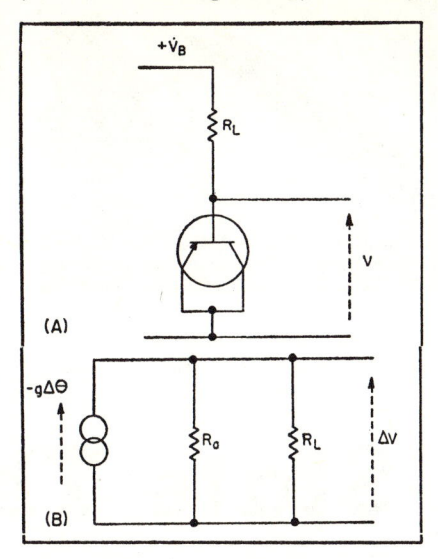

FIG. 1—Circuit for producing a temperature-dependent voltage. Basic circuit configuration to obtain voltage output (A) and equivalent circuit for small current changes (B)

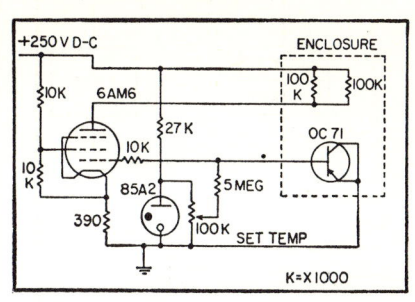

FIG. 2—Circuit of transistorized thermostat. Sensitive temperature controller provides stable and quiet operation

REFERENCES

(1) J. Tellerman, Measuring Transistor Temperature Rise, ELECTRONICS, p 185, April, 1954.
(2) W. Shockley, Transistor Electronics..., *Proc IRE*, p. 1,289, Nov., 1952.
(3) G. E. Valley, and H. Wallman, "Vacuum Tube Amplifiers", p. 418, McGraw-Hill Co., New York, N. Y.

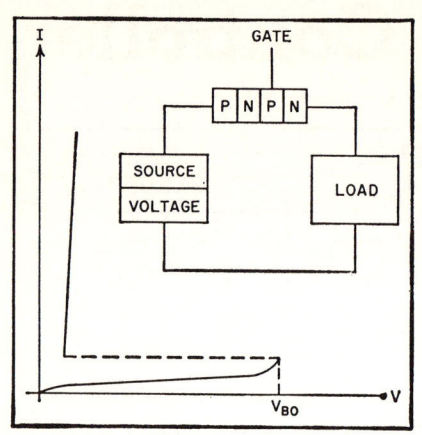

FIG. 1—Forward E-I characteristic of the ZJ-39A silicon controlled rectifier

Silicon controlled rectifier has third or gate lead analogous to grid of thyratron

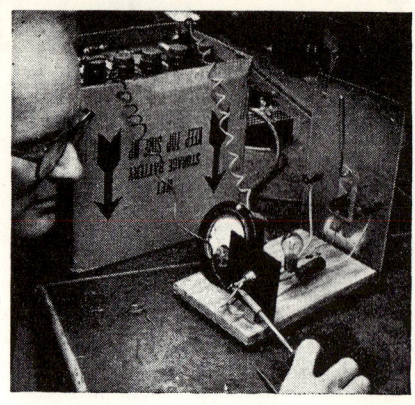

Author soldering connection to gate in demonstration d-c static switching circuit

Solid-State Thyratron

DEVELOPMENT of a solid-state equivalent to the thyratron has been a goal of the semiconductor industry for many years. Such a device needs to switch electronically from an extremely high impedance to a low impedance. And it must do so easily.

The thyratron suffers from low power, slow switching speed, relatively high forward voltage drop, and the usual tube shortcomings. Mercury-arc rectifiers overcome the power disadvantage of thyratrons but retain the high forward voltage drop characteristic.

The silicon controlled rectifier is one answer to the problem. It is neither a transistor nor a rectifier but combines features of each. It opens up new fields of application for semiconductors, some of which are outlined in the latter part of this article.

Switching Mechanism

Design objectives in development of the silicon controlled rectifier were: current ratings comparable to thyratrons, blocking voltages useful in industrial circuits, complete control of current turn-on without complicated circuitry, switching speeds of the same order as small-signal transistors, efficiency equal to similarly rated silicon rectifiers and construction conducive to high-quality mass production at reasonable costs. Result of the developmental program is the ZJ-39A silicon controlled rectifier. Its specifications are available from General Electric.

To understand the rectifier's operation as a two-terminal switch, Fig. 1 is helpful. The rectifier will block current flow in either direction until a critical forward break-

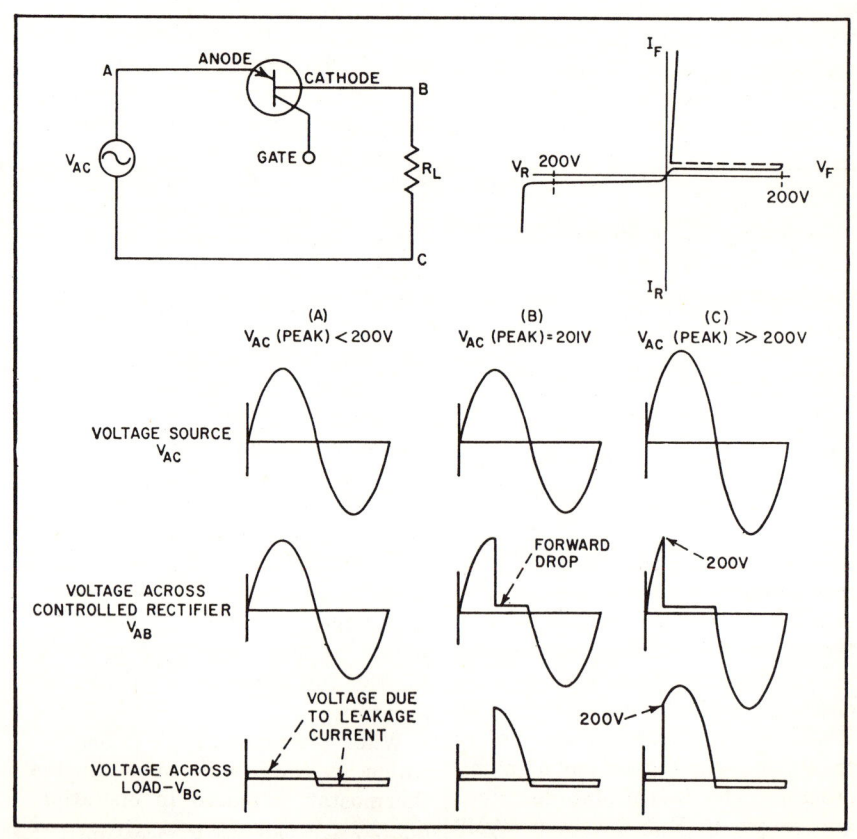

FIG. 2—Simple half-wave circuit showing operation of controlled rectifier without gate signal. Also shown are the voltage-current characteristic and pertinent waveforms. Losses due to forward voltage drop during conduction and forward and reverse leakage during blocking have been exaggerated for clarity

Applications for the silicon controlled rectifier, a recent addition to the growing list of semiconductor switches, include replacement of relays, thyratrons, magnetic amplifiers, power transistors, and conventional rectifiers of all types. Typical circuits presented and discussed are static switches, synchronized inverters, d-c to d-c converters, regulated d-c power supplies, dynamic braking, surge-voltage suppression and power flip-flop

By **R. P. FRENZEL**, Product Planner and **F. W. GUTZWILLER** Application Engineer,
Semiconductor Products Department, General Electric Co., Syracuse and Clyde, New York

Switches Kilowatts

over voltage, V_{BO}, is exceeded. At this voltage, the center pn junction begins to avalanche. Current through the device increases rapidly until the current gain—the sum of the current gains of the overlapping pnp and npn structures—exceeds unity. This current level is relatively low. When reached and exceeded, it effectively reverses the bias of the center pn junction. Voltage across the device then becomes low and the current is limited essentially only by the series load impedance.

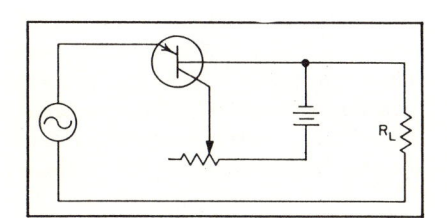

FIG. 3—Half-wave controlled rectifier circuit using simple d-c gate circuit

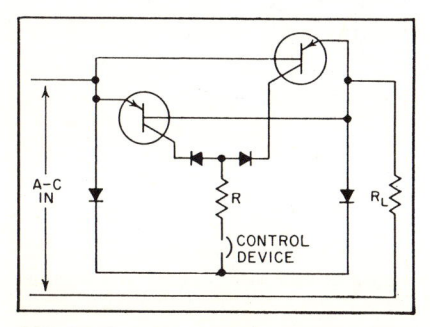

FIG. 4—A-c static switch provides high-speed switching of power loads

A third electrode called a gate, which is an ohmic connection to the center p region of the controlled rectifier, switches it from the nonconducting state without the necessity of exceeding the critical breakover voltage.

The device can be fired by pulses of extremely short duration. With pulse switching, the average power control ratio in 60-cps circuits has been found to exceed twenty-five million to one.

Circuit Operation

Operation of the silicon controlled rectifier in a circuit is understood best by comparing it to that of thyratron—the third electrode of the rectifier is comparable to the grid of the thyratron. As with the thyratron, conduction can be achieved by exceeding some critical anode-to-cathode voltage or by applying power to the grid or gate in the presence of positive anode voltage. In a thyratron, this firing power is generally applied as positive grid-to-cathode voltage. In the rectifier, it is positive gate-to-cathode current since the firing mechanism of the rectifier is dependent on current rather than voltage.

Because the gate-to-cathode voltage-current characteristic of the rectifier is essentially that of a forward-biased semiconductor diode,

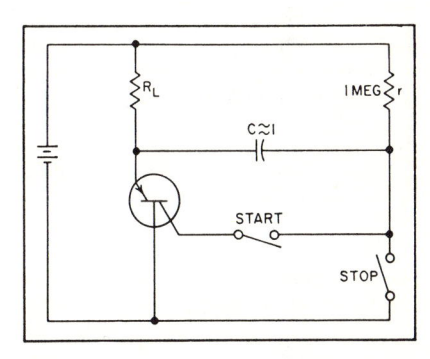

FIG. 5—D-c static switch circuit

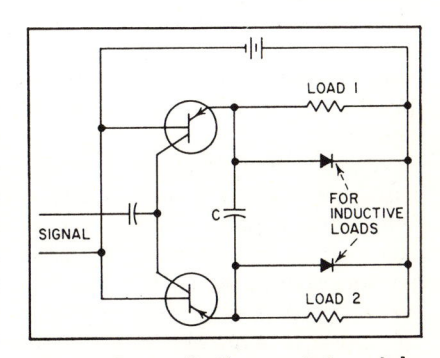

FIG. 6—Power flip-flop or static switching relay switches in one μsec

gate-to-cathode voltage drop for the critical value of firing gate current is about one to two volts.

To illustrate how a controlled rectifier functions, consider it with no applied gate signal and a V_{BO} of 200 v. Assume that the reverse breakdown occurs at a considerably higher voltage. If the peak supply voltage is limited to less than 200

235

A QUICK LOOK AT SEMICONDUCTOR SWITCHES

1. Point-contact transistor — regenerative action with current gains greater than unity.
2. Junction transistor — regenerative if current gains greater than unity are achieved through collector avalanche or charge storage.
3. Field-effect transistor — exhibits negative-resistance region which permits switching from one impedance state to another. (W. Shockley, A Unipolar Field Effect Transistor, Proc. IRE, 40, p 1365, Nov. 1952)
4. Nesistor — improved version of field-effect unit (R. G. Pohl, The Nesistor — A Semiconductor Negative Resistance Device, WESCON paper, Aug. 21 1957)
5. Filamentary transistor action — possible use in flip-flop and counter circuits (W. Shockley, "Electrons and Holes in Semiconductors", D. Van Nostrand Co., Inc., New York, N. Y., 1950, p 81)
6. Unijunction transistor or double-base diode — current multiplication resulting from conductivity modulation (R. F. Shea, "Principles of Transistor Circuits", John Wiley and Sons, Inc., New York, N. Y., 1953, p 467)
7. Semiconductor devices with thyratron-like characteristics (A. W. Berger and R. F. Rutz, A New Transistor with Thyratron-Like Characteristics, AIEE-IRE Electronic Components Conference paper, May 26 1955)
8. Hook collector transistors — three-terminal pnpn devices switched from low-conduction to high-conduction state (W. Shockley, "Electrons and Holes in Semiconductors" D. Van Nostrand Co., Inc., New York, N. Y., 1950, p 112)
9. Four-terminal pnpn hook transistors — feedback to a base electrode necessary to achieve regenerative switching; difficulty in achieving high current and voltage levels (J. J. Ebors, Four Terminal p-n-p-n transistors, Proc. IRE, 40, p 1361, Nov. 1952)
10. Pnpn transistor — regenerative switching achieved through avalanche breakdown of center pn junction (J. L. Moll, M. Tanenbaum, and N. Holonyak, P-N-P-N Transistor Switches, Proc. IRE, 44, p 1174, Sept. 1956; W. Shockley, Unique Properties of the Four-Layer Diode, Electronic Industries, p 53, Aug. 1957)
11. Dynistor — pnp structure with an additional modified junction (A. P. Kruper, The Dynistor Diode, A New Device for Power Control, Machine Tool Electrification Forum paper, Apr. 24, 1957)
12. Silicon controlled rectifier — efficient switching of kilowatts of power at speeds measured in μsec; no feedback circuit necessary; turns on at extremely low power levels; ratios of load power to control power of 100,000 to 1 have been achieved; can be fired by pulses of extremely short duration; average power control ratio in 60-cps circuits exceeds 25,000,000 to 1 with pulse switching.

v, the voltage relationships of the rectifier in a simple a-c half-wave circuit are as shown in Fig. 2A. Here, full a-c forward and reverse half cycles of voltage are across the rectifier. Under these conditions, only the voltage due to leakage current (essentially zero in practical circuits) appears across the load.

Assume next that the peak a-c supply voltage is raised to 201 v. When the supply voltage reaches 200 v on its way toward 201 v, the rectifier will break over in the forward direction, essentially at 90 deg as shown in Fig. 2B. For the remainder of the forward half-cycle, the supply voltage appears across the load except for a one- to two-volt conduction drop across the controlled rectifier.

For supply-voltage peaks considerably higher than 200 v, the rectifier will fire earlier in the cycle, as shown in Fig. 2C. If reverse breakdown voltage is not exceeded, the rectifier will block reverse current flow as does a conventional rectifier. It will also block current flow in the forward direction until the forward breakover voltage is exceeded. Then, it will fire and continue to conduct until the forward voltage is reduced essentially to zero. In reality, a small forward holding current is required to maintain forward conduction.

Consider the rectifier with a signal applied to its gate as shown in Fig. 3. Signal source is usually one of relatively high impedance and several volts strength. The source may be either a-c or d-c. For firing, the gate voltage should normally be positive with respect to the cathode. Gate input impedance of a typical medium-power controlled rectifier is in the 10- to 100-ohm range at the firing point depending upon the firing current required and the specific gate characteristic.

As the gate current is increased, a critical point, I_{GF}, is reached at which the rectifier will break over at any positive anode-to-cathode voltage greater than a few volts. After breakover, impedance of the rectifier is low and supply voltage appears across the load. The gate loses control after breakover and the rectifier can be cut off only by reducing the anode voltage and current to zero. This is analogous to loss of grid control in a thyratron.

The firing scheme shown in Fig. 3 is rudimentary. Firing may be accomplished by as many diverse methods as are used for firing thyratrons.

Circuit Applications

In designing circuits around the controlled rectifier, the engineer should have several factors in mind. (1) As in other semiconductor devices, rated peak inverse voltage and load current should not be exceeded. (2) The controlled-rectifier device should be cooled adequately since satisfactory operation depends largely on maintaining the junction at reasonable temperature levels. (3) To prevent possible damage to the device characteristics, reverse gate current should be limited to low values.

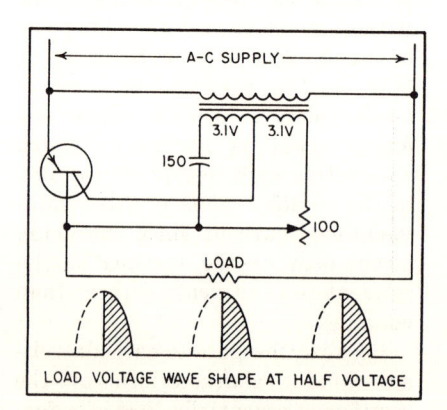

FIG. 7—Half-wave phase-controlled d-c power supply provides uniform output

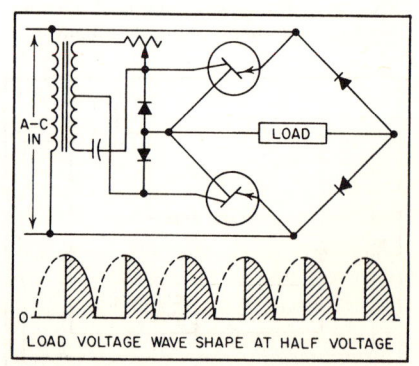

FIG. 8—Full-wave phase-controlled d-c power supply

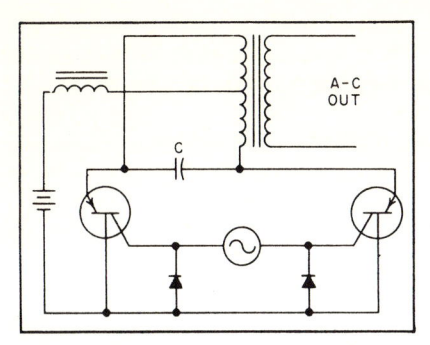

FIG. 9—Synchronized inverter. Feedback makes inverter free running

For this purpose, diodes are used in the gate circuit in several circuit applications. (4) Since the control characteristic varies with junction temperature and anode voltage among individual devices, it is usually desirable to fire the rectifier with a steep wavefront whenever precise timing or phase control is required. Saturable reactor and pulse types of control using such devices as the unijunction transistor are well suited for this purpose. For strictly on-off control, adjustment of gate-current magnitude is satisfactory provided ample excess current is furnished for positive firing.

Examples of well-established and new applications are given in Figs. 4 to 11. Circuit configurations and values are suggestions only and are not intended to imply optimized design. Other applications are limited only by the imagination of the circuit designer.

A-C Static Switch

Figure 4 shows a circuit for providing high-speed switching of power loads. It is ideal for applications with a high duty cycle. Contact bounce and mechanical wear as experienced on relays or contactors are eliminated. The control device can be contacts of a thermostat, pressure switch, current relay, or voltage-sensitive device. Signals from magnetic cores, transistors, or tubes can be used to control sizeable blocks of power in the controlled rectifier. Resistor R limits gate current. Its value depends on the magnitude of supply voltage and the current required for firing the rectifier.

Variations of this circuit can be used in connection with conventional d-c rectifier power supplies to provide both switching and rectification with the same device. With contacts of a sensitive current relay in series with the gate current, such a circuit can interrupt fault currents in as little as one-half cycle.

D-C Static Switch

Figure 5 illustrates one way in which the rectifier can be used to switch d-c loads. To close the switch, the gate circuit is energized momentarily from the main d-c supply through some kind of signal device. This device is represented in Fig. 5 as a start pushbutton. As soon as the start button is released, capacitor C charges to essentially the d-c supply voltage through resistor r.

When the stop button is depressed momentarily, the positive terminal of C is connected to ground. This action impresses a negative voltage across the controlled rectifier for the few μsec necessary to return it to the blocking state.

In Fig. 6, voltage transfers from one lead to the other each time a pulse is fed into the gate circuit. Optimized circuits of this type yield switching times in the order of one μsec. Size of capacitor C depends on the load resistance and the energy stored in the rectifier loop. In circuits using the ZJ-39A, five μf are ample for 12-v circuits operating at one to five amperes.

The circuit shown in Fig. 7 uses a potentiometer control scheme that permits shifting the a-c gate-

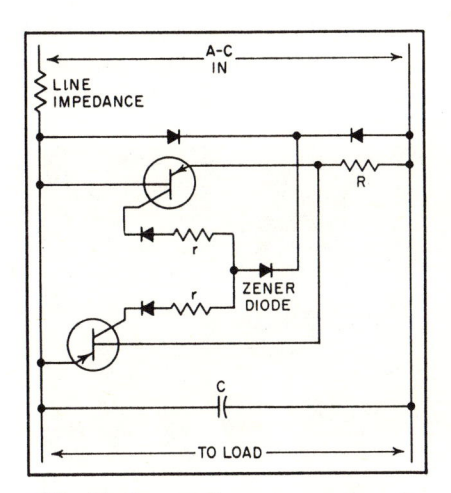

FIG. 10—Surge-voltage suppression circuit. Capacitor prevents overshoot

current signal between 0 and 180 deg with respect to the anode supply voltage. This phase shift regulates the point at which the rectifier fires during each cycle. Average output voltage can be varied uniformly from zero to about 0.45 of the rms supply voltage.

Figure 8 shows a single-phase full-wave bridge using controlled rectifiers in two legs. The rectifiers control the average output voltage from zero to 0.9 of the rms supply voltage. In addition to the R-C circuit shown, excellent phase-shift circuits can be designed around transistors and saturable-core devices. These circuits control output automatically by signal or error currents and voltages. Through this type of gate control, the rectifier lends itself ideally to regulator circuits.

Synchronized Inverter

Figure 9 depicts two rectifiers in a circuit for generating an alternating voltage from a d-c source. The triggering gate signal for this inverter can be a sine wave, a series of pulses, or a square wave. A unijunction transistor in a relaxation-oscillator circuit makes a stable triggering source, assuring reliable starting and constant frequency. Insertion of suitable feedback makes the inverter free-running. By rectifying the output voltage, this circuit makes an efficient d-c to d-c converter.

The type of circuit shown in Fig. 10 is useful in protecting transistor and semiconductor-rectifier circuits from harmful line-voltage surges. When line voltage exceeds a predetermined value, one of the controlled rectifiers fires. It draws enough line current to drop the voltage across the line impedance.

Value of resistor R should be selected to limit anode current to the rating of the rectifier. For surge durations of less than one cycle, R should be selected to limit peak anode current to 150 amp when using this device. Voltage level at which suppression starts is determined by breakdown voltage of the Zener diode. Where temperature variations are not excessive, resistor r can be used instead of the diode to control firing level of the controlled rectifiers.

Here's a time delay relay that is adjustable. It can replace synchronous timers in industrial control circuits. Accurate relay has instant recycling

By **LEON SZMAUZ** and **HAL BAKES**, Engineering Dept., Heinemann Electric Company, Trenton, New Jersey

Time Delay for

AN ADJUSTABLE time delay relay using transistors has been developed for use in various industrial or other processes. Advantages of the relay are timing repeatability and instant recycling. These factors provide for maximum speed in automatic operations and cut production time.

One application of the relay is automatic cycling, where the relay can be used to endurance test other relays and contactors. In other applications the relay can be used to replace synchronous timers, with a consequent saving in space, weight and cost. Delay periods between 1.5 sec and 30 sec are possible. The photograph shows a number of the relays used in an automation setup where a small brass tube goes through the operations of spinning, fluxing, placing solder and brazing. A considerable saving is effected since these operations previously required much handling and personal judgment.

Basic Operation

In a typical application, the user connects 12 volts d-c to the input with an external switch. The circuit is shown in Fig. 1. As soon as switch S closes, transistor Q_1 conducts, charging capacitor C_1 through the normally closed contacts of K_1. Charging rate is controlled by variable resistor R_p. To simplify analysis, other charging resistances will not be considered. Transistor Q_2 is cut off by the action of Q_1 and relay K_1 is held de-energized. When the current in the R_pC_1 network has dropped to approximately 37 percent of its maximum, Q_2 acts as a switch to connect K_1 across the supply. This actuates the relay which remains closed until the supply voltage is broken.

An extra set of contacts on the relay connects R_d across C_1, thus discharging the capacitor preparation for the next cycle. When the 12-volt supply is broken, the circuit is instantly ready for another time

A number of the transistorized time delay relays are used to control a movable table for automatically processing a small brass tube

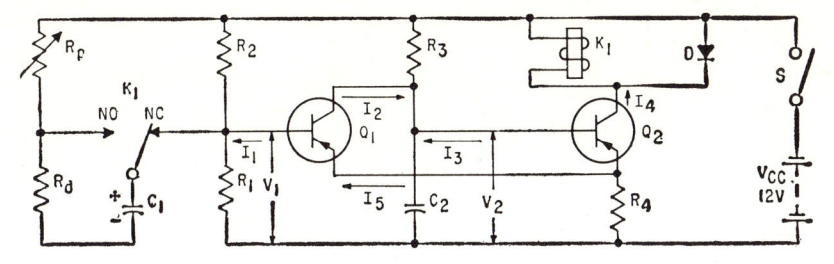

FIG. 1—Time delay begins when S is closed. Charging current of C_1 through R_p and Q_1 controls the circuit. Transistor Q_2 is held cut off until C_1 charges to the time-delay-setting voltage

FIG. 2—Relay current is a function of V_2. Relay turn-on point occurs between V_A and V_B

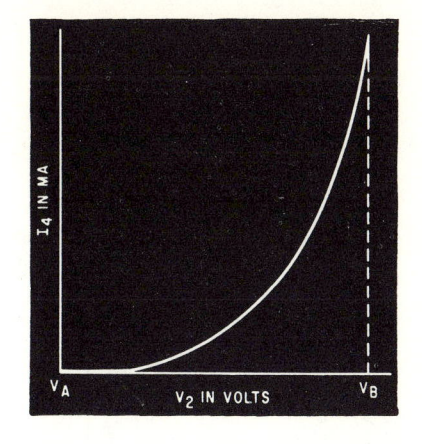

Industrial Control

delay cycle.

The circuit consists of two directly coupled alloy-junction, germanium transistors. Since Q_1 (Fig. 1) serves as a driver and Q_2 as a switch, the collector saturation voltage of Q_1 must be lower than its base to emitter drop. That is, when Q_1 is saturated, voltage V_2 must be sufficiently low that Q_2 is effectively cut off.

Because of direct coupling, the collector voltage of Q_1 is the same as the base voltage of Q_2, as can be seen in Fig. 1. By varying the collector voltage of Q_1, the bias voltage

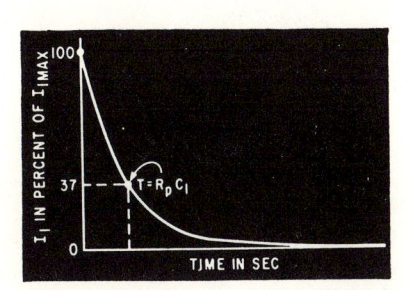

FIG. 3—Charging current of C_1 and thus base current of Q_1 follows typical exponential curve

of Q_2 will vary from V_A to V_B, as shown in Fig. 2. If V_A is the voltage just below or equal to the base cutoff bias, and V_B is the forward bias at saturation, Q_2 will switch the relay ON at some intermediate voltage.

The relay coil and the supply voltage are chosen to give rated relay current $I_4 = V_{cc}/R_L$, where R_L is the d-c resistance of the relay.

Resistor R_3 is determined from saturation conditions. When Q_2 is saturated, maximum current is flowing through the relay. Under these conditions, Q_1 is not conducting since its base current has fallen to zero (except for a small bias current) because C_1 is not in the circuit. The maximum collector current of Q_2 is $I_4 = \beta I_3 = V_{cc}/R_L$. But the maximum value I_3 can have is V_{cc}/R_3. From these equations it can easily be shown that R_3 must be less than βR_L if I_4 is to reach a specified value.

Timing Circuit

The simplified timing circuit consists of R_p in series with C_1. The time delay period begins when external switch S is closed. A charging current begins to flow into C_1 and into the base circuit of Q_1. Transistor Q_1 conducts and holds the base to emitter voltage of Q_2 to a low value. As C_1 charges, the current through Q_1 decreases and the base voltage of Q_2 rises.

Base current of Q_1, ignoring bias current, is given by $I_1 = (V_{cc}/R_p) \epsilon^{-t/R_pC_1}$. The initial base current is $I_{max} = V_{cc}/R_p$. This value of current is much greater than that needed to hold Q_2 at cutoff. Base current I_1 will decrease exponentially as shown in Fig. 3. For fast switching and good timing accuracy, it is desirable that Q_2 turn-on conditions be reached while I_1 is still changing rapidly. This condition is met when the time delay period is approximately equal to the time constant of the network, or $T \cong R_p C_1$.

Because of the variation of parameters of production lot transistors, some experimentation with component values may be required if the best performance of the circuit is to be obtained. The circuit of Fig. 1 is stabilized against minor temperature changes. Thus, for the condition $R_2 \gg R_1 \gg R_4$, the current through R_2 is essentially constant for a given supply voltage. When the emitter current of Q_1 increases because of an increase in temperature, the voltage across R_1 and R_4 increases. The current through R_1 is therefore larger since the voltage drop across it is larger. Since the current through R_2 is constant, base current I_1 must decrease. The decrease in I_1 will then act to decrease I_2 and I_5.

If wide temperature swings are expected, compensation can be obtained with nonlinear elements and temperature sensitive resistors.

A diode is placed across the relay coil to prevent inductive transients which would otherwise appear when switch S is opened. Capacitor C_2 at the base of Q_2 acts as a low pass filter and keeps any sudden changes from being applied to Q_2 and thus accidentally triggering the relay.

Resistor R_d has a low value and is used to discharge C_1 as soon as the relay operates, thus readying the circuit for the next operation.

For appreciable time delays, electrolytic capacitors have to be used. Tantalum electrolytics should be used since they are more stable than the aluminum type.

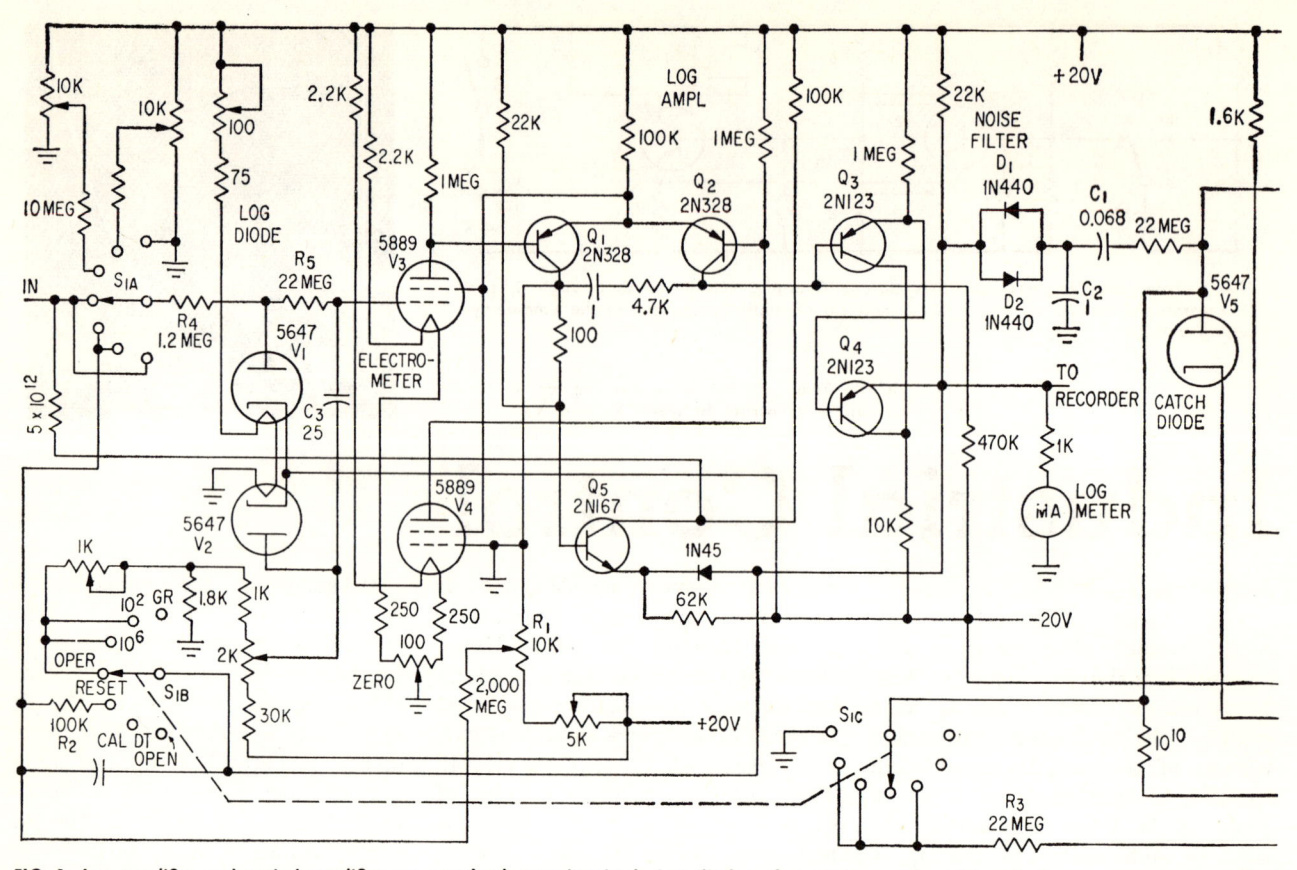

FIG. 1—Log amplifier and period amplifier are completely transistorized. Log diodes, electrometers and catching diode are vacuum tubes. Func-

Amplifiers for Nuclear

Logarithmic and period amplifiers used in nuclear reactor startup ranges are transistorized with exception of log diodes and electrometers. Great saving in size, weight and power consumption is made

LOGARITHMIC AND PERIOD amplifiers are indispensable to the operation of nuclear reactors because of their wide indicating range without switching.

The log diode V_1 shown in Fig. 1 is a nonlinear element whose characteristics are sensitive to changes in cathode temperature.

This effect can be compensated by using two diodes to balance out effects of temperature and power supply variations. When V_1 and V_2 are connected in series back to back, a constant current flows through balance diode V_2 which is large compared with the maximum current to be measured. Input flows through both diodes but the potential change across the balance diode is negligible and is included in calibration.

The log diode can be considered as a variable resistance that may vary greatly over the amplifier operating range.

Log Amplifier

As transistors with sufficiently high input impedance are not available, low-current electrometer tubes are used. Balanced electrometer tubes V_3 and V_4 drive a differential stage consisting of Q_1 and Q_2. The silicon transistors have low I_{co} and minimum temperature effects. A stable operating level is obtained by connecting the electrometers as tetrodes and supplying their screen grids from the common emitters of Q_1 and Q_2. Changes in d-c level are highly degenerated without reduction in signal gain. The current output of the differential stage is amplified by two cascaded emitter followers Q_3 and Q_4.

Calibration of the log amplifier is done in the conventional manner. For period calibration, log diode V_1 is biased nonconducting by potentiometer R_1 and the log amplifier is connected as an integrating amplifier to generate a linear ramp voltage. To discharge the circuit capacitance quickly, switch S_1 is placed in the reset position and resistor R_2 is connected around the

tion switch is used for different modes of operation

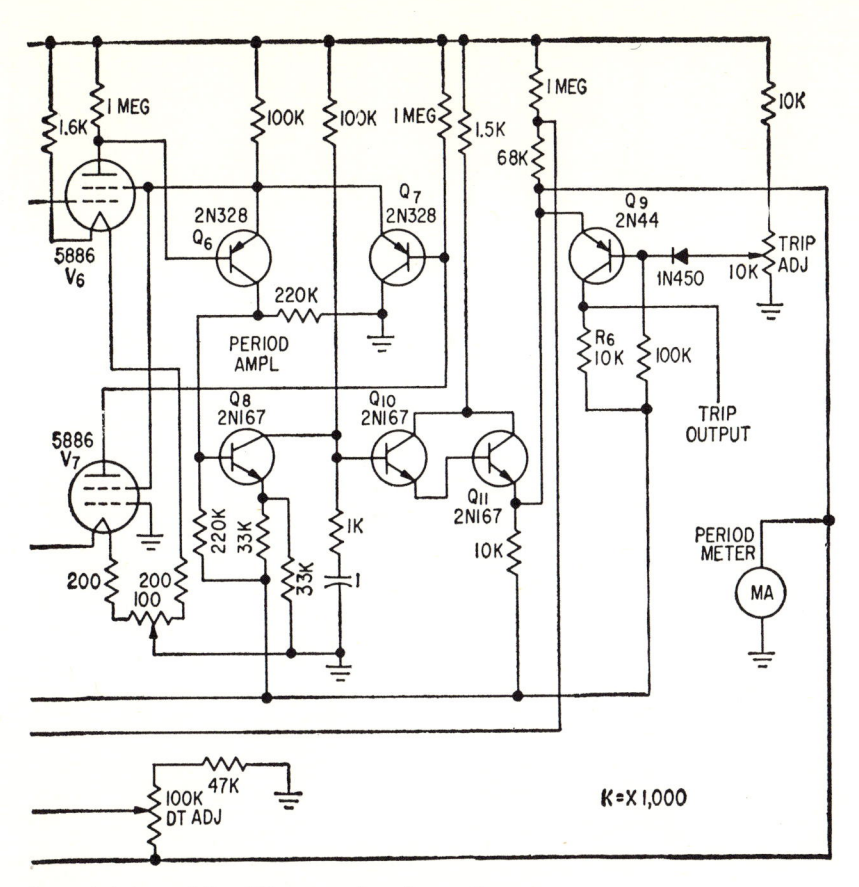

Reactor Control

By E. J. WADE and D. S. DAVIDSON*
Knolls Atomic Power Laboratory, General Electric Co., Schenectady, N. Y.

log amplifier and R_3 is connected around the period amplifier. These resistors discharge the ramp generator in 0.1 second and the differentiating capacitor C_1 in 1 second.

Period Amplifier

This amplifier is a feedback-type differentiating circuit. The factors that determine the input current are the log amplifier output per decade, the period to be measured and the value of capacitor C_1. The output voltage depends upon the capacitor C_1 discharging resistance and the time constant of the amplifier gain.

* Now with Technical Measurements Corporation, New Haven, Conn.

When the amplifier is used to initiate the operation of protective circuits, it is desirable to provide a time delay before generating the trip signal. This delay is a function of the period and is met by choosing the correct time constant of the differentiating circuit. Calibration is made by applying a simulated period from the log amplifier.

To reduce noise, a nonlinear filter consisting of D_1, D_2 and C_2 is placed between the log amplifier output and the period amplifier input. The silicon diodes have high resistance at low voltage and are connected in parallel with reversed polarities. In conjunction with C_2 the filter has long time constant at low voltage decreasing until it is negligible above 0.5 v and has little effect on the tripping time.

Input resistor R_4 is in series with log diode V_1 preventing noise transients from being rectified by the diode. The filter consisting of R_5 and C_3 at the input electrometer grid further reduces the high-frequency gain.

Catching Circuit

When the log amplifier is operating at low current, diode V_1 may either not be in its logarithmic range or cut off due to transient or grid current. When the input current increases under this condition the amplifier output is no longer logarithmic but is linear. This simulates a much shorter period until the log diode reaches its operating range and can cause tripping during reactor startup.

To prevent this tripping, negative feedback is applied from an auxiliary amplifier which operates only when the signal output is slightly reversed. This circuit has negligible effect during normal operation. If the output signal reverses, indicating a diode current less than 10^{-12} amp, the feedback maintains the log diode current at 10^{-12} amp thus preventing the log diode from operating outside its logarithmic range.

The period amplifier uses catching diode V_5 in a low-impedance feedback loop to improve recovery time for input reversals due to switching transients or negative periods.

Trip Output

The trip output circuit consists of transistor Q_9 biased by an adjustable potential applied to its base. When the emitter is driven more positive than the trip setting, the transistor conducts and the output goes positive generating a trip signal. Grounding or opening the trip circuit also causes tripping.

BIBLIOGRAPHY
E. J. Wade, IRE Conv Record, 9, p 79, 1954.
G. F. Wall and M. P. Young, NRL 5025, Sept. 1957.
G. Epprecht (Bern), Tech Mitt P.T.T., p 161, 1951.
W. F. Goodyear, Logarithmic Counting Rate Meter, ELECTRONICS, p 208, July 1951
J. A. DeShong Jr., Logarithmic Amplifier with Fast Response, ELECTRONICS, p 190, March 1954.
E. J. Wade and D. S. Davidson, How Transistor Circuits Protect Atomic Reactors, ELECTRONICS, p 73, July 18, 1958.

Chapter 13
SERVOMECHANISMS

Servo Preamplifiers Using Direct-Coupled Transistors

Two-stage silicon transistor amplifier has adequate d-c stability under conditions of interchanging transistors having a beta range of nearly three to one. Stability over −55 to 125 C range is achieved

By A. N. DESAUTELS*

Senior Development Engineer, Minneapolis-Honeywell Regulator Company, Minneapolis, Minn.

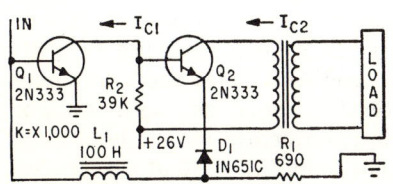

FIG. 1—Direct-coupled amplifier uses Zener diode to provide constant voltage

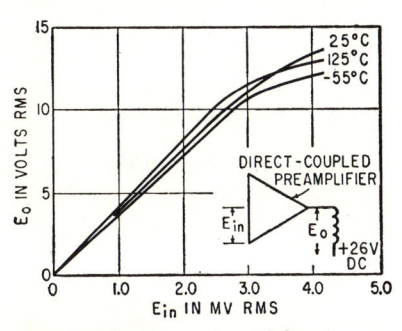

FIG. 2—Adequate gain stability is provided over wide temperature range

VARIATION in characteristics of transistors of the same type has led circuit designers to give only limited consideration to direct coupling when designing preamplifiers for servo applications.

A circuit which emphasizes the desirable characteristics of small size, simplicity and economy associated with direct coupling while still achieving excellent a-c stability is shown in Fig. 1. The circuit uses d-c feedback to achieve d-c stability.

Although it uses no capacitors and has fewer components than conventional stabilized transistor preamplifiers, this preamplifier is d-c stable with interchanging of transistors of the same type. Adequate d-c stability is achieved when interchanging transistors having a beta range of nearly three to one. The preamplifier gives stable gain from −55 C to over +125 C by using silicon transistors.

Zener Diode

The Zener diode, D_1, uses reverse diode characteristics to provide a constant emitter voltage to Q_2. This diode also determines the d-c collector voltage for Q_1. The d-c drop across R_2 establishes the base bias voltage for Q_2.

When interchanging transistors with slightly different d-c characteristics or when collector currents increase with higher temperature, the circuit automatically stabilizes itself. If I_{c1} increases with temperature, the voltage drop across R_2 will rise and cause the base voltage of Q_2 to decrease with respect to ground. Since the Zener diode keeps the emitter voltage of Q_2 constant, a decrease in the base voltage of Q_2 will tend to decrease the collector current I_{c2}. Thus, the drop across R_1 will decrease, tending to reduce the base bias current of Q_1 which reduces I_{c1}. The same theory holds for an increase in I_{c2}.

This d-c stability is obtained by a d-c feedback loop around the two stages. The feedback path is provided by L_1 which has low d-c resistance and high a-c impedance, resulting in tight d-c coupling with practically no a-c feedback.

Performance

Since stabilization is accomplished with negligible a-c degeneration, a-c performance compares favorably with that of conventional transformer or R-C coupled preamplifiers. The gain at various temperatures for the amplifier of Fig. 1, loaded with 20,000 ohms, is shown in Fig. 2. Stability of performance over the temperature range is indicated. Table I shows that there is less than ten-percent variation in d-c collector currents from room temperature to the temperature extremes.

The circuit of Fig. 1 can be adapted to any type junction transistor triode by proper biasing. Typical performance figures which can be obtained by adapting Fig. 1 are: a voltage gain of 20,000, a current gain of 500, a power gain of 70 db and an input impedance of 2,500 ohms.

Table I
Variation of I_c

Temp	25 C	125 C	−55 C
I_{c1} (ma)	0.63	0.75	0.57
I_{c2} (ma)	1.50	1.68	1.38

* Now with Maico Electronics, Inc., Minneapolis, Minn.

Constant-Current Technique

Completely transistorized measuring system detects and responds with 0.1-percent linearity to core displacement in a differential transformer. Low-level, a-c transformer output is converted to 10- to 50-ma d-c transmission signal of one watt maximum by high input impedance feedback amplifier

By LEON H. DULBERGER,
Project Engineer, Research and Development Department, Fischer and Porter Co., Hatboro, Pa.

DIFFERENTIAL TRANSFORMERS have come into wide use in the process industry as the primary element in motion measurement systems. The device discussed here converts the small a-c voltage produced by displacement of a differential transformer core to a d-c voltage large enough to drive a standard high-accuracy measuring instrument or to control a d-c operated control device.

Measurement Techniques

Conventional systems for measuring core displacement in a differential transformer consist of an error amplifier, servo motor and a reference differential transformer. The error amplifier drives the servo motor positioning the reference transformer core to compensate for unbalance caused by core movement in the measuring transformer. Since both transformers are excited in series by the line, voltage shifts introduced by the power source are automatically eliminated. However, response time is poor and care must be taken with phase relationships to preserve accuracy.

It is also possible to rectify the output of the differential transformer directly with diodes. If high-frequency excitation is used, the response time limitation of conventional methods can be overcome; however, the nonlinearity and temperature sensitivity of practical diodes leave much to be desired.

In the system described here diode rectification is used, but only after the signal has been brought to a satisfactory output level by a linear amplifier. Use of a precision exciter and —60 db current feedback from the output rectifier to the amplifier maintain system linearity at 0.1 percent over a 20 to 55 C temperature range. Local a-c and d-c feedback within the amplifier assure stable operation.

System Operation

As shown in Fig. 1, a precision one-kilocycle exciter provides ex-

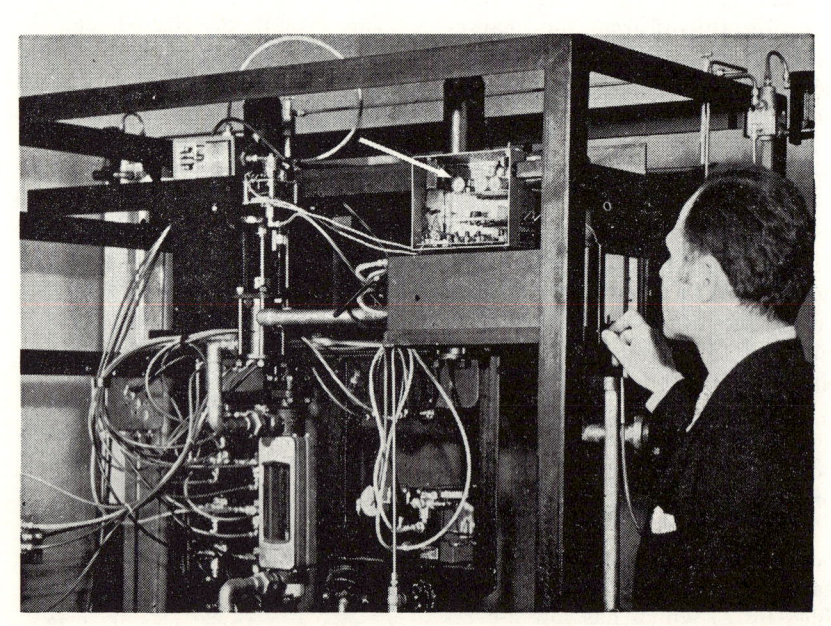

Output current of experimental core displacement measuring unit (arrow) controls an electric-to-pneumatic converter and valve assembly

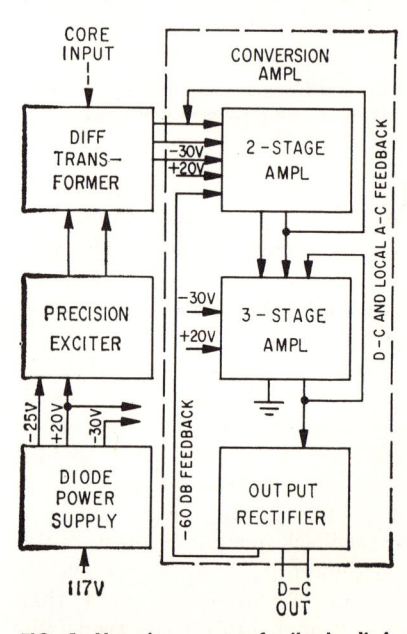

FIG. 1—Negative current feedback eliminates amplifier loading errors resulting from temperature variations

Cuts Servo Response Time

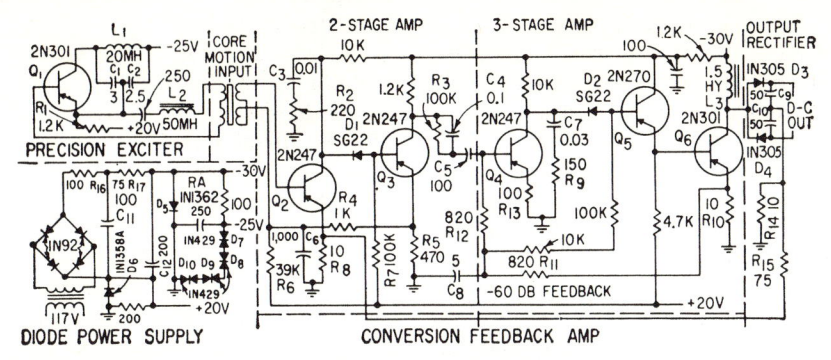

FIG. 2—Operating voltages of measuring unit are regulated by Zener diodes in power supply. Filter network of resistors R_{16} and R_{17} and capacitors C_{11} and C_{12} set regulating point for diodes D_5 and D_6. Zener diodes D_7, D_8, D_9, and D_{10} are temperature compensated

citation for the differential transformer and compensates for d-c resistance changes in the transformer primary resulting from temperature variations. Since the oscillator output is directly dependent on the amplitude of the d-c voltage from the source, temperature compensated Zener diodes used in the power supply are operated at 7.5 ma to provide an accurate and stable reference level.

The voltage produced in the secondary of the differential transformer as the transformer core moves is sensed, amplified and then rectified by the conversion amplifier. System output is made to provide feedback that is proportional to the current in the load, increasing input impedance and providing excellent linearity.

Use of d-c constant current varying from 10 to 50 ma as the transmission medium allows accumulation of d-c resistance in long lines without degradation of accuracy. Filtering can be employed at the input to sensitive readout equipment to eliminate hum and noise picked up on the lines.

A schematic diagram of the core displacement measuring system is shown in Fig. 2.

Precision Exciter

A constant-voltage, one-kilocycle oscillator and a high-Q swamping choke constitute the precision exciter. Selection of the one-kilocycle operating frequency was based on Q and output variations with frequency for the particular differential transformer used.

The Colpitts-type oscillator is formed by transistor Q_1, capacitors C_1, and C_2, and choke L_1. Operating bias established by R_1 produces 5-percent bottoming of the collector waveform shown in Fig. 3A. Since the internal collector resistance of Q_1 is low, the oscillation amplitude is determined by the Q of the L-C components and the accuracy of the negative supply voltage.

Output at the emitter Q_1 is a sine wave as shown in Fig. 3B. Waveform distortion is reduced when driving low-impedance loads by the energy storing action of capacitor C_2.

Compensation Circuits

Swamping choke L_2 provides constant-current output to correct for d-c resistance changes in the differential transformer primary as the temperature varies. High inductance of L_2 at 1 kc as compared to that of the primary winding permits the choke to swamp the current changes in the circuit. The d-c resistance of the swamping choke, which is only 8 ohms at room temperature, is compensated for along with the 22 ohms present in the differential transformer primary.

If the negative supply voltage is held to 0.01-percent variation for each deg C temperature change, the final exciter amplitude stability is 0.3 percent between temperature limits of 25 to 55 C. The exciter produces a 110-mv rms output across the secondary of the differ-

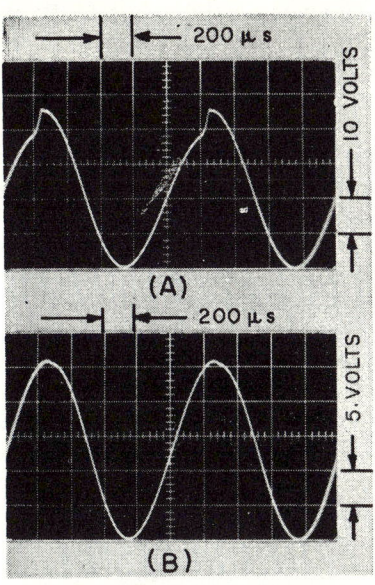

FIG. 3—Bottoming voltage waveform (A) at oscillator collector and excitation voltage waveform (B) at oscillator emitter

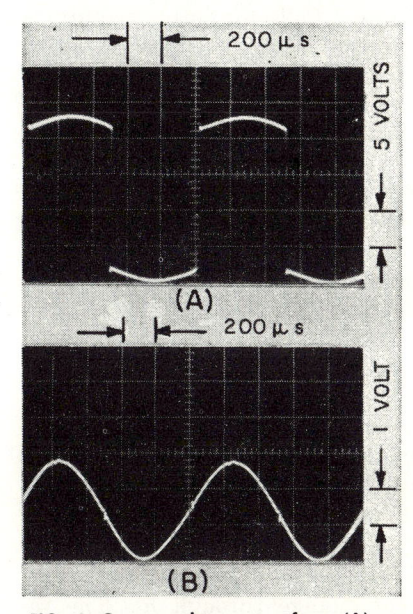

FIG. 4—Output voltage waveform (A) at Q_6 collector and feedback current waveform (B) generated across R_{10}

ential transformer for a core displacement from the null position of 0.088 inch.

Conversion Feedback Amplifier

Loading errors caused by temperature-induced changes in the d-c resistance of the transformer secondary are prevented by providing 625,000 ohms impedance at the input to the two-stage amplifier. This impedance is obtained at the

247

1-kc operating frequency by applying 60 db of negative feedback to the amplifier from the output rectifier.

To avoid amplifier instability resulting from variation in feedback phase with frequency, the amplifier transmission curve is shaped from 5 cps to 220 kc. This is done with local a-c feedback loops and shaping networks, and by using wide-range components.

Good gain stability requires that operating points of all transistors be stabilized. This is done by using a separate d-c feedback loop for each group of d-c coupled transistors.

The two-stage amplifier circuit uses high-frequency drift type transistors for Q_2 and Q_3. These transistors have an alpha cutoff frequency of 30 mc. Since a grounded-emitter connection is used in the circuit, the actual high-frequency cutoff is approximately equal to 30 mc divided by the beta of the transistor. Collector load magnitude and the values of C_2 and R_2 shape the high-frequency response for the transistor pair.

A low-frequency step is produced by the C_4-R_3 coupling network to the three-stage amplifier. Capacitor C_5 blocks d-c while maintaining a low reactance relative to the output impedance of Q_3 throughout the passband.

Current Feedback

Current feedback from the emitter of Q_3 is applied through R_4 and the secondary of the differential transformer to the base of Q_2. A change in the collector current of Q_2 is transferred by coupling diode D_1 to Q_3 where it appears as a current change through R_5. The resulting voltage swing is applied to the base of Q_2 at proper polarity to correct for the initial shift in collector current.

Resistors R_6 and R_7 return to a positive bias to establish the operating points of Q_2 and Q_3, respectively. This resistor pair is stable within two percent over the temperature range from 20 to 65 C.

Most of the a-c feedback is removed from the loop by C_6 whose capacitive reactance rises at frequencies below the passband of the amplifier. Unbypassed emitter resistors R_5 and R_8 provide a-c feedback at all frequencies.

Power Amplifier

The three-stage amplifier is a direct-coupled power amplifier. High-frequency response of the stage using drift transistor Q_4 is shaped by C_7 and R_9. This stage is directly coupled to Q_5 by forward-biased diode D_2 which maintains the required voltage difference between stages.

A transistor with a 150-mw collector dissipation rating is used for Q_5 to permit generation of adequate power to drive Q_6. Transistor Q_5 employs a grounded collector connection to more nearly match the low impedance of Q_6. Germanium output transistor Q_6 is biased for

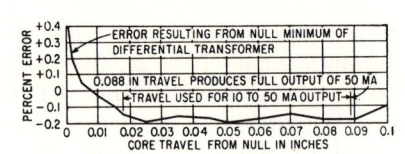

FIG. 5—Linearity curve for system and core displacement measuring unit

linear operation.

The signal applied to the output rectifier is shown in Fig. 4A. Choke L_3 is used instead of the transformer usually required to obtain true a-c feedback through the load. A suitable transformer having the necessary low d-c resistance, high current capacity, low leakage reactance and the like is expensive.

The d-c feedback signal from the emitter of Q_6 shown in Fig. 4B is picked up at emitter resistor R_{10} and applied to the base of Q_4 through R_{11} and R_{12}. This technique holds the operating point within 5 percent over a 20 to 65 C temperature range.

Alternating current feedback below 30 cps resulting from rising reactance in capacitor C_8 compensates for phase shift caused by rapidly falling reactance in choke L_3. The a-c feedback and unbypassed emitter resistor R_{13} determines input impedance of Q_4.

Output Circuit

The output rectifier consists of diodes D_3 and D_4, capacitors C_9 and C_{10}, and current feedback resistor R_{14}. Capacitors C_9 and C_{10} couple the a-c signal to current flowing through the load. Thus, feedback is present around diodes D_3 and D_4 to correct for changes in their characteristics with ambient temperature variations.

Resistor R_{15} couples the feedback voltage to the emitter resistor of Q_2. This system generates a total of -60 db of current feedback at 1 kc. Final power gain of the entire amplifier with all feedback circuits in operation is 73 db.

Output d-c is true constant current for load values from zero to 400 ohms. Maximum output power of one watt is developed into a 400-ohm load.

With reduced input levels to the amplifier, loads up to 5,000 ohms can be used before distortion becomes pronounced. To limit errors resulting from leakage in C_9 and C_{10}, high-quality aluminum foil capacitors are used.

Performance

For full d-c output of 20 volts across a 400-ohm load, the a-c ripple is 82-mv rms. Response time for 0.1-percent accuracy is under 20 millisec with filtering action of C_9 and C_{10} introducing most of the delay. Stability checks made on the laboratory model indicate a repeatability of 0.05 percent for an input signal from 10 to 110-mv rms.

Maximum output of 50 ma is provided for 0.088 in. of core travel from null. A minimum transmission signal of 10 ma which corresponds to 0.0176 in. core travel establishes transmission zero thereby voiding null problems.

Linearity curve for the experimental model, shown in Fig. 5, was checked using highly accurate core positioners. Error given is in percent of core setting and is -0.1 percent maximum down to residual null voltage. Linearity is maintained within runs taken at any temperature in range of 20 to 55 C.

The author acknowledges the assistance of J. Franklin who built the experimental models and obtained the performance data.

BIBLIOGRAPHY

L. H. Dulberger, Transistor Oscillator Supplies Stable Signal, ELECTRONICS, p. 43, Jan. 31, 1958.

V. Learned, Corrective Networks for Feedback Circuits, *Proc IRE*, p 403, July 1944.

F. E. Terman, "Electronic and Radio Engineering", McGraw-Hill Book Co., New York, 1955.

Transistorized relay servo has transistor amplifier (left), step-function potentiometer (right rear), reversible motor (right) and differential relay (front)

Reducing Relay Servo Size

Relay servo system simulates on-off control device by using step-function potentiometer to provide on-off characteristic of the null detector. Easily adjusted damping is applied through differential relay contacts to eliminate oscillations; fast response to small angle displacements assures close following

By SAUL SHENFELD, U. S. Underwater Sound Laboratory, Fort Trumbull, New London, Connecticut

DESIRE FOR COMPACT and efficient servo devices has grown with the postwar expansion of servo applications. Originally great efforts were expended on the design of linear components, but currently servo systems using less expensive nonlinear components are in demand.

This article describes a transistorized relay servo system which illustrates the simplicity possible in design of the nonlinear type.

Positional Units

Nonlinearities existing in the servo system shown in Fig. 1A are approximated by linear transfer characteristics. When the servo device is used as a positional system,

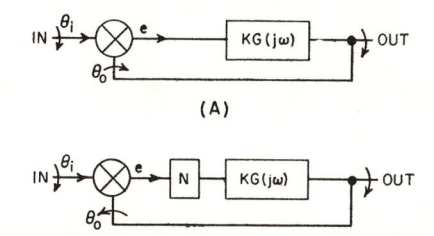

FIG. 1—Linear (A) and relay (B) servo systems can both be used as positional units in antenna systems and the like

the block $KG(j\omega)$ represents an electronic amplifier, a motor and a gear train. Except for errors resulting from nonlinearities such as static friction and backlash, the positional error is zero. A high-gain amplifier provides desirable performance, and compensating networks prevent oscillation.

The relay-type system shown in Fig. 1B has a full on or off voltage applied to the servo motor. In this system, a high degree of nonlinearity exists between the input error and the drive force applied to the servo motor.

Relay-type servo systems have

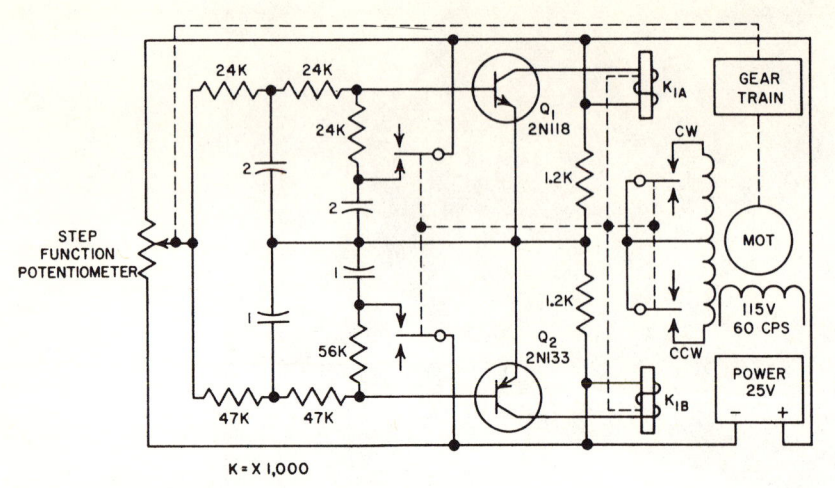

FIG. 2—Tapping positive voltage off step-function potentiometer causes current flow in Q_1 transistor, operation of relay K_{1A} and upward movement of arms. Negative voltage moves arms downward reversing direction of motor travel

the error signal available in both magnitude and sign, or in sign only. Block N of Fig. 1B, the nonlinearity in the system, may be in one or more places including the error-detector system.

Stabilization

Oscillation cycles whose amplitude and frequency depend on the parameters of the system exist in the relay servo. Damping devices that increase the natural frequency of oscillation to the point where the amplitude of the cycle is negligibly small stabilize the relay servos.[1] These devices anticipate the point of correspondence between the input and the output shaft and apply a breaking torque to the motor prior to this point. By introducing a small dead zone into the system, continuous hunting is eliminated.

Reduction in size and weight of relay servo systems is achieved with efficient and compact relays which control power output in watts with microwatts of input. Use of nonlinear servo systems is mandatory where the available null detector is nonlinear outside of a narrow region at the null position. Since the system is insensitive to variations in gain, the gain parameter may vary over wide limits without affecting response characteristics.

Description of Equipment

The transistorized relay servo system shown in Fig. 2 has a step-function potentiometer with the characteristics shown in Fig. 3. The potentiometer simulates the on-off characteristic of the null detector. The dead center or null position of the servo system corresponds to the center of the linear range of the potentiometer.

The potentiometer has 360 deg of mechanical rotation and 357 deg of electrical contact. During a 3-deg segment of the potentiometer rotation, a linear variation from zero to full output is obtained. The R-C time lag from the potentiometer to the base of the transistors simulates the time constant of the actual mechanical and motor system. Time constant variation produces undamped oscillations at the frequency of the actual system.

To equalize the sensitivity of the *npn* and *pnp* transistors, the series base resistors of the *npn* transistor are halved and the value of the capacitor associated with the smaller resistor doubled to compensate for the unequal time constant.

Compatible Performance

Transistors and relays are extremely compatible for on-off servo applications. Current drain for operation of the system shown in Fig. 2 is about 14 ma at 25 v. Since 10 ma is required as bleeder current for a center-tapped supply, actually only 4 ma at 25 v is required by the relay servo system.

The transistors are connected across a split 25-v power supply. Connection of the arm of the potentiometer to the positive side of the supply causes 4 ma, sufficient to operate the relay, to flow in transistor Q_1. Except for reverse collector current, there is no current in Q_2.

With the potentiometer arm connected to the negative supply, 4 ma flows through the other coil of the

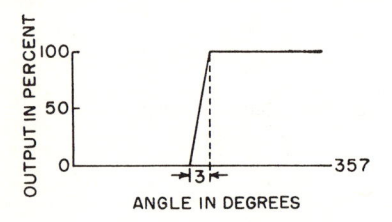

FIG. 3—Characteristic of step-function potentiometer has 3-deg segment during which there is linear variation from zero to full output current

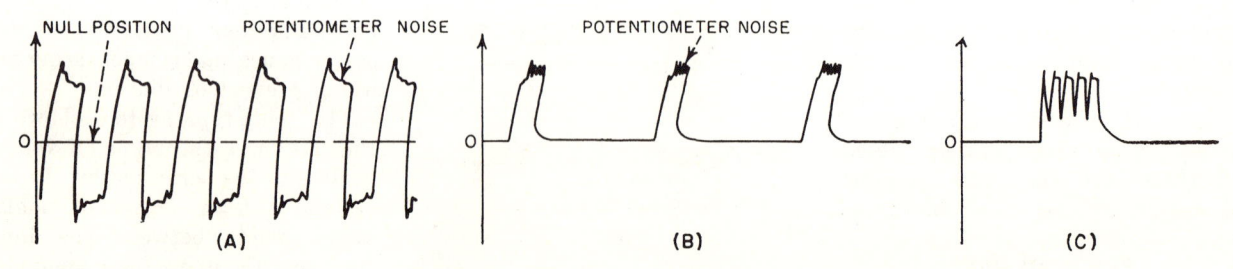

FIG. 4—Slider voltage of step-function potentiometer has undamped oscillations (A) and response to three step displacements of output shaft (B). Damping voltage (C) applied to base of off transistor has exponential character

differential relay and actuates the relay in the opposite direction. Since relay-system contacts are connected to the control winding of a reversible, shaded-pole motor, closing either set of contacts drives the motor toward the central position of the potentiometer. When there is no current in either coil, the relay is in its neutral position and the motor is not excited.

Damping

For a step displacement of the motor shaft, the slider voltage of the step-function potentiometer is shown in Fig. 4A where damping is not used. The voltage waveform illustrates the oscillatory nature of the output shaft position. The amplitude of the displacement corresponds to several deg on each side of the null position.

When base current is supplied to the off transistor as the relay closes, 2 ma of collector current, sufficient to return the relay to its neutral position, is produced in that transistor. By adjusting the value of the capacitance in the base circuit the amount of damping is changed for the desired response. The slider voltage obtained for three step displacements of the output shaft is shown in Fig. 4B. Potentiometer noise produces the variation of the correction voltage appearing at the top of the waveform.

The damping voltage shown in Fig. 4C and applied to the base of the off transistor decelerates the motor by tripping the relay off. This feedback voltage decays exponentially. Feedback prevents the motor from running at top speed and overshooting after it reaches the null position.

Position and Velocity

A sketch of the position and velocity of the output shaft superimposed on the phase-plane plot is shown for the damped and undamped cases in Fig. 5. The curve for the undamped case corresponds to one set of initial conditions. A family of such curves is obtained for various initial shaft displacements or initial shaft velocities. In the undamped case the phase-plane plot consists of parabolic segments when the motor's self-damping is small.[2] When damping forces on the motor are neglected, the equation of motor torque and acceleration is

$$T = I\,(d^2\theta_o/dt^2) \quad (1)$$

where T is motor torque for full excitation, I is rotational inertia and $d^2\theta_o/dt^2$ is acceleration. Then

$$\omega_o(t) = (T/I)\,t + C_1 \quad (2)$$

where ω_o is output shaft angular velocity and constant C_1 is shaft angular velocity at $t = 0$. Furthermore,

$$\theta_o(t) = (T/2I)\,t^2 + C_1 t + C_2 \quad (3)$$

where constant C_2 is the angular shaft position at $t = 0$.

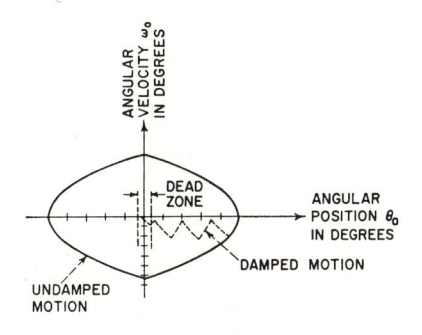

FIG. 5—Position and velocity of output shaft is superimposed on phase-plane sketch for damped and undamped cases

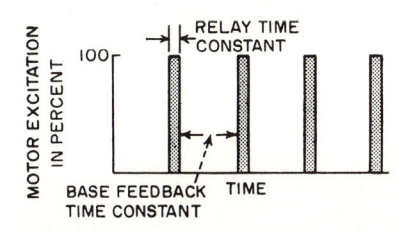

FIG. 6—Graph shows motor excitation time as function of relay time constant, and braking time as a function of feedback time constant

Equations 2 and 3 are the parametric equations of the shaft motion in terms of time. When plotted on the phase plane for initial shaft displacements or velocities, the parabolic curves of Fig. 5 are obtained. With on-off control the point of motor reversal is

$$-\omega_o^2 + (2T/I)\,\theta_o = 0 \quad (4)$$

or

$$\omega_o^2 + (2T/I)\,\theta_o = 0. \quad (5)$$

The quadrant of operation determines the appropriate equation to be used.

Equations 4 and 5 indicate the points on the reference plane for reversing the motor torque and bringing the output shaft to the null position in minimum time. This defines an optimum system in terms of minimum response time and zero overshoot. Equation 1 is only applicable for time intervals which are short with respect to the time constant of the servo drive motor.

System Advantages

Base-current damping does not give optimum system speed response but has the valuable features of simplicity and ease of adjustment. The degree of damping is adjusted by varying the time constant of the base feedback voltage. Essentially, impulse excitation of the motor control winding restores the shaft to the null position with the spacing of the impulses determined by the feedback time constant. Figure 6 shows the excitation time as controlled by the relay time constant, and the braking time by the feedback time constant.

With this system, the shaft is returned to the null position with a low energy storage assuring small amplitude of the stable oscillation. By introducing a dead zone and a center-position relay into the system, the oscillation is reduced to zero. Where a velocity input must be followed, base-current damping is desired, since the speed of response is fast for small displacement angles.

The system also has high damping. For small angles, operation approaches that indicated by Eq. 4 and 5. The maximum average velocity followed by the system is reduced by the impulse excitation, but within the velocity follow-up limits, satisfactory performance is obtained.

REFERENCES

(1) A. Tustin, "Automatic and Manual Control," Academic Press, New York, 1952.

(2) D. MacDonald, Nonlinear Techniques for Improving Servo Performance, *Proc NEC*, **6**, p. 400, 1950.

Controlled Rectifiers Drive A-C and D-C Motors

Saturable magnetic core firing circuits can be used with controlled rectifiers to drive both a-c and d-c motors. Specific applications covered include half-wave and full-wave control circuits

By **W. R. SEEGMILLER**, Magnetic Controls Engineer, General Electric Co., Schenectady, N. Y.

CONTROLLED RECTIFIERS are rapidly gaining acceptance as an effective means of reducing size and weight and increasing efficiency in power stages of control equipment. Because the controlled rectifier is similar to the thyratron, it can be used in many applications where thyratrons, magnetic amplifiers, power transistors, relays and switching devices are now used. These applications include drives for d-c torquers, a-c servo motors, d-c power supplies and power oscillators.

Characteristics

The gate firing characteristics (gate voltage versus gate current) of a controlled rectifier are shown in Fig. 1. These characteristics are important in the design of firing circuits for the controlled rectifier. The rectifier will fire at some gate current-voltage point within the shaded area. Reverse gate current out of the gate terminal should be limited by a diode in series with the gate lead if reverse voltage appears in the gate circuit.

Another plot of the gate firing characteristic of a typical controlled rectifier is shown in Fig. 2. This is a plot of the gate current required for firing as a function of temperature and applied forward voltage. Although these curves will vary somewhat from unit to unit, they clearly indicate the temperature dependency of the firing characteristic and the resulting importance of this dependency should be considered when designing firing circuits.

To provide for firing of the controlled rectifier at a definite point in the supply voltage cycle, it is necessary to apply a current pulse with a steep wavefront. Circuits utilizing saturable magnetic cores in a manner similar to magnetic amplifiers are a convenient means of generating this steep wavefront. Although the controlled rectifier actually requires only a very short pulse of a few microseconds duration to fire, the gate current may be continued throughout the remainder of the firing half cycle without adversely affecting the rectifier.

A typical half-wave circuit is shown in Fig. 3. The firing circuit consists of the saturable magnetic core with gate winding N_G, control winding N_c, current limiting resistor R_1, diode D_1, shunt resistor R_2, and supply voltage e_{s1}. Diode D_1 prevents reverse voltage from appearing on the gate terminal of the controlled rectifier.

Current in the control winding determines the time in the cycle at which saturation occurs. The saturation angle, thus, can be varied from zero to 180 electrical degrees

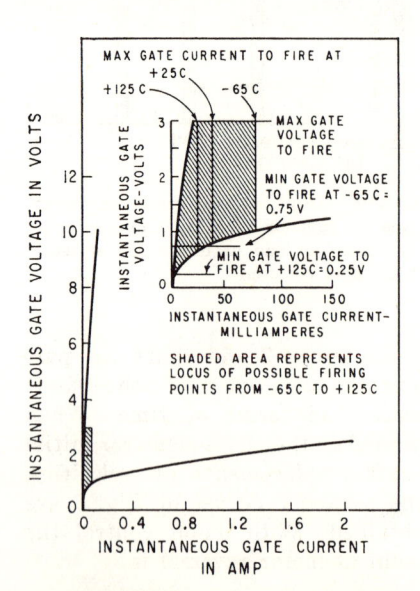

FIG. 1—Gate firing characteristics of C35 controlled rectifier show variation of gate firing point with temperature. This temperature dependency must be considered by the engineer when he is designing firing circuits for controlled rectifiers

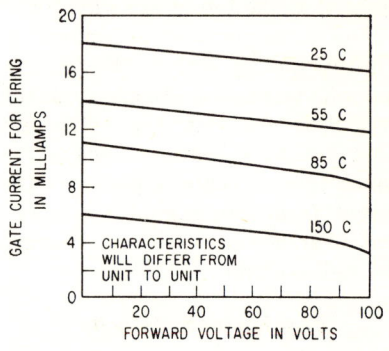

FIG. 2—Temperature variation of characteristics are shown

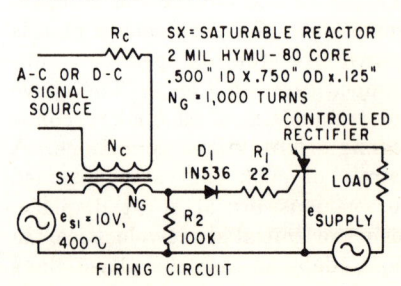

FIG. 3—Basic circuit for motor control

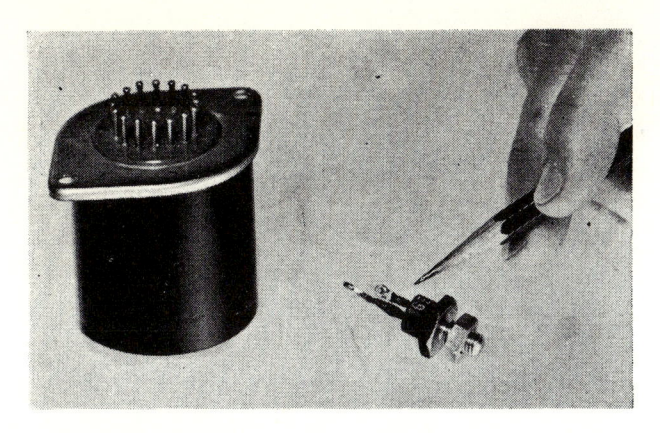

Controlled rectifier replaces magnetic amplifier

FIG. 4—Drive circuit for reversible d-c shunt motor

in a manner similar to magnetic amplifiers.

Firing circuit design values given in Fig. 3 are for a 400-cycle supply frequency. The circuit is designed for a peak gate current of 200 milliamps. With this firing circuit the controlled rectifier fires over a range from 30 to 150 electrical degrees at minus 65C or from 10 to 170 electrical degrees at normal room temperature.

For complete 0 to 180 electrical

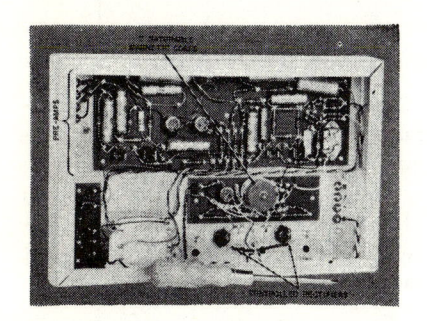

Series actuator of flight control system uses controlled rectifiers to drive solenoids. Circuit can also be used to drive d-c split-series motor

degree coverage under all conditions, a square wave of voltage can be used as the firing circuit supply voltage, e_{s1}. The firing circuit will then operate along a single load line independent of the firing angle. Another alternative is to increase the magnitude of the sinusoidal supply and use a Zener diode from gate to cathode to limit the maximum voltage. However, in most control applications this complete coverage refinement is not necessary.

Saturation Current

Another point to be considered in the design of the firing circuit is the magnitude of the saturable reactor exciting current before saturation. Design values given in Fig. 3 keep the maximum exciting current over the normal control range below 0.5 ma.

Since the magnitude of the exciting current will rise with large negative values of control current it may be necessary, in some applications, to limit the maximum value of control ampere-turns. Where the magnetic core firing circuit is being driven by a preamplifier, this limiting will be accomplished by saturation of the preamplifier output. Since the rectifier will not fire below 0.25 v (Fig. 1), a safety margin can be obtained by adding a 100-ohm resistor shunted from the gate terminal to cathode terminal of the rectifier. This insures that more than 2.5 milliamps of exciting current is required before the gate reaches 0.25v—the minimum firing voltage.

In summary, the saturable magnetic core firing circuit has the following features:

(1) The steep wavefront of the firing pulse overcomes the temperature dependency of the firing characteristic.

(2) Signal circuits are isolated from each other and from the power source.

(3) Signal windings can be connected to respond to either a-c or d-c signal sources or preamplifier outputs.

Half-Wave Push-Pull Circuit

An application of controlled rectifiers to control the armature of a d-c shunt motor or d-c torquer is shown in Fig. 4. This circuit is for applications requiring a push-pull output for a reversible drive. To keep the number of control components to a minimum, a half-wave output circuit is used. The firing circuit is of the type previously discussed. The saturable reactor control windings are wound over both cores together. The bias winding is connected to the a-c supply in the proper polarity to retard the firing angle on each core.

At standby, the firing angle on each controlled rectifier is set with the bias adjustment potentiometer R_1 to approximately 160 electrical degrees.

There is a small a-c component of current through the armature at standby. Power dissipated in the armature by this a-c component is kept small by adjustment of the standby firing angle.

The signal winding is connected to a d-c control source and the gate winding polarity is such that a d-c signal retards the firing angle on one core and advances the firing angle on the other core.

Recordings of the armature current acceleration and reversing are shown in Fig. 5. Armature current

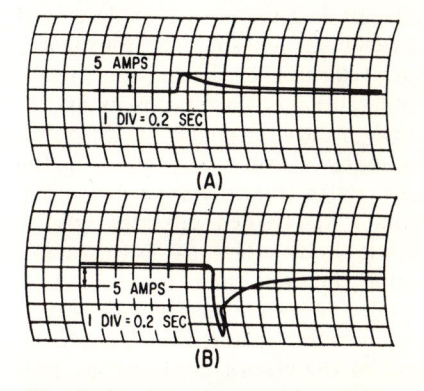

FIG. 5—Armature current zero to top speed (A) and during reversal from top speed in one direction to top speed in the other (B)

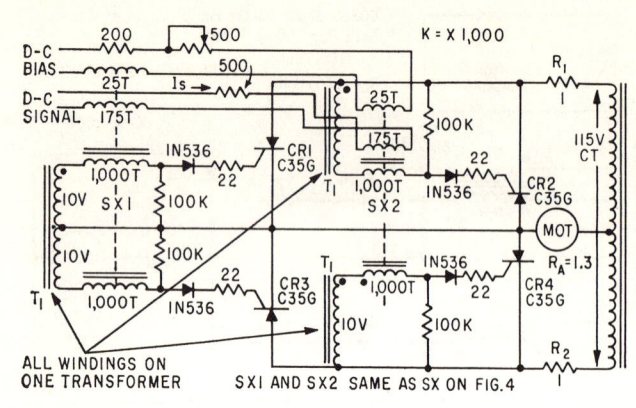

FIG. 6—Full-wave circuit is used for d-c shunt motor or d-c torquer

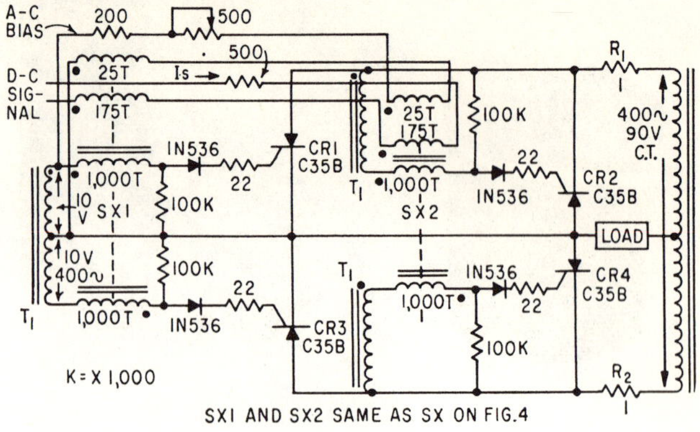

FIG. 7—Drive for a-c motor is very similar to that for d-c

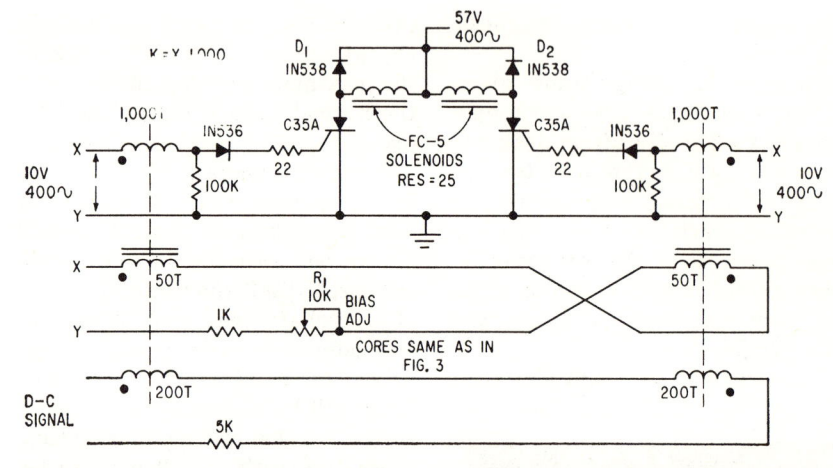

FIG. 8—Solenoid drive circuit can also be used to drive d-c split-series motors

during an acceleration from standstill to top speed is shown in Fig. 5A. Maximum current during a reversal from top speed in one direction to top speed in the other direction is approximately 20 amp as shown in Fig. 5B. The current drops to 10 amp in less than 0.1 sec, well within the capability of the C35 controlled rectifier.

With high inertia loads the high armature current during reversals is extended for longer periods of time, depending on the torque to inertia ratio. In some applications it may then be necessary to limit the armature current to protect the rectifier from these high currents during reversals. The maximum values of current that can be tolerated will depend upon ambient temperature, type of heat sink and duty cycle of the particular application.

Full-Wave Push-Pull Circuit

A circuit for full-wave push-pull operation of a d-c shunt motor, or d-c torquer, is shown in Fig. 6. This circuit requires four controlled rectifiers and a center-tapped transformer. The magnetic core firing circuit is similar to that described previously, except that four cores are required for the full-wave push-pull action. Limiting resistors R_1 and R_2 are required to limit the standby current when two of the controlled rectifiers, such as CR_1 and CR_4, (which operates on the same half-cycle) are both turned on at the same time.

The full-wave push-pull circuit of an a-c servo motor, Fig. 7, is identical to the circuit used for the full-wave push-pull d-c shunt motor drive except for a different arrangement of the firing circuit. Limiting resistors R_1 and R_2 are again used to limit the standby current.

Figure 8 can replace the half-wave magnetic amplifier stage used to drive the solenoids in the series actuator of a flight control system. The circuit consists of a controlled rectifier in series with each solenoid and a saturable magnetic core firing circuit. Each magnetic core has two control windings, one for adjustment and the other for signal.

Potentiometer R_1 provides for adjustment of the standby firing angle and standby current in the solenoids. Increasing R_1 will increase the standby current in each solenoid.

The signal windings are connected in the proper polarity to increase the current in one solenoid and decrease the current in the other solenoid. Free-wheeling rectifiers D_1 and D_2 provide a path for continuous flow of current in the inductive load during that portion of the cycle when the controlled rectifier is not conducting. Response time of the saturable magnetic core firing circuit is similar to a half-wave magnetic amplifier; approximately one-half cycle of supply frequency.

The circuit shown in Fig. 8 can also be used to drive a d-c split-series motor. The windings of the motor replace the solenoids shown.

Applications

Because controlled rectifiers offer a significant reduction in size and weight in the power stages of control equipment, the application possibilities for these rectifiers are almost unlimited. This is especially true in applications requiring load powers of 10 watts and upward, which are presently being supplied by magnetic amplifier output stages.

BIBLIOGRAPHY

H. F. Storm, "Magnetic Amplifiers", p 260, John Wiley and Sons, Inc., New York, N. Y., 1955.
R. P. Frenzel and F. W. Gutzwiller, Solid-State Thyratron Switches Kilowatts, ELECTRONICS, Mar. 28, 1958.
Silicon Controlled Rectifier (Specification) ECG-371-A, G.E. Semiconductor Products Dept., Syracuse, N. Y.
Application Notes on Silicon Controlled Rectifier ECG-371-1, G. E. Semiconductor Products Dept., Syracuse, N. Y.
Proceeding of National Electronics Conference, Oct. 15, 1958.

Chapter 14
INDUSTRIAL MEASURING INSTRUMENTS

Wide-band amplifier of millivolt signals

FIG. 1—Bridge circuit at input achieves common mode rejection

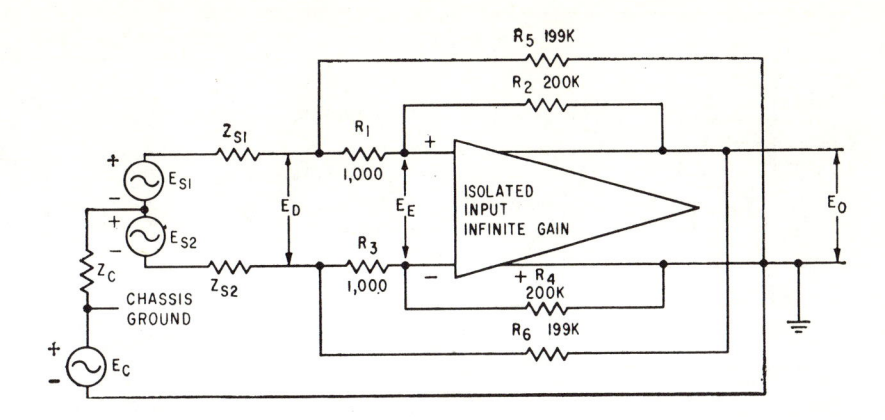

Amplifiers for Strain-Gages And Thermocouples

Positive and negative feedback to a bridge-type transformer-coupled input circuit are used to attain high-impedance differential input in a d-c to 25-kc amplifier. Bridge balances out common mode signals

By **RICHARD S. BURWEN,** Boston Division, Minneapolis-Honeywell Regulator Co., Boston, Mass.

NOISE VOLTAGES produced by grounding at two different points in strain-gage and thermocouple circuits have necessitated the development of amplifiers with high common-mode rejection.

Common-mode voltages arise in data-acquisition systems, where the preamplifier ground is frequently located several hundred feet from the transducer ground.

The amplifier to be described features both a floating input and a floating output, and delivers a low voltage at high current suitable for feeding high-frequency recording galvanometers, f-m recorders, analog-to-digital converters and other equipment, through long lines if necessary. Direct-current to 25-kc millivolt signals from strain gages and thermocouples are amplified with an accuracy of ± 15 µv instantaneous peak equivalent input on a wide-band basis, and ± 7 µv from d-c to one kc.

Equivalent input d-c drift of less than ± 0.5 µv peak during 40 hours is achieved by chopper stabilization and a unique way of cancelling transistor drift using a feedback amplifier. The input impedance is infinite when used with balanced sources. Common-mode rejection, the ratio of common-mode voltage to the equivalent differential input it produces, is 2×10^6 at d-c and up to 1×10^6 at 60 cps with balanced sources.

Common-mode rejection is achieved by a resistance bridge at the input which passes differential signals while balancing out common-mode signals. The differential output signal from the bridge then divides into high- and low-frequency channels. Frequencies are converted to 400-cycle square waves.

Input Signal

Analysis of the bridge circuit by the principle of superposition allows the effects of several voltage generators to be considered one at a time while the others are turned off and shorted out. In Fig. 1, the input signal consists of two components E_{S1} and E_{S2} generated with respect to an input ground which is at a common mode voltage E_c with respect to the output ground. Resistors R_1, R_2, R_3 and R_4 constitute the bridge. Resistors R_5 and R_6, each 199,000 ohms, equalize the currents through the differential input terminals. In effect, the current through R_6 develops a positive feedback voltage across the impedance Z_{S2} in the negative side of

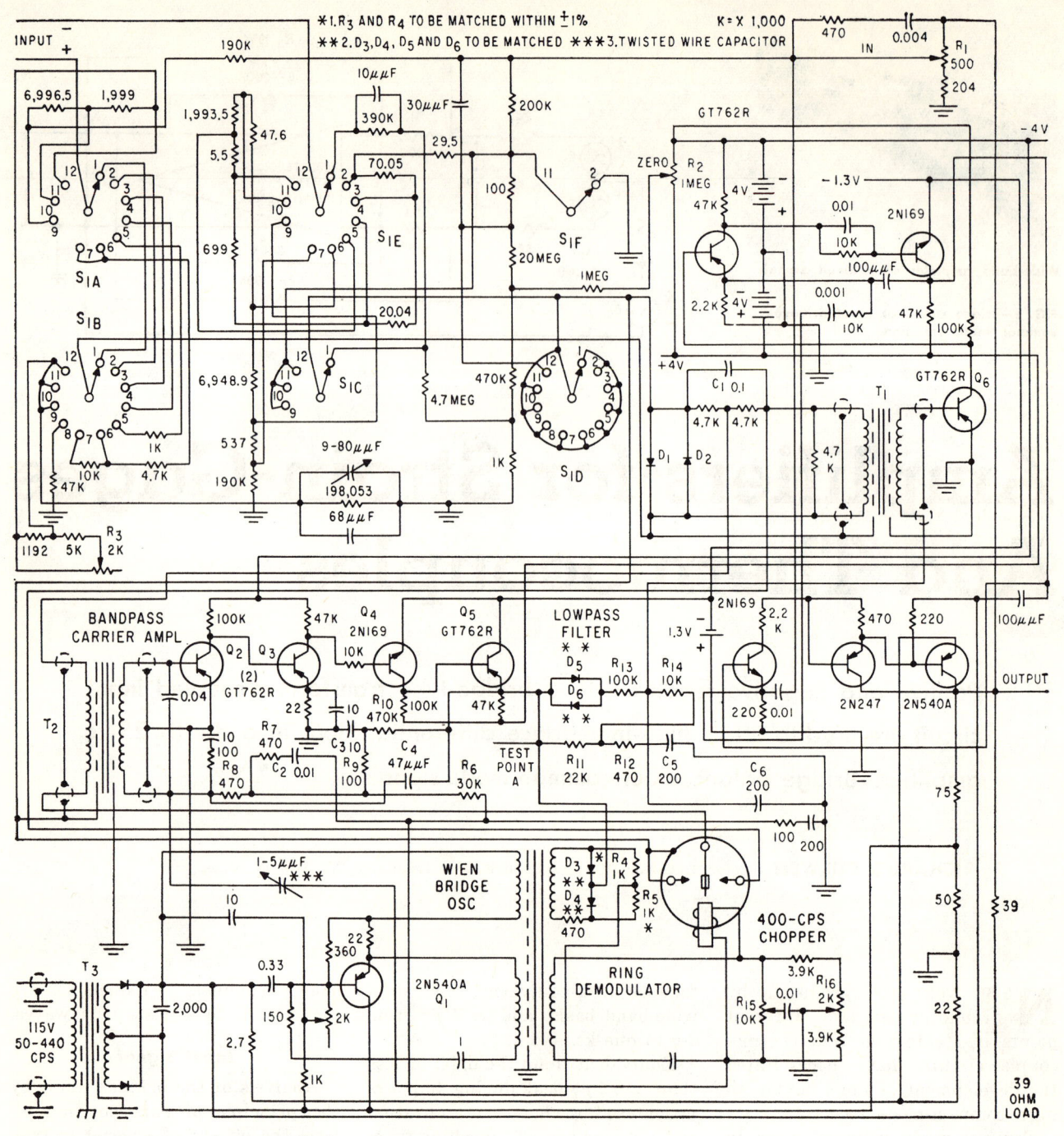

FIG. 2—Differential-input chopper-stabilized transistor amplifier

the source, while that through R_2 develops a negative feedback voltage across the impedance Z_{s1} in the positive side of the source. Resistor R_5 completes the symmetry of the circuit so that the common-mode voltage E_c produces no differential input voltage E_d when equal source impedances Z_{s1} and Z_{s2} are added.

In the schematic, Fig. 2, the eleven-position switch S_1 changes the gain in three-to-one steps by varying R_1 thru R_6 of Fig. 1 to maintain the proper relationships for infinite input impedance and common mode rejection.

Continuous Coverage

In Fig. 2, potentiometer R_1 in the feedback path varies the gain to cover intermediate points. These controls provide continuous coverage of full-scale inputs of ± 100 μv to ± 100 mv single ended and ± 3 mv to 100 mv differential, plus zero check, and an open loop position which permits the use of external feedback networks when special characteristics are needed. Adjustment R_2 zeros the no-signal output and R_3 compensates for tolerances in the bridge resistors and unbalances in the source up to ± 5 ohms.

The error voltage appears across diodes D_1 and D_2 which protect the first stage against damage from excessive input signals. At this point the error voltage is divided into two channels. Direct current to 25-cycle components are developed across C_1 while the high-frequency components pass directly through T_1 to a six-stage direct-coupled wide-band amplifier.

The chopper, which converts the

low-frequency components to a 400-cycle carrier, is excited by a Wien-bridge oscillator using a single power transistor, Q_1. The carrier produced by the chopper passes through T_2 and is then amplified by the four-stage bandpass carrier amplifier Q_2 through Q_5. A ring demodulator, D_3, D_4, R_4 and R_5 synchronously rectifies the carrier, converting it to d-c of the proper polarity. After filtering, the rectified output is added in series with the high-frequency component of the error signal across the secondary of T_1. At d-c, the gain from the input terminals to test point A at the output of the demodulator is 30,000.

Lowering the Gain

Demodulation is accomplished by shunting a network R_6, R_7 and C_2 during alternate half cycles across a portion of the bridged T, 400-cycle selective-feedback filter R_8, R_9, R_{10}, C_3 and C_4. This effectively lowers the gain of the amplifier during alternate half cycles and prevents the overload that would otherwise occur if demodulation were accomplished by grounding the output of Q_5 during this time.

Using this arrangement for demodulation allows the chopper amplifier to serve a second purpose—temperature compensation of Q_6, the transistor primarily responsible for drift.

Since the no-signal d-c output from Q_5 at test point A is referenced to the input of Q_2 through the demodulator and feedback network, the d-c level drifts with temperature to satisfy the bias requirements of Q_2. This variation is the same as needed for temperature compensation of Q_6 which has the same operating point as Q_2. Low-voltage, low-current operation of each of these transistors (collector voltage, 0.3 v; current, 35 μa) minimizes their d-c drift and a-c noise.

Following the chopper amplifier and demodulator is a two-section, nonlinear low-pass filter R_{11} through R_{14}, C_5, C_6, D_5 and D_6. During overloads, diodes D_5 and D_6 convert this filter into a one-stage filter which has less phase shift and allows the amplifier to recover. Without this nonlinear network, the phase shift is so great that the amplifier would continue to oscillate once the amplifier becomes overloaded.

The stability afforded by the nonlinear filter technique permits a substantial increase in chopper amplifier gain with consequent reduction in equivalent input drift.

Total Loop Gain

In the six-stage direct-coupled amplifier, the gain is 10^6. Multiplied by the d-c gain of 30,000 in the chopper amplifier and the loss of three times in the filter, the total open loop d-c gain is 10^{10}. Drift relative to the input due to the transistors is that of Q_6 less the compensation provided by Q_2, all divided by 30,000, about 1.6 μv over the temperature range of 32 to 125 F. This low inherent drift plus cool operation of the low-level input circuit, the use of materials having low thermal voltages with respect to copper, and thermal lagging to minimize temperature differences, results in low drift.

Noise caused by power line voltage variations and serious decoupling problems are eliminated by powering the first transistors with mercury batteries. Batteries operate continuously for a year with no provision for turning them off.

Technician inserts amplifiers in rack-mounting assembly

The h-f channel is reduced to a single 1 : 1 transformer T_1 to eliminate phase shift. An identical transformer, T_2, avoids the necessity of having to float the entire chopper amplifier and its power supply within a separate shielded compartment.

Careful shielding of both primary and secondary and all connecting wires results in a transformer design having a bandwidth of 20 cps to 200 kc with common-mode rejection of more than 25×10^6 at a frequency of 60 cps.

High-frequency stability problems, made more difficult by the presence of T_1 within the feedback loop, are solved through the use of multiple loops which roll off the gain in a slow manner. Since the principal local loop encloses the output stage, distortion and output impedance are reduced to beyond 500 kc.

Stray Capacitances

It is necessary to maintain all the components associated with the error signal at the same potential, to prevent stray capacitances from introducing common-mode signals across the primary of T_2 in the chopper channel. For this purpose the chopper and associated error signal circuitry are all mounted inside an electrically floating aluminum box, placed towards the front of the amplifier. This box is connected to the negative side of the error signal.

Because the frame of the chopper is at error-signal potential, it is capable of picking up 400-cycle noise through capacitance to its coil. This noise is eliminated by grounding the coil through center-tapping potentiometers R_{15} and R_{16} which cancel the electrostatic pickup in both magnitude and phase.

Shielding

Compact construction caused in part by the space taken by the double wall case necessitates good magnetic as well as electrostatic shielding. Transformers T_1 and T_2 have double mu-metal and copper shields and the power transformer T_3 and oscillator transformer have single mu-metal shields.

Construction in a double wall case completely shields the amplifier circuits.

The chopper, a strong source of 400-cycle noise because of its proximity to sensitive circuits, is also double-shielded with mu-metal.

The above considerations together with the low-noise input circuitry, high common-mode rejection, and excellent gain stability and linearity resulting from the high feedback factor, have made possible high-accuracy wide-band amplification of millivolt signals.

Transistorized indicator judges foul or fair in the sport of fencing despite complex rules. Can ingenious electronic devices help umpire other sports?

By **W. R. DURRETT**, Project Engineer, Designers for Industry, Inc., Cleveland, Ohio

Electronic Judging of

USING ELECTRICAL devices to indicate the validity and timing of touches scored in the sport of fencing is not uncommon. However, as ordinary indicators use relays, fencers are occasionally shocked by transients developed by collapsing relay-coil fields. The transistorized touch indicator to be described eliminates the possibility of shock and responds well within the fast-response times required to monitor fencing contests.

Rules

The transistor indicator detects touches in either epee or foil fencing contests. In epee, touches on all parts of the body are fair. In foil contests, only touches on the metallic-thread jackets worn by the fencers are scored. The conductive area of a jacket composes the fair-touch region for a foil. When a foil hits an invalid area, a foul is registered.

A touch must be held long enough to score a point. With either weapon, after a touch is held long enough to score a timing interval begins in which the other fencer may also score. At the end of this interval, which is longer for foil than for epee, no further touches may be scored until the next action.

Swords and Tips

The indicator is connected through two lengths of three-conductor cable to a pair of automatic take-up reels. Each reel holds 50 ft of cable which goes to a connector clipped to the fencer's belt. A body cord runs from this connector to contacts at the tip of the sword; for foil contests, the cord also goes to the metallic jacket. Reel tension does not hamper the fencers.

As shown in Fig. 1, an epee has two wires insulated from the sword blade and running down the blade to a tip assembly that acts as a spst, normally-open switch. A touch closes the contacts between the two wires.

A foil has only one wire, which runs to the tip assembly and is connected permanently to the tip. The foil-tip assembly is a spst, normally-closed switch that shorts the wire to the foil blade and guard. A touch opens this short. The metallic

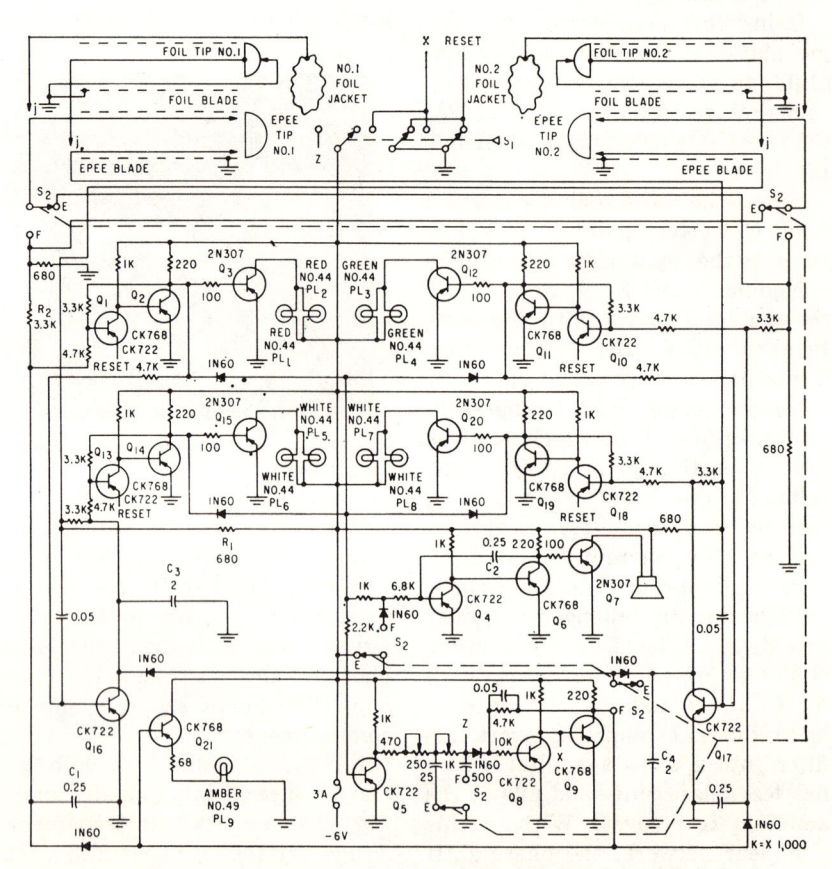

FIG. 1—Transistor fencing indicator is set up for an epee match. For foil, throw switch S_2 to terminal F and connect foils and foil jackets to the j terminals

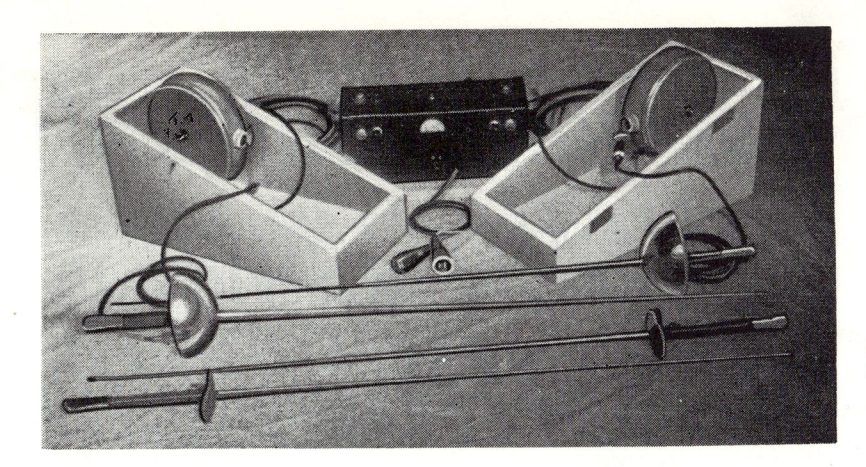

Indicator is in center. Reels are connected to epees, which are behind foils

Fast-Moving Sports Contests

jacket is connected to the third wire of the body cord with a detachable clip.

Scoring With Epee

Considering Fig. 1, assume that *epee No.* 2 scores a touch. The tip contacts close, applying a negative voltage to the input of transistor Q_1. Transistors Q_1 and Q_2 form a bistable multivibrator that is called a valid multivibrator. At the start of a fencing action Q_2 conducts, keeping Q_1 and transistor amplifier Q_3 nonconducting. Integrating capacitor C_1 bypasses transients so that the touch must be held until C_1 charges through R_1 and R_2. When the charge level reaches the critical value of about -1.5 v, the multivibrator flops over lighting red-indicating lamps PL_1 and PL_2 and transmitting a signal to Q_4 and Q_5.

Transistor Q_4 is the input transistor of a multivibrator oscillator which drives a 4-in. p-m speaker. Capacitor C_2 provides the feedback path; if C_2 is 0.25 µf, the oscillator tone is about 1,000 cps.

Transistor Q_5 is the input transistor of a timer. With the application of an input signal, the timer input r-c network begins to discharge toward zero v. The timer is set for a given discharge rate. After 0.04 to 0.05 sec the timer flops over, switching on Q_9. This transistor brings the input stages of all lamp circuit inputs to a potential near ground. This action does not affect the red indication but it prevents any subsequent touch by *epee* No. 1 from being scored on the green lights, PL_3 and PL_4. The indicating condition is thus frozen, indicating a touch by *epee* No. 2 and sounding tone until switch S_1 is reset. *Epee No.* 1 scores touches in the same way when it applies a signal to Q_{10}. Transistors Q_{11} and Q_{12} correspond to Q_2 and Q_3.

Scoring With Foil

When fencing with foils, switch S_2 is at position F and the foils are are connected to the circuit junctions indicated by j in Fig. 1. Scoring action begins when the closed circuit in the foil tip between tip and blade opens. Assume that *foil No.* 2 touches a foul, or invalid, area. A touch opens the tip contacts, applying a negative voltage to input transistor Q_{13} of the invalid-indicating bistable multivibrator. Integrating capacitor C_3 bypasses transients so the touch, to indicate, must be held until C_3 charges to the point of triggering the invalid multivibrator. The multivibrator flops over, lighting the white lamps, PL_5 and PL_6 and transistor Q_{15} sends an initiating signal to the timer and tone oscillator, as in epee. The timer, after a 2-sec discharge period, sends a disabling signal that prevents further scoring. This signal also goes to the tone oscillator to silence the audible signal.

The 2-µf value of C_3 requires a touch duration of about 0.004 sec. This requirement protects against a false invalid signal that could be caused by poor reel contacts and momentary openings of the foil tip.

If a touch is made in the valid area, opening of the tip contacts generates a signal to an invalid multivibrator and to a valid multivibrator. As integrating capacitor C_1 is smaller than C_3, the valid multivibrator flops over before the invalid multivibrator after *foil No.* 2 makes a fair touch. The valid multivibrator flops in about 0.002 sec. Actuation of valid multivibrator Q_1-Q_2 sends signals to the tone oscillator and timer as well as lighting PL_1 and PL_2.

Simultaneously a signal goes to transistor Q_{16}, which has the job of inhibiting (disabling) the invalid circuitry when a valid touch is scored. Conduction of inhibit transistor Q_{16} grounds the input of invalid multivibrator Q_{13}-Q_{14}, bleeding off the charge building up in C_3.

If a valid touch arrives and then glances to an invalid target the valid indication is the only one to register, since transistor Q_{16} has disabled the invalid multivibrator. However, if an invalid touch comes first and then glances to the valid area, both valid and invalid indications will show as long as the valid touch arrives before the timer disables all circuits. This action does not extinguish the previously illuminated invalid light; by the time a signal comes from transistor Q_{16}, the invalid multivibrator has flopped.

No touch registers from a touch made on the blade of either an epee or foil.

261

Balloon-Borne Circuits Sort

Transistorized circuits operate effectively over wide temperature and pressure range for high-altitude study of cosmic rays. Telemeter coding system is designed for reception through noise

By DONALD ENEMARK, Physics Department, State University of Iowa, Iowa City, Iowa

BALLOON-BORNE APPARATUS has proven to be invaluable in the measurement of cosmic ray phenomena. This high-altitude observation technique opens many new areas of study to the geophysicist, but it presents a serious problem to the electronics engineer. Any circuits designed for such use must be able to withstand wide ranges of temperature and pressure.

The system shown in Fig. 1 has been tested for temperature stability from 38 C to -45 C. Using inexpensive germanium transistors, the circuit measures cosmic ray energy and intensity. It has already been used successfully at altitudes of 130,000 ft.

General Description

The system uses a sodium iodide scintillation crystal optically coupled to a multiplier phototube. Radiation falling on the crystal produces light scintillations that excite the multiplier phototube to produce electrical output pulses. The amplitude of each pulse is proportional to the radiation energy. The pulse sizes are separated into increments by the discriminator and sent to scalers for counting. The outputs of the four scaling channels are coded and applied to the telemetry system.

An automatic integrating ion chamber acts as a second detector. It triggers a scaler at a rate dependent on the ionizing power of the radiation. The output of the scaler is applied to the telemetry. The plastic scintillators act as a third detector. They are attached to multiplier phototubes, and the axes of the tubes are lined up vertically. The two amplifier-discriminator channels are connected in a coincidence circuit. Whenever a coincident pulse occurs, it is gated through to a height-to-time converter, and the output of the converter is applied to a subcarrier oscillator in the telemetry system. The Olland barometric altimeter provides the pressure information used to determine balloon altitude.

Scalers

The transistorized scalers used for counting are shown in Fig. 2. The input trigger circuit requires about .5-v negative pulse to operate the scaler.

The simplest detector channel in the instrument is the one using the ion chamber. Each time the electrometer in the chamber recharges, it produces a negative output pulse of several volts. The signal is connected to the input of a scaler, and the scaler output is used to modulate the telemetry system.

Four-Channel Discriminator

The four-channel discriminator is composed of a common amplifier and four individual amplifiers which drive four trigger circuits. The four trigger outputs are used to drive four channels of scalers. The individual channels require either 256 or 512 counts to produce one output to the telemetry system.

The detector that operates with this circuit is the sodium iodide scintillator-multiplier phototube combination. The plate supply for the multiplier phototube is 900 v. The individual dynodes are supplied from taps to various points within the batteries. The output appears as negative pulses of various amplitudes across R_1 in Fig. 3. The gain of the 6199 multiplier phototube is in the order of one million, but there are rather large differences between the gains of individ-

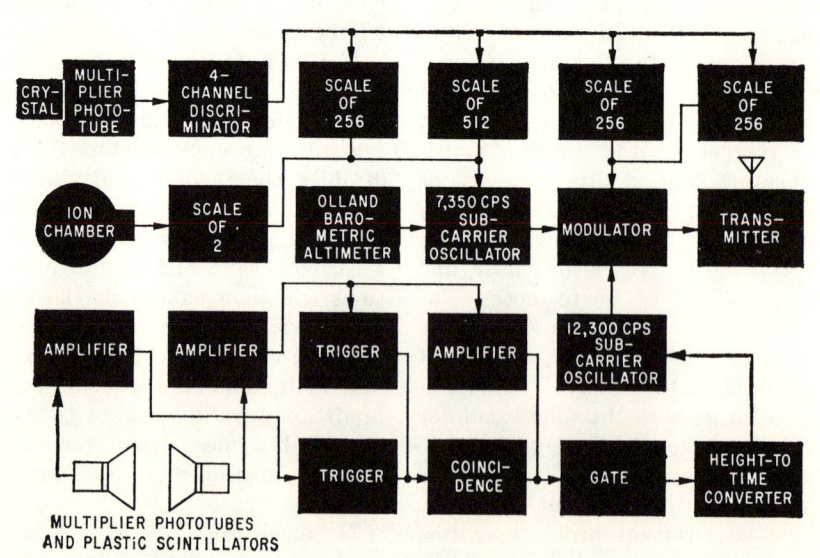

FIG. 1—Block diagram of cosmic ray measuring system is designed for high temperature and pressure stability

High-Altitude Cosmic Rays

ual tubes. The input network of R_1, R_2 and R_3 is used for attenuation, and in practice the value of R_2 is selected to provide the required pulse size at the base of Q_1.

The first two stages of the common amplifier operate together to supply a high input impedance, a gain of one, and a low output impedance. The first stage Q_1 is an emitter follower with some output across R_7 to drive Q_2. Transistor Q_2 amplifies this signal and feeds it back across the emitter load R_5 in the proper phase to insure a gain of one. The output of the first two stages, which act as an impedance transformer, is fed to Q_3 where it is amplified and applied to the bus which feeds the four individual discriminator channels. Resistor R_6 and d-c feedback through R_7 improve gain stability. The voltage gain from the input of Q_1 to the output of Q_3 is about four. The β of Q_3 is chosen to be about 30 thus allowing a maximum signal of about 7 v on the common bus.

The bus supplies the signal to four potentiometers, the settings of which determine the discrimination point of that particular channel. The four channels are identical, and only one is shown in the diagram. The input amplifier Q_4 is similar to the preceding stage and has a voltage gain of about four. The trigger circuit is a Schmitt type adapted to transistors. Feedback for regenerative action is applied across R_8 in the emitter circuit. The output pulse is about 10 μsec long. Accelerating capacitor C_1 has a small effect on the output pulse width. The thermistor network is necessary to keep the discrimination point constant with temperature changes. It has some shunting effect on the feedback resistor R_8, and, as a result, the triggering point is not as distinct as with the vacuum tube version.

The gain of the amplifier decreases slightly as the temperature decreases, but the thermistor network overcompensates slightly making the circuit insensitive to temperature.

Energy-Loss Telescope

The multiplier phototubes for the scintillation detectors in the energy-loss telescope receive their plate and dynode potentials from the same 900-v battery as the single scintillation detector. The output of each multiplier phototube is fed through an impedance transformer, an amplifier, and a discriminator, and finally the outputs of the two channels are compared for coincidence. The signal in one channel is also sent through another amplifier, a gate circuit, and finally to a height-to-time converter.

The gate is an AND gate which only allows the pulse through to the converter if there is a pulse from the coincidence circuit. In order to get an output from the height-to-time converter there must be a signal in each channel large enough to pass through its discriminator, and the signals in the two channels must be coincident.

The amplifiers in all three channels are almost the same as the one in the four-channel discriminator. The trigger circuit is also similar, except that the transistors are replaced by npn types to get positive input and output pulses, and some component values are changed. The most significant change is the replacement of the thermistor network and emitter resistor with a

Balloon is made ready for launching of cosmic ray measuring apparatus

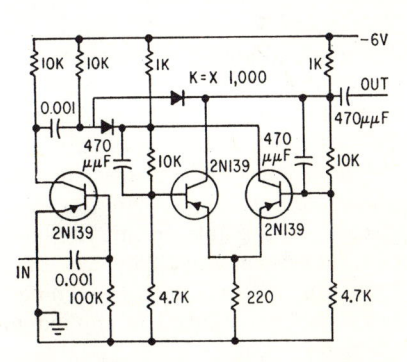

FIG. 2—Each scaler stage is a bistable circuit with trigger amplifiers between each two stages

single Sensistor as shown in Fig. 4. There is no longer a shunting effect on the feedback, and the regenerative triggering circuit compares well with the vacuum tube version. The trigger output pulse has a minimum length of about 3 μsec.

The coincidence circuit is a transistorized version of a common

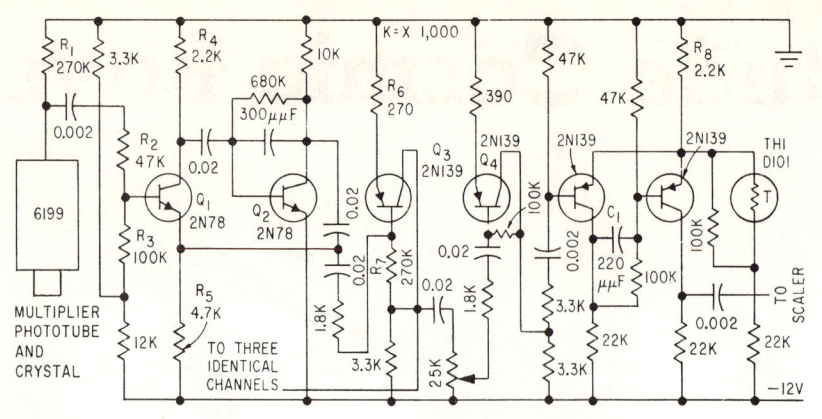

FIG. 3—In this four-channel discriminator, the first two stages make up the impedance transformer

tube circuit. In the quiescent state the base resistors of both Q_6 and Q_7 keep the transistors conducting heavily, and the potential across R_1 is nearly the full supply voltage. When a positive pulse appears at the base of Q_6, it is cut off momentarily, but Q_7 remains in conduction, and thus the potential across R_1 changes only slightly. If a pulse appears at Q_7, but not at Q_6, the condition is the same as before; however, if a pulse appears at both Q_6 and Q_7 at the same time, then a negative output pulse is sent as a gate pulse to Q_2.

Gate Circuit

The height-to-time channel gets its signal from the divider R_4, R_5. The pulse is delayed slightly by R_2 and C_1 to get it in time with the gate pulse which must go through several stages and thus is delayed. The delayed signal goes through the impedance transformer, is amplified, and is applied to the gate circuit. In the gate circuit the base of Q_2 is returned to 1.25 v, and thus it is in a condition to conduct heavily if a potential is applied to the collector.

As long as there is no gate pulse at the base of Q_2, the positive pulses from Q_3 are developed across R_6 by the shunting action of Q_2. If a negative gate pulse is applied to the base of Q_2 at the same time as a positive signal pulse appears at the collector, the signal will not be attenuated. When this condition is satisfied, the positive pulse charges C_2 through D_1.

The time constant of C_2, R_6, the resistance of the diode and the resistance of Q_3 in conduction is approximately 1 µsec. since the rise time of the pulse at this point is about 1 µsec, the charging capacitor C_2 is charged nearly to the peak pulse voltage. The capacitor is allowed to discharge through R_7, and the exponential voltage is used to cut off V_1. This amplifier is saturated by the large input voltage, and its output is nearly square. The voltage to which C_2 is charged is a function of the energy lost in the scintillator.

The time at which V_1 is cut off is a function of the voltage on C_2; hence the width of the output pulse of Q_4 is a function of the energy loss and can be calibrated to measure that energy. At the time of calibration the ratios of the divider circuits in the emitters of Q_1 and Q_8 are adjusted to compensate for the variation in the gain of the multiplier phototubes.

Subcarrier Oscillators

To complete the instrumentation, a stable subcarrier oscillator is needed. A Colpitts oscillator designed for 7,350 cps and 12,300 cps is used with the frequency modulation by the reactance method.[1]

Transistor Q_3 as shown in Fig. 5 is connected in a Colpitts configuration with C_3 and C_4 in parallel with C_5 acting as the feedback divider. Capacitor C_5 is also used for tuning. The output is taken from the emitter circuit because it has a rather low impedance and also has a waveform which is quite simple to filter to a near sine wave.

The signal from the other oscillator comes in through another RC filter, and the mixed output is taken across R_1. The emitter current of Q_2 is a combination of d-c from the first stage and the a-c fed back and shifted in phase by C_2. The collector current drawn through L_1 by Q_2 is not exactly in quadrature with the current drawn by the oscillator. The result is a change in the apparent inductance in the circuit, which causes a change in the oscillator frequency. The frequency is determined by the emitter cur-

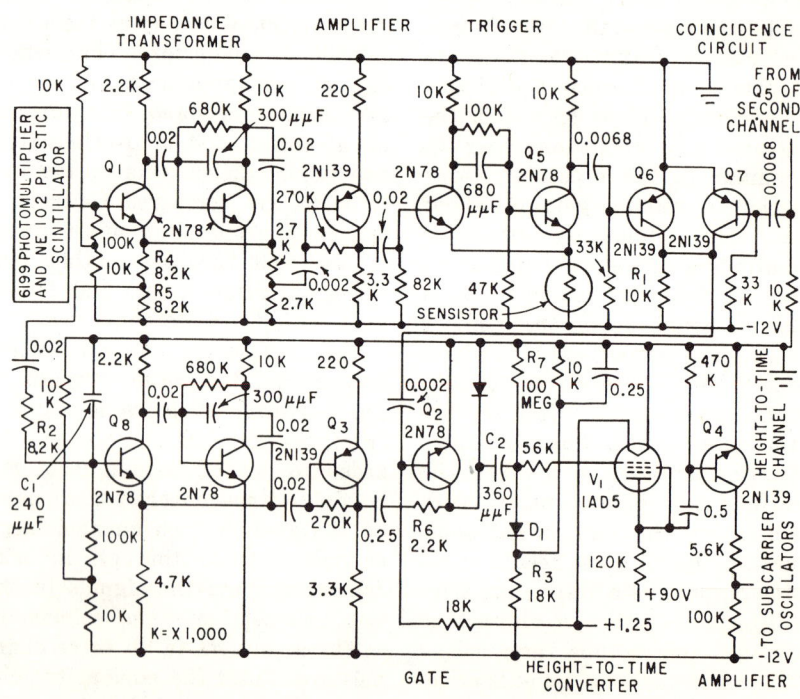

FIG. 4—Energy-loss telescope uses Sensistors to help compensate for temperature effects. Circuit normally employs two identical channels for the two phototubes

rent of Q_2. The input impedance of Q_2 is very low, and the first stage is added to make it easier to modulate the oscillator. To avoid temperature stability problems, Q_1 and Q_2 are connected in different configurations so that the drift of one tends to nullify the drift in the other, but the compensation is not complete. The d-c feedback gives the additional stability required.

In this instrumentation only pulse, not analog, information is fed to the sub carriers, so there is no need for either absolute frequency stability or absolute deviation linearity.

The input pulse to the 12,300-cps oscillator is a square wave from the height-to-time converter of the telescope, and it is allowed to modulate the oscillator directly. The pulses from the scalers going to the 7,350-cps oscillator, however, must be modified in some way so that they can be identified later. The input networks differentiate and rectify the square waves from the scaler. The diodes are connected in opposite directions to allow pulses of opposite polarity in the two channels.

Capacitors C_1 and C_6 are necessary across the diodes to prevent rectification and resultant biasing from the r-f carrier.

The input to the subcarrier from the ion chamber is through a large capacitor and no diode, so the square wave is converted to a very wide, differentiated pulse of both polarities. The input from the barometer-altimeter is through resistor R_2. As the drum in the Olland cycle rotates, it grounds one end of R_2 at keyed intervals. For any pressure there is a unique arrangement of pulses.

Modulator and Transmitter

The modulator and transmitter used in this instrumentation is described in detail in ELECTRONICS.[2] The transmitter is a quarter-watt, 92-mc, transistorized, f-m unit. The modulating signals are mixed at the input to the modulator, and the composite signal is applied to the f-m transmitter. Two scaler channels are sent to the subcarrier oscillator, and the other two are differentiated and clipped by networks similar to the ones used with the subcarrier and applied directly to the transmitter modulator.

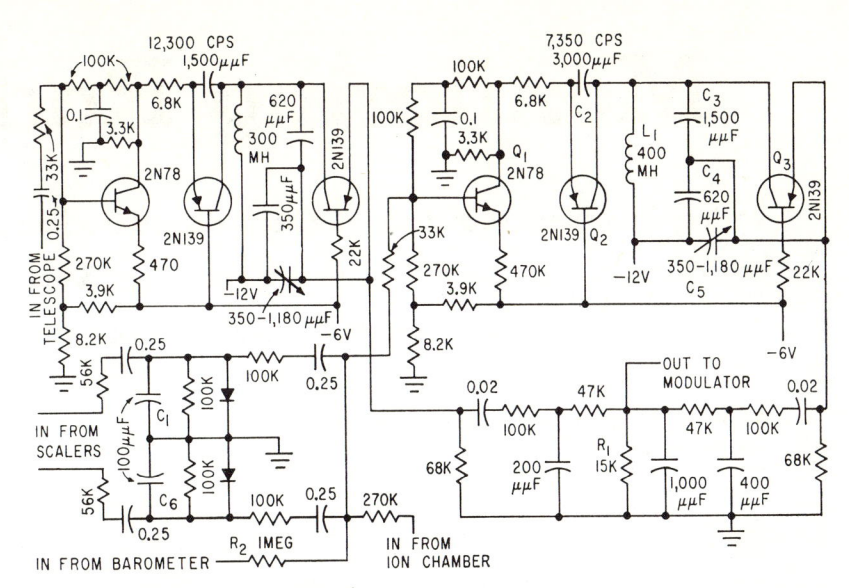

FIG. 5—Stabilization of the input stage of the subcarrier oscillators against temperature effects is obtained with d-c feedback

Coding of Modulation

With seven channels of information to be telemetered with one carrier and two subcarriers, some type of coding is necessary for channel identification. The 7,350-cps subcarrier has four channels of information impressed on it. One scaler channel registers as a positive spike, another scaler channel as a negative spike, the ion chamber scaler as a very wide pulse of both polarities, and the Olland cycle information appears as a shift in the base line.

Since the information from the energy-loss telescope may come at high rates, and the measurement is carried in the pulse width, it is the only channel on the 12,300-cps subcarrier. The pulse widths are measured directly by a time-interval meter, and they are recorded immediately by an automatic digital printer. The last two scaler channels appear as positive and negative spikes directly on the r-f carrier.

The information from the scaled channels is all transmitted with uniformly shaped pulses, and can be received through considerable noise. The height-to-time channel, however, must have a high signal to noise ratio if the pulse width is to be measured accurately; consequently, a large part of the bandwidth is assigned to that channel. The modulation of each channel is adjusted by varying the size of the mixing resistors at the input to the modulator.

In actual operation nickel-cadmium storage batteries have been used for power. They are capable of operating through the cold night hours without any appreciable lowering of performance.

Many of the individual circuits such as the subcarrier oscillator or the trigger circuit may be applied profitably to uses other than ballooning.

This project was done under the direction of Kinsey A. Anderson, Physics Department, State University of Iowa. The work was supported by funds from the Office of Naval Research.

REFERENCES

(1) F. M. Riddle, California Institute of Technology, Patent No. 2,728,049.
(2) Donald Enemark, Transistors Improve Telemeter Transmitter, ELECTRONICS, Mar. 13, 1959.

BIBLIOGRAPHY

J. L. W. Churchill and S. C. Curran, Pulse Amplitude Analysis, "Advances in Electronics and Electron Physics," Academic Press Inc., New York, 1956.
RCA Laboratories, "Scintillation Counters for Radiation Instrumentation," Navy Department, BuShips, Electronic Division, 1951.
G. H. Ludwig, "The Instrumentation in Earth Satellite 1958 Gamma," Physics Department, State University of Iowa, 1959.
H. V. Neher, Review of Scientific Instruments, No. 24, p 99, 1953.
L. E. Peterson, R. L. Howard, and J. E. Winckler, Balloon Gear Monitors Cosmic Radiation, ELECTRONICS, p 76, Nov. 7, 1959.
D. C. Enemark, Transistors Improve Telemeter Transmitter, ELECTRONICS, p 136, Mar. 13, 1959.
M. A. Korff and H. Kallman, "Electron and Nuclear Counters," D. Van Nostrand Co., Inc., New York, 1955.

Monitoring Radioisotope

Battery-operated radiation monitors record radioactivity level of flowing liquids or gases under field conditions. Unit described has battery life in excess of 300 hours and uses economical circular chart recorder

By F. E. ARMSTRONG and E. A. PAVELKA,
Div. of Petroleum, Bureau of Mines, U. S. Dept. of Interior, Bartlesville, Oklahoma

USE OF RADIOACTIVE ISOTOPES as tracers in the petroleum industry often involves monitoring the radioactivity level of flowing liquids or gases over relatively long periods of time. In many instances, location of monitoring points is such that reliable power is not available. Often no housing is available for the monitoring instrument and it is subjected to wide extremes of temperature and humidity. The proximity of heavy moving machinery may impose severe shock and vibration problems. As operation by untrained personnel is an occasional necessity, a minimum of controls or adjustments is desirable. The instrument described in this article was designed for field use under such conditions.

Recorder

Choice of a recorder was dictated primarily by cost considerations. A strip-chart recorder would have been satisfactory from an operational standpoint, but the cost of such units was more than twice that of the circular recorder used. The clock-stopping type of instrument[1] has a number of inherent faults. If an ordinary 12-hour clock is used, some method of eliminating ambiguity of indications is necessary, unless the device is attended at least twice during each 24-hour period. More important is the lack of quantitative data, particularly with an exposed and unshielded radiation unit. Variation in background count due to fallout, radon buildup or other phenomena can cause excursions of more than 3 to 10 times a normal reading. This in turn causes false alarms and will deactivate the unit until it is reset. Operations type recorders[2] eliminate this difficulty but still do not present data that would allow analysis of the variation in radioactivity and determination of its cause.

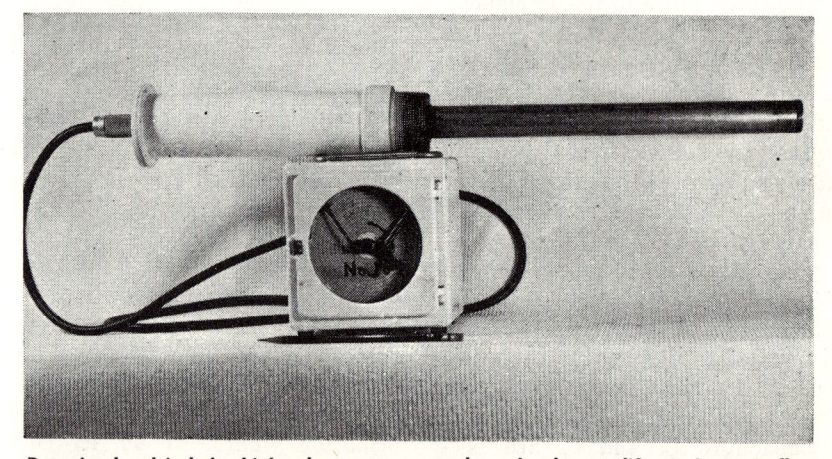

Detection head includes high-voltage power supply and pulse amplifier in hermetically-sealed case. Remainder of circuit and recorder is mounted within housing

Detection Unit

Figure 1 shows the circuit diagram of the detection unit. The Geiger-Muller counter is a brass-wall, bismuth-coated-cathode unit operating at 1,100 v. The normal background counting rate is about 300 cpm. Sensitivity is such that when used as dipping counters in a liquid volume of 7 liters, a concentration of 10^{-10} curies per milliliter (0.1 microcurie per liter) of Iodine[131] yields a counting rate slightly more than 200 cpm above background.

A potted commercial high-voltage unit is used to supply high voltage for the counter. Considerable variation of required input power occurs among different supply units. The average unit requires 100-mw input power to supply 11 mw of high-voltage power. The 100-mw input power is more than half of the total power requirement for the battery recording unit. Improvement of efficiency in this portion of the equipment would yield substantial gains in battery life.

The G-M counter pulse is approximately 5 v in amplitude. To prevent overloading of transistor Q_1, with accompanying pulse-width variations, voltage divider R_1 and R_2 reduces the input pulse to 0.5 v. Diode D_1 across the primary of transformer T_1 damps out any ringing. Transistor Q_1 operates with no bias and draws little or no quiescent current.

The output pulse is 0.75 v in amplitude and 20 μsec wide. The im-

Tracers for Industry

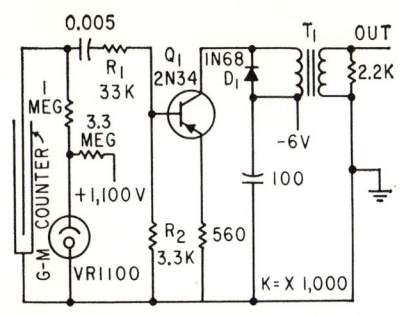

FIG. 1—Geiger-Muller counter receives its high voltage from potted unit

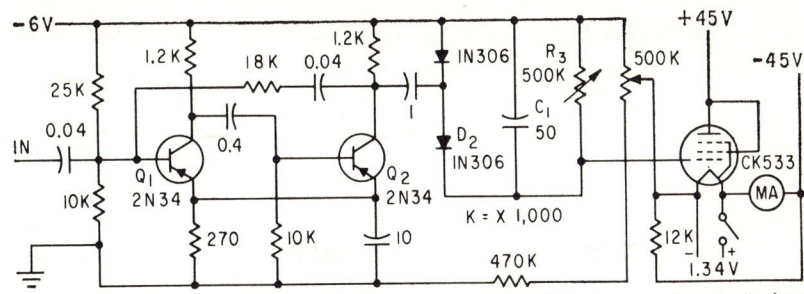

FIG. 2—Two-transistor one-shot multivibrator output signal is integrated and displayed on recording meter

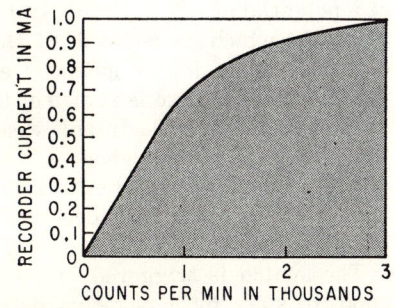

FIG. 3—Recorder response is linear for approximately the first half scale and essentially logarithmic over the remainder

pedance at the output transformer secondary is sufficiently low so that no crosstalk occurs when driving 500-ft lengths of unshielded 4-conductor cables, provided the cables are terminated in 2,000 ohms or less.

Counting-Rate Meter

The counting-rate meter shown in Fig. 2, uses two transistors in an integrating circuit and a vacuum-tube recorder-drive circuit. The vacuum-tube output circuit was used because of the high quiescent current demands of suitable transistor output circuits. The single-ended tube circuit is sufficiently stable so that overall system drift is less than 10 percent.

The counting-rate circuit consists of transistors Q_1 and Q_2 in a one-shot multivibrator circuit whose output is a 4.5-v, 260-μsec square pulse. This pulse charges integrating capacitor C_1 through diode D_2. The multivibrator requires a 0.1-v pulse for reliable triggering. Since several times this voltage is available from the detection unit, no input amplifier is required.

Linear response in a counting-rate integrating circuit requires that the charging-pulse amplitude be very large with respect to the range of voltage over which the integrating capacitor voltage will swing. In this application, non-linearity in the higher portion of the instruments range is desired; therefore the integrating capacitor is allowed to charge nearly to the maximum pulse voltage at full scale. The resultant response curve is shown in Fig. 3. The instrument is linear over a little more than the first half of the scale and essentially logarithmic over the remainder. This allows small variations of interest to be seen without losing the record if an unexpected large amount of radioactivity appears. Because only changes in radioactivity level rather than absolute measurement are of interest, the rate meter is not calibrated. Variable-discharge resistor R_3 allows the instrument range to be continuously adjusted to permit setting the recorder to $\frac{1}{4}$ to $\frac{1}{3}$ full scale regardless of background variations at various locations. When absolute readings are required, the instrument may be calibrated with either a standard source or a pulse generator.

A mercury cell supplies filament current to the triode-connected output stage. With a plate supply of 45 v, the E_g/I_p curve is linear over a wide range although tube ratings are exceeded by a factor of 2. Five mercury batteries supply the plate voltage. The batteries will operate the unit almost 1,000 hours under normal operating conditions except for the battery powering the transistor multivibrator and the high-voltage supply. This battery, which is mounted on the outside of the assembly for ease of replacement, is good for 250 to 300 hours of operation.

Performance

The instrument has been operated continuously for as long as three weeks and the overall drift did not exceed 10 percent. Half of this drift is caused by battery aging and the remainder is thermal drift. Most of the drift could be corrected by adding a zener diode to the 6-v circuit. This would take an additional 20 to 40 mw of power. Range changes of approximately 10 to 1 may be made with the adjustable discharge resistor across the integration capacitor. Changes of larger magnitude than this require that the duration of the charging pulse output from the multivibrator be shortened by changing the R-C feedback elements.

Although this instrument was designed specifically for use in the application of radioisotope tracers to petroleum production, it is adaptable to many other field problems which require continuous recording of similar information.

REFERENCES

(1) R. Farmer and O. Reiner, Jr., Determining Arrival Time of Radioactive Fallout, ELECTRONICS, 31, p 76, Aug. 1, 1958.
(2) R. Rhody, T. Evans and C. Ford, Inexpensive, Rapid, Automatic Recorder, Nucleonics, 14, May, 1956.

Radioactive Tracers Find

Simultaneous gating of oscillator and radiation detector permits recording of flow rate of jet fuel containing radioactive tracer. Reliable transistor circuit can be used for other time-interval measurements

By J. D. KEYS and G. E. ALEXANDER, Department of Mines and Technical Surveys, Ottawa, Ontario, Canada

DURING THE COURSE of investigations using radioactive tracers to measure flow rates of liquids, a circuit was required to measure time intervals of about 25 milliseconds. The particular application involved is the measurement of the rate of flow of fuel to a jet engine. Previous methods used to measure flow rates[1,2] are considered unsuitable either from the instrumentation point of view, or, in the case of the total-count method, the flow rates encountered are too great.

The circuit developed for this purpose uses a Geiger tube as the detecting element. A transistor switch circuit is operated by an initiating pulse, and it controls gate circuits for the oscillator and Geiger amplifier circuits.

The first gate permits output from an oscillator to be recorded, and the second gate permits output from the Geiger tube to pass to a scaling circuit. After a predetermined number of pulses, the scaling circuit feeds back its output to the switch circuit to close the two gates. Time interval is derived from the recorder based on the known oscillator frequency.

Operating Principle

A block diagram of the overall circuit is shown in Fig. 1. The operating cycle is initiated by a trigger pulse that is amplified by the trigger amplifier and applied to the switch circuit. The switch circuit performs two simultaneous functions. The first is to open the gate for the crystal-controlled oscillator, permitting oscillator output to be applied to the recorder. The second is to open the gate controlling the detector amplifier, permitting output from the Geiger tube to pass into the scaling circuit, which is of conventional design.[3]

Output from the scaling circuit is fed back to the switch circuit, which closes the two gates previously opened. The time elapsed between the initiating trigger and the closing pulse from the scaling circuit is read out from the recording device. Thus, the time interval depends upon the pulse rate delivered by the Geiger tube, which depends in turn on the fuel flow rate.

The scaling circuit acts as a discriminator against random background. Depending on the time interval to be measured, scales of 2, 4 or 8 may be used to prevent spurious radiation from closing the gates prematurely. When the radioactive tracer passes the detector, the rapidly increased count rate is sufficient to create an output pulse from the last binary stage of the scaler. This output actuates the switch circuit.

A manual reset is incorporated into the circuit for use in applications where the time interval is terminated mechanically rather than by pulses from a Geiger tube.

Switch and Gates

The switch circuit consists of a bistable flip-flop circuit, which is shown in Fig. 2. With the circuit in the ready condition, no current flows to the collector of Q_1, which is at a potential of -5 v. In this state, point A, which is coupled to the detector and oscillator gates, is at -2.6 v. This voltage is sufficient to keep the gates closed. In this case, the gates are closed when Q_3 and Q_5 are conducting and shunting the signals applied to their collectors to the zero line.

The switch is triggered into the opposite state by a negative pulse of at least -2.8 v applied to the collector of Q_1. When transistor Q_1 is

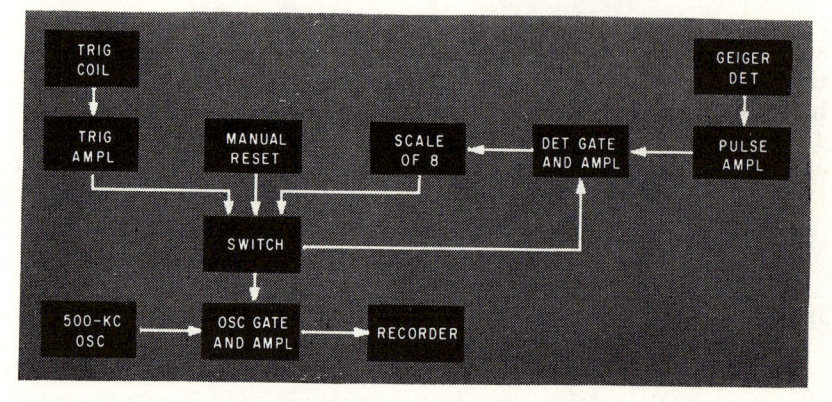

FIG. 1—Switch opens gates for detector and oscillator outputs

Jet Fuel Flow Rates

conducting, the potential on its collector rises to −0.5 v, and the collector of Q_2 is cut off.

With the switch in this state, the potential at point A falls to −4.8 v. This drop in potential is applied to the bases of gate transistors Q_3 and Q_5, cutting them off.

The circuit remains in this condition until a positive pulse from the scaling circuit is applied to the base of transistor Q_1. This increase in the potential on the base of Q_1 cuts this transistor off again. Collector potential falls to −5 v and transistor Q_2 again conducts. The drop in collector potential is again coupled to the bases of gate transistors Q_3 and Q_5, causing them to conduct which effectively closes the gates again. Since the scaling circuit gate is closed, pulses from the Geiger tube are shunted through Q_5 and no longer arrive at the scaling circuit. Therefore, the scaling circuit remains in the ready or reset condition.

The switch circuit is very stable and its operation is relatively independent of the pulse shape with one exception. The leading edge of the pulse must be sharp. Some overshoot can be tolerated because a positive pulse appearing on the collector of transistor Q_1 has no effect on the operation of the circuit.

The circuits of the oscillator and detector gates are also shown in Fig. 2. In the ready condition, the base of transistor Q_3 is at a potential of +0.53 v. Because transistor Q_3 is conducting under these conditions, the potential on its collector is zero volts and output from the oscillator remains shunted to the zero line.

When the switch circuit is triggered to its ON state, voltage at the base of Q_3 falls to +0.50 v. This slight drop in base voltage is sufficient to cut Q_3 off. Voltage on the collector of Q_3 rises to +1.8 v, raising voltage on the base of Q_4 to the same level. Therefore, oscillator output appearing at the base of Q_4 is amplified and fed to the recorder.

The pulse transformer in the collector circuit of Q_4 serves two purposes. The first function is to increase the output to the level necessary to actuate the recording instrument, which is a commercial scaling unit with a 1-µs input strip. The second purpose served by the pulse transformer is to provide isolation for the recorder.

The operation of the detector gate is exactly the same as that of the oscillator gate. However, an additional feature is incorporated in the detector gate—the insertion of a 10,000-ohm tapped resistor from the collector of gate transistor Q_5 to the zero line, rather than the fixed resistor used with the oscillator gate. Use of a tapped resistor permits some control over pulse height appearing at the base of detector amplifier Q_6.

Performance

The circuit described has been in operation for several months, both in and out of the laboratory. Its performance has proved to be very reliable. The particular application in connection with measuring flow rates of jet fuel with radioactive tracers is only one of many for which the circuit is suitable.

The contribution of G. G. Eichholz, in the form of many discussions during development of the circuit, is gratefully acknowledged.

REFERENCES

(1) W. H. Gauvin, I. S. Pasternak, L. B. Torobin and L. Yaffe, A Radioactive Tracer Technique for Particle Velocity Measurement in Solid-Gas Systems, Technical Report Series, Pulp and Paper Research Institute of Canada, Feb. 1959.
(2) D. E. Hull, The Total-Count Technique: A New Principle in Flow Measurements, *Int J Appl Rad and Isotopes*, 4, No. ½, 1, 1958.
(3) T. A. Prugh, Junction Transistor Switching Circuits, ELECTRONICS, 28, No. 1, p 168, 1955.

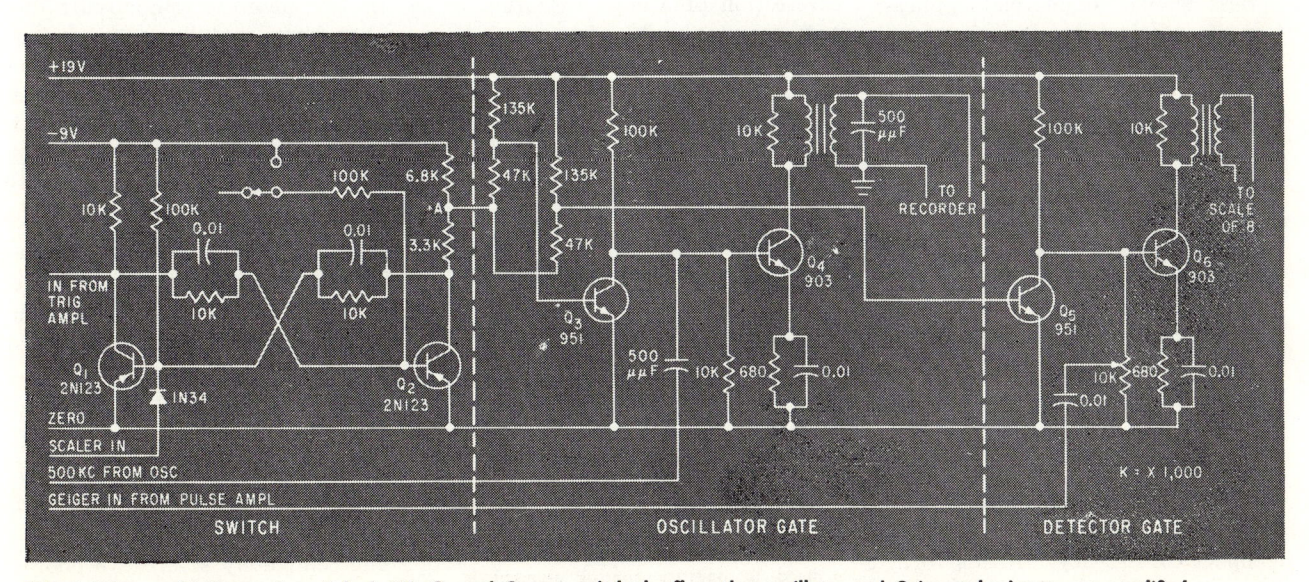

FIG. 2—When multivibrator is switched ON, Q_3 and Q_5 are switched off so that oscillator and Geiger tube inputs are amplified

Interruption of high-energy beam from radioactive source changes resistance across cadmium-sulphide detector. Transistor amplifier converts variation into signal capable of actuating limit switches and positional devices. Photocell system is unaffected by most environmental conditions, is small, inexpensive and reliable. Source-detector separation must be less than two in. to insure consistent detection

By PAUL WEISMAN, and STANLEY L. RUBY,
New Products Engineering Dept., Materials Engineering Dept., Westinghouse Electric Corp., Pittsburgh, Pa.

Solid-State Photocell

WEAK LINK in conventional photoelectric systems — the light source—is characterized by easily obscured light beams and lamps which require a power supply. To overcome these shortcomings, the photoelectric system discussed here uses a radioisotope source to furnish high-energy radiation and a solid-state photocell to detect the radiation. The beam formed cannot be obscured by hazy atmospheres; components are unaffected by temperature and shock and do not require a power supply.

Best known commercial application of radioisotopes is in beta-ray thickness gages used for measuring and controlling fast-moving, continuous sheet materials to within one percent of sheet thickness. The isotope used in most of these gages is strontium-90. This material retains half its radioactive energy after 25 years—a convenient decay rate for the present application.

Radioactive material is produced at a pile operated by the Atomic Energy Commission and is in the form of an aqueous solution. Chemical processing can be used to concentrate the material to a solid, compact consistency. In this form, the radioactive material can be permanently sealed into stainless steel containers which have one face thin enough to pass most of the emitted radiation.

A ten millicurie, strontium-90 source is used. It is cylindrically shaped and approximately $\frac{1}{8}$-in. thick and $\frac{1}{2}$-in. in diameter. If ultra-high resolution is desired, concentration of the source energy using radiochemical techniques could reduce the radioactive material to the size of a pinhead.

Photocell Detector

Detection of atomic radiation is most commonly done on an individual basis; that is, each disintegration is counted singly. This method gives great sensitivity but requires delicate, sensitive and expensive detection equipment.

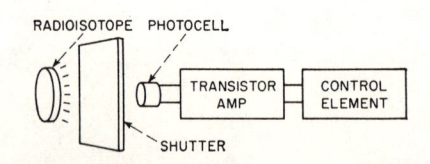

FIG. 1—Physical arrangement of photoelectric system. Radioisotope strontium-90 source generates high-energy beam which is detected by cadmium-sulphide photocell. Detected signal, after being amplified by bistable transistor amplifier, is used to operate control element in machines, elevator stops and the like

Scintillation counters, proportional chambers and the like are not, therefore, readily adaptable for use as small, low cost, rugged photoelectric devices.

A simpler method of detection is to measure the total ionization produced by the radiation absorbed in a volume of material. In this case the ions are collected and averaged into a d-c current representing the average ionization rate. The material generally used is a low pressure gas in which ions are free to move.

In the photoelectric circuit described here, the change in resistance caused by ionization in an appropriate solid is used to detect the high-energy beam. The solid used is a cadmium-sulphide photocell which features compactness, low cost and reliability. Several types of commercially available cadmium-sulphide photocells are sensitive not only to visible light but also to atomic radiations.

The particular cadmium-sulphide photocell used is cylindrically shaped and approximately $\frac{1}{2}$-in. long and $\frac{1}{4}$-in. in diameter. Size of the active sensing area is less than $\frac{1}{16}$ sq in.

The completely solid-state photoelectric system consists of the

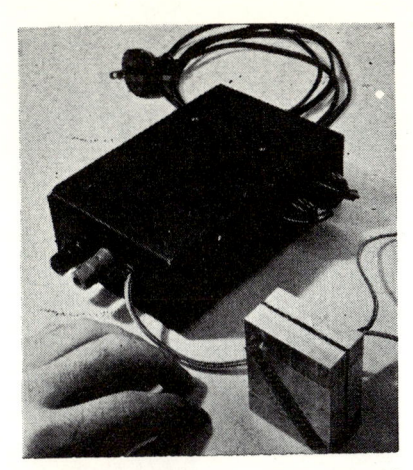

Complete photoelectric system. Bistable transistor amplifier is contained in chassis at top of photo. Slotted block holds radioisotope and shutter

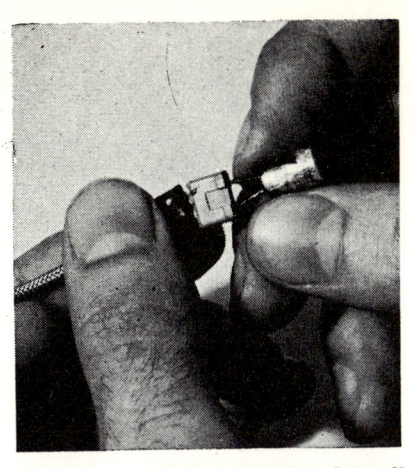

Closeup of cadmium-sulphide photocell mounted in transistor socket. This detector features compactness, low cost and reliability. Several types are available

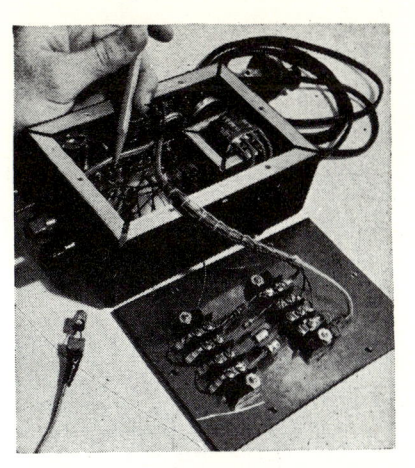

Internal construction of transistor amplifier. Pencil is pointing to 2N138A transistor. In left foreground are cadmium-sulphide photocell and cable

Sees Through Haze

radio-isotope source, the cadmium-sulphide detector and a bistable transistor amplifier. Physical arrangement of these components is shown in Fig. 1.

The radioisotope strontium-90 source generates a high-energy beam which is detected by the cadmium-sulphide photocell. When the shutter, which is made of $\frac{3}{32}$-inch aluminum and simulates the physical interruption of the beam, is inserted between the source and detector, the beam is interrupted and the resistance of the photocell increases.

The output state of the bistable transistor amplifier depends on the photocell resistance. When this resistance rises above a certain level, the output of the amplifier changes in a fraction of a millisecond from off to on. The resulting output signal is used to trigger alarms, actuate control elements or the like.

Bistable Circuit Operation

A power supply filter network formed by resistor R_1 and capacitors C_1 and C_2 provides a 100-v d-c supply across both the photocell-potentiometer R_6 and resistor R_4-R_5 voltage dividing networks. The off state of the circuit is caused by an absence of voltage across load resistor R_3, which is in series with transistor Q_2. Transistors Q_1 and Q_2 are a-c coupled to the photocell-potentiometer R_6 dividing network. When the photocell resistance is increased by interruption of the beam, a voltage is impressed across load resistor R_3 energizing a 20,000-ohm relay coil in the control device and converting the circuit to the on state. A slight amount of d-c coupling is provided by leakage through C_3 and C_4 allowing the output to follow slow shutter closures.

Many photoelectric applications such as assembly line counting or high-speed limit indicating require fast time response. In this respect, the strontium-90 isotope source and cadmium-sulphide photocell combination has its severest limitation.

The cadmium-sulphide photocell has a slow change of resistance after sudden application or removal of radiation.

The transistor circuitry can be adjusted to switch states at a photocell resistance only slightly below the dark value, giving a fast off-to-on response but a slow on-to-off response, or conversely.

Change in circuitry characteristics with temperature and age makes it inadvisable to balance too close to either extreme of the photocell resistance range. A reasonable compromise gives response times of about 0.05 sec and counting rate of about five pieces per second.

Maximum source-to-detector distance for consistent detection is about 4 in. A separation of 2 in. requires a source strength one-quarter that required at 4 in. With strontium-90 it takes over 50 years to reach this level.

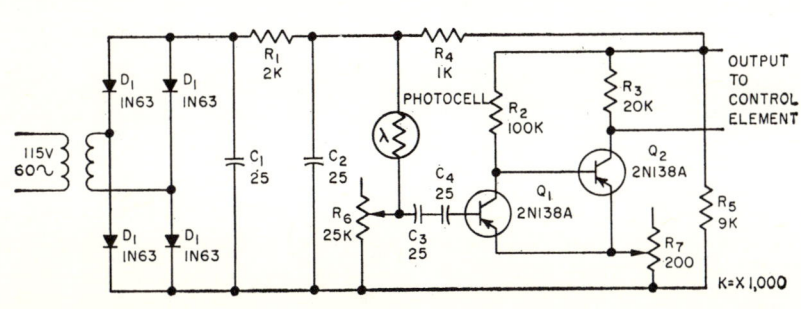

FIG. 2—Bistable transistor amplifier circuit. Isolating transformer is used in power line. Capacitors C_1 and C_2 are electrolytic; C_3 and C_5 are sintered electrolytic

Chapter 15
MEDICAL INSTRUMENTS

Circuit Substitutes for Larynx

EXPERIMENTAL artificial larynx has been announced by Bell Laboratories for persons who have lost their voices through surgical removal or paralysis of their vocal cords. Great impetus to development of the device was given by the National Hospital for Speech Disorders.

With limited difficulty and training, patients are said to be able to use the electronic larynx to speak conversationally. It is especially effective when conversing over the telephone.

Using a finger-operated combination push-to-talk switch and inflection control, the user can easily control pitch of his artificial voice, giving his speech a natural sounding quality previously unobtainable.

Operating Principle

The underlying principle of the new artificial larynx is a transducer held against the throat. Self-contained and cylindrically shaped, it measures only 1¾ inches in diameter by 3¼ inches long—sufficiently unobtrusive for the user.

Included in the package is a modified telephone receiver serving as the throat vibrator or transducer, a transistorized pulse generator with pitch control and a battery power supply. To miniaturize the artificial larynx, experimental units were built using modular techniques. However, by using printed-circuit techniques it is anticipated that an even more compact unit can be built.

In operation, the user presses the vibrator against his throat. Switching on the pulse generator with his finger, he transforms vibrations transmitted into his throat cavities into speech sounds by normal use of the articulatory mechanisms—throat cavity or pharynx, tongue, mouth, teeth and lips.

Output speech volume is equal to that of a person speaking at a normal conversational level, though the sound is a bit buzzy and mechanical. Nevertheless, the frequency spectra of vowel sounds show that the frequency range transmitted into the person's throat is sufficient for satisfactory production of such sounds. Users of the new device can achieve a sentence intelligibility of 97 percent or more, depending on their experience.

Circuit

Because the artificial larynx requires an economical, self-contained power source, parameters had to be adjusted to yield maximum acoustic output with minimum current drain. Accordingly, two transistors are used in a relaxation oscillator whose frequency is controlled by a variable resistor and whose pulse width is determined by a feedback network, as shown in Fig. 1.

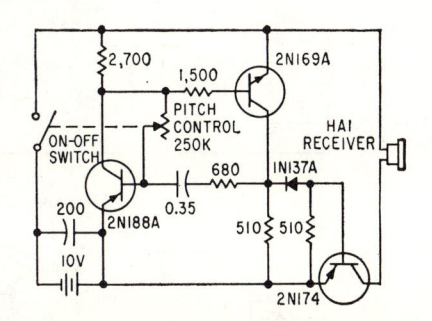

FIG. 1—Multivibrator and single-ended output amplifier provide good compromise between maximum output power and minimum current drain

Output is a negative pulse which occurs at a frequency of about 100 cps. This repetition frequency may be varied from about 100 to 200 cps by a rheostat which the user operates by pressure on the push-to-talk switch—or inflection control—while speaking, thus changing the pitch of his voice. For use by women, the frequency range is adjusted to 200 to 400 cps, to correspond with the normal range of a woman's voice.

A third transistor acts as a single-ended power output stage that amplifies the pulses applied to it from the relaxation oscillator. A diode isolates the multivibrator from the power amplifier input impedance during the period between pulses and is necessary for stable operation. Because a large pulse is required for sufficient acoustic power output at low frequencies, the oscillator drive circuit has heavy current requirements.

Two 5.2-v mercury cells in series provide the power necessary to operate the device continuously for a period of about 12 hours. These batteries have a 250-ma hour rating with maximum permissible current drain of 25 ma. With push-to-talk operation such as the patient requires, 12 hours of continuous operation should be equivalent to several days or even weeks of normal talking.

An alternative to using the self-contained mercury cells for powering the artificial larynx is a small a-c power supply which can be fed from a normal wall outlet at home or in the office. When the artificial larynx is plugged into the power supply, its batteries are disconnected from the circuit.

Instrument Monitors

Continuous indication of blood pressure, with better than 3-percent full-scale accuracy, is obtained by using variable-reactance pressure transducer mounted in 5-cc syringe. Transistorized excitation supply amplifier and power-supply circuits permit packaging entire instrument in 8 by 10 by 10 inch unit

By O. Z. ROY and J. R. CHARBONNEAU
Electromedical Project, National Research Council, Ottawa, Canada

RELIABLE AND ACCURATE methods of continuously measuring a patient's blood pressure during an operation have been provided for many years by commercial instruments. However, because operating room space is at such a premium it was felt that many of the bulky commercial models could be replaced by a transistorized monitor.

The instrument to be described has three ranges, 0 to 75, 0 to 150 and 0 to 300 mm of Hg; the mean blood pressure is indicated on a panel meter, while an additional output permits continuous recording of systolic and diastolic pressure variations. The complete instrument measures 8 by 10 by 10 in. and has a full-scale accuracy of better than 3 percent.

Basic Principles

By definition the maximum intra-arterial pressure during contraction of the heart, or the systole phase, is called systolic and the minimum pressure between relaxation and the start of the next heart contraction is called diastolic. The mean pressure is usually given as half the sum of the values for the systolic and diastolic pressures. A pressure introduced at the transducer effects an electrical relationship which is exactly proportional to the applied pressure.

The transducer, which is activated by a needle inserted directly into a patient's artery, is a commercially available variable-reactance unit that replaces the plunger

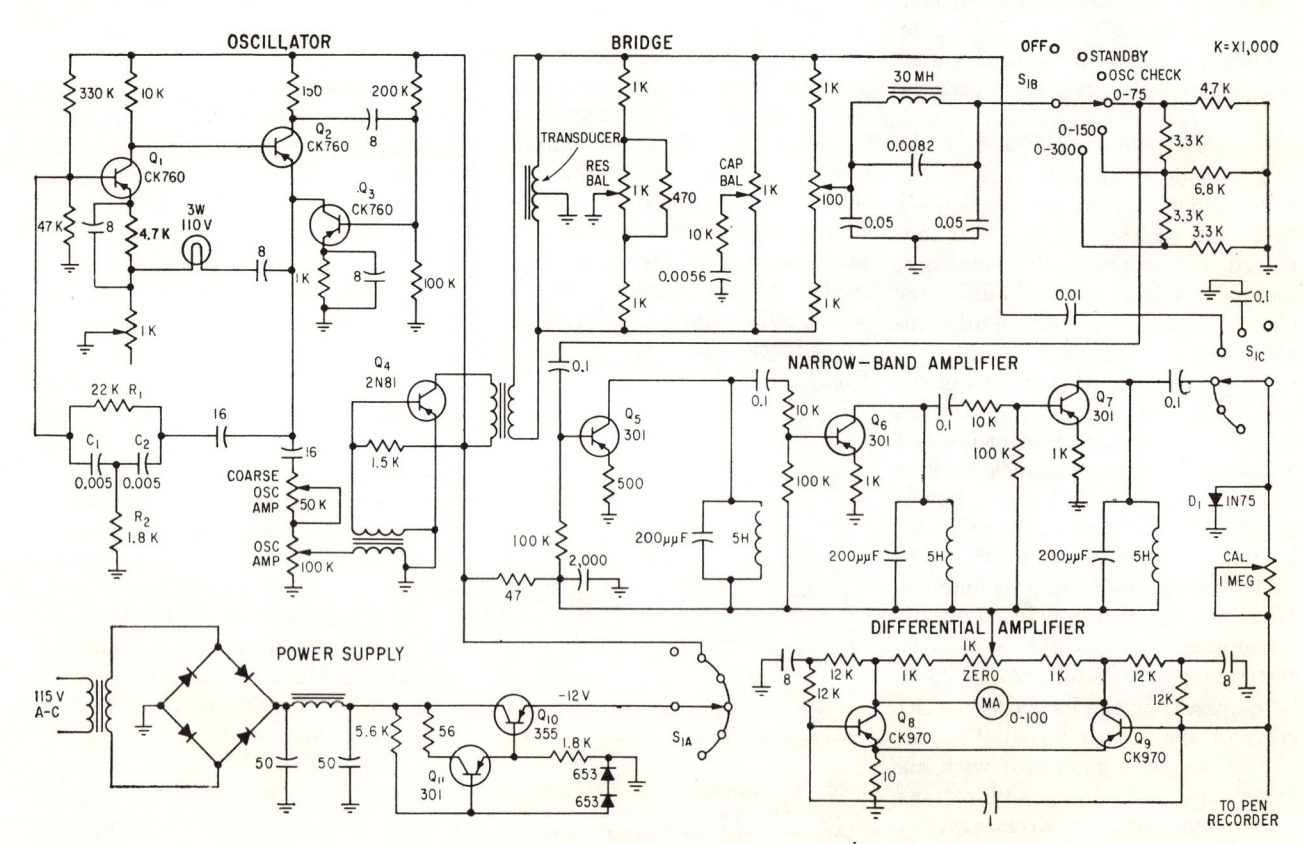

FIG. 1—Accuracy and repeatability of readings obtained with this circuit are limited only by characteristics of transducer used

Blood Pressure

in a 5-cc syringe. This assembly facilitates sterilization by solution or autoclaving. To prevent blood clotting, a three-way stopcock is used between the needle and the syringe for the introduction of anticoagulant solution.

Circuit

The transducer is excited by a low-distortion sine wave produced by the oscillator shown in Fig. 1. This circuit is the counterpart of the vacuum-tube Wien-bridge oscillator. Positive and negative feedback circuits generate a 5-kc, 1-v rms signal.

Frequency of oscillation is determined by the bridge circuit (R_1, R_2, C_1, C_2) in the negative-feedback loop. The amplitude of oscillation is stabilized by the lamp filament resistance in the positive feedback circuit. Power output stage Q_4 couples the oscillator to the low-impedance transducer bridge circuit.

With the bridge parameters shown and proper balancing procedure, the null potential can be made as low as 0.1 mv. The null point, although not absolute zero, is low enough to be negligible and serves as a reference for the output readings.

The signal from the bridge is fed through a range switch into a three-stage 5-kc amplifier comprising Q_5, Q_6 and Q_7; bandwidth and gain are shown in Fig. 2. Provision of sufficient negative feedback throughout the amplifier allows variations between transistors and provides good thermal stability.

The amplified pressure signal is rectified by D_1 and applied to the bases of Q_8 and Q_9. A microammeter is connected between the collectors. Silicon transistors are used because of their greater stability with temperature variations.

To obtain a true mean-pressure indication on the meter, the ripple voltage produced by the systolic and diastolic pressure variations is fed in phase to both sides of the differential amplifier through a large capacitor and thus does not affect the meter reading. However, a pen-recorder output is incorporated to obtain a record of systolic/diastolic pressure changes.

The constant-voltage transistor-regulated power supply produces −12 v at the load with better than 1-percent regulation for line variations of ±10 percent, and better than 5-percent regulation variations from zero to 100 ma.

Compactness, accuracy and dependability of unit make it ideal for operating-room service. Complete instrument measures only 8 by 10 by 10 in.

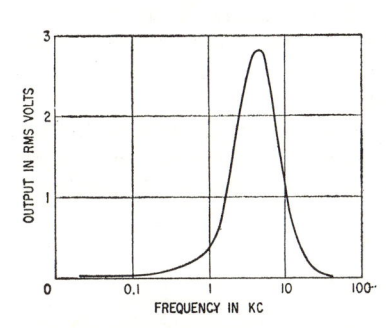

FIG. 2—Tuned amplifier response showing bandwidth and gain

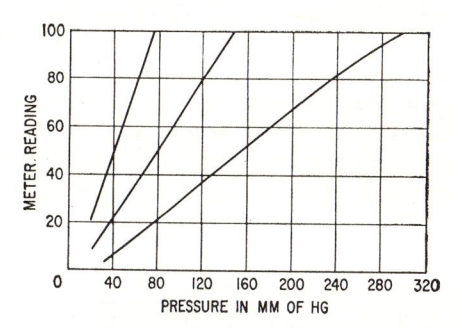

FIG. 3—Calibration curves for different operating ranges

Calibration

The instrument is calibrated by applying a known pressure on the transducer through a cuff manometer. With the range switch on the 0 to 75 mm range and a static pressure of 75 mm of Hg set by the cuff manometer, a full-scale reading is obtained by adjusting the current flowing into the base of Q_8 with the calibration potentiometer. Typical calibration curves for the instrument are shown in Fig. 3.

Accuracy and repeatability of readings are determined chiefly by the transducer used. With a Crescent type MPQ6 (0 to 300 mm Hg) transducer in the circuit, an accuracy of better than 3-percent full scale was obtained. The overall base-line drift, after a warmup period of two hours is less than 1 percent of the full-scale reading per hour.

277

Impedance Measurements

Transistorized device measures changes in impedance of living tissue resulting from nonrhythmic fluctuations of blood content. Changes of as little as 0.1 percent of the total tissue impedance can be detected by a resistance bridge and phase-sensitive detector that scans the bridge unbalance

By SAM BAGNO and FREDERICK M. LIEBMAN, D.D.S.

Walter Kidde Co.,
Newark, N. J.

Dept. of Physiology,
Dental College, New York University, New York, N. Y.

RHYTHMIC FLUCTUATIONS of blood content cause body tissue during the cardiac cycle to expand or contract with consequent change in tissue impedance. A device which detects and records these impedance variations is known as a plethysmograph. The plethysmograph to be described will detect changes in blood volume caused by nonrhythmic smooth-muscle activity in the walls of the blood vessels (vasoconstriction or dilation), as well as changes in extravascular fluid content and ionic changes of the tissue. These three causes of impedance change can be distinguished from one another only to the extent that their occurrence can be controlled under experimental conditions.

Operation

To measure impedance changes caused by vasomotor activity, a 50-kc signal is modulated by the unbalance of an impedance bridge. A pulsating unbalance results from the rhythmic changes in blood volume of the tissue being measured. These are the volume-pulse changes that coincide with the cardiac cycle.

However, nonpulsating change in resistance unbalances the bridge and requires a change in setting of the 25,000-ohm potentiometer for rebalance. The potentiometer is calibrated in ohms making the steady-state change in resistance equal to the difference between the new and previous potentiometer setting.

A schematic of the circuit is shown in Fig. 1. Transistor Q_1 in conjunction with T_0 constitutes an oscillator that generates approximately 50 kc. The output of T_0 is fed to emitter follower Q_2 which isolates the oscillator from the rest of the circuit.

Transformer T_1 supplies R_1, R_2, R_3, the 25,000-ohm potentiometer and the tissue segment which form a bridge circuit. The bridge permits measurement of resistances independently of the contact resistances. This is accomplished by using a four terminal connection, I_1 and I_2 to supply the current input and E_1 and E_2 to detect the voltage drop across the resistance being measured. The voltage drop across E_1 and E_2 is counteracted by the voltage from T_2 to form a null circuit.

The bridge also balances out noise fluctuations. The variation in impedance with the cardiac cycle is measured by the fluctuations it imparts to an electrical potential. Since the impedance fluctuations are generally less than 0.1 percent of the total impedance, their order of magnitude approaches the noise fluctuations in the voltage used to measure them. Use of the bridge to minimize noise insures that the modulation of the output of the

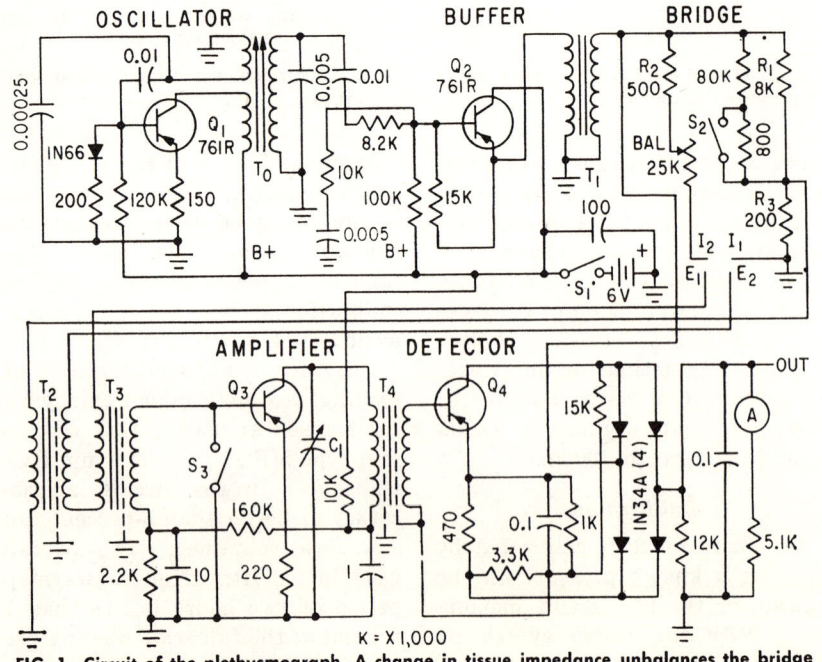

FIG. 1—Circuit of the plethysmograph. A change in tissue impedance unbalances the bridge

# of Living Tissue

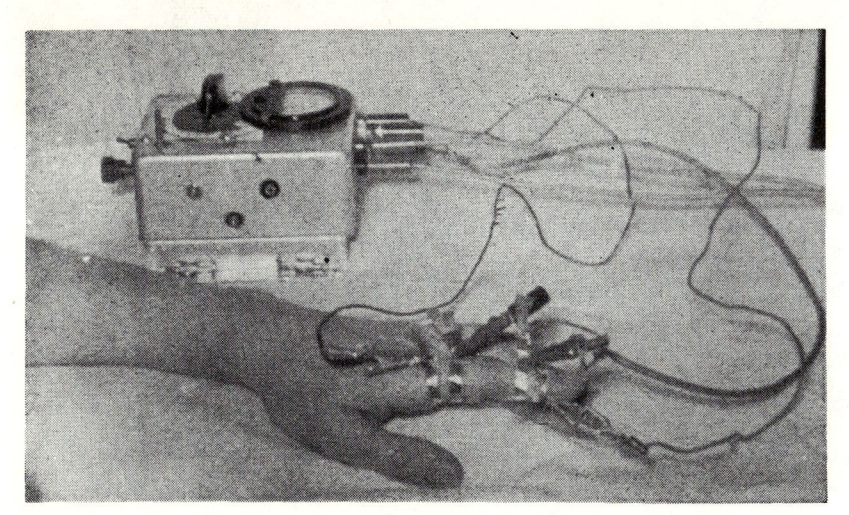

Electrode arrangement for living tissue impedance measurement

bridge is basically due to the variations in the resistance of the tissue segment.

The voltage picked off across E_1 and E_2, which is the voltage drop across the tissue segment, is subtracted from the voltage drop across R_3 to establish the null. This is accomplished by the use of 1:1 coupling transformer T_2 which transfers the voltage drop across R_3 so it is in series and bucking the E_1-E_2 potential. Once the bridge is balanced any subsequent difference in potential is impressed across the high-impedance primary of T_3. The secondary of step-down transformer T_3 matches the input impedance of a grounded-emitter stage Q_3 while the primary presents a high impedance so that practically no current flows between points E_1 and E_2. Common-emitter amplifier Q_3 amplifies the unbalanced potential of the bridge. Impedance matching transformer T_4 couples the amplifier to a phase detector. Variable capacitor C_1, in conjunction with the core of T_0, adjusts the phase of the unbalanced bridge signal to a reference coming directly from the generator. Both signals feed the phase-sensitive detector. Closing switch S_3, after placing the electrodes and activating the circuit, eliminates any signal from the bridge and permits phase adjustment through T_0 so that the galvanometer reading is initially zero. It also permits a check on phase adjustment at any time during a recording session.

Determining Null

Since the instrument measures the pulsating variation of the absolute value of impedance, the bridge null adjustment is simplified by using a phase-sensitive detector to scan bridge unbalance. Therefore, small unbalanced quadrature components of the impedance do not affect the null of the detector and the null measuring circuit is relatively independent of small variations in phase. Thus the bridge can be balanced simply by the 25,000-ohm potentiometer.

In operation the vascular volume-pulse modulates the 50-kc signal so that the output of the bridge fluctuates on either side of a null. The phase detector demodulates the 50-kc unbalanced signal without distortion. This is true even when the bridge is so well balanced that the 50-kc carrier of the modulated signal coming from the bridge is completely eliminated and only the sidebands remain. The phase detector thus permits the use of a simple adjustment.

The unit is powered by four 1.5-v pencil batteries. Because of the low voltage and high frequency of the signal, switches with gold plated contacts are used.

In previous electrical impedance circuits the current was kept constant and the output of T_2 was balanced by a potentiometer against the potential detected along the segment of tissue being measured. In this circuit the voltage is held constant. The output of T_1 is in series with the potentiometer and the current output of T_1 is adjusted so that the voltage drop along the test segment is balanced. This modification is important because of the low-voltage power supply and the noise level of the transistors. If the voltage were allowed to vary in proportion to the impedance of the tissue the voltage drop could approach the upper limit of the batteries and the voltage drop could be so low that the noise level of the transistors would be reached.

Detection

The modulated signal due to the unbalance of the bridge is amplified and drives the base of the detector Q_4. The detector circuit is similar to one described in the literature.[1]

If the resistance of the tissue remains constant there is no pulsating or other unbalance of the bridge and the signal from the bridge is unmodulated or zero. Under these conditions the signal driving the base of Q_4 is the constant bias voltage derived from the voltage divider and filter capacitor in the emitter-collector circuit of Q_4. The resultant collector voltage is proportional only to the rectified reference signal, thus its adjacent half waves will have the same amplitude. As a result, the current on the a-c side of the rectifier will be a sinusoidal current without a d-c component, hence a d-c pulse will not be detected.

To obtain the d-c component, both sets of half waves pass through detector Q_4. Any alteration in the sensitivity of the detector cannot affect the balance.

REFERENCE
(1) S. Bagno and J. Fasal, Intruder Alarm Uses Phase-Sensitive Detector, p 106, ELECTRONICS, Feb. 14, 1958.

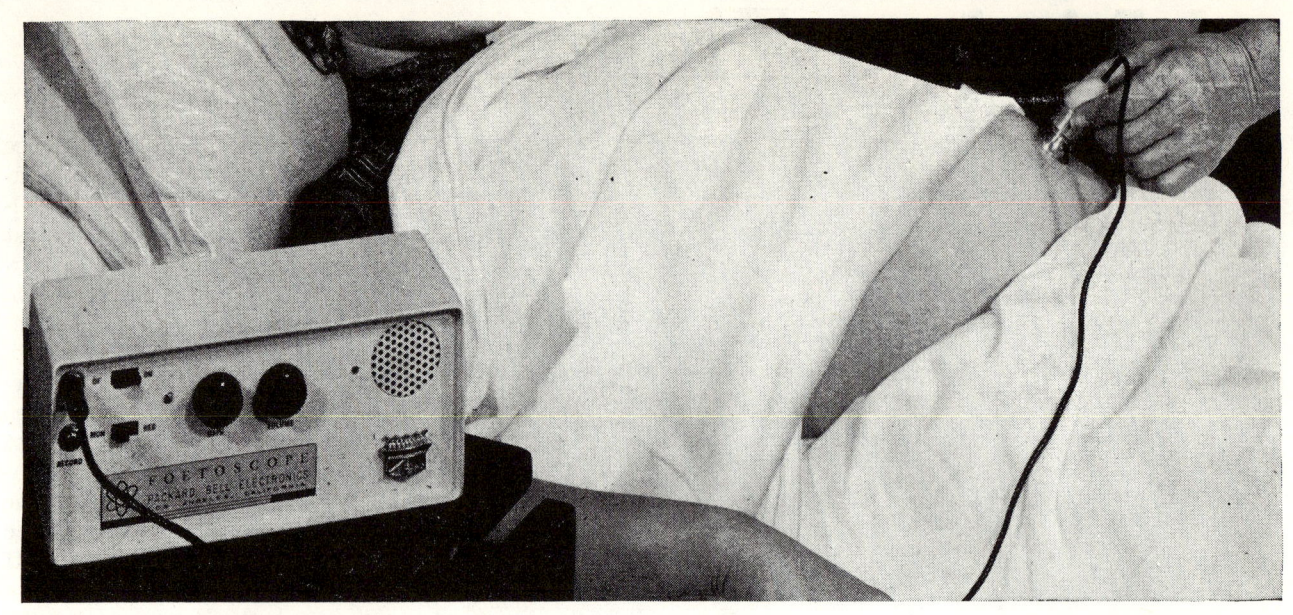

Maternity patient receives prenatal examination. Used with a recorder, instrument also analyzes adult heart action

Detecting Foetal

Amplified 2-to 3-cps signal from foetal heart modulates transistor oscillator operating between 800 and 1,200 cps. Frequency modulation technique overcomes poor low freqency response of human ear and loudspeakers. Device has additional cardiograph applications when used with recorder

By T. I. HUMPHREYS,
Assistant Chief Development Engineer, Packard-Bell Electronics Corp., Los Angeles, California

DETECTING FOETAL HEART sounds and amplifying them so that they are readily usable has been the subject of experimental work since 1906. This article reports another approach to the electronic problems involved and describes the equipment which resulted from a recent project.

The foetal heart beats approximately 125 to 180 times per minute. Thus, the fundamental frequency of the sound from a given foetus is somewhere in the range of two to three cps. The source of this sound lies inside the maternal abdominal cavity. The sound generated by the foetal heart must be conducted through a portion of the foetus and surrounding media to the external abdominal wall. This path attenuates low and high frequencies by different amounts.

Normal Techniques

The vibration that does get to the outer wall is normally picked up by the obstetrician with his stethoscope. This sound is conducted by actuating a column of air which directly connects to the eardrum, so that any change in the pressure of this column gives an audible sensation. This closed column of air does not exist when the sound is picked up by a microphone, amplified and converted to an audible acoustic signal. The coupling of a speaker to the air at low frequencies is slight since the air displacement at frequencies of a few cps is negligible. The spreading of the signal in the air further reduces the signal intensity so that by the time it reaches the ear it has been greatly attenuated. In addition, most ears will not detect an acoustical signal below 16 cps, making it almost impossible to use the technique of direct amplification unless the higher frequency com-

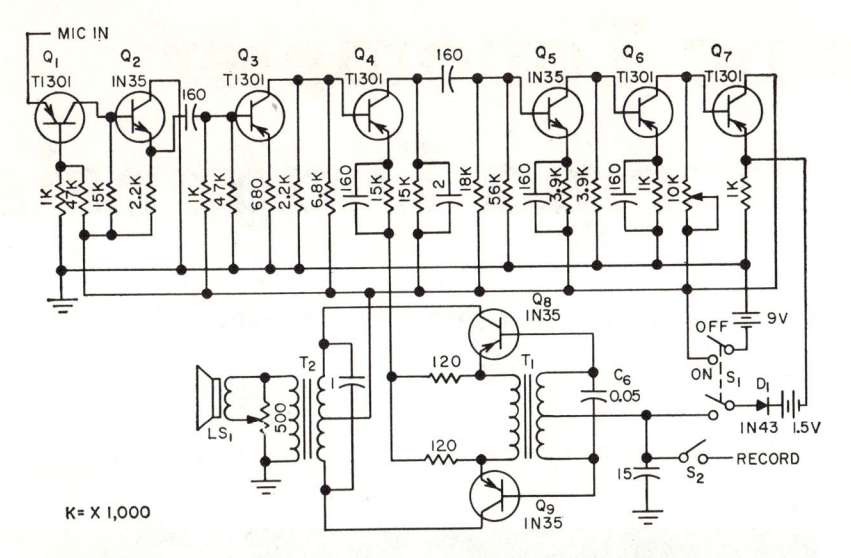

FIG. 1—Circuit schematic of foetal heartbeat detector. Amplifier low-frequency response is enhanced by large time constants in a-c coupled stages to insure that the low-frequency foetal heart signals are amplified

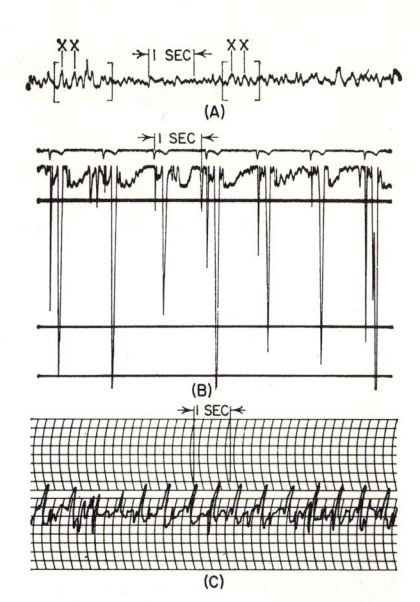

FIG. 2—Oscillograms of (A) 5-month foetus, (B) adult and (C) 4 year old boy

Heart Sounds

ponents alone are employed. These components occur at the same relative time as the foetal heartbeat, but may not coincide with the actual sound.

Circuit Description

The Foetoscope utilizes a carbon microphone, several transistor stages of amplification, an oscillator and a small speaker. The circuit diagram for the unit is shown in Fig. 1.

The transistorized amplifier uses direct-coupled stages wherever possible to pass the low-frequency signals. In those stages that are a-c coupled, large time constants are used. A potentiometer in the base lead of Q_7 provides gain adjustment. The output of this stage is fed through a battery and crystal diode to the center-tap of the oscillator coil T_1. The signal produces current changes in the coil and corresponding inductance changes in the secondary coil cause frequency modulation of the oscillator. By use of the frequency modulation technique, the low frequency component of the heartbeat can be heard in the form of an audible change in oscillator frequency which is set in the range of 800 to 1,200 cps. In this region the ear is quite sensitive and can readily detect small changes in frequency.

Figure 2A is a phonocardiograph of a five and a half month foetus, taken using this equipment. The sound from the foetal heart is indicated by the points marked X. This trace shows a repeating complex of signals, indicated by the brackets. The form of this particular complex indicates the possibility of the presence of more than one foetus.

Other Uses

The possibility of using the instrument for other than foetal heart sounds is suggested by recordings made using a young man and a small boy as subjects. Figure 2B shows the recording of heart sounds of the young man. Here the positioning of the microphone was found to have a considerable effect on the waveforms obtained. Figure 2C shows the recording of the heart sounds of a four year old boy.

The sensitivity of the unit is great enough to provide for the distinguishing of the several heart sounds as various valve and muscle motions occur.

Interior view of chassis shows construction details. Printed circuit and automatic insertion techniques are evident in compact unit

Electronic Tonometer For Glaucoma Diagnosis

Glaucoma can be detected by a probe that measures pressure within the eyeball. Early diagnosis makes ultimate cure possible

By R. STUART MACKAY and ELWIN MARG, U. of California Medical Center, San Francisco, Cal.

OF all people over forty, it is estimated that two percent are going blind from what is known as simple glaucoma. In this malady an increase in pressure, if not detected and corrected in time, causes irreparable damage to the optic nerve. Its onset is diagnosed earliest by measuring the pressure within the eye using devices called tonometers, after which the pressure can be relieved by drugs or surgery. A new electronic tonometer has been conceived that is so fast and gentle that it does not require anesthetics and yet it is more accurate because it does not respond to extraneous factors that lead to uncertain readings in the classic devices.

Conventional tonometers measure pressure in one of two ways. Either a plunger is placed upon the front surface of the eye in a vertical position and the indentation due to loading with a known weight measured, or else the area flattened by pressing a transparent plate against the eye with a given force is measured optically. Bending of the cornea introduces an uncertainty into the reading because of its stiffness or rigidity. The difficulty is increased in an astigmatic eye where the curvature may be different in different directions. Even the surface tension of tears introduces an uncertain component of force.

Operational Principles

Figure 1 shows the arrangement of the new tonometer. The eye is momentarily flattened beyond the pressure sensitive region. Since the bending takes place at the periphery of the probe, the central plunger is not acted upon by bending forces. Any tension in the tissues is a centrifugal force that does not act on the pressure sensitive area. The probe is a small handheld device that is momentarily touched to the eye. As long as the front surface of the probe remains approximately flat, essentially the only variable that will be recorded is the intraocular pressure.

In the present device flatness, as well as freedom from drift caused by changes in amplifier gain or changes in resiliency of the mechanical components, is assured by a feedback mechanism. Any tendency for the plunger to be deflected inwards is sensed by a sensitive motion transducer and is counteracted by the resulting change in

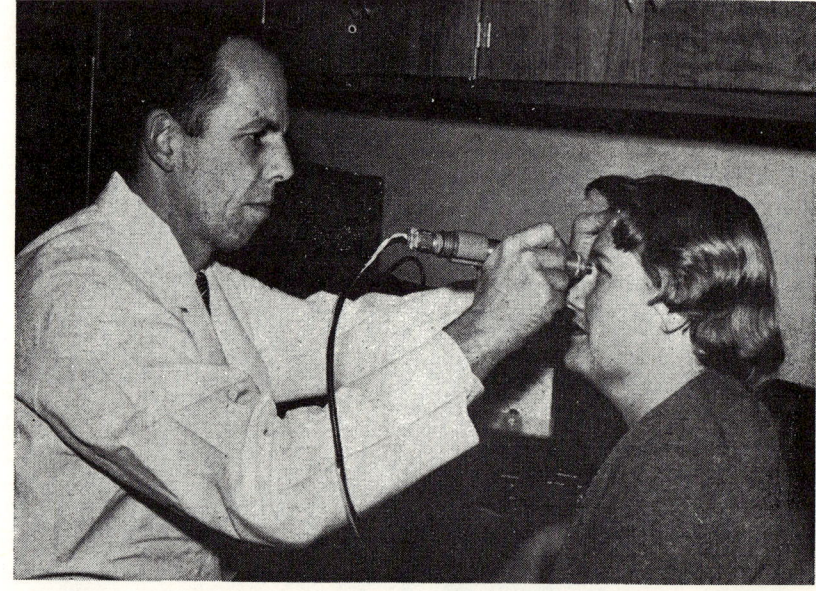

Eye pressure measurement can be completed in less than one second

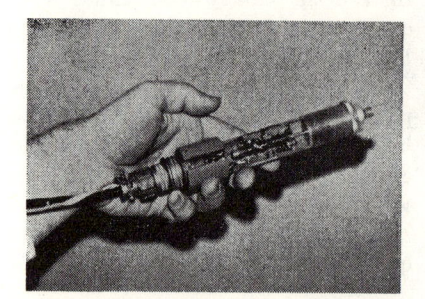

Probe is shown with shield removed

Close-up of central unit and probe

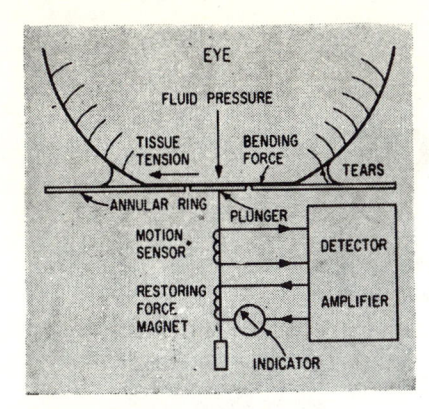

FIG. 1—Magnetic feedback is phased so that plunger resists eyeball pressure and remains approximately co-planar with surrounding annular ring

current in a direct coupled magnetic actuator. A measure of the pressure is then obtained by recording the current to the feedback magnet. The scale is linear and free from involved calibration because the plunger never moves appreciably and thus changes in transducer sensitivity with deflection, or variations in restoring force field, are not introduced.

In using the device the probe is momentarily pressed against the eye. As the force of contact increases, the recorded reading will increase until the plunger, which is about two millimeters in diameter, is covered. The further increase in force will not change the reading until the pressure within the eye begins to be raised by the external force. Thus the reading, which can be recorded on a penwriter, displays a plateau whose elevation is a measure of intraocular pressure (Fig. 2).

Circuit Operation

The requirements of the motion-transducer are twofold. It must (1) be extremely sensitive to small displacements and (2) very little force should either be required by the transducer, or reflected back into the system by it. The transducer employed made use of the motion of a ferrite core which altered the inductance of an adjacent coil. The change in inductance is measured by a circuit resembling some types of frequency-modulation detectors. The circuit is shown in Fig. 3.

Changes in position of the moving ferrite core cause a signal to be developed across capacitor C_1, with signal polarity dependent upon the displacement direction. The

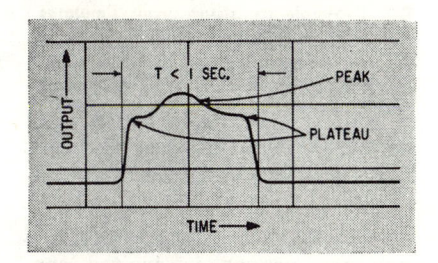

FIG. 2—Recorded response as probe is applied to, and taken away from, the eyeball. The peak indicates extra pressure caused by probe

signal (voltage) is fed to the d-c amplifier via the shielded probe-cable and the output from the d-c amplifier feeds the transistor-pair, Q_1 and Q_2. The amplified signal drives the restoring coil which forces the probe into the plane of its surrounding annular plate. The actuator consists of the voice coil and magnet from a small loudspeaker.

The signal from the displacement sensor is about 0.1 volt per micron of movement of the plunger. The feedback system applies a force of about 0.8 gram to the eye for a normal intraocular pressure of 16 millimeters of mercury. The deflection of the system is 0.6 microns for a pressure of 40 millimeters of mercury. A change in oscillator frequency from 5 megacycles down to 100 kilocycles decreases the sensitivity to uselessness.

In Fig. 3, T_1 and T_2 are Miller transformers, type 1467, the primary of T_1 being L_1. Coupled to T_1 and T_2 are L_2 and L_3, which each have 16 turns in two layers of No. 22 wire. The primary and secondary of T_2 are labeled L_4 and L_5 respectively. The probe is shielded and the shield has an axial slit so that it does not act as a shorted turn. The probe shield is grounded to the shield of the connecting cable. This connecting cable carries both radio frequency to the probe and the detected signal from it.

The plunger diameter is approximately 2 millimeters and the diameter of that part of the eye flattened by the surrounding plate is approximately 3 millimeters. The mass of the moving parts in the device is kept to a minimum so that accelerometer or seismograph effects are minimized in the recording as the probe is moved. Problems of friction were minimized in the design of the probe but their remaining interference was removed by including in the feedback loop a small 200 cycle voltage signal (fed in by T_3). The resulting small amplitude motion introduces no noticeable signal but eliminates the effects of static friction.

The help of Mr. Raymond Oechsli in perfecting the circuits and the probe is gratefully acknowledged.

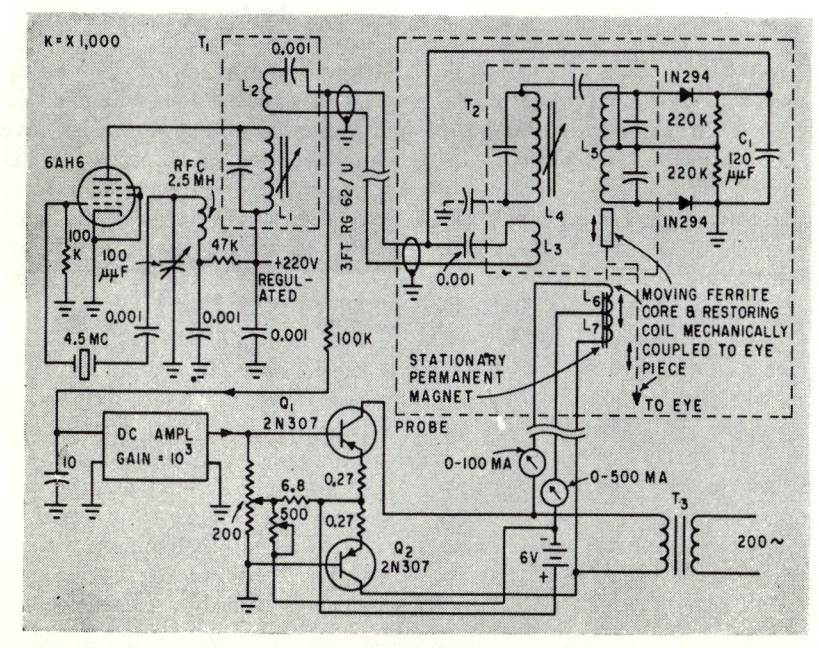

FIG. 3—Circuit diagram shows interconnection of oscillator, probe, and d-c amplifier

Artificial Neuron Uses Transistors

Elements of electronic nerve cell are mounted on printed-circuit card for ease of handling. Network of such cells can be seen in background

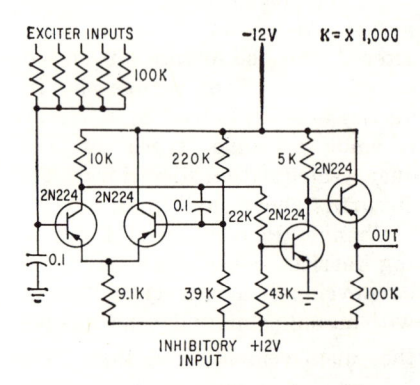

FIG. 1—Simple transistor circuit simulates many of the functions of nerve cells of the eye and ear. Groups of the cells can be combined to form simple nerve systems

ELECTRONIC circuit, developed at Bell Telephone Labs, simulates some functions of individual biological nerve cells (neurons). Numbers of the artificial cells are being combined into experimental networks that are roughly analogous to the nerve systems of the eye and ear.

Scientists are especially interested in discovering how visual and auditory nerves function and how their signals are interpreted by the brain.

The circuit shown in Fig. 1 delivers 6-millisecond pulses, considerably longer than the biological cell, but they can be shortened. If the circuit is driven by a constant stimulus, simulating receptor cells of the eye or ear, trains of pulses are emitted. Higher intensity excitation increases frequency; and when the neuron is excited continuously, frequency can be made to decrease with time, exhibiting accomodation as a living nerve cell does.

Input Threshold

Input must, as in a biological cell, surpass a threshold value, and the cell will integrate two or more input pulses below threshold value to cause firing. A particular input connection can also, while energized, inhibit firing of the neuron by other inputs. Similarly, immediately after firing, the electronic neuron's threshold rises to infinity and for a few milliseconds no input signal can fire the neuron again.

The cell has an integrating time constant of two milliseconds and a refractory time constant of about ten milliseconds, approximating time constants of the biological neuron. Because the electronic inputs and outputs are compatible, the cells can be assembled into chains and networks.

Electronic neurons can be combined with photo-resistive cells to simulate simple functions of nerves in the retina. Some receptors (on receptors) fire only when light intensity is increasing, off receptors fire only when light is decreasing and during receptors fire while light is steady.

Flicker-fusion phenomena have also been produced. In the human eye, these cause a sequence of flashes to be seen as continuous illumination.

Mutual inhibition of cells in an array has been demonstrated experimentally. Some animals have been observed to possess this arrangement, in which a cell receiving greater light inhibits firing of nearby cells that receive less light. Result is local sharpening of image boundry detail.

Chapter 16
MISSILE AND SATELLITE TELEMETRY

Data Conversion Circuits
For Earth Satellite Telemetry

Transistorized pulse height-to-time converters can be included in satellite payload. Output width is proportional to input height

By D. N. CARSON, Bell Telephone Laboratories, Murray Hill, N. J.
and S. K. DHAWAN, Dept. of Electrical Engineering, Columbia University, N. Y.

AT PRESENT, EARTH satellite experiments measuring radiation require that the amplitude distribution of pulses from radiation detectors be determined by tedious manual reduction of telemetered data. This time consuming work can be eliminated by including a complete pulse height analyzer as part of the satellite payload. Thus, the telemetered data would directly contain the amplitude distribution of detected radiation over a given period of time. Poor signal-to-noise and other problems preclude the use of an analyzer on the ground.

Key portion of the analyzer is a circuit which converts pulse heights to widths or time. The time pulse controls an oscillator, which gives a number of cycles of output directly proportional to the height of the input pulse. Output information may then be stored in some type of memory so that the number of pulses of each amplitude occurring in a given period of time may be determined. Such an analyzer (using vacuum tubes and mechanical registers) was first described by D. H. Wilkinson in 1950.[1]

Linearity of the analyzer depends upon the linearity of height-to-time conversion and the oscillator stability. The oscillator can be easily stabilized. However, linear conversion of height-to-time over a reasonable range of input pulse amplitude is not easily accomplished. Furthermore, when using semiconductor devices, as dictated by satellite requirements, the leakage current is particularly troublesome in the height-to-time portion of the analyzer. Two simple circuits which provide linear conversion are discussed.

The schematic diagram of what will be referred to as the constant current charge circuit[2] (CCC circuit) is shown in Fig. 1. This circuit accepts negative input pulses of up to 40 v in amplitude. Transistor Q_1 is used to charge capacitor C_1 at constant current. Transistor Q_2 and associated diodes make up the comparator.

It should be noted that the voltage comparison is made only at the instant C_1 is recharged, thus reducing leakage current problems. Transistor, Q_3, is merely an emitter follower output, which may be eliminated if a low output impedance is not required. Thus, the circuit performs the required function with just two transistors, Q_1 and Q_2. Direct connection between stages greatly simplifies the biasing and eliminates components.

CCC Operation

Under quiescent conditions, the approximately 200 μamp of current available from Q_1 is used to supply the base current of Q_2. The collector load resistance of Q_2 is selected so that the collector is saturated under these conditions. Upon application of a negative input pulse, C_1 is charged to the peak value of the pulse through diode D_1. In order for C_1 to charge faithfully, the rise time of the input pulses must be longer than the charging time constant of D_1 and C_1.[1] As soon as C_1 has been discharged (charged negatively) by the input pulse, diode D_2 is reverse biased, cutting off the base current to Q_2. After the input pulse has disappeared, the 200 μamp of current begins to recharge C_1. Diode D_1 isolates C_1 from the pulse source during the ramp generation.

Since the base current of Q_2 is reduced to zero by the reverse bias on D_2, the collector current of Q_2 cuts off and the collector rises to the full supply voltage of +25 v. When the ramp amplitude reaches

Table 1—Specifications for Constant Current Charge Circuit

Output pulse amplitude: 8 v (25 v without D_4).
Output pulse rise time: less than 4 μsec.
Output pulse fall time: less than 2 μsec.
Output pulse width: 415 μsec with 40 v input pulse.
Linearity: within 3.23 percent for input pulse amplitudes of 1 to 40 v and within about 1 percent for input pulse amplitudes of 1 to 30 v.
Temperature stability: stabilized from −30 C to +100 C.
Voltage Supply: +25 v and −4 v.

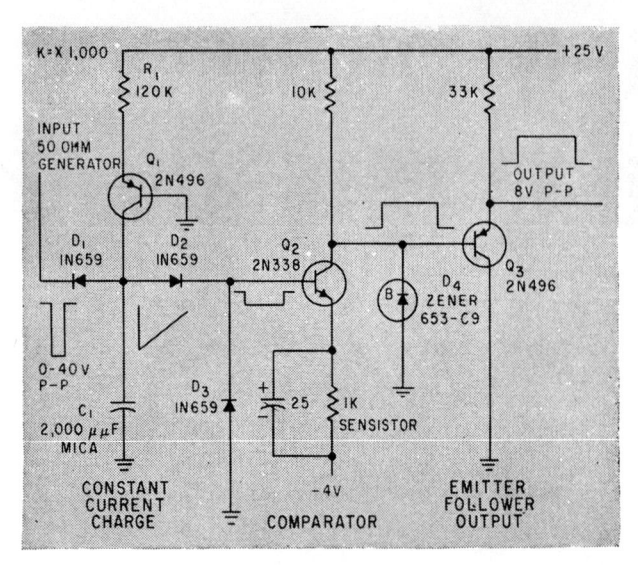

FIG. 1—Last stage of constant current charge circuit can be eliminated if output impedance can be high

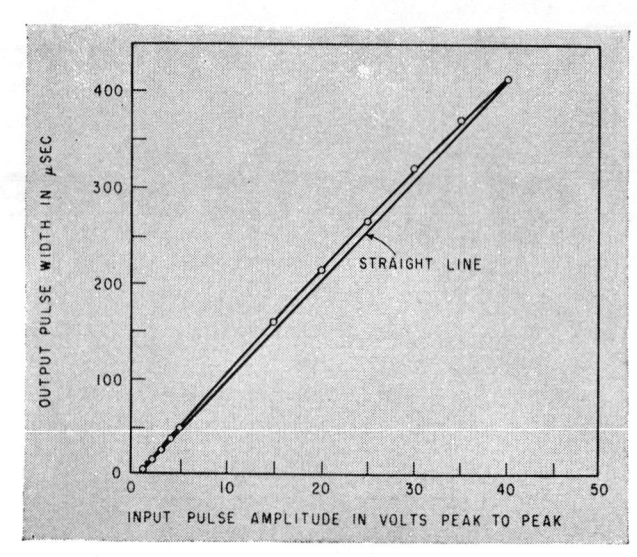

FIG. 2—Comparison with straight line shows a maximum departure from linearity of 3.2 percent for CCC circuit

the point where D_2 is once more conducting, the collector of Q_2 is again saturated.

A large amplitude (25 v p-p) rectangular pulse appears on the collector of Q_2. The width of this pulse is linearly related to the input pulse amplitude.

Zener diode D_4 is used to clip the output pulse to about 8 v p-p. In many applications it might be desirable to differentiate the output pulse for triggering purposes; in this case, D_4 can probably be eliminated. Diode D_3 may also be eliminated since its only function is to make it easier to adjust the nominal −4 v supply to the optimum value for biasing D_2.

Linearity

The linearity of the circuit is shown in Fig. 2. A straight line has been drawn between the end points to show the departure from exact linearity over the complete range of input pulse amplitudes. Maximum departure from linearity is 3.2 percent. If the input pulse amplitude range is restricted to 1-30 v, the linearity improves to about 1 percent. Obtaining input amplitudes this large with transistors presents problems. Unfortunately, the linearity below 1-v input could not be measured accurately. Improved methods of measuring the linearity, particularly at small input pulse amplitudes, are required.

Temperature stabilization of the CCC circuit was accomplished by using silicon diodes and transistors. In addition, a Sensistor is used in the emitter of Q_2 (Fig. 1). With these simple expedients, the upper limit of operation was between 100 C and 120 C, while the lower limit was between −30 C and −40 C.

Effect of temperature on linearity was not investigated because it was felt that as long as the output pulse width was stable for a fixed-amplitude input pulse, the linearity probably was not affected. Attempts at stabilization by control of R_1 were not examined.

Pulse Stretcher

Another approach[3] to the problem of providing linear conversion of height-to-time is shown in Fig. 3 and 4. In this circuit, which accepts positive pulses up to 7 v in amplitude, D_1, D_2 and C_1 stretch the input pulse for a fixed period of time while preserving the amplitude. Transistors Q_1 and Q_2 form a monostable multivibrator which

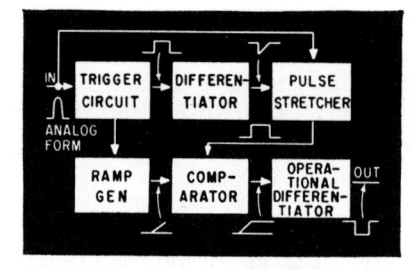

FIG. 3—Block diagram of stretcher shows waveform at various points

is used for delay and triggering.

Diode D_3 is used as a voltage comparator, the stretched input pulse being compared to the ramp from the bootstrap ramp generator, Q_3. Comparison is made only when the magnitude of the ramp reaches the stretched pulse amplitude. Transistor Q_5, an operational differentiator, provides a rectangular output pulse, the width of which is directly proportional to input pulse amplitude.

Stretcher Operation

The positive input pulse triggers the monostable multivibrator (Q_1 and Q_2), which gives a positive-going pulse of 20 v at the collector of Q_2 and a negative-going pulse of 4 v at the collector of Q_1. Time duration of these pulses (68 μsec) is several μsec greater than the widest output pulse required. The extra delay gives better comparator output pulse shape. The input pulse also charges storage capacitor C_1 through D_2 to its full voltage, minus the 0.3 volt drop across D_2. During the stretching period of 68 μsec, D_1, D_2 and D_3 are reverse biased.

The 20-v positive pulse from the collector of Q_2 is differentiated by C_2, C_3, R_1 and R_2. A low impedance path is provided for positive pulses by D_4, and the negative pulse discharges C_1 after the 68-μsec delay.

The ramp generator is gated by the 4-v negative pulse from the collector of Q_1. Normally, Q_3 and Q_4 are conducting, grounding C_4. When the negative pulse is applied to its

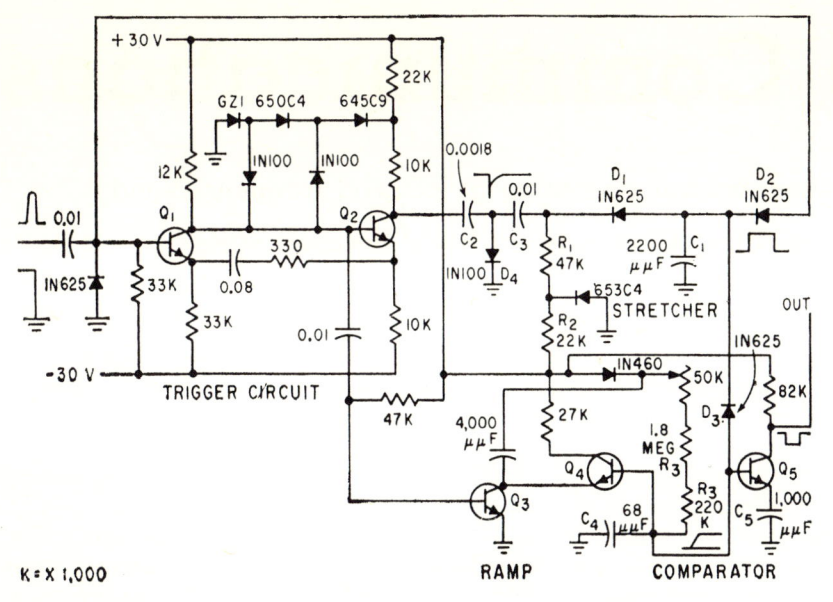

FIG. 4—Output of pulse stretcher circuit is used to control an oscillator

base, Q_3 cuts off. Capacitor C_4 starts charging at constant current because of the bootstrap action. The ramp is applied to the anode of D_3, while the stretched pulse is applied to its cathode. When the amplitude of the ramp is 0.3 v greater than the stretched pulse, D_3 conducts, thus bypassing most of the charging current of C_4.

The comparator waveform is differentiated in the emitter follower differentiator Q_5. Because of the emitter follower arrangement, the emitter follows the base. Because of C_5, the current that flows in the emitter of Q_5 (hence in the collector of Q_5) is the derivative of the base voltage. The negative output pulse appears at the collector of Q_5 and may be differentiated for triggering purposes.

Linearity and Stability

Silicon transistors and diodes are employed in the pulse stretcher circuit. Preliminary measurements indicated that the linearity of height-to-time conversion is of the order of 0.1 percent over a temperature range of −40 C to +50 C.

Nonlinearity introduced at elevated temperatures is caused by the ramp generator and diodes D_1, D_2 and D_3. Diodes D_1 and D_2 do not affect the stretcher voltage waveform up to +75 C, but the comparator works over a more limited range of −40 C to +50 C. It is not necessary to compensate the ramp generator since temperature-dependent changes in the current through R_3 are balanced by opposite changes in the leakage current. Thus, the charging current of C_4 is kept constant over the temperature range mentioned.

Amplitude sag in the stretcher itself (because of a finite leakage time constant) does not seriously affect the operation of this circuit. Voltage across C_1 is a linear function of time during the stretching period, and, therefore, does not affect the linearity of the height-to-time conversion.[3]

Applications

Both circuits are designed to provide linear conversion of pulse height to time within the limitations of their specifications. Specifications are shown in Tables I and II. These circuits were designed to

Table II—Specifications For Pulse Stretcher Circuit

Output pulse amplitude: about 25 v.
Output pulse rise time: 2 μsec.
Output pulse fall time: 1 μsec.
Output pulse width: 64 μsec with 7 v input pulse.
Linearity: 0.1 percent for input pulse amplitudes up to 7 v.
Temperature stability: stabilized from −40 C to +50 C.
Voltage supply: ±30 v.

accept input pulses with rise times of 1 to 2 μsec. For other rise times the charging time constant for the storage capacitors (C_1 in Fig. 1 and 4) have to be changed.

Possible applications include pulse width modulation systems, as well as pulse height analyzers as considered here.

Since the height-to-time converter is normally synchronized so that the oscillator controls the start of the digital output, the leading edge of the time pulse is not important in this application.

In the CCC circuit (Fig. 1) the collector-to-emitter voltage rating of the 2N496 is exceeded when large amplitude pulses are applied. The short duty cycle apparently protects this transistor from damage, but use of another type transistor is desirable for critical applications.

In general, the circuit shown in Fig. 1 would be most useful where the higher amplitudes required would not present a problem. It offers simplicity, good linearity, and good temperature stabilization. The pulse stretcher circuit of Fig. 4 would prove useful where lower amplitude pulses must be converted to time information. This circuit offers somewhat increased complexity, but provides very good linearity and reasonable temperature stability.

In both circuits, particularly the CCC circuit, improved methods of linearity and jitter measurements are desirable.

This work, performed in the Cosmic Ray Laboratory of the State University of Iowa, Department of Physics, was supported in part by the Office of Naval Research. The authors gratefully acknowledge this aid and that of the staff members of the Physics and Electrical Engineering Departments.

REFERENCES

(1) D. H. Wilkinson, A Stable Ninety-Nine Channel Pulse Amplitude Analyzer for Slow Counting, *Proceedings of the Cambridge Philosophical Society*, 46, p 508, July 1950.
(2) D. N. Carson, Transistorized Pulse Height-to-Time Conversion Circuits for use in Multichannel Pulse Height Analyzers Employing Digital Techniques, State University of Iowa, Physics Dept., No. 59-19, Iowa City, Iowa, August 1959.
(3) S. K. Dhawan, A Transistorized Analog to Digital Converter Circuit for Use in Multichannel Pulse Height Analyzers Employing Digital Techniques, State University of Iowa, Physics Dept., No. 59-20, Iowa City, Iowa, August 1959.

Infrared Communications

Experimental communications receiver operating at infrared frequencies may be used for future space vehicles. Circuits reject solar infrared noise

By W. E. OSBORNE,
President and Director of Engineering, Twenty-First Century Electronics, Inc., Riverside, Calif.

AS A MEANS OF COMMUNICATIONS, infrared will undoubtedly come into greater prominence over the next few years. At present at least two systems are in military use. Other IR systems, usually of a passive type, have been in use for some time. The Sidewinder missile is a good example of a simple but effective unit.

For communications purposes within the earth's atmosphere, radio scores heavily over infrared in terms of range. This situation changes rapidly with altitude and a sensitive, relatively simple infrared receiver at an altitude of 20,000 ft possesses an average range of several hundred miles in reasonably good weather. Outside the atmosphere, infrared is on equal terms with radio. Its range is virtually unlimited. Infrared also has spectrum space for literally millions of channels. The use of semiconductor detection, amplifying and oscillator components coupled with the highly directional characteristics that such short wavelengths provide are also advantageous.

An experimental fully transistorized infrared receiver is shown in Fig. 1.

Circuit

The detection elements may be a pair of matched lead sulphide (Pbs) or indium antimonide cells. The latter type require no cooling whereas the Pbs cells perform much better inside a CO_2 filled jacket. This is by no means mandatory. Cells of PbTe may be substituted but require a case cooled with liquid nitrogen.

Since the time constant of many photoconductive cells is now in the microsecond range, pulsed communication is possible. Lead sulphide cells are still relatively slow.

Cell D_2 is used as a reference while cell D_1 looks at the target or transmitter and detects any changes from the IR level of the reference cell. Selectivity is obtained by either an optical or electronic filter. The output or difference signal from the detection cell is first amplified by Q_1 and Q_2 and then passed through emitter follower Q_3 for low-impedance transmission to the main unit.

At this point the signal is electronically chopped or modulated at a frequency of approximately 1,000 cps. Chopping originates in the circuit of oscillator Q_{10}, which triggers multivibrator Q_{11} and Q_{12}. After amplification by buffer Q_{13} the square-wave from the oscillator is transformer coupled to transistors Q_{14} and Q_{15}. The synchronized rectifying action then chops the incoming IR signal at the pre-set chopping frequency. It is probable that such chopping (which allows a-c amplification at the chopping frequency) will be rendered unnecessary in the near future by the use of amplifiers which operate at the signal frequency.

The modulated component (which we assume to be a message pulse-modulated at a frequency much higher than the chopping frequency) is now applied to a tuned amplifier consisting of Q_4, Q_5, and Q_6, assisted by feedback transistor Q_8. The pass-band is narrowed to as low a point as the pulse-modulated message will allow, providing discrimination against unwanted noise frequencies. To increase the signal/noise ratio, the gated output is used. It is the electronic equivalent of former mechanical types of synchronous rectifiers used in IR receivers. In a system designed solely for communications purposes, the signal from the tuned amplifier would be boosted by Q_7, and then could be taken directly to transformer T_1. Operation of the latter is controlled by Q_9, which is in turn triggered by the multivibrators. The output of the amplifier is thus synchronized with the input. Transformer T_1 would then pass the message for discrimination and operation of a recorder or display device. These pulses would ride on the top of the waveform produced by the chopping circuitry.

Figure 1 also shows an arrangement whereby the overriding modulation or message is taken through a discriminator and thence to an audio amplifier and recorder. This then allows the unit to operate as a communications receiver.

The output signal is also fed to four diodes in a balanced output circuit which provides a rectified signal level which is indicated on a meter. This converts the receiver from a communications set to a radiation intensity measuring device.

While the transmitter used to operate this experimental receiver was of a Xenon Mercury type, solid-component power oscillators at IR frequencies are under development for military purposes. Synchronizing pulses from the transmitter (or blank periods of no transmission) will be used to hold the receiver in line, and may be adjusted in the cir-

Receiver for Space Vehicles

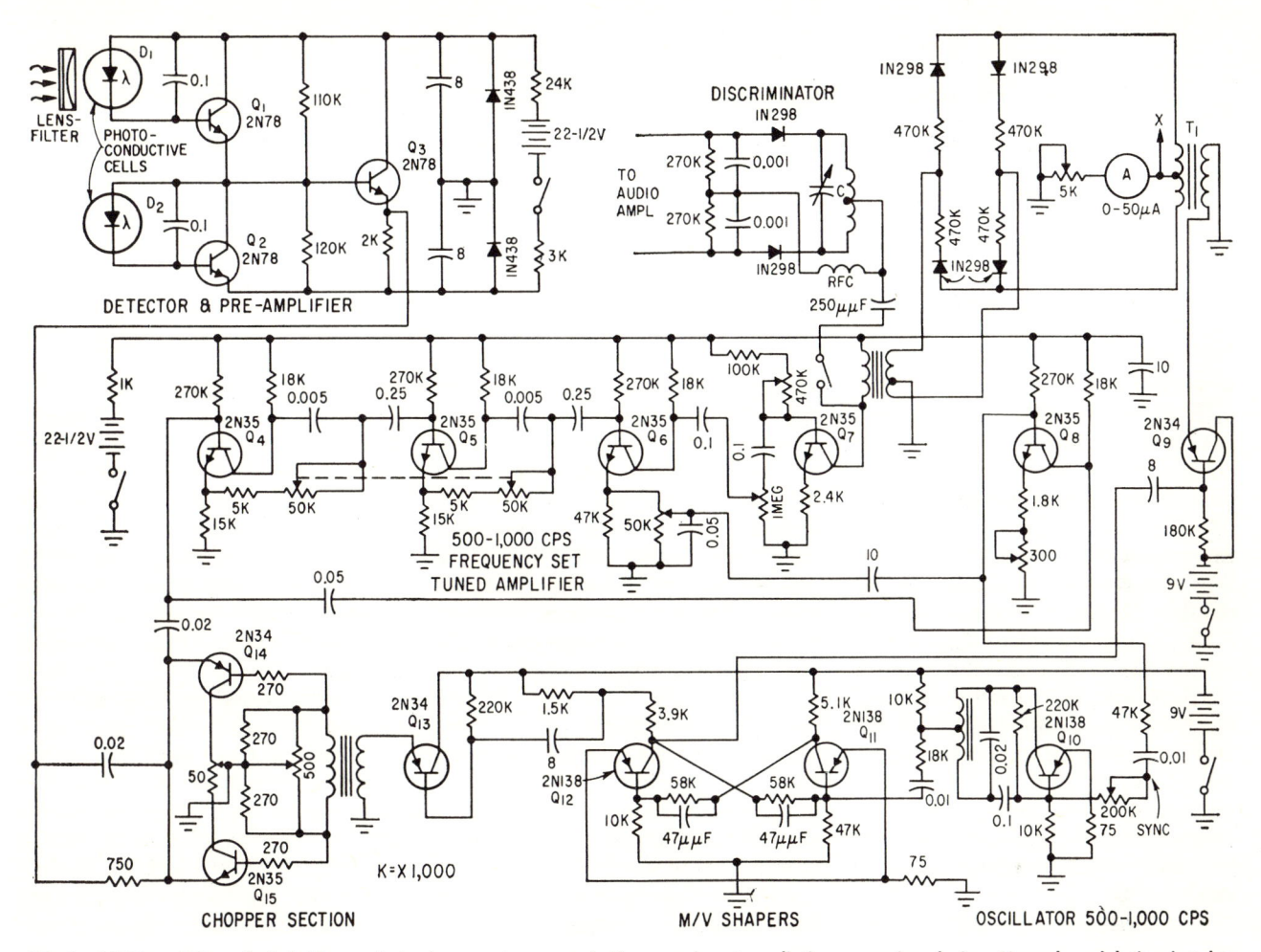

FIG. 1—Addition of four-diode bridge and circuit converts communications receiver to radiation-measuring device. Normal modulation is taken from conventional diode discriminator

cuit of oscillators Q_{10}.

Filters

Electronic filters are also under development for use in the intermediate portion of the IR spectrum (1-12μ). One of these, capable of selecting a 10-angstrom-unit portion of this spectrum, was recently tested successfully. As one micron, or 10,000 A is equivalent to three hundred thousand billion cps it is obvious that 10 A still represents an immense pass-band.

If such a communication system were used in space, it may be argued that the tremendous IR signal from the sun would completely prevent any reception. However, the receiver sync control allows it to trigger only on a modulated signal of a chosen type, and the extremely narrow angular limits of the modulated carrier combine with this to permit useful operation. In addition, the direct IR from the sun may be rejected in a great majority of cases by simple rate circuitry which measures the closing rate (in terms of emission) between sun and vehicle, and vehicle-to-vehicle. The sun's rate is infinitesimal while the vehicle-to-vehicle rate changes (relatively) rapidly and is accepted.

Radar could be used for the combined purpose of target location and simultaneous communication. However we are faced with the problem of an efficient power oscillator to replace the present magnetron or klystron.

The use of millimeter wavelengths would to some extent relieve the overcrowding problem but the time will certainly come when even this portion of the spectrum will be fully occupied.

Infrared possesses a peculiar advantage over radio. Taking the intermediate portion of the infrared spectrum, from one to 12 microns, during certain portions of the space journey no transmitter would be necessary as solar heat or the heat generated within the vessel for propulsion purposes could be used as the carrier. A modulator would be required like that of a radio or radar transmitter.

Present infrared detectors remain sensitive under direct sunlight for only a few days. This defect will eventually be eliminated and the cells may also require no cooling. Developments now under way show promising results.

Multiplexing Techniques for

Explorer VII satellite, now orbiting, is using this transistorized 10-channel multiplex system that accepts conventional as well as random pulse inputs. A portion of the system accepts variable impedance inputs

By OLIN B. KING, Army Ballistic Missile Agency, Redstone Arsenal, Huntsville, Alabama

SIGNAL-TO-NOISE RATIO is of prime importance in any data link. This ratio is affected by many factors such as bandwidth, transmitter output power, antenna gains and space loss. One of the few parameters which the satellite-link designer may manipulate to obtain improved signal-to-noise ratio is bandwidth. The transmitted bandwidth is a function of the information bandwidth as well as of the type modulation employed. The effective information bandwidth may be reduced appreciably by time-division multiplexing of a number of data channels with the resulting overall system improvements.

A multiplexer must be simple and reliable and moderate accuracy must be maintained over a wide range of environmental conditions. Size, weight and power consumption must be minimized. A 10-channel multiplexer designed for an Army IGC satellite is typical. Eight channels are available for information inputs. Fixed levels are applied to the other two channels. One of these levels corresponds to zero input and the other to 110 percent of information full scale, thus providing for ease of frame identification. The high-level pulse may be used as a sync signal for automatic ground demultiplex if desired. In addition the pulses provide endpoint calibration levels each frame which may be used to correct for the effects of drift in subsequent portions of the link. A timing oscillator and its associated amplifier drive a decade ring counter continuously. An analog gate is controlled by each of the counter stages except the second. (This omission of a gate allows the zero pulse each frame). A voltage from a reference source is applied to the first gate to provide the other fixed level previously mentioned. Information inputs are applied to the remainder of the gates. The gate outputs are connected in parallel to provide the multiplexed output.

Counter

The basic counter circuit has been used in a number of applications[1]. The counter, shown in Fig 1, consists of ten modified bistable multivibrator stages, Q_1 through Q_{20}, coupled in the usual ring manner. Each individual stage employs a pnp and npn transistor. Base, load and cross-coupling resistors function as in a conventional Eccles-Jordan circuit. The use of complementary transistors permits two stable states. In one state both transistors are ON or conducting and in the other state both transistors are OFF or nonconducting. A common emitter resistor is used for the npn transistors. The pnp transistors also have a common emitter resistor. Only one counter stage is normally ON. The emitter currents of this stage build up emitter voltages of the proper polarity to reverse bias the OFF stages. Advantages of such a scheme are obvious. In an n stage counter the current drain is reduced by a factor of $n/2$. The number of emitter resistors required is divided by n as compared to conventional circuits. Stability is also improved. The emitter holdoff voltages permit reliable operation to much higher temperatures than possible with other circuits. Ability to accept loading is greatly increased. If the collectors are loaded heavily it is necessary only to increase the size of the appropriate emitter resistor to regain the unloaded stability factor. In such a case relatively equal loads should be placed on every stage to prevent unbalance.

As an accepted practice, power is applied through an R-C network. A series resistor and capacitor couple a differentiated pulse to the first stage upon application of power. This is done to provide an initial set with the first stage ON and all others OFF. The emitter bus presents an ideal input for the count pulses. Count action is initiated by applying a negative pulse to the emitters of the pnp transistors. Such a pulse does not affect the OFF stages but turns the ON stage OFF. The resultant positive step at the npn collector of the turning OFF stage is differentiated and coupled as a pulse to the base of the next stage npn transistor. The time constant of the coupling network should be longer than the duration of the count pulse. This allows the transfer pulse to be present when the count pulse disappears. The transfer pulse turns the next-stage npn transistor ON and the cross coupling network turns the pnp transistor of the same pair ON. Thus the ON stage is shifted one stage to the right. The same action could be obtained by applying a positive count pulse to the npn emitters. It should be noted that no steering circuits whatever are required as in conventional counters.

Satellite Applications

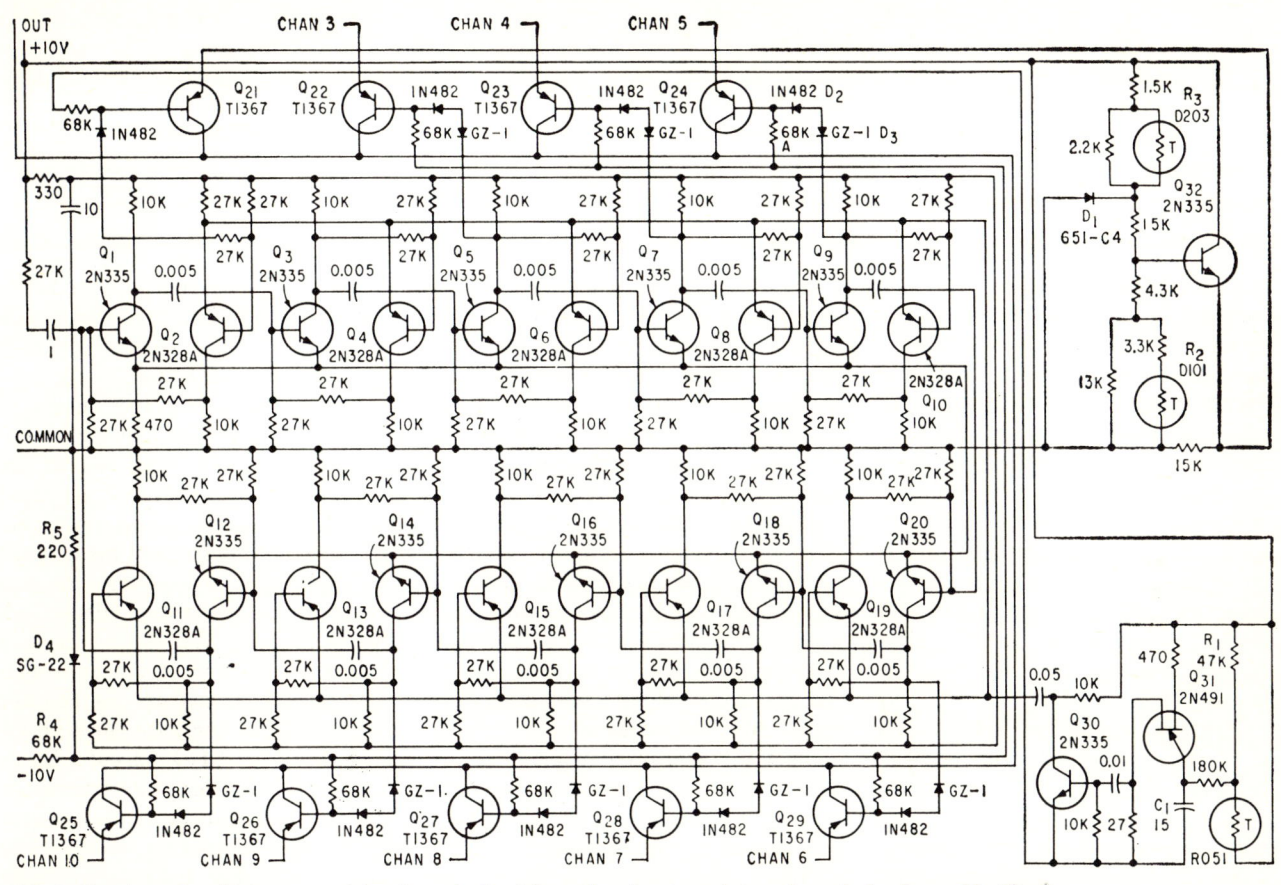

FIG. 1—Ten-channel multiplex uses eight channels for information inputs and two channels for frame identification

This results in saving one or possibly two diodes per stage. Although reset is not required in this instance, this too is easily and economically accomplished with a counter of this type.

Oscillator and Amplifier

The timing oscillator is a relaxation type with unijunction transistor Q_{31} used as the active element[2]. Upon application of supply voltage, current flows through the resistor R_1 and builds up a voltage across capacitor C_1. When the peak point potential of the transistor is reached, it conducts and discharges the capacitor. During conduction, current flows through the base one resistor and develops a positive pulse across it. This cycle repeats itself as long as the supply voltage is present. The frequency is determined by the RC time constants of the unijunction emitter circuit. The thermistor is included to compensate for changes of the capacitor and transistor due to temperature.

Amplifier stage Q_{30} is included to insure sufficient drive to the counter, even after considerable component deterioration. The pulse output is derived from the collector waveform of the amplifier transistor.

The oscillator pulse repetition rate determines the sampling rate of the multiplexer. This rate must either be several times higher than the highest sampled frequency or several times lower. The type of information to be multiplexed must be considered before making the decision. If, as in most satellite cases, the data varies at a slow rate and evaluation of transients is not important, the sampling rate may be low with respect to the modulating frequency. In such a case the multiplexing process does not increase the system bandwidth.

As was previously mentioned, a voltage reference is included and is fed to one channel. This voltage is developed by series regulator Q_{32} controlled by breakdown diode D_1. The major difficulty is obtaining the desired stability within the power limitations. A breakdown diode must be used as the basic element and such a device inherently requires current for satisfactory operation. Unfortunately, diodes in the voltage range where low-temperature coefficient is exhibited have soft knees, that is, considerable current must be passed to drive the diode into the low impedance region where good regulation is obtained. If the allowable current is small, a voltage region must be selected where a sharp knee is found. The higher temperature coefficient must be externally compensated in such a case. Only by such a compromise may satisfactory regulation be ob-

293

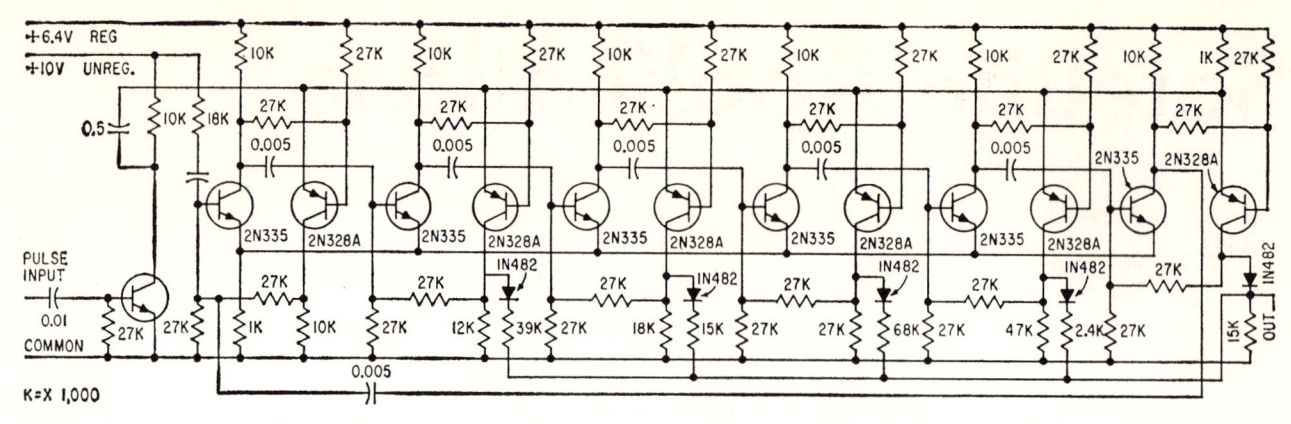

FIG. 2—Pulse converter transforms random information into analog form suitable for multiplexing and provides memory between events

tained under conditions of widely varying supply voltage and temperature while keeping power dissipation low. The resistive network in parallel with diode D_1 serves to divide the diode voltage to the value required at the transistor base. Thermistor R_2 in this network introduces the compensation necessary to correct for variations of diode breakdown voltage with temperature. It also corrects for temperature variation of the base-emitter voltage of the regulator transistor.

Current is supplied to the diode through a resistor-thermistor network as shown. Thermistor R_3 is used to prevent the current drain from increasing at cold temperatures. The diode breakdown voltage decreases in such a case and the voltage across the resistor network increases. The network values are chosen to provide the diode with only the required current regardless of temperature.

The overall circuit provides a 5.5-v reference to transistor Q_{21} with a tolerance better than 1-percent for worst conditions of ±10 percent supply voltage and temperature from −55 C to +100 C. This is accomplished with a total power drain of approximately 20 mw.

Gate

Perhaps the most important circuit in any multiplexer is the gate itself. A wide variety of gate circuits may be envisioned and the choice of a particular one is dependent upon many factors. Items such as sampling and nonsampling input impedance, trans-gate drop, input and output leakage currents, ease of drive and circuit complexity must be considered. In the case in point, overall multiplexing accuracy of the order of 1-percent is adequate and sampling impedance need not exceed 50,000 ohms. This permits the choice of a simple, easily-driven gate circuit. The circuit selected is of the inhibited common-base transistor type. If point A of Fig 1 was connected to ground, a large positive voltage applied to the emitter of Q_{24} (for example) will forward bias the emitter-base diode and cause current to flow in the base resistor. If the resistor is small enough to allow sufficient base current flow, the transistor will be driven into saturation, and the input voltage will appear at the output terminal minus the saturation voltage of the transistor.

Thus, the power to drive the switch ON is supplied by the signal source itself. This places a limit on the input impedance of the gate during sampling. However, transistors are available which meet all other requirements and allow saturation voltages less than 15 mv with a circuit input impedance as high as 100,000 ohms. If lower saturation voltage is desired it may be obtained by inverting the transistors [3,4]. However, the effective beta of the inverted transistor is lower with available transistors and the input impedance must be lowered accordingly. Six-mv saturation voltage and 20,000 ohms input impedance would be typical values for such a connection. For the satellite application, the normal configuration was chosen.

It is evident that for an input signal of less magnitude than the breakdown voltage of the emitter-base diode, the gate proposed would be inoperable. If, however, point A were removed from ground and returned to a negative bias equal to the transistor V_{bc}, the gate would function for any positive voltage greater than a few mv.

The gate may be placed in a nonconducting state by placing an inhibiting voltage on the base of the gate transistor. This voltage must be larger than the maximum expected gate input voltage to insure that the transistor diodes are reverse biased. In the multiplexer described, the inhibiting voltage is derived from the collector of the appropriate npn transistor of the counter and applied to the gate through diodes D_2 and D_3 of Q_{24} circuit. Diode D_2 serves to isolate the gate from the shunting resistances of the counter during the conducting period. The requirement for breakdown diode D_3 may be understood by recalling that the collectors of the npn counter transistors are not clamped to ground during the on time, but rather to the voltage developed across the emitter resistor. This voltage would be sufficient to inhibit the gate for input voltages less than a volt. If, however, the control voltage is coupled to the gate through a breakdown diode, whose conduction voltage is greater than the counter emitter voltage, this source of difficulty is eliminated. The transistors employed in the gates are of the alloyed-silicon type. An alloyed device is used to obtain low saturation voltages. Silicon is mandatory in order that leakage currents be negligible. I_{EO} and I_{CO} for the devices used are typically 5×10^{-10} amperes at ten volts and room temperature. Such leakages allow the gates to have resistances of several hundred megohms even at elevated temperatures. This is more than adequate for the application.

294

The base resistor of the gate transistors must be returned to a negative bias approximately equal to the transistor emitter-base voltage. This voltage is temperature sensitive and for proper gate operation the bias voltage must track the transistor voltage. A negative voltage is applied through dropping resistor R_4 to forward bias silicon diode D_1. The diode acts as a regulator to maintain a bias voltage approximately equal to the diode-breakdown voltage of the gate transistors regardless of temperature.

All of the gates are connected to a single diode regulator. To compensate for the fact that the diode and transistor forward characteristics do not match exactly, resistor R_5 is placed in series with the bias diode. The potential across this resistor plus the diode potential provides the bias voltage required within a few mv over the range of supply voltage and temperature.

Pulse Converter

Time-division multiplexing as described above is compatible with many data formats. One notable exception exists, however, and it is found frequently in satellite applications. This is the case where information is in the form of random binary or pulse functions. Such is the output of radiation counters or some types of micrometeorite detectors. Obviously a random pulse cannot be reliably detected over a channel which is connected for only a small percentage of the time. Thus a converter must be added to transform the binary information into an analog form suitable for multiplexing and also provide memory between events. This may be done in a number of ways. The simplest methods involve some type of integration and reactive storage. Such an approach destroys the digital nature of the data and if the information rate varies widely over a period of time (as is usually the case) storage element leakage introduces considerable errors. However, use of binary techniques makes possible the conversion of random pulses to quantized analog voltages which retain the digital character of the original input. These binary techniques also provide for infinite memory if required. Simple circuits allow this more sophisticated approach to compete with the cruder methods in all aspects.

Figure 2 shows the schematic of the converter. The input pulses are connected to an amplifier whose output is applied to a ring counter. The outputs of individual counter stages are combined through an adding network to provide an analog voltage as the converter output. With suitable weighting in the adding network, the output voltage relates the state of the counter. It is apparent that the output is quantized into the same number of levels as the counter has stages. So long as the quantum steps are larger than the resolution capability of the data link the digital character of the input information is retained. The input rate, number of quantum levels, and the rate at which the output is sampled must be related in order to prevent ambiguity. The counter must not complete one cycle between output samplings. If the input rate and multiplexer frame rate are fixed, the number of counter stages must be chosen to satisfy this condition. The upper limit on the number of stages is fixed by the resolution of the data link as mentioned above. Should this limit be below that required by the input and sampling rates, the input rate might be reduced by binary division. If the loss of time resolution thus incurred were too great, the only alternative would be to increase the multiplexer frame rate. In the example described, six stages were considered optimum. To facilitate readout, the counter may be reset at the end of each sampling. However, for simplicity the counter may be allowed to operate in a continuous ring manner. In this case the count during any multiplexer frame period may be obtained by subtraction from counts of adjacent frames. The latter choice was made in the example.

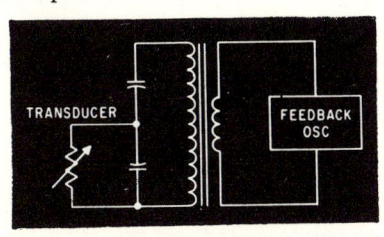

FIG. 3—Resonant frequency varies with impedance of transducer

The basic counter and pulse amplifier circuits used in the converter are the same as in the multiplexer and have been previously described. The adding network may be recognized in Fig. 2. The output from each counter stage is taken from the collector of its pnp transistor. A portion of the collector current of the ON stage flows through a weighting resistor and develops a voltage across a common output resistor. The weighting resistors are proportioned to provide the proper quantum level for each stage. The diode in series with each output isolates the OFF from the ON stages. The collector load resistors are chosen so that the total collector current of each stage is the same.

Impedance Multiplexer

The basic multiplexing idea may be applied to a special case commonly found in satellite instrumentation to achieve even greater advantages. This occurs when the transducers used to detect physical phenomenon have variable impedance outputs[5]. This is the case with thermistors, micrometeorite detectors, and photo cells for example. In the general case the output of such transducers would be connected to suitable conditioning circuitry which would transform the information to an acceptable format, usually a voltage. This conditioning circuitry might be relatively complex with the resulting increase in size and weight and possible decrease in reliability. One of the problems in this area is the difficulty in supplying the reference voltages required in such an application. As previously mentioned, any means of deriving a stable voltage for a long period of time requires a relatively large amount of power. A number of these problems might be eliminated if it were possible to multiplex the transducer outputs directly with no type of conversion.

In most situations a subcarrier oscillator is used between the multiplexer and the transmitter to enable frequency division multiplexing of additional data or to provide wide band gain. A number of impedance controlled subcarrier oscillators have been designed[6] and might be

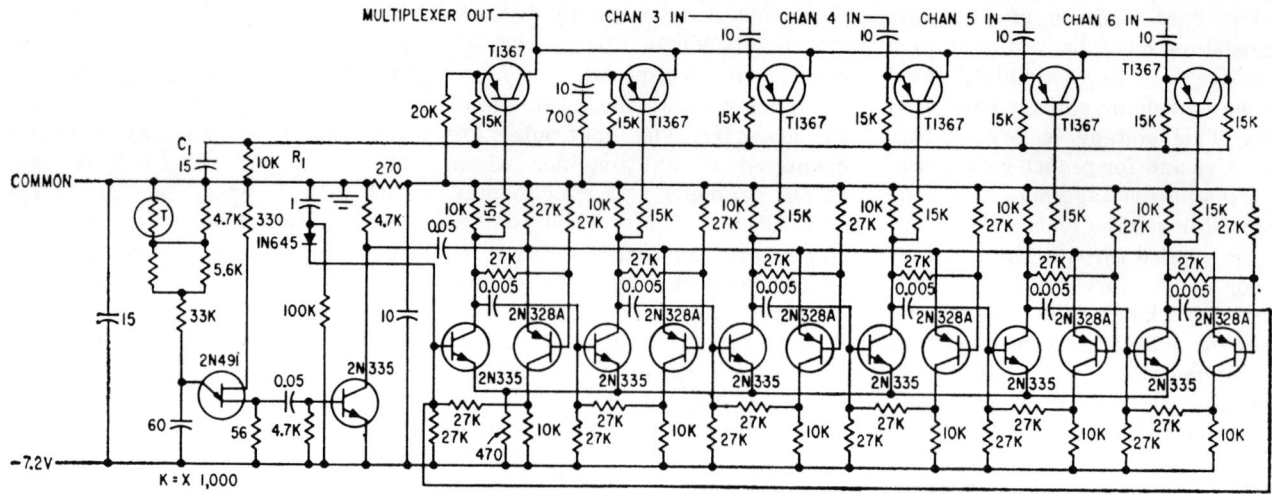

FIG. 4—Impedance multiplexer accepts variable impedance inputs. System inaccuracies introduced by impedance multiplexing are insignificant in a practical system

used with a multiplexer if compatibility were achieved. In order to analyze the functions required of the multiplexer, the oscillator itself must be examined.

The oscillator demands may be met by a number of circuits. However, a design of the general type described by Riddle[6] seems very well suited for satellite applications. The general method of operation may be understood by considering Fig. 3. The capacitive branch of the tuned tank of a feedback oscillator is split and a transducer is connected across the lower capacitor. Non-rigorously the mechanics of functioning may be followed by observing that as the impedance of the transducer varies, the effective admittance of the capacitive branch varies also. This change in admittance causes the tank resonant frequency to vary and results in a proportionate change in output frequency.

The actual effect of the transducer is to control the current flowing in the capacitive branch. Thus, the quantity which actually must be switched by an impedance multiplexer is an alternating current and the gates used must be bilateral.

Gates

If the transistors are to be used to capitalize upon their inherent advantages as switches, both diodes must be reverse biased during the OFF period to prevent feed-through. A method for accomplishing this is illustrated in Fig. 4. The transducer input signals to the gates are capacitively coupled to prevent any d-c current from flowing through the transducers from the biasing circuitry. The transistor bases are connected to ground through resistor R_1, across which the driving signal is applied at this appropriate time. The input terminal of each gate transistor is returned to ground resistively. A portion of this resistance is made common to all gates as shown. The output terminals of all gates are connected in parallel and also connected by a resistor to the common resistor.

A gate is turned on by applying a negative voltage to its base and bias current flows through both transistor junctions. Thus, both transistor diodes are forward biased and a vary low impedance exists between the input and the output terminals of the gate transistor.

Any a-c current or voltage applied to either terminal will appear at the other, provided the peak value of the current is somewhat smaller than the bias current. The bias current flowing in both branches of the ON transistor is drawn through the resistor R_1 to all stages. The voltage across this resistor is of the proper polarity to reverse bias both junctions of the other transistor and satisfy the OFF condition. Thus, the ON stage holds all additional stages OFF. Filter capacitor C_1 serves to maintain the reverse bias voltage during transient switching periods. Offsets due to saturation voltage and leakage current are rendered unimportant by capacitive coupling. The impedance of the ON switch is less than ten ohms and is negligible with respect to the load. The impedance of the OFF switches is in excess of 100 megohms even at elevated temperatures and consequently crosstalk is also negligible.

A schematic of a six channel design is shown in Fig. 4. The first two channels are terminated in fixed resistors to provide the reference functions described in connection with the basic multiplexer. It should be noted that the values of the gate components are flexible and may be chosen to provide an input impedance which provides the desired characteristic when using nonlinear transducers.

Performance

All of the circuits operate from −55 C to +100 C. The data errors introduced are less than 1 percent. Power drain of the basic multiplexer is less than 60 mw, including the reference supply. Power consumption of the complete converter is only 8 mw and that of the impedance multiplexer 18 mw. The equipments weigh only a few ounces each and withstand 25 g's vibration to 2,000 cps and accelerations beyond 100 g's.

REFERENCES

(1) O. B. King, A Transistorized Calibrator for Missile Telemetry, *Proc IRE*, 1958 National Symposium on Telemetering, Sept. 1958.
(2) T. P. Sylvan, Design Fundamentals of Unijunction Transistor Relaxation Oscillators, *Electronic Equipment*, p 20, Dec. 1957.
(3) J. J. Ebers and J. H. Moll, Large Signal Behavior of Junction Transistors, *Proc. IRE*, v 1,761, Dec. 1954.
(4) R. Bright, Junction Transistors Used as Switches, *Communications and Electronics, AIEE*, p 111, March 1955.
(5) O. B. King, A Data Transmission System for Use With Impedance Transducers, *Proc Amer Rocket Soc*, Semi-Annual Meeting, June 1959.
(6) F. M. Riddle, Satellite Environment Via One-Milliwatt Oscillators, *Proc IRE*, 1958 Nat Symp on Telemetering, Sept. 1958.

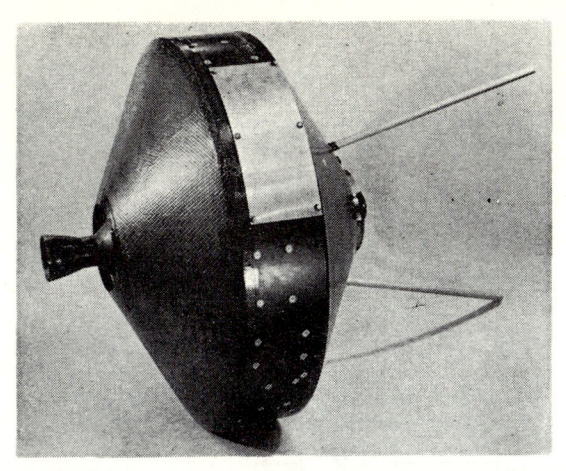

Pioneer lunar probe vehicle. Retro rocket to place the vehicle in lunar orbit protrudes from shell. Antenna is at rear

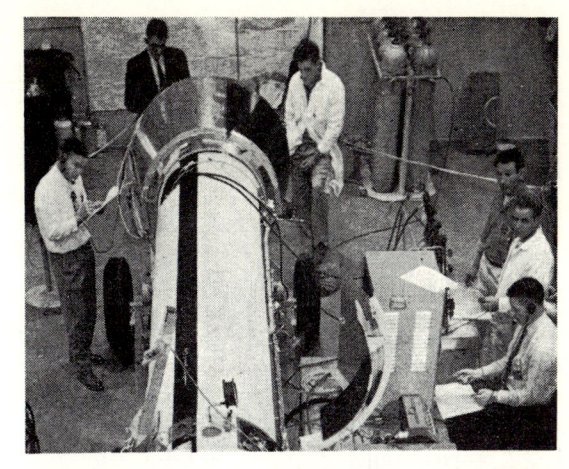

Ground testing portion of missile to check proper operation of second, third and fourth stages

Circuits for Space Probes

They measure magnetic fields, sense radiation level and transmit tv pictures back to earth. Circuit design allows use of vehicle as space station relay for long-distance vhf communications

By R. R. BENNETT, G. J. GLEGHORN, L. A. HOFFMAN, M. G. McLEOD and Y. SHIBUYA,
Space Technology Labs., Inc., Los Angeles, Calif.

ON OCTOBER 11, 1958, the United States launched the first space probe using the Pioneer vehicle. This highly-instrumented vehicle travelled over 70,000 miles from the earth before it returned to destruction in the earth's atmosphere. Portions of the electronic circuits that composed the fourth or payload stage will be covered in this article.

MISSILE — The first stage was a conventional Thor IRBM missile generating 150,000 lbs of thrust. The normally-used guidance system was removed to save weight. The second stage was a special vehicle which had been successfully flown previously, and had a thrust of 7,500 lbs.

Control about the pitch and yaw axes was provided by gimballing the engine (thrust chamber) with hydraulic servos. The pitch and yaw channels were identical except for the addition of a gyro torquing signal used to program the missile in pitch.

Roll control was provided by a set of helium jets located about the periphery of the missile. The jets used the main propulsion system helium supply. The roll jet helium supply was controlled by solenoid-operated valves. The jets operated in pairs and provided 7.5 lbs of thrust each.

Certain critical second stage airframe, control, and propulsion functions were telemetered by a three-band f-m/p-m system that radiated 2 w of r-f. Pulse-amplitude modulation was employed on one subcarrier to multiplex a number of information channels.

An electronic commutator capable of time-division multiplexing 14 channels of information, inserting a reference level between information pulses and supplying a synchronizing pulse to identify each frame was used.

The Pioneer's third stage used a solid-propellant rocket motor. The fourth stage was to be used as

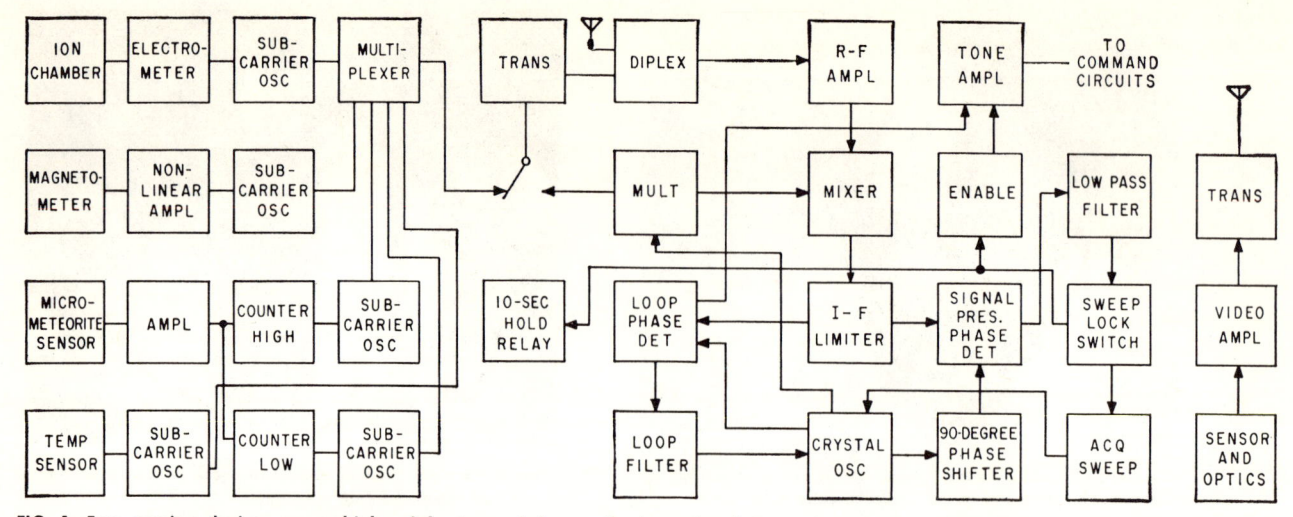

FIG. 1—Four sensing devices are multiplexed for transmission as f-m/p-m signal. Transmitter can be switched for use as Doppler transponder and as receiver-transmitter repeater system

a retro rocket. It's solid-propellant motor was mounted within the payload. The retro rocket was intended to be fired when the payload approached lunar orbit and guide the payload toward the moon.

PAYLOAD—The total weight of the payload was 83 lb of which 34 lb was in electronic components. Power was obtained from a combination of mercury cells and silver-zinc hydroxide batteries. A honeycomb structure made of epoxy-fiberglass housed the electronic components. The exterior of the shell was coated with a special blue paint pattern with proportion of the coated to the uncoated portion of the shell providing temperature control within the vehicle.

A block diagram of the payload is shown in Fig. 1. The telemeter was an f-m/p-m system. Five frequency-modulated subcarrier oscillators operated in the five lowest bands normally associated with frequency-division telemetering; 400, 560, 730, 960 and 1,300 cps. The subcarrier oscillator outputs are resistively summed and amplified to become the complex modulation for the transistorized crystal-controlled transmitter. Each tone phase-modulates the r-f carrier approximately 0.3 radian to form a frequency spectrum where a few percent of the total r-f power is carried in each of the five sets of symmetrical sideband pairs about the carrier. This low degree of modulation insures that a major portion of the total power always remains in the carrier so that a locked condition can be maintained in the microlock receiving system used at ground tracking and receiving stations.

The five subcarrier channels were employed to transmit ion density, two levels of micrometeorite particle impacts, magnetic field strength and compartment temperature.

MAGNETOMETER—Typical of the circuits is the magnetometer used to measure the earth's magnetic

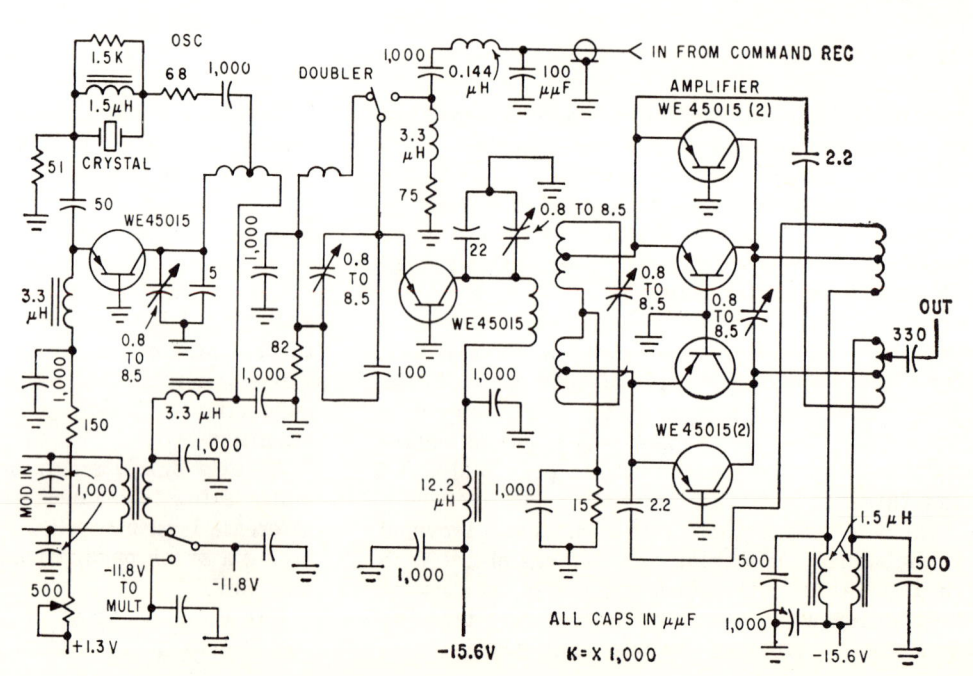

FIG. 2—Pioneer's all-transistor transmitter uses a push-pull parallel output stage. It will accept inputs from either the telemeter system or signals from the command receiver

field at great distances. It consists of 30,000 turns of fine wire wrapped around a ferromagnetic core. This core is mounted against the inner circumference of the payload. As the payload spins about its axis at 2 cps, the output of the magnetometer coil consists of a sine wave with a frequency equal to the payload spin rate and an amplitude proportional to that component of the earth's magnetic field vector perpendicular to the spin axis. The magnetometer is in effect a simple a-c generator in which the magnetic field strength in space takes the place of the field windings and the coil fastened to the spinning payload is the armature.

The output of the magnetometer is amplified by a nonlinear amplifier that provides gain and compresses a large input range to a small output range to make wide-range magnetic field measurements possible.

The output of the nonlinear amplifier is fed to an f-m subcarrier oscillator to produce a signal frequency modulated at the spin rate. The frequency deviation is proportional to the amplitude of the amplifier output which in turn is a function of the magnetic field strength. This subcarrier output is mixed with the other subcarriers of the telemeter system and the resultant signal is used to phase modulate the transmitter.

The transmitter may be coupled to the receiver to form a phase-coherent transponder system to provide measurements of Doppler velocity and position from earth tracking stations. The receiver also furnishes output commands to operate certain payload equipment sequences. Unless interrogated for command or Doppler purposes, the system continuously transmits telemetry data.

The electrical schematic of the transmitter is shown in Fig. 2. The output stage uses four transistors in a push-pull parallel combination. The transmitter has an output of 400 mw with a frequency stability of 2 parts in 10^7 per deg C, an efficiency of 53 percent and weighs 10 oz.

RECEIVER—The receiver uses 62 transistors, has a gain of 20 db and a noise figure of 8 db. The i-f bandwidth is centered about 6 mc and is determined by a crystal-lattice filter that is 10 kc wide at the 6 db point.

The receiver sweeps in frequency until phase lock is accomplished. The sweep circuit uses a voltage-variable semiconductor capacitor effectively shunting the crystal of a crystal-controlled oscillator. The receiver has a sensitivity of −130 dbm, consumes less than 270 mw and weighs 5.7 lb including batteries for a life of 120 hours.

MISSILE TESTING AND CHECKOUT—After the various subsystems were installed in the missile, exhaustive checks were performed to check proper operation of the equipment as a missile system. The propulsion system was exercised by pressurizing gas fed through the ground console. All systems were operated in as realistic a time scale as possible.

GROUND STATIONS—To track the Pioneer and re-

FIG. 3—Aerial photograph of the Hawaii tracking station showing the parabolic receiving antenna and the multiple helix arrays arranged as an interferometer system

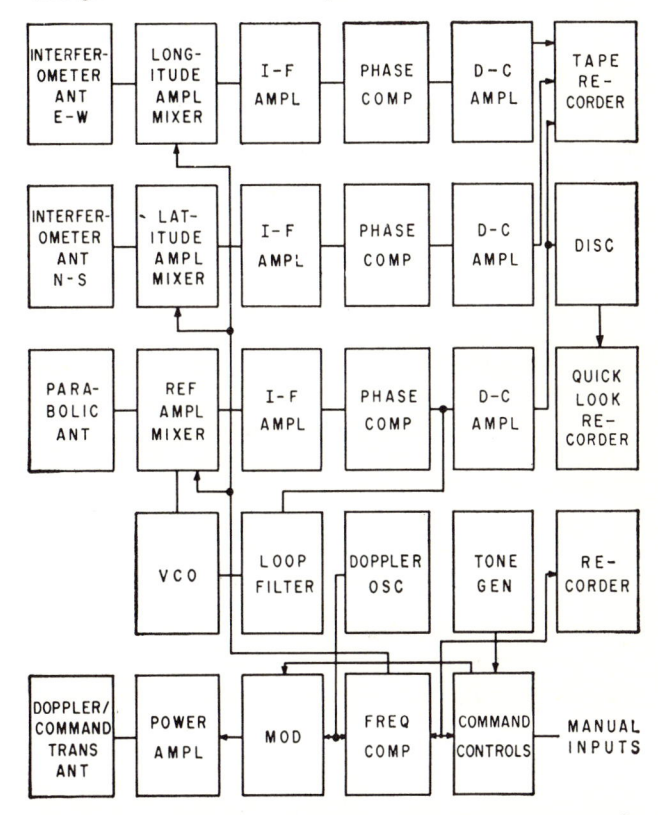

FIG. 4—Main elements of the Hawaii tracking station showing the microlock receiving system and the command transmitter

ceive the telemetry signals, a network of ground stations was established. Two of these, at Cape Canaveral and Hawaii, included provisions for sending commands to the payload. Figure 3 is an aerial view of the Hawaiian station showing the 60-ft parabola and multiple-helix arrays. The latter are used in an interferometer arrangement. The block diagram of the Hawaii station including the microlock receivers and command transmitter is shown in Fig. 4.

Because the Pioneer contained a receiver-transmitter repeater system, it was possible to communicate through the payload from one ground station to another.

On October 12, 1958, a vhf relay was accomplished between Florida and England and between Hawaii and Florida. This portion of the experiment highlights the use of satellites for intercontinental communications.

Chapter 17
MISSILE AND AIRCRAFT GUIDANCE

Solid-State Guidance For Able-Series Rockets

Here are the transistorized circuits for guiding the Able series of probes into space. Lightweight, simple, reliable, these circuits may be applicable to many types of control

By ROBERT E. KING and HENRY LOW,
Members of the Technical Staff, Space Technology Laboratories, Inc., Los Angeles, Cal.

THE ABLE SERIES of missiles is being used for space exploration experiments. An Able missile consists of a Thor or Atlas first stage, a second-stage liquid rocket, a solid-propellant third stage, a solid- or liquid-propellant fourth stage, and an experiment package or payload.

The control system to be described is for the second stage. It makes the missile follow a predetermined course in response to commands generated by an internal programmer, or in response to guidance commands from the ground.

One of the primary design criteria for the second-stage attitude-control system was light weight without compromising the ability of the equipment to survive and operate in a high-vibration environment. Thus, this system uses transistor and magnetic-amplifier circuits instead of vacuum tubes. A novel modular design uses stacked etched circuit board. Weight of similar equipment in the earlier Able missiles was 120 lb. The corresponding control equipment described here weighs about 30 lb. Power requirements were reduced from 200 to 18 v-a.

Figure 1 shows the attitude-control system. The command converter normally converts signals from a command receiver; it can also establish an attitude program.

Receiver commands are converted to gyro torquing currents. Gyro output signals command pitch, yaw and roll channels. Pitch and yaw channels, which are identical, control missile attitude in these axes by gimbaling the second-stage rocket engine with hydraulic actuators. The roll channel operates pneumatic jets located on the periphery of the airframe.

A stop command from the receiver interrupts the gyro torquing currents. After being servoed to its new attitudes, the missile follows a constant course until the next gyro-torquing commands arrive.

Pitch and yaw channels have more than 10 db of gain margin and 30 deg of phase margin at all times during flight, without the complexity inherent in the use of rate gyros. A lead network with a break frequency of 2.5 radians per second provides rate damping, and an additional lag term acts to stabilize for missile bending.

The roll control channel is a discontinuous on-or-off servo system which exerts a corrective torque on the missile for roll angles greater than 3 deg. This torque is obtained by the action of four fixed pneumatic jets which are controlled in pairs by solenoid-operated valves. Working fluid for the jets is helium, which is supplied from the

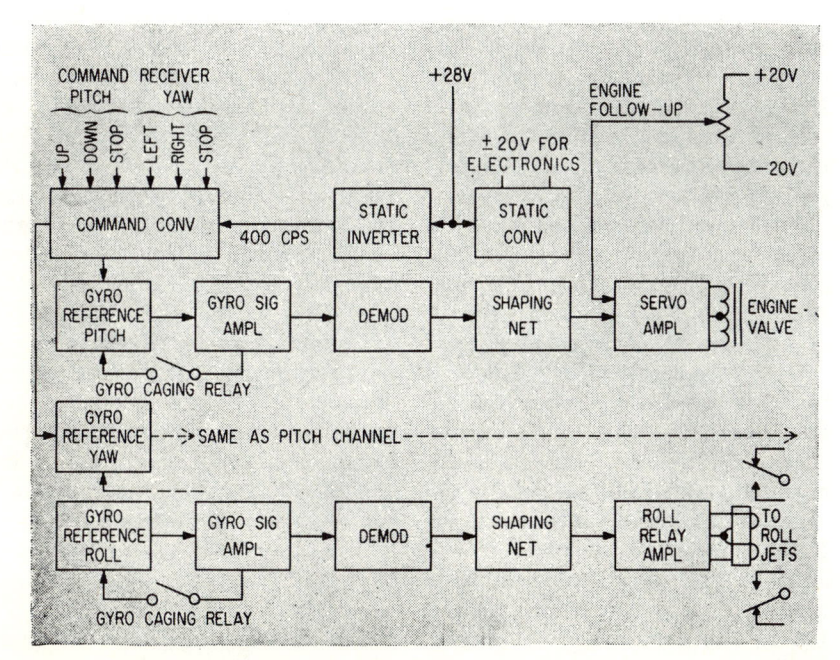

FIG. 1—Attitude-control system. Not shown here is the provision for internal programming in the command converter

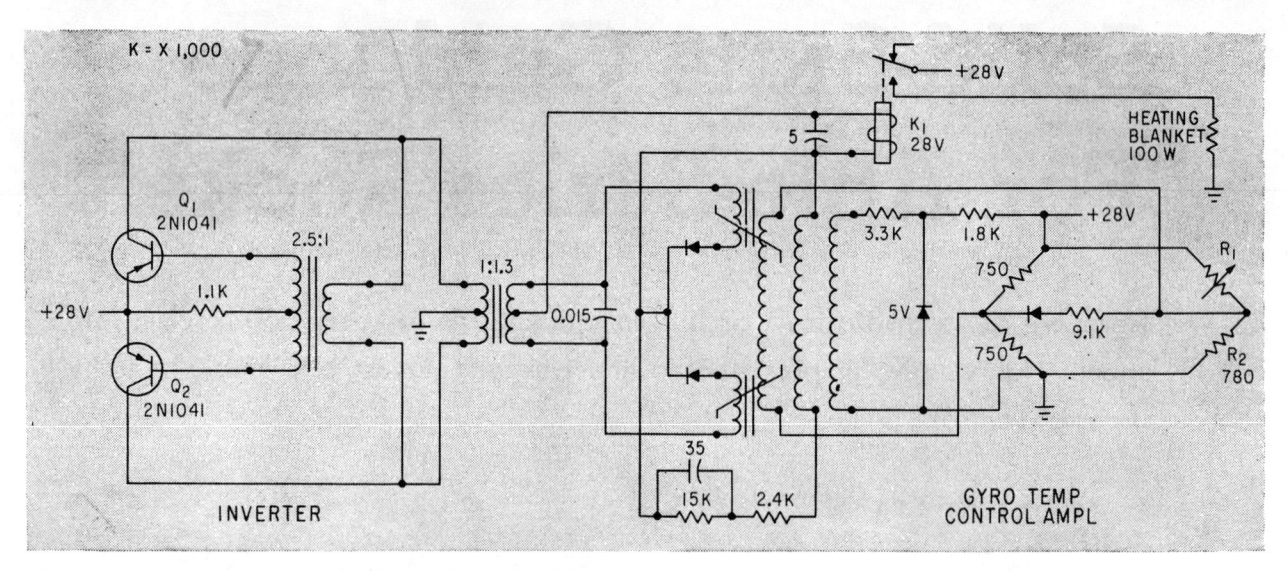

FIG. 2—Gyro temperature control connects and disconnects heater

propellant gas-tank pressurization system. Each jet can exert 15 lb of thrust. In flight, the missile roll attitude oscillates continuously at a fixed amplitude between the plus and minus 3-deg limits. Instead of a rate gyro, a lead network provides rate damping to minimize the pneumatic impulse requirements. A low-pass filter attenuates demodulator ripple. This filter has negligible effect on system dynamics.

Power from a 28-v silver-zinc battery is converted to 115 v, three phase, at 400 cps, which is then transformed to the distribution voltage of 10 v. A regulated plus and minus 20 v d-c for the control circuitry is produced by a transistorized static-converter voltage regulator.

Command Converter

Some of the commands that come from the receiver are: pitch up, pitch down, pitch stop, yaw left, yaw right, and yaw stop. These commands are received in the form of a relay contact closure which persists for 5 seconds. An arrangement of relays in the command converter causes these momentary contact closures to supply and maintain a constant a-c of the proper phase to the gyros, thus commanding the desired missile attitude change. Transmission of a stop command in a particular channel causes the current to be interrupted, and the missile will then follow a constant heading.

The command converter also has a provision for establishing an attitude program consisting of a constant pitch turning rate for a specified time. Turning rate is established by setting the current with a series rheostat, and time is established by a transistorized time-delay relay operating from a regulated d-c supply.

The command converter also contains a three-phase transformer which converts the 115-v output of the static inverter to the 10 v required by the gyro spin motors and demodulator reference transformers.

Gyro References

Each channel of the gyro reference assembly has a body-mounted hermetic integrating gyro of the HIG-4 type, a gyro temperature-control amplifier, circuits to excite the gyro, and spin-motor monitoring circuits.

Each gyro is mounted with its input axis parallel to either the pitch, roll or yaw axis of the missile; hence, each is sensitive to attitude deviations in only one axis.

Gyro temperature is regulated to $\pm \frac{1}{2}$ F by a temperature-control amplifier (Fig. 2). The amplifier senses differences between gyro-mounted, temperature-sensitive resistance R_1 and fixed resistance R_2, both connected in a bridge circuit. If the gyro temperature is higher or lower than the desired 165 F, the bridge has an output. This voltage goes to the control winding of a high-gain single-stage magnetic amplifier, which de-energizes or energizes relay K_1; the closed relay contacts connect a heater element within the gyro to the 28-v d-c power source. Operating frequency of the magnetic amplifier is 1 kc, permitting use of relatively small cores. This h-f power is generated in a two-transistor (Q_1, Q_2) low-power static inverter.

Gyro heat losses and heating power requirements were reduced by mounting the gyros in metal rings which are thermally isolated from the case by a dimensionally stable epoxy-glass mounting board. Each gyro is contained in a gold-plated, thin metal can, which reduces heat loss due to radiation. Each can is, in turn, encased in a thick molded-in-place shell of foamed polyurethane plastic. The foamed plastic adds structural rigidity to the entire package, and its thermal insulation property reduces warm-up time; the reduction in heater cycling increases the useful life of the control relays.

Current transformers are incorporated in one leg of each gyro spin motor. Before launching, the output of these current transformers is monitored on a recorder in the blockhouse. The current level indicates performance of the gyro spin motor.

Pitch and Yaw Channels

The pitch and yaw channels (Fig. 3) are identical, each containing a gyro signal amplifier, demodulator, shaping network, and servo amplifier. The gyro signal amplifier uses silicon transistors Q_1, Q_2 and Q_3 to provide a voltage gain of 1.5 at 400 cps with a power output of approximately 100 mw. An amplifier phase shift of 50 deg lagging, which compensates for the phase shift of the gyro signal, is obtained by an R-C filter at the input of Q_1. In order to avoid loading the gyro signal generator, the necessary high input impedance was obtained by a large resistance (R_1, R_2) in the series arm of the filter. To prevent the gyro signal amplifier from loading the input filter, the first stage has an input impedance of approximately 100,000 ohms. The three amplifier stages are direct coupled, and base bias for the first stage is obtained from the voltage developed across resistor R_3. This feed-back arrangement assures stability of the operating point of all three transistors despite variations of temperature and of transistor parameters.

The amplifier gain is obtained in the first stage, which is operated as a heavily degenerated amplifier.

Gyro signal amplifier coupling to demodulator transformer T_1 uses capacitor C_1 to avoid saturation of the transformer core by the d-c component of the amplifier output. The demodulator is a full-wave sum-and-difference type using four silicon diodes. Input to the demodulator from T_1 is a 400 cps signal; demodulator output is a d-c or l-f signal whose amplitude is proportional to the amplitude of the input signal and whose polarity is dependent on the phase of the input signal. A 400-cps reference signal is fed to the demodulator by T_2, an 80-v center-tapped transformer. Output circuits of the demodulator include an R-C low-pass filter to attenuate ripple components resulting from the rectification process and a balance potentiometer to permit setting the output to zero when no input signal is present.

The demodulator has a voltage gain of 9.4 and provides a maximum output of ± 60-v with a linearity of 3 percent of full scale.

The shaping network which follows each demodulator is a lead-lag resistance and capacitance network which provides the desired anticipatory system response.

The servo amplifier is a direct-coupled differential amplifier with a gain of 5 differential ma per volt of input. Output differential current goes to a dual-coil hydraulic valve. Maximum linear output is 10 differential ma into a dual 1,000-ohm-coil valve.

The amplifier has two inputs: one from the shaping network which represents attitude error of the missile, and one from the engine follow-up potentiometer which represents engine gimbal angle. In operation, the amplifier causes a current to flow in the hydraulic valve which causes engine motion in such a direction that the output of the engine follow-up potentiometer R_3 will equal and cancel the output of the shaping network.

To reduce drift effects, seven silicon transistors are used in a balanced configuration with their base-emitter voltages matched. Use of Sensistors further reduces drift. The last stage acts as a differential current amplifier, which uses transistor Q_4 instead of a large common emitter resistor to minimize the common-mode effect. A range of temperatures from + 20 C to +65 C causes a maximum drift of about 2 percent of full-scale output. Emitter followers in the servo amplifier's input stages provide high input impedance and avoid loading effects on the shaping networks.

Roll Channel

The gyro signal amplifier and demodulator of the roll channel are identical to those of the pitch and yaw channels. As shown in Fig. 4, a signal from the roll-channel demodulator goes to its roll shaping network, which applies it to the switching amplifier.

The roll switching amplifier is composed of two sections. The section whose input transistor is Q_1 operates for positive error signals and the section whose input transistor is Q_2 operates for negative error signals.

Trasistors Q_1 and Q_3 operate as a cascaded emitter follower to ob-

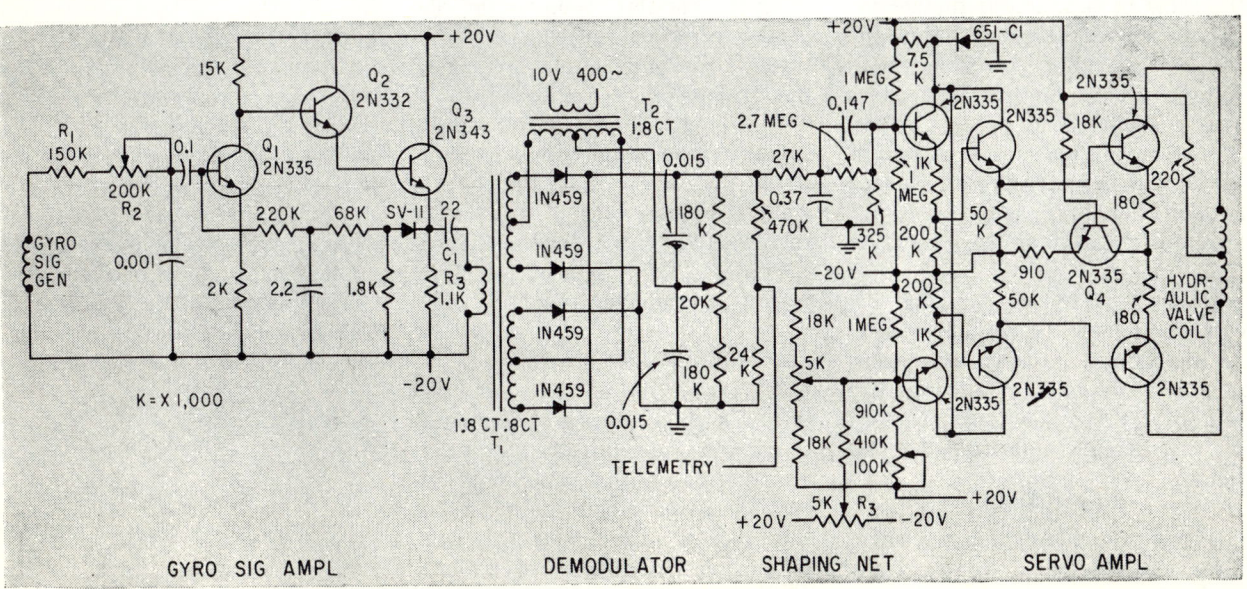

FIG. 3—This is the schematic of either the pitch or the yaw channel; channels are identical

tain a high input impedance. An input impedance of one megohm is necessary to prevent loading the shaping network and changing frequency-response characteristics of this network. Output of the emitter followers activates a Schmitt trigger circuit, and a 10,000-ohm potentiometer sets the input level at which the trigger circuit will operate. This potentiometer is normally set so that the trigger actuates at a shaping-network voltage output corresponding to 3 deg of missile-attitude error; this voltage is approximately + 2.5 v.

The Schmitt trigger is arranged so that in the OFF state transistor Q_4 conducts heavily and Q_5 is effectively cut off. When the input signal reaches the proper value Q_5 starts to conduct, turning off Q_4. When Q_4 turns off, the voltage at its collector rises toward the + 20 v and in so doing sufficient current is supplied to the base of Q_6 to turn it on. Transistor Q_6 now causes Q_7 to conduct, thereby energizing pilot relay K_1 and actuating the clockwise solenoid valve. Since relay K_1 requires approximately 100 ma to actuate and the saturation resistance of Q_7 is 5 ohms, only 50 mw of power is dissipated in the transistor in the ON state. Silicon diode D_1 is connected across the relay coil to protect the 2N389 driving transistor from inductive transietents.

The other section of the roll switching amplifier is essentially the same in function as the section just described. However, this circuit is energized by a negative voltage input and *pnp* transistors are used in place of *npn* types. Since the output circuit uses *npn* types it was necessary to include two additional stages to change the level and polarity of the base drive to Q_8.

In both of these switching circuits the hysteresis was adjusted to be less than 1 percent for reasons of missile dynamic stability and to minimize the amount of gas used in correcting roll-attitude deviations.

Power Supply

The power supply for the channels consists of a commercially available transistorized static con-

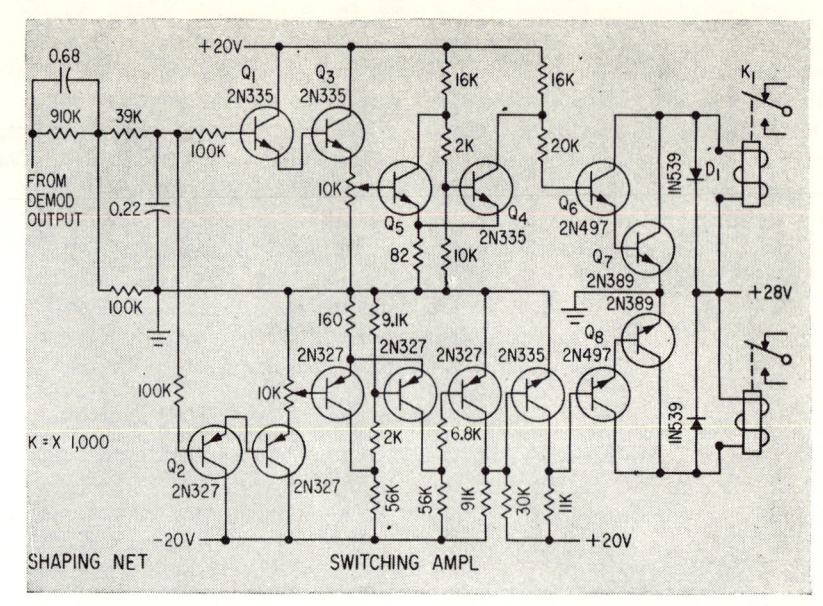

FIG. 4—Roll channel. Since gyro signal amplifier and demodulator are same as those of pitch and yaw channels, they are not shown

verter-regulator which provides ± 20 v d-c regulated to ± 1 percent. Prime power is obtained from the 28-v missile battery.

The static inverter consists of an oscillator, semiconductor power switches, control circuits, and output filters.[2]

The open loop gyro torquing technique used to program missile maneuvers imposes strict limits on voltage and frequency regulation. Total angle through which the gyros are torqued is a function of the gyro's angular momentum, the torquer current, and time; thus it is necessary to hold each of these factors to close tolerance in order to achieve a precise program.

Since angular momentum depends on wheel speed, which in turn is dependent on frequency, the frequency is regulated to ± 0.25 percent. Since gyro torquing current depends on supply voltage, it is necessary that the output voltage of the static inverter be regulated to ± 0.5 percent. These tolerances were determined from trajectory considerations.

Environmental Testing

On the basis of data obtained from Vanguard, Jupiter, Redstone, Thor, and other missile flights, it was estimated that maximum vibration levels would be near 3-g rms over a 5 to 2,000-cps range for the second-stage powered flight, with a burst of somewhat higher vibration levels at second-stage cutoff.

The equipment described was designed to withstand environmental specifications more severe than field or flight conditions in order to provide better assurance of locating design faults. These test environments were not intended to be severe enough to exceed reasonable safety margins or to excite unrealistic modes of failure.

Packaging

Electronic packaging is unusual in that individual circuits are assembled on separate etched wiring boards which are stacked vertically. Interconnections between boards are made by flat-tape cables, which permit the whole assembly to be unfolded for servicing and checkout. Size and weight are reduced and reliability is increased by eliminating plugs and keeping the number of soldered connections to a minimum.

A separate test plug is provided on each package for checkout of the control system after it is installed. Calibration and testing of the system is done with simple equipment.

REFERENCES

(1) R. R. Bennett, G. J. Gleghorn, L. A. Hoffman, M. G McLeod and Y. Shibuya, Circuits for Space Probes, ELECTRONICS, June 19, 1959.

(2) D. W. Moore and R. D. Gates, Special Report on the Solid Inverter, *Western Aviation*, Nov. 1958.

Systems engineer checks out automatic-pilot computer and amplifier as part of overall system test

Designing Safety Into Automatic Pilot Systems

Systems can be monitored by protection circuits that give alarm or shut equipment off when either control or safety circuit breaks down. Typical torque limiting and modulating circuits for control surfaces are cited

By C. W. McWILLIAMS, Chief Engineer, Autopilot Systems
Eclipse-Pioneer Div. of Bendix Aviation Corp., Teterboro, N. J.

SAFETY REQUIREMENTS for aircraft surface-actuator systems require that they never produce hazardous loads on the aircraft or create dangerous deviations in the flight path either during normal operation or in case of malfunction.

One of the most direct ways of achieving protection in main control-surface actuating systems is by limiting the torque that can be applied to the control surface or its control tab. As the margin between the torque required to properly fly the aircraft and the maximum torque permitted is small, accurate and reliable control of torque is necessary.

In aircraft whose automatic pilot provides automatic pitch trim for steady-state flight, an important safety requirement is that the automatic trim system operate the trimming surface. This is an auxiliary control surface on the elevator whose function is to trim the elevator for a steady-state flight, an important must be operated continuously during autopilot operation to relieve the main pitch-surface actuator (elevator servo) of holding a fixed surface deflection against a load.

The trim servo is normally set into operation when the voltage applied to the elevator servo indicates a sizable torque is being applied to the elevator. The trim surface deflection should be in a direction to minimize the elevator torque required to hold the aircraft in the trim condition.

The surface-actuator safety systems to be described use circuits that are typical of those required in modern flight control systems.

Pitch Axis System

A simple, but typical, servo system for the pitch axis of an aircraft

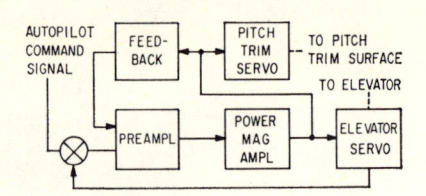

FIG. 1—Pitch-axis servo channel

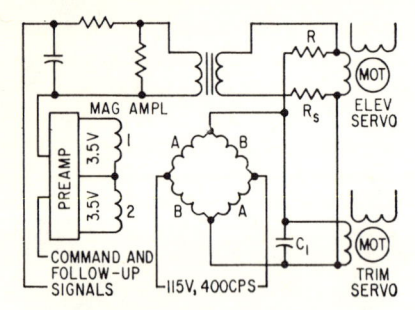

FIG. 2—Detailed circuit of block representation of Fig. 1

is shown in Figs. 1 and 2.

Automatic pilot command signals are fed through a transistor preamplifier (Fig. 3) to a power magnetic amplifier that drives the elevator servo motor. The motor operates the elevator surface through the gear train and engage clutch. Rate generator and follow-up signals are fed back into the input circuit in opposition to command signals. Degenerative feedback improves amplifier operation and stabilizes gain. The trim servo receives power commands from the power magnetic amplifier in parallel with the elevator servo motor.

Power Amplifier

The power magnetic amplifier is of the bridge type whose output voltage and phase are controlled by d-c signals in its control coils. As shown in Fig. 2, 115-v 400-cps power is applied across two points of the bridge and output power is fed to motors from the other two points. Capacitor C_1 provides impedance matching and phasing of the variable-phase voltage with respect to fixed phase.

When high current is present in control coil 1 and low current is present in 2, the bridge arms A are saturated, forming low impedance paths for the a-c power. This results in most of the excitation voltage being applied across the motor. When control currents are reversed, a similar voltage of opposite phase is applied to the motor producing opposite output torque. In the same way, power is applied concurrently to the trim servo motor, which is electrically in parallel with the elevator motor.

The magnetic amplifier and the two variable-phase motor fields are designed for parallel connection of the trim and elevator servos. Such operation results in a trim system possessing a high degree of safety since practically all of the trim system circuitry is common with the elevator system.

Electrical faults occurring before the parallel connection points that result in loss of signal cause both servos to be inoperative, which is a fail-safe condition. Any faults that result in a bias or apparent command signal will be evident to the pilot through abrupt elevator response, which is quite rapid compared with slow moving trim system.

Short circuits or opens at the trim motor or in cabling to it will result in nonoperation of the motor. No single failure can cause the trim servo to run away or operate independently of the elevator channel so as to build up a large counteracting sustained elevator deflection. Hence, the danger of large transient maneuvers occurring at autopilot disengagement is greatly minimized.

Torque Limiter

The element which limits the power applied to the servo motors, hence the torque, is the bridge magnetic amplifier. This amplifier is designed to supply the highest voltage and largest amount of power required for the application.

Independent of the command-signal strength, output voltage is sharply limited to what the magnetic amplifier is inherently capable of supplying, or approximately 90 percent of excitation voltage. No single fault can cause any appreciable increase in voltage applied to motors. For this reason the power magnetic amplifier can be relied upon as a safe type of voltage limiter.

Further protection is afforded by the servo variable-phase voltage being high compared with other voltages present in aircraft wiring. As a result, there is no chance of a higher voltage being applied to the servo motor because of a short circuit in the cabling.

The elevator servo is critical regarding maximum torque; this is usually not true of the trim servo. The maximum voltage applied to the elevator servo can be controlled by R in Fig. 2. Since the bridge magnetic amplifier is a fixed limit, variation in R results in variation of torque limit.

The overall feedback network starts at the elevator servo motor leads and provides about 20 db of negative feedback. This amount permits sizable changes in the value of R without any appreciable change in overall gain. Therefore, maximum permissible torque can be varied without appreciably changing gain or torque gradient in the linear amplifier region that is important to autopilot performance.

Torque Modulation

The torque limit may be a fixed value on each axis of a specific aircraft or it may be changed during flights. When wide variations are required a method of torque variation or modulation is available that

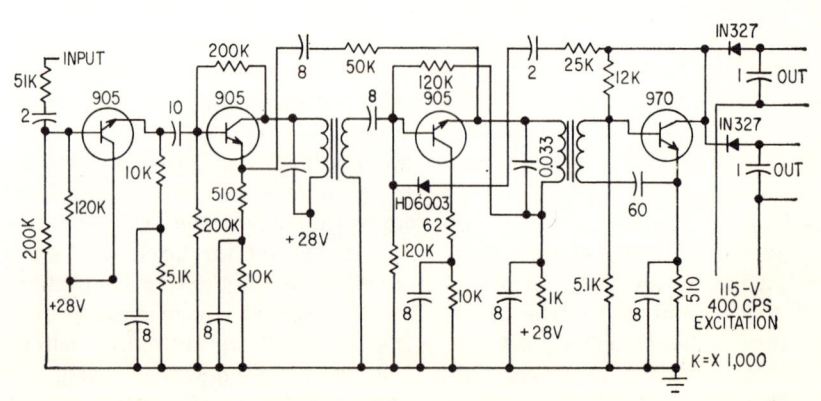

FIG. 3—Preamplifier for pitch-axis channel employs four npn silicon transistors

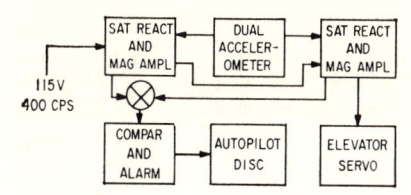

FIG. 4—Torque-modulation system has dual accelerometer that measures acceleration in pitch plane of motion. Servomotor torque is proportionately decreased with increasing acceleration

is actually a self-adjusting system wherein normal and angular accelerations in the pitch axis are measured and cause a proportional reduction in torque as normal g loading on the aircraft increases. In this manner the elevator servo always exerts the torques necessary for good performance under different flight conditions, but within safe limits. Since such a system is protecting the aircraft from structural damage or a dangerous maneuver, it in itself must be fail safe.

A block diagram illustrating this type of system is shown in Fig. 4.

The sensing element is a dual accelerometer located in the aircraft to measure acceleration in the pitch plane of motion. Both accelerometers are identical and feed signals to separate and identical magnetic amplifiers which control saturable-reactor coils. Both coils are in series with the a-c power source and the fixed phase of the elevator servo motor.

As acceleration signals increase from zero, independent of sign, the impedance of the reactors is increased to proportionally reduce motor torque. Each accelerometer, amplifier and saturable reactor channel contributes to half the voltage drop needed for the torque reduction required for a given acceleration.

Use of dual channels instead of one provides a high degree of safety since a comparison type of monitoring system can be used.

Comparator Circuit

As shown in Fig. 4, a comparator-alarm circuit measures and compares each of the voltage drops across the two reactors. When both channels are operating properly, the voltage across each reactor coil will be equal within tolerance limits and the comparator will not activate an alarm or disengage the automatic pilot. In the event of an appreciable difference, however, an alarm will be activated and the automatic pilot disengaged. The circuit, which includes both semiconductors and magnetic amplifier elements, is shown in Fig. 5.

Mixing

The two input voltages are applied to the input winding of the mixing transformer whose output, a voltage representing the arithmetic difference of the two inputs, is applied to the base of Q_1 through the R-C and diode network. The positive emitter voltage back biases Q_1 to permit normal I_{ceo} current flow in the collector circuit through control coils A and B when R is properly adjusted and no input signal is present.

When the a-c input level just rises to a point where the transistor is cut off, the base-collector junction is biased in reverse and no collector current flows. Any further increase in the a-c signal will produce a negative collector current. The parallel diode and resistor network prevent a-c voltage feedback to Q_1 because of unbalance between the two halves of the magnetic amplifier.

The control currents in coils A and B are always of the same polarity so the ampere-turns produced are always additive. The d-c normally present in A and B contributes to maintaining the core of the magnetic amplifier in a normally saturated condition.

Normally gate windings C and D are low impedance as a result of core saturation by the high current generated by the excitation voltage and the current in the control coils. As a result, a unipolar voltage normally appears across points X and Y, energizing the relay through the Zener diode. Excitation of the relay results in continuity of its contacts, which is a no-alarm condition.

When the a-c input to Q_1 reaches or rises above a value which causes cutoff and the collector current is either zero or negative, the flux in the core of the magnetic amplifier is reduced below saturation. The output impedance of the two windings becomes high and most of the excitation voltage drop is across the gate windings. This results in a sharp reduction of d-c output across points X and Y. When this voltage is reduced, the current through the Zener diode is sharply reduced causing the relay contacts to open and an alarm indication results. The alarm circuit may be used to auto-

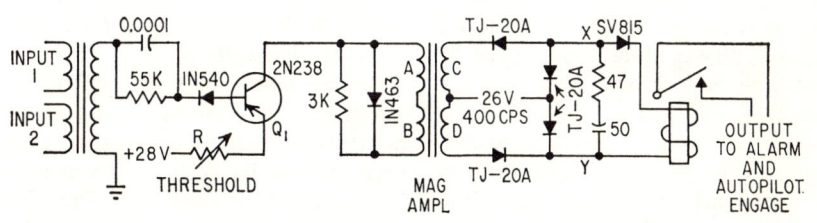

FIG. 5—Comparator-alarm circuit for system of Fig. 4 activates alarm or disengages automatic pilot if signals from dual accelerometers differ appreciably

matically disengage the automatic pilot.

Current flows during the normal or safe condition throughout the circuit. This practice not only insures that an alarm will be given in the event of power failure, but that open circuit faults that interrupt or diminish appreciably the current flow will also result in an alarm.

The Zener diode insures that the XY voltage required to drop out the relay is held above a minimum value compatible with fail-safe operation. Failure of any of the resistors, diodes, saturable reactors or the transistor, mixer transformer and connecting wiring that would result in open circuits incapacitating the monitor will cause the relay to revert to its deenergized position and give an alarm indication. Short circuits and breakdowns in magnetic components and in other vulnerable components that affect monitor operation also result in an alarm.

Switching Circuits for

Production lines and industrial processes as well as missile count-downs, can be stopped if any of a number of variables is out of tolerance. Transistorized level-sensitive switch uses nonlinear negative feedback to provide stable operation over wide temperature ranges

By DONALD W. BOENSEL,*Jet Propulsion Laboratory, California Institute of Technology, Pasadena, Calif.

SYSTEMS for absolute or tolerance measurement often rely on go/no-go indications designed to halt a missile firing count-down, pull a faulty module from a production line or perform some similar function. In such operations, a number of variables, such as power-supply voltage, signal level or bias current, are monitored continuously, sequentially or on command. Each variable is ordinarily compared with a predetermined reference. Both the reference and its tolerance are often programmed to vary with some parameter (such as time).

Usually a process is stopped if one or a combination of variables are out of tolerance. More sophisticated systems may include feedback control techniques to provide automatic correction for out-of-tolerance variables.

An a-c level-sensitive switch has been developed as a reliable comparator. It is stable to ±1.5 percent over a 100-deg C temperature range. The circuit has a switching gain of about 40 db and uses standard components throughout.

Unlike many switches which are stabilized by compensating temperature-sensitive elements, this design derives its stability from nonlinear negative feedback. This arrangement results in characteristics that are not susceptible to either active component changes or temperature changes.

Requirements

An input-output characteristic similar to that in Fig. 1 must be obtained from the switch. Depending on whether the measurement is absolute or a differential tolerance, drift limits on the switch or on the out-of-tolerance point, E_s, can range from ±0.5 percent to as much as ±10 percent of E_s. In military applications, it may be necessary to maintain these limits not only with component substitutions but over severe ambient temperature ranges. These requirements indicate the desirability of including overall negative feedback to supplement the positive feedback necessary for a regenerative change of state at the switch point. Both types of feedback must be nonlinear to allow the switch to operate in a bistable fashion.

Most high-accuracy systems either process a-c signals or modulate d-c signals and use a-c amplification. An a-c level-sensitive switch is therefore compatible with many critical applications.

Circuit Operation

The basic comparator consists of a somewhat unconventional monostable multivibrator followed by a rectifying transistor and filter. For signals above the trigger level, the circuit is periodically switched into its transient state.

The more important waveforms

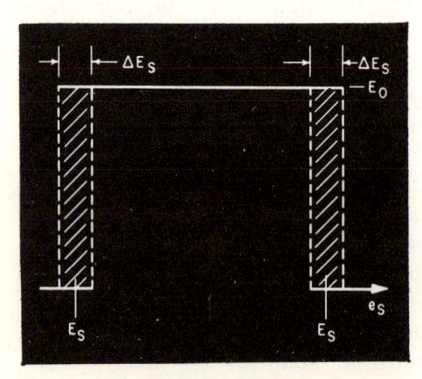

FIG. 1—Level-sensitive switch must operate at E_s despite wide temperature changes and component substitutions

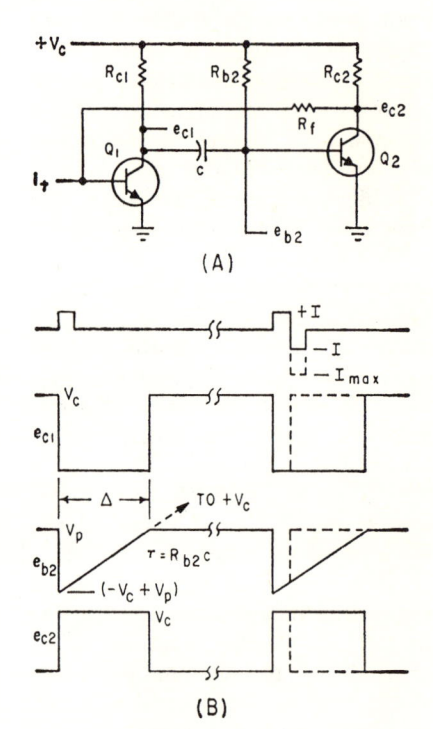

FIG. 2—Conventional multivibrator at (A) provides waveforms at (B). Large negative pulse following positive trigger pulse would affect waveforms as shown in dotted lines

* Now with Space Electronics Corp., Glendale, Calif.

Missile Countdowns

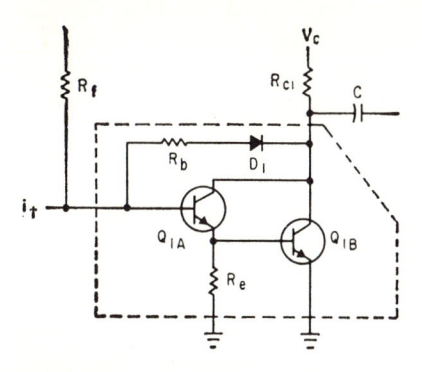

FIG. 3—Portion of circuit in dotted lines replaces Q_1 in Fig. 2A to get required gain and stability

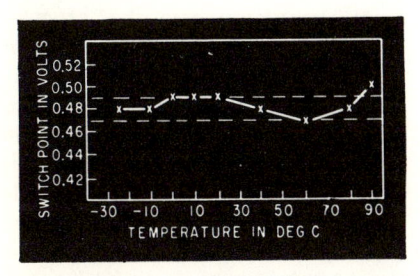

FIG. 4—Switch point stability with temperature indicates a variation between dashed lines of about ±1.5 percent

involved in the operation of the conventional one-shot multivibrator in Fig. 2A are shown in Fig. 2B. The positive feedback loop consists of Q_1, C, Q_2 and R_f. Duration of the transient state[1,2] in which Q_1 is saturated is $\Delta = \tau \ln (2V_c - V_p)/(V_c - V_p)$, where $\tau = R_{b2}C$ and where V_p is emitter-base contact potential of Q_2. If $V_c \gg V_p$, transient duration is $\Delta \cong \tau \ln 2$.

To use this circuit as a level-sensitive switch requires knowledge of the minimum trigger level. If duration of the input trigger is much greater than time required for switching, minimum triggering level can be calculated as below.

If R_f and R_{b2} are assumed to be much greater than R_{c2} and R_{c1}, respectively, minimum capacitive current necessary to bring Q_2 out of saturation is $i_{c\,min} = V_c [(1/R_{b2}) - (1/\beta_2 R_{c2})]$, where β_2 is common-emitter current gain of Q_2. Consequently, the minimum positive trigger pulse with an input resistance of Q_2 much less than the load impedance of Q_1 is $I_{min} = (V_c/\beta_1)[(1/R_{b2}) - (1/\beta_2 R_{c2})]$, where β_1 is corresponding current gain of Q_1.

The dotted waveforms at the right of Fig. 2B indicate the possibility of improper circuit operation if the positive trigger pulse is followed immediately by a large negative pulse. The maximum permissible value of such a negative pulse is calculated in the same way as that for the minimum positive trigger pulse. Thus, $I_{max} \cong V_c [(1/R_f) - (1/\beta_1 R_{c1})]$.

If $R_{b2} \cong R_f$, $R_{c1} \cong R_{c2}$, and $\beta_1 \cong \beta_2 \cong \beta$, there is no possibility of operational failure with reasonably symmetrical triggering. Under such conditions, $I_{max} \cong \beta I_{min}$. Therefore, triggering the switch with a symmetrical pulse train, rather than a unilateral pulse, is desirable. Although rectangular trigger pulses have been used in the above calculations, in a broad sense, results apply equally to sinusoidal triggering.

Since I_{min} corresponds to the trigger point of the switch, it must be stabilized with respect to temperature and component substitution. With present-day transistors, it is virtually impossible to obtain a trigger point stability better than ±10 percent with the circuit in Fig. 2A because of manufacturers tolerances in current gain and temperature drifts in current gain and emitter-base contact potential. These problems arise primarily in considering the parameters of input transistor Q_1.

If $\beta_2 R_{c2} \gg R_{b2}$, minimum positive current is $I_{min} \cong V_c/\beta_1 R_{b2}$, which is dependent only on current gain of Q_1. Although trigger source impedance must be high for proper regenerative switching it cannot be made infinite, so that trigger level is also dependent on input characteristics of Q_1 (in particular, emitter-base contact potential).

If Q_1 is biased and feedback is stabilized in a linear operating range, rather than allowed to remain in its normal cutoff condition, trigger level can be made practically independent of temperature and component change; effects of drift in V_p can be removed by capacitive input coupling techniques.

In practice, gain-temperature variation of standard small-signal silicon transistors is about 1.5 or 2.5 to 1 over a temperature range of 100 C. It is therefore impossible to adequately stabilize a single transistor performing the function of Q_1 and maintain sufficient gain. Therefore an amplifier configuration that is compatible with requirements for gain and stabilization must replace Q_1.

The transistor combination in Fig. 3 was developed to replace Q_1 in Fig. 2A. Both Q_{1A} and Q_{1B} in Fig. 3 have nominal current gains of 100, so that open loop gain of the cascaded pair is about 10,000. The feedback network consisting of avalanche diode D_1 and resistor R_b not only stabilizes d-c operating point but also a-c gain.

If $\beta_1 R_{c1} \gg R_b$, d-c gain with feedback is $A_{d-c} = R_{b1}/R_{c1}$. Similarly, if $\beta_1(R_{c1}//X_c) \gg R_b$, a-c gain is $|A|_{a-c} = R_b/|R_{c1}//X_c|$, where $X_c =$

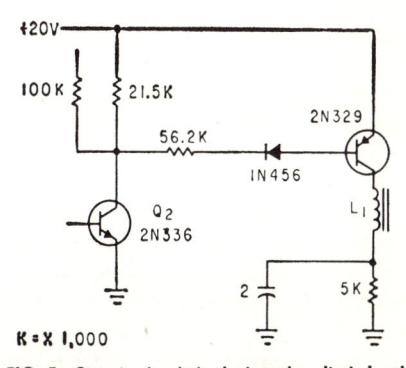

FIG. 5—Output circuit is designed to limit load effects on level-sensitive switch

$1/\omega C$ and ω is sinusoidal triggering frequency.

In this case, effective β_1 of the amplifier is 10,000; furthermore, R_b can be made relatively small because of the voltage drop afforded by D_1, so that the above approximation is applicable.

With values chosen for the experimental comparator ($R_b = 147{,}000$ ohms, $R_{c1} = 10{,}000$ ohms, $R_e = 14{,}700$ ohms, $C = 0.03$ μf, $\beta_{1A} \cong \beta_{1B} \cong 100$ and D_1 breakdown voltage $= 9$ v), d-c gain is about 15 and a-c gain is about 150 at 6 kc.

In this way, operating point was stabilized to about ±0.2 percent and a-c gain to ±2 percent for a temperature range or component changes which produce a 4:1 change in β_1. It is possible to stabilize a-c gain to as little as ±0.5 percent under the same conditions and maintain the same sensitivity that would be obtained if Q_1 had a nominal β_1 of about 40.

Feedback Path

This simply involves reducing a-c gain to 40 by placing an a-c feedback path in parallel with R_b and D_1 or by reducing R_b by a factor of 4. Reduction of R_b produces no significant change in collector operating point of the cascaded pair, since this point is largely controlled by the breakdown voltage of D_1.

The nonlinear negative feedback loop is effective only when the switch is in its stable state. Application of a trigger current sufficient to bring Q_2 out of saturation produces a current that is fed back to the base of Q_1 and that is large enough to open D_1 and drive Q_1 into saturation.

Two other factors that affect trigger point stability are related to variations in the parameters of Q_2. In practice, switching speed places an upper limit on R_{c2} of 20,000 to 50,000 ohms. For β_2 nominally 100, R_{b2} must be below 200,000 ohms for the relationship $\beta_2 R_{c2}$ always $\gtrless 10 R_{b2}$ to be even partially satisfied. In any event, for typical variations in β_2, it is necessary that $\beta_2 R_{c2}$ be at least $20 R_{b2}$ to assure that the error in triggering level from this source be less than ±2 percent. Thus, these values should be $R_{b2} = 100,000$ ohms, $R_{c2} = 21,500$ ohms and $\beta_2 = 100$.

The second factor is associated with the change in input resistance of Q_2 with temperature during saturation. It has been pointed out by Ebers and Moll that incremental input resistance of a common emitter transistor decreases significantly when it saturates. This phenomenon results not only because there is no feedback associated with the resistance of the emitter region, but also because of an order of magnitude decrease in base spreading resistance. The decrease in base spreading resistance has been attributed to a shift in injection coordinates as well as carrier injection into the base region caused by forward bias on the collector-base junction.

Incremental input resistances during saturation of about 10 to 15 ohms can be expected for the 2N336 transistor used as Q_2. If C is chosen to make $\Delta = 20 \, (\tau/4)$, such an input impedance represents about 1 percent of total a-c load impedance on Q_1. A simple calculation shows that $C = 0.03 \, \mu f$ ($X_c = 1,000$ ohms at 6 kc). Since the temperature coefficient of silicon is about −0.7 percent/deg C, the change in trigger level contributed by this variation is about ±0.4 percent over a 100-deg C temperature range.

Under the worst possible accumulation of these errors, stability of the switch would be about ±4.4 percent over a 100-deg C temperature range. However, there is direction as well as magnitude associated with these drift errors. The effects of similar drifts in β_1 and β_2 on trigger level tend to cancel each other. A more precise estimate of switch stability would therefore be ±0.5 to ±1.0 percent. Experimental data, a sample of which is shown in Fig. 4, has indicated that this figure is probably closer to ±1.0 to ±1.5 percent.

Output Circuit

The switch output must be converted to d-c. Since filtering at the switch output would interfere with monostable operation, an isolating transistor is introduced between the sensing switch and filter.

This circuit should not cause appreciable loading of the level-sensitive switch. It should furnish a drift-free, low-ripple d-c of prescribed magnitude to the load when the monostable switch is in its stable state. It should also furnish an adequate change in current to the load when the one-shot is triggered.

The circuit in Fig. 5 was developed to satisfy these requirements. The 5,000-ohm load resistor and the 56,200-ohm base resistor were chosen to assure that the output transistor is saturated when Q_2 in Fig. 2A is in its stable saturated state. Inductor L_1 must have sufficient inductance at maximum load current to provide adequate filtering during switching. An inductance of 10 henrys at 4 ma for L_1 and a capacitance of 2 μf reduced ripple to about 20 μamp peak-to-peak low-current state or about 0.5 percent of maximum current.

Although load current does not drop to zero in the unstable state, a characteristic like that shown in Fig. 1 can be obtained by coupling the actual load through an avalanche diode. For example, voltage across the 5,000-ohm resistor in Fig. 5 drops from 19 to 6 v as the monostable switch is triggered into recurrent oscillation. A 9-v coupling diode (similar to D_1) can therefore be used to obtain the desired characteristic. A schematic of the complete switch and output circuit is shown in Fig. 6.

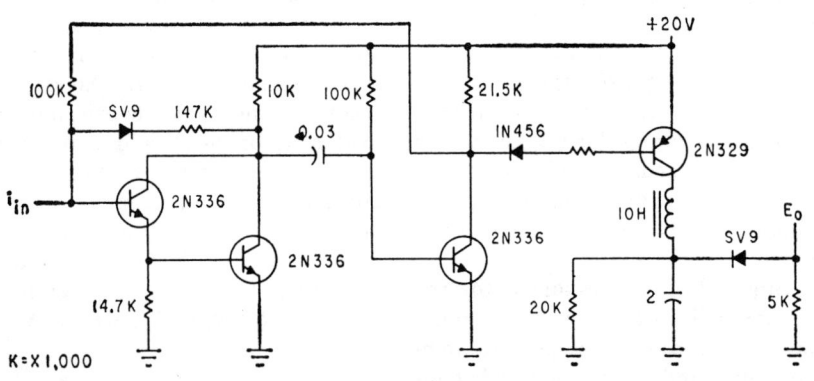

FIG. 6—Final circuit with nonlinear negative feedback is stable within ±1.5 percent over 100-deg C temperature range

References

(1) J. Millman and H. Taub, "Pulse and Digital Circuits," McGraw-Hill Book Co., Inc., 1956.

(2) R. B. Hurley, Designing Transistor Circuits, *Electronic Equipment Engineering*, Feb. 1959.

Eleven meter superregenerative receiver (left) controls the decoding and servo circuits (right)

Simplified Controls for Target Drones

Pulse symmetry and repetition rate control servos which drive rudder and elevator. Pulses modulate the transmitted carrier, which is picked up and detected by a superregenerative receiver on the plane

By G. B. HERZOG, R. C. A. Laboratories, Princeton, N. J.

IN SIMPLE FORMS of pilotless aircraft control, two separate and continuous control channels are sufficient, one for the rudder and one for the elevator. The control system described here (Fig. 1) transmits two completely separate and continuous pieces of information over one radio link. A third bit of information is transmitted by momentarily interrupting the transmission, thus operating a digital control.

One continuous channel of information is conveyed by varying the symmetry of a pulse waveform and the other channel of information is conveyed by varying the repetition rate of the waveform. Advantages of this form of transmission are that nonlinearities in the transmitting and receiving equipment are unimportant and a constant signal amplitude can be obtained by clipping. Furthermore, selective filtering or synchronization between receiver and the transmitter is not necessary as with more complicated forms of frequency or time multiplex transmission.

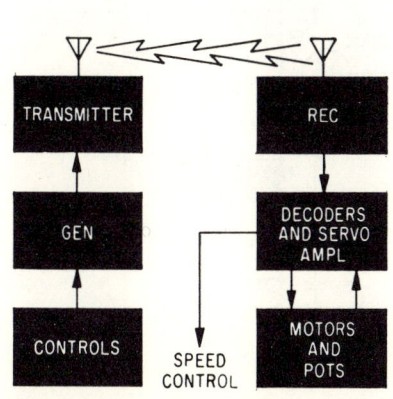

FIG. 1—By controlling the modulation of the carrier, an operator guides the plane

The control signal is obtained by generating a sawtooth waveform of variable repetition rate for the frequency controlled. This wave is then clipped to a desired nonsymmetrical waveform for the symmetry-controlled channel.

Signal Generation

A sawtooth voltage is generated across capacitor C_1 by blocking oscillator Q_1, Fig. 2. The oscillator time constant is in the emitter circuit of Q_1, whose charging capacitor, C_1, is fed by transistor Q_2. Frequency is varied directly by rotating linear potentiometer R_1, which feeds Q_2.

Since the collector impedance of Q_2 makes Q_2 essentially a constant-current source, the sawtooth is linear. This reduces any possibility of the frequency affecting the sym-

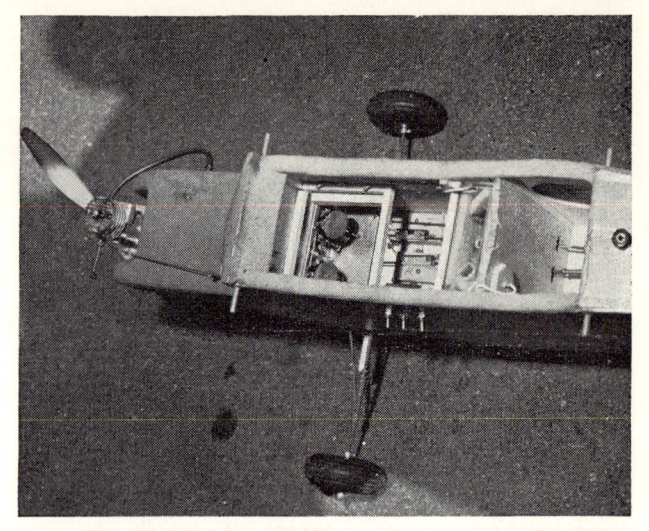

Closeup shows motors, gear box, linkage and potentiometers. In both photos the servo circuit has been removed to show the servo motors

metry channel.

The sawtooth is amplified by d-c amplifier Q_3, which has sufficient emitter degeneration to minimize loading on the sawtooth generator.

The following stage, Q_4, has a variable emitter bias which clips the waveform to the desired symmetry. The bias is controlled by potentiometer R_2.

Since the waveform is d-c coupled from the sawtooth generator, the frequency of the sawtooth will not affect the point on the wave at which the clipping transistor begins conduction. Therefore, variation of the bias point sends a second bit of information which is completely independent of the first piece of information carried by the frequency of the waveform. Because a linear sawtooth is clipped, the resulting nonsymmetrical waveform is a linear function of the bias point, hence linearly related to the potentiometer setting.

Stage Q_5 further shapes the wave, squaring the sawtooth portion of the input wave at the point that this wave crosses the clipping level.

The output, Fig. 3, whose peak is 13.5 v, modulates the transmitter carrier wave.

Signal Reception

The transmitted signal is received by a logarithmic mode (self-quenching type) superregenerative receiver (Fig. 4). A stage in the audio section limits the signal to a constant level regardless of reception conditions. By operating in the logarithmic mode, the receiver rejects brush noise interference from the servo motors. The audio

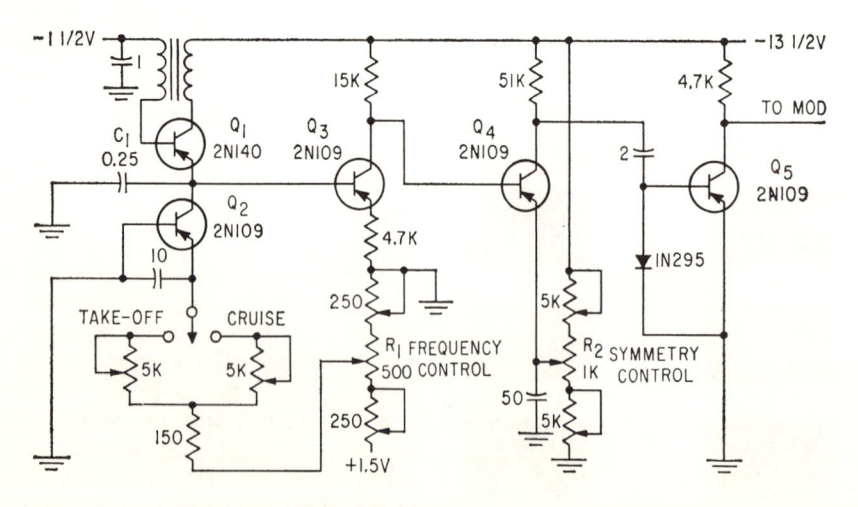

FIG. 2—Potentiometer R_1 sets the frequency of the sawtooth and potentiometer R_2 adjusts the symmetry of the output to the modulator grid of the transmitter

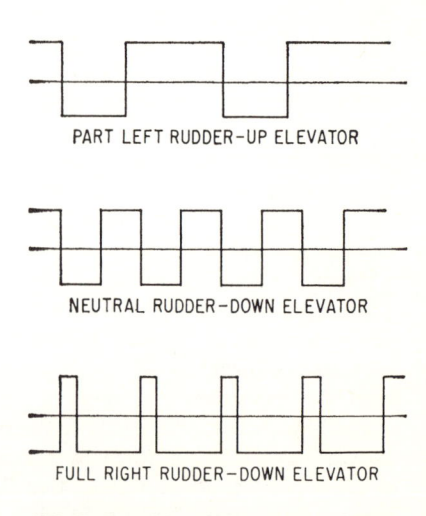

FIG. 3—Repetition rate positions elevator. Pulse shape positions rudder

Robot plane ready for takeoff

part of the receiver consists of amplifier Q_1, clipper amplifier Q_2 and complementary-symmetry emitter followers Q_3 and Q_4, which drive the signal decoding circuits.

Signal Decoding

The two continuous channels are separately demodulated by frequency and symmetry detectors which are unaffected by the information in the opposite channel.

The symmetry detector consists of average voltage detectors, D_1 and D_2. When the pulse shape deviates from a symmetrical form, the detectors change the voltages that they apply to the divider that contains R_1. The tap of R_1 signals transistor Q_5 of the amplifier that drives servo motor No. 1. This motor rotates the tap of potentiometer R_1 until the tap finds the reference voltage. When a signal is absent, the reference potential brings the servo to its center position.

Diodes D_3 to D_6 form a balanced pulse counter detector which demodulates the frequency channel information. A change in the repetition rate changes the voltages applied to the divider that contains R_2. The tap of R_2 signals transistor Q_{11}, which drives motor No. 2 until the tap finds the reference potential. The motor is centered when a signal is absent.

Servo Amplifiers

Since the symmetry and frequency servo amplifiers are identical, only the symmetry servo amplifier will be described. Transistors Q_5 and Q_6 are biased so that zero voltage appears at the bases of transistors Q_7 and Q_8 when R_1 is at its reference voltage point. A change in the symmetry of the transmitted pulse changes the reference point of R_1. Transistors Q_7 and Q_8 form a complementary-symmetry input arrangement which drives transistors Q_9 and Q_{10}. Motor No. 1 is a miniature p-m field type which requires a low driving current.

Third Channel Information

A third bit of information, engine speed control, is transmitted by momentarily interrupting modulation. Signal interruption removes the detected voltage across R_2, which counteracted the forward bias of Q_{17}. Transistor Q_{17} conducts, switching Q_{18} on, thus pulsing the step actuator. The actuator advances the engine speed by one step each time the modulating signal is interrupted. Momentary interruption of the modulation signal does not interfere with the information conveyed by the other two channels. Should the receiver fail to receive a signal continuously, the servos center and the engine control advances to a stop position.

Application Data

Tests have shown that the transistors impose a ceiling of 140 F for safe operating temperature.

The transmission system might be used to control the autopilot of a target-drone airplane.

FIG. 4—Receiver output transistors, Q_3 and Q_4, simultaneously signal the symmetry and frequency decoders and the detectors that control the step actuator

Chapter 18
RADAR, SONAR, AND BEACONS

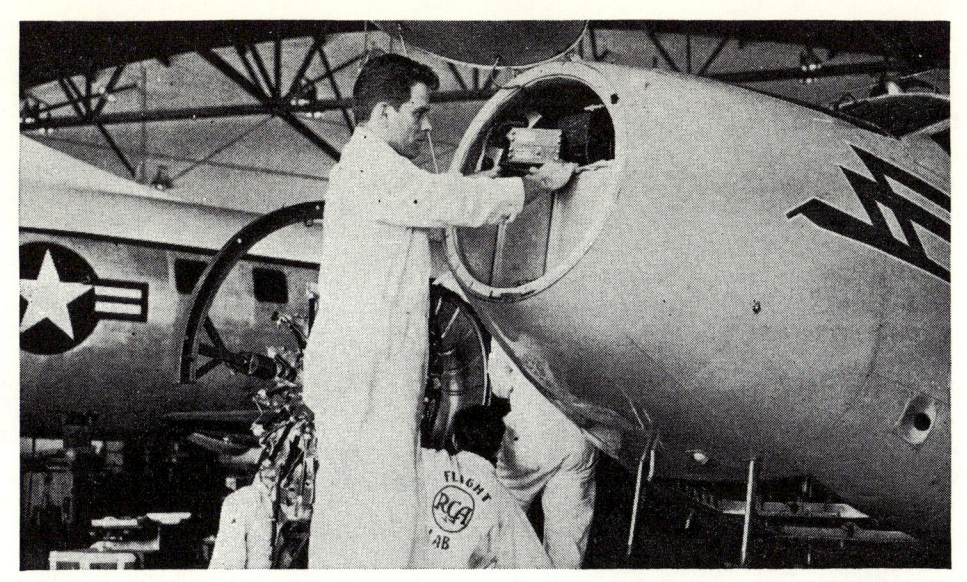

Three-indicator-light adapter is mounted on one-indicator-light receiver to form small, light package

Dual Conversion for Marker-Beacon Receivers

Airborne marker beacon receiver has high first i-f for good image rejection and lower second i-f for stable gain. With same sized transistorized adapter, one-indicator-light receiver is converted to three-indicator-light receiver weighing about two pounds

By **RICHARD G. ERDMANN,** Airborne Communications Engineer, Radio Corporation of America, Camden, N. J.

JUNCTION TRANSISTORS and diodes completely replace vacuum tubes and relays in a dual-conversion superheterodyne marker beacon receiver. The one-indicator-light receiver with transistorized adapter converts to a three-indicator-light marker beacon receiver that weighs less than two pounds and draws less than two watts.

The transmitted 75-mc carrier in marker beacon systems is amplitude modulated to denote marker function (airways, 3,000 cps; outer runway, 400 cps; middle runway, 1,300 cps). Two receiving system types are in general use. The one-light receiver responds to any of three modulating frequencies with identification being aural. The three-light receiver separates the frequencies to operate three color-coded lights.

When the aircraft passes over a marker beacon transmitter, the tone is heard. At a predetermined signal level, an indicator lights to establish aircraft position.

One-Light Receiver

The dual-conversion receiver in Fig. 1 has a high-frequency first i-f for good image rejection and a low-frequency second i-f to provide stable gain. The first i-f circuits include only passive tuned circuits.

With dual conversion, the number of i-f stages could be reduced because of the increased gain per stage. A single-conversion receiver with a high i-f would have required that the i-f transistors operate at a frequency where gain would not be high. Tests of the dual-conversion receiver showed a signal-to-noise ratio of 20 db or more. Input to the receiver during these tests was 500 μv at 75 mc modulated 30 percent.

The 75-mc filter in Fig. 1 consists of four tuned circuits. The tuned circuits feed a mixer whose output is fed to a triple-tuned filter tuned to the first i-f of 4.2 mc. This signal is converted to the second i-f of 520 kc by a 4.72-mc converter.

The broadband second i-f stages

feed a diode that detects the audio and agc signals. Part of the output is rectified and turns on the indicator light when the received signal exceeds a predetermined level.

Stray ground-current coupling between the preselector tuned circuits was encountered at 75 mc. This coupling, resulting from lack of a good ground plane with the printed circuit, made the receiver susceptible to spurious responses. Therefore, a one-can assembly was made that includes T_1 through T_4, first i-f coil T_5 and oscillator coil T_8.

Zero-temperature-coefficient ceramic capacitors are used in the tuned circuits of T_1 through T_4 and in T_8 for stable preselector performance. Near critical coupling allows maximum nose bandwidth with good skirt selectivity. It also simplifies tuning of the filter.

First I-F

A drift transistor, Q_1, is used in the oscillator in a grounded-base configuration with a fifth-overtone crystal. The circuit is stabilized in relation to d-c operating point by a large emitter resistor and a relatively low d-c impedance provided by base-biasing network R_1 and R_2.

Frequency stability is better than ±10 kc. Using a transistor and a permeability-tuned oscillator coil make the oscillator much less susceptible to vibration and shock.

Diode D_1 is supplied with injection power from Q_1. Forward bias on D_1 minimizes variations in diode impedance caused by supply-voltage and temperature changes and their effect on oscillator power and receiver gain. Since frequency output is a fundamental rather than a harmonic, spurious responses through the receiver preselector fall at 70.8-mc intervals and are easier to control.

I-F Selectivity

Near-critical coupling is used in the first i-f filter T_5, T_6 and T_7 for the same reason as in the preselector. This i-f selectivity also contributes to the skirt selectivity of the receiver passband.

The first i-f is fed to crystal-controlled converter Q_2 which oscillates at 4.72 mc producing a 520-kc second i-f. Frequency error contributed to the receiver by this conversion is only a few hundred cycles maximum, and conversion gain variations with temperature and

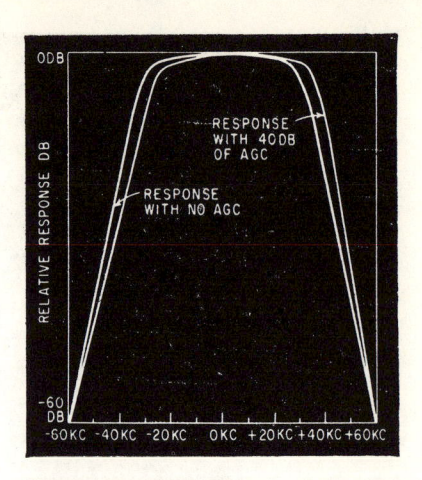

FIG. 2—Composite selectivity of first and second i-f filters is influenced only slightly by parameter shift of the transistors

supply-voltage changes are comparable to a straight i-f amplifier.

The drift transistor (2N247) produces enough power gain at the oscillator frequency to eliminate need for a separate oscillator inductance, eliminating one tuned circuit.

Output impedance of the converter is high enough to permit coupling the converter collector directly to the top of the second i-f filter T_9 through T_{11}. The filter is used rather than distributed selectivity to eliminate effects of transistor parameter variations with temperature and supply-voltage changes, transistor replacement and agc action. The filter is critically coupled and operates at medium Q, yielding the required broad-nose selectivity of 80 kc to the 6-db points.

Composite selectivity of the first and second i-f filters shown in Fig. 2 is influenced only slightly by parameter shift of the transistors.

The second i-f amplifier Q_3, Q_4, Q_5 and T_{12}, T_{13}, T_{14} is completely broadbanded and requires no tuning adjustments. Since all receiver selectivity is before these circuits, broadbanding does not have some of the usual undesirable effects such as susceptibility to cross modulation. Power gain of the second i-f stages is better than 20 db including conversion loss. The drift transistors in the second i-f amplifier operate well inside the flat gain region of the power-gain versus frequency curve and have high d-c betas for good agc action.

These transistors are stabilized

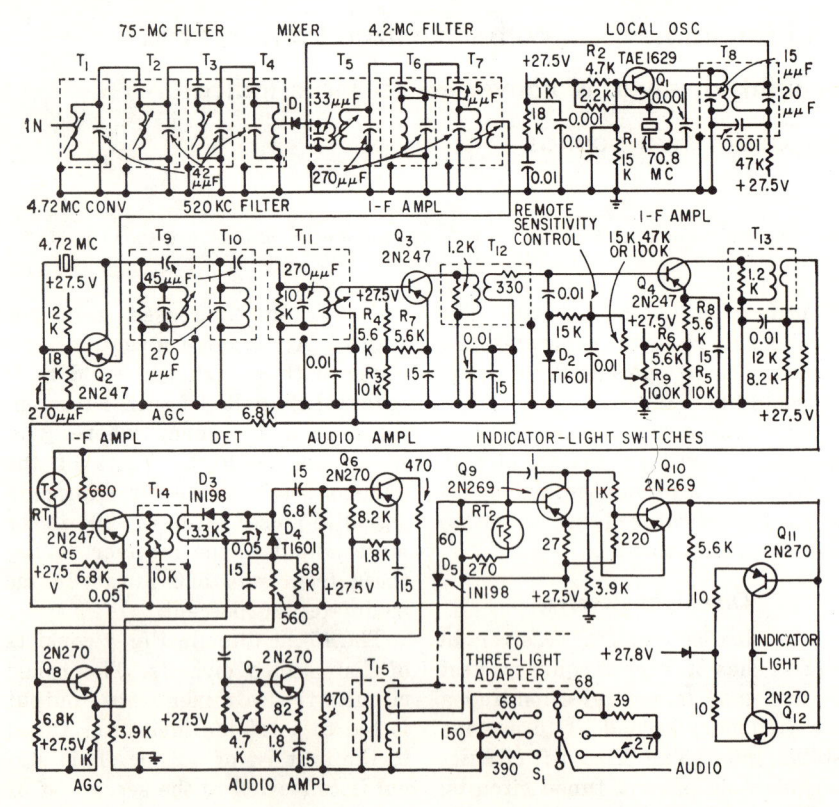

FIG. 1—Schematic of one-indicator-light receiver shows first i-f filter circuits which consist of only passive tuned elements. Second i-f stages use three drift transistors

in relation to operating point and interchangeability by large emitter resistances and a relatively low d-c impedance base-bias network. Emitter resistors of Q_3 and Q_4 are tied to a voltage divider network (R_3 and R_4 for Q_3, R_5 and R_6 for Q_4). These networks provide fixed emitter potentials that cut off collector current rapidly when base voltage approaches agc threshold. Below agc threshold, resistors R_7 and R_8 serve as emitter-stabilizing resistors for Q_3 and Q_4, respectively.

Thermistor RT_1 compensates all gain variation with temperature incurred in the receiver from antenna to audio output.

The circuit between T_{12} and Q_4 forms an r-f voltage divider for remotely adjusting receiver sensitivity. Since agc lowers collector current and power gain of this stage, the network permits changing gain without disturbing d-c parameters.

Radio-frequency impedance through diode D_2 is inversely proportional to d-c through it. Since R_9 controls d-c through D_2, it controls r-f impedance of the diode. Remote control of current through D_2 is easily accomplished, since it involves a d-c of 1 ma or less. Gain adjustment of approximately 20 db can be provided in this manner.

The third i-f amplifier, Q_5, must be stabilized at a high output level to prevent i-f clipping and provide sufficient power for detection.

AGC

Agc amplification is necessary because available d-c power from detector D_3 is low. Agc amplifier Q_8 receives rectified power from D_3 and reduces forward bias on Q_3 and Q_4 as signal strength increases.

The base of Q_8 is isolated from second detector D_3 by D_4, so that Q_8 can be stabilized for temperature

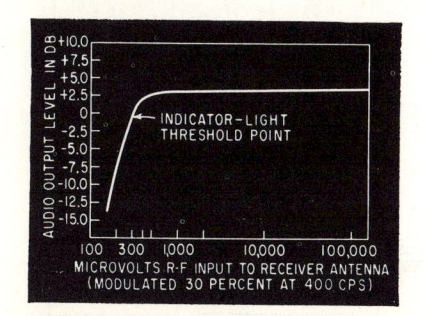

FIG. 3—Automatic-gain-control characteristics of receiver at 25 C

when developed bias on D_3 is below the agc threshold level. The agc is delayed for maximum receiver sensitivity with D_4 providing the delay. No coupling between detector D_3 and the agc line occurs until forward voltage across D_4 reaches about 0.5 volt. The receiver agc curve is shown in Fig. 3.

First audio amplifier Q_6 provides audio gain and serves as an impedance-matching stage between detector and audio amplifier Q_7. Output from Q_7 is passed to the audio output terminals from a secondary winding of T_{15} through attenuator S_1.

Attenuator S_1 maintains constant load on the output winding while permitting four steps of audio attenuation. A constant load is required to keep a-c gain of Q_7 constant. Failure to do so would result in variation of the indicator-light threshold not only with audio output level but with r-f input signal level.

Indicator-Light Circuits

The indicator-light circuits provide relay-type switching when receiver input exceeds a threshold level. Transistors rated in milliwatts of collector dissipation control almost three watts of indicator-light power.

Diode D_5 rectifies audio and supplies forward bias directly to the base of Q_9. Transistor Q_9 is normally cut off and Q_{10} is normally conducting. Transistor Q_9 conducts when a rectified signal from D_5 places forward bias on its base. Transistor Q_{10} is cut off and the potential on its collector drops. This drop appears on the bases of Q_{11} and Q_{12}, permitting collector current to flow out to the light circuit.

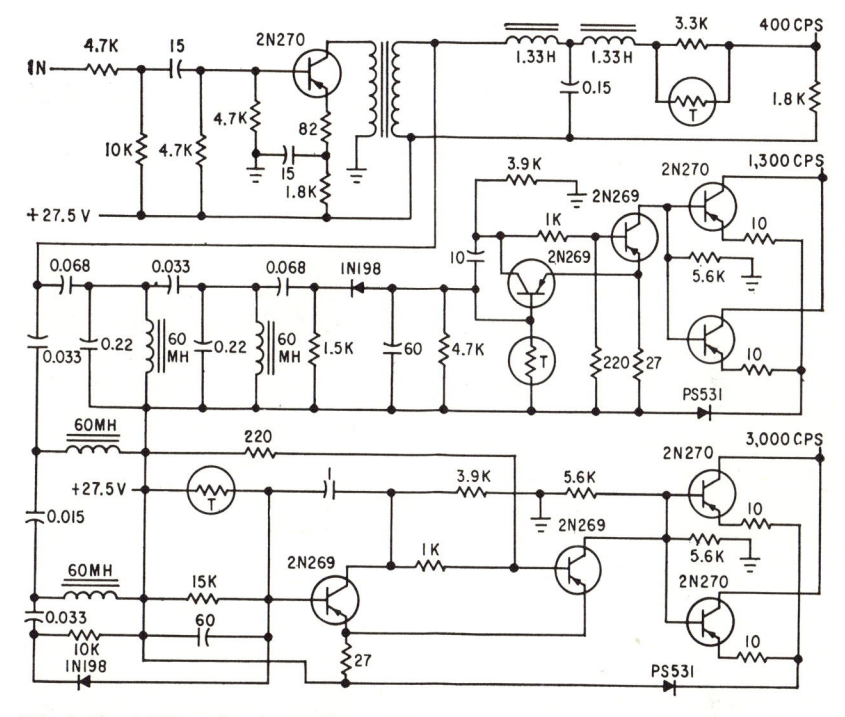

FIG. 4—The 1,300-cps band-pass filter is the only one in the three-indicator-light adapter that requires a high degree of stability

The switching action of Q_9 and Q_{10} occurs suddenly, causing the switching of Q_{11} and Q_{12} to occur in microseconds. Since this time is much less than the thermal time constant of the base-collector junction of Q_{11} and Q_{12}, time of peak-power dissipation is so short that it does not cause unsafe transistor junction temperatures.

Three-Light Adapter

The transistorized three-light adapter requires only two more electronic switches, in addition to a loss amplifier and filters. The schematic in Fig. 4 shows the filter network, which consists of low-pass, band-pass and high-pass filters. Component tolerances and temperature stabilization of the filters are not critical. The 1,300-cps band-pass filter is the only one that requires a high degree of stability.

Low-Power Sweep Circuits

High duty factor ppi sweep circuit features efficient and simple gated clamp, extremely linear push-pull d-c amplifier and novel use of silicon transistors to prevent thermal runaway. Total power consumption is less than 35 watts

By **CHARLES E. VEAZIE,** Electronics Department, The Martin Co., Baltimore, Md.

TRANSISTORIZED CRT DEFLECTION CIRCUITS for use in radar and television sets can be made much smaller and require far less power than their vacuum-tube counterparts. The unit to be described was designed for a 50-deg deflection radar crt and uses transistors with recovery times of approximately 100 μsec. Power consumption is less than 35 w, (not including crt) and provisions are included to blank the crt during flyback time. Figure 1 shows a block diagram of the sweep system.

The circuit shown in Fig. 2 makes use of emitter-follower Q_4 between triggered transistor Q_3 and the width-controlling R-C network to assure a fast rise and fall time.

Sweep Generator

The sweep generator shown in Fig. 3 generates a voltage that rises at a constant rate during the off time of the monostable multivibrator, and is held at zero voltage during the on time. A portion of the output voltage of this circuit is fed back through R_1 to the center tap of two-section integrating capacitor C_1, to permit control of the sweep voltage linearity. The output of this circuit is coupled to the sweep amplifier by two cascaded emitter-followers Q_7 and Q_8 to match the impedances of the two circuits.

The resolver is driven by power transistor Q_{11}. As power transistors have a large I_{co} and are subject to thermal runaway, silicon transistor Q_{10} is used as a directly connected emitter-follower to drive transistor Q_{11}. The amplified I_{co} of silicon transistor Q_{10} is comparable to the I_{co} of transistor Q_{11}. When they are connected in this configuration, the I_{co} of Q_{11} is cancelled, thus preventing thermal runaway at normal temperatures.

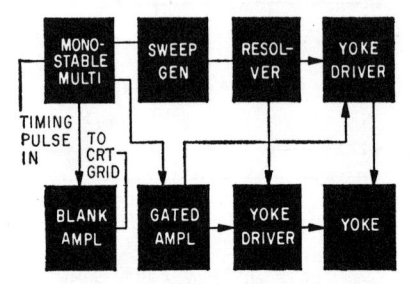

FIG. 1—Trigger pulse initiates the sweep. Resolver is driven by antenna rotation to generate rotating ppi sweep. Crt is blanked during flyback time

Yoke Driver

In a high duty factor ppi sweep, the spot starts at the center of the ppi screen and moves radially outward to the edg of the screen, then rapidly flies back. For accurate range measurements this sweep must always start from the exact center of the ppi screen. It is necessary for the sweep current to change from a fixed reference value, and return rapidly to this reference value at the end of the sweep. This condition requires a yoke with a fast recovery time and an efficient clamp circuit.

The recovery time of a yoke is the time required for the voltage induced by the rapidly changing yoke current during the beam fly-back time to collapse to zero.

The requirements for an efficient clamping circuit are a short time constant charging source and a long time constant discharging circuit. The short time constant charging source is obtained by push-pull emitter-followers Q_{12} and Q_{13} of Fig. 4 connected between the resolver secondary circuits and the push-pull clamping circuit. The long time constant discharging circuit consists of a high resistance voltage dividing network.

To present a high impedance load to the voltage dividing network, two-stage, common collector d-c amplifier Q_{16} and Q_{17} drives the push-pull power amplifiers Q_{18} and Q_{19} to furnish the sweep current to the deflection yoke. To prevent thermal run-away of the power amplifiers, transistors Q_{14} and Q_{15}, which follow the clamping circuit voltage divider, are silicon npn units.

The other winding of the deflec-

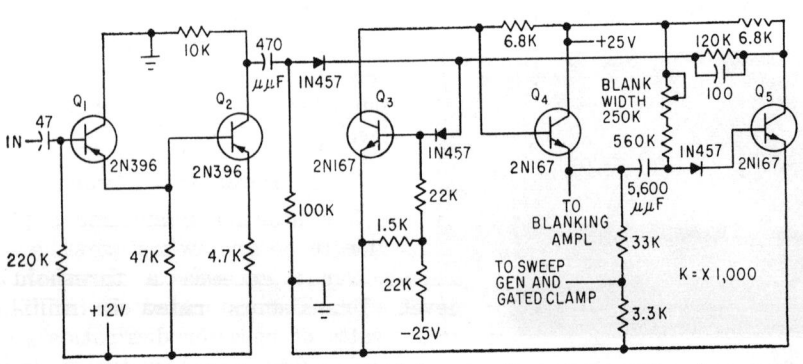

FIG. 2—Monostable multivibrator converts timing signal to narrow pulse of accurately controlled width

for Radar Indicators

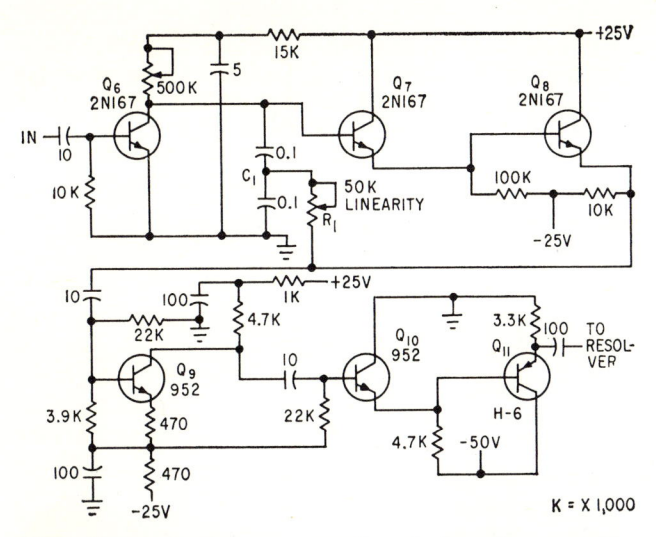

FIG. 3—The sweep generator accepts the pulse from the monostable multivibrator and generates the signal for the sweep resolver

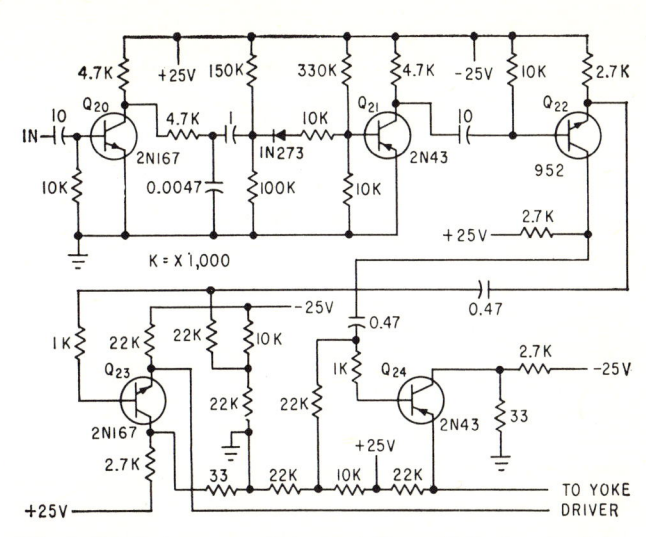

FIG. 5—Gated clamp uses the monostable multivibrator signal to generate reference level for the yoke driver

tion yoke is driven by a similar circuit, excited by the sweep voltage from the other secondary winding of the resolver.

Gated Clamp

The voltage applied to a ppi deflection yoke driver circuit is a function of sin θ, or cos θ, where θ varies from 0 to 180 deg. A single polarity clamp cannot be used so it is necessary to clamp both polarities during the clamping period and remove the clamp during the sweep period.

The gated clamp circuit is used to clamp the sweep signal voltage to a reference voltage during the clamping time at the end of the sweep, and to remove the clamp during the sweep.

The clamp circuit uses a pair of diodes D_1-D_2 and D_3-D_4, connected in opposite polarity to each signal line. During the sweep time these diodes are reverse biased by a pair of transistors Q_{23} and Q_{24} of Fig. 5, and they cannot conduct. During the clamping period, transistors Q_{23} and Q_{24} are driven into saturation by the control gate from the monostable multivibrator. This removes the reverse bias from the clamping diodes, clamping the signal to the reference voltage.

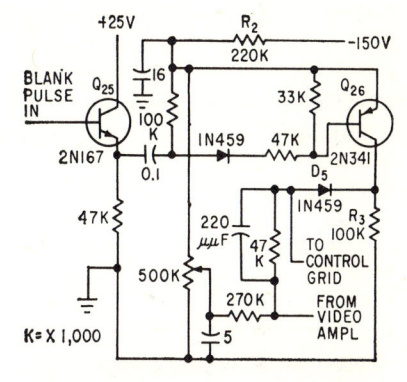

FIG. 6—Blanking amplifier blanks the crt during flyback time

Blanking Amplifier

The blanking amplifier shown in Fig. 6 amplifies the blanking pulse from the monostable multivibrator to the level required to blank the crt screen during fly-back time. This circuit uses high-voltage transistor Q_{26}. The emitter is connected to -150 v through resistance R_2. The collector is connected to the crt control grid through diode D_5, and to ground through resistance R_3. Transistor Q_{26} is normally nonconducting, and is driven into saturation by the blanking pulse thus applying a high negative voltage to the crt control grid and cutting off the beam.

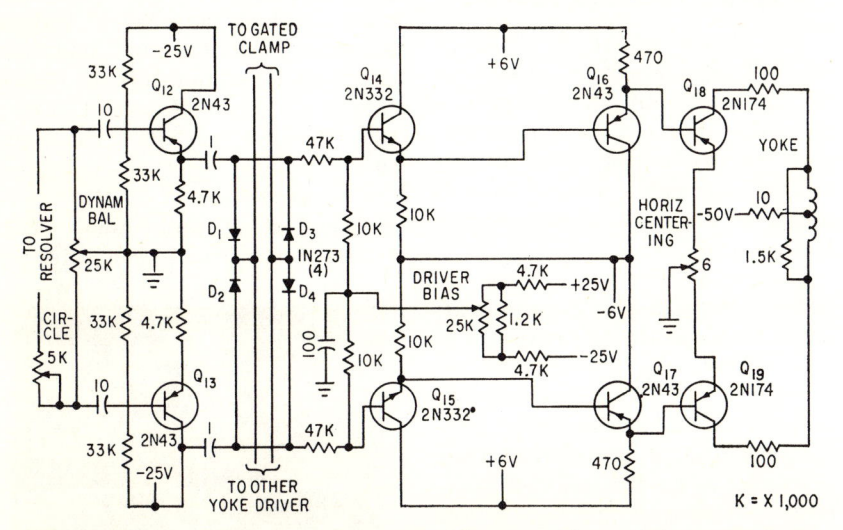

FIG. 4—Yoke driver accepts sine or cosine output of sweep resolver and clamps to reference level from gated clamp. Resolver signal drives deflection yoke

323

Using Magnetic Circuits

Size and weight reduction are obtained in design of pulse generator by using transistors and new core materials. Reliability is also increased

By ARTHUR KRINITZ, Department of Electrical Engineering, Servomechanisms Laboratory, Massachusetts Institute of Technology, Cambridge, Mass.

RECENT ADVANCES IN MAGNETIC MATERIALS are put to good use in this radar pulser. Transistor drive and high-permeability, high-saturation flux-density cores are used to obtain a small, light weight design. Applications might be airport surveillance or boat radars, but other uses are also possible.

Magnetic discharge and pulse shaping networks are used instead of thyratrons or vacuum tube amplifiers. Transistorizing the drive circuit further increases reliability.

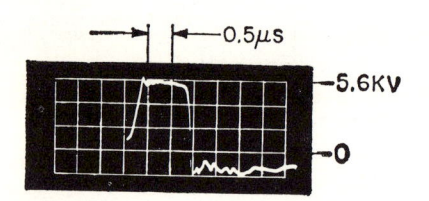

FIG. 1—Generator output pulse. Peak power of the 1-μsec pulse is 24-kw

The output waveform and the output of a magnetron when driven with this pulse are shown in Figs. 1 and 2.

Transistor Driving Circuit

The pulse generator consists of a square wave oscillator, driver circuit, two-section magnetic discharge circuit and a two-section pulse forming network. These functions are indicated on the schematic diagram, Fig. 3.

Two medium power transistors and a magnetic core are used in the square wave oscillator. Similar to d-c to d-c converters, the 1,700 cps oscillator is highly efficient and its square wave output is well suited to switching the power transistor of the driver circuit. The circuit of transistor Q_1 determines the operating voltage of the oscillator transistors and thus controls frequency. This control is adjusted for best operation.

Driver transistor Q_4 is a low voltage, high current device with switching time of 15 to 30 μsec. With square wave drive it operates as a switch, being either full ON or full OFF. Since current is low in the OFF state, and voltage across the transistor is low in the ON state, dissipation is low and occurs mainly during the brief switching time. Transistor limitations are maximum allowable current in the ON state and maximum allowable voltage in the OFF state.

Magnetic Discharge Network

The magnetic circuit takes the low voltage, low power pulse from the driver circuit and compresses the pulse into a high power output. Energy is not added during this process but pulse duration is decreased from about 300 μsec to about 1 μsec.

Figure 4 shows the magnetic paths of the saturable elements during the pulse forming process. The bias circuit has brought the reactors to the zero points on the B-H curves. At t_0 the driver transistor of Fig. 3 is switched from OFF to ON and a voltage step is applied to charging inductor L_1 and the primary of T_1. The reactors then travel path 0-1-2 (Fig. 4),

Breadboard generator showing reactor oil bath and magnetron load. Metal panel is 19 in. long

with T_1 acting like a transformer and T_2 being driven further into the saturation region. The circuit beyond T_2 is effectively decoupled and T_3 is not affected by the actions of T_1 and T_2. Capacitor C_1 of Fig. 3 is reflected into the primary of T_1 and the resulting transient is the step response of a high-Q RLC network. The effect of the first interval is to charge C_1.

At t_3 the driver transistor is turned OFF. The voltage on C_1 now takes T_1 along path 2-3-4. During the same interval the excursions of T_2 and T_3 act to discharge C_1 and charge C_2. The tank circuit during this time is essentially a short circuit. The next interval—t_4-t_5-t_6—is the discharge time of C_2. Reactor T_3 operates along path 4-5-6 (Fig. 4C), and the load is connected by

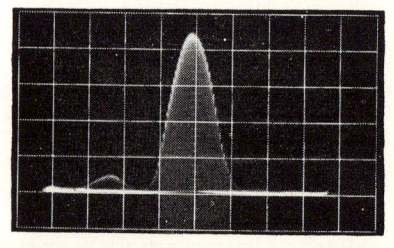

FIG. 2—Output spectrum from 2J42 magnetron

To Pulse Radar Sets

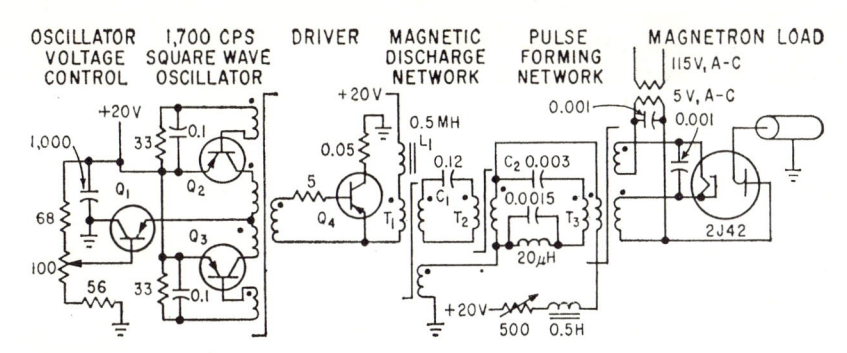

FIG. 3—Square loop magnetic cores are used with transistor drive to generate and shape pulses

transformer action.

The magnetic discharge circuit forces C_1 to discharge more rapidly than it charges; similarly for C_2. The pulse therefore becomes taller and narrower as it passes from stage to stage.

The pulse forming network is in two sections. One section is the tank circuit of Fig. 3 and the other consists of C_2 and the saturated inductance of T_2. The combination is a two-stage, linear, pulse forming network of the Guillemin type.

Core Reset and Bias

Following each pulse, the bias circuit resets the reactors to the zeros of Fig. 4, preparing the network for the next pulse. During the reset operation the reactors are not saturated and therefore act as transformers. Transformed voltages thus appear across the corresponding capacitors and it is these voltages which actually reset the cores.

When the charging time of any reactor is short compared with the reset time available, the voltage developed on the capacitor during reset will be small and thus have little effect on the charging cycle. But the charging time of C_1 is approximately equal to the reset time of T_1. Voltage induced on C_1 therefore has an important effect on the charging cycle. This problem is analyzed in detail in Appendix I of reference (1).

The bias inductor is used to keep bias current constant. Its impedance must be large compared to any impedance in the associated mesh. Bias winding ampere-turns must be large enough to reset the core and also overcome coercive force. Magnetic network design can be based on core volume and hysteresis loss considerations or on eddy current loss considerations. Theoretical work shows that using more than two to four stages would not generally be worthwhile.[1]

Experimental Generator

A breadboard version of the generator is shown in the photograph and its characteristics are listed in Table I.

Power dissipation is the transistors is low and, with typical heat sinks, they can operate in air without special cooling. The driver transistor, for example, operates at about two-thirds its rating. Dissipation in the magnetic networks is higher, especially in T_2 and T_3. These two cores are mounted on heat sinks and placed in an oil bath. The oil acts as both heat sink and insulator.

Main losses in the generator are hysteresis loss, eddy current loss, copper loss, and dielectric loss in the capacitors. Eddy current loss can be reduced by using thinner magnetic materials. One- and two-mil laminations were used in the generator but ½-mil and ¼-mil materials are available. An additional five percent improvement in efficiency can be obtained from a better designed oscillator and bias circuit. Ultimately, an efficiency of 60 to 65 percent should be possible.

Jitter in the generator output pulse is less than 0.05 μsec when a storage battery is used for the power supply. Any ripple in the supply will increase the jitter.

Output power can be increased by using several power transistors in the driver circuit. Characteristics of recently developed solid-state thyratron-like rectifiers indicate these could be used instead of—or in conjunction with—the power transistor of the driver circuit.

Research was supported by

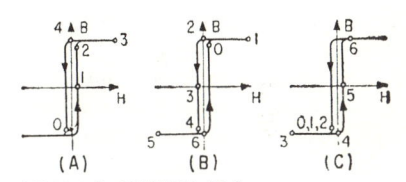

PATH 0-1-2 : CHARGING OF C_1
PATH 2-3-4 : DISCHARGE OF C_1 AND CHARGE OF C_2
PATH 4-5-6 : DISCHARGE OF C_2

FIG. 4—Path of operation of T_1 (A), T_2 (B) & T_3 (C)

M. I. T. Servomechanisms Lab and by USAF, Weapons Guidance Lab, Wright Air Development Center, under Contract No. AF 33(616)-5489, Task No. 50688.

Table I—Experimental Generator Characteristics

Repetition frequency	1,700 cps
Peak output power	24 kw
Peak output voltage	5.6 kv
Pulse width	1 μsec
Average output power	40 watts
Efficiency (including d-c adjuster circuits, oscillator and bias)	50 percent
Weight (including heat sinks, oil bath, filament transformer and 2 lb mounting panel)	10 lb

REFERENCES

(1) A. Krinitz, "Transistor-Magnetic Pulse Generator For Radar Modulator Applications", Report 7848-R-1, Servomech Lab, MIT, Sept., 1958.

BIBLIOGRAPHY

W. S. Melville, "The Use of Saturable Reactors and Discharge Devices for Pulse Generators", The Inst of Elec Engrs, (London), Vol 98, p 185-204, Feb, 1951.
G. N. Glasoe and J. V. Lebacqz, "Pulse Generators", MIT Radiation Lab Series, Vol V, McGraw-Hill Co., New York, 1948.
E. J. Smith, J. Antin, and K. T. Lian, "Magnetic Modulators for Radar Applications", Report R-419-55, PIB-35, Poly Institute Of Brooklyn, Microwave Research Institute.

Portable Depth Finder For Small Boats

Instrument locates fish and measures depth to 120 feet. Here is the circuit and explanation of a portable device in which a mechanical arrangement removes the need for a cathode-ray tube

By HERBERT C. SINGLE, Design Engineer, Commercial Apparatus and Systems Div., Raytheon Co., Waltham, Mass.

DEPTH INDICATORS ARE BASICALLY time-measuring devices. They measure the time taken for a transmitted sonic pulse to travel down through the water, reflect from the bottom and return to the instrument. The instrument then converts the elapsed time into feet of water below the transducer. If there are any objects, such as a school of fish, between the transducer and the bottom, a lower-intensity echo will arrive at the transducer before the stronger bottom echo.

The transistorized depth indicator to be described is a portable unit using either an external 12-v battery or an internal 7.5-v battery and has an indicating range of 120 feet.

The depth indicator is a neon lamp rotated at the end of an arm driven by a constant-speed motor. A magnet is also located on the revolving arm and is used to trigger the transmitter when the neon lamp is approaching the zero feet indication. When the transmitter operates, the generated sonic pulse is passed simultaneously to the transducer and the receiver causing the neon lamp to glow brightly as it passes the zero feet indication of the depth scale. The neon lamp glows again when the echo pulse is received and the angular position at the second glow can be read in terms of water depth on the calibrated scale.

The transducer is a single disk of barium titanate operating in the thickness mode. It has a beamwidth of 6 degrees and is potted in a salt-water resistant plastic.

Circuit

Figure 1 shows the circuit of the five-transistor instrument. Transmitter transistor Q_1 is connected in a normally-quiescent Hartley-type oscillator circuit with its base circuit loaded by R_1. When the magnet, rotated by the constant-speed motor, passes close to inductor L_1, a negative pulse is generated and passed to the transistor through C_1 and the pulse-shaping network R_2, C_2 and R_1.

Transmitter Q_1 is driven into oscillation for a period of 300 to 500 μsec. Tank circuit L_2 and C_3 determines the transmitted frequency.

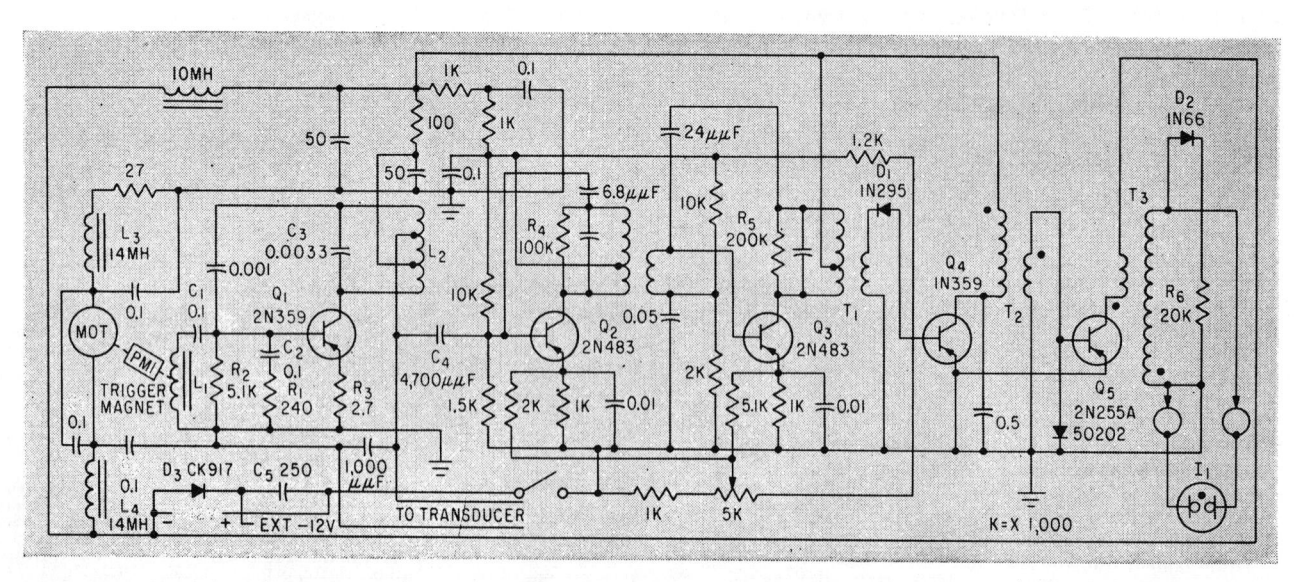

FIG. 1—Constant-speed motor simultaneously drives triggering magnet past L_1 and rotates neon lamp I_1 around transparent depth scale

THE FRONT COVER—Likely fishing spots for sportsmen are indicated by presence of low-intensity echoes

Transducer (left) is connected to the fathometer depth sounder with a twelve-foot cable

The transmitter operates 1,200 times per minute at a frequency of approximately 200 kc. Using L_2 as an autotransformer holds the coupling losses to a minimum while matching the transmitter to the transducer.

The returning echo is coupled through C_4 to the first r-f amplifier Q_2. The conventional r-f amplifiers Q_2 and Q_3 have their bandwidths increased by damping resistors R_4 and R_5. The amplifiers are also slightly stagger tuned to further broaden their bandwidth to be compatible with the slight differences between transducers.

Gain Control

Since the received pulses can vary in amplitude between 75 μv and 2 v and overloading the r-f amplifiers can cause incorrect operation of the following stages, a large dynamic range of control is necessary.

Amplifier gain is determined by varying the base voltage applied to Q_2 and Q_3. With this method of varying gain, the impedance of the transistors and their operating point is changed and slight detuning and bandshifting occur.

The peak of the bandpass will shift about 1.5 kc between maximum and minimum gain. By correctly neutralizing the r-f amplifiers at full gain, maximum stability at lower gain is assured.

Output of amplifier output transformer T_1 is detected by diode D_1 which is directly connected to the base of amplifier Q_4. High-power transistor Q_4 is a grounded-emitter unbiased amplifier which will conduct only when a negative-going pulse is applied to its base.

Lamp Operation

The detected negative-going pulse drives Q_4 into saturation and the resulting pulse in the collector circuit is transformer-coupled to output amplifier Q_5. Amplifier Q_5 is also a grounded-emitter unbiased power amplifier. The pulse from transformer T_2 drives Q_5 into saturation. When Q_5 conducts, approximately 1.5 amperes peak current flows through the primary of transformer T_3.

The voltage pulse developed

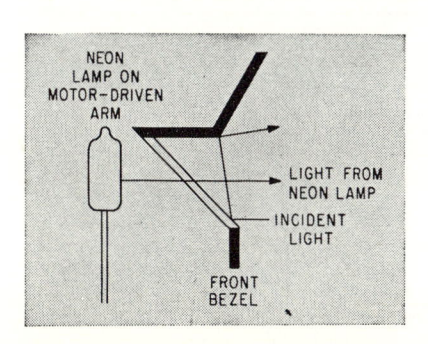

FIG. 2—Design of depth scale bezel permits viewing at high light levels

across the secondary of T_3 is sufficient to ignite (through slip rings) neon lamp I_1. When the voltage pulse is over, the neon lamp will extinguish. When the flux in transformer T_3 collapses, a reverse-polarity voltage is generated that will re-ignite I_1. To prevent this, diode D_2 in series with R_6 conducts to remove the voltage.

Power is supplied by internal mercury cells or external batteries. Diode D_3 is used to protect the circuit from accidental reverse polarity battery installation. Capacitor C_5 is used to provide a low-impedance power source.

Reflections Reduced

To reduce reflections on the indicating face of the depth scale, the portion directly over the indicating scale is sloped backwards as shown in Fig. 2. Incident light is reflected to a black, light-absorbent ring and the high-intensity red light from the neon lamp can shine directly through the transparent part.

In common with many other types of electronic equipment, the depth sounder and associated cables should be kept as far away from motor ignition leads as practicable.

Ignition interference shows up as stray flashes of light from the depth indicating lamp. Shielding or bypassing of offending leads may reduce the trouble.

Receiver For

Efficient use of transistors and the elimination of relays produces a compact, economical marker beacon receiver suitable for light aircraft

By F. PATTERSON SMITH, National Aeronautical Corporation, Fort Washington, Pennsylvania

AN AIRCRAFT MARKER BEACON receiver furnishes the pilot of a plane with information as to his position as he flies over marker beacon transmitters. The transmitters are located along the main air lanes and are coded by modulation and pulsing.

Beacons that are primarily route markers are amplitude modulated at 3,000 cps and are sometimes pulse coded for further identification. At airports, the beacons are used in instrument landing systems. A beacon located about five miles from the airport is called the outer marker. It is modulated at 400 cps and pulsed to give two dashes per second. Passage over the beacon is the signal to begin the descent at a specified rate.

Another beacon, called the middle marker, is mounted about ½ mile from touchdown. Modulation is 1,300 cps and pulsing is in alternate dots and dashes. As the plane passes over this station, procedure requires the pilot to make a new approach if the runway is not visible.

The transmitters all operate at a fixed 75 mc and modulation is 95 percent. Transmission is directed upward in a relatively small angle cone. These points are illustrated in Fig. 1.

Low-Cost Receiver

In the light aircraft field—private and small commercial planes—equipment cost, size and weight are important factors. Marker beacon receivers for this use have been primarily of the trf type; typically, they provide the pilot with aural signals only. Airline type receivers present information with colored lights: white for enroute beacons, blue for the outer marker and amber for the middle marker.

Three-lamp indication can be designed for a trf receiver but spurious signals, primarily from tv broadcasting stations, can cause a lamp to light. When visual indication is provided, this type of interference cannot be tolerated. With only aural indication, the interfering signal will not be confused with the tone from the marker beacon.

To meet the needs of the light plane field, an inexpensive superheterodyne receiver has been developed that provides both aural signals and airline-type colored lights. The unit is shown in the photographs.

Light output from the lamps is adequate for bright daylight operation and dimming is available for night instrument landings. Antenna sensitivity is 1,000 microvolts and is based on antenna efficiency and field strength of the radiated signal. Higher sensitivity would

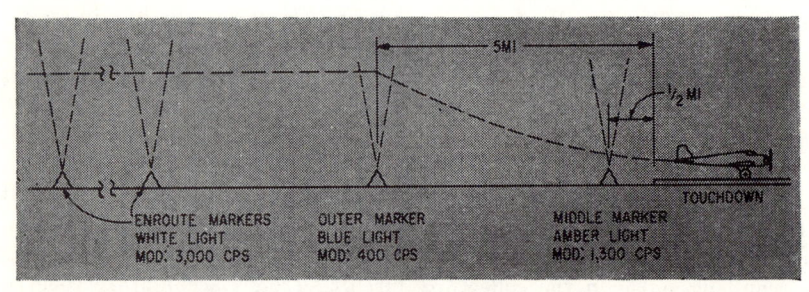

FIG. 1—Enroute beacons are sometimes pulse coded for further identification. The outer marker is pulsed at two dashes per second and the middle marker with alternate dots and dashes

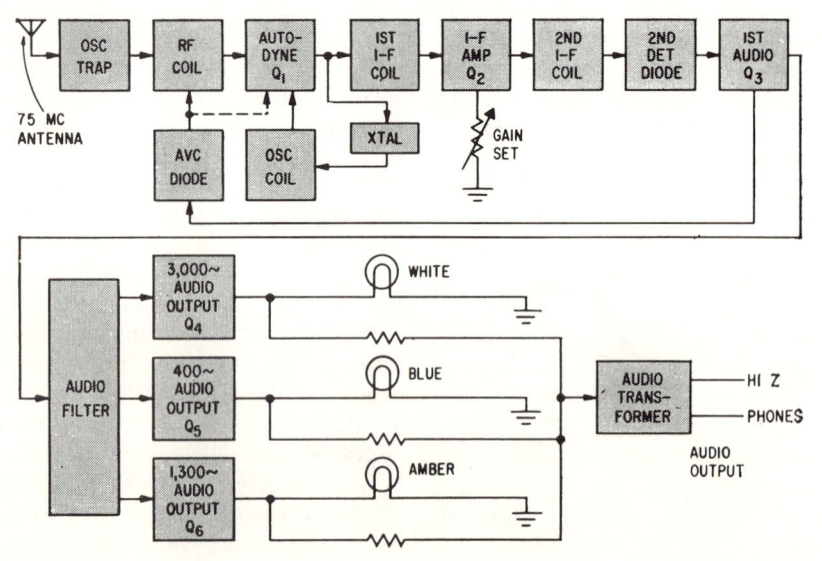

FIG. 2—Crystal-controlled autodyne oscillator provides maximum gain with minimum number of parts. Output transistors replace commonly used relays

Marker-Beacon Use

Pilot adjusts brightness of indicator on marker beacon receiver. Reverse side of front panel is specially labeled to permit vertical mounting

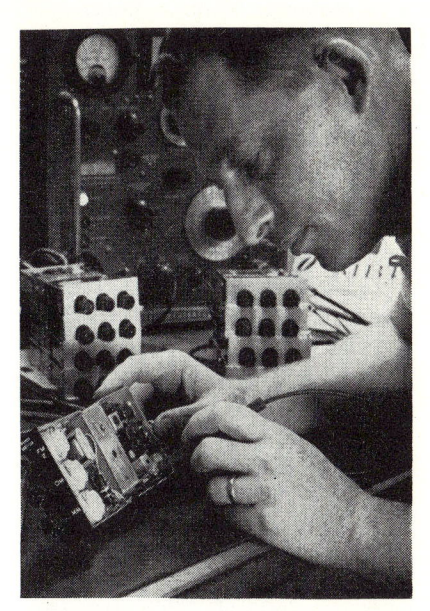

Receiver dimensions are 1 3/32 × 3 13/32 × 7 5/8 inches. Weight is 18 ounces

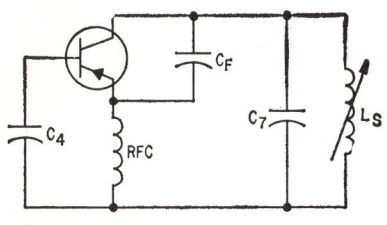

FIG. 3—Basic vhf oscillator circuit showing signal paths only

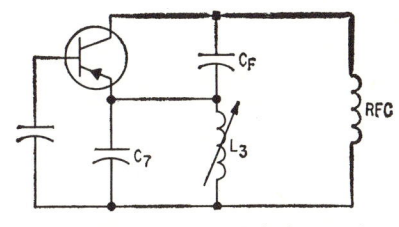

FIG. 4—Tank circuit is shifted to emitter load

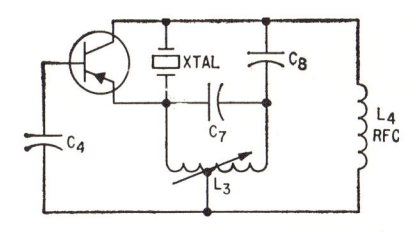

FIG. 5—A crystal and a neutralization circuit are added to the oscillator of Fig. 4

provide a broader fix; lower sensitivity might be inadequate at high altitudes.

Crystal Oscillator

A block diagram of the single-conversion, all-transistor receiver is shown in Fig. 2. It uses an autodyne front end for maximum gain with a minimum of components. The conversion gain of an autodyne may be higher than an r-f amplifier for the same transistor. The autodyne also gives less variation in gain for a given parameter spread and thus a relatively inexpensive transistor can be used. The local oscillator frequency was carefully chosen to keep spurious responses at frequencies where there is little interference. This approach permits low-insertion loss circuits to be placed ahead of the autodyne.

Oscillator frequency is 68.75 mc and i-f frequency is 6.25 mc, thus image response is at 62.5 mc which is 1.25 mc above the picture carrier in tv channel 3.

The crystal-controlled vhf autodyne circuit evolved from the basic vhf oscillator circuit shown in Fig. 3. The transistor operates in the common base mode using a tank circuit and feedback through C_f. This circuit was revised, Fig. 4, to allow both input leads, base and emitter, to be a-c grounded at other than the oscillator frequency.

Crystal control was instituted by replacing C_f with an overtone crystal in the feedback path. Neutralization is necessary and is accomplished by C_s in combination with the phase reversal of coil L_3 as shown in Fig. 5. The signal circuits were then added to the basic oscillator circuit to give the final circuit indicated in Fig. 6.

The problem of local oscillator radiation from the receiver was solved by placing a series tuned trap, L_1C_1, in the antenna circuit, tuned to the oscillator frequency. This ensures that the base circuit is well grounded at the oscillator frequency, improves the image ratio, and introduces little insertion loss. Maximum 75-mc gain is achieved with the trap properly tuned.

The i-f amplifier is standard and is reasonably neutralized to prevent regeneration. The i-f coils are designed for matched gain and have an insertion loss of 3 db each. Resistor R_6, in combination with R_5 and R_8, establishes the bias on this stage for maximum gain. After alignment, R_6 is bridged with a selected resistor R_7 to reduce the receiver gain to the required value. A control range of approximately 20 db is obtained by this method, without appreciably deteriorating the avc characteristic.

Automatic Volume Control

When the voltage drop across the audio d-c collector load resistor R_{10} exceeds the value established by the voltage divider consisting of R_2 and R_3, avc diode D_1 goes into conduction. This diode loads antenna coil L_2 to reduce receiver gain. If the dynamic range of this circuit is exceeded, further current drawn through D_1 will reduce the collector current of autodyne transistor Q_1 by increasing the drop across its emitter resistor R_1. This gives approximately 10 db additional control range. If emitter current control of the autodyne were carried too far, it could result in oscillator "squegging".

Filter Design

The audio filter to separate the three channels is designed for constant loaded input impedance, matched gain and maximum loaded Q. Filter output is matched so that receiver sensitivity is the same for all three channels. This requires matching audio output transistors Q_4, Q_5 and Q_6. The nonlinear load presented by the output transistors required an empirical filter design.

Audio output transistors Q_4, Q_5 and Q_6 function as a combined rectifier and single-ended class-B power amplifier. Collector current is a half sinusoid until the stage is over driven. As the signal level increases, clipping occurs until, at medium signal levels, the output waveform is essentially a square wave.

Light output is not a linear function of lamp current. The steep slope of the curve improves the low-level noise rejection. The slightly non-flat top of the lamp input to light output curve, however, permits more accurate location of the center of the transmitted field than is possible with relays.

The three outputs are matrixed into a common audio lead and applied to transformer T_1. The transformer provides a power match between the matrix resistor source and the headphone load, providing about 15 mw to 600-ohm earphones. Another winding gives about 4 volts rms for a cabin amplifier. Though audio distortion is high, tonal quality is adequate for identification purposes, which is all that is required.

Nominal operating voltage is 13.75. With no signal, current drain is 5 ma; with maximum signal, 195 ma. Thus a separate power switch is not provided. Satisfactory operation is obtained with supply voltages from 11 to 16 volts and power supply transients of the type normally encountered in aircraft radio installations will not damage the receiver. Transients will normally be inaudible. With minor modifications, nominal 27-volt operation is possible.

As noted earlier, the receiver image frequency is 62.5 mc within the band of tv channel 3. Measurements with 400-cps modulation show a typical image rejection of the r-f system alone of 25 db. Rather than simulate an accurate tv signal, flight tests were used to check the design.

Flights were made in the vicinity of the Empire State Building, from which channels 4 and 5 are broadcast, and in the Philadelphia area which uses channels 3 and 6. Within one half mile of the radiating antenna, at an altitude close to maximum radiation, no light output or audible buzz was detected.

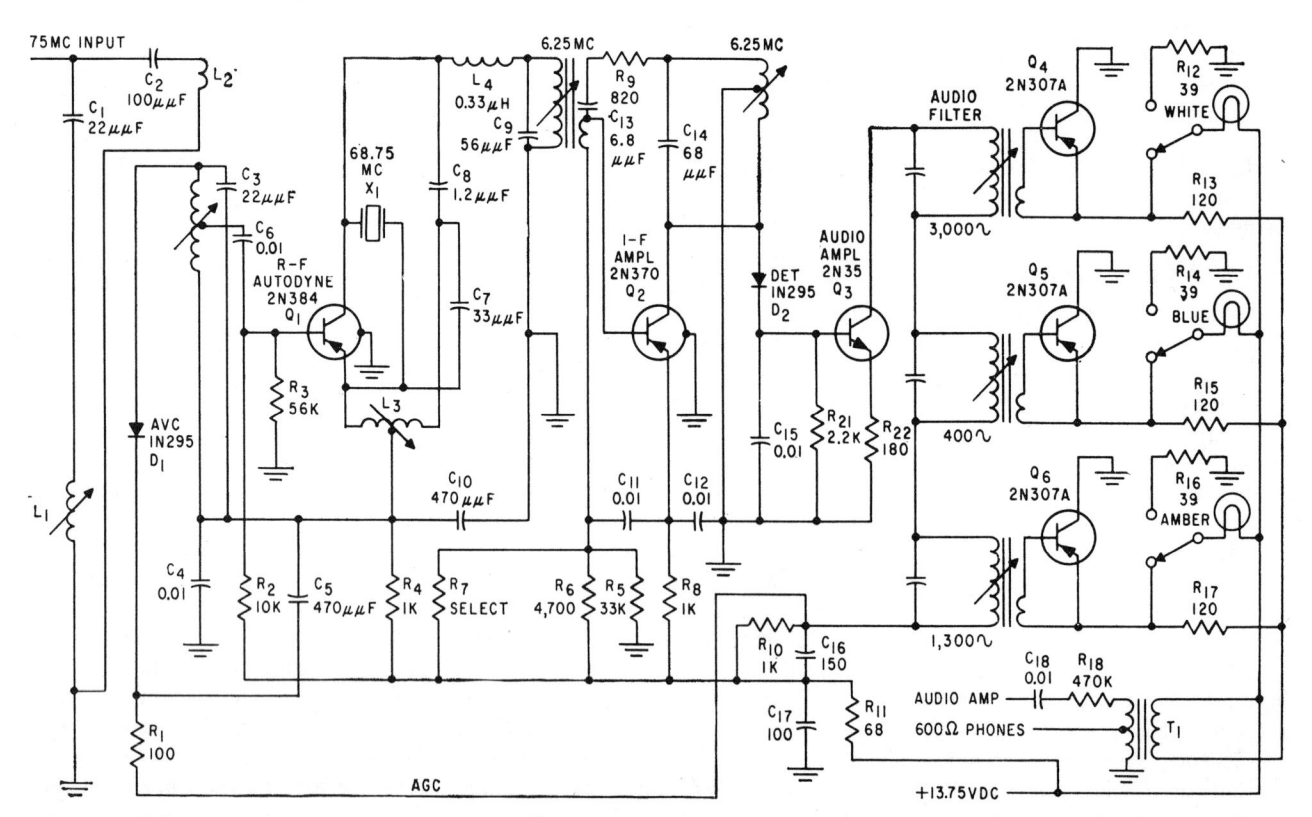

FIG. 6—Transistorized marker beacon receiver gives audio output as well as the colored light presentation typical of airline type units. Careful, efficient design gives performance comparable to more complex equipment

Radio Beacon Helps Locate Aircraft Crashes

This radio beacon, designed to withstand high g's and extreme environments, flies free of a crashing aircraft and then automatically transmits a distress signal

By **DAVID M. MAKOW,** Radio and Electrical Engineering Div., National Research Council of Canada, Ottawa

DESIGNING A RADIO BEACON which will survive an aircraft crash and then radiate a useful distress signal poses several problems not encountered in conventional equipment. Such a beacon must be safely separated from the crashing aircraft as early as possible—no present equipment can withstand the tremendous forces and temperatures encountered in a direct crash. The beacon must somehow be transferred to a safe and operational position not too far from the wreck; it then should transmit a useful signal for about 100 hours and over a wide range of temperatures, even if it lands in swamp, in water or in vegetated country.

The Crash Position Indicator,[1,2] is a promising solution. It consists of an enclosure shaped somewhat like a short section of an aircraft wing inside which, potted in plastic foam, is a specially designed battery-operated radio beacon. Two units have been developed: a large beacon and an enclosure, called the tumbling aerofoil, with the total weight of 11.4 lb, for large aircraft (to the left in Fig. 1); and a small unit with the total weight of 5.7 lb, for small aircraft, shown to the right.

Operation

The indicator is placed on the tail of the aircraft and is held by a slim metal ribbon passing through a spring-loaded knife (see Fig. 2). The knife is operated by a change in tension of the tightly stretched trigger wires connected to the extremities of the aircraft structure. Any abnormal structural change in the aircraft as a result of a crash or other emergency is transmitted through the wires and releases the aerofoil, switching on the beacon, which is then quickly rotated into the airstream and rapidly pulled away by the aerodynamic lift and drag forces. These cause it to curve away on an arc of about 100-foot radius and slow down by half every 35 feet until it reaches its terminal velocity of 20 to 25 mph at sea level. Its safe landing speed is about 40 mph. The unit will also float about 85 percent out of water and will act as a snowshoe because of its small wing loading.

To operate as outlined, the indicator must have all parts including the transmitting antenna placed inside the aerofoil, and the ratio of weight to aerofoil area must be small. A fractional wavelength capacitor antenna could be designed to fulfill the first requirement. A beacon system with great battery economy has also been obtained using a pulsed-carrier, pulse-filament transmitter in connection with the Sarah search receiver which is available commercially[3]. This has been chosen in preference to a c-w signal where crystal control and several frequency multiplier stages would have to be provided in the beacon so that the communication receivers available on aircraft could be employed during a search.

Antenna Characteristics

Choice of carrier frequency was influenced by the antenna size,

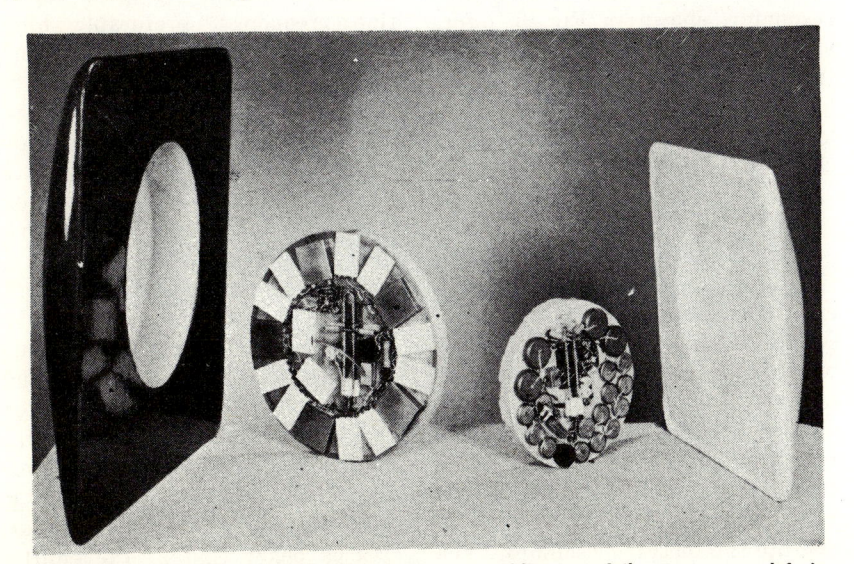

FIG. 1—Radio beacons ready for installation in tumbling aerofoils. Beacon on left is 14 inches in diameter; its aerofoil measures 2 ft × 2 ft × 5 inches. Right beacon is 10 inches in diameter; its aerofoil, 20 inches × 20 inches × 4.5 inches

FIG. 2—Crash position indicator is held on aircraft tail by a slim metal ribbon passing through a spring-loaded knife. Extreme tension in trigger wire operates knife

FIG. 3—Complete circuit of the radio beacon for the crash position indicator

which was restricted by the dimensions of the aerofoil. The frequency of 243 mc, used in distress signaling, was considered the best compromise, as at this frequency a relatively efficient low-Q capacitor antenna[4] could be realized and compatibility with the Sarah receiver was possible. The antenna consists of two aluminum or copper disks which form the capacitor plates. In the large unit the disk diameter is 14 inches and disk spacing 1½ inches. In the small unit the corresponding values are 10 and 1¼. The capacitor dielectric is plastic foam with a small dissipation factor. The equivalent circuit of the antenna is shown in Fig. 3. The elements in the dashed rectangle represent a series resonant circuit consisting of the capacitor antenna, an inductive lead and the radiation resistance. The adjustable capacitors C_1 and C_2 are used to obtain an optimum matching condition at the operating frequency. Measurements of the large antenna at 243 mc indictate a value of Q = 50 and an efficiency of 80 percent. The Q of the small antenna is estimated to be 120 and the efficiency is 60 percent. The measured radiation pattern, when the antenna is placed on the ground, is shown in Fig. 4; grazing angle depends on the conductivity of terrain.

Transmitter

Figure 3 shows the circuit of the beacon transmitter. Conversion of the available d-c power into r-f power could be efficiently carried out employing pulse plate modulation of the transmitter tube. An unusually high peak power output has been obtained with the filamentary oxide-coated cathode, CK 5971 vhf triode by using short pulse width and a relatively long pulse interval[5,6] and by exceeding the maximum anode voltage considerably. This tube, in a push-pull oscillator, consumes 89 ma at 1.3 v for the filament, has a peak anode current of 16 ma, and delivers 2-3 watts peak r-f power for a 9-microsecond pulse of 300-400 v applied to the plate. This is about 10 times the power available at the rated value of 90 v d-c. The parallel line resonator, shown in Fig. 3, offers satisfactory mechanical and electrical stability. The antenna has been coupled with a small coupling loop placed over the shorted end of the parallel line.

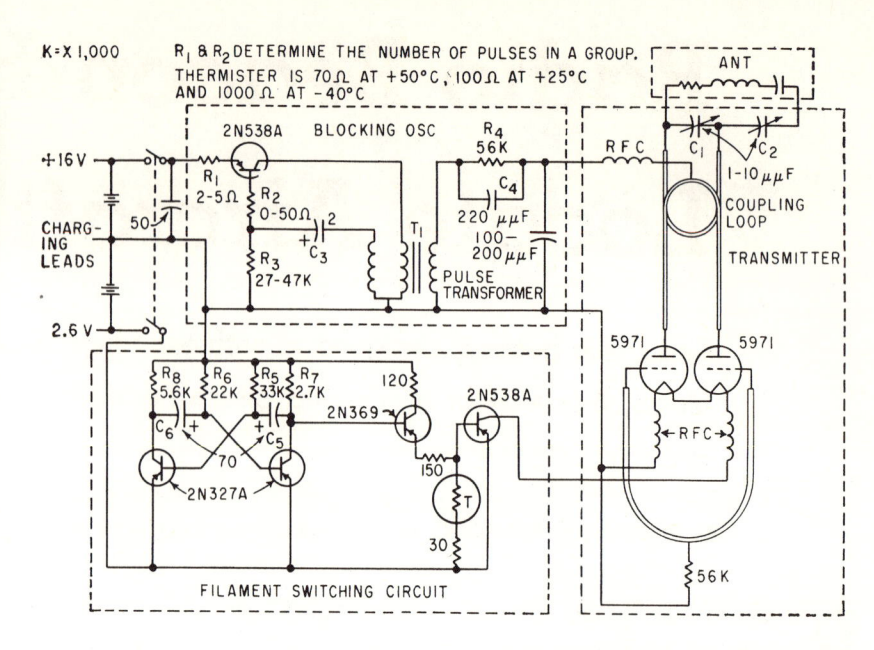

Modulation Waveform

Plate pulse modulation, in addition to the above-mentioned property, requires no separate plate supply, as the high voltage pulses which are generated in a transistorized blocking oscillator (Fig. 3) are applied directly through a step-up transformer to the plates of the transmitter oscillator tubes. A low voltage supply can then be used to operate the blocking oscillator. The modulation waveform consists of groups of pulses with a repetition frequency of about 65 groups per second. Four pulses in a group, 9 microseconds wide and spaced by about 75 microseconds, were chosen. The first pulse of each group triggers the sweep of the Sarah receiver, thus permitting the remaining three pulses to be displayed on the screen of this receiver. As a result of synchronization of the time base, these pulses appear stationary, improving the rejection against the unstationary noise and interference signals. The pulse width was determined by the properties of the receiver, and the number of pulses in a group was chosen from considerations of reliability of reception and effectiveness of display. The group repetition frequency, preferably high, was limited to 65 groups a second, as the highest values permitted by the capacity of the batteries.

The process of pulse group formation in the blocking oscillator is, in its simplest representation, similar to the squegging (generation of several frequencies simultaneously) operation of an r-f oscillator. At the end of the fourth pulse the capacitor C_3 (see Fig. 3) has charged to a voltage value which exceeds the potential of the emitter, and the transistor cuts off. During the OFF period, the charge accumulated in C_3 leaks away through R_3 until the potential of the base decreases below that of the emitter and the operation is restored again. The number of pulses in a group is

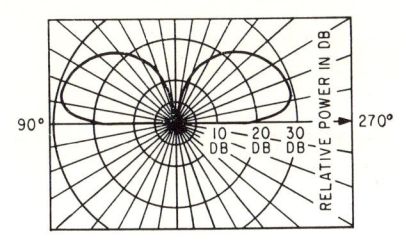

FIG. 4—Radiation pattern of the parallel plate capacitor antenna when the beacon is on the ground

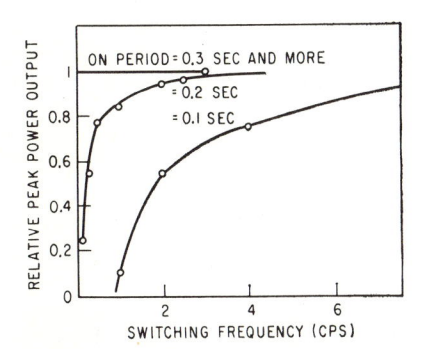

FIG. 5—Peak power output as a function of the filament switching frequency

determined by the charging time constant and the group repetition frequency by the discharging time constant of this capacitor. The pulse spacing is related to the primary inductance of the pulse transformer T_1 and the pulse width can be modified with the pulse slicing circuit $R_4C_4^2$.

Filament Switching

In addition to pulse plate modulation of the transmitter, further economy in battery supply can be achieved by periodically switching ON and OFF the filaments of the tubes[7]. The relationship between the r-f output power and the switching frequency for three values of ON periods is shown in Fig. 5. For filament ON periods equal to 0.3 sec or more, the maximum r-f output is obtained, as the filament has reached the operating temperature. If the ON period is reduced, the power output decreases at low switching frequencies but approaches the full value at high switching frequencies as a result of an increase in the average filament temperature. The filament is switched by a transistor operated through a grounded collector stage from a transistorized multivibrator as shown in Fig. 3. The ON and OFF periods have been adjusted to 0.8 sec and 1.2 sec respectively, since this ratio has been found to result in favorable performance. Total power consumption of the filament-switching circuit is about 5-10 mw, which is a small fraction of the total filament power. Although, in general, switching of the filament reduces the life of a tube, a large number of tests have indicated that the CK 5971 operates satisfactorily for at least 500 hours, which is 5 times the required period of operation.

Power Supply

For its power source, the beacon uses Saft Ni-Cd cells, trickle-charged from the 28-v aircraft battery. The respective voltages are obtained then with dropping resistors in series with diodes which prevent reverse discharge. In the large unit, where the filament is continuously operated and not switched ON and OFF periodically, a 1.3-v filament supply with a 160 ma drain is provided by five 4-amp hr VO4 Saft Ni-Cd cells, totaling 2 lb. In addition, a 16-v supply with a drain of 6.5 ma is made up by twelve 0.8 amp hr VO8 Saft Ni-Cd cells, totaling 1.3 lb. Total weight of the beacon supply is 3.3 lb. This supply will power the beacon for 125 hours at room temperature and 106 hours at −40 C.

In the small unit, filament switching has been used permitting a reduction of the battery weight. Here, four Saft Ni-Cd 2 amp hr button cells form a 2.6 v supply with a drain a 85 ma and weight of 0.94 lb. Fourteen 0.5 amp hr button cells form a 2.6-v supply with a drain of 7 ma and weight of 0.66 lb. The total weight is then 1.6 lb. Using a filament ON period of 0.8 sec. and an OFF period of 1.2 sec this supply will last 80 hours at room temperature and slightly less at −40 C.

Experimental Results

The radio beacon potted in the tumbling aerofoil has undergone several tests simulating the conditions of an actual aircraft crash and subsequent search. Aircraft crashes were simulated by a rocket sled speeding into a cliff at a known velocity. The unit, suitably mounted on the sled, was designed to be released when the nose of the rocket hit the obstacle. In the tests, one at 120 mph and the other at 230 mph, the unit upon release was lifted high in an arc and landed safely away from the point of impact. Normal flight operation of a Beechcraft Expeditor with the unit on its tail has been proven in two flight tests.

Laboratory shock tests were made on a spring-mounted platform shocked by heavy hammer blows about three different axes. The unit tested survived, with no change in performance, a total of 18 tests up to 1,100 g; the batteries and electronic components received shock up to 700 g.

In numerous flight range tests at altitudes of 9,000 ft, using Sarah search equipment, ranges of 20-35 miles were obtained when the unit was located in an open exposed site, and ranges of 5-12 miles were obtained when it was buried in snow in a narrow valley.

Appreciation is expressed to colleagues in the Royal Canadian Air Force, the Defence Research Board, Canadian Army Signals and the divisions of Applied Chemistry, Mechanical Engineering, and Radio and Electrical Engineering of the National Research Council of Canada, for their cooperation and assistance. In particular, the author wishes to thank H. T. Stevinson of the Flight Research Section, who was in charge of the mechanical and aerodynamical development of this project, H. Ross Smyth, head of the Navigational Aids Section, and W. A. Cumming of the Microwave Section who designed the antenna.

REFERENCES

(1) H. T. Stevinson and D. M. Makow, A Tumbling Aerofoil Radio Distress Beacon System for Locating Crashed Aircraft, Report No. MR-22, National Research Council of Canada, Ottawa, July 1958.
(2) D. M. Makow, H. R. Smyth, S. K. Keays, and R. R. Real, A Low-Drain Distress Beacon for a Crash Position Indicator, The Brit IRE, 19, No. 3, March 1959.
(3) D. Kerr, S.A.R.A.H., A UHF (243MC) Pulse Coded Air/Sea Rescue System, J Brit IRE, 17, p 669, Dec. 1957.
(4) H. A. Wheeler, Fundamental Limitations of Small Antennae, Proc IRE, 35, p 1479, Dec. 1947.
(5) R. Loosjes, H. J. Vink and C. G. J. Jansen, Thermionic Emitters Under Pulsed Operation, Philips Tech Rev, 13, p 337, June 1952.
(6) Pulse Operation of Transmitting Valves, Mullard News Letter, p 4, Jan. 1956.
(7) D. Makow, S. Keays and S. Mak, Filament Pulsing Cuts Power Needs, Electronic Engineering of Canada, p 26, Aug. 1957.

Distress Transmitter Uses Hybrid Circuit

By HENRY B. WEISBECKER Project Engineer, Simonds Aerocessories, Inc.
Tarrytown, N. Y.

TRANSISTORS and tubes are combined to get required power output from minimum power input in a distress transmitter. The usual hybrid design was reversed in that tubes are used in the power output and driver stages.

The system operates on both 500 kc and on the 8.326-mc distress frequency. A code-wheel operated photoelectric flip-flop accomplishes band switching automatically. The same wheel is used to key the transmitter in a predetermined code.

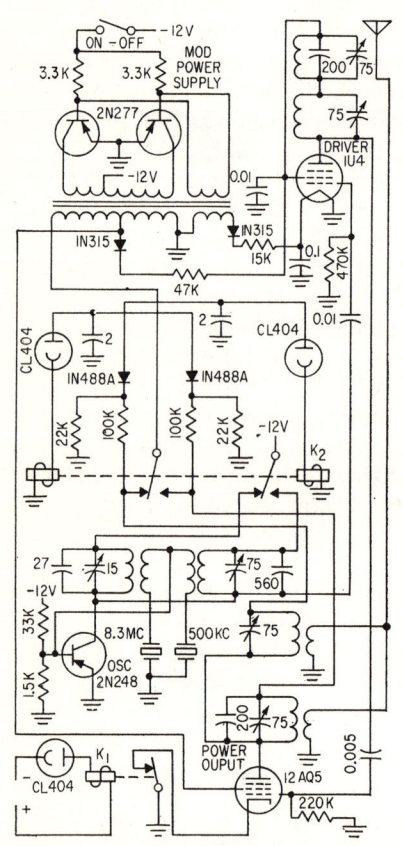

The transmitter operates from a 12-volt battery and requires no cranking of a generator by the operator. It provides about 5 watts output with a current consumption of 1.5 amp from the battery.

The modulator-power supply shown in Fig. 1 uses a pair of 2N277 transistors in push-pull. It generates a square wave of 1,000 cps. Output from a tap on the secondary of a toroidal step-up transformer is rectified for screen and plate voltage for the 1U4 driver. A-c from this tap is applied directly to the screen of the 12AQ5 output tube. High-voltage a-c is applied by the band-switching circuit through the appropriate tank circuit to the plate of the 12AQ5. Applying the 1,000-cps signal directly to the plate and screen of the output tube assures 100-percent tone modulation of the transmitter. It also eliminates the need for producing a high d-c voltage for the tube.

Output from a separate winding of the transformer is rectified to furnish 1.5 volts d-c for the driver filament.

A 2N248 transistor is used in the oscillator, which is crystal controlled at both frequencies. The tank circuit for the oscillator is switched by the photoelectric band-switching circuit. Oscillator output at one of the two operating frequencies is coupled to the driver. The tank circuits in the driver for each of these frequencies are connected in series.

A motor-driven code wheel, a bulb and three photocells key the transmitter and switch frequency once a minute. To key the transmitter, the bulb produces light which falls on a photocell through the code wheel. The wheel is translucent or opaque, as called for by the coded information. The light on the photocell actuates relay K_1. The relay contacts, which are in the cathode circuit of the output stage, key the transmitter.

For band switching, a special slit in the code wheel passes light only once per code-wheel revolution. The photocells operate relay K_2 as a flip-flop. With K_2 at one of its two positions, light falls on both photocells through the slit. Current flows in the opposite coil, pulling the arms of the relay toward that position. When the originally closed contacts

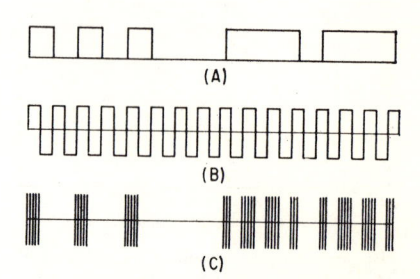

FIG. 2—Portion of S-O-S produced by code wheel is shown at (A), 1,000-cps square wave generated by modulator (B) and modulated output signal (C)

are opened, current in the coil would normally cease flowing, but this is prevented by a 2-microfarad capacitor, which, in discharging, keeps current flowing long enough for the arms to reach the opposite side.

When this is accomplished, light is no longer present and the relay remains in the new position. On the next light pulse, a similar ac-

FIG. 1—Complete transmitter shows how relay K_1 through code wheel encodes output and relay K_2 switches frequency

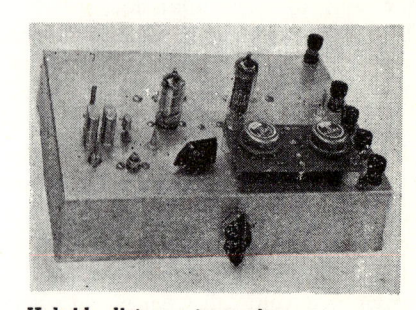

Hybrid distress transmitter uses two tubes and three transistors to transmit coded output at two frequencies

tion returns the relay to the original position. Thus the relay is thrown to its opposite position on each light pulse. Its contacts switch 12 volts d-c to the appropriate tank circuit in the oscillator and high voltage a-c to the appropriate tank in the output stage.

Chapter 19
DIGITAL COMPUTER SWITCHING CIRCUITS

Frequency Control of Magnetic Multivibrators

New magnetic-coupled multivibrator circuits permit wide-range step-by-step and continuous control of output frequency without adversely affecting waveform. Circuit variations also permit independent control of output voltage

By **WILLIAM A. GEYGER** U S. Naval Ordnance Laboratory, Silver Spring, Md.

MAGNETIC-COUPLED transistor d-c to a-c converters usually provide square-wave outputs with frequencies dependent on d-c input voltage. Output frequency of one type of magnetic-coupled multivibrator is controlled by a d-c bias, but the shape of the output square wave is adversely affected.

A basic circuit has been developed in which output frequency can be controlled step-by-step and/or continuously over wide frequency ranges. Variations of the circuit also provide step-by-step and/or continuous control of output amplitude.

A low-power (10-watt) prototype multivibrator circuit has been developed as a square-wave generator for magnetic-amplifier transient analyzers[1] and magnetic-switch B-H loop tracers[2]. These and similar applications require a clean output waveform for proper operation of the silicon diode chopper circuits.

Because output frequency of the multivibrator can be made directly proportional to angular position of a variable transformer slider shaft, the new circuit may be used as a high precision transducer to convert angular (or linear) displacement to a corresponding frequency shift. Movement or rotation of the shaft may be controlled remotely with a conventional servo system. These characteristics suggest application of the circuit to telemetering systems.

Frequency Control

Earlier transistor d-c to a-c converters use a saturable transformer with rectangular hysteresis loop core material[3,4]. They are finding increasing applications as square-wave power supplies for magnetic amplifiers, induction motors, gyros and torque relays.

These magnetic-coupled multivibrators are widely used in d-c power-supply devices, where low-voltage d-c is converted to a square-wave a-c, transformed to a high voltage level and rectified.

The linear relationship between multivibrator output frequency (f) and d-c input voltage (E_{d-c}) is desirable for telemetering-transducer applications. However, in designing test circuits for magnetic amplifiers, it is necessary to vary frequency independently of supply voltage.

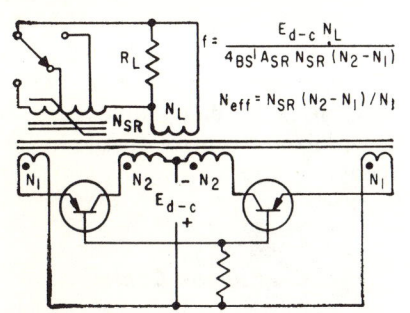

$$f = \frac{E_{d-c}}{4 B_S{}^! A_{SR}} \frac{N_L}{N_{SR}(N_2 - N_1)}$$

$$N_{eff} = N_{SR}(N_2 - N_1)/N_L$$

FIG. 1—Saturable reactor in parallel with load permits step-by-step control of output frequency

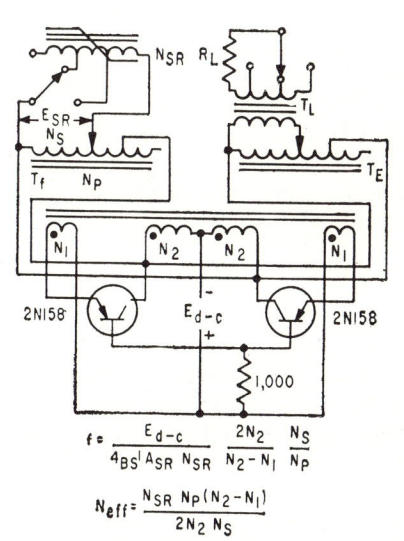

$$f = \frac{E_{d-c}}{4 B_S{}^! A_{SR}} \frac{2N_2}{N_{SR}} \frac{N_S}{(N_2 - N_1)} \frac{N_S}{N_P}$$

$$N_{eff} = \frac{N_{SR} N_P (N_2 - N_1)}{2 N_2 N_S}$$

FIG. 2—Variable transformers across collector windings provide linear control of frequency and output voltage

In the variable-frequency magnetic-coupled multivibrator developed by Van Allen[5], frequency is controlled by varying a d-c voltage applied to additional windings of a twin-core arrangement similar to that of self-saturating magnetic amplifiers. Core flux, instead of swinging between saturation levels as in earlier converters, is reset to an unsaturated value on alternate half cycles. However, d-c bias magnetization of the cores has an adverse effect on the output wave shape.

Frequency of the magnetic-coupled multivibrator shown in Fig. 1 can be controlled without impairing output waveform by varying

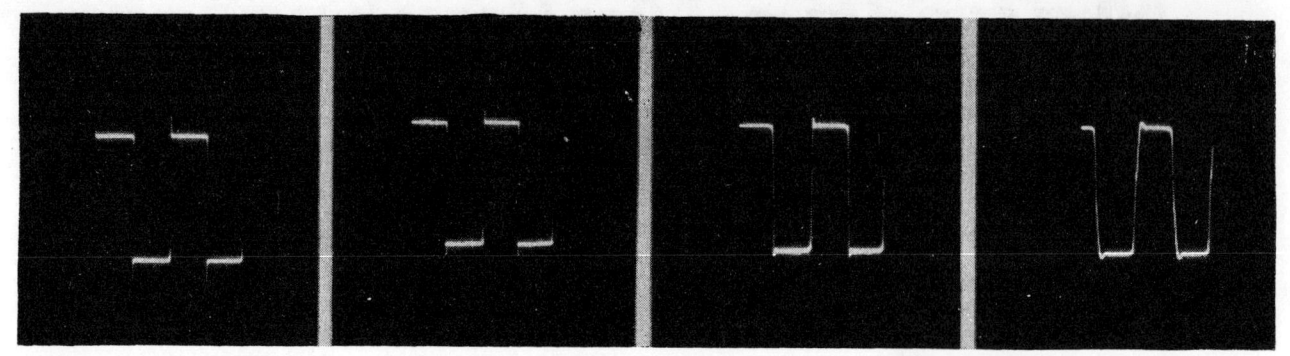

Prototype of circuit in Fig. 2 produced these waveforms. Beginning at left, frequencies are: 55, 500, 3,000 and 20,000 cps

the effective number of turns of the transformer windings. Effective number of turns (N_{eff}) determines frequency in accordance with the basic relationship $N_{eff} = E_{d-c}/4 B_s A_T f$ where B_s is saturation flux density of transformer core material in gausses and A_T is transformer core cross-sectional area in cm.

Operating Principle

It is impractical to vary the actual number of turns of the four transformer windings that operate in conjunction with the two switching transistors in Fig. 1. Therefore, additional components are used to vary effective number of turns through tapped windings and/or continuously variable transformers.

In working with the Campling circuit, the author discovered that frequency of a differential multivibrator can be varied within wide limits without changing magnitude of output voltage. To control frequency, a saturable reactor with rectangular hysteresis loop core material is connected in parallel with the load terminals, as in Fig. 1. The effective number of turns of the saturable reactor are then varied to make step-by-step changes in frequency.

Further investigations revealed that the loading saturable reactor may be placed across an addition transformer winding or the collector terminals, either directly or with additional transformer components. If the product $B_s' A_{SR} N_{SR}$ of the saturable reactor, shunted across both collector windings, is smaller than the product $B_s A_T (N_2 - N_1)$ of the transformer, output frequency is solely determined by the saturable reactor. The transformer then operates with the two switching transistors merely as an unsaturated isolation transformer with correspondingly reduced core losses.

With the saturable reactor in parallel with the load, frequency is inversely proportional to N_{SR} and directly proportional to N_L. By varying the actual number of turns of a secondary winding in Fig. 1 with properly distributed taps, it is possible to control frequency directly.

Continuous Control

Figure 2 illustrates a convenient method for controlling frequency with a standard low-power variable transformer, T_f. This circuit arrangement permits wide-range step-by-step frequency control by varying the taps of reactor N_{SR} and continuous frequency control by moving the slider of variable transformer T_f without changing magnitude of output voltage. A second variable transformer, T_E, added to the circuit makes it possible to vary output voltage without affecting frequency.

Output frequency is directly pro-

FIG. 3—Multiple-control multivibrator has coarse and fine control of both frequency and output voltage

portional to N_s (actual angular shaft position of transformer T_f). Similarly, output voltage can be varied by changing the actual turns ratio of the second variable transformer, T_E. Transformer T_L isolates the load from the d-c source and permits selection of the range of output voltage (0-6, 0-30 and 0-150 volts).

Multiple Control

The multivibrator in Fig. 3 has coarse and fine adjusting transformers T_f, T_f' and T_E, T_E'. The large primary voltage for T_f and T_E (coarse adjustment) is the voltage across both collector windings. The small primary voltage for T_f' and T_E' (fine adjustment) is derived from two separate secondary windings.

With $N_T/2N_2 = N_L/2N_2 = 1/10$, fine adjustment of frequency and voltage output can be achieved by varying the slider positions of T_f' and T_E', respectively.

Design Problems

In designing the new magnetic-coupled multivibrator circuits, the fundamental principle for wide range frequency control consists in loading the original differential-type circuit with a square-wave loop core saturable reactor. Because the windings of the reactor should have low copper resistance, it is desirable to provide several groups of equally rated (preferably bifilar) windings. They may be either series or parallel connected to vary frequency step-by-step over wide ranges.

For optimum switching performance and to minimize spikes in the output square wave, magnetic leakage effects must be minimized. When using the turns ratio $N_1/N_2 = \frac{1}{2}$, the transformer can be considered to have $N_1 + 2N_1 + 2N_1 + N_1 = 6 N_1$ turns acting as collector and emitter windings. Six twisted wires may be hand wound simultaneously around the core of the transformer to limit magnetic leakage between windings.

Although the core of the multivibrator transformer will always be unsaturated when operated at higher frequencies, it will operate with the lowest frequency between saturation levels in the same manner as conventional multivibrator circuits. It is therefore desirable to use rectangular hysteresis loop core material in the multivibrator transformer if the lower part of the frequency characteristic (Fig. 4) is to be used. However, when working only with higher frequencies, this transformer will never be saturated and will have correspondingly reduced core losses. For operation at higher frequencies, therefore, lower grade core material may be used.

The continuously variable transformer providing linear frequency control will never be saturated, so commercially available components may be used. However, to minimize magnetizing current requirements of the variable transformer, it is

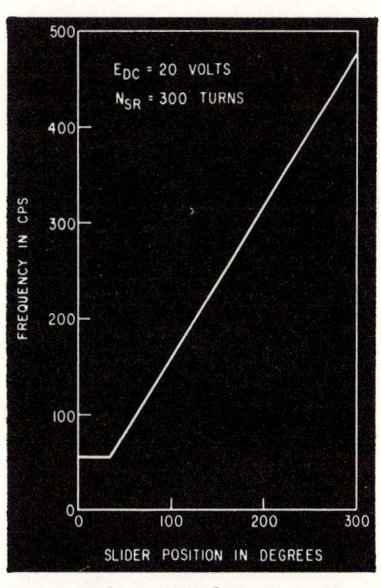

FIG. 4—Performance of circuit in Fig. 2 shows effect of reactor above 55 cps

necessary to use standard low-power design. The scale of this transformer may be calibrated in cps for various values of d-c supply voltage and number of turns of the saturable reactor.

Construction

The multivibrator transformer with Orthonol 2-mil tape core (inside diameter, 1.5 in.; outside diameter, 2.5 in.; tape width, 1.0 in.) has 6 x 275 turns of No. 24 (B & S) wire (6 twisted wires, hand wound, 6 x 2.5 ohms). It is connected so that N_1 has 275 turns and N_2 has 550 turns.

The saturable reactor with Orthonol 2-mil tape core (inside diameter, 1¼ in.; outside diameter, 1⅝ in.; tape width, 1.0 in.) has 6 x 75 turns of No. 22 (B & S) wire (6 twisted wires, hand wound, 6 x 0.4 ohm). A common-base resistor of about 1,000 ohms may be used.

Performance

Measured frequency as a function of the variable transformer slider position (actual angular deviation of the slider from its initial position in which N_s equals zero) is shown in Fig. 4. Within the initial range from zero to 30 deg, the saturable reactor remains unsaturated. The multivibrator transformer operates between saturation levels, as with conventional multivibrator circuits. Hence, the transformer alone determines frequency (in this case, 55 cps).

At higher frequencies (with the slider moved toward higher values of N_s), the saturable reactor starts to operate between its own saturation levels, while the transformer starts to become unsaturated. The saturable reactor takes over and becomes the frequency-determining factor from frequencies of 55 to 500 cps. Thus, with an input of 20 volts d-c and 300 turns on the reactor, smooth linear frequency control is possible from 55 to 500 cps. Higher frequencies (55 to 1,000 or 55 to 3,000 cps) may be obtained by reducing reactor turns (to ½ or ⅛ of the original number).

The spikes that appear in the waveform photographs of the output voltage as it appears across the load resistor are sufficiently small for the present applications. If they were objectionable, they could be eliminated by connecting small capacitors (0.05μf) between collector and emitter of each transistor.

REFERENCES

(1) W. A. Geyger, Magnetic-Switch Transient Analyzer, ELECTRONICS, Jan. 1956.
(2) W. A. Geyger, Magnetic-Switch B-H Loop Tracer, ELECTRONICS, Oct. 1956.
(3) G. H. Royer, A Switching Transistor D-C to A-C Converter Having an Output Frequency Proportional to the D-C Input Voltage, *AIEE Trans*, Part I, July 1955.
(4) C. H. R. Campling, Magnetic Inverter Uses Tubes or Transistors, ELECTRONICS, Mar. 14, 1958.
(5) R. L. Van Allen, A Variable Frequency Magnetic-Coupled Multivibrator, *AIEE Trans*, Part I, July, 1955.

Computer Switching With

How to select the right power transistor and switching circuit to obtain required switching speed, gain and current-carrying capacity

By JAMES S. RONNE, Military Systems Studies Dept., Bell Telephone Laboratories, Whippany, New Jersey

HIGH-SPEED COMPUTERS utilize power transistors in increasing numbers. In selecting one of these units for a particular switching application, it is necessary to consider three requirements: switching speed, pulse gain, and current-carrying capacity. Unfortunately, these parameters are strongly dependent upon circuits and operating conditions. Furthermore, it is difficult to specify standards of comparison which would be appropriate for all types of transistors. As a result, these data are not directly available from published specifications.

It is not difficult, however, to determine the approximate capabilities of a power transistor on the basis of the published data sheet. The method by which a transistor can be evaluated relative to switching speed, pulse gain and current will be discussed. In addition, the considerations involved in realizing the full capabilities of a transistor in a practical circuit will be examined and a design example will be given.

Switching Speed

Pulse rise time, t_r, of a transistor is closely associated with frequency cut-off and collector capacitance[1,2,3]. The expression for rise time with the emitter grounded is approximately given by[1,2,3] (see Fig. 1 and Table I):

$$t_r = t_k \ln [1/(1 - 0.9A/h_{FE})]$$

Typically the rise time will lie in the range, $0.1 t_k < t_r < 3 t_k$, depending upon the circuit gain.

Turn-off time is given approximately by:

$$t_f = t_k \ln [1/(0.1 + 0.9 h_{FE}/(h_{FE} + A_R))]$$

The fall time will be similarly found to lie in the range:

$$0.1 t_k < t_f < 2 t_k$$

depending upon the magnitude of the reverse base current, I_{b2}.

Unfortunately, storage time (t_s) cannot be calculated from normally available small signal parameters. A typical curve illustrating the dependence of storage time on base current is given in Fig. 2. Current I_{b1} refers to the base current which flows during saturation and current I_{b2} refers to the reverse base current which flows during the storage and turn-off times. It should be emphasized that to pull a transistor out of saturation quickly, a reverse base current must flow which is in a direction opposite to that of the turn-on current. This reverse current can be obtained in several ways.

A simple scheme is to ground the base as shown in Fig. 3A. The base-emitter junction of the transistor resembles a battery of voltage V_j throughout the storage time. The reverse base current is

$$I_{b2} = V_j (R_2 + r_b)$$

where V_j and r_b are the junction voltage and base resistance, respectively, during saturation. The junction voltage can be found from the knee of an I_{b1}, V_b curve or from the expression $V_j = V_b - I_{b1} r_b$ where V_b is the d-c base-to-emitter voltage during saturation. In Fig. 3B the equivalent-circuit switch and limiting resistor have been replaced by transistor Q_2. The storage time can be reduced by selecting a transistor for Q_2 which has an effective series resistance (R_2) that is as small as possible. Other methods of reducing storage time will be discussed when the circuit design example is considered.

The maximum available pulse gain is determined by the large-

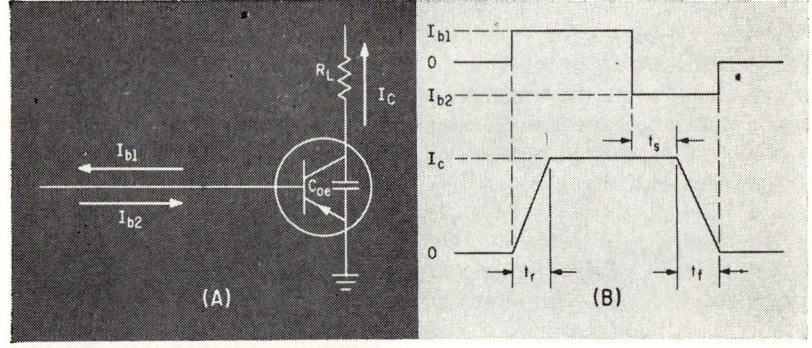

FIG. 1—Transistor capacitance C_{oe} (A) is one of parameters affecting rise time of I_c (B)

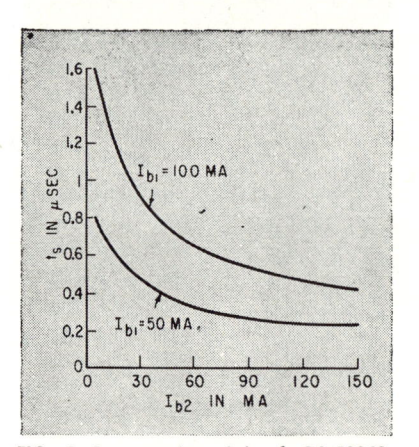

FIG. 2—Storage time (t_s) of GA-53242 transistor is function of I_{b1} and I_{b2}

High-Power Transistors

signal gain of the device. However, the circuit gain is determined by the degree of saturation required. Reducing the circuit gain substantially below the large-signal gain by overdriving will reduce rise time, but unfortunately this method has the adverse effect of increasing the storage time. Furthermore, since some power transistors have low gain at high currents, use of excessive overdrive would sacrifice too much gain. In general, a circuit is designed to have the maximum possible pulse gain consistent with the required switching speed.

Current-Carrying Capacity

Data sheets for switching transistors normally list maximum power dissipation, maximum collector current, and collector saturation resistance (or voltage). However, it is not immediately obvious what the current-carrying capacity of the unit is for a particular switching application. Furthermore, it is difficult to compare the capabilities of transistors on the basis of these specifications.

To evaluate transistors, using their published specifications, a simple expression will be derived which relates pulse width to maximum collector current for various repetition rates. Consider a pulse train (Fig. 4) at the collector with repetition rate, f (in Mc), switching time, t ($t_r + t_f$ in μsec), pulse width t_p (in μsec), with amplitudes E_c and I_c (in volts and amperes). The total average collector dissipation, P_c, is equal to the sum of the

Table I—Parameters

t_r = rise time (0 to 90 percent)
$t_k = 1/\omega a_e + R_L C_{oe}$
$A = I_c/I_{b1}$ = circuit forward gain
$A = I_c/I_{b2}$ = circuit reverse gain,
ωI_e = cutoff frequency, common emitter
h_{FE} = d-c current gain, common emitter
C_{ob} = collector-to-base capacitance
$C_{oe} = h_{FE}C_{ob}$ = collector-to-emitter capacitance

average power dissipated during current saturation (P_1), the average power dissipated during switching (P_2), and the average power dissipated during voltage saturation (P_3).

$$P_1 = I_c^2 r_{ce} t_p f$$
$$P_2 = (I_c E_c/4) t f$$
$$P_3 = E_c h_{FE} I_{CBR} t_0 f$$

where the product, $h_{FE}I_{CBR}$, represents the maximum collector current during voltage saturation under conditions of maximum junction temperature and r_{ce} is the collector saturation resistance. Power term P_3 becomes important only at high ambient temperature or when the transistor is operated near the maximum power capabilities.

The average total power dissipation is:

$$P_c = P_1 + P_2, \quad \text{if } P_3 \ll P_c$$
$$P_c = (I_c)^2 r_{ce} t_p f + 0.25 I_c E_c t f$$

Solving for the pulse width, we get:

$$t_p = \frac{P_\tau - 0.25 I_c E_c f t}{(I_c)^2 r_{ce} f}, \text{ with } t_p \leq \frac{1}{f} - t$$

and where P_c has been replaced with P_τ, the maximum power the transistor is capable of handling. In this way the capabilities of the transistor can be determined for various combinations of t_p, f, and I_c.

The values for switching time (t), collector voltage (E_c), and saturation resistance (r_{ce}) are obtained from transistor specifications and circuit conditions. The maximum power-handling capabilities (P_τ) of the transistor in a particular application are not immediately obvious from most published data sheets. The collector current should never be allowed to exceed the maximum value recommended by the manufacturer. The equation for t_p does not take into account this important rating.

Transistor Power Rating

No industry-wide standard has been adopted for transistor power-dissipation rating. Some published ratings are contingent upon maintaining the entire transistor case at room temperature while others represent average values for specified operating conditions. Since the actual power-handling ability of a transistor is strongly dependent upon application and environment, a better scheme would be to provide sufficient engineering data to permit the calculation of the power-handling capability for individual circumstances. These data would include the maximum allowable collector-junction temperature (T_c), the thermal resistance from collector junction to free-air ambient with no external heat sink, (θ_{c-A}) and the thermal resistance from collector junction to mounting surface (θ_{c-c}).

Temperature drop from the col-

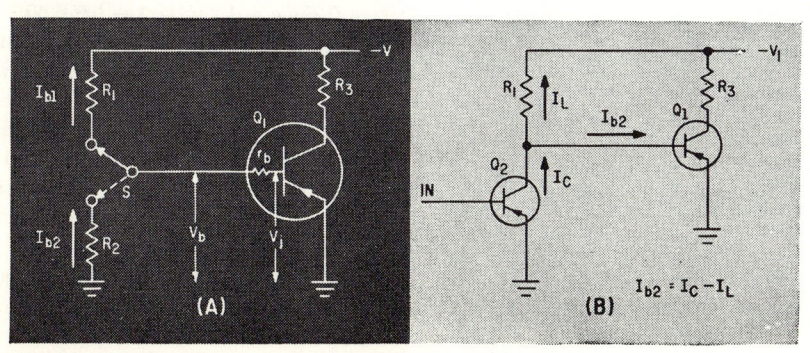

FIG. 3—Switch S is circuit equivalent for transistor Q_2 (B)

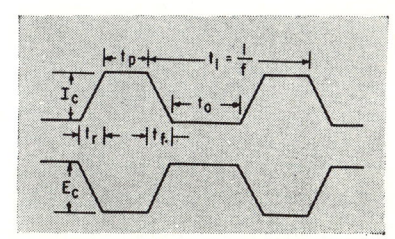

FIG. 4—Waveforms of I_c and E_c

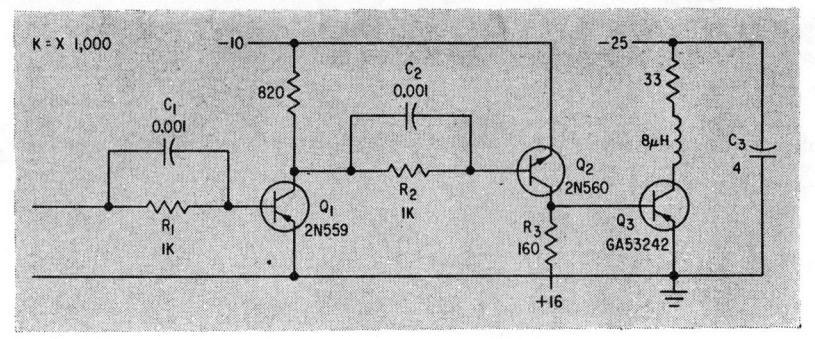

FIG. 5—All transistors of this driver amplifier are initially off

lector junction to the surrounding air is analogous to the voltage drop between two potentials. The thermal resistance from collector junc to mounting surface, θ_{c-c}, is in series with the thermal resistance of the heat sink, θ_{hs}. Thermal resistance θ_{hs} includes the thermal resistance of the transistor case in parallel with that of any external heat sink. Increasing the size of the external heat sink reduces θ_{hs}, thus allowing a greater temperature drop within the transistor and permitting more power to be dissipated. Thermal resistance has the dimensions of deg C/watt.

An external heat sink can be designed based on the rule-of-thumb thermal resistance figure of (125 C/w)/A for surface areas greater than 5 in.², where A is the total exposed area of the heat sink in in.². This area includes both sides of a flat plate but only the outside surface of a transistor case. For small heat sinks (less than 5 in.²) the value (62 C/w)/A should be used. These values are determined from power dissipation based on heat transfer as a result of radiation and convection from a shiny surface. A properly blackened heat sink can reduce the thermal resistance by as much as 25 percent. Simply painting the area is not recommended, however, since the insulating properties of the paint may increase the thermal resistance.

Obviously, mounting arrangements and the degree of confinement can alter these values considerably in actual circumstances. For this reason, the thermal resistance values given here can be considered to be within a factor of two of the true value. Based on these numbers, however, a quick calcula-

Table II—Typical Characteristics of W E GA-53242

I_c	≤3 a
$BV_{CBO, EBO, CEO}$	≧40 v
r_{ce}	0.5 ohm
I_{CBO} (25C, $V_{CB} = -40v$)	5μa
h_{FE} ($I_c = 800$ ma)	60
C_{ob} ($V_c = -4.5v$)	≤50 pf
$f\alpha_e$	100 Kc
θ_{c-c} (int. thermal resistance)	24 C/w
T_j (max. junction temp.)	85 C

tion will determine the approximate capabilities and/or requirements for a given application.

Commonly, only one of the two thermal resistance figures is specified in the published transistor data sheets. Given one, a simple approximation will allow the calculation of the other. This procedure entails a guess as to the effective heat dissipating area of the transistor cases.

As an example, consider the WE 2N560. This transistor has a θ_{c-A} of 250 C/w and a maximum junction temperature (T_c) of 150 C.

The internal thermal resistance, θ_{c-c}, is found from $\theta_{c-c} = \theta_{c-A} - \theta_{hs}$. The external area of the case (built-in heat sink) is 0.56 in.² Thus, internal temperature rise is

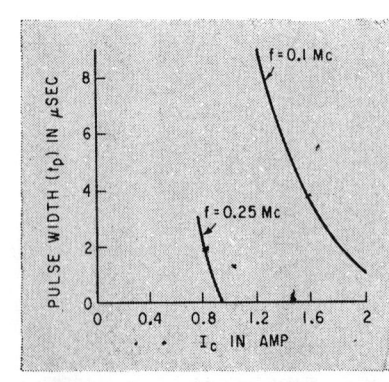

FIG. 6—Current-handling capability of the GA 53242

$\theta_{c-c} = 250$ C/w $- 62$ C/0.56w $= 140$ C/w

Similarly, given θ_{c-c} we can find θ_{c-A}. The addition of a 3½ x 3½ in. heat sink will reduce θ_{hs} to:

$$\theta_{hs} = \frac{125 \text{ C/w}}{2 \times (3½)^2} = \frac{5 \text{ C}}{\text{w}}$$

The 2 in the denominator takes into account both sides of the heat sink. The area of the transistor case is now insignificant and need not be included. The thermal resistance of transistor plus heat sink, θ_t, is $\theta_{c-c} + \theta_{hs}$, or 145 C/w. (The specification sheet gives the figure as 150 C/w.)

These values can now be combined with the maximum junction temperature to get the maximum power dissipation. For example the 2N560 operating with a 3½ x 3½ in. heat sink in 75 C free-air ambient (T_A) can dissipate $P_T = (T_c - T_A)/\theta_t$. Thus

$$p_T = (150 - 75)/(150/w) = 0.5 \text{ w}$$

Circuit-Design Example

An experimental solid-state memory requires a drive amplifier which will supply an 8-μh load with a current pulse (I_c) of 750 ma, $t_p = 1$ μsec, and ($t + t_p$) equal to or less than 2 μsec. (See Fig. 1B.) Repetition rate, f, is 0.25 Mc.

Consider the transistor characteristics listed in Table II. This transistor exhibits low saturation resistance (0.5 ohm) and high gain (60 at 800 ma).

Based upon these data

$$t_k = \frac{10^{-6}}{(0.1)(2\pi)} + (33)(50 \times 10^{-12})(60)$$

$= 1.7$ μsec

for a load resistance (R_L) of 33 ohms (Fig. 5).

If we design the circuit to have a rise time (t_r) not greater than 0.3 μsec,

$t_r = 0.3$ μsec $= 1.7 \ln[1/(1 - 0.9A/60)]$

Solving for A we see that A must be less than 11. The base drive (I_{b1}) required must therefore be greater than $I_c/A = 750$ ma/11 $= 68$ ma. The addition of an 8-μh inductance in the collector circuit will add approximately 0.3 μsec to the rise time.

Similarly, a fall time (t_f) not greater than 0.3 μsec is calculated from

$t_f = 1.7 \ln\{1/[0.1 + 0.9 \times 60/(60 + A_R)]\}$

Solving for A_R we get $A_R < 13.1$; thus $I_{b2} > 57$ ma. Referring to Fig. 2, we see that a reverse base current of perhaps 100 ma would be desirable to reduce storage time. In the absence of storage-time information, use as large a reverse base current as can be conveniently supplied, in turning off high-speed high-current pulse amplifiers.

A 2 x 2 in. heat sink has a thermal resistance of 16 C/w. The thermal resistance of heat sink plus transistor become $\theta_t = 16$ C/w + 24 C/w = 40 C/w. The maximum power-handling capability of the device at an ambient temperature of 30 C is

$$P_T = (85C - 30C)/(40C/w) = 1.4w$$

The current-handling capability of the unit is

$$t_p = (2.8 - 12 I_c f)/(I_c^2)f$$

where switching time t is 1 μsec and E_c is 25 v (Fig. 4).

This expression is plotted in Fig. 6 for $f = 0.25$ Mc and $f = 0.1$ Mc. As indicated, the power requirements are within the capabilities of the transistor. However, the curves of Fig. 6 represent an absolute maximum. Operating a transistor at this value is inviting trouble, since a slight increase in supply voltage or ambient temperature could permanently damage the transistor. Furthermore, the power dissipation term, P_3, would no longer be insignificant. In most instances the safety margin of a border-line transistor can be increased by reducing the supply voltage or increasing the area of the heat sink or doing both.

To obtain the high reverse base current (I_{b2}) required to cancel storage time and turn the transistor off quickly, a special circuit is required. The objective is to establish the maximum possible I_{b2} that is consistent with the circuit and the safe operating margins of the transistor. Some of the coupling schemes which could be used to establish a reverse base current are shown in Fig. 7.

In Fig. 7A, a negative pulse to Q_1 switches it on, grounding the base of Q_2 through R_1, which has a small resistance. In Fig. 7B, a negative pulse switches on Q_1, causing

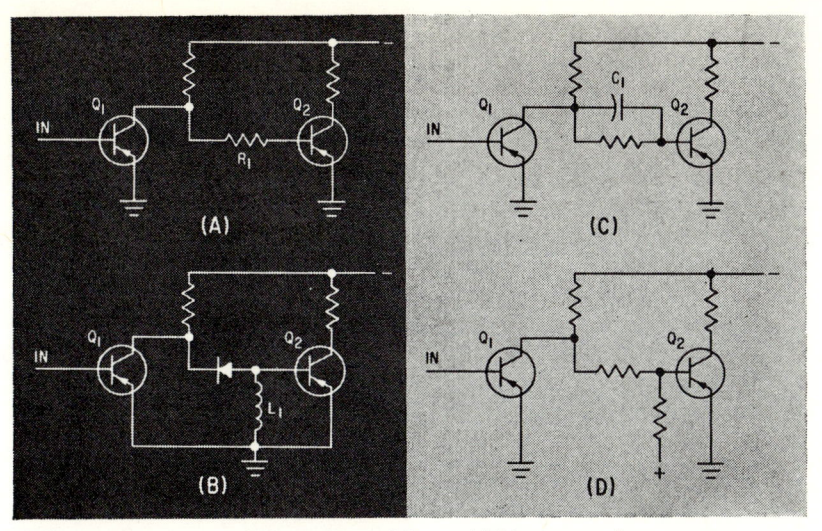

FIG. 7—Coupling circuits which can be used for establishing reverse base current

L_1 to impress a turn-off voltage on the base of Q_2.

In Fig. 7C, a negative pulse switches on Q_1, causing C_1 to apply a turn-off voltage to the base of Q_2.

A variation of the method illustrated in Figure 7D is used to turn off transistor Q_3 of Fig. 5. In Fig. 7D, the reverse voltage is applied by means of a voltage-divider network. The resulting coupling network introduces a substantial loss in power. In Fig. 5, a positive turn-off voltage is automatically applied with no added loss in gain or power. This is accomplished by driving pnp transistor Q_3 with an npn type Q_2. Transistor Q_1 is a 150-mw diffused-base germanium transistor which allows the amplifier to operate from direct-coupled low-level circuits. Switching time in the stages Q_1 and Q_2 of the amplifier is reduced by incorporating the coupling scheme shown in Fig. 7C.

It will be noted that there is no phase reversal between transistors (they are all on or off at the same time). This condition has the effect of reducing the demands placed on Q_2 since it will have the same duty cycle as Q_3. Under no circumstances would transistor Q_2 be subjected to long periods of "soaking", which tends to increase storage time and power dissipation. However, the condition of having all transistors on at the same time increases the demands placed on the power supply. It was found necessary to by-pass the power-supply leads with capacitor C_3 to reduce the ripple to a tolerable level.

In operation, turning transistor Q_1 on causes the negative 10-v supply to apply a turn-on current to transistor Q_2. Coupling resistor R_1 is by-passed with a capacitor C_1 to speed up switching. Turning on transistor Q_2 effectively connects the base of transistor Q_3 to the negative 10-v supply, which applies a 100-ma turn-on current. As a result, approximately 750 ma will flow in the 8-μh load. The 100-ma base current of Q_3 is determined primarily by the series resistance of the emitter-base junctions of the transistors Q_2 and Q_3. Other transistors may have a lower input impedance and therefore require a lower negative-supply voltage or a series current-limiting resistor or both.

Turning transistor Q_1 off removes the forward base bias of Q_2 thus turning it off. Turning off Q_2 effectively connects the base of Q_3 to the positive 16-v supply, through load resistor R_3. As a result, approximately 100 ma of reverse base current flows, which sweeps the minority carriers from the base region and turns transistor Q_3 off quickly.

The amplifier produces 0.75-amp 1.9-μsec. pulses at a 250-Kc repetition rate in the 8-μh load with a 0.25-ma, 1-μsec. input. Transistor Q_3 remains saturated for approximately 1 μsec.

References

(1) J. J. Ebers and J. L. Moll, Large-Signal Behavoir of Junction Transistors, *Proc IRE*, **42**, p 1,761, Dec. 1954.
(2) J. L. Moll, Large-Signal Transient Response of Junction Transistors, *Proc IRE*, **42**, p 1,773, Dec. 1954.
(3) J. W. Easley, The Effect of Collector Capacity on the Transient Response of Junction Transistors, *IRE Trans on Electron Devices*, **ED-4**, p 6, Jan. 1957.

Alarm Circuit Warns of

Audible alarm system allows central placement and gives distinctive indication of fault location in digital computer and data processing equipment. Simple, flexible circuits simulate horn, collision, battle stations sounds

By STANLEY FIERSTON
Sylvania Electronic Systems Division, Sylvania Electric Products, Inc., Needham, Mass.

FLEXIBLE MONITORING EQUIPMENT is needed to operate and test today's large scale data processing systems. Indicator lamps are normally used for this function but they have several disadvantages. These disadvantages can be severe if a system has many units spread over a wide area or if circuit faults are infrequent. The audible alarms proposed have generally been too complex, too inflexible, or have required voltages not normally available in transistorized circuits.[1] The audible alarm discussed here overcomes these disadvantages and produces distinctive sounds.

Horn and collision signals are generated by the electronic circuits shown in Fig. 1. The two sounds are used to monitor two circuits. Mixing the two signals produces a battle stations sound which is used for a third monitor.

Alarm Circuit Operation

Switch S_5 controls the alarm, selecting the circuit to be monitored and the output signal. With S_5 in position 2, an alarm trigger pulse will activate the one-shot multivibrator. The AND circuit passes the battle stations signal through to the audio amplifier and speaker. The one-shot multivibrator resets after a suitable time and closes the AND circuit, thus silencing the alarm. Positions 3 and 4 of S_5 are similar to position 2 but monitor different circuits, and produce horn and collision signals. No one-shot multivibrator is needed in position 4 since the trigger signal is not a short pulse.

Other positions of S_5 connect to selected points in the monitored system. Pulses with audio frequency components on distinctive pulse trains are listened to directly. These signals give an experienced operator useful information about

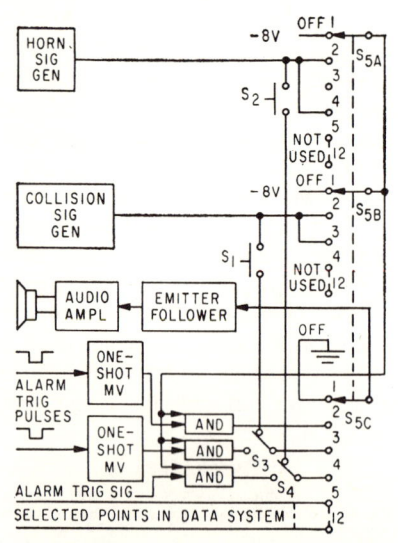

FIG. 1—Alarm system generates sounds electronically

Operator is shown selecting the channel to be monitored. Circuits are identified on cardboard tab

Faults in Digital Systems

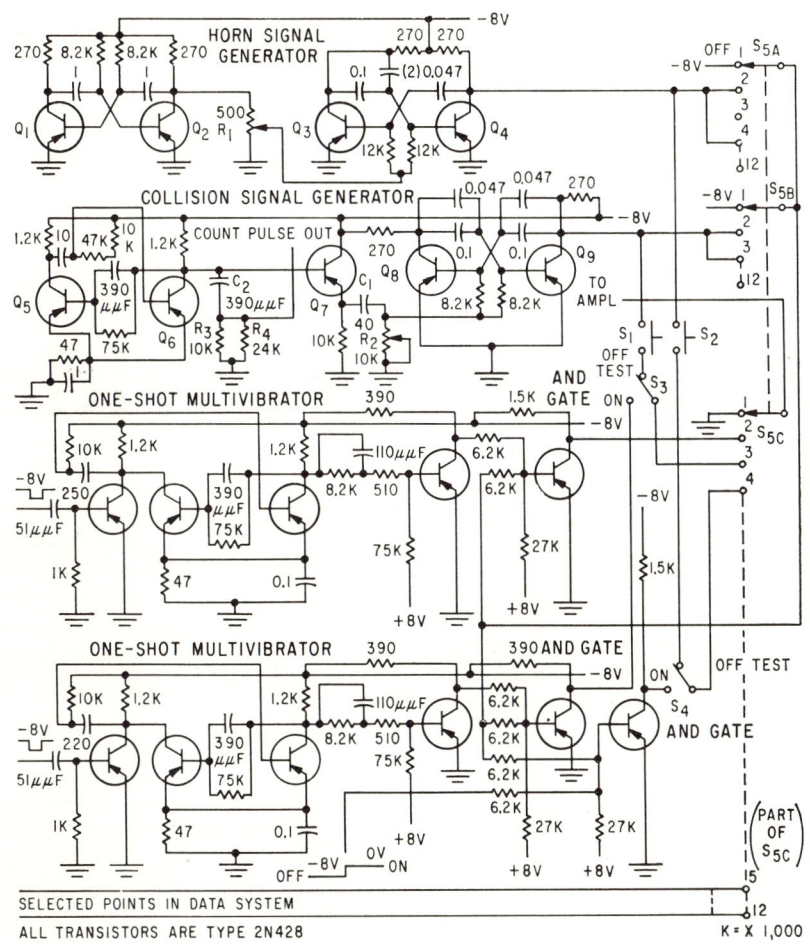

FIG. 2—Horn signal generator uses two multivibrators; collision signal is generated with multivibrator and pulse generator. Mixing the two sounds produces battle stations signal

the system.

The one-shot multivibrators used for positions 2 and 3 could be replaced by flip-flop circuits. The flip-flops maintain the alarm signals until manually reset.

Switches S_1 through S_4 are used for testing the signal generators and for silencing an alarm.

Alarm Signal Generator

Figure 2 is a schematic diagram of the alarm. A low frequency multivibrator Q_1 and Q_2 is combined with a high frequency multivibrator Q_3 and Q_4 to generate the horn signal. The rough sound produced by the square waves is deliberately increased by the unequal timing capacitors used in the Q_3 and Q_4 circuit. Frequency of the Q_3 and Q_4 multivibrator can be controlled by changing the bias voltage. Potentiometer R_1 controls the bias and is set to give the desired sound.

The collision signal is generated in a similar way. High frequency multivibrator Q_8 and Q_9 is similar to Q_3 and Q_4. Circuit Q_5 and Q_6 forms a 2-cps, highly unsymmetrical, free-running pulse generator. Emitter follower Q_7 couples the output pulse to differentiating circuit R_2-C_1.

Output of the differentiator is an exponentially varying signal of 2 cps which controls the bias of the Q_8 and Q_9 circuit and thus its frequency. The actual sound developed by the generator depends on the differentiator and multivibrator time constants and the pulse rate. Component values shown in Fig. 2 tend to produce a birdlike sound. Other values can produce a sound similar to a navy collision signal.

A second differentiating circuit R_3 R_4 C_2 gives a narrow 2-cps pulse for use in other parts of the data system.

Alarm signals are fed to an emitter follower and then to an audio amplifier, both of which are conventional circuits. A variable resistor in the emitter follower provides a convenient volume control. The audio amplifier schematic is shown in Fig. 3.

Applications

Highly specialized circuits are not needed since the alarm uses standard pulse circuits found in most data processing systems. Only a few circuits are required and the alarms can be placed at desired points. Extra connecting cables are not needed since the volume can be set to give the required area coverage. Many other distinctive sounds can be generated by changing the R-C constants of the multivibrators. This flexibility is an obvious advantage over mechanical alarms

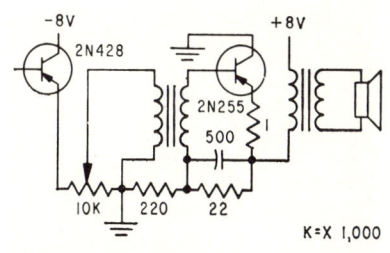

FIG. 3—Audio amplifier uses conventional circuits

such as bells and horns. The various parts of the alarm can be packaged into small plug-in modules.

The author thanks John Malcolmson for his assistance in the construction and testing of the alarm.

REFERENCES

(1) M. B. Freedman and T. E. Rolf, Alarm Signal Generators, *Electronics*, p 204, July, 1952.

Circuits designed to provide clock signals for developing and testing large digital computers use readily available germanium transistors. Except for one oscillator, all transistors operate in either saturated or cutoff state. Other design techniques limit waste of transistor power so that system can provide peak load up to 5 amp

By S. SCHOEN Data Systems, Norden Division, United Aircraft Corp., Gardena, Calif.

Transistors Provide

DEVELOPMENT of several large digital data-processing systems required circuits to provide the basic timing and gating signals. Since the equipment is completely transistorized, use of transistors to generate the computer clock was investigated. Several system and circuit techniques that were used to provide controlling clock signals for digital computers containing several thousand transistors are described.

For this application, transistor switching circuits had to be capable of high speed and also of controlling high peak currents. Although reliability was the major design consideration, other factors included cost, component availability, ease of fabrication and interchangeability. For example, all circuits use commercially available germanium transistors.

Requirements

The clock circuits provide 0.25- to 0.6-μsec pulses with rise and fall times of about 0.1 μsec. Repetition rates are 150 to 750 kc. Peak transient load current is typically 3 to 5 amp. Load is primarily capacitive and may vary appreciably during the pulse.

The load consists mainly of normally nonconducting trigger transistor inputs to bistable circuits. Input impedance of these transistors under transient conditions is considerably influenced by whether they are cut off, active or saturated. Clock pulse amplitude (6 to 8 v) must be kept reasonably constant, since the trigger transistors are operated as voltage-threshold devices.

Two other features are incorporated in the clock circuits to aid design and testing of large computers. One is provision for operation either as a continuous clock or as a gated clock comprised of an arbitrary number of pulses. The second, particularly useful when troubleshooting a large system, is operation at either a slightly higher or a much lower clock repetition rate. The clock circuit can be actuated by an external vfo or by a pushbutton that provides a single clock pulse.

System Design

The clock circuits may be separated into a common pulse generator and a number of individual drivers. In general, multiple drivers are needed because no single high-speed transistor that is readily available is capable of providing peak currents of 3 to 5 amp.

Multiple drivers provide several

Use of commercially available parts and standardized circuits limits cost of computer clock generator

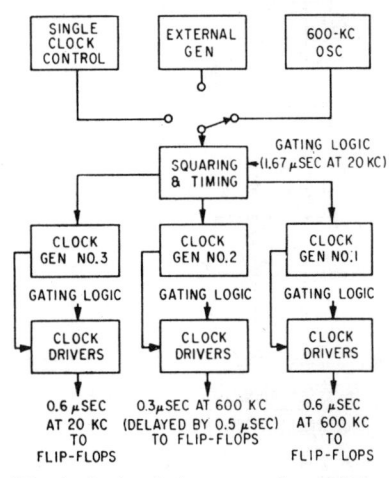

FIG. 1—Basic clock system has 600-kc oscillator but can be operated with an external generator. Provision is also included to generate a single clock pulse

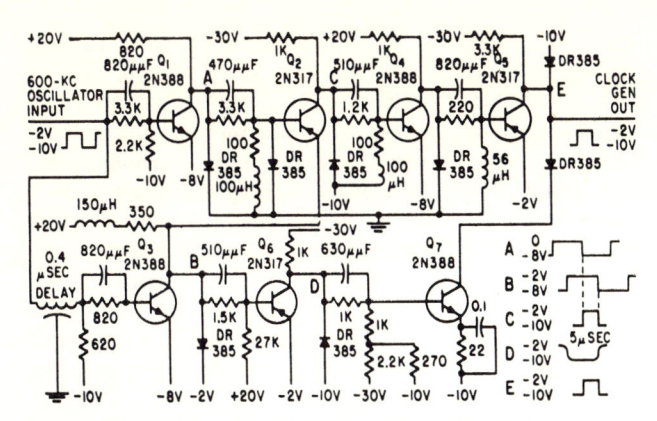

FIG. 2—Input pulse applied through transistor Q₁ switches Q₂ on. Same pulse applied through Q₃ switches Q₂ off after fixed delay

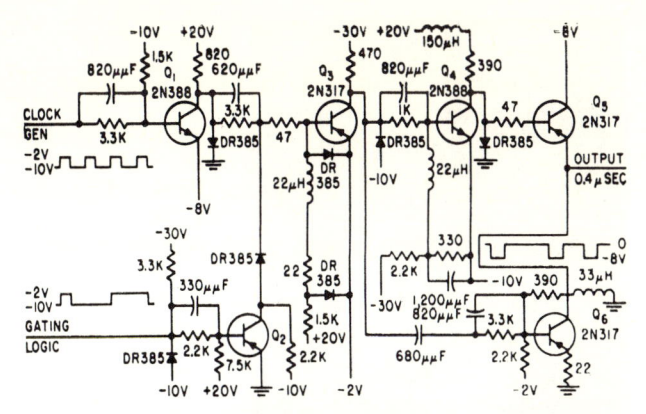

FIG. 3—By replacing conventional emitter-follower load resistor with transistor Q₄, a high impedance is presented to Q₅ during output

Computer Clock Signals

other advantages. The drivers may be individually gated for an arbitrary duration. In fabricating a large system, each clock driver may be placed physically close to the circuit being triggered, reducing ground noises, transmission attenuation and cross coupling. By making all driver circuits nearly identical, per unit fabrication cost is decreased.

A single oscillator and a pulse-forming network generate the basic waveform and repetition rate. Multiple outputs from the generator provide logical gating of entire groups of clock drivers.

A typical clock system is shown in simplified form in Fig. 1. The basic repetition rate of 600 kc is set by a sine-wave oscillator. Frequencies other than that generated by the oscillator may be supplied to the pulse-forming network by an external generator. Single clock operation is provided by a pushbutton controlled flip-flop.

Signals from the oscillator or external generator are partially shaped in the squaring and timing circuit. Delay circuits, which provide for arbitrary phasing of the various clock signals, are included in this network. Also included is provision for a logical gate that in this example provides a 20-kc repetition rate.

Three different clock-generator outputs are provided: 600-kc pulses, 600-kc pulses delayed a fixed amount from the first output, and 20-kc pulses. These outputs are supplied to the clock drivers, which set the actual pulse width used to trigger the flip-flops. In the above example, two widths are required.

Circuit Design

Accurate control of clock repetition rate is obtained with a crystal-controlled oscillator. Distortion-free waveforms are unnecessary, since eventual output is a series of pulses. The oscillator is the only circuit in this completely transistorized clock system that does not use transistors as switches operated either in a saturated or cutoff state.

The squaring and generator circuits provide an output signal whose waveform is relatively independent of input signal waveform. Input may be sine or square waves at any frequency from d-c to one mc. The input signal is amplified by the squaring circuit to provide a relatively fast rise time. Only the leading edge is used in determining pulse width.

Forming Pulses

One technique that may be used in forming the pulse is shown in Fig. 2. The leading edge of the input signal is amplified by Q_1 and applied to the base of normally cut off Q_2, switching it on.

The same leading edge is passed through a delay line and amplified by Q_3. The negative-going, delayed signal is then applied to the emitter of *pnp* transistor Q_2, cutting it off again.

Although there are modifying effects, such as trigger delay and hole storage, pulse width (on time of Q_2) is essentially determined by the delay line. Since the delay line is a passive element, output is relatively unaffected by typical compo-

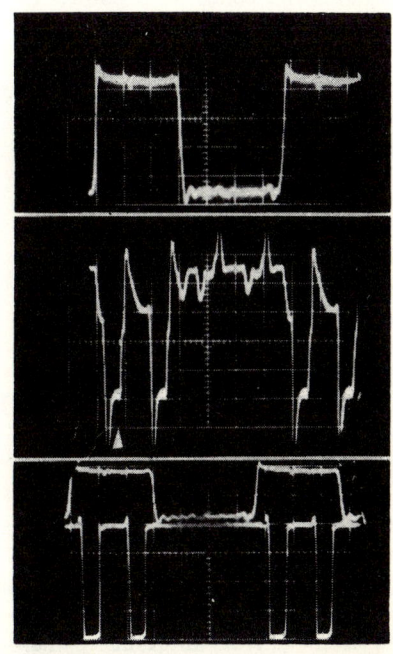

FIG. 4—Gate at top passes input at center to Q_3 in Fig. 3. Bottom shows clock driver output

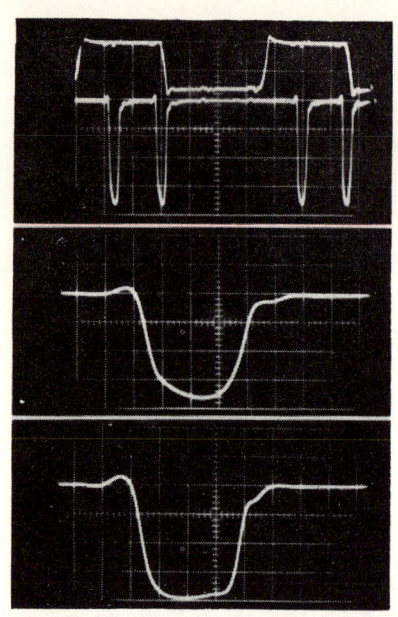

FIG. 5—Gate and resulting clock output are shown at top. Center and bottom waveforms, on expanded time scales, show output when driving thirty and ten flip-flops, respectively

nent and environmental variations or by input signal waveforms. The pulse is amplified to provide an input signal for the clock drivers.

Drivers

When driving a high-current load, it is frequently difficult to maintain a good wave shape at the trailing edge of the pulse.

At the end of the pulse, the output transistor is cut off and load capacitance must be discharged through the collector resistor. Instead of a small collector resistance, which is wasteful of pulse power, transistor Q_7 provides a low-impedance path for discharging circuit capacitances.

This stage, which is normally saturated, is cut off by Q_6 for the duration of the output pulse. At the end of the pulse, Q_7 is switched on and provides the low-impedance path. The 22-ohm resistor in series with the emitter of Q_7 limits steady-state current through it that might result from an inadvertent short circuit between collector and ground during testing.

Individual driver circuits have been designed to trigger thirty flip-flops. In addition to transistor load on the driver, there is a large capacitive load caused by a computer requirement for integrating networks and also by wiring capacitances. Peak current supplied by each driver is 400 ma. Assuming basic pulse width is maintained by the clock generator, additional shorter-duration clock pulses may be obtained by altering reactive time constants of the input networks to the high-frequency switching transistors.

Gating Circuits

Since the same logical signal may be applied to a large number of clock driver circuits, loading of the gating signal by the driver must be limited. Gating should also be done at a low current level to limit load on the gating signal.

The gating circuit used in one application is shown as part of the clock driver in Fig. 3. The clock pulse is amplified by Q_1, and the gate controls the state of Q_2. With Q_2 cut off by the gate, it appears as an essentially open circuit across the output of Q_1. Therefore, the clock pulse is applied, unattenuated, to the base of Q_3.

When the gate saturates Q_2, a low impedance appears across the output of Q_1. If the effective input impedance of Q_3 is high compared to the shunting impedance of Q_2, the clock signal will not exceed cutoff bias of Q_3. The 47-ohm resistor in series with the base of Q_3 keeps the input impedance high compared to the saturated gating transistor.

This circuit provides effective gating action with negligible insertion loss. Typical waveforms are shown in Fig. 4. The upper waveform is an arbitrarily selected gating signal. The center waveform, which is the input signal to Q_3 at the junction of the 47-ohm and 3,300-ohm resistors, illustrates gating action. The lower waveform is the output of the clock driver circuit.

Driver Output

It is difficult for the driver output stage to provide a satisfactory pulse wave shape to the type of dynamic load encountered. An additional complicating factor is the computer requirement for negative-polarity clock pulses, which would normally imply use of npn output transistors. However, the only readily available high-frequency transistors were pnp types.

In one application, transformer coupling was used. For applications with higher clock pulse-repetition rates, transformer coupling became increasingly troublesome. Conventional emitter followers also have limitations. With reactive loads, it is difficult to prevent the emitter follower from being cut off at the end of the pulse. The load and distributed capacitances must then be discharged through the emitted load resistor.

To obtain sufficient transient response, this resistor must be made quite small, which is wasteful of current during the pulse.

An emitter follower circuit that has provided good performance at frequencies approaching one mc is shown in the driver circuit of Fig. 3. The conventional emitter load resistor is replaced by Q_6. This transistor conducts when output transistor Q_5 is cut off.

A positive pulse from the collector of Q_3 applied to the input of Q_6 switches it off. The same pulse, amplified and inverted by Q_4, switches on output transistor Q_5. Since there is one less stage between it and the input signal, Q_6 is switched off slightly before Q_5 is switched on. With the high impedance presented by Q_6, essentially all output current that Q_5 can furnish is provided to the load.

At the end of the pulse, Q_6 is switched on and provides a low-impedance path to discharge the external load and distributed capacitances. As in the generator circuit, the 22-ohm emitter resistor of Q_6 limits steady-state current.

Although this circuit required two transistors, performance is sufficiently better than a conventional emitter follower to warrant its use for high-frequency circuits. Load transistor Q_6 is normally conducting in the absence of a clock pulse. However, average dissipation is negligible, since its return to the power supply is through the normally cut off output transistor, Q_5. Also, since it is normally conducting, Q_6 serves as a low-impedance clamp to maintain output d-c level invariant with respect to changes in repetition rate and load. Typical waveforms are shown in Fig. 5.

Chapter 20
DIGITAL COMPUTER MEMORY CIRCUITS

Digital Recorder Holds Data After Shock

Recorder memorizes instantaneous magnitude of parameters when triggered by a predetermined set of conditions. Data are stored in ferrite cores that can retain the data even after a 6,000-g shock. Subsequent interrogation of the cores releases the stored information for processing

By **CHARLES P. HEDGES,** Santa Barbara Division, Curtiss-Wright Corporation, Santa Barbara, California

MAGNITUDE of any measurable parameter can be recorded quickly and stored indefinitely by a member of a family of recorders termed didmop. This terminology refers to the ability of this type of recorder to record *d*iscontinuously the *i*nstantaneous *d*igital *m*agnitude *o*f *p*arameters.

Magnitudes of parameters are presented in digital form and recorded on command, rather than continuously. When a parameter varies continuously, an analog-to-digital converter is used to supply digital information to the recorder.

Didmop recorders use the coincident current principle to store information in ferrite cores. The magnetic state of a core may be changed by passing half-write current pulses through it in the same direction and time. The information thus stored can be released by pulsing the core to its former magnetic state and reading the resulting pulse delivered by the sense

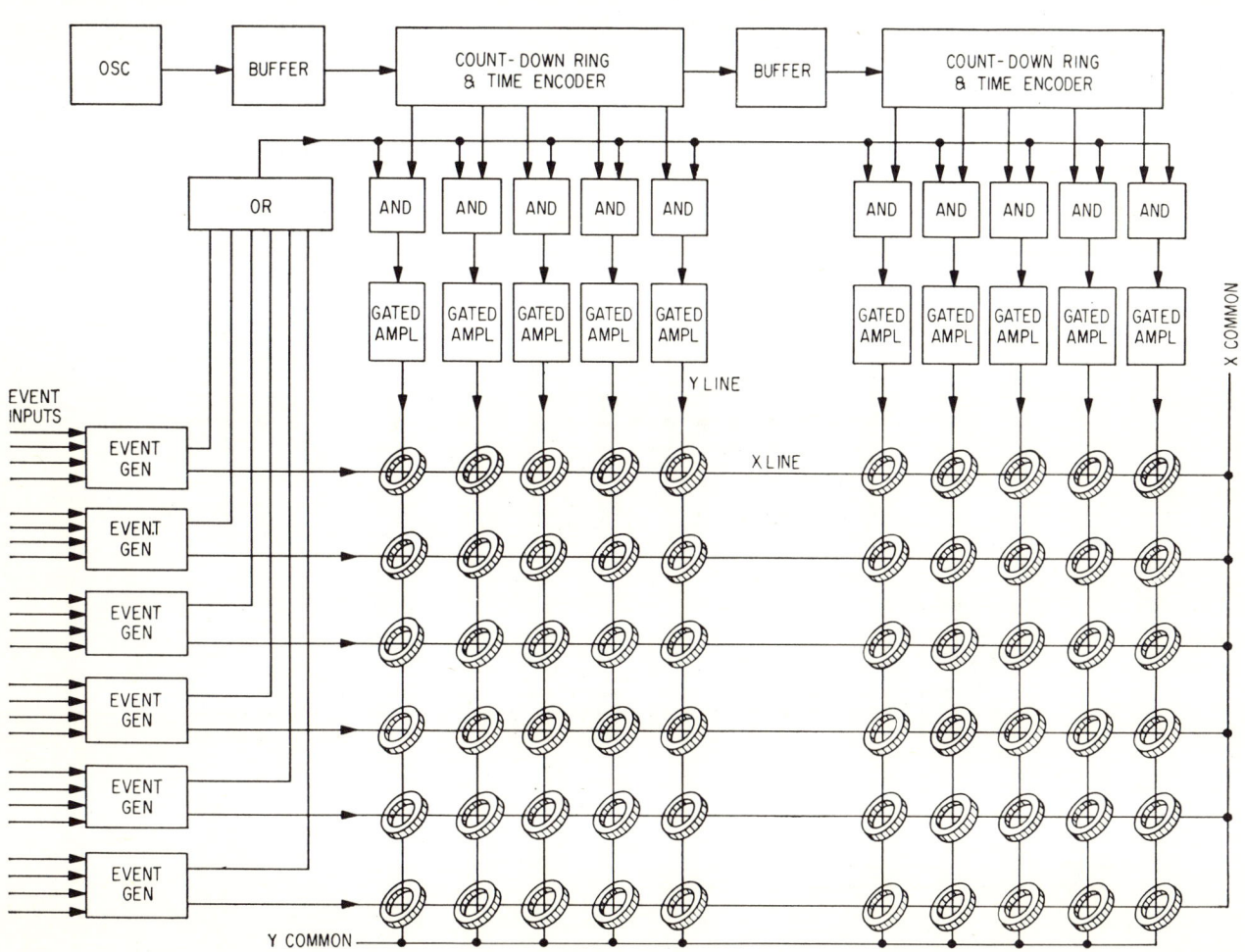

FIG. 1—Time of events recorder. The core windings in each X and Y line of cores are in series. Sense windings are not shown

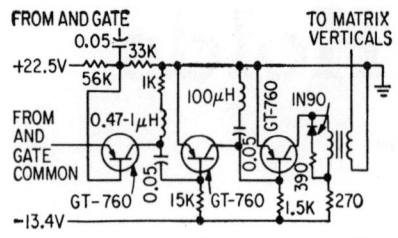

FIG. 2—When pulsed, a gated amplifier drives six toroids

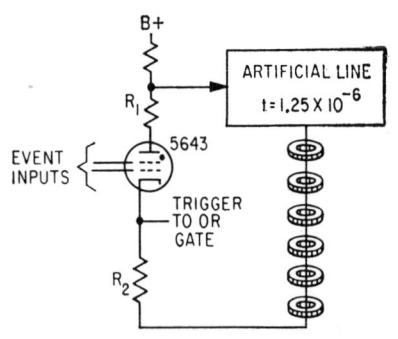

FIG. 3—Only two event inputs to the event generator are indicated. The impedance of the artificial line equals $R_1 + R_2 + e_p/i_p$

The Y inputs of the matrix are at the left, X on top and sense inputs are at right

winding that threads the core.

The time-of-event recorder, shown in Fig. 1, is a member of the didmop family. Before recording, the core matrix is erased by passing a current pulse through the cores which has an opposite polarity to the polarity of the write current. The erase current brings the cores to the ZERO state. To record a digit, a core is pulsed to its ONE state.

The oscillator drives a countdown ring and time encoder. Each output of the encoder supplies one input to an AND gate. The gate opens if a pulse is applied at the same time to its other input.

When all event inputs to an event generator are present, the event generator pulses an X line of tor-

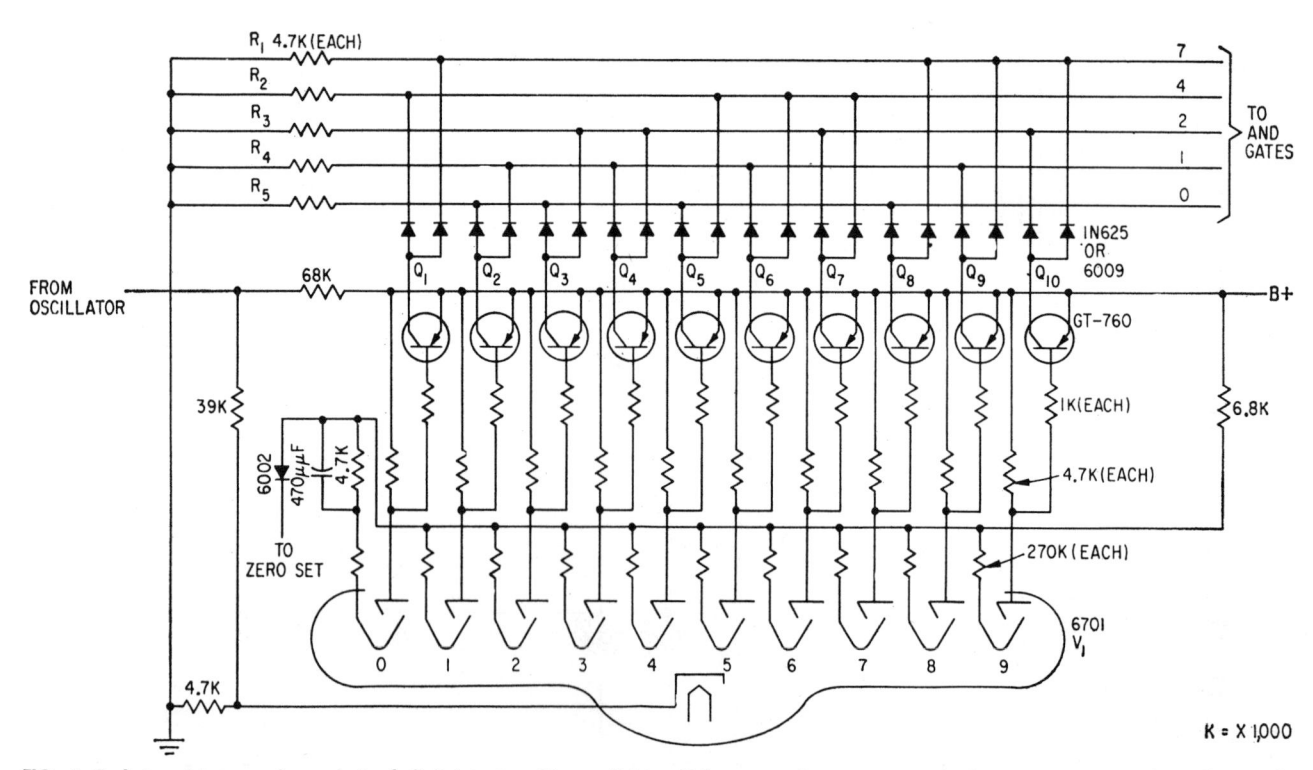

FIG. 4—Each transistor encodes a decimal digit into two binary digits. Voltages at the output terminals represent terminal numbers. The switching grids of V_1 are not shown

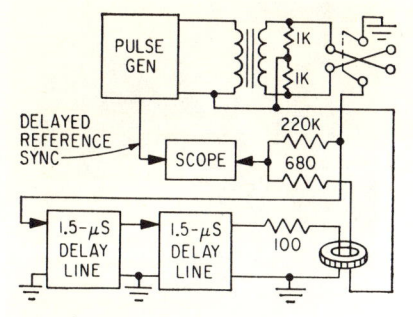

FIG. 5—Read-out circuit. The generator pulses a core with a full-select current. If the core is in the one state, its sense winding pulses the scope

oids, and generates a trigger pulse. The trigger pulse passes through the OR gate to the common input of the AND gates, opening the AND gates that simultaneously receive time encoder pulses.

The AND gate pulses a gated amplifier, Fig. 2, which drives a line of Y toroids. The cores that receive both X and Y half-select currents go to the ONE state. These cores now hold a word that indicates the time corresponding to a specific set of conditions established by the event inputs.

Event Generator

Each event generator, Fig. 3, is a miniaturized version of an artificial-line type of radar modulator.

The event inputs to the generator establish the operating conditions which generate half-current and trigger pulses. These conditions include a suitable cathode temperature, plate supply voltage, screen-grid bias, and an adequate control-grid voltage.

The impedance of the artificial line, resistance of R_1 and R_2, and the supply voltage determine the magnitude of the half-current pulse. The duration of this pulse is 2.5 μsec, twice as long as the minimum core switching time.

Count-Down and Encoding

The oscillator continuously drives tube V_1, a beam-switching decade counter (Fig. 4). Transistors Q_1 to Q_{10} conduct in succession, switching currents through combinations of resistors R_1 to R_5. The voltages at terminals 0, 1, 2, 4 and 7 thus represent one decimal digit.

If the AND gates that receive time voltages from the time encoder are also pulsed by an event generator, they open and pulse their gated amplifiers. The cores that receive half-currents from both the gated amplifiers and an event generator change state and now store the time of an event.

Oscilloscope equipment is used to obtain photographic records of the data. The read-out circuit, Fig. 5, consists of a low impedance pulse generator, an artificial delay line, a core sensing lead and an oscilloscope. The pulse generator triggers the oscilloscope and interrogates a core with a select pulse. An oscilloscope sweep first shows the amplitude and polarity of the core drive signal and then the presence or absence of core response.

Conventional ferrite memory matrices use one sensing wire which threads all of the cores in a plane. Thus the minor changes of core flux produced by half-select currents contribute noise to the sense winding. The time-of-event recorder matrix uses as many sense windings as time digits. Thus the noise produced by other cores in a row that is pulsed does not add noise to the core-turnover-signal sought.

Application

Figure 6 shows a design of a didmop recorder. With auxiliary circuits, this recorder could be used to obtain a record of an airplane's operation during a midair collision. Flight data such as throttle settings, manifold pressure, engine rpm, flap and landing gear positions, aircraft elevation, etc., would be monitored continuously. When various collision phenomena are sensed such as the closing of an impact switch or the opening of a conductor along the skin, logical gating and selecting circuits would then trigger a recording. With careful packaging, the information stored in the cores can and has been recovered after a 6,000-g shock.

The author acknowledges the contributions of P. M. Cruse, V. Alvarez, W. Johnson, K. Wasserman, and R. L. Williams and the cooperation of P. Escher, A. P. Bridges and D. L. Baker.

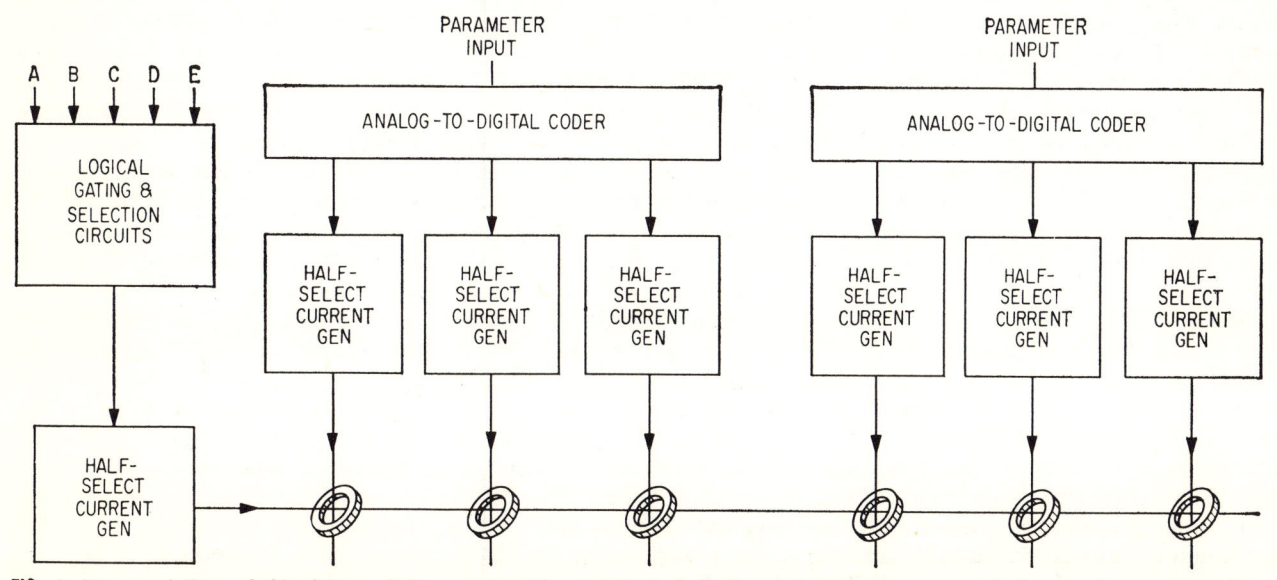

FIG. 6—Many variations of this didmop design are possible. Inputs A, B, C, D and E can be events, magnitudes or other parameters

Coincident-current technique for digital data buffer memories permits compact construction. Capacity can be increased by adding plug-in packages, and systems can provide random and/or sequential access

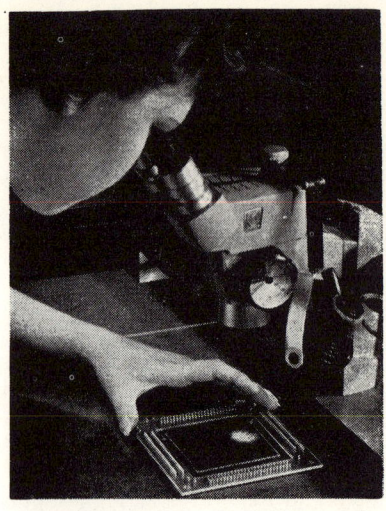

THE FRONT COVER. Production worker examines memory plane using ferrite memory cores for possible flaws

By **DANIEL HAAGENS**,
Engineering Manager, General Ceramics Corp., Keasby, N. J.

Buffer memory in foreground is checked out by engineer. Memory is a 144-character, 4-bit, coincident-current system

Compact Memories Have

COMMON REQUIREMENT of data-handling systems is temporary storage of information transmitted at one rate and processed or received at a different rate. Situations in which this requirement must be met are magnetic tape input to digital computers, transmission of digital data over communications systems and manual keyboard input to digital systems. It is also required for data conversion from one medium, such as punched cards, paper tape and magnetic tape, to another.

Small to medium capacity random-access memories for normal digital data operations are also needed.

Buffering of digital data may be accomplished by such means as magnetic shift registers, small magnetic drums or by a portion of a larger memory if the system has one. The problem can also be solved with coincident-current memory techniques such as described in this article.

The coincident-current approach makes possible unusual flexibility in equipment design, is often more economical and makes possible much smaller memories. These advantages result from the asynchronous nature of coincident-current operation and the rapid access it affords.

Memory System

Modular design of the new Ferramic core coincident-current memory to be described permits storage capacities up to 8,192 bits, with sequential or random access. Transistors are used throughout, and the memory is provided with drive and sense circuits as well as address and output registers. Stable square-loop characteristics and conservative circuit design allow operation over an ambient temperature range from zero to 50 C without external forced cooling. The relay-rack mounted unit is 14 in. deep and has a panel height of 5¼ in. It is less than half the size of comparable systems (about 0.7 cu ft). Density rate is about 2,400 components/cu ft.

The unit is designed and packaged so that the same basic circuit elements could comprise either random or sequential access systems. A wide range of storage capacity was also considered essential, so it was designed to permit convenient expansion of capacity by using a limited number of standard etched circuit cards.

The first unit designed has a storage capacity of 144 characters of four binary digits each. Specifications called for independent loading and unloading operations and a sequential mode with no limitation on the manner in which the two operations were ordered. The combined rate of operations was to be no less than 100,000 cps, allowing 10 μsec/operation. One additional control input permits resetting the address registers of the memory to the initial address within 10 μsec. Another clears the entire memory to

Flexible Capacities

binary zero within 75 μsec.

External control signals set the load and unload flip-flops in the timing generator in Fig. 1, as well as reset and clear one-shot multivibrators, also part of the timing generator. Input information signals actuate the inhibit drivers, one for each memory array. All four arrays are on one memory plane.

Timing Signals

Amplified timing generator signals control timing pulses for the sense amplifiers, the inhibit drivers, the X and Y load and unload address counters, the output register and a clear pulse for the memory planes. The X and Y load and unload address counters provide conditioning levels to the coincidence gates, which are incorporated in the drive circuits for both X and Y loading and unloading. These circuits in turn are coupled to the X and Y transformer matrices, which sequentially drive the X and Y lines of the memory planes with the load and unload current pulses.

The inhibit drive pulse is gated with the input information signals at the inhibit drivers, determining the information pattern to be stored in the selected memory addresses. The stored information is read out during the unload period and amplified by sense amplifiers that are turned on by the unload sense output during the unload cycle. The sense amplifier outputs actuate the flip-flops of the output register, which provide output information levels at the output connector.

Selection System

To simplify the X and Y line selection and drive circuits and to reduce currents in the drive transistors, pulse transformers are used to supply the drive lines with the selection currents on both load and unload cycles.

In the 144-character memory, the memory cores are arranged in four 9 by 16 arrays, with the X and Y drive lines of each connected in series. For example, in the X side, 16 transformers drive the 16 lines. The transformers are arranged in a four by four square array as shown in Fig. 2. Each transformer has one output winding and two input windings. One input is used for load drive, the other for unload drive. The transformers are isolated from each other by diodes and are connected to permit independent selection on load or unload.

Turns ratio of the transformers is 5:1 between either primary and the secondary. Drive current on each X line is 150 ma. The driving transistor therefore must supply 30 ma to the primary. Resistors R_1 and R_2 are selected so voltage at the emitter of Q_1 is slightly negative when the unload drive pulse is absent. Since the bases of transistors Q_2 are clamped to ground, all current passing through R_2 goes through Q_1. When the drive pulse occurs, Q_1 is cut off and the emitters of Q_1 and all Q_2 transistors rise until a Q_2 transistor conducts.

The correct Q_2 was selected by removing the positive input current from its base resistor R_3. Current from R_2 is transferred to the selected X unload drive transformer line. The value of R_3 keeps diode D_1 from being cut off. The base of Q_2 is therefore kept near ground potential and acts as a constant-current source. One of the unload drive return transistors, Q_3, has been selected in a similar manner. As a result, only one transformer has current through its primary. The Y selection is accomplished in a similar manner.

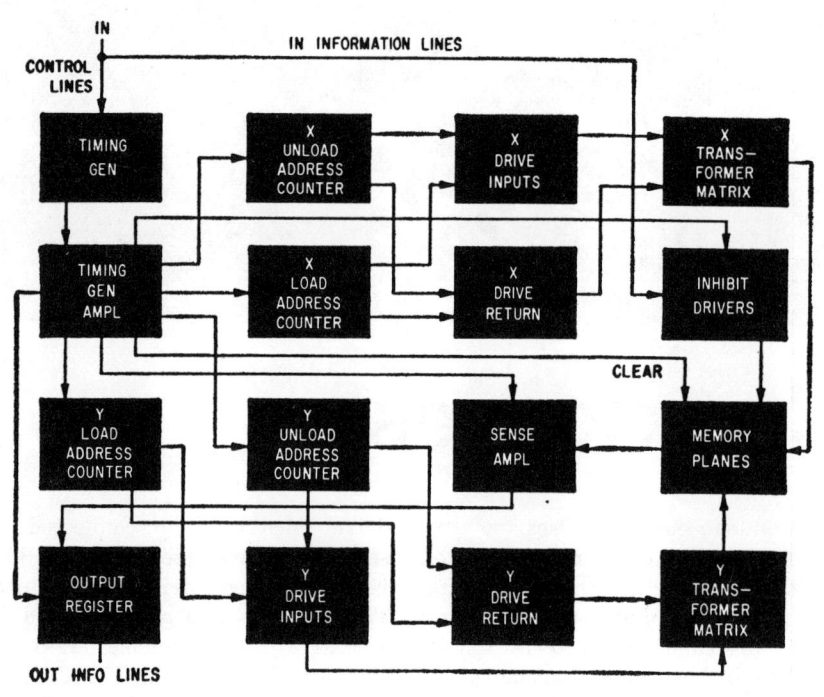

FIG. 1—Input information signals actuate inhibit drivers for each of four memory arrays

This method of drive-line selection permits expansion by varying number of transformer plug-ins, input drive-circuit and return drive-circuit cards. If the address is always the same for load and unload, transformer returns can be connected in common for load and unload, cutting the number of return drive-circuit transistors in half.

Since input current requirements of Q_2 and Q_3 are low, the address decoder inputs can be supplied from diode gates driven directly from the flip-flops in the address registers.

The memory planes are mounted on a plug-in package and have the sense and inhibit windings brought out in the following manner: Each array of 9 by 16 cores has a sense winding brought to two terminals through a coaxial lead and then connected in the same manner to the sense amplifier input. The four inhibit windings are connected in series with their junction points brought out (five leads from the memory plane for the four inhibit windings). Since output of the inhibit drivers is derived from the secondary of a transformer, no ground return is necessary on the inhibit windings, and the inhibit drivers and inhibit windings can be connected in series. This arrangement permits clearing all memory cores with a single long pulse through all the series inhibit windings, and no cycling of the memory address counters is necessary.

Flip-Flops

The transistorized flip-flops in the memory are stabilized against saturation, so that wide variations in h_{fe} can be tolerated.

The one-shot multivibrators are similarly stabilized and they are a variation of the basic bistable circuit.

All flip-flops are arranged to swing between +2 and −8 v, so that it is possible to connect gating chains to the flip-flop outputs maintaining a positive bias on succeeding amplifiers.

In the sense amplifier in Fig. 3, Q_1 is a drift transistor with low output capacitance. Transistor Q_1 amplifies linearly the 20-mv sense output signal from the memory planes. Output of Q_1 is coupled to Q_2 through a pulse transformer, with impedance matched to the input of Q_2.

A full-wave rectifier is necessary at the secondary of the transformer because sense output can be of either polarity. An enabling signal turns on the sense amplifier to permit it to discriminate between an output from the memory cores during an unload and a load cycle. Transistor Q_2 supplies sufficient power to operate a flip-flop in the output register.

Inhibit Driver

The inhibit driver circuit in Fig. 4 uses the same basic approach as used in the load and unload selection circuits. In the quiescent state, Q_2 conducts and its emitter is at a negative potential just below ground. The emitters of Q_1 and Q_2 are tied together, and the base of Q_1 is grounded. Therefore, Q_1 remains cut off until Q_2 is cut off. Quiescent current through Q_2 of about 15 ma flows through one of the primaries of the pulse transformer. With an inhibit signal and a binary ZERO on the data input lead, the base of Q_2 becomes sufficiently positive to cut it off. Emitter voltage rises until the common-based transistor Q_1 starts to conduct, causing current to flow through the other transformer primary. Therefore, the transformer field is reversed.

Secondary transformer current is ten times quiescent primary current. Transistor Q_3 is connected to the output of the transformer secondary and acts as a diode with very low forward drop, allowing the inhibit output to be disconnected from any transformer secondary recovery currents that might otherwise occur during high duty cycles. The input circuit to Q_2 has constant base drive for all conditions when a binary ONE data input is applied and no inhibit action is desired. A 3,600-ohm resistor between ground and the collector of

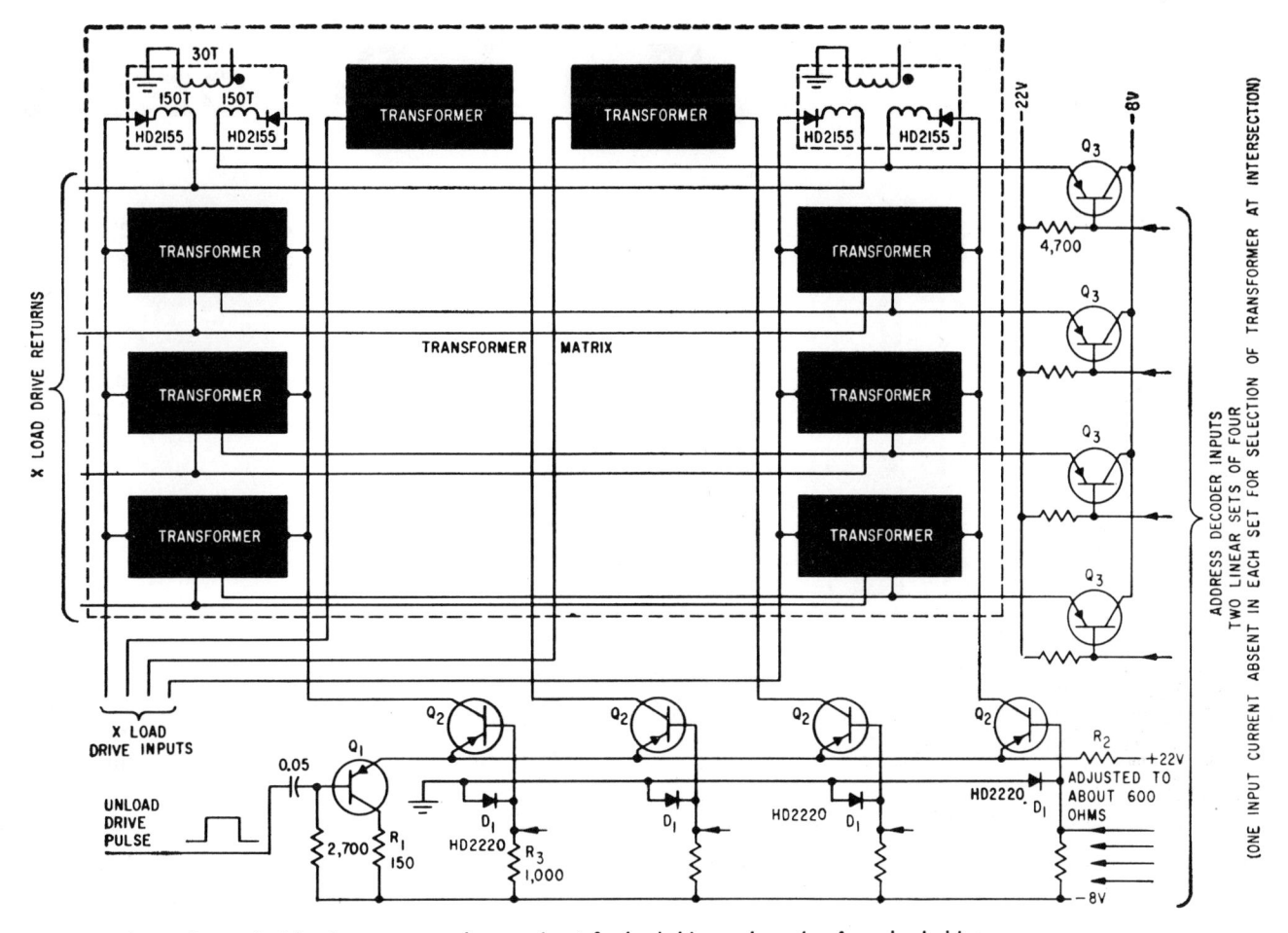

FIG. 2—Each transformer in 4 by 4 square array has one input for load drive and another for unload drive

Q_1 damps excessive transients at the end of the inhibit cycle. Current through Q_2 is adjusted to proper value at the factory.

The transistors in all these circuits are computer-type transistors with carefully controlled cut-off currents held below 6 μamp. All circuits are stabilized for I_{co}.

Using the general configuration of this memory, by adding plug-in packages containing flip-flops, drive circuits, sense amplifiers and inhibit drivers, capacity both in number of words stored and in number of binary digits in each word can be easily expanded or contracted. Random-access memories of similar capacities can readily be constructed. Address registers in such memories would not be coupled together as counters. In general in a random-access memory only one address register is necessary, rather than separate ones for load and unload, since the address for loading and unloading is the same.

Other Designs

Other designs have been constructed including a memory of 32 characters with four bits to the character and a memory of 144 characters with 8 bits to the character. Memories with other capacities are presently reaching the prototype stage.

The sequential-access memory using individual registers for load and unload can replace shift registers in some applications.

It is possible to arrange this type of memory to be reversible in the load or unload mode. This arrangement permits reversal of the time ordering of information in data transmission and processing devices. Combined random and sequential access can be provided.

FIG. 3—Sense amplifier requires full-wave rectifier at input to second stage

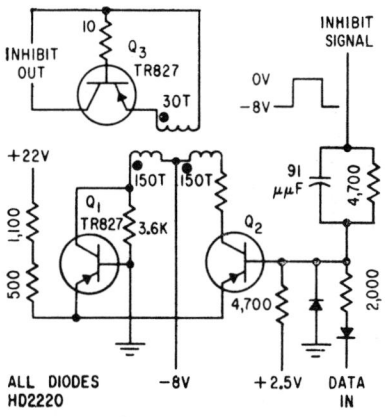

FIG. 4—Inhibit driver uses transistors with carefully controlled cutoff currents

Clock Track Recorder

Card assembly containing two flip-flops

Clock track recorder with the magnetic drum. Complete electronics with control panel are housed in the chassis at left

MAGNETIC DRUMS have found widespread use in data-processing systems as memory storage units. Drums are reliable and can store a great deal of information economically.

The magnetic drum can also serve to generate the basic timing signals for the entire system. At Hughes, a transistorized clock track recorder was developed that can write, automatically, any preselected number of timing bits on one channel of the drum. The recording heads are spaced in proper position around the drum periphery. A timing track on the drum channel synchronizes all the computational cycles and information flow to and from the memory. Signals are then distributed to the rest of the computing system.

Total power consumption of the unit is about 15 watts.

Timing Signals

A block diagram of the transistorized clock track recorder is shown in Fig. 1. The central data source, module 9,999 counter, supplies all of the necessary timing signals used in both head setting and clock track recording. Four module 10 counters, connected in series, produce the total count of 9,999. Standard transistorized two-input flip-flops are used as the basic element, and diode gating accomplishes the proper timing. Since the counter is divided into four identical counters, the logical equations are the same for each, with the exception of an additional AND circuit to gate the following tens counter.

The counter can be reset in two ways: by an *RZ* origin pulse; and by the counter's own final output. The origin pulse is recorded on the drum by passing a transient current through the head while the drum is at a standstill. This records one *RZ* pulse in one spot on the drum.

The origin pulse is fed into the *OP* amplifier. The pulse is transformer-coupled to an a-c grounded-emitter stage. This, in turn, is transformer-coupled to a second a-c grounded-emitter stage. The output stage of the amplifier is transformer-coupled, but the base return of the transformer is returned to a bias point which is 2 volts more negative than the emitter return voltage. This puts the transistor in a cutoff stage until the signal overrides the bias and allows this stage to conduct.

The one-shot multivibrator supplies a gating voltage for the reset flip-flop. The multivibrator is a modified transistorized Eccles-Jordan flip-flop.

Counter Output

The counter output sets up a gating voltage for the reset flip-flop, provides a scope sync and drives the *KK* write flip-flop. The writing amplifier is controlled by a switch, Write MOP-Clock Track. In the MOP position, the output of the *KK* flip-flop is connected to the input of the writing amplifier. In the clock-track position, the write amplifier writes all "1's" on the drum.

The clock-pulse amplifier and digit square-wave amplifier, Fig. 2, serves two functions: it generates clock pulses which drive all of the flip-flops in the system; and it generates digit square wave signals which are used only in the Manchester conversion network of the writing amplifier.

The input of the clock-pulse amplifier and digit square-wave generator comes from either of two sources: an external signal gen-

358

For Memory Drum

All-transistorized unit facilitates writing of timing signals on a magnetic drum. Recording heads, properly spaced around periphery of drum, automatically write any preselected number of timing bits on one channel of the drum, and synchronize all of the memory information

By **A. J. STRASSMAN** Member of Technical Staff, Hughes Aircraft, Los Angeles
and **R. E. KEETER,** Member of Technical Staff, Aeroneutronics, Newport Beach, Calif.

erator provides the desired frequency for writing the time track; and timing signals, previously recorded on the drum, set up the bit spacing between the read and write heads.

The read amplifier and read flip-flop unit is used in conjunction with head space setting of the read and write heads. The input to the read amplifier comes from a reading head on the drum. The read-amplifier output goes to the read flip-flop and can be viewed with an oscilloscope.

Flip-Flops

The JK flip-flop, Fig. 3, is a transistorized Eccles-Jordan switch, with the collectors clamped with diodes to stabilize the operating points. Grounded collector transistors, driven from the switches, provide the current for driving voltage stage gating circuits.

The $K\bar{K}$ flip-flop is really divided into a normal JK flip-flop with increased input capacitance at J, and a pulse-steering network. This network is driven by an npn emitter follower. The normal K-side clock pulse is AND'ed with the emitter follower output and the normal J side pulse is introduced through a 1,200 $\mu\mu f$ capacitor. The pulse reference on the J side becomes essentially the input voltage level and is pulsed only if this level is at -6 v. At the zero-volt level, J pulsing is

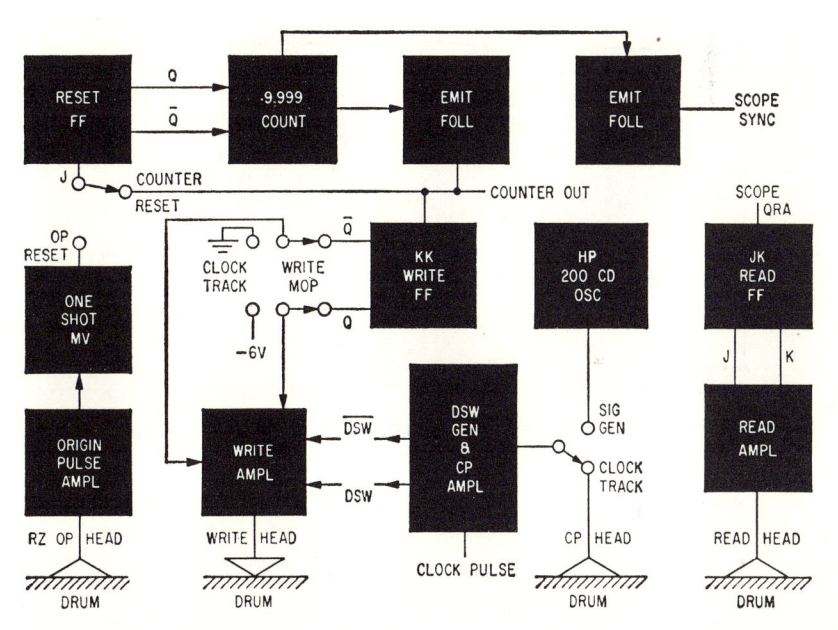

FIG. 1—Block diagram of clock track recorder

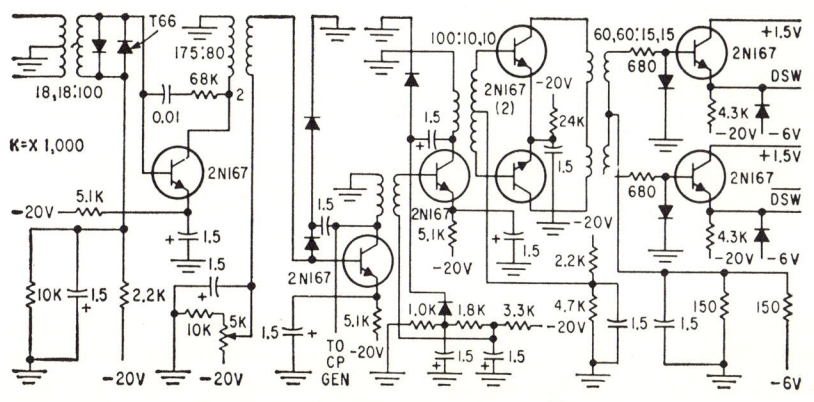

FIG. 2—Clock pulse amplifier and digit square-wave amplifier

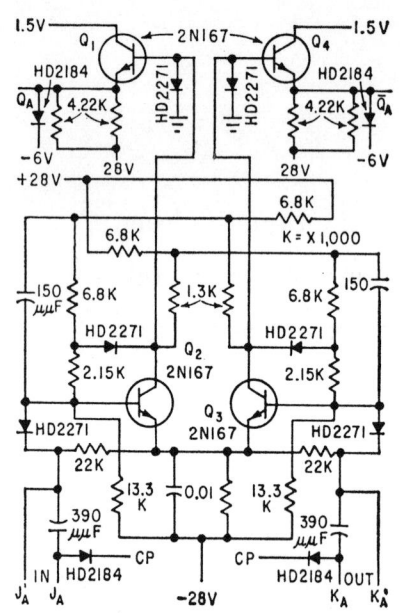

FIG. 3—The flip-flop circuit used with the read amplifier

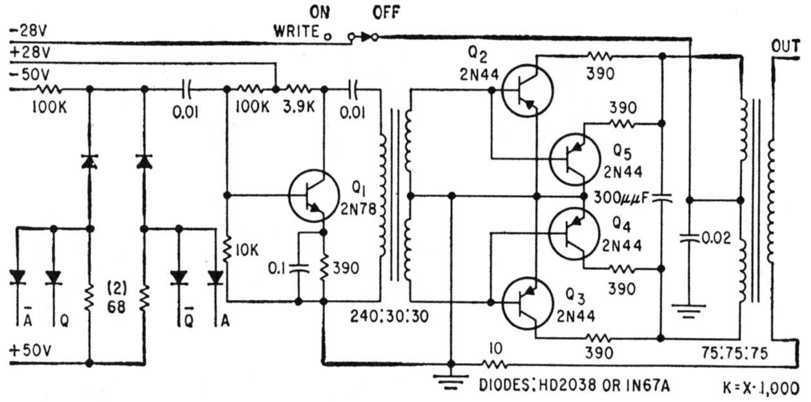

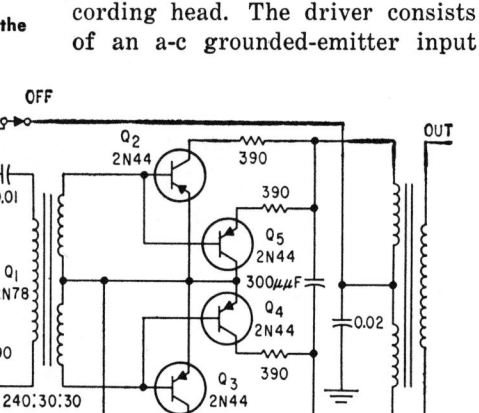

FIG. 4—Write amplifier contains a power amplifier and an impedance-changing device

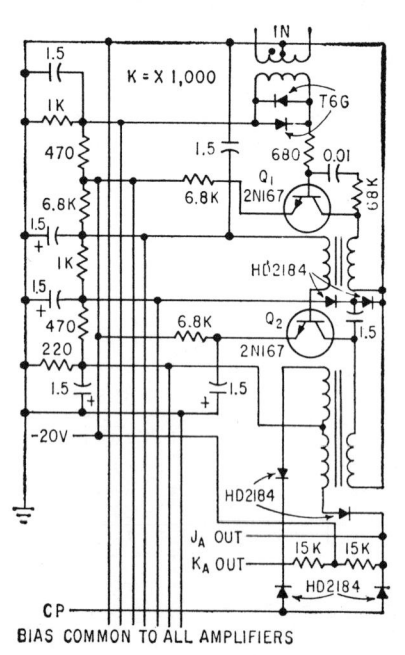

FIG. 5—Read amplifier for the reading head also used to synchronize

prevented by a clamp decoupling diodes.

Write Amplifier

The digit square wave generator receives the square-wave output of the read amplifier and produces two 0 (−6) volt square waves, *DSW* and *DSW*, phase displaced by 180 deg with respect to each other. These waves are utilized in the gating of the write amplifier for the Manchester conversion.

The write amplifier, Fig. 4, contains a power amplifier and impedance-changing device which is arranged to convert the voltage waveform at the output of the flip-flop into a corresponding current waveform in the low-impedance recording head. The driver consists of an a-c grounded-emitter input stage driven through a capacitor by the output stage through a 4 to 1 center-tapped step-down transformer.

The output stage of the write amplifier consists of a push-pull parallel Class B amplifier, using *pnp* transistors. The square-wave signal furnished by the driver's stage is transformer-coupled into the bases of the output stage and serves to switch the two sets of the output transistors alternately on and off, to provide a current square wave in the output transformer.

The write amplifier supplies the proper current waveform for Manchester recording to the write head. A 220 ma peak-to-peak current is required by the write head winding with the head-to-drum spacing of 0.001 inches.

The read amplifier, Fig. 5, is used to amplify and synchronize. Amplification is necessary because of the low read-back voltage achieved with the low-impedance reading head. This voltage is primarily a function of the head-to-drum spacing although it varies inversely with magnetic cell density. The minimum read-back voltage that will be permitted is 0.06 volts, for independence of cross talk and other noise.

The signals from the amplifier output stage are essentially square waves, OR'ed with clock pulses, the outputs of the OR gates are applied to the inputs of the read flip-flop, and trigger it so that the information finally appears as voltage states of QRA and $\overline{QRA}$. These two output states go to the oscilloscope shown in Fig. 1.

The wave form of the read-back signal for the Manchester system of magnetic recording may be considered as a phase-modulated sine wave that is step modulated. Therefore some means of phase detection must recover the stored information.

There are only two frequencies present in the signal received from the reading head—a sine wave of the bit frequency when the reading signal changes from one to zero, or vice versa. This permits use of a bandpass amplifier that must have little more than a two-to-one frequency pass band.

Voltage Feedback

The amplifier portion of the read amplifier has an a-c grounded emitter input stage transformer coupled to a second grounded emitter stage. The first stage uses voltage feedback through the series 0.01 μf, 68-K resistor to increase the band width of the stage. Voltage clamps in the second stage protect the transistor against overswing as well as provide a plateau for the gating.

A 2.5:1:1 stepdown transformer couples the second amplifier stage to the OR gates.

The two diodes connected back-to-back to the secondary of the input transformer produce a symmetrical limiter to insure reliable operation over a wide range of amplitudes of the input signal.

Chapter 21
DATA PROCESSING INPUT-OUTPUT CIRCUITS

Office setup suggests one practical application for monoscope character generator which has been developed at Stanford Research Institute

Character Generator for Digital Computers

Speed with which machine-to-person information links can process information has lagged behind computer capabilities. This monoscope tube, operating in the speed range of computers and memory devices, provides readout directly on a cro or on paper

By **EARLE D. JONES**, Research Engineer, Stanford Research Institute, Menlo Park, California

A COMPACT AND economical monoscope tube apparatus which generates well-formed alphanumeric characters for display at high speeds is described in this article. In this apparatus, the character generation function is completely separated from the character display, thus providing a degree of flexibility not available in some other methods. With repetitive digital information, for instance, an ordinary laboratory oscilloscope may be used to display the characters.

This monoscope character generator may also be used to drive certain types of hard copy printers, either by optical transfer of visually presented information or directly for special types of electrostatic printing tubes.

Input to this monoscope is derived from conventional six-wire parallel binary information. A standardizer at the input supplies uniform and consistent electrical pulses to the generator, permitting the device to be driven by a wide variety of sources.

Sources of Information

Excellent results have been obtained from two sources of input information. First, a 36-position ring counter employing beam switching tubes has synthesized the necessary digital information on six wires. The beam switching tube outputs were sampled off through diodes to six wires.

The device has also worked from information stored in digital form on a six-track magnetic tape. A 360-character message was encoded on the tape and read off at a tape speed of 60 inches per second to give a character rate of approximately 10,000 per second, which is in the speed range of computer and memory devices. (With minimal changes, this rate could be increased to 20,000.) Equally satisfactory performance can be expected from an information source such as a magnetic drum, a paper tape or an electric typewriter.

Information can be displayed on commercially available equipment. If the input information is repeated faster than 20 times per second, the characters may be displayed, without flickering, on any conventional cathode ray tube device. For non-repeating or low-repetition rate information, a direct view storage tube is suitable. Portions of a message can be selected and displayed as long as desired on a tube face. Otherwise, the message can run in its entirety for the observer and can be stored if desired.

Character quality is limited only by the raster frequencies (and therefore, video bandwidth) and

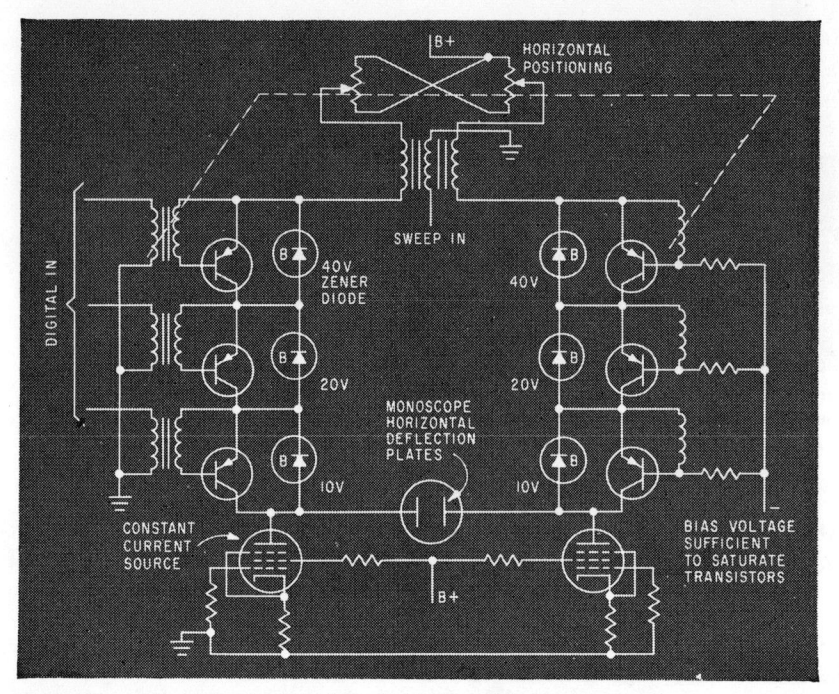

FIG. 1—Simplified schematic of horizontal decoding and deflection circuit

the finite spot size of the monoscope tube. Present techniques utilize about 400 picture elements in the small raster to cover a character and any guard space around it. For a 10 kc character rate (one character each 100 microseconds), about four megacycles of video bandwidth is required. To operate at faster character rates, the switching transistors should be faster and a higher frequency raster should be employed, resulting in wider video bandwidth requirements. Character quality is superior to character display systems utilizing a simple dot matrix for character writing.

Monoscope Construction

Video waveforms necessary to write characters are generated in a monoscope tube—a fundamental cathode ray device. The tube comprises a conventional electron gun and an aluminum target enclosed in a vacuum envelope. All of the desired characters and symbols are on the metal target in printer's ink. A six-wire system makes up to 64 different characters and symbols available, including one for spaces. These are arranged in eight rows of eight characters each on the target.

Preparation of the monoscope target is by common photoengraving and printing techniques. This enables complete freedom in the selection of the style of type, symbols or simple pictures. In the preparation of the target the characters can be located in the matrix in order to conform to six-unit binary codes which vary with machines. Type font or character-selection code can be changed by replacing the monoscope tube (a $100 to $200 item).

Monoscope Operation

As a constant-intensity electron beam is scanned across a character, the secondary emission curent from the target is modulated; aluminum and carbon, the chief ingredient of the ink, exhibit different secondary emission coefficients. The resulting video signal is amplified and used to intensity-modulate the display tube.

The monoscope beam is scanned in a television-like raster which covers only one character on the matrix. The display device is swept in synchronism with the monoscope and modulated with the video signal; thus, the character appears on the face of the tube. The character is positioned to its proper sequence in a word by deflecting the display device.

A character is selected by positioning the small raster to a specified location on the 8 by 8 character matrix. This requires an accurate horizontal and vertical deflection of the beam. This is accomplished in the course of deriving horizontal and vertical deflection voltages from the digital input information. The vertical component of the raster is a 200 kc sinusoidal or sawtooth voltage while the horizontal component is a once-per-character (10-kc) sawtooth. To select a character from an 8 by 8 matrix, eight distinct voltage levels are needed, both in the horizontal and vertical deflection circuitry.

Input information in digital form must be decoded in order to be read as alphanumeric characters. The six-track digital input can be thought of as two three-track sources. From one three-track binary arrangement, eight distinct codes can be derived.

Input information channels 1, 2, and 3 therefore may be decoded to determine one of eight horizontal positions merely by adding binary pulses whose amplitudes are in the ratio 1:2:4. Channels 4, 5, and 6 may be decoded similarly to determine one of eight vertical positions.

The instability and drift nor-

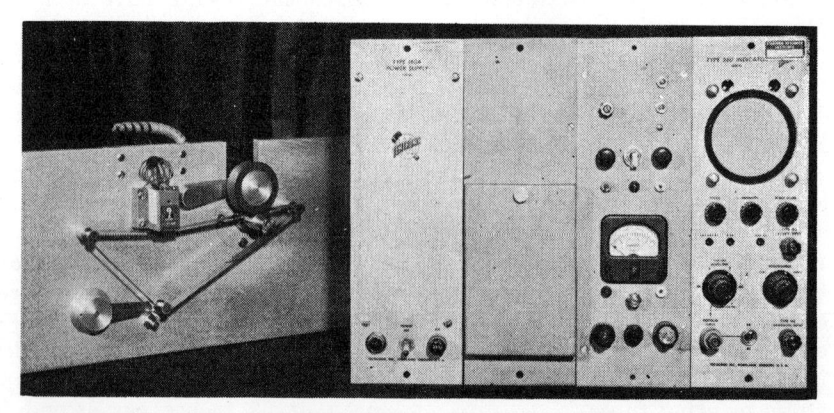

Small loop tape recorder (left) drives system. From left, units shown are: power supply; monoscope chassis and decoding and deflection circuitry; companion unit, providing sweep circuits for both monoscope and display scope, and pulse standardization equipment; and small monitoring display scope

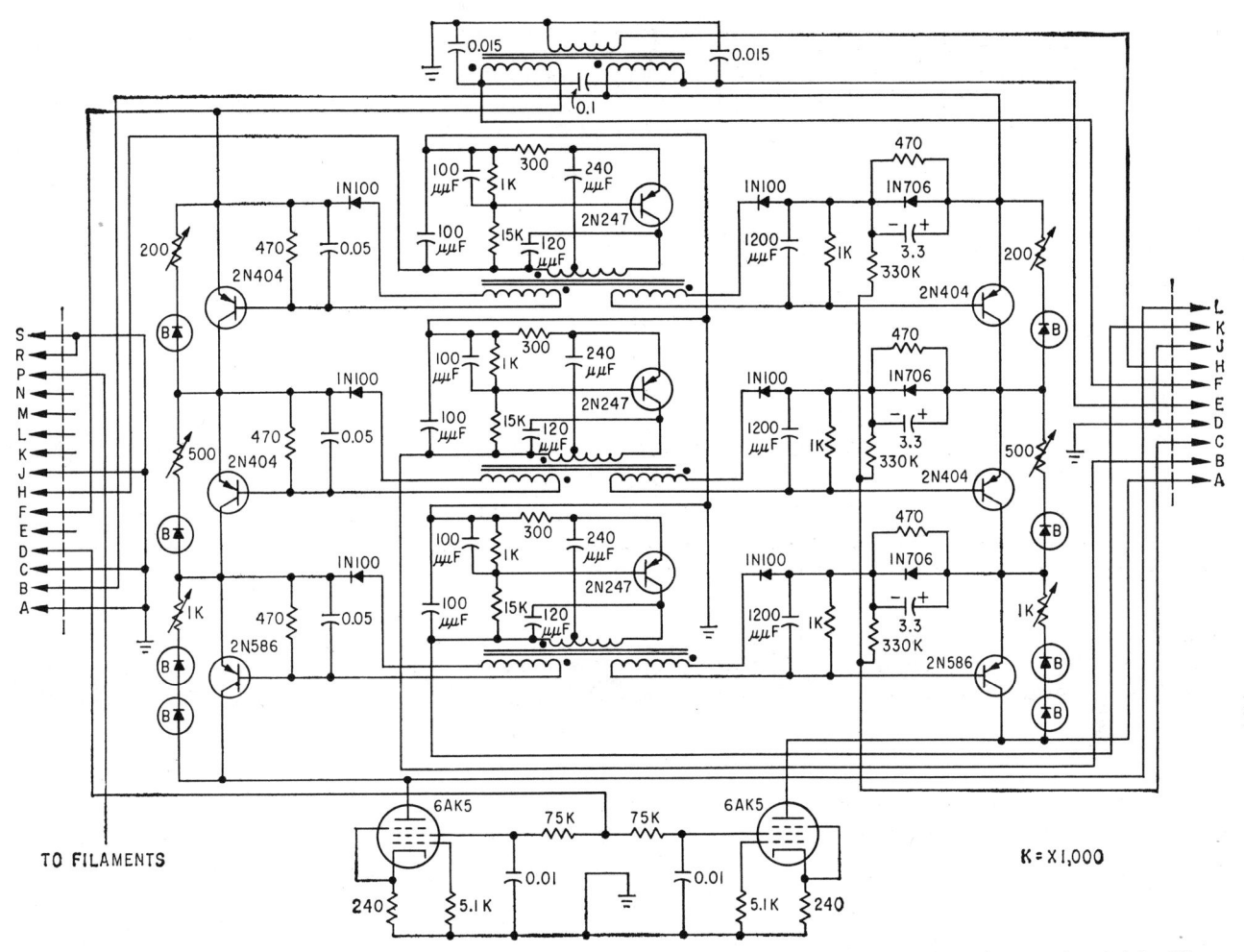

FIG. 2—Complete schematic of horizontal decoding and deflection circuit. Vertical deflection circuitry is identical except for slightly different Zener voltages of the voltage-reference diodes

mally associated with cathode ray tube devices is overcome by a simple but highly effective combination decoding and deflection circuit (see Fig. 1 and 2). The voltages applied to the four monoscope deflection plates are developed across voltage-reference diodes which are driven from constant current sources. This arrangement provides stable d-c levels without amplification.

A series string of three voltage-reference diodes is connected to each of the four monoscope deflection plates (see Fig. 1). The three diodes have breakdown voltages in the ratio 1:2:4. A transistor is connected across each diode and is biased so the constant current may be shunted around the diode or through it, depending on whether or not the transistor is saturated.

The transistors associated with one deflection plate are all normally biased OFF while the transistors associated with the other plate are biased to saturation. In Fig. 1, the right deflection plate will be B+ potential (less the saturation voltage of the three transistors). Because the transistors are biased OFF and the current is through the diodes, the left plate is 70 volts below B+ potential. The transformers which couple incoming pulses to the transistors are polarized to turn off the saturated transistor and saturate the open transistor. The voltage across either diode string can vary from zero to 70 volts in 10-volt increments. As voltage excursions on each deflection plate are equal in amplitude and opposite in polarity, true push-pull deflection results. Thus eight levels of horizontal deflection are established by eight input pulse combinations.

Vertical deflection circuitry is identical to the horizontal except for slightly different Zener voltages of the voltage reference diodes. A higher voltage is required in the vertical direction because of a slightly lower deflection sensitivity.

With the eight voltage levels on each deflection axis, 64 characters are defined by the binary information on the six-wire input.

Results

Excellent stability of deflection has been obtained and is due to three prime factors: the voltage-reference diode is a constant-voltage device with a low incremental impedance; the voltage-reference diodes are driven from constant current sources; and, series-string voltage-reference diodes are used with breakdown voltages near the zero-temperature coefficient point of operation, reducing temperature sensitivity.

The system was developed by W. E. Evans, L. J. Kabell, the author and other members of the Video Systems Laboratory of Stanford Research Institute for the A. B. Dick Company of Chicago as a component of that firm's Videograph equipment.

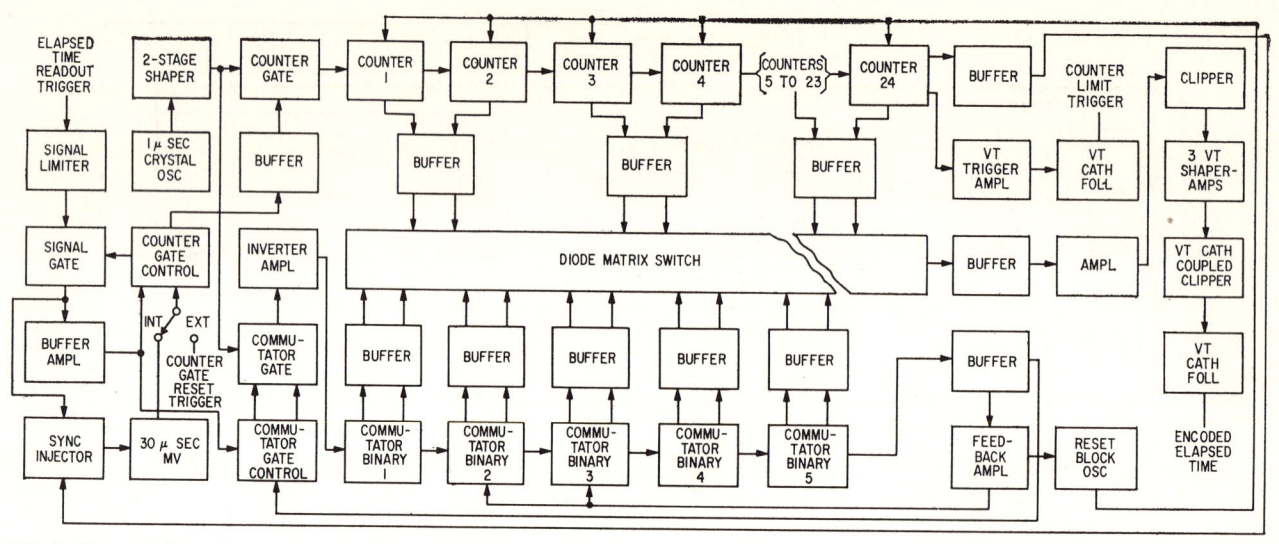

FIG. 1—Size, weight and power consumption of encoder are minimized by using semiconductors in all circuits not identified as VT (vacuum tube). Device will operate reliably over a temperature of −40 to +55 C

Encoder Measures Random-

ANALYSIS AND CONTROL of pulse period jitter in pulse-time and pulse-code modulation communication systems are simplified using the high-resolution random event encoder to be described. The device can also be used to form statistical distributions as in the investigation of variations in period between counts produced by a scintillation counter used for nuclear radiation studies.

A block diagram of the encoder is shown in Fig. 1. A one-mc oscillator is used as a time reference. Output of the oscillator is shaped and used to drive a 24-stage binary counter which stores elapsed time.

When a signal appears at the input indicating the end of the desired storage period, the output of the oscillator is switched from the counter to a commutator circuit. The commutator together with a diode matrix reads out the state of the counters in serial form. Upon completion of readout, a reset pulse is applied to each counter stage returning it to the original state.

Reset

Counter reset technique permits the time required for readout to remain as a stored count. Elapsed time, therefore, is actually the time from the start of the previous readout to the start of the present readout. Thus, the time required for readout (approximately 30 μsec) is not subtracted from the time between the events being studied.

If the period timed is longer than the capacity of the counters (approximately 16 sec), a trigger pulse is generated which initiates the readout process when the limit of the counter is reached. In this instance, only an index pulse is read out. Storage of elapsed time is resumed after the counters have been reset to the time required for this readout.

Output of the diode matrix switch is amplified and shaped for use in intensity modulating the horizontal sweep of a crt. Resulting patterns are then photographed. For a recording of approximately 100 elapsed time readouts, the film is held stationary while successive horizontal sweeps are positioned vertically by a direct-coupled step generator.

Record Analysis

A record made of the jitter in a pulse train generated by a pulse

FIG. 2—Encoded elapsed time record on single-frame 35-mm photograph

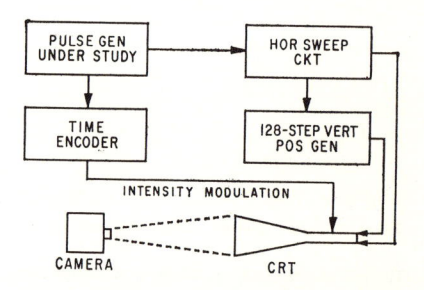

FIG. 3—Circuit used to produce elapsed time record shown in Fig. 2

366

Transistorized encoder stores and reads out elapsed time between consecutive but randomly occurring events. One-mc oscillator triggers 24 elapsed time counters until end of storage period. Oscillator is then switched to an electronic commutator controlling a diode matrix switch. Counter data is read out serially through the switch, converted to traces on a crt screen and recorded photographically

By R. J. KELSO Development Engineer **and J. C. GROCE,** Senior Project Engineer, ITT Laboratories, Nutley, New Jersey

Event Time Intervals

generator operating at approximately 1 kc is shown in Fig. 2. The circuit used to produce the photograph is given in Fig. 3.

Each horizontal line of dots is an elapsed time readout. The first dot at the left hand side is an index dot. Remaining dots are weighted in binary fashion, that is, the first dot to the right of the index represents one μsec (2^0), the second dot two μsec (2^1), and third dot four μsec (2^2), the fourth dot eight μsec (2^3) and so on.

Addition of time represented by the visible dots indicates total elapsed time. As an example note the dots visible on the bottom line: the index, 2^0, 2^2, 2^4, 2^5, 2^7, 2^8 and 2^9. The elapsed time is the sum of these figures or 949 μsec.

Consecutive horizontal sweeps are not in exact vertical alignment because the one-mc oscillator and the external signal trigger pulse that starts the sweep were not synchronized. Readout, therefore, begins within the one μsec period following the initiation of the sweep. Long time readout indicated at the top of the photograph results from the random state of the counters at the start of the testing period. To study a greater number of consecutive periods, the vertical positioning can be held constant while the film is moved at a rate determined by the minimum period expected.

A schematic of the one-mc crystal oscillator and shaper circuits is shown in Fig. 4. An emitter coupled crystal oscillator is

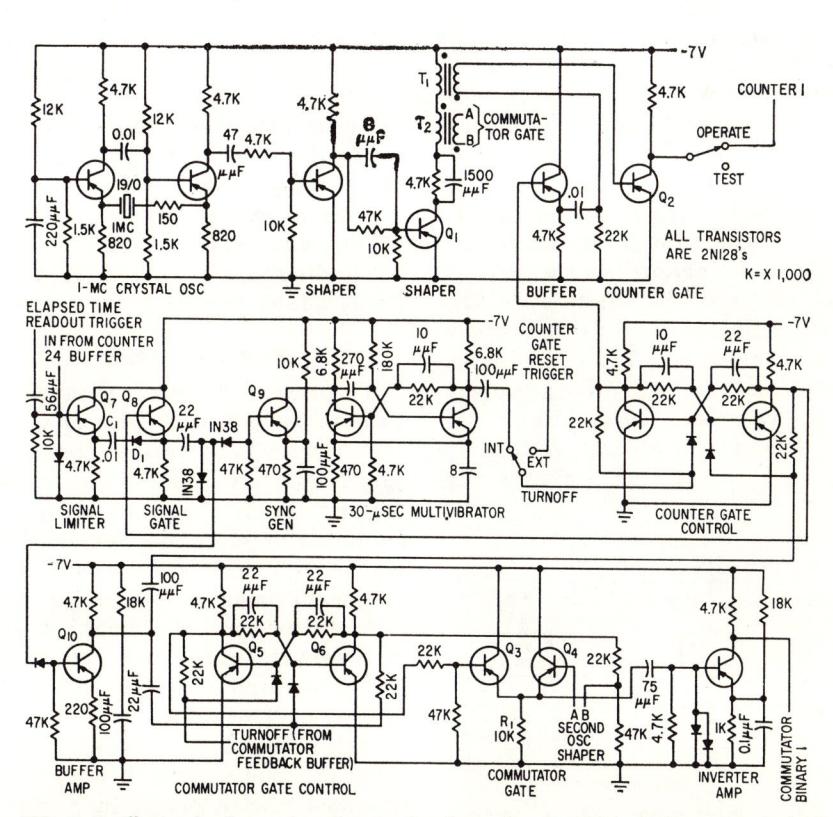

FIG. 4—Oscillator, shaping and gating circuits of encoder. Crystal in oscillator is principal frequency determining element. Gates are designed to maintain amplitude of trigger pulse at one-half the supply voltage or more within —40 to +55 C range

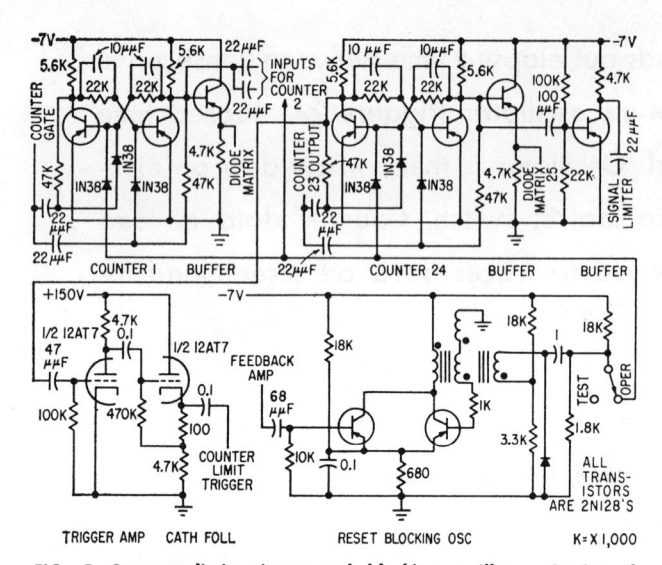

FIG. 5—Counter, limit trigger and blocking oscillator circuits of encoder. Counters 2 through 23 are identical to counter. Similar multivibrators are used in commutators and gate control circuits shown in Fig. 6

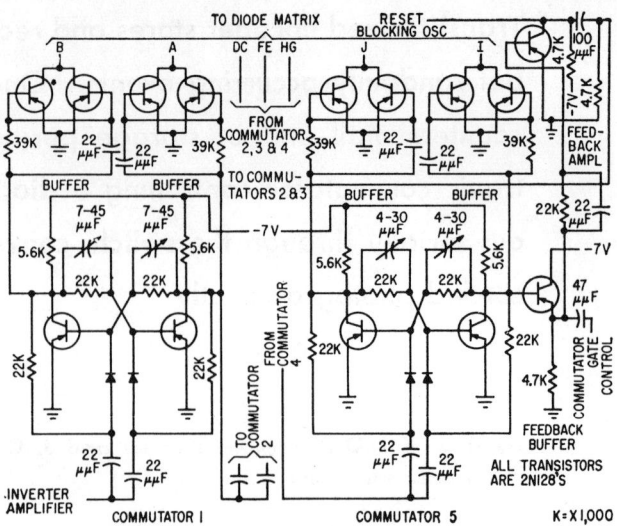

FIG. 6—Commutator, buffer and feedback circuits of encoder. Commutators 2, 3 and 4 are identical to commutator 5 except that value of variable capacitor in commutator 3 associated with output F is 7-45 $\mu\mu$f

used as the time reference. If the crystal ceases to function, oscillations will stop rather than continue at some uncontrolled frequency as might happen in a stage containing additional tuned circuits. Output of the oscillator is amplified and shaped to produce synchronizing pulses or triggers.

Gating Circuits

Gating circuits used are shown in Fig. 4. Ground reference is removed from the triggers by pulse transformers T_1 and T_2 in the collector of Q_1. The secondary windings of the transformers are applied to the inputs of the counter and commutator gates respectively.

The gating circuits are designed to minimize the effects of leakage and gain resulting from temperature variations. These circuits operate satisfactorily over a temperature range of -40 to $+55$ C.

The gate which controls the triggers to the counter blocks the triggers only during the 30 μsec readout period. This gating action makes it possible to a-c couple the control waveform from the counter gate control multivibrator thereby removing the d-c reference level. The control waveforms are then permitted to go positive with respect to ground and effectively bias off gate transistor Q_2. Thus, trigger pulses present in the base circuit are prevented from appearing at the collector circuit and triggering the counters.

The commutator gate is normally off and is turned on to pass triggers only during the readout period. If a pedestal appears at the collector of this gate when it is turned on, the leading edge of the pedestal produces a false trigger in the output.

Push-Pull Gate

To prevent the generation of a pedestal, a push-pull gate is used which consists of two emitter followers Q_3 and Q_4 and common emitter resistor R_1. The bases are connected to opposite collectors of the commutator gate control multivibrator. Trigger pulses present in pulse transformer T_2 do not appear at the emitter of Q_4 if sufficient base current is supplied to Q_3 by the gate control multivibrator.

Base current is supplied to Q_3 when the multivibrator is triggered to place transistor Q_5 in a nonconducting state. When this occurs, the voltage developed at the emitter is sufficient to bias Q_4 to a point where the negative pulses present in the base circuit do not appear at the emitter. When the multivibrator is triggered to place Q_6 in a nonconducting state, the d-c level at the emitters of Q_3 and Q_4 does not

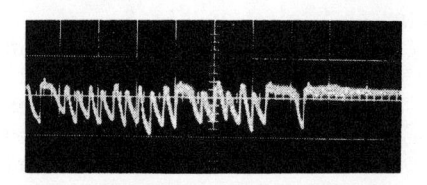

FIG. 8—Encoded elapsed time at diode matrix output

change appreciably, but the negative triggers present at the base of Q_4 now appear at the emitter. These triggers are amplified and inverted to drive the commutator.

The external trigger that is supplied to the encoder to start the elapsed time readout passes through a gate before it reaches the 30 μsec multivibrator and the two gate-control multivibrators. Thus, triggers that may appear during the readout period are prevented from affecting the timing of the 30 μsec multivibrator. The gate consists of two emitter followers Q_7 and Q_8 coupled by a capacitor in series with a diode. Emitter follower Q_7 acts as a clipper and limits the external trigger pulse amplitude to approximately -7 v. When this gate is off, the base and, therefore, the emitter

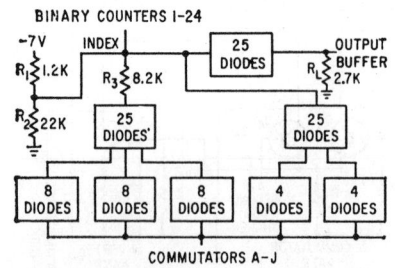

FIG. 7—Diode matrix switch of encoder. Gold-bonded germanium 122G diodes are used throughout

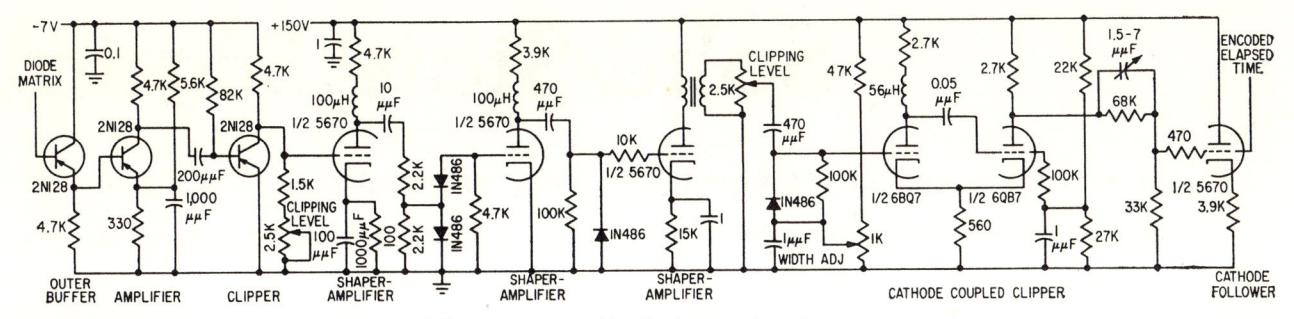

FIG. 9—Output circuits of encoder. Cathode follower stage provides low output impedance

of Q_8 are at approximately -7 v because of the state of the counter gate control multivibrator.

Back resistance of diode D_1 limits the charging of capacitor C_1. The time constant formed by the diode back resistance and the capacitance is made large in comparison with the gate off time of 30 μsec. If a trigger pulse appears during this time, the pulse amplitude appearing at the emitter of Q_7 is limited to the voltage present at the emitter of Q_8 resulting in no signal current passing through capacitor C_1. This action prevents the trigger pulse from reaching the sync injector Q_9 and the buffer amplifier Q_{10}. When the gate is on, no current is supplied to the base of Q_8, therefore, any trigger pulse appearing at the emitter of Q_7 also appears at the emitter of Q_8 with practically no attenuation.

Counters

Counter and associated circuits are shown in Fig. 5. The counters consist of 24 identical bistable multivibrator stages. Capacity of the counter is 2^{24} or 16,777,216 μsec.

In theory the operation of a stage is closely analogous to that of a vacuum-tube multivibrator. One transistor is supplied with sufficient base current to cause saturation which causes the voltage drop across the resistor in the collector circuit to become almost equal to the supply voltage.

Since the voltage at the collector is practically zero, negligible current is supplied to the base of the opposite transistor which is, therefore, in a nonconducting state. Voltage at the collector of the opposite transistor is equal to the supply voltage minus the small voltage drop resulting from the leakage current and plus any small base current supplied by the saturated transistor.

Interchange of the state of the two transistors can be done by supplying a positive trigger to the base circuit through triggering diodes. The diode that is connected to the base of the nonconducting transistor is back-biased by the voltage present at the collector. A positive trigger will therefore act only to turn off the transistor that is conducting. As the conducting transistor is turned off, its rising collector voltage acts to turn on the opposite transistor thereby permitting the use of positive going triggers to change the state of the multivibrators.

Readout Switch

The commutator shown in Fig. 6 when used with a diode matrix shown in Fig. 7 forms a single-pole, 26 position switch to obtain serial readout of the count stored in the counters. Normally, this switch is in the off position. When the commutator gate opens allowing the commutator to be triggered, a readout is obtained of an index pulse, produced by the voltage divider formed by resistors R_1 and R_2 shown in Fig. 7, followed in sequence by the state of counters 1 through 24. After readout, the switch returns to the off position.

The commutator is a five stage binary counter that employs two feedback paths to obtain a count of 25^1. With the switch in the off position, all buffers connected to the commutator are supplied with sufficient base current to cause saturation. Shunting effect of the collector saturation resistance acts to prevent the voltage present at the switch input terminals from appearing across output resistor R_L of the diode matrix. Saturation resistance of 2N128 transistors is less than 100 ohms. When the commutator is driven by trigger pulses, the shunting effect of the buffers is sequentially removed from the 25 diode matrix inputs allowing a portion of the voltage, determined by the voltage divider action of R_s and R_L, to appear across R_L.

Readout Waveforms

A typical digital readout waveform appearing at the output of the switch is shown in Fig. 8. A rapid decrease in voltage occurs at the output after the readout of each on counter, but the rate at which the output voltage increases because of an on counter is limited by the minority carrier storage time of the buffer transistors. Results of this minority carrier storage action are desirable because the slow on and fast off action produces a separation in the readout between adjacent on counters. In certain positions, it is necessary to supplement the minority carrier storage effect by adding integrating capacitors to further reduce the counter on transition time.

The output circuits used to obtain the desired output waveform are shown in Fig. 9.

The authors wish to acknowledge the assistance of John D'Aiuto in designing and testing the circuits and of Arthur Walter, John VanderHorn and Norman Tirpak in laying out the circuits and constructing the encoder.

REFERENCES

(1) B. R. Gossich, Predetermined Electronic Counter, p 813, 37, *Proc IRE*, July 1949.
(2) D. R. Brown and N. Rochester, Rectifier Networks for Multiposition Switching, p 139, 37, *Proc IRE*, Feb. 1949.

Solid-State Digital

Up to 13 bits can be converted from Gray to straight binary with this reliable, simple, code converter. Basic building block is a circuit composed of a magnetic core, a single junction transistor and an RLC delay

By **REUBEN WASSERMAN**, Associate Division Head, **and WILLIAM NUTTING**, Technical Staff Member, Digital Systems Division, Hermes Electronics Co., Cambridge, Mass.

IN SOME SYSTEMS it is advantageous to handle information in both Gray and binary code form. Gray code, which derives its usefulness from its property of changing one and only one digit in proceeding to a next higher or next lower number, is often used to prevent ambiguities in readings and then, in a final stage, is converted by some means to straight binary form.

The Gray-to-binary code converter shown in Fig. 1 accepts up to 13 bits of Gray information from an analog to digital shaft encoder, converting it to straight binary code. Binary results are presented on an array of gas tube display bulbs.

Elements of this converter such as dynamic flip-flops, logical gates, adders and shift registers are all constructed from a basic circuit composed of a magnetic core, a single junction transistor, and an RLC delay. Theory and operation of this converter is best understood, therefore, by first considering the functioning of the basic magnetic core logic circuit.

Magnetic Core Logic

The small toroidal ferromagnetic cores used are characterized by rectangular hysteresis loop and microsecond switching time. Information storage in these cores is based on the fact that a core with a rectangular hysteresis loop is capable of storing one binary bit in the form of its flux. Figure 2A shows how a binary bit ONE is stored as a positive residual flux ($+B_r$) and binary digit ZERO is stored as a negative residual flux ($-B_r$) in the opposite direction.

A magnetic storage core consists of several windings on a toroidal magnetic core (Fig. 2B). If a negative pulse is applied to the core through the shift winding, a voltage is induced in the output winding if the digit stored is ONE ($+B_r$), and a negligible voltage is induced if it is a ZERO ($-B_r$). The induced voltage is large enough to magnetize another core of identical construction. Binary digits can thus be transferred from one core to another.

However, in applying the magnetic core to a shift register in which all information is to be stepped along simultaneously, temporary storage or delay at the output of each core is required in order that the core can transfer its in-

Table I—Gray and Binary Codes

Decimal	Gray	Binary
0	0000	0000
1	0001	0001
2	0011	0010
3	0010	0011
4	0110	0100
5	0111	0101
6	0101	0110
7	0100	0111
8	1100	1000
9	1101	1001
100	1010110	1100100
1000	1000011100	1111101000

Table II—Sequence of Conversion After Each Shift Pulse

				Cores					
Time	10	11	12	13	C_2	C_1	1	2	3	4
t_0	0	1	1	0	0	0	—	—	—	—
t_1	—	0	1	1	0	0	0	—	—	—
t_2	—	—	0	1	1	0	0	0	—	—
t_3	—	—	—	0	0	0*	1	0	0	—
t_4	—	—	—	—	0	0	0	1	0	0
t_5	—	—	—	—	—	—	0	0	1	0

*Binary bit is complemented

formation before receiving new information from the preceding core. Therefore, the core storage element comprises two essential units—the magnetic core for permanent storage and an RLC delay circuit for temporary storage.

Other uses for magnetic cores, in addition to delay and storage, are in power amplifying and AND, OR, and INHIBIT circuits. Power amplification can be derived by adding a transistor which then serves as the power driver. The core lends itself to two elementary logical functions of OR and INHIBIT, from which it is possible to synthesize all other digital logical functions. Therefore, a circuit composed of a magnetic core, a transistor, and an RLC delay forms an ideal computer building block in constructing dynamic flip-flops, logical gates, adders, and shift registers.

Basic Circuit

The circuit of Fig. 3, which serves as the basic building block in the Gray-to-binary converter, uses a *pnp* junction transistor, and therefore is wired in such a fashion that the circuit resembles a blocking oscillator. If the core is in either of two saturated states, ($+B_r$) or ($-B_r$), the permeability of the core is low and the gain around the feedback loop is less than unity. On the vertical slope of the B-H curve the permeability is high so that loop gain is well above unity and regeneration is possible.

To analyze the Fig. 3 circuit, consider the transistor connected as a grounded emitter and the core initially in the ONE state. The transistor is normally cut off. A small trigger pulse is applied which is sufficient

Code-to-Code Converter

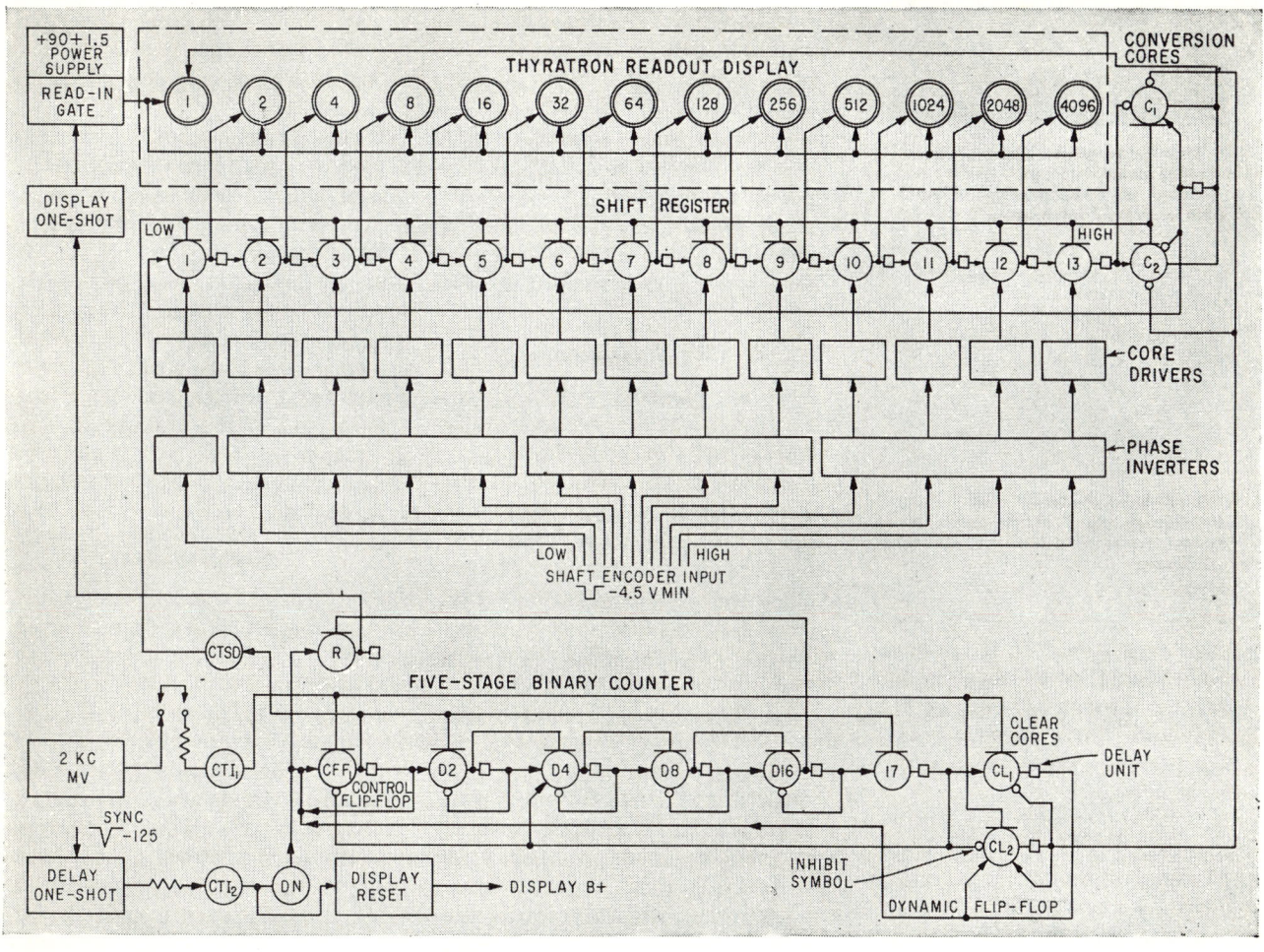

FIG. 1—Core logic diagram of complete thirteen bit Gray-to-binary converter

to change the state of the B-H curve. The resulting flux change induces a negative voltage on the base of the transistor and causes collector current to flow. This current flows in the same direction as the trigger current pulse and therefore shifts the core further, inducing a larger negative voltage on the base. This in turn will further increase the collector current. The cycle continues until the core is completely shifted, at which time the feedback-loop gain falls below unity, and the transistor ceases to conduct. The result is that the core has been completely shifted and a large pulse of current has passed through the collector and supplied power to the load.

If, however, the core was initially in the ZERO state and the same trigger pulse was applied, there would be a negligible voltage developed on the base since the gain is sufficiently below unity, and regeneration would not take place. The core can then be reset to the ONE state by feeding current in the proper direction through the insert winding. While the core is being reset, a positive voltage appears on the base of the transistor which tends to drive it further into cutoff, preventing any output at this time. If, however, as the current pulse is applied to the insert winding an equal magnitude of current is applied simultaneously to the inhibit winding, the effect of the reset pulse will be cancelled and the core will remain in the ZERO state.

Successive stages of the Fig. 3 basic circuit form the shift register. All the stages with ONES stored change their state to ZEROS after being triggered, and during regeneration supply energy to the output. The output current pulse is delayed before it resets the following stage to the ONE state. In this manner, information is advanced one stage after each shift pulse.

Dynamic Flip-Flop

The dynamic flip-flop (see Fig. 1) is actually a one-stage shift register in a closed loop circuit which can exist in two dynamic states. If, for example, the state is initially ONE, the trigger pulse will switch the core to ZERO and, after a time delay, the core's own output will reset the core back to ONE. The state of the flip-flop can be changed to ZERO by applying a signal to the inhibit winding while the delayed output of the core is trying to reset itself, thereby preventing the core

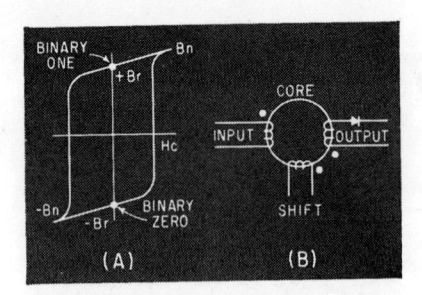

FIG. 2—Rectangular hysteresis loop (A) allows binary storage. Pulses through windings on toroidal magnetic cores (B) transfer binary information

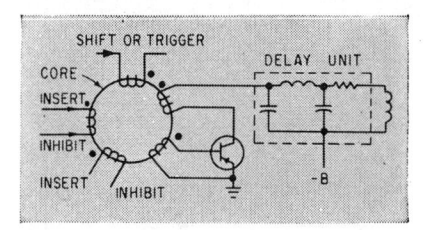

FIG. 3—Basic building block circuit for constructing many of the converter elements uses core, transistor and RLC delay unit

from being reset. In the ZERO state the next trigger will have no effect and the flip-flop will remain in the ZERO state until such time as the core is reset to ONE by an external insert pulse.

Binary Counter

To analyze the binary counter circuit which appears in Fig. 1, assume that the first stage ($D2$) is ZERO. A trigger pulse will then have no effect as there will be no regeneration. However, if the first stage is ONE and a trigger pulse appears, one of two possibilities will occur. If the second stage was in the ZERO state, the output current passing through its trigger winding will have no effect. However, the delayed output of the first core will reset the second core to the ONE state, or to use a simple terminology, will write a ONE in the second stage. If this second stage was already ONE, it will be switched to ZERO by the collector current or undelayed output of the first stage. The delayed output of the second stage passing through its own inhibit winding will prevent the delayed output of the first stage from writing a ONE into the second stage. It then becomes evident that this is a bistable device which changes its state every time an input pulse is applied, and puts out a pulse identical to the input pulse for every second input pulse. This essentially is the operation of a binary counter.

Timing Generation

Circulation rate of the converter register is determined by the internal clock which runs at a 2-kc rate. Clock output is fed into a core-transistor input amplifier (CTI) which is similar to a Schmitt Trigger in the sense that it triggers on a minimum voltage input pulse. Its output is a current pulse for shifting or triggering, inserting or inhibiting.

Shift Operation

The magnetic core shift register, consisting of cores 1-13, initially stores all thirteen bits of the Gray coded word to be converted. After a short delay, allowing all the inputs to be read into their respective cores, the information is shifted along into the two converter cores, C_1 and C_2, which in operation resemble a combination of the dynamic flip-flop described above and an exclusive OR gate. Table I shows the relationship between the Gray and binary codes.

Operation of cores C_1 and C_2 can be traced in Fig. 1. Initially assume that the Gray code number 0110 (represents decimal 4) is stored in cores 10 to 13, with high order bit in core 13. Cores C_1 and C_2 are both in the binary ZERO state. Table II shows the sequence of conversion after each time or shift pulse.

At time t_2, core C_2 stores a binary ONE, which is the first high order binary ONE of the Gray code number. At time t_3, core C_2 emits the binary ONE which complements the next binary bit in sequence from core 13, or, in other words, converts the binary ONE to binary ZERO. At t_4 and t_5 the information is shifted two places to the right and the converted information is now stored in cores 1-4.

Output of cores C_1 and C_2 is connected to core 1 to make a circulating register in order that all thirteen bits are stepped along to the right (high order first), and the converted code is read back into core 1. The register is shifted thirteen positions and the original input information is converted to straight binary form. The register is shifted once more and the outputs of core C_1 and cores 1-12 are sensed or looked at simultaneously by the gas tube display.

Control Counter

In order that the register be circulated the proper number of shift positions a control counter, which consists of a control flip-flop, CFF_1, and a five-stage counter, $D2$, $D4$, $D8$, $D16$, and 17 is incorporated (see Fig. 1). The control flip-flop is a dynamic flip-flop which is turned on by a pulse from CTI_2. CTI_2 gives an output pulse whenever any new input information is inserted. Since CFF_1 is triggered continuously by CTI_1, once reset to ONE it will produce output pulses at a 2 kc rate. The output of CFF_1 is fed into $CTSD$ which provides the proper shift pulse current to all the register cores as well as the two converter cores. After the fourteen

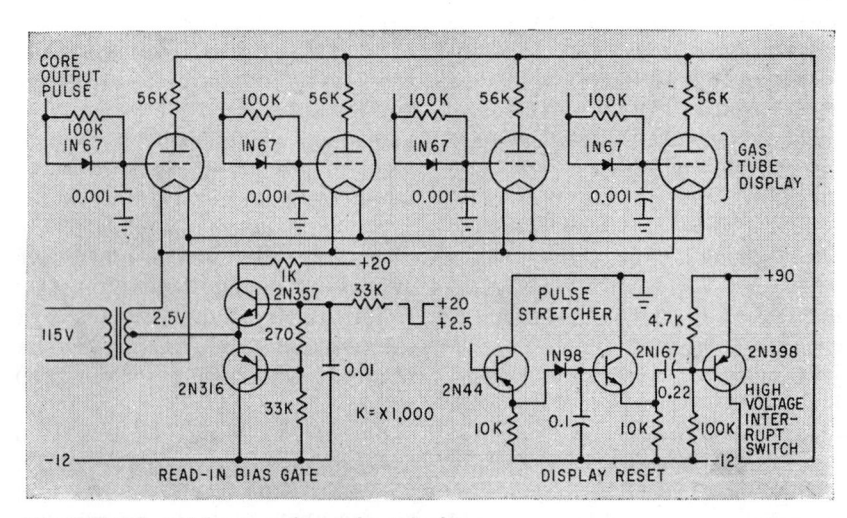

FIG. 4—Circuit used for gas tube read-out display

output pulses, CFF_1 is inhibited by the output of the counter, core 17 in particular. The counter is basically the binary counter described earlier, arranged in cascaded stages and with a feedback connection from core 17 to core $D4$.

Clear Cores

Cores CL_1 and CL_2 (Fig. 1) clear the register and converter cores of any ONES stored in them following a word conversion and thus prepare for accepting new input information. This is done by resetting the control flip-flop so that it will generate fourteen more shift pulses and at the same time inhibit both converter cores so that any ONES entering from the register will be canceled. This means fourteen inhibit pulses are required as well as the shift pulses.

Clear cores CL_1 and CL_2 are connected much like the converter cores in a dynamic binary counter. A single pulse from core 17 is used to write a ONE into CL_1. The next trigger pulse from CTI_1 will cause an output from CL_1 which writes a ONE into CL_2 as well as into CFF_1. With CFF_1 reset it will produce a train of pulses; CL_2 will also until the second output pulse from core 17 shuts it off. CL_2 output then inhibits both C_1 and C_2.

Readout Display

After thirteen shifts all the information is converted except the last bit and that bit is written into the converter cores. During the following, or fourteenth shift, all the input information is converted and is shifted out of core C_1 and cores 1-12. At the thirteenth shift there is an output from the counter core $D16$ which shifts the Read (R) core which was reset to ONE by CTI_2. The output of R triggers a static one-shot which gates the level of the filament center-tap of the thyratron display bulbs so that any ONES shifted out of cores C_1, 1-12 at the time of the fourteenth shift pulse will cause the associated thyratrons to ignite.

These display bulbs (Kip "Memolites", manufactured by Transistor Electronics) remain on until the next input sync pulse occurs. When the next sync pulse occurs, a static delay one-shot circuit is triggered, and its output triggers CTI_2. The undelayed output of CTI_2 triggers the static display circuit shown in Fig. 4, which extinguishes the gas display bulbs by dropping their plate voltage below the ionization point.

The bulbs are extinguished only when new input information is to be received. Therefore, the single pulse from CTI_2 which triggers indirectly from the input sync pulse and directly from the input-delay one-shot, activates the display reset. This pulse, however, is short. Therefore, two npn transistors in a cascaded emitter follower circuit, with a pulse stretcher between them (see Fig. 4), control a high-voltage

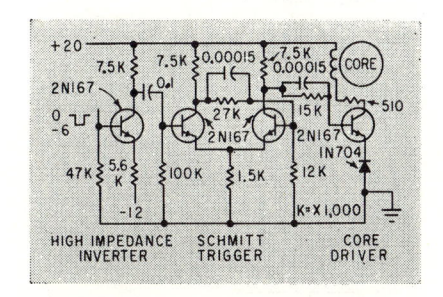

FIG. 5—Phase inverter, in addition to phase reversal, provides isolation between the negative input pulse and the core driver

pnp transistor in series with the thyratron plate voltage. The pnp transistor, normally conducting with only a few volts drop across it, is shut off by the stretched pulse, causing the full plate supply of the thyratrons to be dropped across it. The transistor stays shut off for 500 to 600 microseconds.

Phase Inverter

The phase inverter (Fig. 5) is a single npn transistor which in addition to phase reversal of the input signal provides isolation between the negative input pulse and the core driver. The output of the analog to digital shaft encoder is directly coupled to the base of the transistor and requires that the external input be at ground level or be capacitively coupled into the converter. The output of the phase inverter is capacitively coupled to the input of the core driver. There are thirteen phase inverters and core driver circuits, one for each input bit.

Each core driver uses two npn transistors connected in a Schmitt trigger circuit with discrimination level including the phase inverters of -4.5 v ± 10 percent.

Any signal below that level is disregarded.

However, any signal in range from 6 to 12 v will cause a pulse output from the Schmitt, the duration of which is dependent on the duration of the input signal. This output is direct-coupled to an npn transistor in a grounded emitter circuit. This transistor is normally cut-off and uses a Zener diode in its emitter circuit to insure its cut-off until the output pulse of the Schmitt trigger exceeds the Zener breakdown voltage. This is necessary to prevent false read-in of the cores if the transistor is not fully cutoff. The insert winding of a register core and a series limiting resistor make up the collector load of this transistor.

When the Schmitt trigger emits a pulse, this driver supplies a 30 ma current pulse, writing a ONE into its associated core.

The minimum duration pulse for writing a ONE results from a 6 microsecond input pulse to the phase inverter.

Read-In Gate

The read-in gate, which appears in Fig. 4, is a two transistor amplifier employing complementary symmetry to provide a low-impedance path for the filament return of the thyratron display bulbs at either of two operating levels. The input to the read-in gate is the output of the display one-shot which is directly coupled through a series resistor to the bases of the npn and pnp transistors. Another resistor, from the bases to the negative supply, forms a voltage divider which determines the normal output operating point. When the negative gate occurs, the read-in gate output, which is connected directly to the filament center-tap of the thyratron display bulbs drops to -5 v allowing the output of the register cores to ignite their respective bulbs at the fourteenth shift pulse. Before the next pulse the level is returned to $+5$ v (normal) and prevents any further core outputs from triggering the bulbs.

Pulse-Height-to-Digital

Transistorized analog-to-digital converter provides 7-digit binary output for an input of 0 to 2 v at a maximum sampling rate of 13,000 pps

By W. W. GRANNEMANN, C. D. LONGEROT, R. D. JONES, D. ENDSLEY, T. SUMMERS, T. LOMMASSON, A. POPE and D. SMITH,
University of New Mexico Engineering Experiment Station, Albuquerque, New Mexico

PULSE-HEIGHT data reduction can be speeded with pulse-height-to-digital converters. Moreover, tedious arithmetical computation can be avoided by using the converters with a computer. This article describes a transistorized pulse-height-to-digital unit developed for a radar. The counter-type converter provides 7-digit binary output for input signal of 0 to 2 v at maximum sampling rate of 13,000 pps.

Converter Specifications

Function of the converter is to digitize radar return pulses for processing by a datum-reduction system. The unit converts pulsed analog voltage to a full-cycle pulse sequence representing a binary number.

The converter requires two input signals of positive polarity, on separate lines, and produces outputs on two other lines. Input signals are a pulse which initiates a timing sequence in the converter, and an analog-datum pulse following the timing pulse by 5 to 10 μsec. The converter produces a synchronizing pulse on one output line, a control pulse followed by the binary representation on another line. Output represents amplitude of the input-datum pulse to less than one percent.

FIG. 1—Block diagram of converter function

Timing pulses have width of one μsec, rise-and-fall time of 0.1 μsec, amplitude of 15 v. Data pulses have amplitude of 0 to 2 v, width of 0.25 to 5 μsec, rise-and-fall time of 0.08 μsec, maximum pps of 13,000. Synchronizing output pulses have width of 1 μsec, amplitude of 2 v. The first pulse, representing the most significant figure of the binary output, follows the control pulse by 18-μsec; appearance of a 4.5-μsec pulse in any other interval depends on amplitude of the data input pulse.

Principles of Operation

Data and timing pulses are received on separate lines, Fig. 1 and 2. Data pulses pass through height-to-width conversion and the new pulses are used to gate clock pulses, proportional to the amplitude of data pulses. After clock pulses are stored by a 7-digit binary counter, the state of the binaries represents the datum-pulse amplitude and the binaries are ready for readout. Timing pulses, leading data pulses by 5 to 10 μsec, trigger the timing sequence, generating the necessary output pulses at the proper time and delaying readout of the 7-digit binary counter until it has time to count the gated clock pulses.

Operating at 13,000 cps, the converter will complete one cycle in 76.9 μsec. A full-cycle representation of seven digits at 4.5 μsec/digit accounts for 31.5 μsec. The data pulse, maximum duration 5 μsec, follows the timing pulse by maximum of 10 μsec, accounting for another 15 μsec. Thirty microsec-

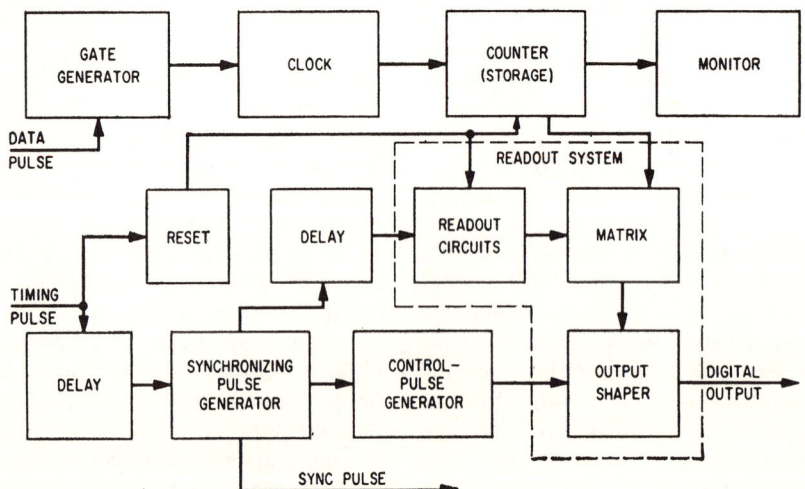

FIG. 2—Block diagram of counter-type converter

Signal Converter

onds is then left for the counting operation.

To achieve one-percent accuracy, requiring a minimum of 100 counts for the maximum input signal, a 7-stage binary counter is required (2^7 or 128). The 30 μsec remaining for counting, when combined with 2^7, indicates a minimum clock crystal frequency of approximately 4.27 mc; therefore 5.5 mc was selected. This allows rundown of 23.3 μsec for maximum data input signal of 2 v.

Figure 3 shows time relationship between the several converter events. With the 5.5-mc oscillator running continuously, the sequence starts at arrival of the timing pulse that triggers a 21-μsec delay (line 6). This provides part of readout delay until the state of the binaries represents the data signal amplitude. After 21 μsec delay, a 1-μsec pulse generator is triggered—this pulse synchronizes the sweep of a recording scope. The trailing edge of the synchronizing pulse is used to trigger two other stages: the control pulse generator producing a 4.5-μsec pulse for readout reference; and the 18-μsec delay preventing the first binary readout from occurring until 18 μsec after the control or reference pulse. After 18-μsec the trailing edge of the delay pulse triggers a 31.5-μsec gate operating the readout circuits. This gate allows full-cycle representation of the 7-digit counter.

The data pulse follows the timing pulse by 5 to 10 μsec and starts another sequence concurrent with that just described. However, the data pulse is supplied on a separate line and its trailing edge triggers the gate generator which makes the pulse-height-to-pulse-width conversion. Gate width, directly proportional to data pulse amplitude, is applied to the clock circuits to allow clock pulses to pass to the 7-digit binary counter. A maximum of 38.3 μsec is required for the 7-digit counter to collect the information necessary to represent the analog input signal. Readout occurs after 40 μsec has passed.

Conversion of the 0- to 2-volt pulse to a gate is accomplished by a gate generator, Fig. 4. Action of the gate is dependent on a constant-current generator discharging a capacitor. The capacitor is charged through a diode and peaking inductance so that after the pulse has been applied the voltage on the capacitor is of such polarity that the diode is reverse-biased.

Since the capacitor is connected to the collector of the npn grounded-base transistor, it starts to discharge after the input pulse has dropped to zero. High impedance in the emitter of the grounded-base 2N332 transistor limits emitter current to a low value, causing the transistor to operate at low collector current. The I_c vs E_c characteristics of this configuration show collector current is constant for wide variation of collector voltage; thus, constant current properties are maintained. The charge on the capacitor is removed at a constant rate; therefore voltage is similarly reduced. With a 7.5-μa constant-current generator rundown is 23.3 μsec for a 2-v input pulse. Temperature control of the critical elements was found essential.

The gate amplifier comprises a differentiator, four amplifiers, and clipping circuits. The gate-generator pulse is differentiated, resulting in a leading-edge spike and gate whose width is proportional to the length of rundown. Following stages clip unwanted spikes and amplify the gate pulse. Output is applied to an oscillator gate, providing control of pulses to be counted.

The oscillator gate, Fig. 5, provides a train of pulses to be read by the 7-digit binary counter. Number of pulses in the train depends on the width of the gate pulse received from the gate amplifier and is proportional to the amplitude of the data input pulse. The oscillator gate is a continuously running crystal-controlled oscillator and gate producing needed pulses at the input to the 7-digit binary counter.

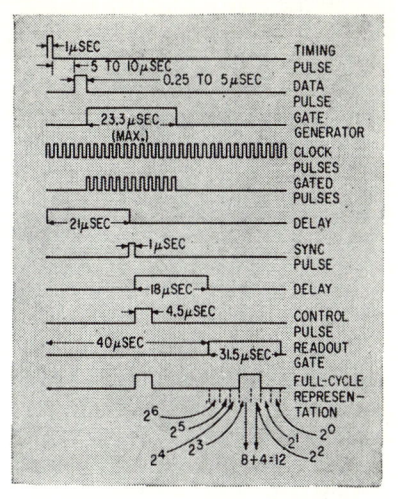

FIG. 3—Timing schedule of converter

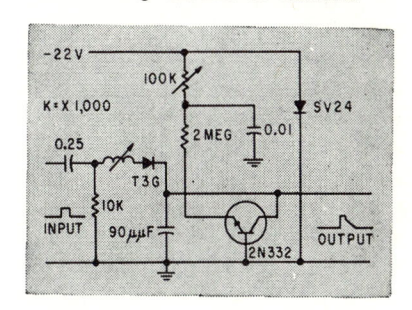

FIG. 4—Schematic diagram of gate generator

Stability is provided by the crystal across the output winding of the three-winding pulse transformer in the free-running blocking oscillator. Gating is accomplished by an emitter-follower and shunt gate. Q_1 normally conducting, shunts output resistor R_1 except when the transistor is cut off by a positive gate signal at its base.

The 7-digit counter, Fig. 6, stores pulses received from the oscillator-gate, the 128th pulse resetting the counter to zero. The binary counter consists of seven cascaded bistable multivibrators, transformer-triggered. Each pulse to the base of trigger Q_3 provides identical pulses to the bases of binary transistors Q_1 and Q_2, causing the multivibrator to change state. D_1 is used to reset the binary to its off condition, in which Q_2 is near cut off and Q_1 near saturation. Hence a positive pulse applied at A will reset the binary to its off state if it is not already in that

375

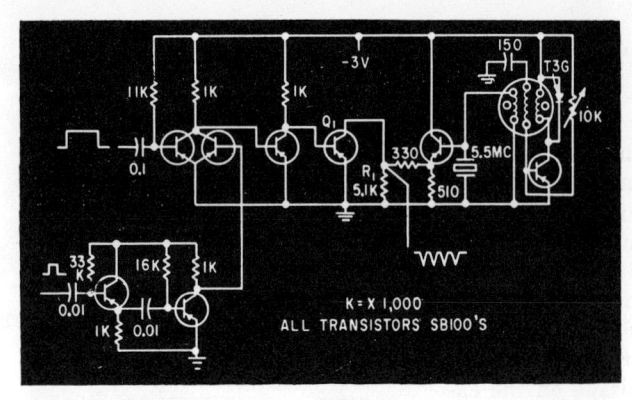

FIG. 5—Continuously running crystal-controlled oscillator gate

FIG. 6—Single stage of 7-digit binary counter

mode. When the binary is in the off condition there can be no digital output from the binary because of associated matrix circuits, discussed later.

The delay stage, Fig. 7, prevents certain circuits from operating until the proper time and generates and shapes required output pulses. The delay, a monostable multivibrator, is triggered by a positive pulse produced by an input differentiating circuit. Circuits are cascaded, one stage operating from the output of another. The second stage starts its delay coincident with the trailing edge of the first delay output pulse. Output of any stage may also be used to trigger several stages, each of which may be used to actuate an independent event. However, all events will have a common time reference in the trailing edge of the output of the first stage. This delay circuit, with simple modifications, is used for pulses and delays ranging from 0.25 to 32 μsec and can be used for considerably longer delays.

The readout circuit is composed of an 8-pulse train generator, a 3-digit binary counter, and an AND matrix. Readout is accomplished by reading the state of each binary in the 7-digit counter.

The 8-pulse train, Fig. 8, consists of the previously described delay circuit triggered by the 18-μsec delay circuit. The delay generates a 32-μsec negative gate to pulse a blocking oscillator having a period of 4.5 μsec. Following the blocking oscillator, the 32-μsec gate

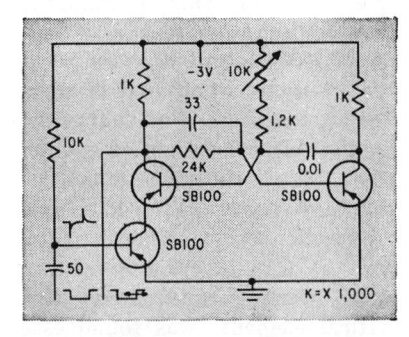

FIG. 7—Schematic diagram of delay stage

produces eight pulses at 4.5-μsec intervals at the output of the emitter follower.

These pulses put a 3-digit binary counter, consisting of three binary circuits similar to those in the 7-digit counter, through one counting cycle. At the collectors of the six transistors in the 3-digit counter appear the output signals, which in conjunction with the states of the binaries of the 7-digit counter operate an AND matrix. This matrix produces pulses, at proper interval, to form a pulse representing amplitude of the data input pulse which has been stored in the 7-digit counter.

The AND matrix is composed of seven AND circuits which read out the states of the binaries in the 7-digit counter. The AND circuits have four inputs, operating on positive signals taken from the six transistors in the 3-digit counter and the seven output transistors in the 7-digit counter. To obtain an output signal of 4.5 μsec duration, all inputs to a particular AND circuit must be positive and coincide in time for a period of 4.5 μsec.

Output Circuits

Output section consists of an OR gate and associated circuits. The OR circuit produces an output for any one or another combination of more than one input. It has eight inputs, seven from the AND matrix and one from a 4.5-μsec control-pulse generator. The OR gate thus allows the serial output in full-cycle representation.

Output consists of the control pulse followed 18 μsec later by a train of 4.5-μsec pulses, 2 v in height. This train represents the analog input signal amplitude.

Another output, a 1-μsec synchronizing pulse, triggers a display oscilloscope and is provided on a separate line. It precedes the control pulse by 1 μsec.

The work described in this article was supported by the Naval Research Laboratory.

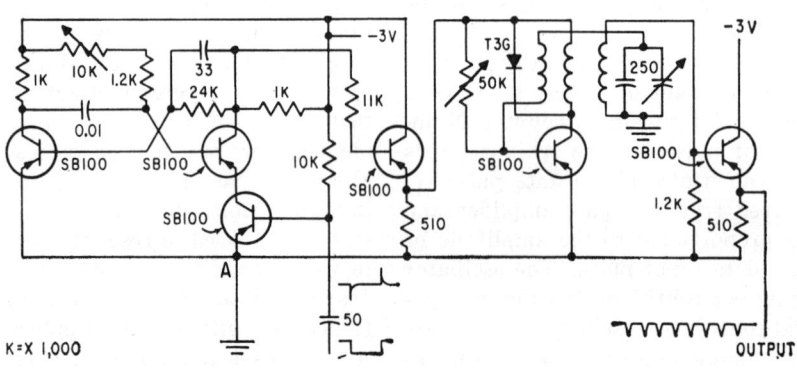

FIG. 8—Eight-pulse train generator

INDEX

A-c motor control (SCR), 254
A-c preamplifier, 74
A-c static switch (SCR), 235
Acceptors, 3
Active filter theory, 77
Agc amplifier, 321
Agc circuit, 164
Alloy junctions, 14
Alpha-cutoff frequency, measurement, 228
Amplifier, agc, 321
   audio, 60, 61, 203, 208
      high-power, 56
   bandpass, 80
   bistable, 271
   blanking, 323
   carrier, 197
   chopper, 72, 258
   class-A, push-pull, 202
      sliding, 55
   clock-pulse, 359
   common-base, 12
   common-collector, 12
   common-emitter, 13
   comparison, 105
   d-c, 41, 283
      operational, 72
      portable, 75
   differential, 121, 258, 276
   direct-coupled, 31, 258
   driver, 342
   error, 246
   feedback, 246
      nonlinear, 312
   filter, 78
   forced-feedback, 133
   gyro signal, 305
   horizontal, 218
   i-f, 40
   integrating, 73
   isolation, 311
   logarithmic, 240
   low-frequency, 281
   narrow-band, 277
   operational, 72, 74
   period, 241
   phototube, 262
   power, 57, 59, 74
   pulse, 215
   push-pull, 41
      class-A, 202
      one transistor, 55
      parallel, 298
      single-ended, 69
   quasicomplementary and series, 190
   read, 360
   recorder, 75
   repeater, 196
   r-f, 173
   selective, 26
   sense, 357
   servo, 315
   servo-power, 248

Amplifier, square-wave, 359
   strain-gage, 75, 257
   summing, 72
   switching, 306
   television, 189
   temperature control, 304
   tetrode, 122, 197
   thermocouple, 257
   trigger, 217
   tuned, 291
   unblanking, 217
   video, 65, 67
Analyzer, pulse-height, 287
AND gate, 226
Anticipatory response network, 305
Artificial larynx, 275
Artificial neuron, 284
Audio amplifier, 60, 61, 203, 208
   high-power, 56
   tunnel-diode, 40
Audio-modulated oscillator, tunnel diode, 40
Audio volume compressor, 79
Autodyne converter, 160, 162
   with crystal oscillator, 330
Automatic beta (h_{fe}) checker, 220
Automatic hunting system, 200
Automatic pilot, 307
Automobile generator regulator, 118
Automobile receiver, 162
Avalanche effect, 4
AVC for autodyne detector, 330
AVC overload diode, 161

Backward diode, 37
Balanced mixer, 197
Balanced modulator, 197
Balanced negative feedback, 61
Bandpass amplifier, 80
Base-driven autodyne converter, 160
Beam-switching tube counter, 352
Beta (h_{fe}), automatic measurement, 220
Bias, 13
   power-amplifier, 59
   slope detector, 188
Binary counter, 141, 376
   magnetic, 372
   reversible, 138
Bistable amplifier, 271
Bistable cathode follower, 28
Blanking amplifier for crt, 323
Blocking oscillator, 97, 145, 182, 215, 226, 332
Blood pressure monitor, 276
Bootstrapped collector, 32
Breakdown voltage, 57
Broadcast and long-wave receiver, 159
Broadcast receiver, 159, 161
Building-block assembly, 138

Capacitor-charge timing circuit, 239
Carrier transmission, 195

Cathode follower, bistable, 28
Character generator, 363
Charge-carrier recombination, 7
Chopper, 290
   amplifier, 72, 258
Clamp, gated, 323
Clapp oscillator, 225
Class-A amplifier, push-pull, 202
   sliding, 55
C-L-C and C-L-L feedback network, 85
Clock-pulse amplifier, 359
Clock signals, computer, 346
Clock source for counters, 225
Closed-circuit television, 195
Code-to-code converter, 371
Coincidence circuit, 264
Cold-cathode decade counter, 144
Collector-base resistor for negative feedback, 62
Collector capacitance, 10
Collector characteristics, 11
Collector multiplication, 6
Collector potential, effects on high frequency response, 10
Collision-signal generator, 345
Colpitts oscillator, 247, 264
Common-base amplifier, 12
Common-collector amplifier, 12
Common-emitter amplifier, 13
Common-mode rejection, 76, 257
Commutator gate, 368
Comparator alarm circuit, 309
Comparison amplifier, 105
Complementary phase shifter, 58
Computer clock signals, 346
Computer switching, 340
Constant-current charge circuit, 288
Constant-current-coupled power supply, 114
Control, frequency, for magnetic multivibrator, 337
   gas tube read-out, 372
   gate, 369
   live-voltage, 105
   motor, a-c, 254
      d-c shunt, 254
      d-c split-series, 254
      full-wave push-pull, 253
      half-wave push-pull, 254
      reversible d-c shunt (SCR), 252, 253
      using silicon-controlled rectifiers, 252
   nuclear reactor, 240
   target-drone, 313
   temperature, 108, 233
Controlled rectifier, 17
   for motor control, 252
Converter, autodyne, 160, 162
   with crystal oscillator, 330
   code-to-code, 371
   d-c/a-c, 124
   power, 102, 127
   pulse-to-analog, 294
   pulse-height-to-digital, 374
Core driver, 143, 357, 373

377

Core selection for oscillator-type power supplies, 110
Counter, beam-switching tube, 352
   binary, 141, 376
     magnetic, 372
     reversible, 138
   clock source, 225
   cold-cathode, 144
   flip-flop, 141, 368
   glow-tube, 136
   reversible, 131, 138
   ring, 28, 141, 292
Counting-rate meter, 267
Coupling circuits for reverse base current, 343
Covalent bonds, 3
Crash-position indicator, 331
Crystal lattice, 3
Crystal oscillator, 87, 196, 201, 225, 334, 347, 367, 376
Crystal reference for multichannel communications, 199
Current amplification, 6
Current regulator, 113, 115
Curve tracer, tunnel diode, 38
Cutoff frequency, measurement of, 228
Cutout, silicon rectifier, 120

Damping in relay servo, 251
D-c amplifier, 283
   drift-current compensation, 41
   operational, 72
   portable, 75
D-c feedback in servo preamplifier, 245
D-c motor control (SCR), shunt, 254
   split-series, 254
D-c static switch (SCR), 235
D-c/a-c converter, 124
Decay time of common-emitter switch, 52
Decoder, character-generator, 365
   frequency, 315
   symmetry, 315
Decommutator, battery-powered, 142
Decoupling, video-amplifier, 67
Deflection, horizontal, character generator, 365
   crt, 182
Demodulator, ppm, 206
Demultiplexer, synchronizing, ppm, 206
Depletion layer, 5
Depth finder, small boats, 326
Detector, diode, 165
   f-m, 186, 283
   foetal heart-sound, 280
   nuclear radiation, 270
   null, 278
   ratio, 167
   r-f, tunnel-diode, 39
   synchronizer, ppm, 206
Diagonally symmetrical power-supply control, 101
Differential amplifier, 121, 258, 276
Differentiating network, 137
Differentiator, 143, 226
Diffused junctions, 14
Diffusion, 4
Diffusion capacitance, 10
Digital recorder, 351
Diode, double-base, 17
   forward-biased, 5
   reverse-biased, 4
   tunnel, 16, 36
Diode acceptance, 5
Diode action, 4

Diode compensation for d-c amplifier drift current, 41
Diode detector, 165
Diode feedback circuit, 230
Diode harmonic multiplier, 88
Diode-matrix encoder, 368
Diode modulator, 197
Direct-coupled amplifier, 31, 258
Direct-coupled servo preamplifiers, 245
Distress transmitter, 334
Donors, 3
Double-base diode, 17
Double-bootstrap sweep circuit, 229
Double doping, 14
Drift, 4
Drift-current compensation for d-c amplifier, 41
Drift transistors in video amplifiers, 65
Driver, crt yoke, 323
   ferrite core, 143, 352, 357, 373
   horizontal, 183
   inhibit, 357
   magnetic-network, 325
   oscillating slope detector, 188
   solenoid, 254
Driver amplifier, 342
Dry-cell stabilization, 113
Dual-conversion superheterodyne receiver, 319
Dynamic flip-flop, 371

Electrochemical etching, 15
Electrometer tubes, log amplifier input, 240
Electronic switch (see Switch)
Emitter bypassing in video amplifiers, 67
Emitter-driven autodyne converter, 160
Emitter efficiency, 6
Emitter follower, flip-flop cross coupling, 148
Emitter-resistor compensated bias, 13
Encoder, diode-matrix, 368
   random-event, 366
Envelope detection in reflexed radios, 155
Equalization circuit, 60
Equivalent-T circuit, 9
Error amplifier, differential-transformer, 246
Excess-three binary code, 131

Falling sawtooth generator, 229
Feedback, crystal oscillator, 201
   negative (see Negative feedback)
   positive, 25
Feedback amplifier, high input impedance, 246
   nonlinear, 312
Feedback network, oscillator, 85
Fencing indicator, 260
Ferrite core, triple-wound, 142
Ferrite-core driver, 352
Ferrite-core memory, 351, 354
Ferrite-core sense amplifier, 357
Ferrite-core transformer, repeater amplifier, 196
Filament switching, 333
Filter, noise, low-frequency, 191
   nonlinear, 241
   notch, 206
   subaudio, 76
Filter amplifier, 78
Flip-flop, 143
   complementary, 131
   dynamic, 371

Flip-flop, feedback stabilized, 148
   nonsaturating, 206
   photoelectric, 334
   power, 140, 235
   read amplifier, 360
Flip-flop counter, 368
   $pnpn$ trigger, 141
Flip-flop demodulator, ppm, 206
Fluorescent lamp power supply, 116
Flyback oscillator, 127
F-m detector, 186
   oscillator locking, 187
F-m modulator, 83, 168, 208
F-m oscillator, 83
F-m transducer, 207
F-m transmitter, 207, 208
F-m tuner, 166
Foetal heart-sound detector, 280
Forced-feedback amplifier, 133
Forward-biased diode, 5
Forward current transfer rates, 9
Four-stage binary counter ($pnpn$ trigger), 141
Four-terminal network, 8
Free-running multivibrator, 275
   hybrid, 30
   tunnel diode, 40
Frequency control, magnetic multivibrators, 337
Frequency decoder, 315
Frequency divider, 97
Frequency-selective calling equipment, 26
Frequency sweep generator, 83
Full-wave push-pull motor control (SCR), 254

Gain, power and voltage, 11
Gas tube read-out control, 372
Gaseous diffusion, 15
Gate, AND, 226
   commutator, 368
   multiplexer, 294
   push-pull, 368
Gate control circuit, 369
Gate generator, 375
Gated clamp, 323
Gating circuit, 269, 347, 368
Geiger-Müller power supply, 267
Generator, character, 363
   collision-signal, 345
   falling sawtooth, 229
   frequency sweep, 83
   gate, 375
   harmonic, 200
   horn-signal, 345
   microwave, 88
   pulse, 140, 150, 223, 376
   regulator, automobile, 118
   rising sawtooth, 229
   sawtooth, 222, 229, 230
     variable, 314
   square-wave, 229
   sweep, 322
     ppm modulator, 204
   tone, 96
   triangle, 229
   variable delay, 134
   variable pulse width, 213
   wavetrain, with zero-crossing sync, 226
Germanium tunnel diode, 36
Glow-tube counter, 136
Gray code, 370
Grown junctions, 14
Guidance system for rockets, 303

Gyro signal amplifier, 305
Gyro temperature control, 304

Half-wave controlled rectifier (SCR), 234
Half-wave phase-controlled power supply (SCR), 236
Half-wave push-pull motor control (SCR), 253
Harmonic distortion in power amplifiers, 57
Harmonic generator, 200
 using capacitance diode, 88
Hartley oscillator, 326
 high-frequency, 213
Heat dissipation, switching transistors, 50
Heat drift, silicon transistors, 42
Heat sink, copper sleeve, 203
H-f and broadcast-band receiver, 158
High-frequency current gain, 8
High-frequency cutoff, 45
High-frequency equivalent-T circuit, 10
High-frequency oscillator, 90
 Hartley, 213
High-power oscillator, 90
High-power switching, 341
High-voltage hybrid power supply, 219
Horizontal amplifier, 218
Horizontal decoding and deflection, character generator, 365
Horizontal deflection, 182
Horizontal driver, 183
Horn-signal generator, 345
Hunting system, automatic, 200
Hybrid-matrix parameters (table), 21
 conversion to common-emitter values, 46
 measurement, 8

I-f amplifier, tunnel diode, 40
Impedance multiplexer, 296
Impedance transformer phototubes, 264
Indicator, crash-position, 331
 fencing, 260
 touch, 260
Indicator lamp switching, 34
Inductive feedback, 67
Industrial control relay, 238
Infrared communications receiver, 290
Inhibit driver, 357
Input circuit, tv tuner, 177
Input impedance, 8
 how to raise, 32
Integrated semiconductor devices, 18
Integrating amplifier, 73
Interpolation oscillator, 199
Interstage coupling, tv tuner, 178
 video amplifier, 67
Intrinsic semiconductors, 3
Inverter, for fluorescent lamp, 116
 synchronized, 236
 using linear circuits, 121
Isolation amplifier, 311

Jet-fuel flow-rate measurement, 268
Junction breakdown, 7
Junction temperature, 48

L-pad attenuator, 79
Larynx, artificial, 275
L-C pulse generator (*pnpn* trigger), 140
L-C-C feedback network, 85

Line-voltage control, 105
Load lines, 12
Logarithmic amplifier, reactor control, 240
Logarithmic diode, log amplifier, 240
Logic, magnetic-core, 370
 nines complement, 131
Low-frequency amplifier, 281
Low-frequency Hartley oscillator, 213
Low-frequency noise filter, 191
Low-level modulator, 146
Low-noise preamplifier, 191

Magnetic amplifier preamplifier, 304, 308
Magnetic binary counter, 372
Magnetic-core logic, 370
Magnetic-core pulse generator, 150
Magnetic-drum clock-track recorder, 358
Magnetic multivibrator, 94, 337
Magnetic network driver, 325
Magnetic pulser, radar, 324
Marker-beacon receiver, 319
 light aircraft, 328
Matrix inversion, 21
Maximum oscillation frequency, 45
Measuring, $f_{\alpha co}$ (alpha cutoff), 228
 h_{fe} (beta), 220
 fuel-flow rates, 269
Mechanical construction of video amplifier, 66
Melt-quench, 14
Memory, ferrite-core, 351, 354
 magnetic-drum, 358
Memory circuit (*pnpn* trigger), 141
Mesa transistor, 15
Meter, counting-rate, 267
 living tissue impedance, 278
Microalloy diffused-base transistor, 15
Microwave generator, 88
Microwave ppm modulator, 204
Miller effect, 83
Miller sweep generator, 217
Missile countdown switching, 310
Mixer, balanced, 200
 tv tuner, 180
 series-resonant, 174
Mobile radio, 198
Modulator, balanced, 197
 f-m, 83, 168, 208
 low-level, 146
 ppm, 204
 sweep generator, 204
Monitor, alarm circuit, 344
 blood pressure, 276
 radioisotope, 266
 television, 189
Monostable multivibrator, 134, 137
Motor control (*see* Control),
Motorboating in reflexed radios, 155
Multifunction circuit: r-f oscillator, modulator, audio amplifier, 208
Multiple feedback loop, 70
Multiplexer, impedance, 296
 ppm, 204
 satellite, 292
 telephone communications, 204
Multivibrator, free-running, 275
 hybrid, 30
 tunnel diode, 40
 magnetic, 94, 337
 monostable, 134, 137
 one-shot, 134, 137, 226, 267, 310, 322
 hybrid, 30
 pnpn trigger, 140
 Schmitt trigger, 217, 226, 373

Multivibrator, sweep-gating, 217
 time-delay, 147
 trigger, 217
Musical timbre demonstrator, 96

Narrow-band amplifier, 277
Negative feedback, 62
 balanced, 61
 in power amplifiers, 59
Negative resistance, tunnel diode, 17
Negative-resistance oscillator, 166
Network, anticipatory response, 305
 differentiating, 137
 magnetic, 325
 oscillator feedback, 85
 pulse-stretching, 137
Neuron, artificial, 284
Nines complement logic, 131
Noise filter, low-frequency, 191
 nonlinear, 241
Nonlinear feedback amplifier, 312
Nonlinear noise filter, 241
Nonsaturating flip-flop, 206
Notch filter for ring-down, 206
Nuclear radiation detector, 270
Nuclear reactor control, 240
Null detector, 278

One-mc crystal oscillator, 367
One-shot multivibrator, 134, 137, 226, 267, 310, 322
 hybrid, 30
 pnpn trigger, 140
Operating point, tv tuner, 177
Operational amplifier, 72, 74
Oscillating slope detector for tv sound, 186
Oscillator, blocking, 97, 145, 182, 215, 226, 332
 carrier, using tetrodes, 196
 Clapp, 225
 Colpitts, 247, 264
 crystal, 87, 196, 201, 225, 334, 347, 367, 376
 flyback, 127
 f-m, 83
 graphical design, 84
 Hartley, 213, 326
 high-frequency and high-power, 90
 interpolation, 199
 negative resistance, 166
 Pierce, 225
 power supply, 110, 111
 relaxation, saturable-core, 124
 tunnel-diode, 37
 unijunction transistor, 293
 r-f, 208
 ringing-choke, 127
 square-wave, 325
 subcarrier, 264
 television tuner, 175, 180
 transfluxor, 94
 tuning-fork, 121
 tunnel-diode, 39
 audio-modulated, 40
 ultrasonic, 276
 vhf, 90
 frequency modulated, 168
 voltage-controlled, 94
 Wien-bridge, 277
Oscilloscope, battery-operated, 218
Output impedance, 9
Output power amplifier, computers, 74
Output transformer, audio amplifier, 61

379

Overload protector, 31, 128

Period amplifier for reactor control, 241
Phase discriminator, 200
Phase inverter, 373
   split-load, 56
Phase shifter, complementary, 58
Phonograph wireless pickup, 207
Photocell detector, nuclear radiation, 270
Photocell preamplifier, 291
Photoelectric flip-flop, 334
Photoflash power converter, 127
Phototube amplifier, 262
Pierce oscillator, 225
Pilot, automatic, 307
Planar transistor, 15
Plethysmograph, 276
p-n junction, 4
pnpn trigger device, 140
Portable radio, 154
Positive feedback, 25
Potential barrier, 4
Power amplifier, 57, 59, 74
   servo, 248
Power converter, 102
   photoflash, 127
Power flip-flop, *pnpn* trigger, 140
   SCR, 235
Power gain, 11
Power supply, constant-current-coupled, 114
   design equations, 110
   fluorescent-lamp, 116
   full-wave phase-controlled, 236
   gas-tube auxiliary, 115
   Geiger tube, 267
   half-wave phase-controlled, 236
   high-voltage, 219
   operational amplifier, 74
   oscillator-type, 110
   positive and negative, 101
   repeater amplifier, 197
   series regulated, 114, 219
   series stabilized, 106
   short-circuit-proof, 108
   shunt stabilized, 107
   temperature controlled, 108
   transistor stabilized, 106
Power transistor, thermal design, 48
Preamplifier, 60
   analog computer, 74
   magnetic amplifier, 304, 308
   photocell, 291
   regulated power supply, 107
   servo, 245
   tv monitor, 191
Prebiasing technique, 35
Pulse amplifier, 215
Pulse converter, analog output, 294
Pulse-delay circuit, 376
Pulse discriminator, 262
Pulse-forming circuit, 214, 347
Pulse generator, 223, 376
   pnpn trigger, 140
   single transistor, 150
Pulse-height analyzer, 287
Pulse-height-to-digital converter, 374
Pulse-position-modulation multiplexer, 204
Pulse shaper, video, 205
Pulse sorter, 142
Pulse stretcher, 137, 289
Pulse-train generator, 376
Push-pull amplifier, class-A, 202
   diode compensation, 41

Push-pull amplifier, one transistor, 55
   parallel, 298
   single-ended output, 69
Push-pull gate, 368

Q-multiplier, 25, 83
Quasicomplementary and series amplifier, 190
Quasicomplementary symmetry, 56

Radio beacon, 331
Radio receiver (*see* Receiver)
Radioactive tracer recorder, 268
Radioisotope monitor, 266
Random-event encoder, 366
Rate growing, 14
Ratio detector, 167
Read amplifier, magnetic drum, 360
Receiver, automobile, 162
   broadcast, 159, 161
   frame in eyeglass, 153
   h-f, 158
   infrared, 290
   long-wave, 159
   marker-beacon, 319, 328
      dual-conversion, 319
   mobile, 198
   reflexed, 154
   single-channel, 153
   superregenerative, 314, 315
   vhf, 157
Recorder, digital, 351
   magnetic-drum clock-track, 358
   radioactive tracer, 268
Recorder amplifier, 75
Rectifier, half-wave controlled, 234
   silicon controlled, 17
   voltmeter, tunnel-diode, 39
Reflex circuit, 161
Reflexed radio receiver, 154
Regulator, automobile generator, 118
   current, 113, 115
Relaxation oscillator, saturable-core, 124
   tunnel-diode, 37
   unijunction transistor, 293
Relay, industrial control, 238
   servo, 249
   static-switching, 235
   time-delay, 238
Repeater amplifier, 196
Reverse base current, switch turn-off, 343
Reverse-biased diode, 4
Reverse-biasing to raise breakdown point, 34
Reverse voltage feedback ratio, 9
Reversible counter, 131, 138
R-f detector, tunnel-diode, 39
R-f television amplifier, 173
Ring counter, four-stage, 28
   pnpn trigger, 141
   ten-channel, 292
Ringing-choke oscillator, 127
Rise time, common-emitter switches, 52
Rising sawtooth generator, 229
Rocket guidance system, 303

Safety factor, power transistors, 48
Satellite multiplex, 292
Saturable-core relaxation oscillator, 124
Sawtooth generator, 222, 229, 230
   falling, 229
   rising, 229
   sweep, 230

Sawtooth generator, variable, 314
Scaler, binary, 263
Schmitt trigger multivibrator, 217, **373**
Schmitt trigger slicer, 226
Selective amplifier, 26
Semiconductor materials, 3
Semiconductor solid circuits, 19
Sensitor, temperature compensation, 264
Sensitive switch, 146
Series feedback loop, 70
Series and quasicomplementary amplifier, 190
Series power-supply regulator, 219
Series regulated power supply, 114
Series-resonant mixer, 174
Series stabilized power supply, 106
Servo amplifier, 315
Servo preamplifier, 245
Shielding, video amplifier, 67
Shift register, 18
   pnpn trigger, 141
Short-circuit-proof power supply, 108
Shunt feedback loop, 70
Shunt stabilized power supply, 107
Silicon controlled rectifier, 17
   applications, 234
Silicon transistor heat drift, 42
Single-channel radio receiver, 153
Single-ended push-pull amplifier, 68
Single-pole, single-throw switch, 146
Sliding class-A amplifier, 55
Solenoid drive circuit (SCR), 254
Sound detector, tv, 186
Sound movie amplifier, 202
Space-probe circuits, 297
Split-load phase inverter, 56
Square-wave amplifier, 359
Square-wave generator, 229
Square-wave oscillator, 325
Stability factor, 14
Starved transistors, 32
Static-switching relay (SCR), 235
Steering circuit, 131
Step-function potentiometer, 250
Strain-gage amplifier, 75, 257
Subaudio filter, 76
Subcarrier oscillator, 264
Summing amplifier, 72
Superregenerative receiver, 315
Suppressor, surge-voltage, 237
   transient, 239
Surface-barrier transistor, 15
Surge-voltage suppressor (SCR), 237
Sweep circuit, 322
   tunnel-diode, 38
Sweep-gating multivibrator, 217
Sweep generator, 322
   ppm modulator, 204
Switch, 227
   common-emitter, 52
   flip-flop, 269
   high-power, 341
   high-voltage, 37
   indicator-lamp, 34
   relay, 321
   reverse base current turnoff, 343
   rise and decay time, 52
   sensitive, 146
   static, a-c, 235
      d-c, 235
Switching amplifier, 306
Switching circuit, missile count-downs, 310
Switching transistor, power dissipation, 50
Symmetrical high-voltage switch, 37

Symmetry decoder, 315
Synchronized inverter (SCR), 236
Synchronizer detector, ppm, 206

T-pad gain control, 75
Target-drone control, 313
Television, closed-circuit, 195
Television monitor amplifier, 189
Television sound detector, 186
Television tuner, MADT transistors, 173
　mesa transistors, 176
Temperature coefficient, 20
Temperature control, stabilized power supply, 108
Temperature control amplifier, 304
Temperature controller, 233
Temperature-sensing transistor, 233
Tetrode amplifier, 197
　variable-gain, 122
Tetrode oscillator, 196
Tetrode transistors, 16
Thermal coupling, power transistors, 56
Thermal design, power-transistor circuits, 48
Thermistor, 19
Thermistor-compensated bias, 13
Thermocouple amplifier, 257
Thyristor, 18
　in waveform generator, 224
Time-delay multivibrator, 147
Time-delay relay, 238
Timer, capacitor charge, 239
Tissue impedance meter, 278
Tone generator, 96
Tonometer, glaucoma diagnosis, 282
Touch indicator, 260
Transfluxor oscillator, 94
Transformer, ferrite, 196
　impedance, 264
　power converter, 102

Transformer coupling, audio amplifier, 60
Transformer design, blocking oscillator, 184
　oscillator-type power supply, 111
Transient catching circuit, 241
Transient suppressor, 239
Transistor, mesa, 15
　microalloy diffused-base, 15
　modulator, 169
　planar, 15
　surface-barrier, 15
　tetrode, 16
　unijunction, 17
　unipolar, 18
Transistor action, 6
Transistor gain, 11
Transistor-stabilized power supply, 106
Transit time, 7
Transmitter, beacon, 331
　distress, 334
　f-m, 207, 208
　mobile, 198
　space-probe, 298
Transport factor, 6
Triangle waveform generator, 229
Trigger amplifier, 217
Trigger circuit, 347
Trigger multivibrator, 217
Triggered bistable circuit, 140
Trip output, reactor control, 241
Tube-transistor hybrid, 28
Tuned amplifier, 291
Tuned-radio-frequency receiver, 153
Tuning-fork oscillator, 121
Tunnel diode, 16
　germanium, 36
Tunnel-diode circuits and applications, 36

Ultrasonic oscillator, 276

Unblanking amplifier, 217
Unbypassed emitter negative feedback, 62
Unijunction relaxation oscillator, 293
Unijunction transistor, 17
Unipolar transistor, 18
Unneutralized i-f amplifier, 162
Untuned r-f amplifier, broadcast receiver, 161

Variable capacitance harmonic generator, 88
Variable delay generator, 134
Variable pulse-width generator, 213
Variable sawtooth generator, 314
Vertical amplifier, 216
Vhf oscillator, 90, 168
Vhf receiver, 157
Video amplifier, 65, 67
Video pulse shaper, 205
Voltage-controlled oscillator, 94
Voltage gain, 11
Voltage regulator, 116
Voltmeter rectifier, tunnel-diode, 39
Volume compressor, 79
Volume control, 165

Wavetrain generator, zero-crossing sync, 226
Wien-bridge oscillator, 277
Write amplifier, magnetic drum, 360

Yoke driver (crt), 323

Zener diode, line-voltage regulator, 105
　multivibrator control, 147
　reference element, 123
Zener effect, 5